Handbook of Liquefied Natural Gas

Handbook of Liquefied Natural Gas

Saeid Mokhatab
Gas Processing Consultant, Canada

John Y. Mak
Technical Director and Senior Fellow, Fluor, USA

Jaleel V. Valappil
Senior Engineering Specialist, Bechtel OG&C, USA

David A. Wood
Principal Consultant, DWA Energy Limited, UK

AMSTERDAM • BOSTON • HEIDELBERG • LONDON
NEW YORK • OXFORD • PARIS • SAN DIEGO
SAN FRANCISCO • SINGAPORE • SYDNEY • TOKYO
Gulf Professional Publishing is an imprint of Elsevier

Gulf Professional is an imprint of Elsevier
The Boulevard, Langford Lane, Oxford, Kidlington, Oxford, OX5 1GB

First edition 2014

British Library Cataloguing in Publication Data
A catalogue record for this book is available from the British Library

Library of Congress Catalog Number
A catalog record for this book is available from the Library of Congress

ISBN–13: 978-0-12-809996-4

For information on all Gulf Professional publications
visit our website at books.elsevier.com

Printed and bound in the United States

14 15 16 17 18 10 9 8 7 6 5 4 3 2 1

Dedication

We are truly fortunate to have collaborated with so many expert contributors from all over the world in the development of this handbook. Their insight taking "theory to practice" is much needed in the academic as well as the business worlds. Thanks are due to the many companies, institutions, and universities in the support of this work, many of whom are identified in the text and the back of each chapter as sources. We would particularly like to acknowledge the generous support of Fluor Corp., Bechtel Corp., DWA Energy Limited, Baker Engineering & Risk Consultants Inc., University of Illinois at Urbana-Champaign, University of Maryland, the Mexican Petroleum Institute, and the Bulgarian Academy of Sciences. We also recognize the support of our families, friends, and significant others; without their patience, encouragement, and inspiration throughout the process, this wouldn't have been such an enjoyable endeavor. Appreciation is also due to our various mentors over the years for the encouragement of the development of technologies and innovations. Thank you, all.

To live for a time close to great minds is the best kind of education.

John Buchanan

Contents

Authors' Biographical Sketches

Saeid Mokhatab is one of the most instantly recognizable names in the natural gas industry with dedicated technical expertise in the midstream sector. He has been on the international advisory board of a number of petroleum/energy consulting firms around the world and has been actively involved in several large-scale gas-field development projects, concentrating on the design, precommissioning, and startup of processing plants. He has over 200 technical publications to his credit, including Elsevier's *Handbook of Natural Gas Transmission & Processing*, which has been very well received by professionals in the natural gas industry and academia worldwide. He founded the Elsevier peer-reviewed *Journal of Natural Gas Science & Engineering*; has held editorial positions in several prestigious journals/book publishing companies for the hydrocarbon processing/LNG industry; and served as a member of technical committee for a number of professional societies and famous gas-processing conferences worldwide. As a result of his outstanding work in the natural gas industry, he has received a number of international awards including the Einstein Gold Medal of Honor, Chilingar Gold Medal of Scientists Without Borders, Kapitsa Gold Medal of Honor, and the Golomb-Chilingar "Giants of Science and Engineering" Gold Medal of Honor; and has been listed in several international biographical directories.

John Y. Mak is a Senior Fellow and Technical Director at Fluor, USA, and leads the technology and design development for the chemical and energy sectors. He has made significant contributions to the technologies in gas treating, emission controls, NGL recoveries, carbon capture, natural gas liquefaction and regasification, power generation, and coal gasification. He has over 50 publications on these subjects and is also listed as the primary inventor on over 80 patent and patent-pending processes. Throughout his career, he has developed over 100 oil and gas and energy projects in various locations and applied his design innovations to improve energy efficiency, lower emissions, and reduce cost for many of his clients. He is pioneering on the use of LNG cold in power plant integration that increases overall efficiency and reduces emissions at LNG receiving terminals. He is also on the advisory board of the China Coal Forum and several technology companies in China.

Jaleel V. Valappil is a senior engineering specialist with Bechtel Oil, Gas & Chemicals's Advanced Simulation Group in Houston, TX, USA. He has several years of experience in process simulation, advanced process control, and optimization of various processes including LNG. Before joining Bechtel, he was a senior consulting engineer with the Advanced Control Services Group of Aspen Technology Inc., where he was responsible for developing and implementing advanced control and optimization solutions for a variety of processes. He has worked extensively on developing and deploying process simulation models for applications in design, engineering, and operations of LNG plants built by Bechtel in collaboration with Conoco Phillips over the years. He identified the benefits of life cycle modeling for LNG plants and its use for plant design, operability/controllability studies, startup simulations and Operator Training Simulators, multitrain LNG facility debottlenecking, and operational troubleshooting. Dr. Valappil has also developed and deployed Advanced Process Control technology for LNG processes, for plants in the engineering stage and for plants that are already operating. He has published several papers on LNG plant simulation and control in both industry conferences and journals and holds patents related to LNG plant control and turbo-machinery operation.

David A. Wood has more than thirty years of international oil and gas experience spanning technical and commercial exploration and production operations, midstream and downstream projects, contract evaluation, and senior corporate management. His early energy industry experience includes Phillips Petroleum, Amoco, Lundin Oil, and other independents working around the world. He is now an independent international consultant, training provider, and expert witness. He has published an extensive body of work on diverse energy related topics including the international energy markets, oil and gas fiscal designs, LNG, GTL, gas storage, and gas supply. He publishes regular reports on global LNG industry activities and developments for World Oil. He acts as an adviser to governments and companies on technical and commercial aspects of the LNG industry through his consultancy DWA Energy Limited. Dr. Wood is actively involved in research, publication, and training programs and is the assistant editor-in-chief of the *Journal of Natural Gas Science & Engineering*. He has developed and participates in online oil and gas MBA programs with a focus on LNG and global gas supply chains.

Contributors

Hanfei Tuo
University of Illinois at Urbana-Champaign, USA
Author of Chapter 4, Energy and Exergy Analyses of Refrigeration/Liquefaction Cycles.

Yunho Hwang, Abdullah Al-AbulKarem, Amir Mortazavi and Reinhard Radermacher
University of Maryland, USA
Authors of Chapter 5, Natural Gas Liquefaction Cycle Enhancements and Optimization.

John L. Woodward
Baker Engineering & Risk Consultants Inc, USA
Author of Chapter 9, LNG Safety and Security Aspects.

Blanca E. García-Flores, Jacinto Águila-Hernández, Fernando García-Sánchez (Mexican Petroleum Institute, Mexico) and Roumiana P. Stateva (Institute of Chemical Engineering, Bulgarian Academy of Sciences, Bulgaria)
Authors of Appendix 2, Modelling the Phase Behaviour of LNG Systems with Equations of State.

Preface

Natural gas is definitely a viable option in bridging our energy gap to the next century of renewable energy. Natural gas is recognized as a safe and environmental responsible fuel and has reduced emissions in many parts of the world. Liquefied natural gas (LNG), being a denser fluid, is an economical way to transport and store natural gas, whether via LNG cargo ships or trucks from remote locations to the consumers. LNG continues to increase its share of year-on-year growth in the global natural gas trade and remains one of the fastest growing sectors of the energy industry.

The LNG supply chain extends from upstream production, LNG production plant, shipping, storage, and regasification to supply to sales gas pipelines and power plant users. LNG production is capital intensive and the recent costs have deterred the commitment of most investors, and any future LNG production plant owners must reevaluate the current technologies for a "fit-for-purpose" design to reduce the life cycle costs.

This book is written for the complete LNG supply chain, from liquefaction to regasification, for LNG plant designers, engineers, and operators, as well as LNG project developers and managers. This book provides an overview of the LNG industry' fundamentals, engineering and design principles. It can be used as a textbook for students in petroleum and chemical engineering curricula or as a reference guide to engineering and operating companies. The first chapter addresses the fundamentals of LNG from liquefaction to regasification and LNG commerce and the later chapters include more advanced topics, such as liquefaction cycle optimization, control and automation, safety and security, and LNG innovations. We believe this book has assembled a "first of a kind" collection of materials that address the full spectrum of the technologies in the LNG supply chain.

The materials used in this book are obtained from different sources including distinguished papers and publications in recent years, standards and recommendations published by several research institutions, and our own research papers. These references are listed in the back of each chapter for those readers interested in learning more on the subjects. We wish to thank chapter authors for their contributions. We also appreciate the patience and assistance of the editorial staff at Elsevier Science & Technology. Finally, we wish to express gratitude to our families and friends for their support and patience during the preparation of this book.

We understand that LNG technologies are evolving and the LNG business is changing. We accept the contents of this work may not be perfect, but we believe that the publication of this book is timely as natural gas is a favorite topic in today's news. We also hope that the materials of this book will be updated with newer and better technologies and innovations as the LNG industry progresses to the next millennium.

Saeid Mokhatab
John Y. Mak
Jaleel V. Valappil
David A. Wood

Endorsements

"This book is a great primer on all aspects of the LNG Industry. I highly recommend it to anyone interested in a comprehensive overview of the industry. It contains the key information we all search for in a single location and it does an excellent job of highlighting the key considerations that must be taken into account for any LNG project in development."

Tom Phalen

Vice President of Upstream Project Operations, Fluor, USA

"This book will be a welcome addition to libraries for anyone in the LNG Business from commercial, technical or operational. Even a brief review of the Table of Contents leaves one with a WOW and wondering where to start as it covers all the areas that one would want to either review or dive into something new. I highly recommend this valuable resource."

David Messersmith

Bechtel Fellow and Manager of LNG Technology Group, Bechtel OG&C, USA

"One thing that makes this book a valuable addition to anyone's technical library is its completeness in the coverage of all aspects of the LNG industry inclusive of contract negotiations, design, EPC execution, facilities startup, and operation. As well as covering some topics rarely discussed and hard to find in the literature, the complex elements of the LNG industry are fully addressed in a straightforward fashion, which makes the book appealing to all parties who are involved in the global LNG business."

Philip Hunter

Senior LNG Consultant, Bechtel, UK

"This is the first book that has filled all the gaps in the treatment of complete LNG supply chain from conceptual to commissioning and beyond. It is an essential addition to the bookshelf of any LNG industry professional regardless of which sector of the industry that individual is involved. I will highly recommend it as a textbook for any Natural Gas graduate degree program anywhere."

Dr. Suresh C. Sharma

ONEOK Chair Professor and Director of Natural Gas Engineering and Management
University of Oklahoma, USA

"This well-balanced handbook is the only book of its kind, covering all aspects of the LNG supply chain in more detail. I highly recommend it as an excellent reference for all professionals, engineers, and scientists working in the LNG industry, and as a textbook for graduate students in the gas engineering curriculum."

Dr. Brian F. Towler

Professor of Chemical and Petroleum Engineering and CEAS Fellow for
Hydrocarbon Energy Resources, University of Wyoming, USA

"This book is an important contribution to the professional literature in the crucial area of the emerging energy scene, offering a complete coverage of key topics in the LNG supply chain. The complete, accurate, and easy-to-use description of issues is also extended to cross-cut controversial topics such as LNG safety. The book provides an excellent access to a number of design guidelines and operating procedures, which makes it a standalone reference for LNG industry professionals and a state-of-the-art textbook for graduate programs on the subject."

Dr. Valerio Cozzani

Professor of Chemical Engineering and Director of Post-Graduate Program on Oil & Gas Process Design, University of Bologna, Italy

"It is the first book on LNG supply chain management and gives an accurate picture of where the LNG industry stands today. This high-quality, comprehensive book provides a better understanding of LNG plant design and operational considerations and covers subject areas missed by other references in these areas. I believe it is a valuable addition to the literature and will serve as a desk reference for practicing engineers and technologists, and as an excellent source of teaching and learning in undergraduate and graduate programs on the subject area."

Dr. Faisal Khan

Professor and Vale Research Chair of Safety and Risk Engineering Memorial University, Canada

"Given the fact that LNG is the fastest growing energy carrier in the world and the trend is toward a wider range of applications for LNG technologies, a handbook like this one covering the entire LNG supply chain with issues of design, operation, and safety will be highly appreciated by plant designers, engineers, and operators in the LNG industry. Of course, parts of the book also provide excellent textbook or reference material for graduate courses in the gas processing field."

Dr. Truls Gundersen

Professor of Energy and Process Engineering Norwegian University of Science and Technology, Norway

"This comprehensive handbook provides practical guidance for all professionals in the LNG industry while maintaining as much rigor as possible. I believe academic institutions that have courses in natural gas engineering should also consider this handbook as a textbook. Congratulations to the authors for producing a useful, quality handbook."

Dr. Kenneth R. Hall

Jack E. & Frances Brown Chair and Professor of Chemical Engineering, Texas A&M University, USA & Associate Dean, Research & Graduate Studies, Texas A&M University at Qatar, Qatar

LNG Fundamentals

1

1.1 Introduction

Natural gas has remained the fastest growing energy resource in most regions of the world for more than two decades, driven by the low greenhouse gas emissions as well as high conversion efficiency in power generation. For almost a century, natural gas has been transported safely, reliably, and economically via pipelines. Pipelines proved to be ideally suited to the supply and market conditions of the twentieth century, when large reservoirs of conventional natural gas are found in accessible locations. Pipelines provided the stability and security of supply and continue to do so where large accessible gas reserves remain. However, over the past decades it has become clear that significant quantities of new gas reserves are not so conveniently located. Attention has shifted to more isolated large gas reservoirs that were previously thought to be too remote, or technically too difficult and costly to develop. A number of solutions for exploiting stranded gas reserves are currently being developed and considered for commercialization. On the other hand, over the past three decades, only the liquefied natural gas (LNG) industry has successfully brought many large remote gas fields to the gas markets that are unreachable by pipeline (e.g., Japan, South Korea). Today, the LNG supply chains have diversified and introduced competition into markets previously "captured" by pipeline gas suppliers, and have improved the security of energy supply of many consuming nations and reduced the geopolitical and political constraints on global gas supply.

This chapter briefly summarizes the components of the LNG supply chain—the steps and industrial processes used in producing, storing, and delivering LNG to commercial and residential customers. For those readers new to the LNG industry and unfamiliar about what it is, how it is used, and its basic safety and environmental track record, Appendix 1 is included.

1.2 Monetizing stranded gas

Despite being one of the most abundant energy sources on the planet, more than one-third of global conventional natural gas reserves remain stranded (Thackeray and Leckie, 2002) and undeveloped. Growing global energy demand, diminishing oil resources, higher oil prices, the no-flaring regulations, and the benefits of lower greenhouse gas emissions from the burning of natural gas are leading to the urgency in the quest for commercially viable technologies for transporting stranded gas over long distances (Mokhatab and Wood, 2007a; Wood et al., 2008a).

Over the last two decades different technologies have been developed and proposed for monetizing hitherto remote gas reserves (see Figure 1-1). However, most of them are not fully matured to the

Handbook of Liquefied Natural Gas.

extent for commercialization, and would require improvements in order to compete with existing technologies. On the other hand, selection of existing technologies is dependent on the distance to consumer markets and the gas field production rates (see Figure 1-2). Stranded gas is currently transported to markets primarily via two long-established technologies: some 70 percent of gas traded internationally is exported by pipeline; the remaining 30 percent by liquefied natural gas.

Figure 1-3 shows that LNG cost becomes competitive to pipeline gas for long-distance transport. As can be seen from Figure 1-3, for short distances, gas pipelines are usually more economical. LNG is competitive for long-distance routes, particularly those crossing oceans or long stretches of water, since construction of undersea pipelines is cost prohibitive (Mokhatab et al., 2006). For offshore stranded gas, LNG can be competitive when the offshore pipeline is less than 700 miles. For onshore pipelines, the breakeven point is about 2,200 miles (Mokhatab and Purewal, 2006).

When natural gas is cooled to approximately –162°C or –259°F at atmospheric pressure, the condensed liquid is LNG. The volume reduction is about 1/600th the volume of natural gas at the burner tip. The more condensed form of LNG allows transport using cargo ships or trucks. For long-distance transport, the gas pipeline option would require large diameter pipes and gas recompression facilities to overcome the transmission pressure drop, which are very costly.

The LNG option offers economics, flexibility, and security of supply advantages over gas pipeline and other technology alternatives. Many countries in Europe and Asia are embracing LNG for these reasons, even though large volumes of gas from Russia, North Africa, and the Middle East could satisfy their domestic energy demands through existing and planned pipeline connections. Today, with the increase in gas supply from the nonconventional gases, export from North American to the Asian

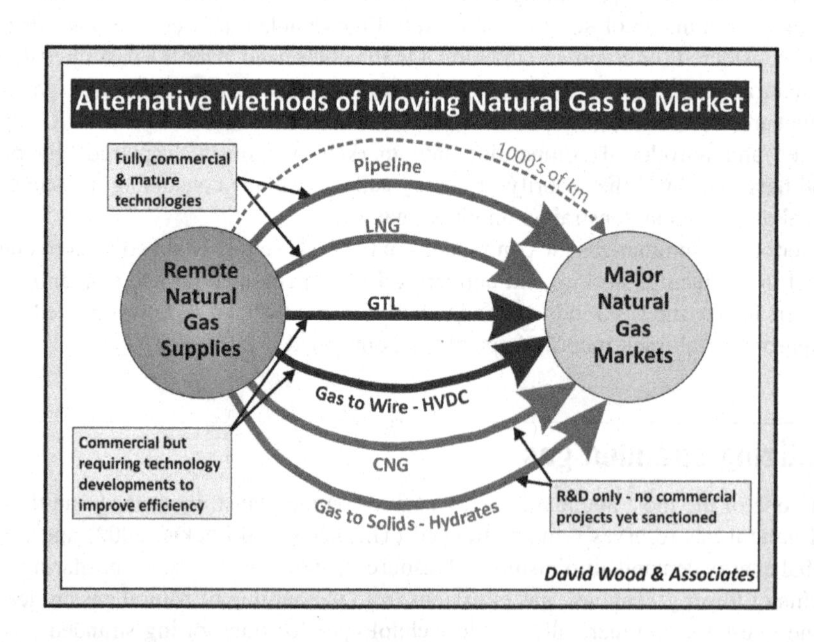

FIGURE 1-1

Technologies available to transport natural gas long distances.

(Wood and Mokhatab, 2008a)

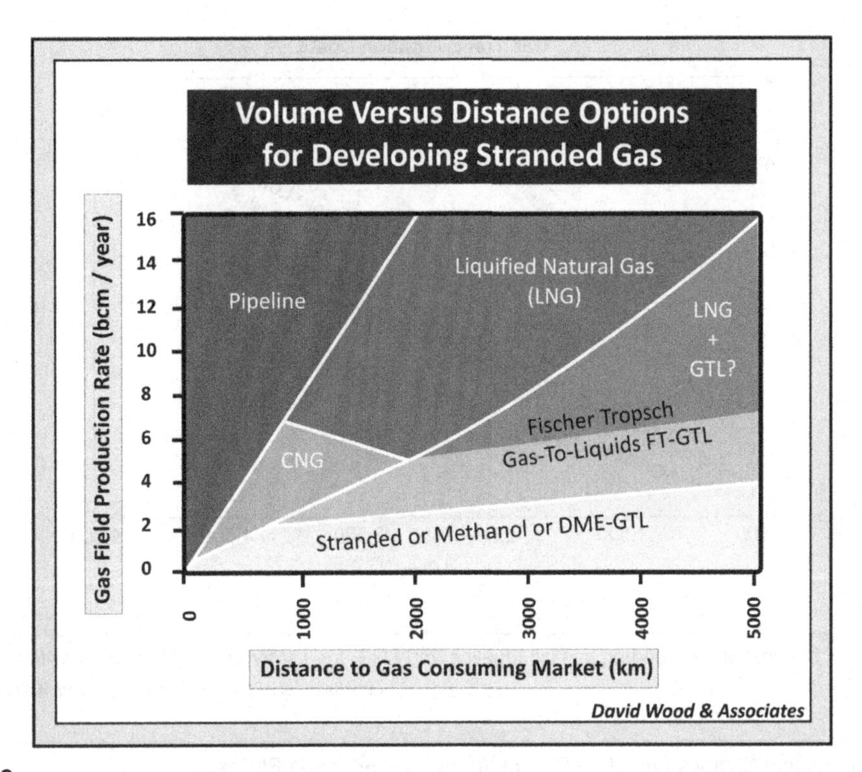

FIGURE 1-2

Production volume versus distance to market framework for gas technologies.

(Wood & Mokhatab, 2008a; Wood et al., 2008a)

countries is imminent. The share of the international natural gas trade as LNG is expected to increase in the coming decades.

1.3 LNG characteristics

1.3.1 Basic properties

A basic knowledge of LNG must begin with an examination of its chemical and physical properties, which is a prerequisite for accurately assessing potential LNG safety hazards and risks.

The properties of LNG vary with its composition, which depends on the reservoir source of the original gas and its processing/fractionation history. While LNG is predominately methane (about 87 mole % to 99 mole %), its composition also includes other higher hydrocarbons, typically, the C_2 to C_4 and heavier, nitrogen and trace amounts of sulfur (less than 4 ppmv), and CO_2 (50 ppmv; see Table 1-1).

LNG is an odorless, colorless, and noncorrosive cryogenic liquid at normal atmospheric pressure. When LNG is vaporized and used as natural gas fuel, it generates very low particle emissions and

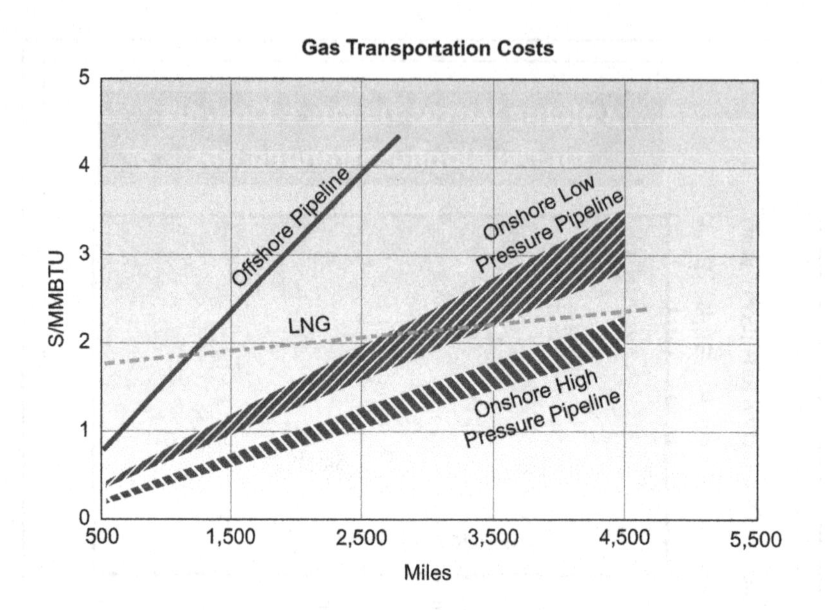

FIGURE 1-3

Comparison of the cost of transporting gas via pipeline and LNG; for 1 tcf/yr and including regasification costs.

(Mokhatab et al., 2006; Economides and Mokhatab, 2007)

Table 1-1 Typical Composition of LNG from Various Liquefaction Plants

Component, mole %	Nigeria LNG	Arun LNG	Brunei LNG	Oman LNG	Atlantic LNG	Kenai LNG
Methane	87.9	88.48	89.4	90	95	99.8
Ethane	5.5	8.36	6.3	6.35	4.6	0.1
Propane	4	1.56	2.8	0.15	0.38	0
Butane	2.5	1.56	1.3	2.5	0	0
Nitrogen	0.1	0.04	0.2	1	0.02	0.1

(ILEX Energy Consulting, 2003)

significantly lower carbon emissions than other hydrocarbon fuels. The combustion products from LNG contain almost no sulfur oxides and a low level of nitrogen oxides, which makes LNG a clean source of energy.

LNG is nontoxic. However, as with any gaseous materials, natural gas release from LNG can cause asphyxiation due to lack of oxygen in an unventilated, confined area, and can be ignited if mixed with the right concentrations of air.

The boiling point of LNG varies with its composition, typically –162°C (–259°F). The density of LNG typically falls between 430 kg/m^{3}and 470 kg/m^3 (3.5 to 4 lb/US gal), which is less than half the density of water. LNG, if spilled on water, floats on top and vaporizes rapidly because it is much lighter than water. Initially LNG vapors are heavier than air and will stay near ground level. However as LNG vapors begin to be warmed up by the surroundings and reach temperatures of approximately –166°F, the density of the vapors is lighter than air and the vapors become buoyant. Cold LNG vapors (below –166°F) are more likely to accumulate in low areas until the vapors warm up. A release of LNG in an enclosed space or low spot will tend to displace air, making the area hazardous for breathing.

Vapors released from LNG, if not contained, will mix with surrounding air and will be carried downwind, which may create a vapor cloud that may become flammable and explosive. The flammability limits are 5 percent and 15 percent by volume in air. Outside of this range, the methane/air mixture is not flammable.

When fuel concentration exceeds its upper flammability limit, it cannot burn because too little oxygen is present. This situation exists, for example, in a closed, secure storage tank where the vapor concentration is approximately 100 percent methane. When fuel concentration is below the lower flammability limit, it cannot burn because too little methane is present. An example is leakage of small quantities of LNG in a well-ventilated area. In this situation, the LNG vapor will rapidly mix with air and dissipate to less than 5 percent concentration (Foss et al., 2003).

Vaporized LNG has the same thermal characteristics as natural gas. In well-ventilated areas, natural gas burns with a low laminar burning velocity and has high ignition energy relative to other hydrocarbon fuels. Natural gas vapor in open areas has not produced unconfined vapor cloud explosions (UVCE), which are more prevalent with higher hydrocarbons. Whether any flammable vapor cloud merely burns back to the vapor source or undergoes an explosion depends on many factors: the chemical structure of the vapor molecules, the size and concentration of the vapor cloud, the strength of the ignition source, and the degree of confinement of the vapor cloud. The conditions needed to produce an unconfined vapor cloud explosion of natural gas are generally not present in an LNG facility, so such explosions should not be considered as potential hazards.

1.3.2 Thermodynamic properties

Knowledge of the LNG phase behavior and thermodynamic properties is required for the successful design and operation of LNG production plants and handling facilities. Therefore, it is important to apply appropriate models within a thermodynamic modeling framework to predict, describe, and validate the complex phase behavior of LNG mixtures. In this case, equations of state, which can both qualitatively and quantitatively predict the complex types of phase behavior found in real systems, are usually the primary choice. Phase behavior prediction and modeling of LNG systems with equations of state will be described in Appendix 2.

1.3.3 LNG safety

The LNG industry has an excellent safety record, as a result of several factors. First, the industry has been developed to ensure safe and secure operations, from engineering to technical competency of personnel. Second, the physical and chemical properties of LNG are well understood and the plant designs are well proven through many years of operation. Third, the standards, codes, and regulations that have been developed for the LNG industry ensure safety and are continuously evolving and improving.

Chapter 9 explores hazards associated with and safety features designed for the unusual characteristics of LNG.

1.3.4 Units and conversion factors

The metric ton is commonly used to account for quantities of LNG. LNG production is commonly expressed in million metric tons per year. The units often used are MMTPA, MTPA, tpy, or a more SI conforming expression Mt/a, which stands for mega but can also be read as million tons per annum.

The various conversion factors for common units in the LNG business are given in Table 1-2. These are the thumb rules of conversion of LNG; the actual factors would vary based on the composition of LNG.

1.4 Traditional LNG supply chain

To make LNG available for use in a country, energy companies must invest in a number of different facilities that are highly linked and dependent upon one another. The major components of the traditional LNG supply chain, including pipeline connections between the stages, include the following as shown in Figure 1-4.

Table 1-2 LNG Conversion Factors

Frequently Used Conversions for Natural Gas and LNG

To:	Billion Cubic Meters of Natural Gas	Billion Cubic Feet of Natural Gas	Million Tons of LNG	Trillion Btu
From:		Multiply By		
1 Billion Cubic Meters of Natural Gas	1	35.315	0.76	38.847
1 Billion Cubic Feet of Natural Gas	0.028	1	0.022	1.100
1 Million Tons of LNG	1.136	46.467	1	51.114
1 Trillion Btu	0.026	0.909	0.02	1

Typical Liquid-Vapor Natural Gas and LNG Conversions*

	Liquid Measures			Vapor Measures		Heat Measure
To:	Metric Ton LNG	Cubic Meter LNG	Cubic Foot LNG	Cubic Meter Natural Gas	Cubic Foot Natural Gas	Btu*
From:				Multiply By		
1 Metric Ton LNG	1	2.193	77.445	1,316	46,467	51,113,806
1 Cubic Meter LNG	0.456	1	35.315	600	21,189	23,307,900
1 Cubic Foot LNG	0.0129	0.0283	1	16.99	600	660,000
1 Cubic Meter Natural Gas	0.00076	0.001667	0.058858	1	35.315	38,847
1 Cubic Foot Natural Gas	0.000022	0.000047	0.001667	0.02832	1	1,100

Conversion Factors

1million metric tons/year = 1.316 billion cubic meters/year (gas) = 127.3 million cubic feet/day (gas)

1billion cubic meters/year (gas) = 0.760 million metric tons/year (LNG or gas) = 96.8 mcf/day (gas)

1million cubic feet/day (gas) = 10.34 million cubic meters/year (gas) = 7,855 metric tons/year (LNG or gas)(US DOE/FE-0489, 2005)

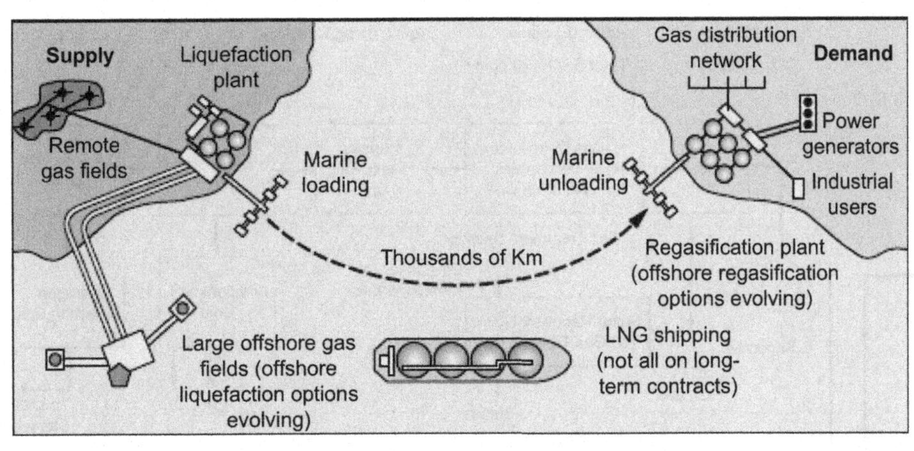

FIGURE 1-4

Key elements of traditional LNG supply chain.

1.4.1 LNG production plant

A typical scheme for an LNG production plant designed to produce LNG from a sour gas feed is shown in Figure 1-5. The process and utility requirement design depends on site conditions, feed gas conditions, compositions, and products specifications.

1.4.1.1 Feed gas conditioning

In a typical scheme, field production upon arrival at the processing plant is first separated in a slug catcher, which removes the liquids and routes the gas to a high pressure (HP) separator. The liquids are flashed to a medium pressure separator, where the hydrocarbon liquids are further separated and fed to the condensate stabilizer. The liquids are fractionated in the stabilizer, producing a bottom condensate product consisting of the C_5 and heavier hydrocarbons. The condensate is stripped with steam to remove its H_2S content and has to meet an RVP (Reid Vapor Pressure) specification of 12 psia, which are required for transport and storage.

Vapors from the medium separator and stabilizer column overhead are compressed and recycled back to the HP separator. The vapors then flow to the Gas Sweetening Unit, GSU (also called Acid Gas Removal Unit, AGRU), in which H_2S and CO_2 are removed. H_2S is removed by an amine solvent to meet the total sulfur product specification, typically 4 ppmv. CO_2 is removed to 50 ppmv to avoid CO_2 freezing in the main exchangers in the liquefaction plant. Carbonyl sulfide (COS) and mercaptans (R-SH) contribute to the sulfur contaminants and must also be removed. The acid gas from the regeneration section is sent to the Sulfur Recovery Unit (SRU), typically consisting of a Claus unit and a Tail Gas Treating Unit (TGTU). The off-gas from the TGTU absorber is incinerated.

The sweet gas from GSU is required to be dried in a dehydration unit utilizing molecular sieves technology to below 0.1 ppmv to avoid hydrate formation in the NGL recovery unit. The sweet gas is saturated with water, and in hot climate operation the water content can be significant. It is more energy efficient and cost effective to chill the sweet gas first, removing the bulk of water before passing

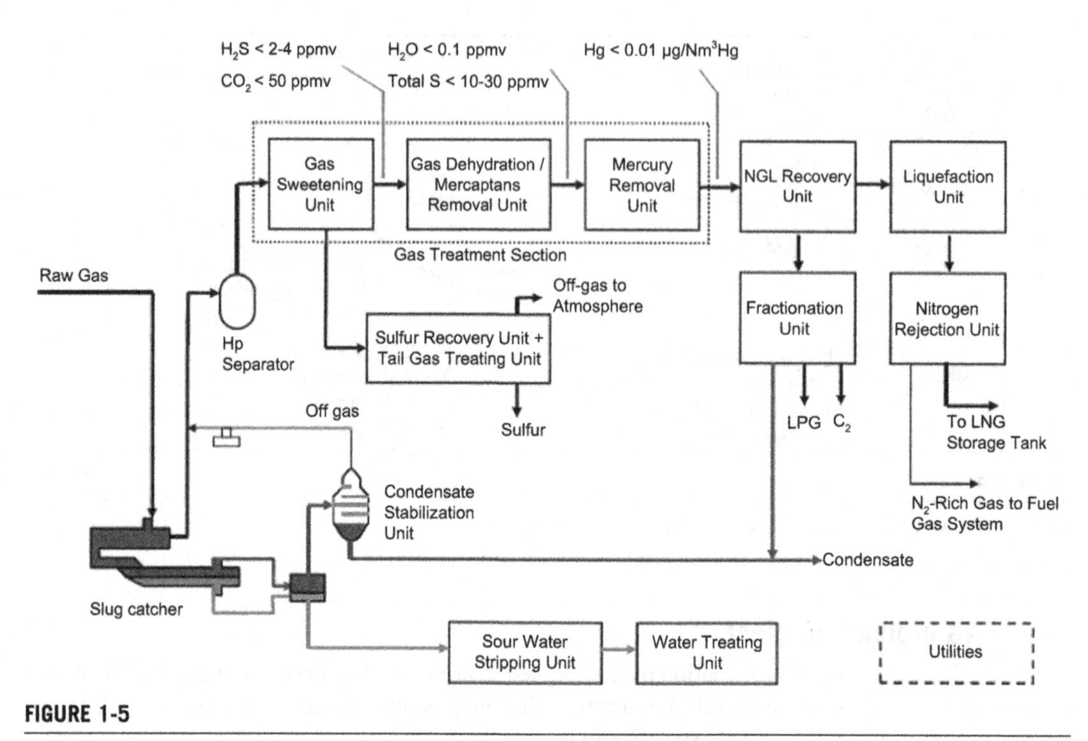

FIGURE 1-5

Typical illustration of LNG production plant: sequence and requirements.

(Mokhatab, 2010)

to the molecular sieve units. The molecular sieve can also be designed to remove mercaptans from the dried gas to meet the sulfur specification.

Typically, traces of mercury are present in the feed gas, which must be removed by mercury removal beds to less than 10 nanograms per cubic meter to avoid mercury corrosion in the downstream cryogenic exchanger.

1.4.1.2 NGL recovery

The dried gas is routed to a Natural Gas Liquid (NGL) recovery unit, which is designed to remove and recover the C_2+ or C_3+ hydrocarbons and produce a lean gas for liquefaction. Removal of the NGL components would eliminate the need for a scrubber column in the liquefaction plant, which typically is used to remove aromatics and heavy hydrocarbons to avoid waxing in the main exchanger. The NGL recovery unit can be designed for ethane recovery to produce an ethane product that can be fed to an ethylene cracker in a petrochemical complex.

The NGL components, the C_3 to C_5+ liquids, are valuable marketable products. The NGL components are fractionated into individual products for sales. Propane and butane are exported as separate products, or a combined propane-butane mix product. C_5+ and heavier components can be exported for gasoline blending. If mercaptans were present in the feed gas, they would show up in the C_5+ liquids, which must be treated to meet the liquid's sulfur specification.

1.4.1.3 Liquefaction

The lean gas leaving the NGL recovery unit enters the liquefaction unit that chills and liquefies the gas in a refrigeration process. Liquefaction technology is based on the principle of a refrigeration cycle, where a refrigerant by means of successive expansion and compression removes the heat content of a gas stream, by rejecting to the ambient air or cooling water. The refrigerant may be part of the natural gas feed (open-cycle process) or a separate fluid continuously recirculated through the liquefier (closed-cycle process). After liquefaction of the natural gas, a nitrogen rejection unit is required if the nitrogen content is above the commercial specification of LNG, typically at 1 mole %. The low nitrogen content in the LNG product is required to avoid low liquefaction temperatures, to reduce the nitrogen content in the boil-off such that it can be used as fuel gas, and to reduce the risk of rollover in the storage tanks in the terminals when delivered to the customers. Rollover occurs in a weathered tank, where there is rapid mixing of stratified layers within an LNG tank causing sudden release of very high levels of boil-off gas within a short period.

Typically, LNG from the liquefaction plant is let down in pressure in a flash drum to close to the LNG storage pressure, and nitrogen, being the lighter component, is flashed off and removed. The nitrogen-rich vapor is compressed and recovered as fuel gas. The flashed liquid is pumped into the storage tanks for export. This end-flash process is well suited to feed gases with nitrogen content of up to 2 mole % (Vovard et al., 2011). However, with a high nitrogen feed gas, the simple end-flash process is not sufficient and an additional fractionation step is required. If not removed, the high nitrogen content would lower the liquefaction temperature, and would increase the power consumption by the refrigeration unit. In addition, the nitrogen content of the flash gas and boil-off gas would be higher, which may not meet the heating value specification of fuel gas. Therefore, there is a justification to remove the nitrogen content either before or during liquefaction.

Nitrogen removal by cryogenic separation process is the proven nitrogen removal process for LNG production. The other alternatives such as pressure swing adsorption or membrane technology are not cost competitive in meeting the very low nitrogen specification (Finn, 2007; Garcel, 2008).

Feed gas conditioning, NGL recovery, liquefaction, and nitrogen removal technologies will be addressed in more detail in subsequent chapters.

1.4.2 Liquefaction plant types

Natural gas liquefaction plants can be classified into large base load, peak-shaving, and small- to medium-scale plants depending on their sizes and functions.

Most of the base load plants are located in large gas reservoirs in Asia, Australia, the Middle East, and West Africa and are mega projects. These base load plants supply natural gas as LNG from the natural gas producers to the consumer nations. Base load plants typically consist of one or multiple trains. The liquefaction train size has steadily increased over the last 40 years with capacities of over 4 million tons per annum now being conventional. Single trains of 7.8 MTPA are now operational in Qatar.

Peak-shaving plants, which are commonly used to balance the fluctuation in natural gas supply and demand during summer and winter months, are small in size, typically up to 0.1 MTPA. They are used to liquefy and store excess gas production and to provide extra capacity during peak demand periods.

As the large reservoirs are slowly being depleted, smaller reservoirs in remote locations are now being explored. These opportunities have prompted some market players to evaluate mid-scale LNG technology applications. Mid-scale LNG plants, ranging from 0.3 to 1.5 MTPA, are applicable for medium-sized onshore and offshore gas fields (Finn et al., 2000).

Small-scale LNG plants, with capacities as low as 0.01 MTPA, are economically viable when excess capacity of gas pipeline is available. LNG can be distributed by LNG trucks to customers that are located in remote areas inaccessible by pipelines or as an emergency fuel backup. LNG can also be used to replace diesel fuel in refuge trucks, public transports, and drilling rigs operation. The use of LNG instead of diesel fuel would significantly reduce emissions and lower fuel costs. The small-scale LNG units have also been installed recently on large LNG carriers to reliquefy boil-off gas and reduce shrinkage of the LNG during the shipping voyage.

1.4.3 LNG train size

Base load LNG plants are designed along the train concept. The train concept allows the plants to continue production when one of the trains is down for maintenance or unexpected shutdown. The train concept has evolved to meet market demands, delivery flexibility, and shipping logistics. Train size is determined by limitations on critical equipment, mainly proven gas turbine drivers' sizes and liquefaction heat exchangers' capacity (Smaal, 2003). However, for the most part, equipment developments have kept pace with the increases in train capacity.

The trend within the LNG industry has been to push the limits of LNG production capacity of a single train. In fact, many base load liquefaction plants built since 2004 have been plagued by high capital cost inflation. In spite of these cost obstacles the large IOCs, NOCs, and technology providers in the industry continue to increase the size of single liquefaction trains. Larger single train size tends to lower the unit cost of production, helping to make it competitive in the marketplace (Wood and Mokhatab, 2007a).

One of the main cost components in a liquefaction plant is the refrigeration compressor turbine driver. With the advances in larger and more efficient gas turbine drivers, producing more LNG in a single train can justify the incremental increase in capital cost. The cost of the support facilities, such as utility and offsite, are not much affected by the larger train size. The larger train would require more frequent shipment, but overall, this approach will improve the overall project economics (Durr et al., 2005). However, to fully realize the cost advantages of the larger trains, a high plant reliability and availability must be maintained; design must be robust and must be provided with sufficient spare equipment in case of equipment shutdown.

LNG technology has undergone a significant improvement in the past decade versus its 50 year history (see Figure 1-6). It took 30 years for the industry to progress from less than 0.5 MTPA capacity trains in Arzew, Algeria, to 3 MTPA trains in the late 1990s in Bonny, Nigeria. In the past decade, several trains in the 4+ MTPA range came on-stream such as the Damietta LNG plant in Egypt producing about 4.8 MTPA of LNG in 2004 and the 7.8 MTPA trains of the Qatargas II plant in 2009. This steady increase in train capacity has enabled the industry to achieve economy of scale benefits that at least have managed to offset the rising capital costs the industry has experienced over the past decade.

In the past few years liquefaction plants in the following countries have come on-stream with liquefaction train capacities shown in brackets: Australia NWS train 5 (4.4 MTPA—2008); Russia Sakhalin (4.8 MTPA—2009); Indonesia Tangguh (3.8 MTPA—2009); Yemen (3.4 MTPA—2009); Peru (4.4 MTPA—2010); Australia Pluto (4.3 MTPA—2012); Angola (5.2 MTPA—2013). Several large liquefaction plants are under construction in Australia (and Papua New Guinea) and are due to come on-stream in the 2015 to 2017 period. In addition several liquefaction plants are now in the

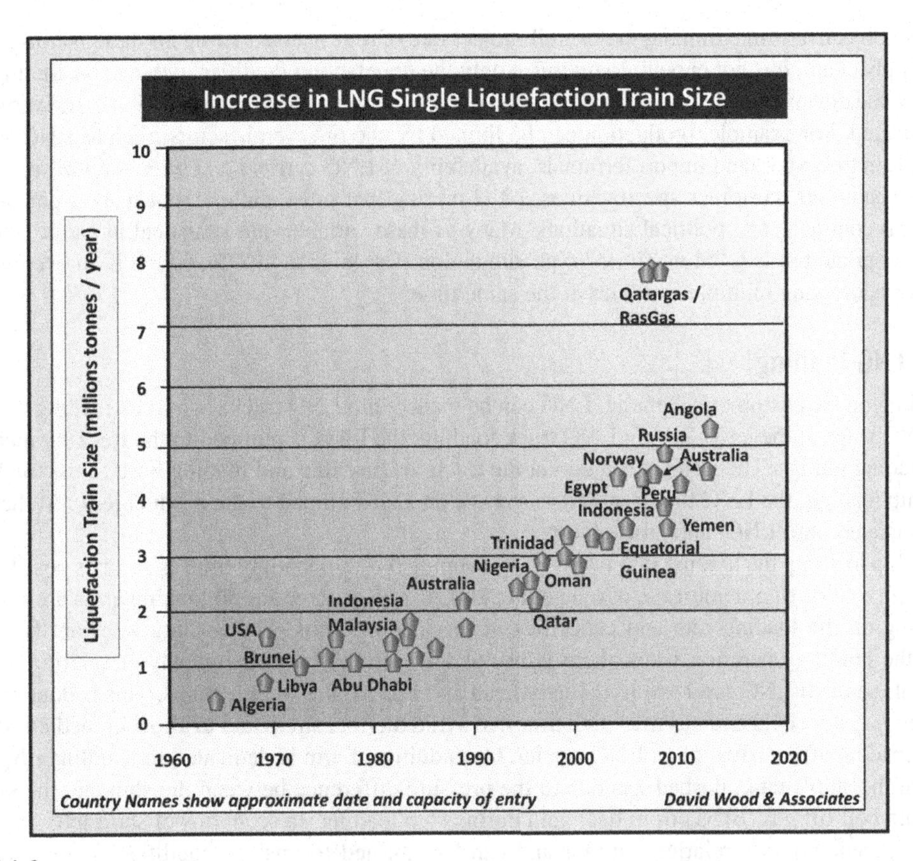

FIGURE 1-6

Liquefaction single train size capacity growth history updated in 2012. Original diagram published and discussed in Wood and Mokhatab (2007a) and Wood (2012).

advanced planning stages in North America (United States and Canada), planning to export vast resources of shale gas unlocked by recent advances in fracture stimulation technology. The vast natural gas resources found recently on offshore East Africa are also attracting interest from LNG suppliers and buyers for potential new projects. The natural gas industry therefore seems destined to continue its significant growth over the coming decades.

During development of an LNG business, there has been a continuing debate on an optimum size of an LNG train (Liu et al., 1992; Hunter et al., 2004). Many factors should be considered when choosing an optimal train size for an LNG project (Avidan et al., 2001), such as:

- Gas deliverability from the gas field
- Market demands and LNG delivery buildup profile
- Overall production, storage, and shipping logistics
- Commercially proven equipment sizes
- Capital and operating costs
- Operational flexibility and reliability.

The main objective is maximizing the overall project rate of return, considering all these factors. For an LNG supply chain, it is not enough to optimize only the liquefaction facilities without considering both upstream and downstream facilities. Optimization of these parameters, from production to delivery, must be considered. For example, production can be limited by one or several factors, such as LNG storage capacities in the export and import terminals, availability of LNG carriers and sizes, ship travel speeds, weather conditions, customer specifications, LNG pricing that must address spot market demands or long terms contracts, and political situations. Many of these variables are statistical in nature, thus the most appropriate too is a "Mont Carlo" type simulation (Coyle et al., 1995), which is an event-driven simulator addressing multiple variables at the same time.

1.4.4 LNG loading

Depending on the customer's demand, LNG can be loaded into LNG trucks at a truck loading bay and/or to LNG ships at the jetty. During LNG truck loading, the LNG is pumped to the truck by means of truck loading pumps. The LNG is routed via the truck loading line and loading hose to the truck. For LNG ship loading, the LNG that is stored in storage tanks is pumped to the product jetty. At the jetty, LNG is loaded onto LNG ships for export.[1]

For ship loading, the loading rate is driven by keeping the ship loading time as short as possible due to the high cost of ship demurrage (Coyle et al., 2003). Two or three liquid loading arms are required depending on the loading rate and capacities of the loading arms. The loading lines are kept cold during the holding operation when there is no loading activity by continuously circulating a small stream of the plant LNG rundown to the jetty head and back to the storage tanks. This is done to keep the loading system cold and gas free at all times, to avoid thermal stress and to allow immediate startup of ship loading after arrival of an LNG carrier. One additional arm is dedicated to handling displaced gas from the ship's tank, flashed gas due to the pressure difference between the ship and the storage tanks, and boil-off gas (BOG) from heat gain during ship loading. In scenarios of short jetty lines, the ship BOG generation is relatively minor and can be returned to onshore facilities where it can be recovered. However, at some locations the jetty needs to be extended several kilometers to reach ship berths at shallow ports or coastlines (Kotzot, 2003). In this scenario, BOG management poses some challenges. Long jetties, which imply long LNG loading lines, result in higher BOG generation rate due to higher pumping energy and heat leakage through piping and vessels. Moving large volumes of low-pressure BOG across long jetties is fairly costly. Therefore, it is generally uneconomical to recover BOG from long jetties. Historically some BOG was flared during LNG ship loading operations. However, as environmental regulations become more stringent, flaring of BOG during ship loading is not permitted. There are various BOG recovery options suitable for long jetties. These may include compressing the BOG back to LNG plant for recovery, in-situ reliquefaction, and in-situ power generation. In each LNG project, the feasibility of the alternatives should be considered from perspectives of environmental impacts, operational safety and efficiency as well as marine operational limitations (Huang et al., 2007).

Note that the right choice of insulating the pipeline from storage to ship can minimize heat gain and boil-off gas. Many insulation choices have come to the market in recent years for use in the cryogenic LNG piping application. The materials/techniques most commonly used are mechanical insulation such

[1]The time period during which the ship is berthed and the loading operation is under way is termed the loading mode.

as glass foam and polyisocyanurate, "powder" insulations such as aerogel, perlite, izoflex, and high vacuum insulation. However, many factors need to be considered for each LNG pipeline application and installation to determine the most efficient insulation that will provide the best solution (Kitzel, 2008).

1.4.5 LNG transportation

The next step in the LNG supply chain is transporting the liquefied natural gas to the regasification facilities. Primary modes of transport are by ships and trucks.

1.4.5.1 By ship

LNG is transported by specialized ships, LNG carriers, with insulated double-hulled tanks, designed to contain the cargo slightly above atmospheric pressure at a cryogenic temperature of approximately –259°F (–169°C). Typically, the storage tanks operate at 0.3 barg with a design pressure of 0.7 barg.

The tank design ensures integrity of the hull system and provides insulation for LNG storage. Because the insulation cannot prevent all external heat from reaching the LNG, some of the liquid boils off during the voyage. LNG vaporization is not homogenous: components with the lowest boiling point (nitrogen and methane) tend to evaporate more readily than heavier components. This phenomenon is called ageing or weathering and its consequence is that the LNG composition becomes heavier, and the heating value and Wobbe Index of LNG increases over time.

The boil-off gas, typically at the rate of about 0.10% to 0.15% of the ship volume per day, must be removed to keep the ship's tanks at a constant pressure. The boil-off gas can be used as fuel in the ship's dual fuel engines or burned in the boilers to produce steam or reliquefied and returned to the cargo tanks, depending on the design of the vessel. BOG reliquefaction can eliminate LNG shrinkage during long voyages and maintain the cargo's composition.

Containment systems for LNG carriers

The fundamental difference between LNG carriers and other tankers is the cargo containment and handling system. There are four LNG containment systems; two freestanding solid type structures, and two nonfreestanding (membrane) type designs.

Freestanding tanks. The freestanding or independent tanks are self-contained, usually spherical (developed by Moss Maritime of Norway) or prismatic in shape (designed by Conch International Methane Ltd), and made of aluminum alloy or 9% nickel steel with layers of insulation on the outside (Figures 1-7 and 1-8). Independent tanks are completely self-supporting and do not form part of the ship's hull structure. Moreover, they do not contribute to the hull strength of a ship (McGuire and White, 2000). The tanks are welded to cylindrical skirts or otherwise tied to supporters that are welded to the ship structure.

As defined in the IGC Code (International Code for the Construction and Equipment of Ships Carrying Liquefied Gases in Bulk), and depending mainly on the design pressure, there are three different types of independent tanks for gas carriers. These are ones built according to standard oil tank design (Type A), others that are of pressure vessel design (Type C), and, finally, tanks that are neither of the first two types (Type B). All LNG tanks are Type B from the Coast Guard perspective, because Type B tanks must be designed without any general assumptions that go into designing the other tank types. The self-supporting, prismatic Type B (SPB) tank is independent of the ship's structure and has the advantage over its spherical version of making maximum use of the available cargo space. However, they contribute significantly to weight and cost due to the fact that free-standing prismatic

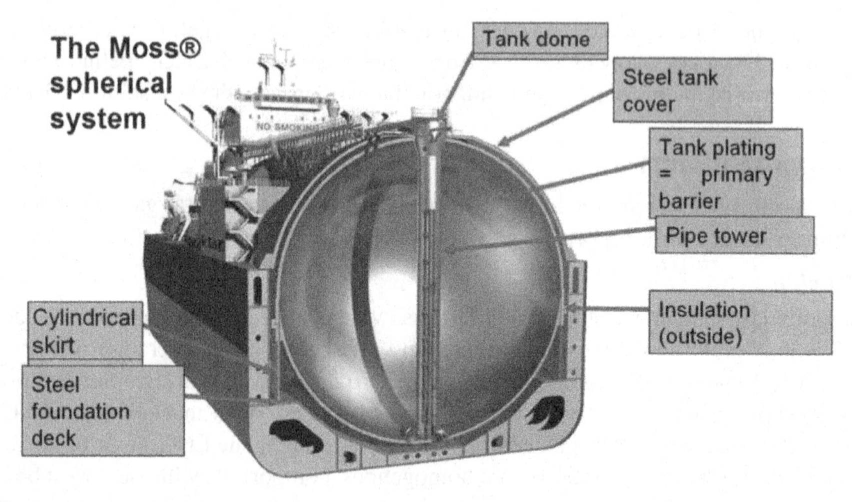

FIGURE 1-7

Free-standing spherical LNG tank.

(Courtesy of Moss Maritime)

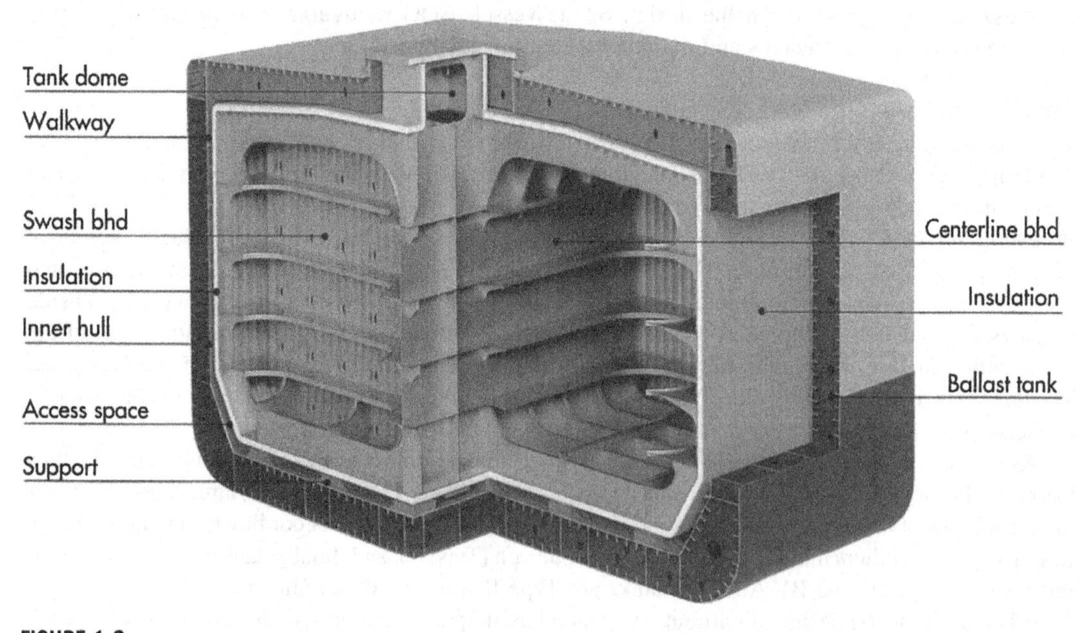

FIGURE 1-8

Typical illustration of free-standing prismatic type B LNG tank.

(McGuire and White, 2000; courtesy of IHI Corporation)

tanks include heavy plates and a considerable amount of bracing to keep the plates from distorting under hydrostatic load.

Membrane containment systems. Membrane tanks are non-self-supported cargo tanks surrounded by a complete double hull ship structure. The membrane containment tanks consist of a thin layer of metal (primary barrier), insulation, secondary membrane barrier, and further insulation in a sandwich construction (Figure 1-9). The membrane is designed in such a way that thermal and other expansion or contraction is compensated for without undue stressing of the membrane.

With the membrane design, the ship's hull, in effect, becomes the outer tank. Insulation is installed thereon, and a membrane placed on the inside to retain the liquid. The inner surface of this "double hull" is either high nickel content (36%) steel (Invar), offered by Gaz Transport, or 18% chrome/8% nickel stainless steel, offered by Technigaz (Figures 1-10 and 1-11).

The Gaz Transport membrane containment system (GT NO96) consists of a grillage structure made of plywood and filled with perlite in order to maintain tightness and insulation. On the other hand, the Technigaz membrane system (TG MARK III) consists of two layers of reinforced polyurethane foam separated by a material called triplex in order to configure an insulation system. Gaz Transport and Technigaz are now one company, whose latest containment system (Combined System Number One, CS1) incorporates features from the existing GT No 96 and TG Mark III systems. CS1 uses reinforced polyurethane foam insulation and two membranes, the first one 0.7 mm thick made of Invar, the second made of a composite aluminum-glass fiber called triplex. The system has been rationalized to make assembly easier and is prefabricated, allowing quick assembly onboard. However, this design has suffered from secondary membrane leakage problems in some vessels, but established shipyards have decided to maintain production of the GT NO96 & TG Mark III.

Selection of containment systems. Table 1-3 gives the comparative characteristics of different LNG containment systems. As all cargo tank system designs have proven safe and reliable in service, the choice of cargo tank design is primarily based on economics—price, delivery schedule, and shipyard

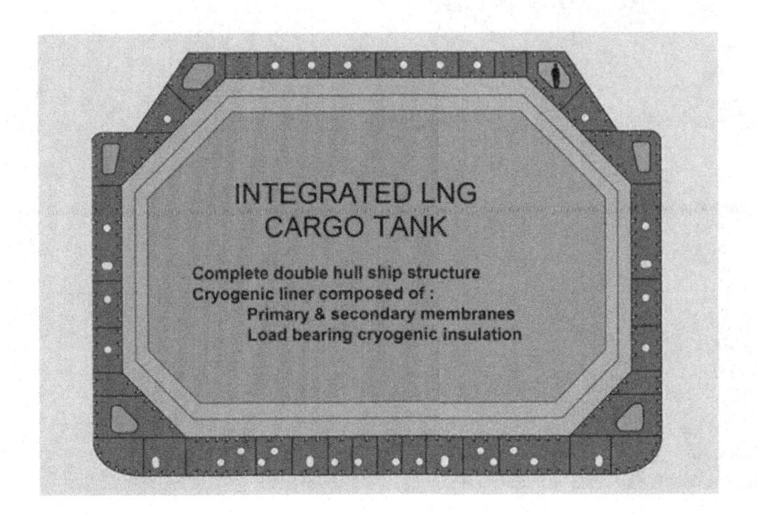

FIGURE 1-9

Generic membrane tank.

(DNV, 2011)

FIGURE 1-10

GTT NO96.

(Courtesy of GTT)

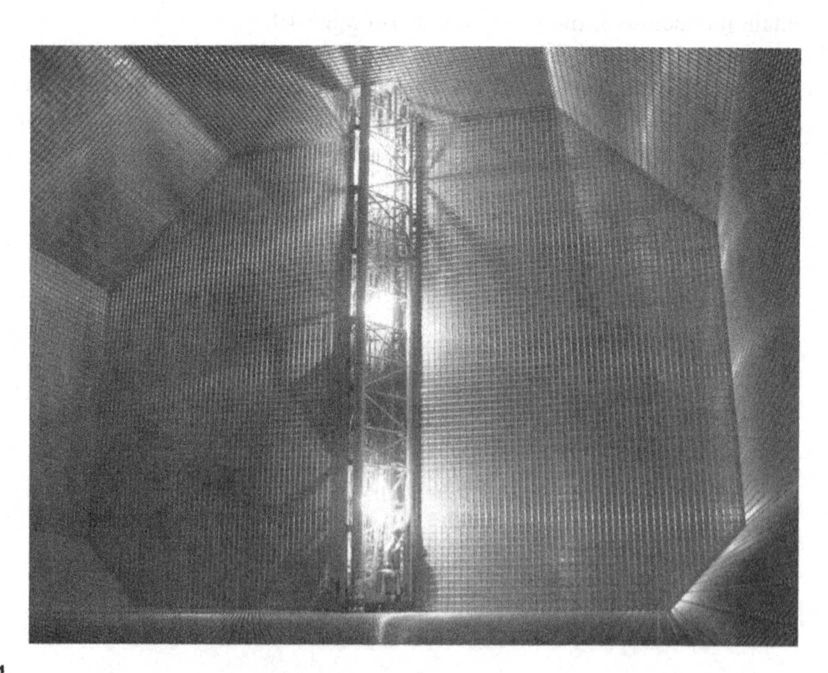

FIGURE 1-11

GTT Mark III.

(Courtesy of GTT)

Table 1-3 Comparative Characteristics of LNG Containment Systems

Characteristics	Freestanding Tanks		Membrane Tanks
	Prismatic	Spherical	
Safety in event of vessel grounding/collision or other emergency	Compared with membrane system less likelihood of hull damage being transmitted to cargo tanks. More efficient use of cubic space.	Safest system in event of grounding or collision-tank structure independent of hull and most void space between vessel hull and cargo tanks. Spherical tanks can be pressurized for emergency discharge in case of cargo pump failure.	Damage to hull of vessel may be more easily transmitted to tank structure than with freestanding tanks. Membrane systems are also more liable to damage or puncture due to causes such as surging of cargo in tank and entry of tank for inspection or repair.
Reliability of containment system	Most ship years operating experience and most experience without primary barrier failure. Structure can be analyzed and risk of fatigue failures minimized. Tanks can be constructed and 100% inspected prior to installation in vessel.	Tank system easiest to analyze structurally; therefore can be made most reliable.	Structure cannot easily be analyzed and therefore difficult to assure absence of fatigue failures. This could potentially lead to costly off-hire and repair time over the project life.

(OTA, 1977)

availability—rather than technical or performance criteria. In the last several years there has been a clear move toward membrane-type carriers, because membrane tanks utilize the hull shape more efficiently and thus have less void space between the cargo tanks and ballast tanks. More than three-quarters of the new LNG ships constructed in the decade 2001 to 2011 were of the membrane design due to their cargo capacity and capital cost advantages. However, self-supporting tanks are more robust and have greater resistance to sloshing forces,[2] which is an important design consideration for offshore storage.

LNG carrier cargo capacities and ship dimensions

LNG ships vary in size from less than 30,000 m^3 to some 265,000 m^3 cargo capacity but the majority of modern vessels are between 125,000 m^3 and 140,000 m^3 capacity (58,000 to 65,000 tonnes). The

[2]Sloshing is liquid motion within a tank that can be produced by wave-induced ship motions (e.g., ships at sea). Sloshing may resonate with structural frequencies and can impose stress loads on the internal tank structure, which may result in damage of the membrane structure.

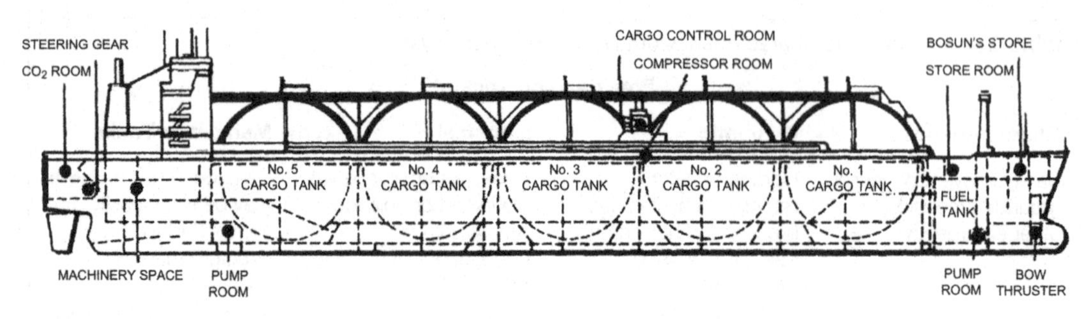

FIGURE 1-12

Inboard profile of LNG tanker.

(OTA, 1977)

industry standard some decades ago became the 125,000 m^3 LNG ship, which usually had five cargo tanks, each with a capacity of about 25,000 m^3 (Figure 1-12).

New larger LNG ships called Q-Flex (with capacity of about 216,000 $m^3$of LNG) and Q-Max (with capacity of up to approximately 265,000 $m^3$of LNG) were built to service long-distance supply chains from the large liquefaction trains commissioned in Qatar between 2009 and 2011.[3] These LNG carriers are propelled by slow-speed diesel engines, which are more efficient, easier to maintain and operate, and more environmentally friendly than traditional steam turbine drivers. These ships are also equipped with an on-board reliquefaction system that prevents losses of boil-off gas. While the Q-Flex and Q-Max LNG vessels are beyond the capacity of some ports, nearly half of the world's existing LNG terminals can, or with modifications could, accommodate these ships.

Typical dimensions of LNG ships quoted by Lee et al. (2008) are:

- Cargo Volume 138,000 to 173,000 cubic meters**:** length overall 277 to 290 meters; depth 26.0 to 26.5 meters, breadth 43.3 to 45.8 meters, number of tanks 4; single or twin propeller systems
- Cargo Volume 210,000 cubic meters (Q-flex): length overall 315 meters; depth 27 meters, breadth 50 meters, number of tanks 5; twin propeller systems
- Cargo Volume 263,000 cubic meters (Q-max): length overall 345 meters; depth 27 meters, breadth 55 meters, number of tanks 4; twin propeller systems

Small LNG ship

The recent LNG logistic chain research indicates a new market segment that requires the use of smaller LNG ships and smaller LNG receiving terminals. The shortage of clean fuel promotes the utilization of LNG for small- and medium-scale applications in developing countries (Mak et al., 2013). Two small LNG carriers with 2,500 m^3 capacity have been in service since 2004 for coastal LNG transportation in Japan. Currently, a Japanese shipyard is building an LNG carrier with 19,000 m^3 capacity. In Norway, coastal transportation of LNG consisting of small LNG ships with 1,000 m^3 capacity is being used to supply fuel to areas lacking gas pipelines.

[3]Q-Flex stands for Qatar and flexibility—a design enabling delivery to about 65% of existing LNG ports; Q-Max stands for Qatar and the maximum vessel size that could dock at the Qatar port facility—these may have access to approximately 50% of current LNG ports.

Unlike the larger vessels, the containment systems in smaller ships are designed with the independent Type C tanks, according to pressure vessel codes. The vessels were originally developed for ethylene carriers and can be upgraded to meet the LNG terminal requirements in an offshore environment. The type C tanks are typically not economical for large vessels due to the high cost due to wall thickness. On the other hand, the type C tanks are easy to fabricate and do not require a secondary barrier because of the pressure vessel design. The tanks can be fabricated outside the shipyard, making them more cost competitive.

A conceptual design of a 30,000m^3 LNG carrier using the IMO Type C tanks design is illustrated by TGE in Figure 1-13. The Bilobe tank arrangement is shown in Figure 1-14.

The containment can be designed for higher design pressures of 4 to 8 barg, which can be used to suppress BOG generation. In fact, some of the small carriers are not equipped with a BOG handling system, which is the simplest way to manage BOG by allowing pressure to increase during the journey. However, storage tanks in the receiving terminals must also be designed for high-pressure storage compatible with the ship, in order to avoid BOG losses during unloading.

The LNG industry is currently experiencing increasing activities for spot LNG cargoes. The introduction of small LNG ships will provide the flexibility and efficiency to accept and deliver smaller LNG cargoes expediently.

LNG ship propulsion systems

Traditionally LNG carriers use a boiler and steam turbine propulsion system that can consume the boil-off gas during transport. On the other hand, most of the ocean-going freight vessels are fitted with high efficiency two-stroke slow speed diesel (SSD) engines. Many new LNG ships built in the past decade have moved away from the traditional less efficient steam propulsion system. Recent high fuel costs may have made alternative propulsion systems more attractive (Lee et al., 2008).

- **Dual fuel diesel electric (DFDE) propulsion system:** Offers high fuel efficiency and freedom of fuel choice between fuel oil and BOG from the LNG cargo tanks. The DFDE system is typically

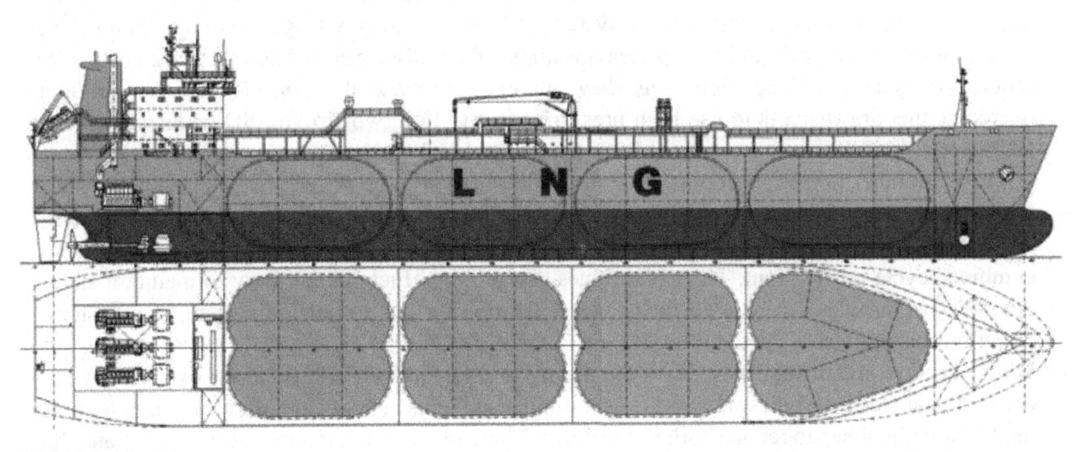

FIGURE 1-13

30,000 m^3 LNG carrier with IMO Type C tanks.

(Courtesy of TGE Marine Gas Engineering)

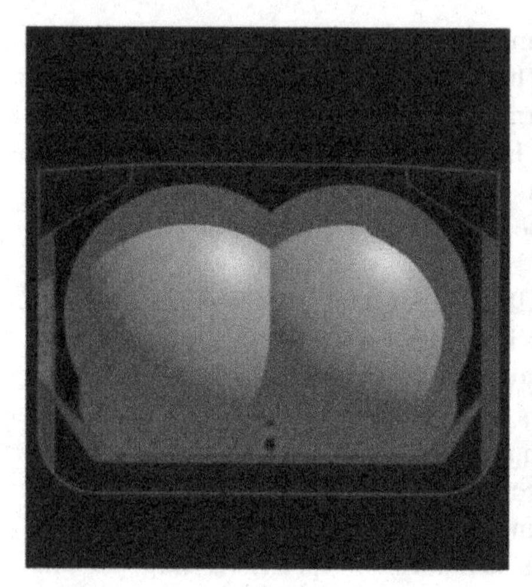

FIGURE 1-14

Bilobe tank arrangement for LNG carrier.

(Courtesy of TGE Marine Gas Engineering)

implemented with electric propulsion systems requiring high capacity electric switch gear, frequency converters, and electric motors, which require a crew with special skills to maintain. Compared to steam engines, the maintenance costs of the many cylinders involved in the typical four-stroke engine are higher.

For the large fleet of LNG carriers built for Qatar's Nakilat with more than 210,000 m^3 capacity, two-stroke slow speed diesel engines and onboard BOG reliquefaction system are used. The diesel engines are the same as those of most of commercial ships, with proven performance and reliability. The onboard reliquefaction system is ideal for long voyages avoiding boil-off gas losses. However, when liquid fuel prices are higher than LNG prices, the operating cost of this propulsion system will be higher, as they are unable to use the boil-off gas. An option to overcome this drawback is to use high-pressure gas injection two-stroke SSD engines. However, such propulsion systems have not yet been applied to LNG carriers.

There are other potential propulsion systems that are yet to be proven in LNG carriers.

- **Gas turbine propulsion systems using onboard BOG:** Proven drivers for refrigeration system and for propulsion. They are energy efficient especially in combination with steam turbines in a combined cycle power plant. The gas turbines can use dual fuel and are easy to maintain and operate. Gas turbines can be used for power generation for the propulsion system, or used for direct mechanical drive for the refrigeration compressors in FLNG vessels.
- **High pressure steam turbines:** Have the ability to improve the fuel efficiency over the conventional steam turbines. However, operating a high pressure steam system requires high quality water makeup to reduce boiler blowdown. The cost of the demineralization system and the weight of the boiler system add costs to the ship system. The fuel efficiency of the high pressure steam system, unless it is a supercritical pressure design, is no more than the diesel engines or the

gas turbines. It is unlikely that a high pressure steam system will be adopted in an LNG carrier propulsion system.

All propulsion systems include a gas combustion unit (GCU) for the disposal of excess BOG to avoid pressure buildup in the cargo tanks. The use of the GCU is typically required during the ship loading and unloading operations when the fuel demand is low or during the ship cool-down operation when excess boil-off gas is generated.

LNG ships for arctic service

The expected increase in the exploitation of gas fields in arctic, subarctic regions of Russia, and other Northern areas will precipitate the development of arctic LNG shipping. Ships navigating in ice areas perform quite differently due to ice resistance as compared to their open sea operations.[4] Therefore, it is a challenge for ship designers to find a design solution that not only optimizes the propulsion performance in open sea but also provides the ship with good ice performance (Lee, 2008).

Over the last five years there has been significant technical development in the ice-class tanker design with potential application to future Arctic LNG ships (Tustin, 2005). The first ice-class LNG vessels are about to enter service for the Sakhalin-II project in eastern Russia. Five new LNG ships will service the liquefaction terminal at Prigorodnoye in Aniva Bay: three built in Japan with the Moss-type independent tank and hulls designed to Finnish-Swedish ice class 1B standard; two ships built in South Korea, each with different membrane tank design. All five ships have their propeller and line shafting built to Russian Maritime Register of Shipping (RMRS) ice class LU2 standard and the membrane containment ships also have their ice-strengthened hulls built to that standard (Tustin, 2006). The performance of these vessels will provide an indication of the standards required for a more extensive Arctic LNG carrier fleet to withstand sea ice seasons of 100 days and more.

Plans to build regasification terminals along the St Lawrence River in Canada a few years ago suggest that some LNG ships may have to operate eventually in ice at both ends of their routes. The power installed and the ice class of vessels plying the more challenging Arctic routes, such as to the western Arctic coastline of Russia, will need to be higher unless dedicated ice breaker vessels are commissioned to assist them. Winterization features include low-temperature-proof materials for the deck equipment on the vessels themselves and for the loading and unloading facilities. The ships will have to withstand severe wave conditions and persistent cold environments. Carriers of membrane containment design will need reinforced tank supports to avoid cargo sloshing damage. Indeed membrane designs will need to prove their reliability in such challenging conditions for arctic service. LNG ships built for dedicated service to the Snohvit LNG facility in Northern Norway (ice free all year) are all of the Moss-type design.

The challenges associated with first-year ice navigation and those with multiyear ice navigation are very different. Multiyear ice is prevalent in the Kara Sea and for year-round navigation requires ice breaker assistance with typical hull structure designed for level ice sheet thickness varying from 120 cm to 170 cm thickness in the summer and autumn seasons and 170 cm to 320 cm thickness (with hummocks) in the winter and spring seasons (Tustin, 2006). Movement in such winter conditions requires very powerful engines (85 to 120 MW), narrower beams, and strong propulsion equipment pushing ice breaking hulls moving at times as slowly as two nautical miles per hour (Scherz, 2006).

[4]The presence of first year and multiyear ice imposes additional loads on the hull, propulsion system, and appendages.

FIGURE 1-15

Arctic LNG Shuttle Hoegh LNG.

(Courtesy of Höegh LNG)

Even the highest ice-classed LNG vessels will need to have ice breaker assistance. The vessels and support services will not only be expensive, but the periodic slow speeds along the most challenging parts of their routes will require more tankers to transport similar contract quantities than for ice-free supply chains. Shuttle-tanker methodologies might make sense in some cases; for example, ice-classed tankers to move cargoes past the ice edge either to trans-shipment ports or for ship-to-ship transfer. The reality is that each port and shipping route will probably pose its own challenges and require tailored vessel design solutions (Wood and Mokhatab, 2009). Figure 1-15 shows a schematic image of the Arctic LNG Shuttle (ALS) concept that is under development by Hoegh LNG in collaboration with LMG Marin.

Note, low temperatures impact the ship, and the cold, lack of light, and visibility affect the crew. To help address these challenges, the American Bureau of Shipping (ABS) has produced a guide for vessels operating in low temperature environments. Guidance is provided for the preparation of vessels and other marine structures and their crew for operation in harsh environments (Conachey, 2006).

1.4.5.2 By truck

Where consumers are located in coastal areas, LNG can be delivered by ship. But when they are located inland, the only viable method is by trucking with mobile equipment such as road trailers (Figure 1-16), ISO intermodal cryogenic containers, or smaller delivery units.

On the production side, liquefaction of small scattered sources of natural gas or biogases in remote areas would not be economical where there are no close-by users. While the reservoir capacity and life may not justify construction of long pipelines, liquefaction and trucking of LNG presents an economically viable solution.

FIGURE 1-16

Typical LNG truck.

On the consumer side, the price of natural gas remains historically low, due to the rapid development of shale gas in North America, which makes natural gas a clean and low cost fuel. LNG trucking is a mature industry. Using specialized, double-skinned vacuum insulated tank trucks, liquefied natural gas can be reliably and safely delivered to LNG refilling stations. At the satellite stations, LNG is unloaded into insulated pressurized storage tanks. Under normal conditions, the vacuum insulated tank can store LNG for long periods without venting (Chrz and Emmer, 2007).

For domestic gas consumers, LNG is pumped, vaporized with ambient air vaporizers, odorized, metered, and delivered to the local pipelines.

Commercial truck fleet operators are now making the switch to the cheaper natural gas from the high priced diesel. Conversion of a diesel engine to use natural gas is simple using a bifuel conversion kit. Such conversions are being done on drilling rigs for shale oil production and can save fuel costs and reduce emissions. Globally, there is a move toward building LNG refilling station infrastructure to reduce energy cost.

The main markets for LNG refueling stations are to serve vehicle fleets, operating directly on LNG onboard. Heavy-duty trucks, busses, refuse vehicles, and frequently operated fleet vehicles such as taxis are typical vehicles with great potential for LNG fueling. Compared to much more traditional CNG systems, LNG fuel has the advantages of high fuel efficiency, lower vehicle dead weight and longer runs because of its high density and low pressure. Centrifugal pumps ensure the delivery of LNG into vehicles engine with cleaner burning and higher efficiency than conventional engines.

LCNG stations can use LNG to support the operation of existing CNG vehicles. High-pressure piston pumps deliver LNG into atmospheric vaporizers and buffers so that CNG vehicles can be refueled quickly. LNG/LCNG stations are mostly combined for fueling both versions of vehicles. Underground tanks are a typical option for urban areas to conserve real estate requirements and to meet regulations.

Currently in the US, there are several companies building infrastructure to support the LNG fuel markets, including Clean Energy, BLU and Shell to take advantage of the low natural gas prices. The truck fleets using LNG rather than diesel can significantly reduce fuel costs and reduce air emissions by over 25%. The LNG units can produce 100,000 to 250,000 gallons of LNG per day, or about 60,000 to 1,500,000 gallons of diesel per day.

1.4.6 LNG receiving terminals

LNG carriers deliver LNG to a receiving terminal, which then returns LNG to a gaseous state. The natural gas is transmitted to the natural gas customers by means of distribution pipelines.

Historically, onshore LNG terminals are close to densely populated areas and industrial areas, where a diverse range of customers are located. However, large tracts of land with adequate marine access for LNG ships are difficult to locate in densely populated areas. Building terminals close to dense population would raise the environmental and safety/security concerns of local communities. Protracted planning and regulatory approval for a new LNG terminal permit are time consuming and very costly. Offshore LNG receiving terminals are an alternative to circumvent these difficulties.

The decision to go offshore with respect to LNG receiving terminal design and operation hinges on a number of considerations. Although offshore/floating LNG terminals may appear to offer many advantages over onshore terminals, they also introduce new complexities, risks, and questions about feasibility, where only a few offshore LNG terminals have actually been achieved to date. However, as technologies are maturing, technical and economic uncertainty will be reduced and some of the resulting risks mitigated.

1.4.6.1 Offshore LNG terminals

An offshore LNG terminal receives LNG from ocean-going vessels, regasifies the LNG either immediately or subsequent to being stored, and delivers the LNG to the onshore customers through a subsea pipeline.

There are basically two fundamental concepts for offshore LNG terminals: Gravity Base Structures (GBS), and Floating Storage and Regasification Units (FSRU). The design selection depends on the site conditions, such as depth of water, subsea soil, sea state, and the sendout capacities (Kulish et al., 2005).

A GBS is a fixed concrete structure laying on the sea floor. The structure is installed with the LNG storage tanks and regasification equipment. The Adriatic LNG terminal located 9 miles (14 km) offshore near Rovigo, Italy in the northern Adriatic was the world's first offshore gravity-based structure LNG regasification facility. It is designed around a large concrete structure in 95 ft of water, which houses two LNG storage tanks, a regasification plant, and facilities for mooring and unloading LNG vessels. The terminal is operated by Qatar Terminal Limited, a subsidiary of Qatar Petroleum, ExxonMobil Italiana Gas, and Edison. The terminal is 47 meters (154 ft) high, 88 meters (289 ft) wide, and 180 meters (590 ft) long.

An FSRU is an LNG ship that can be a custom design or an existing carrier modified to include the regasification facility. They are floating structures, either moored to the seabed via a turret mooring system or tethered to a jetty in a port area. The GBS design is more permanent and takes longer to build, and the capital cost is higher. However, the GBS design can be expanded for higher throughput while FSRU is limited by the real estate on the ship.

1.4.6.2 Onshore LNG terminals

A typical onshore LNG receiving terminal process schematic is shown in Figure 1-17. Figure 1-18 also shows the process units and the plant layout of a typical onshore LNG receiving terminal. As can be seen from Figure 1-17, LNG is unloaded by means of the ship pumps to the unloading arms on the Jetty and then to the storage tank through the unloading lines. It is then pumped to high pressure through various terminal components where it is warmed in a controlled environment. The LNG can be heated by several methods including direct-fired heaters, seawater, heated water, or air. Once regasified, the natural gas is delivered into the distribution pipelines to the different uses or power generation stations.

The process of an LNG receiving terminal is further described in the following.

LNG ship unloading

In most LNG terminals, ship unloading usually takes about 12 hours for a 145,000 m^3 LNG carrier with an average unloading rate of 12,000 m^3/hr. LNG is typically unloaded into a single tank. There are other activities before and after unloading, such as turning basins, berthing, preparation for unloading, and departure. The total time that the ship stays at port is about 24 hours.

Three LNG unloading arms plus one vapor return arm and a hydraulic arm package are provided on the jetty. One of the LNG arms can also be used as a vapor return arm during an emergency.

A single large unloading line and a small recirculation line are typically installed from the jetty to the storage area. If necessary, the recirculation line can also be used for unloading LNG. During the

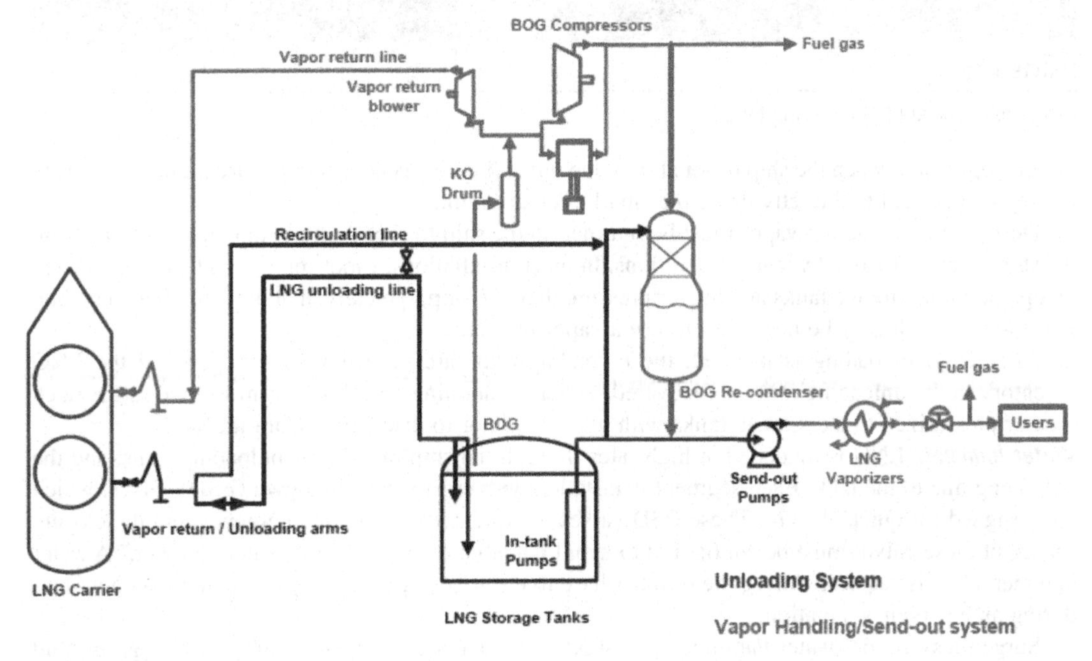

FIGURE 1-17

Typical process scheme of an LNG receiving terminal.

(Lemmers, 2009)

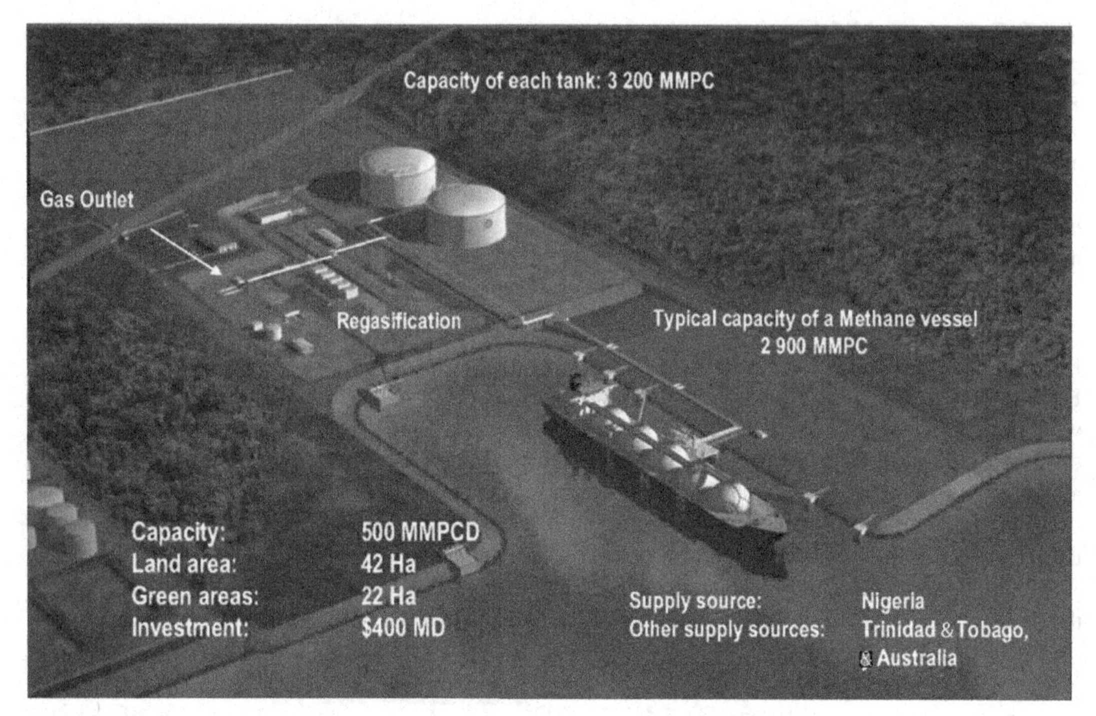

FIGURE 1-18

Model of a typical LNG receiving terminal.

holding operation, when the ship is not at port, the unloading line is kept cold by circulating LNG from the sendout system to the jetty using the small circulation line.

During ship unloading, vapor must be returned to the ship to replenish the unloaded volume from the ship, in order to avoid vacuum conditions. In most installations, vapor return can be accomplished by operating the storage tanks at a higher pressure than the ship, typically at about 8 to 10 KPa higher. For a long jetty, it may be necessary to use a vapor blower.

After LNG unloading is finished, the unloading arms are purged with nitrogen and the LNG inventory in the unloading arms is recovered, either by draining to a drain drum located in the dock area or pressurized to the storage tanks with nitrogen, prior to disconnect from the ship.

Water hammer. LNG is unloaded at high rates through the ship manifolds, unloading arms, and the unloading line to the tank. Each segment is installed with emergency shutdown (ESD) valves, which are designed for quick closure. These ESD valves are initiated during an emergency and the closure timing of these valves must be configured to avoid potential damages from "water hammer." A water hammer study is required during the design phase to ensure design integrity of the unloading system during an emergency situation.

Surge pressure, or "water hammer," is a short-term increase in pressure due to a change in fluid velocity in a pipeline. In a water hammer study, it is assumed that fluid hammer occurs when a valve is suddenly closed and the fluid continues to move away from that valve because of its momentum. A vapor cavity will appear on the downstream side of the valve, and when the fluid runs out of momentum, the vapor cavity will collapse, causing water hammer. Most frequently in a receiving terminal, water hammer is caused by halting the LNG flow by closure of the tank inlet valve. The valve closure sends a pressure wave that can generate a "water hammer" incident.

The water hammer scenarios can be analyzed by transient simulation tools, using software such as AFT Impulse. With this tool, the piping design and ESD valve design can be configured to avoid the potential damage from water hammer.

LNG storage

LNG is stored at atmospheric pressure in double-walled, insulated tanks that are designed for storing the liquids at cryogenic temperatures. The insulation is designed to minimize heat gain and reduce product losses due to boil-off. The boil-off rate from a typical tank is about 0.05 volume % per day.

The tank capacity is typically 160,000 m^3, matching the size of an average LNG carrier. Large tank capacity with 200,000 m^3 or higher capacity is being built and may become the norm for new terminals in order to match today's large LNG carriers. Typical LNG tank operating conditions are shown in Table 1-4.

LNG storage tanks of various designs are employed around the world. Selection of LNG tanks is project specific. They should address site conditions, design criteria, safety, geological considerations, environmental requirements, and applicable design, codes, and regulations.

There are two main types of LNG storage tanks: in-ground storage tanks and above ground storage tanks.

In-ground tank. An in-ground tank consists of a stainless steel membrane, supported by rigid polyurethane foam insulation (Figure 1-19). This, in turn, is supported within a reinforced concrete caisson. The roof consists of a dome-shaped carbon steel structure supporting a suspended deck with glass wool insulation.

In-ground tanks are less visible in their surroundings and more secure from an antiterrorist standpoint. There is no risk of spillage with this high-integrity storage design.

It is also more earthquake-proof as the seismic motion is not amplified in the underground tanks compared to the aboveground counterpart. They are common at receiving terminals located at seismically sensitive areas like Japan, South Korea, and Taiwan. With the earth berm, the tanks can be located close to one another, which is an advantage where land and space are limited.

In-ground tanks were developed by Tokyo Gas Engineering (TGE) in the early 1970s, based on earlier designs in the United Kingdom, United States, and Algeria, and subsequently used by other Japanese companies. As of 2005, there were 61 in-ground storage tanks in Japan. The record for the largest LNG tank in the world was first set by an in-ground tank (200,000 m^3), although several aboveground tanks have recently been built with a similar capacity. These tanks are more expensive

Table 1-4 Typical LNG Storage Tank Design and Operating Pressures

Maximum Design Pressure	30 KPag
Vacuum Design Pressure	−1.5 KPag
Normal Operating Pressure	10 KPag
Minimum Operating Pressure	2.5 KPag
Maximum Operating Pressure	25 KPag

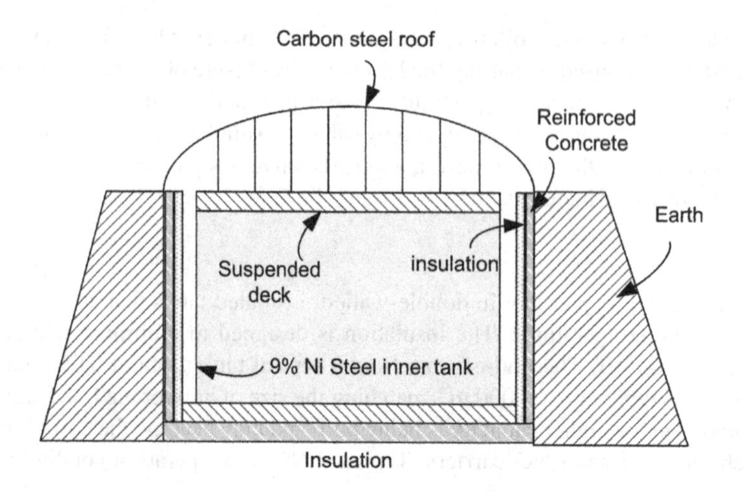

FIGURE 1-19

Typical earth berm LNG tank.

and take longer to build than aboveground tanks—about 4 to 5 years compared to 3 years for a tank built above ground.

Aboveground tanks. Aboveground LNG tanks have two layers of containment. The primary containment is provided by the inner tank, which holds the LNG. Secondary containment is provided either by the use of dykes, berms, and impoundment dams around storage tanks, or by building a second tank around the primary storage tank to contain the LNG, which will protect against failure in the primary tank. All LNG storage tanks are constructed with thermal insulation to minimize heat transfer, reduce boil-off vapors, and protect the carbon steel materials from reaching cryogenic temperatures.

The containment system is designed in compliance with LNG codes and standards, which provide guidelines for material selection and design requirements for LNG storage tanks and other equipment at LNG facilities.

There are basically three tank types used for onshore terminals:

- Single containment tank
- Double containment tank
- Full containment tank.

Single containment tank A single containment tank is composed of a self-supporting inner cylindrical container made of 9% nickel steel (Figure 1-20). This inner tank is surrounded by an outer tank made of carbon steel, which holds an insulation material (perlite) in the annular space. The carbon steel outer tank is not capable of containing cryogenic materials; thus the inner tank provides the only containment for the cryogenic liquid. However, single containment tanks are surrounded by a bund wall (constructed of prestressed concrete) or dyke external to the tank, which provides the secondary containment in the event of inner tank failure, although vapor would not be contained. In case of failure, vapor dispersion and flame radiation, if ignited, could result in serious damage to surrounding equipment and structures. The land requirements are greater than for other tanks because of the separation distance between the tank and the bund wall.

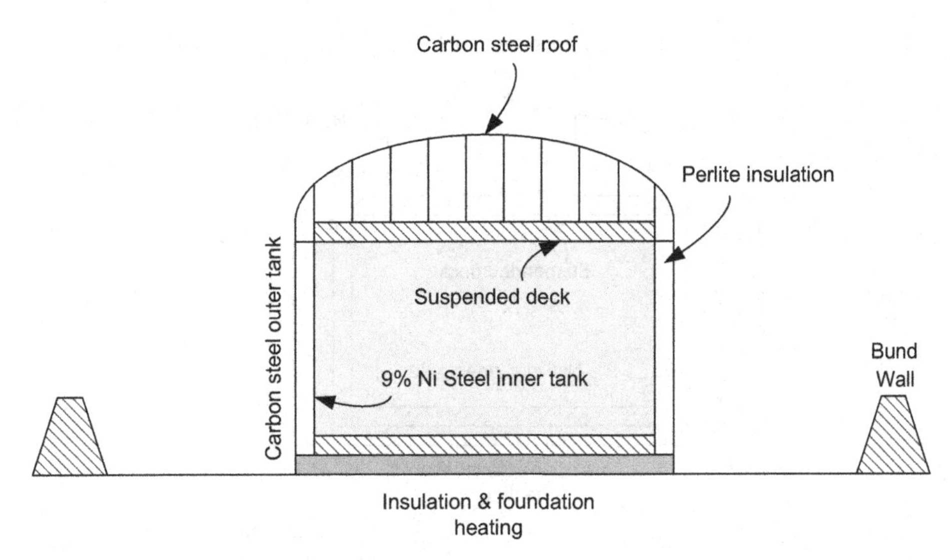

FIGURE 1-20

Typical single containment LNG tank.

This type of tank is the lowest cost option, which has been successfully used in the past. However, because it is more prone to external hazards than other types, insurance premiums are typically higher than the full containment, which penalizes the cost advantages.

Double containment tank The double containment tank is similar to a single containment tank, with the addition of constructed walls as the secondary containment instead of a containment dyke (Figure 1-21). Therefore, if the inner tank fails, the secondary container is designed to contain the cryogenic liquid. The outer wall also limits dispersion of the LNG vapor. The outer concrete wall increases the cost of the tank, but less space is required because the containment dyke is no longer necessary.

Full containment tank A full containment tank is a double containment tank in which the annular gap between the outer and inner tanks is sealed (Figure 1-22). The majority of LNG storage tanks built in the last 10 years worldwide have been designed as full containment tanks. In this tank, the secondary container is liquid and vapor tight. In case of inner tank failure, the secondary container remains LNG tight. The secondary container wall is generally made of post-stressed concrete and the roof is usually reinforced concrete.

The full containment tank has provisions for top connections only. There are no connections for pipelines or instruments penetrating the sides or bottom of the tank. All connections, such as the top fill line, pump wells, pump discharge, and vapor recovery, are permitted only through the inner tank. Top filling is done via a splash plate device supported from the roof of the tank to obtain good flashing and mixing of the LNG into the tank. This ensures that any superheat associated with the incoming LNG is released by flashing at the tank pressure during filling.

Full containment tanks cost from 10 to 20% more than single containment tanks (Kulish et al., 2005). However, this type of storage has the advantage of an additional layer of safety against external elements such as fire, blasts, and atmospheric impacts. The full containment tank design is very

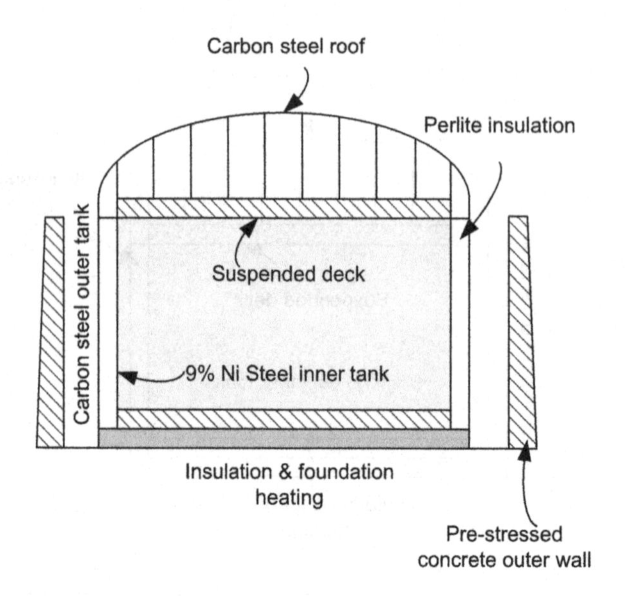

FIGURE 1-21

Typical double containment LNG tank.

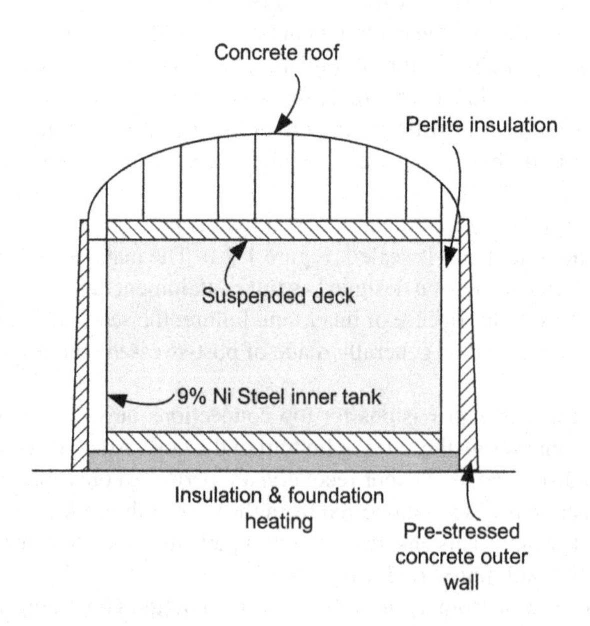

FIGURE 1-22

Typical full containment LNG tank.

compact and is currently the acceptable selection for most projects where land availability, location, local regulations and/or security do not permit the use of a single-containment design.

Due to the high costs and long schedule requirements of constructing traditional storage tanks, new LNG storage designs continue to be developed. Two of the advanced solutions, which potentially can reduce construction time and costs, are the All-concrete LNG (ACLNG) tank developed by Arup Energy (Powell and Thomas, 2008) and the concrete/concrete (C/C) LNG tank presented by Statoil Hydro (Skovholt, 2008).

Storage tank operation. The storage tanks are filled with top connections during LNG unloading. Internal piping is configured to permit top and bottom loading. The top loading operation is typically carried out via a spray device/splash plate to promote flashing and mixing of the unloaded LNG with the LNG inventory in the tank. The bottom loading operation is carried using a standpipe that directs the liquid to the bottom of the tank. In general, a lighter LNG is unloaded into the bottom of the storage tank and a heavier LNG is unloaded from the top.

Care must be taken to avoid subcooling the vapor space in the storage tank. Overcooling may result in a vacuum inside the tank, causing damage to the tank structure. The tank temperature and pressure must be monitored during the tank cooling operation. Tank cool-down is usually carried out before ship unloading to minimize excess vapor boil-off during unloading. The tank must be cooled with the spray and the vapor space temperature kept below –130°C before LNG can be unloaded into the tank.

LNG tank rollover. Top-filling with LNG of varying densities could result, in exceptional circumstances, in stratification of the LNG inside the tank, which may eventually lead to tank rollover. In another instance, when LNG inventory is stored for a long duration, the top layer of liquid is slowly weathered by vaporization in the form of boil-off gas, and its composition becomes richer and heavier than the lower layer. As shown in Figure 1-23, there may be two or more cells of liquid formed with

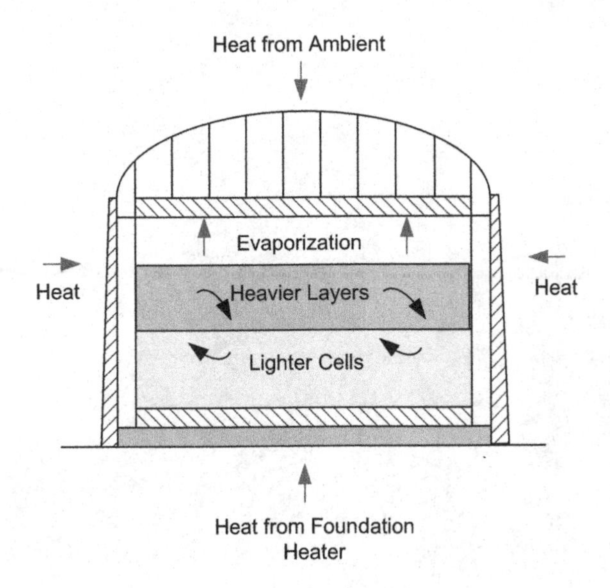

FIGURE 1-23

Heat transfer inside LNG tank.

different densities, and when the top layer is heavier than the lower layer, and under a stagnant environment, rollover can occur. Rollover, the spontaneous mixing of these two layers, will result in the release of a significant amount of vapor in a short time, which may result in overpressure that can lift tank relief valves. More detailed discussions on the rollover phenomenon can be found in Chapter 9 (Section 9.2.5).

Some of the measures to prevent tank rollover are:

- Store liquids of different density in different storage tanks.
- Load storage tanks through nozzles or jets to promote mixing.
- Use filling pipework at an appropriate level in the storage tanks.
- Monitor the LNG shipment cargo density and temperature before unloading.
- Monitor liquid density and temperature profile in the tanks.
- Monitor storage tank boil-off rates.
- Monitor storage tank pressure.
- Transfer contents to other tanks.
- Recirculate the content within the storage tanks using the LP sendout pump.

LNG regasification

LNG arms. LNG arms are used for LNG transfer between LNG ship and onshore tanks (i.e., marine vessel unloading). Each arm consists of the riser pipe, the inboard arm, and the outboard arm. The diameter of the arm varies between 16 and 24 in (Figure 1-24). Typically an LNG swivel joint is installed between each pipe segment, which is used to accommodate the changing positions between the onshore and the ship components. Emergency release systems, quick connect/disconnect coupling (QCDC), and position

FIGURE 1-24

Typical LNG arms.

monitoring systems are provided to ensure safe operation of the arms. A nitrogen purge is also provided for each arm for draining the content prior to disconnection from the ship.

The capacity of a loading arm varies between 4,000 m^3/hr and 6,000 m^3/hr. Typically, there are two or three liquid loading arms, a vapor return arm, and a common spare arm for liquid/vapor.

LNG pumping. The submerged electric motor pump (SEMP) has been used exclusively in most of the LNG applications. The pump motor is submerged along with the pump, since LNG is a dielectric fluid, electrical cables and the motor can be safely submerged in the liquid. The other safety advantages of a submerged pump over external motor pumps are that there is no penetration required in the tank and mechanical seals are not needed. Several pump vendors are available for this service, such as Ebara, Atlas Copco, Nikkiso, Hitachi, and Shinko.

LP sendout pump The low pressure (LP) sendout pump (Figure 1-25) is installed in a vertical pump well inside the tank. The pumps are retractable and are installed from the top of the LNG storage tanks. The pump weight opens a spring-loaded suction valve, which is used for pump removal and installation. The pump column is purged with nitrogen to ensure safety.

The LP sendout pump typically is designed to discharge at 8 to 10 barg pressure. A typical pump curve is shown in Figure 1-26. The pump should be operating at close to the design flow for maximum pump efficiency. When the pump flow is operated under turndown, pump efficiency will drop, and the temperature of the discharge liquid will increase, which may cause problems in the BOG recondenser.

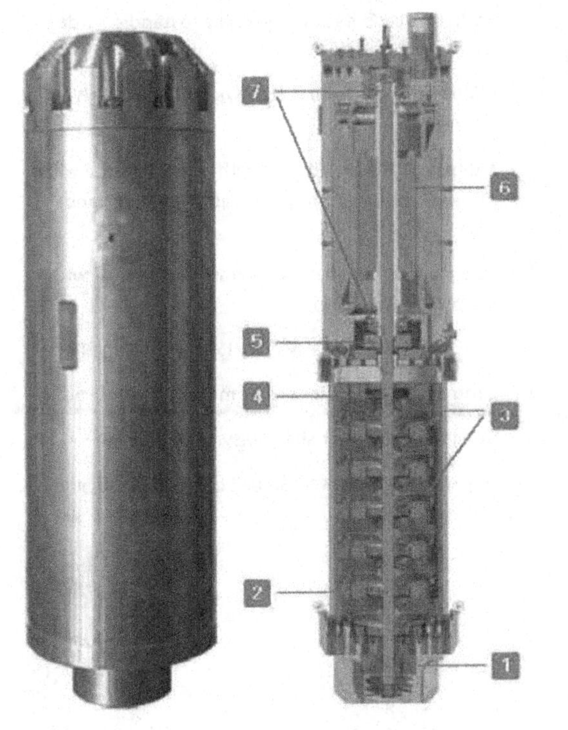

1. Atlas Copco JC Carter Pumps patented HyPerInducer® has the capability to handle fluids with high vapor fractions
2. Multistage radial diffusers: wide, stable operating range
3. Multistage impeller design with collets, bronze wear rings and bushings: high pressure, high reliability
4. One piece pump shaft: a constant diameter provides dimensional stability
5. Active thrust balance system: longer bearing life
6. Robust hollow rotor shaft: easy maintenance
7. Upper/lower AISI 440C SS ball bearing: reliability

FIGURE 1-25

In-tank and suction mounted pump.

(Courtesy: Atlas Copco JC Carter web site)

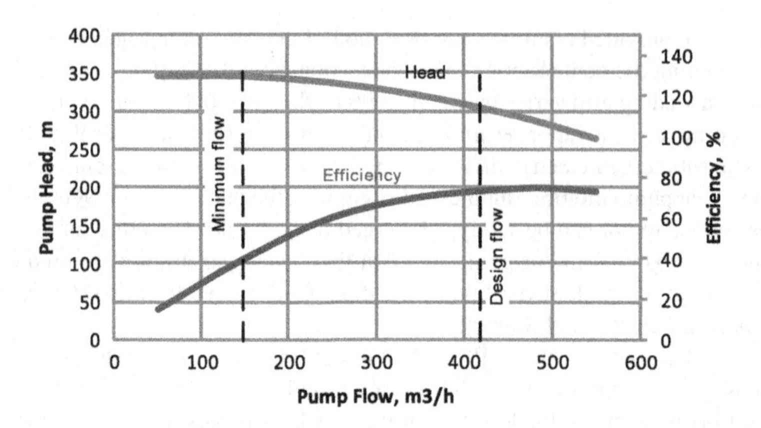

FIGURE 1-26

Typical LP sendout pump curve.

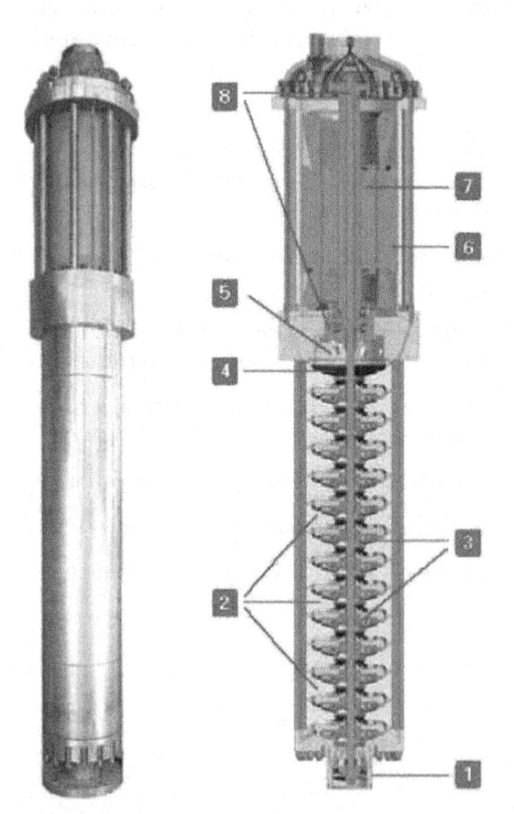

1. Atlas Copco JC Carter Pumps patented HyPerInducer® has the capability to handle fluids with high vapor fractions
2. Multistage continuous crossover diffusers: high efficiency
3. Multistage impeller design with collets, bronze wear rings and bushings: high pressure, high efficiency, high reliability
4. One piece pump shaft: a constant diameter provides dimensional stability
5. Active thrust balance system: longer bearing life
6. Long aspect ratio/slim profile motors: high efficiency
7. Robust hollow rotor shaft: easy maintenance
8. Upper/lower AISI 440C SS ball bearing: reliability

FIGURE 1-27

Third-generation high-pressure, pot-mounted pump.

(Courtesy: Atlas Copco JC Carter brochure)

During turndown there may not be sufficient LNG Cold for BOG recondensation, which may result in BOG carryover to the HP sendout pump, causing a pump cavitation problem.

During pump startup, the minimum flow valve will be open to the tank and the pump heat will generate additional boil-off gas that must be recondensed by the BOG recondenser.

In addition to supplying flow to the recondenser and feeding the HP pump, the LP sendout pumps are also used to circulate cold LNG to the jetty to maintain the cold unloading lines during the holding operation. The pumps can also be used for mixing LNG in the tanks to avoid tank rollover.

HP sendout pump The high pressure (HP) sendout pumps take suction from the BOG recondenser typically at about 8 barg pressure and increase the pressure to meet the sendout pressure requirements, typically 80 to 120 barg. The HP sendout pumps (Figure 1-27) are vertical canned type pumps, with submerged motor that drives the closed-coupled, multistage pump. The pump is very compact and is equipped with an inlet inducer to reduce the net positive suction head (NPSH) requirement for operation. A typical pump curve is shown in Figure 1-28.

Heat generated by the motor is mostly removed by the sensible heat of the inlet LNG. A small portion is vaporized, which must be vented from the pump casing to avoid the pump from being vapor locked. The vent vapor typically is sent to the BOG recondenser. Because of the fact that the BOG condenser and the HP sendout pumps are closely coupled, and that the BOG recondenser liquid is at saturation, the design of the piping system, including elevation of the BOG recondenser, must be carefully analyzed to ensure sufficient NPSH available to operate the pumps. This system is critical in the regasification terminal operation and must be designed for the different pump operations, including startup and shutdown, ship unloading, and holding operations.

For a base load terminal, the HP pump is a multistage high-head pump driven by a large motor typically with 1500 to 2000 kW rating. Startup and shutdown of this pump requires the minimum flow bypass on the pump discharge to open to prevent the pump from overheating. The pump minimum flow typically is returned to the BOG recondenser, which must be designed with sufficient surge to hold the minimum flow volume while maintaining the vapor recondenser operation.

Boil-off gas recondenser system. The vapor flows from the LNG tanks vary significantly between ship unloading and holding operations. These vapors must be compressed, removed, and condensed in the BOG recondenser to maintain the storage tank at a low pressure.

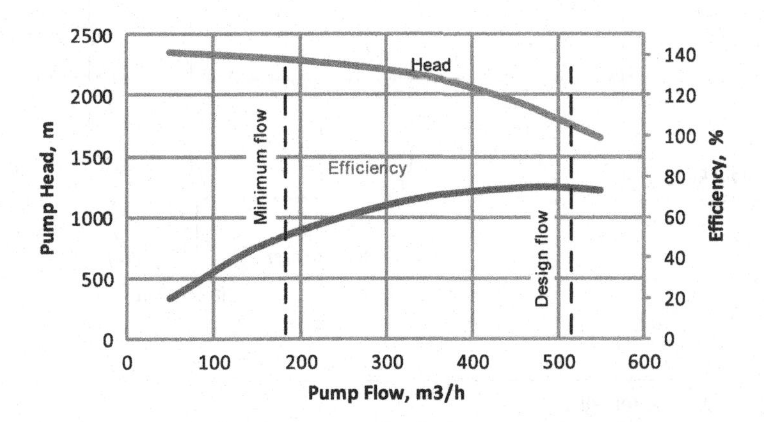

FIGURE 1-28

Typical HP sendout pump curve.

Vapor flow during ship unloading is contributed to by the heat input from the ship unloading pumps, heat gain in the unloading lines, heat leak to the storage tanks, and volume displaced from the tank by the unloaded LNG. A portion of the vapor is returned to the ship, and the majority is recondensed in the BOG recondenser.

During the holding operation, when the ship is not at port, vapor flow is significantly lower. BOG is generated mainly due to heat leak to the storage tanks, typically 0.05 volume % per day. When the plant operates at high sendout, the BOG flow is further reduced and can become negative such that makeup vapor from the natural gas sendout system may be required.

Typically, the vapor rate varies from 4 to 40 MMscfd between holding and ship unloading operations. These vapor flows are compressed by the BOG compressor and condensed in the BOG recondenser using a slip stream of the cold sendout LNG.

The BOG recondenser serves the following functions:

- Condensing the BOG, maintaining the storage tanks at a low pressure
- Serving as a mixing drum and suction drum to the HP sendout pump
- Maintaining a constant suction pressure and providing sufficient NPSH to operate the HP sendout pump.

A typical BOG recondenser system is shown in Figure 1-29. BOG from the storage tank is first compressed by a BOG compressor to about 8 barg. The compressor can be a centrifugal or a

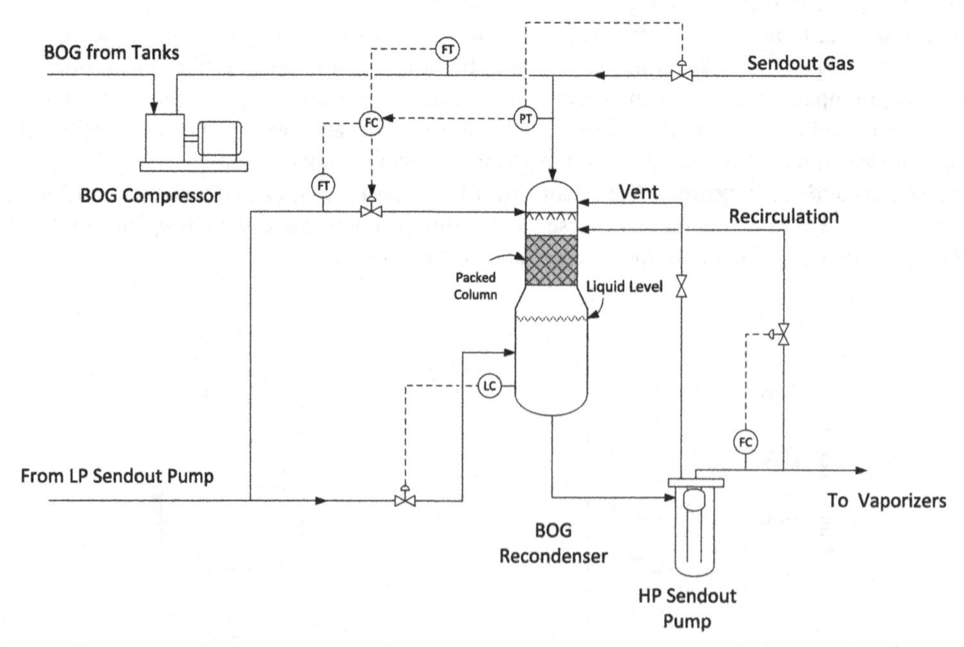

FIGURE 1-29

BOG recondenser system schematic.

(Courtesy of Fluor)

reciprocating compressor. Typically two or more reciprocating compressors equipped with automatic unloaders are used to meet the BOG turndown requirements.

The BOG recondenser is designed with a packed section for vapor and liquid contact. Sufficient LNG must be supplied to condense the BOG to produce a saturated liquid. The balance of the sendout flow is mixed with the condensate from the packed column in the lower section of the recondenser.

The process design and control system of the BOG recondenser/HP sendout pump system must ensure a stable flow with sufficient subcooling and NPSH for the HP sendout pump operation. Some of the key control parameters are:

- Liquid flow rate to the packed column is controlled by a flow ratio controller based on the BOG flow rate. The flow ratio is reset by the pressure controller on the recondenser.
- The main LNG sendout is fed to the lower section of the recondenser under level control on the recondenser. This ensures sufficient liquid is available to the HP sendout pump.
- The liquid level in the recondenser should be kept below the packed section, allowing all the packing to be available for heat transfer.
- The vent from the HP sendout pump is routed to the top of the recondenser, and must be free drained back to the pump suction, to avoid liquid buildup in the vent header.
- HP sendout pump recirculation flow is returned to the recondenser, which must be designed with surge volume.
- Elevation difference between the recondenser and the pump must be sufficient to meet the NPSHR of the HP sendout pump even during turndown operation.

Vaporizers. The optimum choice of an LNG vaporization system is determined by the terminal's site selection, the environmental conditions, regulatory limitations, and operability considerations. It has to comply with the LNG industry's requirements for minimizing life cycle costs. The selection should be based on an economic analysis in maximizing the NPV of the project and meeting the emissions requirements.

Today, the base load regasification terminals use two types of proven vaporizers: Open Rack Vaporizer (ORV) at about 70% and Submerged Combustion Vaporizer (SCV) at about 20%. In addition to these vaporizers, several other types of vaporizers have been used. These include the Ambient Air Vaporizers (AAV), the Shell and Tube Exchange Vaporizers (STV), and the Intermediate Fluid Vaporizers (IFV).

While the ambient air vaporizers option appears to be desirable in terms of operating costs and environmental considerations, it is also the more expensive option, requiring more real estate, and is more sensitive to ambient condition changes. The vaporizer selection is project- and site-specific and must be evaluated on a case-by-case basis.

Open rack vaporizer An open rack vaporizer (ORV) is a heat exchanger that utilizes water as the source of heat. The source of water for these units is dependent on the location of the terminal and the quantity of water available. LNG receiving terminals generally are located close to the open sea. Therefore, seawater is the most commonly used source of heat for open rack vaporizers.

These units generally are constructed from finned aluminum alloy tubes, providing the mechanical strength for the cold operating temperature of LNG. Corrosion protection coating is used on surfaces that come into contact with seawater that is sprayed on the outside of the finned tubes.

The mechanical construction of these units is simple. The tubes are arranged in panels, connected through the LNG inlet and the regasified product outlet piping manifolds and hung from a rack (Figures 1-30 and 1-31). This feature provides ease of access for maintenance purposes.

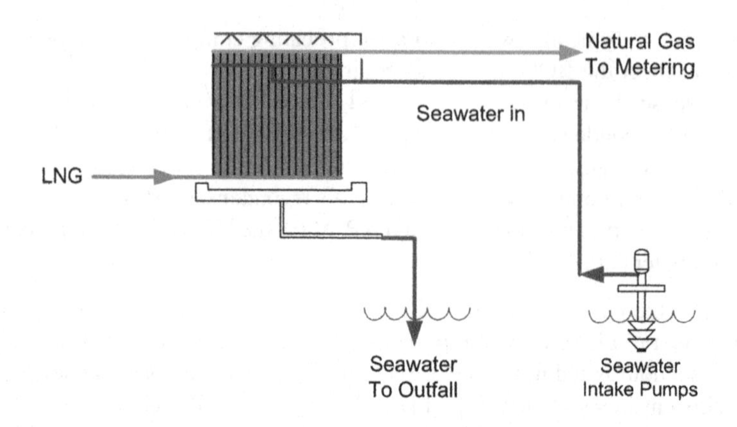

FIGURE 1-30

Open rack vaporizer schematic.

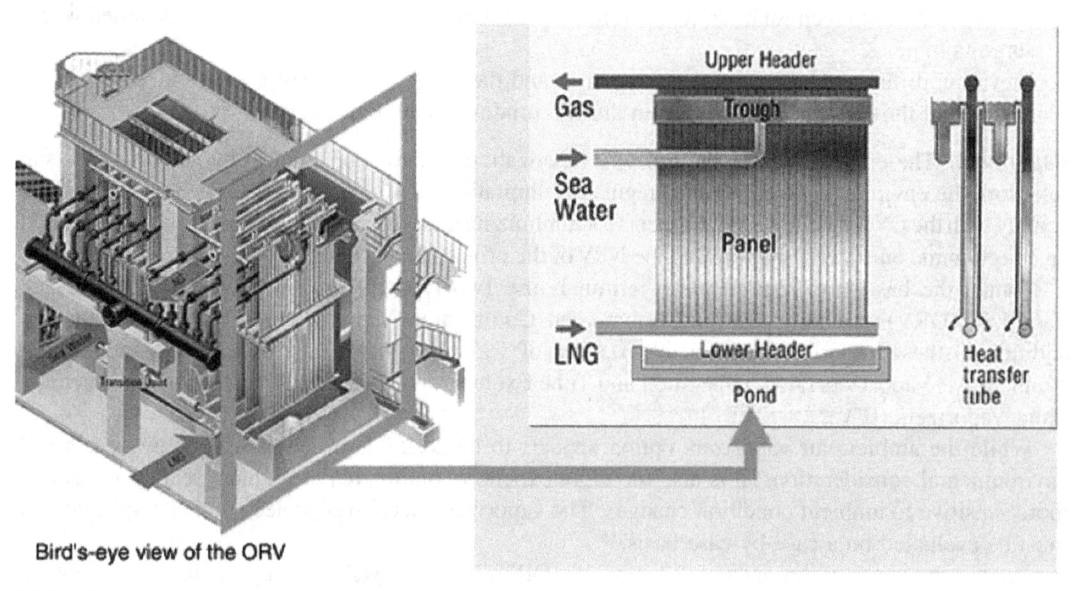

FIGURE 1-31

Open rack vaporizer cross section view.

(Courtesy of Tokyo Gas Co., Ltd.)

The maintenance of these units is also simple, since there are no moving parts and very little instrumentation is involved. The unit can be started or stopped from a remote control station, without physical intervention of an operator. However, maintenance is more frequent than other types of vaporizers as the tubes should be cleaned regularly.

The operating load on these units can be adjusted by varying the amount of seawater flow to the spraying system and/or the LNG flow through the tubes. Depending on the design of the units, it is possible to isolate sections of the panels and to vary the load on the units.

These units are reliable and have very good safety records. Leakage of gas can be quickly detected and the unit can be safely shut down. There is no danger of explosion, due to the fact that there is no ignition source in the system.

The selection of process design of a seawater heating system must consider the following:

- Is the seawater quality suitable for operating an ORV system?
- Does the seawater contain significant amounts of heavy metal ions? These ions will attack the zinc aluminum alloy coating and will shorten its life.
- Does the seawater contain significant amounts of sand and suspended solids? Excessive sediment will cause jamming of the water trough and the tube panel. A proper seawater intake filtration system must be designed to prevent silts, sands, and sea life from reaching the seawater pumps and exchangers.
- The design must consider the environmental impact of the seawater intake and outfall system, and minimize the destruction of marine life during the construction period and normal plant operation.
- Chlorination of the seawater is necessary to slow down marine growth. However, residual chlorine in the seawater effluent can impact the marine life and the usage must be minimized.
- Seawater discharge temperature must comply with local regulations. The temperature drop of seawater typically is limited to 5°C in most locations.
- Location of the seawater intake and outfall must be studied to avoid cold seawater recirculation.
- If a site is located in a cold climate region, supplementary heating may be necessary to maintain the outlet gas temperature. Boil-off gas from LNG storage tanks can be used as fuel to these heaters.
- Is a backup vaporization system provided? This may be necessary during partial shutdown of the seawater system or during peaking demand operation.
- Is the regasification facility located close to a waste heat source, such as a power plant? Heat integration using waste heat can reduce regasification duty and would minimize the environmental impact.

Shell and tube vaporizers Shell and tube vaporizers (STV) can be operated in an open-loop, closed-loop, or combined mode. In the open-loop configuration, seawater is pumped from the seawater intake to the STVs to vaporize the LNG. The STV operates in a similar fashion as ORV except that seawater is pumped through the shell and tube exchangers instead (Figure 1-32). The exchanger shell is constructed of high pressure stainless steel material for the LNG operation and the exchanger tube is constructed of titanium or other suitable materials for seawater operation. The material cost of STV is expensive, but the size is relatively compact, which may be justified in offshore operation.

Intermediate fluid vaporizers The intermediate fluid vaporizer (IFV) uses an intermediate heat transfer fluid (HTF) in a closed loop to transfer heat from a heat source to the LNG vaporizers. The intermediate fluid can be ethylene glycol or propylene glycol; other low-freezing heat transfer fluids are suitable for the operating temperatures. Heat transfer for LNG vaporization occurs in a shell and tube exchanger.

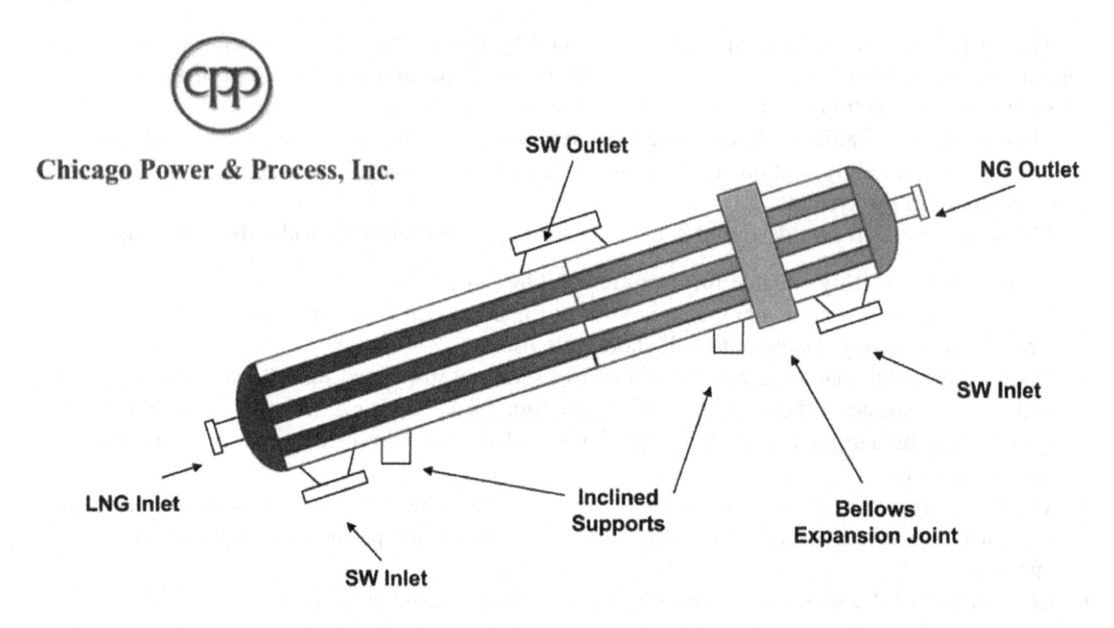

FIGURE 1-32

Shell and tube vaporizers.

(Courtesy of CB&I)

Glycol-Water as Heat Transfer Fluid The warm glycol-water heat transfer fluid is used to heat up the LNG using the intermediate fluid vaporizers (shell and tube exchanger). The cold glycol-water fluid is pumped by the glycol-water pumps. A glycol surge drum upstream of the glycol-water pumps is used to accommodate the system volume changes during shutdown or startup operation.

There are several options to warm the glycol-water solution; for example, waste heat recovery system, fired heater, air heater, plate and frame seawater exchanger, and reverse cooling tower. A general process sketch of this system is shown in Figure 1-33. These glycol-water intermediate fluid vaporizers have a very compact design due to high heat transfer coefficients.

A selective catalytic reduction system can be fitted into the fired heater to reduce the CO and NOx emissions for environmental requirement compliance.

Today, glycol-water intermediate fluid vaporizers account for only a small portion (around 5%) of total vaporization capacity in LNG receiving terminals worldwide.

Hydrocarbon as Heat Transfer Fluid Alternatively, propane or butane can be used as the heat transfer fluid, as shown in Figures 1-34 and 1-35. In this scheme, propane or butane is vaporized on the shell side of a shell and tube exchanger using seawater as the heat source. The vaporized fluid then condenses on the STV that supplies heat to the LNG. The use of a hydrocarbon would avoid the freezing problems encountered with other intermediate fluid.

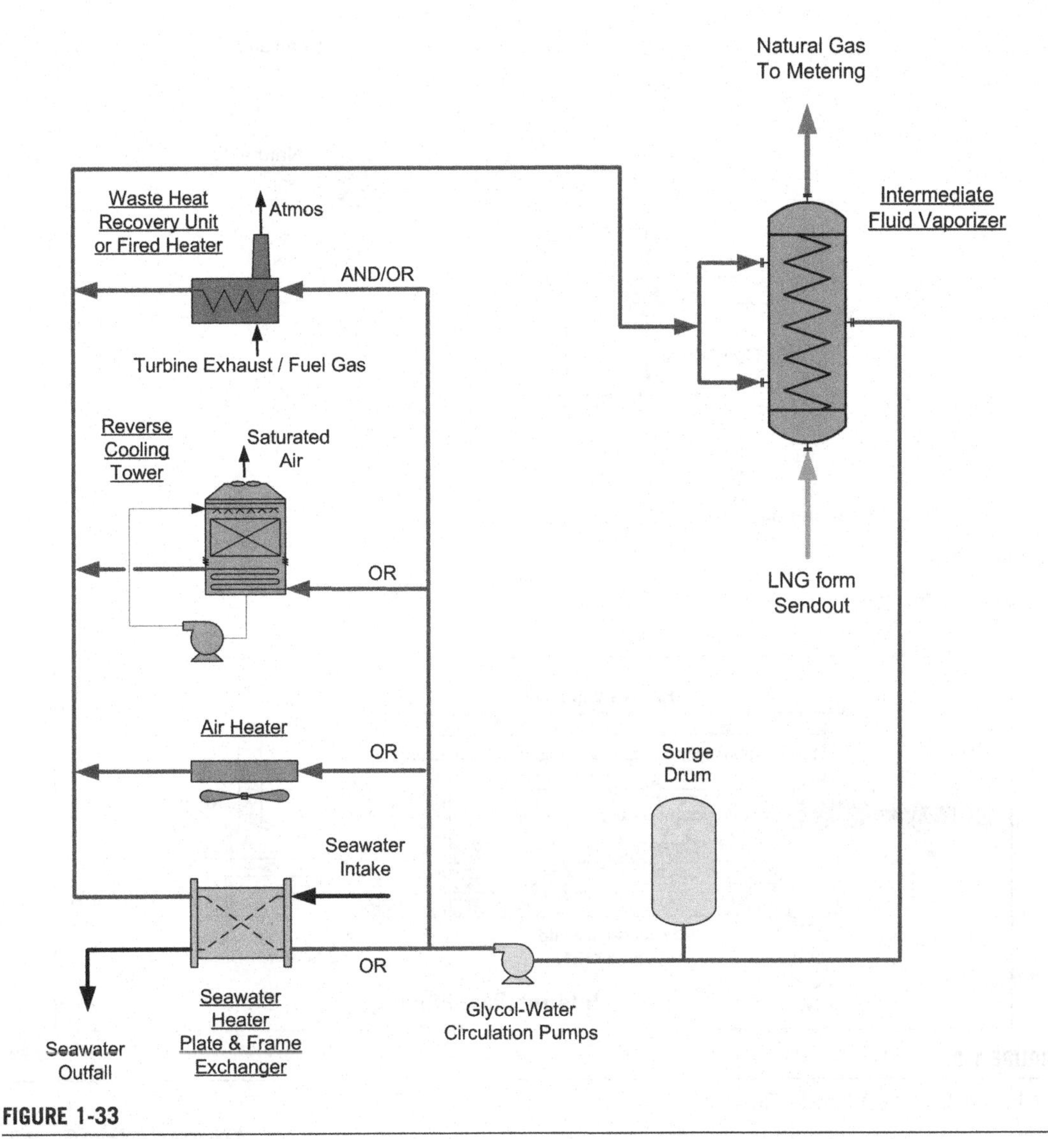

FIGURE 1-33

Intermediate fluid vaporizer process.

(Courtesy of Fluor)

Kobe Steel offers this type of exchanger using propane or butane as the heat transfer fluid in a single shell and tube exchanger called Tri-EX. The design combines the vaporizer and superheater duty in one single shell, split into two sections. The tubes of the Tri-EX are made of titanium that is required for the seawater service, which makes this exchanger relatively expensive. Kobe Steel has built several of these units since 1987.

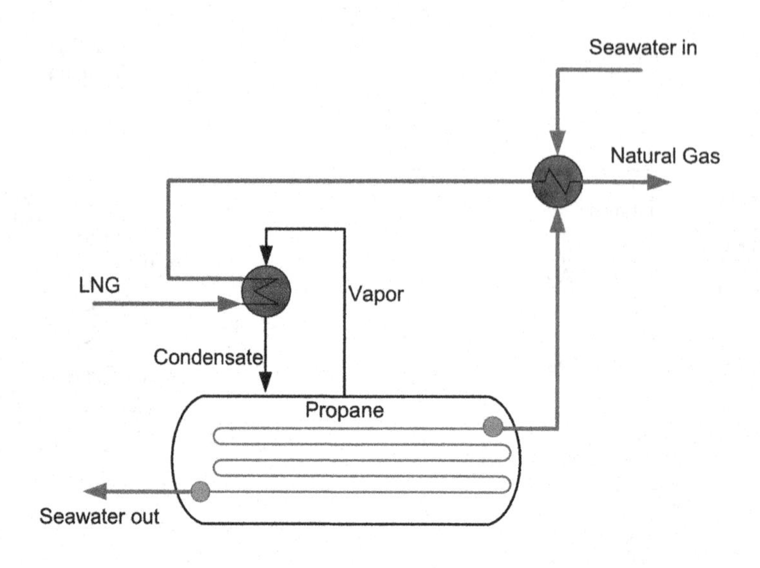

FIGURE 1-34

Propane heating schematic.

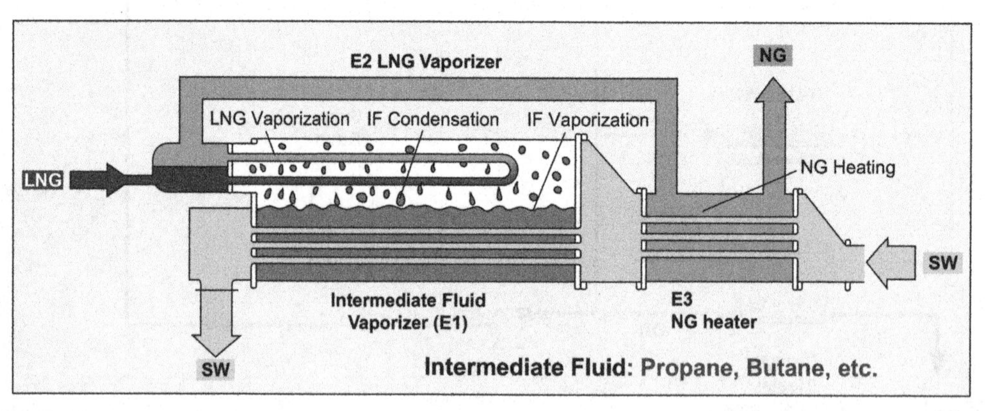

FIGURE 1-35

Hydrocarbon as heat transfer fluid.

(Courtesy of Kobelco Co., Ltd)

Operating with Low Temperature Seawater One advantage with using propane as the heat transfer fluid in IFV is that when seawater drops during winter to a very low temperature, the exchangers can continue to operate but at a reduced rate, as long as the freezing temperature of seawater (typically at −1.5°C) has not been reached. Using the propane as the intermediate fluid as shown in Figure 1-34, performance of the IFV can be maintained even when the seawater temperature drops to 5°C. The unit can continue to operate down to 1°C seawater temperature, but with a much reduced LNG throughput. The reduction in LNG throughput by seawater temperature can be plotted and is almost linear, as shown in Figure 1-36. The exit gas from the IFV exchanger can be trim heated using the standby fired heater or SCVs.

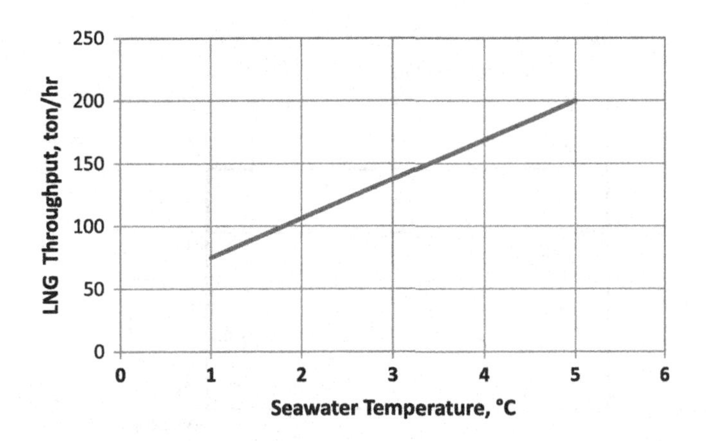

FIGURE 1-36

Impact of seawater temperature on LNG throughput in propane/LNG vaporizer.

There are potential fuel savings by operating with a low seawater temperature in very cold climate regions. This application has to be evaluated with the increase in investment of the additional equipment and operation requirement.

Submerged combustion vaporizer The submerged combustion vaporizer (SCV) system requires approximately 1.5% of the total vaporized LNG as fuel, which adds a significant operating cost to the terminals. For this reason, SCVs are used only where no other free heat source is available. The SCVs can also be designed to utilize boil-off gas.

In a submerged combustion vaporizer (SCV), LNG flows through a tube coil fabricated from stainless steel that is submerged in a water bath (Figures 1-37 and 1-38). Water in the bath is heated by direct contact with hot effluent gases that exit a submerged gas burner. One SCV can be designed to handle a maximum output in the order of 200 MMSCFD. The unit is compact and does not require large tracts of real estate for installation.

Exhaust gases from the burner are sparged into the water through a distributor located under the heat transfer tubes. This causes rapid circulation of water through the tubes resulting in a very high thermal efficiency (over 98%) and high heat transfer rate. Agitation from the sparging action also prevents deposits or scale to be buildup on the heat transfer surface of the tubes.

Since the water bath is always maintained at a constant temperature, the system copes well with load fluctuations and can be quickly started up and shut down. The controls for the submerged combustion vaporizers are more complicated when compared to the open rack vaporizers. The SCV has more pieces of equipment, such as the air blow, sparging piping, and the burner management system.

These units are reliable and have very good safety records. Gas leaks can be quickly detected and the unit can be safely shut down. There is no danger of explosion, due to the fact that the temperature of the water bath stays below the ignition point of natural gas.

The bath water is acidic as the acid gas content in the exhaust gas is condensed. Caustic is added to the bath water to control the pH value to protect the tubes against corrosion. The excess combustion water must be neutralized before being discharged to the open water system.

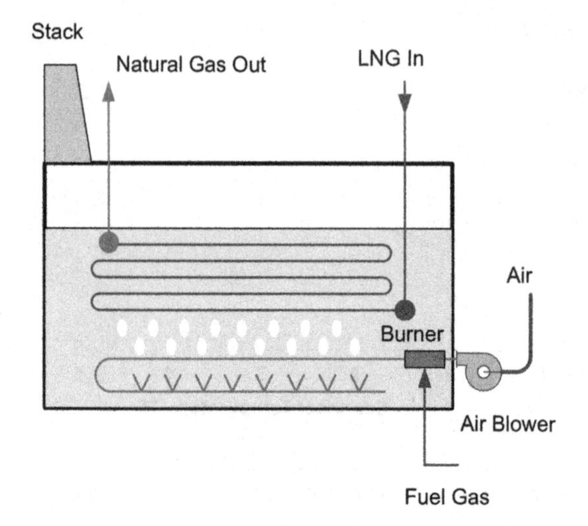

FIGURE 1-37

Submerged combustion vaporizer schematic.

To minimize the NOx emissions, low NOx burners can be used to meet the 40 ppm NOx limit. The NOx level can be further reduced by using a selective catalytic reduction (SCR) system in the flue gas stack to meet the 5 ppm specification if more stringent emission requirements are needed, at a significant cost impact.

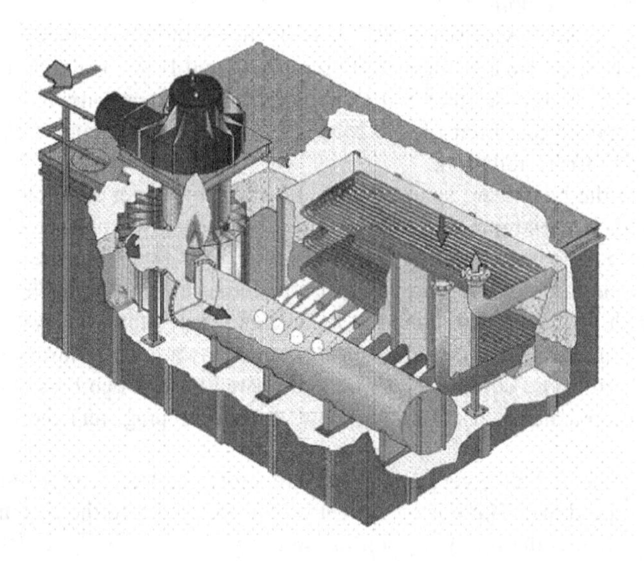

FIGURE 1-38

Submerged combustion vaporizer.

(Courtesy of Sumitomo Precision Products)

Ambient air vaporizer An ambient air vaporizer (AAV) extracts heat from the ambient air for heating, hence avoiding the use of seawater or fuel gas. AAVs are considered the more environmentally friendly solution than ORVs and SCVs, and hence are much easier to permit. The AAV heat exchangers are cost competitive since they can operate on a standalone basis without support of the seawater system, intermediate fluid, and fuel gas system. However, the number of vaporizer units is much higher than other options, which would require a larger plot space. In hot humid locations fog generation may be a problem, which should be evaluated for its impact on the surroundings.

Two types of ambient air vaporizers can be considered for the LNG terminals: direct air/LNG vaporizers and indirect air vaporizers using an intermediate fluid. The air exchanger can be operated on a natural draft or a forced draft mode.

A typical AAV design configuration is shown in Figures 1-39 and 1-40. AAV consists of direct contact, long, vertical heat exchange tubes that facilitate downward air draft. This is due to the warmer and less dense air at the top being lighter than the cold denser air at the bottom. Ambient air vaporizers utilize air in a natural or forced draft vertical arrangement. Water condensation and melting ice can be collected and used as a source of service/potable water.

To avoid dense ice buildup on the surface of the heat exchanger tubes, deicing or defrosting with a 4 to 8 hr cycle typically is required. Long operating cycles lead to dense ice on the exchanger tubes, requiring longer defrosting time. Defrosting can be accomplished by natural draft convection or forced draft air fans. The defrosting time with forced fans is only marginally faster, as heat transfer basically is limited by the ice layers, which act as an insulator. However, the high

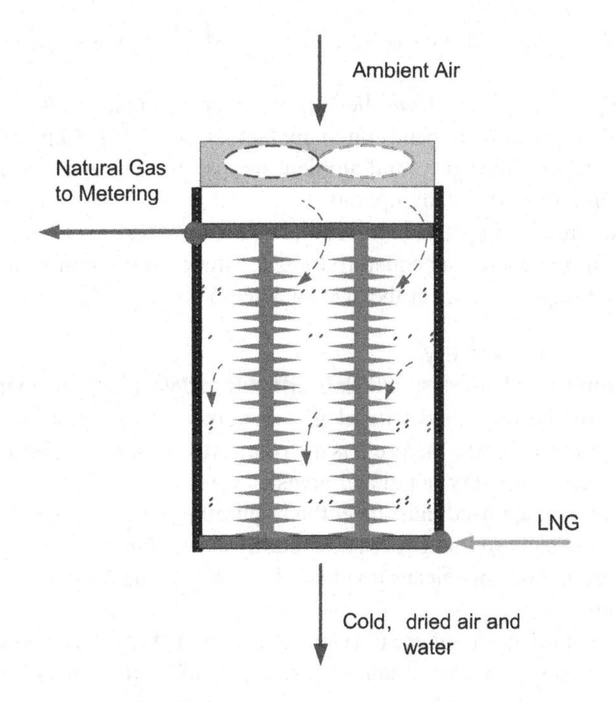

FIGURE 1-39

Ambient air vaporizer schematic.

AAV Side View

AAV Bottom View

FIGURE 1-40

Direct ambient air vaporizers—Plant views.

(Courtesy of Cryoquip Inc.)

air rate using fans helps to disperse the fog generated by the exchanger and may help to improve visibility.

Fog is generated when the cold air from the vaporizers is in contact with the warm and moist air outside. The extent of fog formation depends on many factors, such as the separation distances among units, wind conditions, relative humidity, and ambient temperatures. Dense fog may create visibility problems, which may interfere with daily operation.

An ambient air heater is advantageous in hot climate equatorial regions where ambient temperature is high all year round. In the cooler subequatorial areas, where winter temperature is lower, supplementary heating may be required to meet the sales gas temperature.

Odorization, send-out, and delivery

Since natural gas is colorless and odorless, a leak is difficult to detect without appropriate instruments. Therefore, odorization of the regasified natural gas to make the detection of a gas leak easier is required in many regions and countries before it is distributed to consumers (particularly for residential and commercial customers in densely populated areas).

The properties of any odorant used must meet the requirements of local regulations. Odor strength typically is maintained at a level so that gas may be readily detectable at concentrations of one-fifth of the lower explosive limit of gas. This means that the odor is readily detectable at concentrations of gas in air of 0.9% by volume.

A typical odorant is tetrahydrothiophene (THT) or mercaptan. The odorization station can be in the LNG terminal itself before the send-out of natural gas, or just a few kilometers beyond the terminal on the consumer pipelines. The point at which odorants are added depends on the country. Many terminals that export to a high pressure transmission line do not need to be odorized. Metering, the last step at the terminal, measures the quantity of gas being send out. Natural gas is then delivered by pipeline directly to customers for industrial or residential use.

LNG quality and gas interchangeability

Wobbe index. In the LNG terminal, the quality specification of the delivery gas must comply with the specifications set by pipeline transmission companies and end users to ensure operational safety, reliability, and compliance with environmental regulations. The typical specifications used in most US gas pipeline tariffs are based on meeting the heating value (or gas calorific value) and the gas component composition ranges. However, most combustion devices, such as gas turbines or burners, are designed specifically for a narrow range of gas compositions (or Wobbe Index values) to limit NOx emissions.

Wobbe Index (WI) originally was developed to characterize the similarity of gas mixtures based on the heat release from combustion. It is derived from the fluid dynamic principle, which correlates the heat release from combustion with the heating value and density of the gas. Wobbe Index is used in the gas turbine industry for fuel gas specification and gas turbine performance guarantee. It is also one of the specifications for gas interchangeability required for pipeline distribution. The index is calculated by dividing the higher heating value of the gas by the square root of the gas density or MW (molecular weight) relative to air.

Wobbe Index is internationally the most widely accepted measure of gas interchangeability. Wobbe Index is frequently used as a parameter to determine the upper and lower limits of gas composition specified in gas sales or import contracts.

Historically, each gas market has been isolated from other regions with its own indigenous and gas supply mix, and each market has its own gas quality specifications. In the United States and Western Europe, most of the LPG components are extracted for sales as liquids, and the residual gas is sold to consumer pipeline; consequently, the gas compositions and heating values do not vary very much. However, as shown in Figure 1-41, the LNG compositions and heating values vary significantly among different export terminals, and there are concerns regarding the gas interchangeability with the local gas contents (see Figure 1-41).

From the producers' perspectives, it is necessary to meet the export requirements for the contracted quality, which is typically met by integrating an NGL extraction facility to the liquefaction plant. For the LNG receiving terminals, they must be equipped with a processing facility, especially for spot market cargoes, with provisions that allow them to adjust the heating value or Wobbe Index to meet the pipeline gas specification.

Wobbe index control. The common approach to reduce the Btu value and Wobbe Index of a gas is by diluting the vaporized LNG with nitrogen, up to the pipeline limit for inert content, usually 2 to 3%. Nitrogen is an effective medium for lowering the heating value of the pipeline gas. The addition of nitrogen also increases the molecular weight of the gas mixture that further lowers the Wobbe Index value. For example, for California pipeline gas, the gas specification must meet the 1360 Wobbe Index specification, which is the specification of existing combustion equipment and power generation stations. To meet the California Wobbe Index specification, Alaska LNG is the only gas that meets the low Wobbe Index specification without nitrogen dilution. For the Trinidad LNG, about 1 mole % nitrogen is required. For other LNGs, because of the maximum 3 mole % nitrogen limit, less than 40% of the LNG sources would meet the California specification, as shown in Figure 1-42.

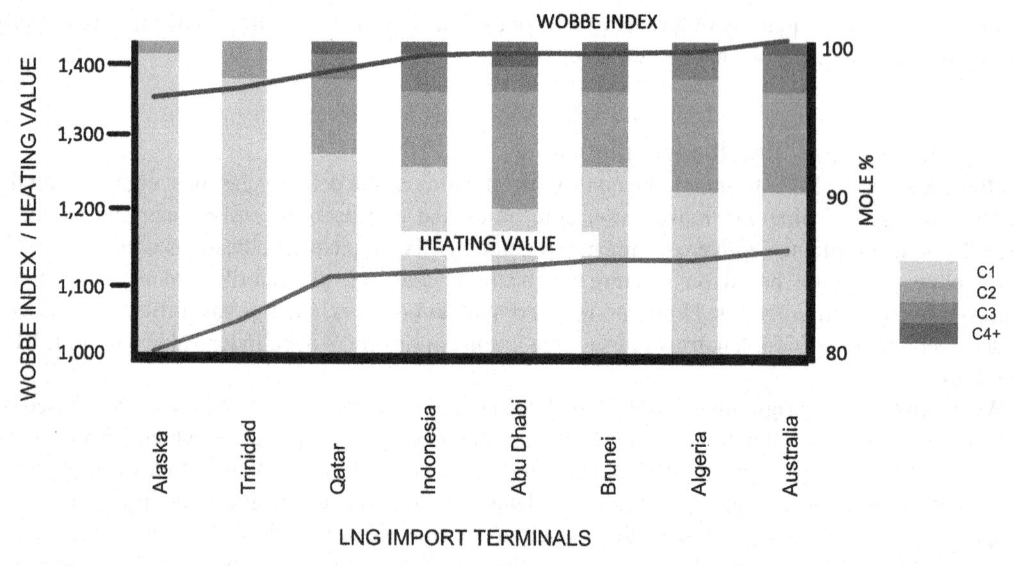

FIGURE 1-41

Compositions and higher heating values of various LNG sources.

(Mak, 2008)

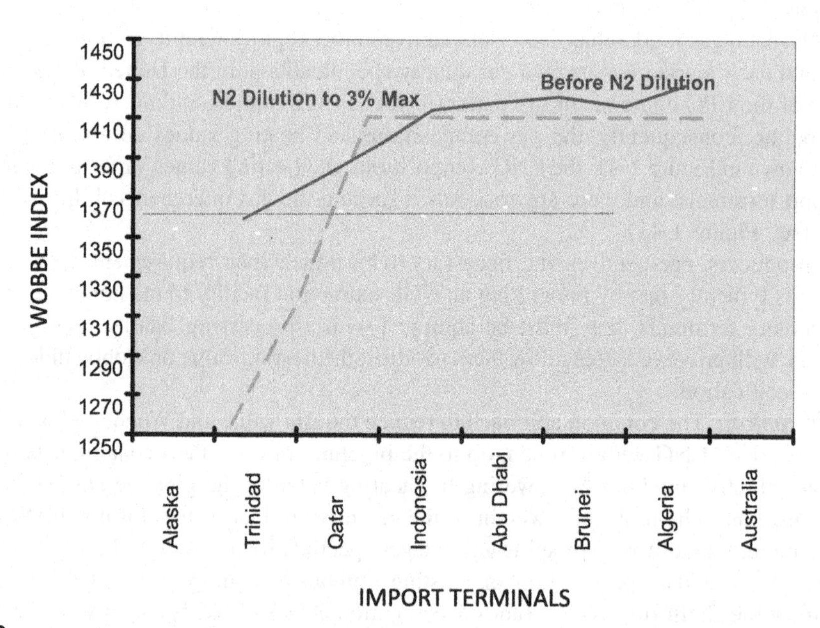

FIGURE 1-42

Nitrogen dilution for different LNG sources.

(Modified after Mak, 2008)

Due to the stringent Wobbe Index specification, weathering of LNG must be monitored in the regasification terminal. The heating value and Wobbe Index will gradually increase due to boil-off of the nitrogen components during transit and storage. For the marginal Wobbe Index LNG, the weathering effect may result in off-specification natural gas product.

Some regasification terminals are equipped with the capability of injecting either nitrogen or LPG to regulate the sendout gas heating value. In North America and the United Kingdom, the pipeline gas typically has a relatively low Btu value, and nitrogen dilution for lowering the Btu value is common. In Japan and most Asian LNG countries, where high Btu gas is desirable, propane can be injected into the gas pipeline to raise its calorific value.

There are various nitrogen injection schemes, which range from simple gaseous pipeline injection to the dual practice of gaseous and liquid nitrogen injections. However, introducing gaseous nitrogen at low pressure by mixing and condensing by the sendout liquid can eliminate vapor compression and is potentially the least costly option. However, it would require sufficient cold LNG sendout available for condensation (Huang et al., 2005; Coyle et al., 2007).

The calorific value reduction by injection of nitrogen is limited by the inert limit of the sendout gas (Figure 1-42). Nitrogen injection alone may not be sufficient for adjusting the quality for the very rich LNG compositions. In cases where nitrogen dilution cannot meet the heating value requirement, NGL components (ethane, propane, and butane and heavier hydrocarbon) need to be removed from the regasified LNG. If the NGL value is at a premium over natural gas, NGL recovery makes economic sense rather than nitrogen dilution. However, if the economics of NGL recovery is not there, then nitrogen injection is the only choice, or by diluting with a leaner LNG source.

There are several NGL recovery configurations that can be used for fractionating the NGL components from the import LNG. The NGL recovery process can be designed to utilize the cold energy from LNG vaporization to provide reflux in fractionation, and is very energy efficient (Mak, 2008). The configuration selection depends on many factors, including the heating value range of the import LNG, the values of the LPG liquid product and ethane product, the capital cost and operating cost, and the feasibility of integrating to an existing regasification terminal (Cuellar et al., 2007).

For LNG that has very high ethane content, and where there is no market for ethane product, an interesting option is using the Catalytic De-Richment (CDR) process (by Johnson Matthey catalysts), which can convert the higher hydrocarbons into methane. The CDR process was originally developed to produce pipeline gas from naphtha, based on steam reforming, CO shifting, and the methanation process to produce methane. Although the CDR option can convert all the higher hydrocarbons into methane, the capital and operating costs of such or similar processes are high and are not economical for LNG import terminals (Carnell et al., 2008). In most markets NGLs are more valuable than methane so this process is not commonly used.

LNG terminal operation flexibility

In addition to the process systems, an LNG terminal requires utility and offsite supports and power supply, and pipeline infrastructure to safely and reliably delivering on-specification gas to the customers.

LNG receiving terminals operate quite differently among different regions of the world. In the United States and Western Europe, LNG receiving terminals are base load plants and are designed to deliver natural gas to the pipelines at a fairly constant rate. These regions have large natural gas surge capacity because of the extensive pipeline network. During low demands when excess gas is produced, pipelines can be packed to store the gas, and underground caverns can also be used for temporary storage. The large surge capacity allows the LNG terminal to operate at a fairly constant rate even during low demand seasons.

However, in other regions of the world, such as in China, where there is a lack of pipeline infrastructure and gas storage capacity, the LNG receiving terminals must adjust their sendout rates to meet the gas consumption. Gas demands typically peak during the day and drop significantly during the night. There are also seasonal variations. Operating these terminals requires frequent startup and shutdown of the sendout pumps and vaporizers. In addition, the large sendout rates, the frequent ship unloading, spot market LNG cargoes, and variations in LNG compositions all make these newer plants more difficult to operate. These LNG import terminals must be designed with more robustness and be provided with flexibility to accommodate these challenging operating parameters.

LNG cold utilization

Approximately 8 to 10% of the energy in natural gas is consumed in the LNG liquefaction process and during LNG transport. In a typical LNG liquefaction plant, it takes about 230 kW to liquefy one MMscfd of natural gas. About 280 MW of power is consumed in a 1,200 MMscfd liquefaction plant. Theoretically, some of the power consumed in LNG liquefaction is recoverable at the LNG receiving terminal if the LNG is used as a refrigerant for industrial use or as a heat sink in power generation.

However, the main difficulty in fully utilizing the "LNG Cold" for cooling is to identify users that can use the different temperature levels of refrigeration released during LNG regasification. The heating curves of LNG at three different pressures are shown in Figure 1-43. The available low temperature refrigeration is higher at 30 barg than at 90 barg. To efficiently utilize the LNG cold, the cooling curves of the users must closely match the heating curve of LNG. This is the same concept used in the design of the refrigeration cycle. Therefore, unless the LNG terminal is located in an industrial complex, it is difficult to find suitable users that can take full advantage of the LNG cold.

From a thermodynamic point of view, the most efficient use of LNG is in cooling and as a refrigerant to provide cooling in refinery and petrochemical complexes, such as in air separation,

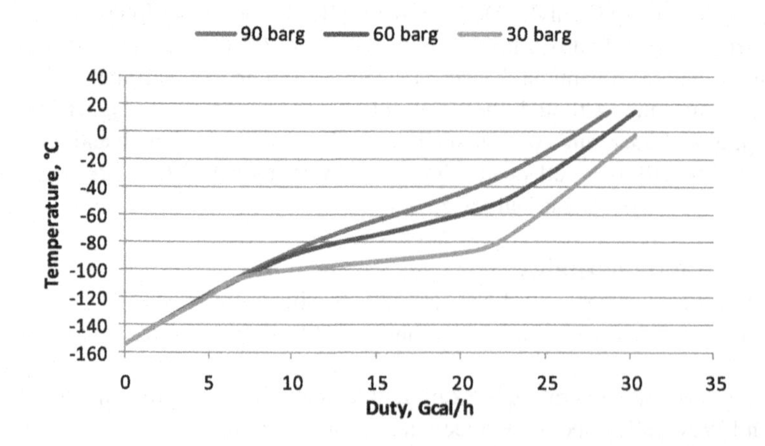

FIGURE 1-43

Typical LNG heat release curves.

carbon dioxide production, vapor recovery, chilled water production, and hydrocarbon fractionation. More discussions on the subject of LNG cold utilization can be found in Chapter 10.

Alternatively, the refrigeration content in LNG can be used as a cold heat sink to generate power. This may not be the most efficient use, but integration to a power plant can eliminate seawater or fuel gas consumption required in regasification since low level waste heat from the power plants are readily available (Mak, 2005). To bring the "LNG Cold" to the users, an intermediate heat transfer fluid is required using a closed circuit in transferring the refrigerant to the users. To minimize the transfer costs, the LNG terminal must be located at the industrial complex, as in Japan's LNG terminals and in the Jurong Island Terminal in Singapore.

Cryogenic power generation. The use of LNG for power plant fuel supply has been proven for many years in Japan. Information relating to some of Japan's gas-fired power plants and their capacities are shown in Table 1-5.

Principle of organic rankine cycle (ORC). The principle for power generation with LNG is based on the Rankine cycle as today's steam-turbine-driven power plants. The ideal Rankine cycle efficiency (or Carnot cycle efficiency) can be defined as (T2-T1)/T2. T2 is defined as the absolute temperature of the heat source and T1 is the absolute temperature of the heat sink. Note that T1 is actually the average vaporization temperature of LNG, which ranges from –250°F to ambient temperature.

Because of the extremely low temperature of LNG, even with a low-level heat such as seawater or turbine exhaust, significant power can be generated using the Rankine cycle concept. In simple terms, the power cycle generation efficiency can be increased almost proportionally to the temperature difference between the heat source (hot side) and the heat sink (cold side).

The Organic Rankine cycle (ORC) operating principal is illustrated in a temperature-entropy (T-S) diagram (Figure 1-44). Typically, a hydrocarbon is used as the working fluid, such as propane or butane or a mixed hydrocarbon. In the cycle, the fluid is liquefied using the refrigeration released during LNG vaporization. The vaporized LNG is piped to the gas consumers while the intermediate fluid is pumped, vaporized, superheated, and then expanded in an expander to a lower pressure, generating power to drive a generator. The low pressure propane vapor is then recondensed by LNG and the cycle is repeated.

Closed rankine cycle. The Organic Rankine cycle can be operated using propane or butane as the working fluid in a closed cycle configuration (Figure 1-45). It basically goes through a typical Rankine cycle of condensing, pumping, heating, and expanding steps.

In addition, since the LNG sendout pump is designed to discharge at 8 to 10 MPag, and the fuel gas pressure to gas turbines is required at 3 to 5 MPag, the high pressure natural gas can be further heated and expanded to recover more power using a gas expander generator.

A combination of the ORC and natural gas expander (Figure 1-46) can significantly improve the overall power generation efficiency. This type of power generation operation has been proven in several applications in Japan's LNG terminals.

Open rankine cycle. LNG itself can also be used as a working fluid in an open cycle configuration as shown in Figure 1-47 (Mak, 2005). In this configuration, only the low temperature portion of the LNG heat curve is used. The T1 (the average temperature at which the heat sink is used) is very low; hence a very high Rankine cycle efficiency can be achieved. With a more efficient cycle, the working fluid circulation can be reduced, lowering the operating cost of the cycle. This process does not use the warmer portion of LNG heat curve, and the residual cold can be further utilized by other cold consumers.

Table 1-5 Japan's Cryogenic Power Plants

Company and Terminal Names	No of Units	Start of Operation	Output (kw)	Type	LNG Consumption (t/h)	Delivery (MPa)
Osaka Gas, Senboku Daini	1	12/1979	1,450	Rankine	60	3.0
Toho Gas, Chita Kyodo	1	12/1981	1,000	Rankine	40	1.4
Osaka Gas, Senboku Daini	1	2/1982	6,000	Rankine/NG direct expansion	150	1.7
Kyushu Electric Power and Nippon Steel, Kitakyushu LNG	1	11/1982	9,400	Rankine/NG direct expansion	150	0.9
Chubu Electric Power, Chita LNG	2	#16/1983	7,200	Rankine/NG direct expansion	150	0.9
		#23/1984	7,200	Rankine/NG direct expansion	150	0.9
Tohoku Electric Power, Nihonkai LNG	1	9/1984	5,600	NG direct expansion	175	0.9
Tokyo Gas, Negishi	1	4/1985	4,000	Mixed refrigerant Rankine	100	2.4
Tokyo Electric Power, Higashi Ogishima	1	#15/1986	3,300	NG direct expansion	100	0.8
Osaka Gas, Himeji	1	3/1987	2,800	Rankine	120	4.0
Tokyo Electric Power, Higashi Ogishima	2	#29/1987	8,800	NG direct expansion	170	0.4
		#3/1991	8,800	NG direct expansion	170	0.4
Osaka Gas, Senboku Daiichi	1	2/1989	2,400	NG direct expansion	83	0.7
Chubu Electric Power, Yokkaichi	1	12/1989	7,000	Rankine/NG direct expansion	150	0.9
Osaka Gas, Himeji	1	3/2000	1,500	NG direct expansion	80	1.5

(Courtesy of Japan Gas Association)

In the Open Rankine cycle, vaporized LNG is used as the working fluid. The working fluid is condensed in the LNG heater/NG condenser at about 5 barg pressure, using the cold portion of the LNG vaporization curve. The working liquid is then pumped by the NG pump, combined with the heated LNG from the NG condenser, and further heated by supplying refrigeration to other users.

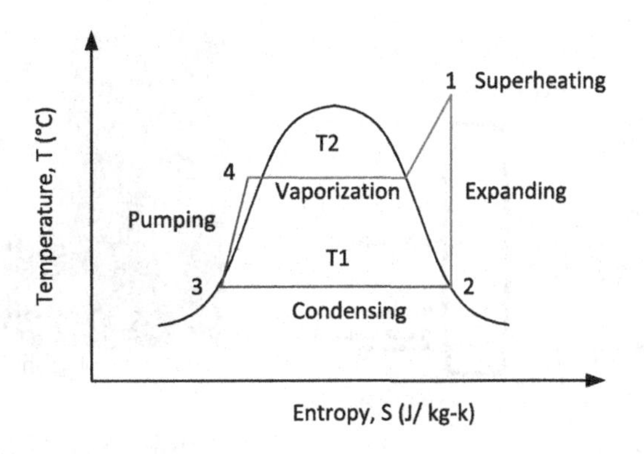

FIGURE 1-44

Typical temperature versus entropy diagram for organic Rankine cycle.

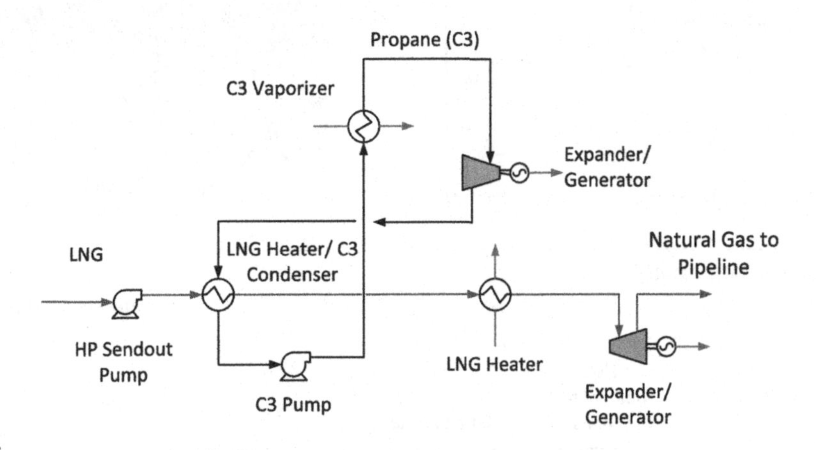

FIGURE 1-45

Typical Closed Organic Rankine cycle.

The LNG working fluid is then split off from the main flow circuit, routed to an exchanger, and heated in a superheater using waste heat prior to being expanded using an expander to generate power. The expanded working fluid is heat exchanged and then condensed in the NG condenser to repeat the cycle.

Although power generation using ORC might seem attractive, it must consider the various operations of the LNG terminals. In a typical LNG terminal, the LNG sendout rate varies with gas and power consumption, and the fluctuation can be quite significant from day to night, from season to season. Therefore while the Organic Rankine cycle is designed for the maximum sendout, it must also be operated at turndown. Subsequently, only a fraction of the installed expander capacity can be utilized during periods of low consumption. The rate of return on investment of power generation must then be based on averaged annual output, which would reduce its economic attractiveness.

Integration with combined cycle power plants. Since a significant amount LNG import today is used to supply fuel gas to operate power generation plants, integration of the regasification facilities to the

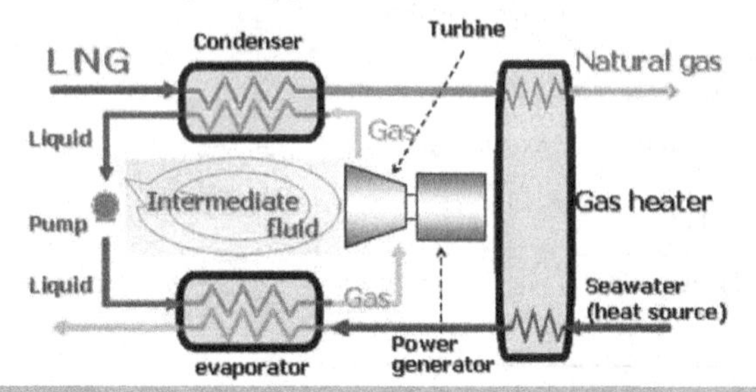

Negishi Terminal

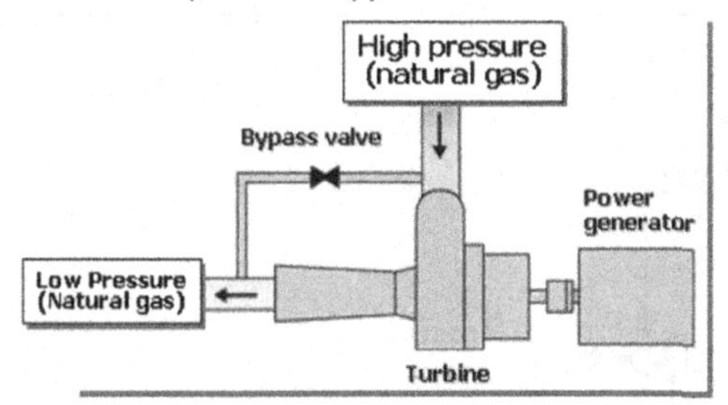

Himeji Terminal

FIGURE 1-46

Japan's cryogenic power plants.

(Courtesy of Japan Gas Association)

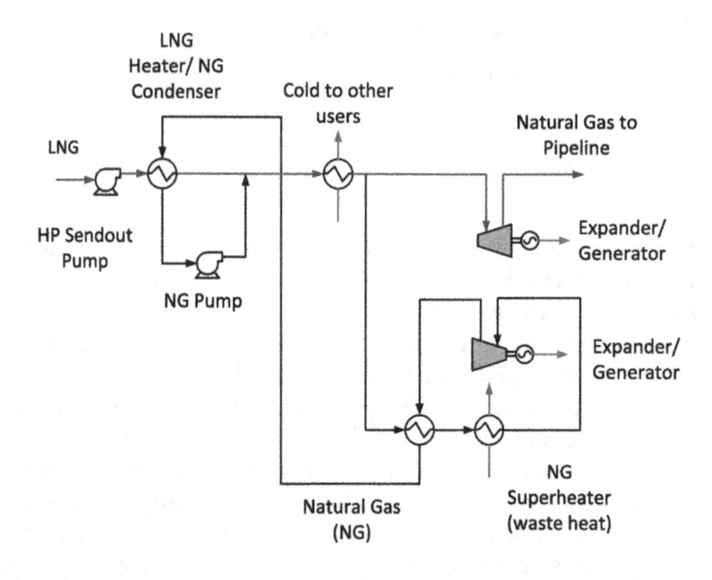

FIGURE 1-47

Open Organic Rankine cycle.

(Mak, 2005)

power plants is a natural progression. With the such integration, the power generation efficiency can be increased using LNG as the cold heat sink. Cooling water consumption in the power plant can then be reduced and the heating duty for LNG vaporization can be supplied by waste heat from the power plant, thus eliminating the use of seawater for LNG heating. More discussions on the integration of LNG plant and power plant approach are discussed in Chapter 10.

1.5 Offshore LNG supply chain

As demand for natural gas continues to grow and the value of natural gas remains high in the major consuming markets, the impetus to monetize offshore gas resources also grows. The potential to unlock offshore gas reserves without investment in capital-intensive pipeline infrastructure, infield platforms, and onshore infrastructure, while minimizing exposure to geopolitical and security risks, makes floating LNG production, storage, and offloading (LNG FPSO and FLNG) concepts worthy of close scrutiny (Barclay and Yang, 2006). Also the restrictions of more stringent no-flaring rules being applied around the world are likely to prompt some existing offshore producers to aggregate gas from several fields and develop large-scale floating liquefaction (>4 MTPA capacity) as an alternative to the costly gas reinjection facilities or long-distance export pipeline. Building an offshore facility requires adapting current technology to an offshore environment, which could be costly and may present a safety risk. But as FLNG technologies on liquefaction and ship-to-ship transfer are being developed and proven, there is increasing confidence that FLNG is a viable option. Although no FLNG facilities currently exist (as of this writing), a facility is under development by Shell for the Prelude offshore gas field in Australia. Once this first unit is operational, it will open up new business opportunities for countries looking to develop their gas resources, bringing more natural gas to market. Shell is the first to go ahead with such a project, Prelude FLNG.

In contrast to the traditional LNG value chains, the offshore LNG value chain will consist of an offshore production facility, floating offshore gas treating and liquefaction facility (FLNG or FPSO), the LNG loading facilities, LNG cargoes transport, the LNG unloading facilities, storage and regasification facility on the FSRU or regasification ship, and finally, tied into a gas distribution pipeline network (see Figure 1-48).

1.5.1 FPSO

A typical LNG FPSO design bases the installation on an LNG carrier hull. The various parts of the process are then located topside and distributed as modules that are installed on the deck. Depending on the intended capacity of the LNG FPSO and the need for treatment of the feed gas composition, the topsides typically may weigh from 20000 tonnes to 50000 tonnes for medium-size units producing

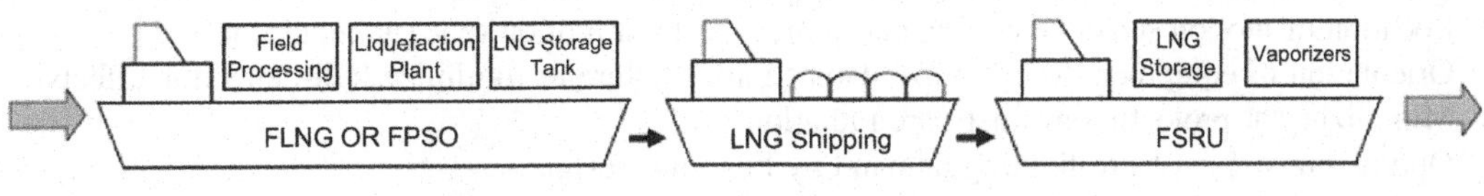

FIGURE 1-48

Floating LNG supply chain.

1.5 to 3 MTPA. For very large scale production units (3–5 MTPA) the topside weight may reach 70000 tonnes or more.

The storage capacity of the FPSO will be related to the processing capacity, the intended offload schedule, and the need to store condensate and LPG, which is dependent on the feed gas composition. Current designs for medium-size units are being proposed with an LNG storage capacity of 180,000 to 190,000 m^3 and LPG and LPG storage capacities of approximately 25,000 m^3 each. For the very large scale units LNG storage capacity of 220,000 m^3 has been proposed together with LPG and Condensate storage in the range of 100,000 m^3 each.

Operating an LNG plant under an offshore setting presents a demanding set of challenges. In terms of the design and construction of the FLNG facility, every element of a conventional LNG facility needs to fit into an area less than one quarter the size of a land base terminal, while maintaining the utmost levels of safety and the flexibility required by LNG production.

The LNG containment systems need to be capable of withstanding the stress that can occur when the sea's wave and current motions cause sloshing in the partly filled tanks. Product transfers have to consider the effects of winds, waves, and currents in the open seas.

Many solutions to reduce the effect of motion and weather are being developed. They are based on adapting current technologies for offshore oil and gas production to LNG production. Notably, the main difference is the cryogenic and volatile nature of LNG and the cargo transfer (ship to ship) operation is recognized as the critical link in the LNG chain (Mokhatab and Wood, 2007b; Mokhatab et al., 2008; Wood and Mokhatab, 2008b).

1.5.1.1 Topside layout

Layout criteria for an FPSO are more stringent than onshore due to the limited footprint, the need for good weight distribution, and the need for personnel refuge and escape routes (Finn, 2008). Safety is the prime consideration for the layout of LNG FPSO. The primary safety concern is the inventory of hazardous, flammable gas, and the consequence of any loss of containment.

A major contributor to safe design is to ensure that initial arrangement and layout are aimed at arriving at a design that meets operational requirements and also will be compliant with regulations. Typically the following principles should be taken into consideration (DNV, March 2011):

- Segregation of accommodation and main working areas from the hydrocarbon processing areas
- Provision of adequate and redundant access and escape ways from normally manned areas
- Ensuring rational flow within the gas and LNG processing system
- Maintenance of availability of lifeboats
- Provision of sufficient access for maintenance and repair and replacement offshore (e.g., LNG pumps)
- Permitting the installation and interconnection of modularized units
- Limiting the amount and areal extent of hydrocarbon and cryogenic spills
- Efficient installation of High Voltage cabling
- Limiting congestion and permitting ventilation to reduce explosion potential and effect
- Location of motion sensitive equipment in areas of least motion (e.g., at centerline)
- Orientation of equipment to minimize damage after failure or minimize hull-deflection effects
- Minimizing or protecting against flare radiation
- Optimizing sizing of fire-fighting demand by fire zone design
- Optimizing vent and blowdown capacity by segregation

- Limiting escalation after fire or explosion by separation distance or physical barriers (fire and blast divisions and deck coamings)
- Ensuring survivability of safety systems following an accidental event (e.g., redundancy and location of power systems, means of evacuation)
- Location of a turret with respect to safety and thrusters demand
- Meeting material handling demands.

For example, Figure 1-49 shows a typical LNG FPSO layout by CB&I, which has considered the following aspects in terms of safety:

- Living quarters, control room, and air intakes are located at the bow, upwind of any potential gas release.
- Power generation with air intakes and exhaust are located away from the NGL and LNG units.
- The internal turret is located forward to provide natural weathervaning capability.
- Utility systems are located just beyond the turret, which helps to maximize the distance between the hydrocarbon source and the living quarters.
- The NGL and LNG plants are located at the stern, providing maximum separation between the living quarters and hydrocarbons.

In terms of storage and transfer, the LPG units will present fewer problems than equivalent LNG installations. However, process plant for LPG production, typically involving tall fractionation columns (deethanizer, debutanizer, and depropanizer), will present some design challenges with regard to accommodating them structurally and reducing their sensitivity to motion.

LPG will potentially cause more severe fire or explosion if an accident occurs during offloading. These will need to be quantified in a safety study. In addition dispersion and detection of any

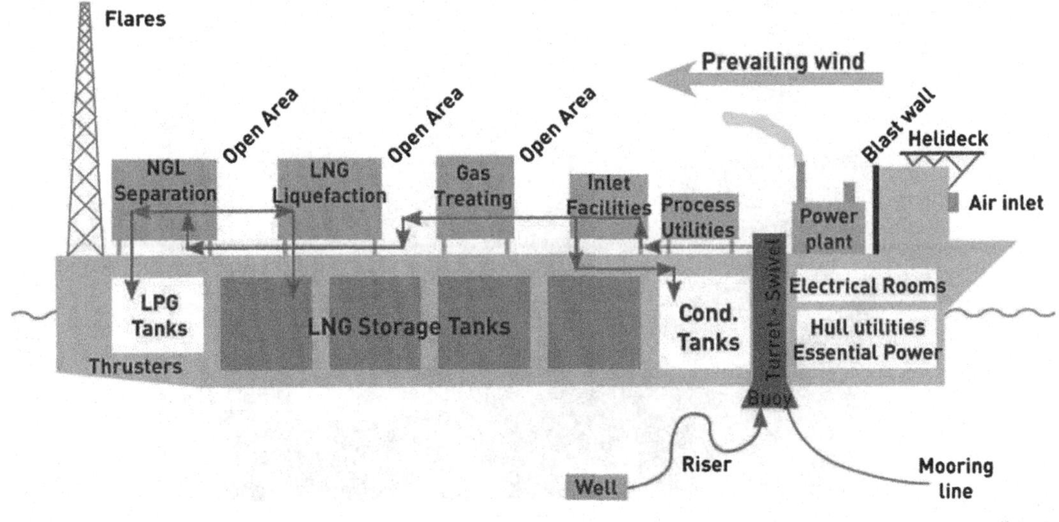

FIGURE 1-49

Typical layout of an LNG FPSO vessel by CB&I.

(Festen et al., 2009)

leakages will need to consider the actual properties of the LPG, being higher in molecular weight than methane.

Shell FLNG

Figure 1-50 shows a rendering of the Shell LNG FPSO design for Northwest Australia. Shell is using its Floating Liquefied Natural Gas (FLNG) technology to develop its Prelude and Concerto gas discoveries, located in the Browse Basin off the northwest coast of Western Australia. The design is being jointly developed by the Technip-Samsung Heavy Industries consortium. The design is planned for other FLNG facilities.

The FLNG vessel will be the world's largest facility afloat when delivered in 2014. It is estimated to be 480 m long and 75 m wide and with 600,000 dwt (Sea Breezes Magazine, December 2009 Issue). The 480 m-long floater, being built at the Samsung Heavy Industries shipyard in Geoje, Korea, will produce an estimated 3.6 MTPA of LNG, 1.3 MTPA of condensate and 0.4 MTPA of LPG from the Browse Basin WA-371-P permit, in 280 m water depths offshore Western Australia (McCulley, 2011).

1.5.1.2 Technical challenges

FPSO vessels have been used to produce oil and gas from deep-water fields since the 1970s, and the industries have proven design and operating experience. Using the same model, the FLNG vessels are custom built to meet the safety requirements of natural gas liquefaction. The facility has to adapt to the gas field's unique characteristics, mainly, the gas reservoir capacity, the gas compositions, acid gas contents, and the local metocean conditions (waves, wind, etc.).

FIGURE 1-50

Rendering of the Shell Prelude FPSO LNG for North West Australia.

(Bradley et al., 2009; Sea Breezes Magazine, December 2009 Issue)

Project economic criteria of offshore plants are different than onshore plants where there are no limits on plot space and weights. Consequently, the design, layout, and efficiencies differ from land-based plants. Some of the key design issues are listed next.

General design

- FLNG operations offshore will impose additional structural loads; for example, arising from topsides loads, sloshing in storage tanks, loads from ship to ship mooring during LNG transfer, and additional design accidental loads arising from activities on board.
- Continuous operation offshore, typically without dry-docking, for the life of the gas field will impose the need for higher design margin and quality in order to avoid the need for in-service repair or replacement, such as from marine environment corrosion and vessel fatigue. The corrosion protection system must be designed to meet a higher standard as the cost of any steel replacements offshore is very high and would require a complete plant downtime.
- FLNG vessel designs require naval architects to coordinate the liquefaction plant design and the FPSO design. It is necessary to optimize the integration of the topsides (i.e., the processing facilities including the major utilities) with the hull and vessel systems (i.e., utilities, control systems, and safety systems). The topsides layout must minimize overall footprint, weight, and cost for specific hull size because space is restricted since process facilities must be located away from the flare, helideck, and accommodation/control room units. Safety considerations for personnel must incorporate clearly designated personnel refuges and escape routes.

Process design

- Due to the space constraint, equipment is designed with "fit-for-purpose" criteria. This reduces the flexibility that may be required for future expansion. If the acid gas content in the feed gas increases, the treating solvent concentration can be increased but ultimately, the plant throughput must be reduced.
- Offshore designs impose limits on equipment dimensions and weights. Compact separation equipment is preferred. For example, lightweight aero-derivative gas turbines are more suitable than the heavyweight industrial gas turbines for offshore installation.
- Multilevel structures are used for equipment layout. Modularization of the process units would reduce installation cost. Modular designs for topside components that utilize standardized equipment can minimize construction costs and logistics. Standardization of plant reduces delivery time and when practical, modules are designed to a 2000 ton limit to facilitate crane availability. Larger modules can be designed if it proves to be more economical.
- The design should address energy efficiency but at the same time, must consider the impact on plot layout. The extra weight and costs of the waste heat recovery and steam generators must be justified by the reduction in emissions and utility consumption.

Operation

- The process equipment must be designed to continue operating despite bad weather conditions. Facility shutdown should be considered only under extreme weather or hazardous conditions.
- The process design must be simple and easy to operate, start up, or shut down. With the limited staffing offshore, the plant must be automated as much as possible, with minimal monitoring

requirements. Equipment design must be robust to handle all expected upset conditions. Equipment maintenance is performed during dry-docking.

Emissions/efficiency

- FLNG vessels are designed for zero hydrocarbon gas flaring under normal operating conditions. There are provisions for the occasional purge flow through the system, which will involve nominal flaring. A high integrity pressure protection system (HIPPs) is required to be installed at the FLNG gas inlet to shut down the gas flow during upset conditions thereby reducing the flaring requirements.
- Gas treating plant wastes, such as amine degradation products, cannot be processed on site. The waste can be stored on site and then transported to an onshore facility for processing.
- Gas turbine waste heat recovery may not be justifiable due to the additional weight and space requirements. High efficiency equipment must be evaluated in consideration of the impacts on space and weight on the FPSO design.
- NOx emissions, mainly from the steam boilers and gas turbines, can be minimized using low NOx burner.
- Liquefaction cycle operation must focus on simplicity and safety. A complex liquefaction cycle that has more equipment counts may not be the optimum choice for offshore. The less efficient but simpler cycle such as the nitrogen expander cycle may be preferable.

Safety

- Equipment spacing must comply with code and standards on safety for offshore installation. Despite congested spacing, egress from the FPSO must be carefully evaluated.
- The design must consider spill and containment of liquid, such as hydrocarbon refrigerants, and safe handling of refrigerants, particularly propane, which is a flammable hazard.
- The plant layout must consider the likelihood of uncontrollable releases of hydrocarbons, the possibility of hazardous gas accumulations, and the probability of ignition and spread of hazardous liquid and gases. The design must consider sufficient separation of hydrocarbons from ignition sources.
- Personnel must be protected from hazardous situations, such as high temperature around the fired equipment and gas turbines, and hazards from gas treating chemical and catalysts.
- The offloading facility must be proven for operation under extreme weather conditions. A quick connect and disconnect system and emergency shutdown system must be included in the design to respond to emergency conditions.

Vessel motions

- An FLNG vessel in operation is subject to the motion of the sea. Stability control systems and vessel design can minimize, but do not completely eliminate, the consequent motion of the process decks. Designing vessel mooring systems that facilitate weathervaning to minimize rolling motions would reduce the negative impacts from plant motion. Process equipment that is sensitive to vessel movement is best located close to a vessel's centerline. It is typically separators and other columns that require phase separation that are the most sensitive to vessel motions (Finn, 2009). As for all offshore facilities, FLNG vessel designs (i.e., integrated topsides and hulls) must be approved by a marine classification society such as Lloyds Register, Det Norske Veritas (DNV), and American Bureau of Shipping (ABS).

- LNG storage tanks must be designed to handle the stresses imposed by sloshing in partly filled tanks, particularly during the ship offloading operation when the liquid inventory in the vessel is changing.

Material of construction

- Topside process equipment is subject to salt water spray, which will cause stress corrosion cracking. Equipment must be protected by coatings or the use of corrosion resistant materials (Bukowski et al., 2011).
- Cold vapor release during emergency may cause metal embrittlement on the top deck. Design provisions, such as the use of a water curtain and suitable material, must be used.

FLNG rule and standards

Each component of the FLNG design must comply with the appropriate rule and standards. The diagram shown in Figure 1-51 identifies the required rules and standards that are to be used as guidelines for the design.

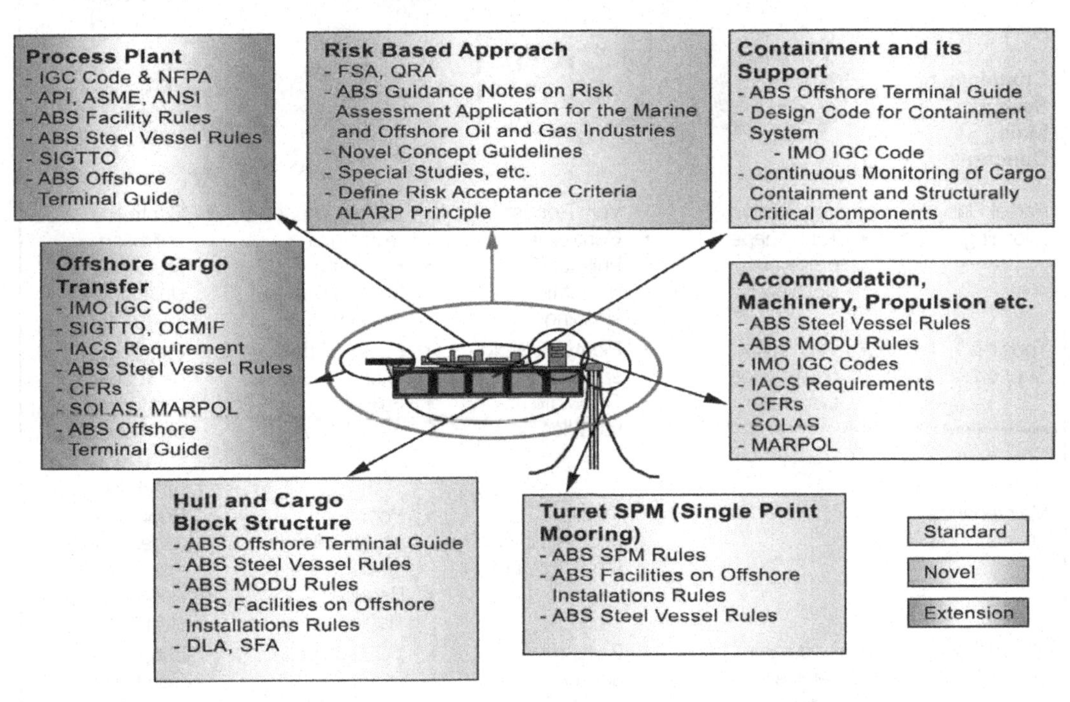

FIGURE 1-51

FLNG rules and regulations.

(Janseens, 2012)

1.5.1.3 Design challenges

There are many design challenges and solutions that may be used in configuring an FPSO design (Wood et al., 2007; Wood, 2009). Some of these design options are described next.

LNG storage

The transport of LNG in marine carriers is well established. However partial fill conditions in an FLNG facility is the prevailing status as the LNG is processed prior to off-take. This may result in sloshing, which is of particular concern in membrane tanks. The consideration of loss of containment must be addressed when considering hull fabrication.

The use of concrete for the hull provides benefits in the storage of cryogenic fluids as it retains its structural integrity when in contact with the LNG but traditional steel ship designs are cheaper to build. The self-supporting, prismatic Type B (SPB) LNG tanks offer some advantages in regards to FLNG

Table 1-6 Cargo Containment System for LNG Carriers

Containment System Main Concerns	MOSS	SPB	Membrane Single Row	Membrane Double Row
Partial Filling (Sloshing)	• Very Robust • Tank shape prevents sloshing	• Very Robust • Centreline bulkhead prevents sloshing	• Limited impact resistance • Full width tanks prone to sloshing	• Centreline bulkhead prevents sloshing
Topside Support	• Very limited deck space available	• Plenty of deck space • Full width modules only	• Plenty of deck space • Full width modules only	• Excellent deck space • Centreline bulkhead can support modules
Maintenance on site		• Excellent accessibility to tank structure	• Poor accessibility inside tanks • Requires staging	• Poor accessibility inside tanks • Requires staging
Other	• >100 ships in service • Mainly Japanese Shipyards	• 2 ships in service • 1 supplier (IHI) • Expensive	• >230 ships in service • Many yards	• Long. Cofferdam heating required

(Janseens, 2012)

over the dome or Moss tanks in that they provide robust tanks but also allow a flat deck space for process equipment.

In any design, a catastrophic tank failure may result in serious structural damage to the offshore facility. Design to avoid such an occurrence and methods to mitigate the problems must be addressed in the offshore designs.

There are generally four different types of cargo containment systems for FLNG. The design details were discussed earlier. These cargo types can be compared in terms of the main concerns on partial filling, sloshing, topside support, and maintenance as given in Table 1-6 (Janseens, 2012).

Marine offloading of LNG cargo storage

Offloading LNG in a marine environment requires bulk LNG carriers to approach and berth either in parallel (i.e., alongside a floating facility) or in tandem. This is routinely achieved for crude oil cargoes in various designs of floating production storage and offloading (FPSO) facilities, but constitutes a major hazard concern for the offshore option. There are hazards associated with potential collisions with approaching LNG carriers. Also the offloading dynamics must be designed to cope with relative motions between the floating structure and the ship that exceed those expected from shore-based jetties.

If offloading is considered with a typical spread-moored configuration such as might be found offshore West Africa, then side-by-side offloading could be considered. This provides the benefit in that typically LNG carriers load at midship, providing more flexibility. However in less benign seas, weathervaning and tandem offloading configurations are more appropriate. To facilitate this, a number of technology suppliers have designed flexible loading arms for transfer of LNG between the production vessel and the tanker such as the SBM soft yoke mooring and offloading (SYMO) system (Sheffield and Mayer, 2001; Faber et al., 2002; Poldervaart and Oomen, 2006).

Side-by-side transfer. Side-by-side transfer involves maneuvering a shuttle carrier alongside the FLNG, temporarily mooring the two vessels together, separated by fenders, conducting the transfer operation via connection to the carrier's midship manifold, then unmooring the vessels. Maneuvering would usually involve assistance of tugs. This operation has similarities with transfer at land-based terminals.

The main side-by-side technologies are:

- Rigid arms with extended envelopes and assisted connection
- Aerial hoses.

FMC has developed a loading arm system for the subsea and topside equipment for Shell's Prelude floating LNG project. The OLAF (Offshore Loading Arm Footless) system will build on FMC's existing Chiksan LNG loading technology, which is based on articulated rigid pipes equipped with Chiksan swivel joints and supported by a separate articulated structure. FMC will supply a total of seven loading systems: four for liquefied natural gas and three for liquefied petroleum gas. The OLAF has no base riser; instead, the articulated assembly is bolted on a turntable at deck level, which allows a 20% reduction in the length of the arm and connection to carriers up to 10 m below Prelude's deck. The advantage is that the OLAF can reach a very low operating envelope to accommodate the large differences in height at deck manifold level that can arise between an LNG shutter tank and FLNG. An artist rendering of the system is shown in Figure 1-52.

FIGURE 1-52

FMC OLAF loading arms.

(McCulley, 2012)

Tandem transfer. Tandem unloading will require LNG carriers purposely built with a bow manifold. The vessels are connected by hawser line and transfer is from stern of FLNG to bow of shuttle tanker. However floating hose solutions may also permit connection to midship manifolds.

Currently, several vendors have certified hoses for LNG service. The system is composed of a flexible cryogenic floating LNG hose and connectors designed to allow the transfer of LNG between vessels in a tandem moored configuration, exactly the same way that traditional floating oil and gas facilities carry out cargo transfers.

Figure 1-53 shows a rendition of the tandem transfer proposed by SBM using SBM's COOL™ hose design. The hose comprises an outer marine hose with an inner composite LNG hose, with the space in between filled with insulating materials. The connector system is a quick connect/disconnect system that handles and connects the hose to the LNG Carrier bow manifold (Malvos, 2012).

Figure 1-54 shows another hose system proposed by Framo Engineering and Aker Solutions. The OCT system is based on adopting the tandem ship-to-ship configuration for crude oil loading operations. The system allows 100 meters separation between the vessels during transfer, thereby maintaining a high level of safety during all operations. The system consists of three flexible pipes—two for LNG transfer and one for vapor return—suspended from movement-free swivels and supported by an A-frame-type crane. When the LNG carrier is in position for connection, the A-frame is lowered and the pipes are simultaneously pulled over to the bow coupling manifold using the carrier's pull-in winches and receptacles. The pipes are then pulled in position one by one and locked to their respective bow coupler (Pusnes News, 2012).

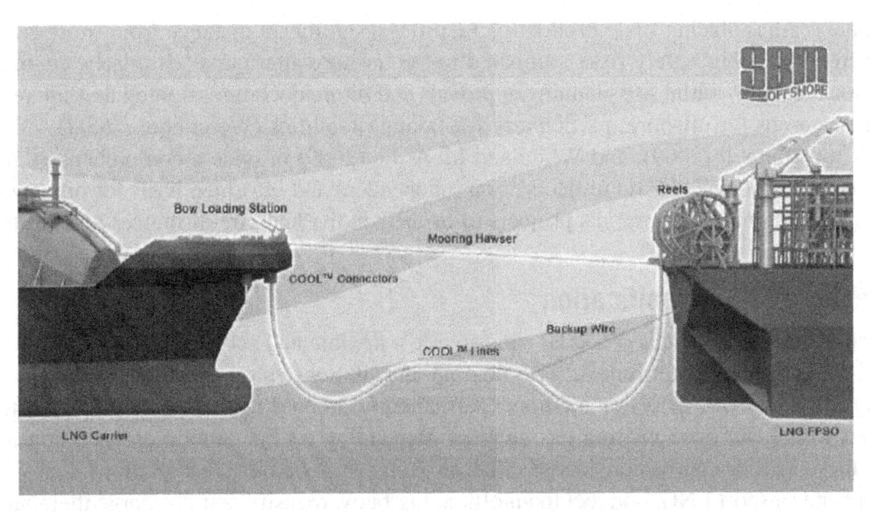

FIGURE 1-53

SBM COOL™ hose design.

(Malvos, 2012)

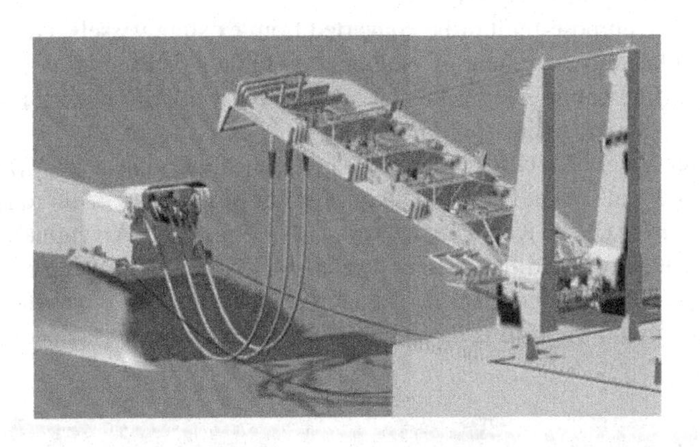

FIGURE 1-54

OCT system.

(Pusnes News, 2012)

1.5.1.4 Geopolitical challenges

The floating LNG (FLNG) technology has taken more than two decades to reach commercial reality. The hurdles are not all technical and operational. Geopolitics represents another formidable stumbling block. Countries that stand to gain the most from floating liquefaction plants, like Nigeria, for example, often insist upon substantial local content in its manufacture, which is almost impossible to accommodate on a commercial basis. Most governments also prefer the guarantees of long-term deployment, local direct and indirect employment that an onshore facility provides. On the other hand, in some regions placing LNG production facilities a significant distance from shore can remove actual or perceived public safety risks compared to the onshore alternative. Similarly, environmental impacts associated with initial site planning approvals and future decommissioning and site restoration become less onerous for offshore, particularly for floating facilities (Wood et al., 2007).

Kerbers and Hartnell (2009), and White and McArdle (2009) provide a comprehensive review of development issues of LNG FPSO projects as an independent and informed basis for organizations to consider as they develop offshore gas projects to fit their individual circumstances and strategies.

1.5.2 Offshore LNG regasification

Floating regasification terminals (FSRUs) can provide a flexible and economic alternative to building land-based LNG receiving terminals. The floating installations may either be located near-shore (e.g., alongside a pier or a jetty) or offshore (permanently moored to the sea floor). Such terminals are supplied with LNG from visiting gas carriers. Depending on their mode of operation and local regulation these FSRUs can be considered either as ships or as offshore installations.

Vessels that transport LNG, connect to an offloading buoy, regasify and discharge their cargo, then disconnect and leave to collect a new cargo are termed Shuttle Regasification Vessels (SRV). These are not normally considered as FSRUs and are readily addressed within maritime classification. SRVs will therefore not normally be engaged in continuous regasification operations and typically will not receive and regasify LNG simultaneously.

FSRUs may either be purpose-built or be converted from existing vessels, typically LNG carriers. Depending on the application the extent of modification to an existing unit will vary; for example, installation of a turret (internal or external) on an offshore moored FSRU or location of loading arms on the FSRU, or on a jetty for a near-shore FSRU.

LNG regasification vessels (LNG RV or shipboard regasification vessels [SRV]) have operated successfully over the past decade around the world (e.g., Gulf of Mexico and offshore New England in the United States, Teeside in the United Kingdom, Kuwait, Dubai, Brazil, and Argentina). An LNG RV has an onboard regasification facility in addition to the conventional LNG carrier's cargo containment tanks and the offshore mooring capability (Figure 1-55). The Gulf Gateway LNG RV was the first offshore LNG facility to be deployed in the world in 2005 and has since been decommissioned. (Lee et al., 2005).

FIGURE 1-55

LNG regasification vessel.

(Kim and Lee, 2005)

The LNG RV offers significant flexibility because of its standardization, interchangeability, and quick deployment to a range of potential locations. The ability to unload its cargo as gas by latching on to an offshore buoy, through a gas manifold through a jetty docked in port, or conventionally as LNG makes these vessels attractive for seasonal or intermittent deliveries to smaller LNG customers. They can avoid the long lead time and high cost of building a land-based regasification plant or can be used as an interim solution while more permanent and larger-capacity land-based systems are built.

The Gulf Gateway, the Energy Bridge regasification ship that was developed by Excelerate, was the first offshore LNG facility to be deployed in 2005 (Lee et al., 2005). This system involves the use of purpose-built LNG tankers incorporating regasification technology and has demonstrated the capability in supplying fuel gas to the Gulf areas during Hurricane Katrina in August 2005.

Höegh LNG and GDF Suez developed the Neptune Deepwater Port (DWP) around two LNG RVs, which became operational in 2010. It is located 10 miles off the coast of Massachusetts and involves an offshore terminal with two buoys and associated pipelines to shore. The two vessels, GDF Suez Neptune (delivered 2009) and GDF Suez Cape Ann (delivered 2010), were built by Samsung Heavy Industries in Korea specifically for the Neptune project. The vessels incorporate the reinforced GTT MKIII cargo containment system with 145,000 m^3 cargo carrying-capacity at 100%. They are fitted with three regasification skids for a total output capacity of approximately 21 million standard m^3 of natural gas per day ($\sim$0.75 bcfd), and are also fully capable of operating as standard LNG carriers.

As of today, there are over 10 offshore regasification vessels built and more than two dozen others are in various stages of development. The underlying enabling technologies such as ship to transfer are well demonstrated and are now almost universally accepted as a viable solution in LNG delivery.

1.5.2.1 Submerged buoy systems

A deepwater port facility usually involves one (or perhaps two) submerged buoys and subsea facilities in one defined location. As the submerged buoy location is known to approaching LNG ships, it is possible for them to locate the mating site with the aid of a differential global positioning system. In the vicinity of the submerged buoy and during the buoy-connection operation an acoustic positioning system is also used to confirm the buoy's exact location and as a self-reliant sensing device. The submerged buoys are fitted with transponders, so the location of the submerged buoy can be detected by a shipboard acoustic transceiver. To approach the mating site and to assist positioning the vessel over the buoy center, the LNG RV vessel is equipped with bow and stern thrusters to provide dynamic positioning.

The LNG RV is moored via its turret system to the submerged buoy, and is free to weathervane around the turret to minimize the drag and ship motion during the regasification operation (e.g., Lee et al., 2005). The vessel motion in the moored position is more restricted than when the vessel is sailing or unattached from the buoy. A swivel system provides a high pressure natural gas connection between the weathervaning LNG RV and the fixed submerged buoy (Kim and Lee, 2005). The normal buoy connection and disconnection takes approximately 1 to 3 hr, but in an emergency situation, the buoy can be disconnected very quickly, in approximately 15 min, and the LNG RV evacuated to a safe location. The vaporized natural gas from the regasification system is connected to the swivel mechanism and submerged buoy through an onboard flexible riser. The turret system typically includes submerged buoy locking mechanism, rope guider, traction winch, swivel-handling system, flexible riser, closed-circuit TV, hydraulic, and control system.

The rest of the subsea system to which the submerged buoy is attached is composed of a subsea riser, pipeline end manifold (PLEM), and subsea pipeline. The buoy typically is located approximately 25 m below the sea surface and moored by a number of spread mooring lines and requires a minimum overall water depth of about 30 m. Shutdown valves and transmitters are located in the PLEM and can be operated by the onboard LNG RV control system. There is usually the requirement for an independent gas metering system to facilitate custody transfer of natural gas dispatched shoreward in the high-pressure gas pipeline.

1.5.2.2 Offshore regasification safety

In addition to the normal shipboard control system, an emergency shutdown (ESD) system is required to shut down the critical valves and the subsea equipment in emergency situations. The integrated LNG RV ESD system needs to coordinate the safety features for the regasification, turret, buoy, and PLEM systems.

During regasification the LNG stored in the vessels' cargo tanks is fed into a suction drum. High pressure pumps take the LNG from the suction drum and send it to the high pressure vaporizers. The LNG is vaporized in the vaporizers and fed to the export gas pipeline network through the turret system. Most LNG RVs are equipped with an alternative gas offloading system for use when at a port jetty, which flows gas through a midship high pressure natural gas manifold.

Protection against LNG spillage and high pressure natural gas leaks are the main concerns during offshore regasification operations onboard an LNG RV. The high pressure LNG lines between the high pressure pumps and the vaporizers typically are kept short to reduce the risk of high pressure LNG spillage. The LNG handling parts of the regasification area typically are protected by stainless steel drip trays and pans to safely collect any LNG spillage. The flange parts of high pressure natural gas lines also typically are protected by shield devices to isolate and deflect jet flames. In addition a deluge system is provided to the regasification area, and sea water is kept flowing to the protection shield and drip pan continuously during regasification operations. Water spray and gas detection systems also typically are provided in the regasification area.

1.5.3 Floating storage and regasification units (FSRU)

The function of the FSRU is similar to that of the LNG RV. The FSRU and the LNG RV can compete with onshore LNG receiving terminals offering speed-of-deployment advantages in the parallel processing of regulatory authority permissions and facility design and construction. FSRUs can be designed and built to include more storage capacity than that of LNG RV, which has the cargo containment size of an LNG carrier (Lee et al., 2008). The storage capacity of FSRUs is typically in the 145,000 m^3 to 350,000 m^3 range and is determined by the cargo tank size and visiting frequency of the shuttle LNG carriers, taking into account storage capacities to compensate for potential supply interruptions caused by weather conditions. In some cases standard LNG carriers are converted to operate as small FSRUs.

The typical regasification capacity for an FSRU is in the range 0.5 and 2.0 bcfd (billion cubic feet per day) and is not only driven by the storage capacity of the FSRU vessel. Regasification capacity can be increased, as necessitated by market demand, provided sufficient space for power generation and regasification equipment is assigned in the design stage. FSRUs can be permanently moored in distant offshore locations utilizing turret systems or moored adjacent to a jetty in port. When necessary, FSRUs can be designed to have thruster-propelled dynamic positioning systems for extreme weather conditions. The regasification is typically by seawater or heat from onboard boilers. In some port areas seawater discharge from vaporizers is not permitted or in areas where seawater temperature is too cold, onboard boilers or other heating sources are used for the regasification operation. However, this adds to the regasification operating costs. In FSRUs, gas turbines usually are used because of the convenient availability of fuel and efficient heat recovery from gas turbine exhausts.

There are a number of projects in the planning and construction stage for LNG RV and FSRU vessels around the world (e.g., Indonesia, Lithuania, and East Mediterranean), which testifies to their growing popularity.

The process units in an FSRU are basically the same as a land base regasification plant. Figure 1-56 shows the different process units in an FSRU.

FSRUs can be used to supply fuel gas for power generation of a medium-size power plant. Generally when used for power generation, the LNG sendout requirement is relatively low. For example, with an efficient combined cycle design, the regasification plant sendout requirement is only about 100 mmscfd in order to generate about 400 to 500 MW electric power.

Compared to land-based terminals, the boil-off rate from an FSRU is much higher, at 0.1 volume % per day, compared to 0.05% of a land-based tank. To avoid loss of natural gas from the BOG, a vapor recovery system should be installed, either with a BOG recondenser, a vapor recovery compressor, or a BOG liquefaction unit.

If the FRSU is to operate in the arctic area, seawater can be used during summer operation. During winter, because of the cold seawater, supplementary heating is necessary using a heat transfer fluid and steam heater, as shown in Figure 1-56.

1.5.4 Gravity-based offshore regasification units

Fixed offshore structures offer an alternative to onshore and floating facilities for regasification of LNG. These could take the form of artificial islands, fixed platforms, or gravity-based structures (GBS). It is the latter of these options that has been developed so far.

The world's first offshore GBS LNG receiving terminal is located in the northern Adriatic Sea 15 km off Porto Levante, Italy, where it is capable of supplying about 10% of Italy's natural gas requirements. The Adriatic LNG terminal is designed to store and regasify LNG to deliver some 775 mmscfd of gas (8 bcma) and it reached that operational capacity in 2009. The facility is majority owned by ExxonMobil and Qatar Petroleum and it receives LNG primarily from Qatar.

The Adriatic LNG terminal was built in Algeciras, Spain and upon completion in 2008 was floated and sailed via a 3000-km trip to its location off Porto Levante. It is situated in a water depth of some 30 m (95 ft) and is connected via pipeline to Italy's natural gas grid. The GBS structure houses in its concrete skirt two LNG storage tanks, a regasification plant, and facilities for mooring and unloading LNG vessels. The concrete used is adapted to handle contact with, and contain, cryogenic liquid in the event of a spill. Also included in the facility are an accommodation unit for its operating crew, gas metering units, an emergency flare system, power, and other utility systems.

GBS structures are generally more expensive and time-consuming to build and install than LNG RV and FSRU, but offer more permanent facilities that can be expanded in phases over time, if necessary. They do pose more significant decommissioning costs than floating alternatives.

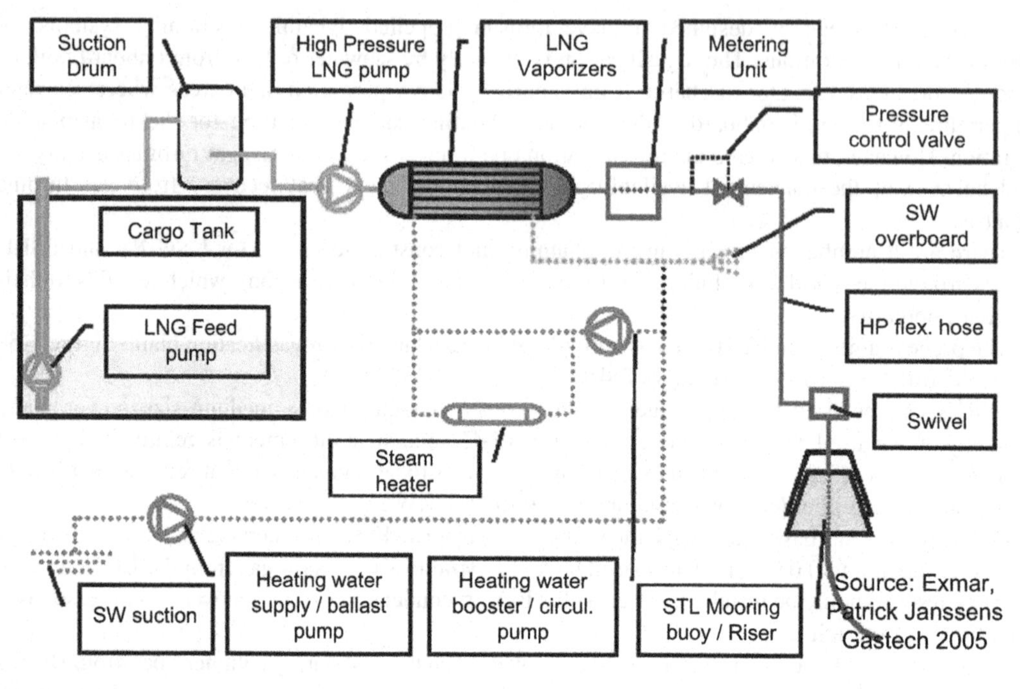

FIGURE 1-56

Schematic of regasification system.

(Janseens, 2012)

1.5.5 Heating value control in offshore LNG regasification plants

When rich LNG with high heating values is being imported, they may not meet the local gas specifications in terms of Wobbe Index and heating value. In such a situation, the propane and heavier components can be separated from the vaporized LNG using fractionation. One of the processes that can be used for managing the heating value of the vaporized product is a patented process (Mak, 2012) as shown in Figure 1-57. The conceptual process consists of an offshore regasification plant and an onshore LPG extraction plant.

The offshore plant uses the Rankine power cycle to supply heat for LNG regasification while generating power for the offshore facility. The Rankine power cycle can produce sufficient power for the offshore consumption without the use of fuel.

As shown in the flow diagram, LNG from the storage tank is pumped by the low pressure LNG pump to an intermediate pressure such as 100 psig, and then pumped by the high pressure LNG pump to supercritical pressure, typically 1650 psig or higher. LNG is then heated and partially regasified to about 0°F using the condensation duty from Rankine power cycle (see Chapter 10).

It is important to note that the LNG is only partially regasified to 0°F instead of completely regasified to 40°F. Also the operating pressure of 1600 psig is higher than the final gas pipeline pressure requirement. The high pressure (supercritical) and low temperature of the semi-regasified

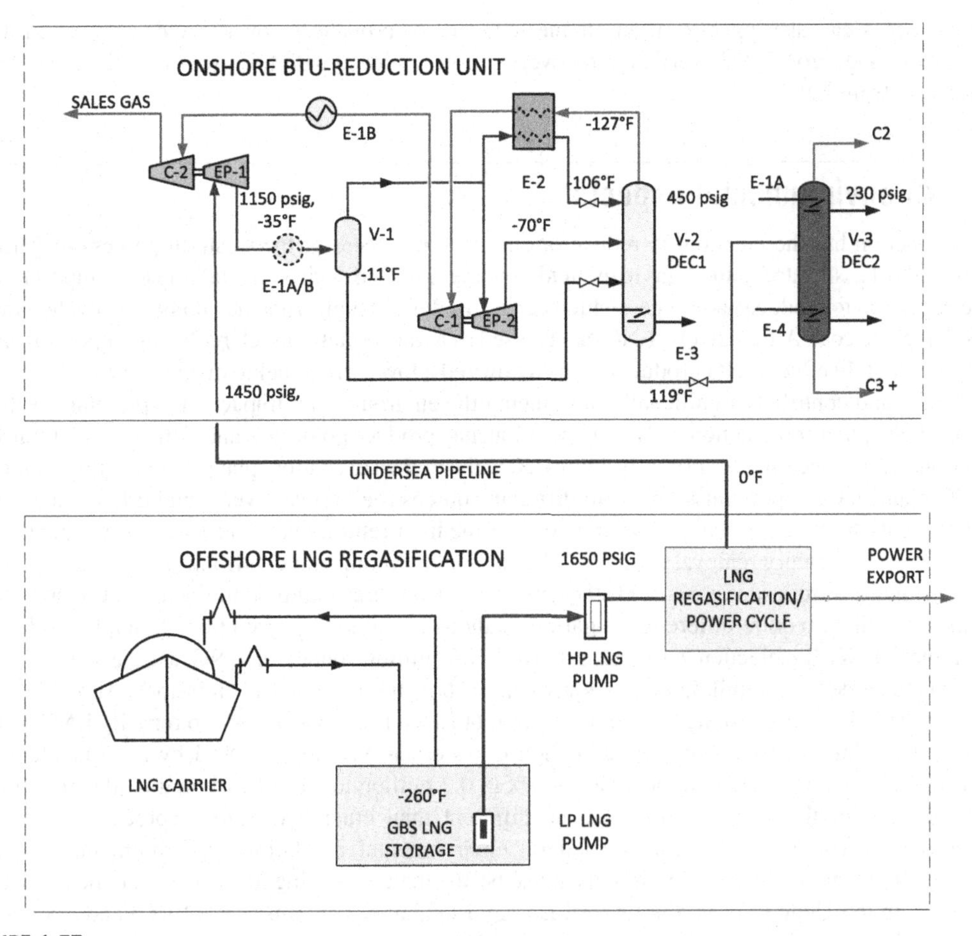

FIGURE 1-57

Offshore NGL recovery.

(Mak, 2012)

LNG effectively preserves the refrigeration content in LNG that is needed in the offshore fractionation facility. Also the regasification duty at 1650 psig and 0°F is significantly less than that of the typical lower regasification design, hence reducing the heating duty in the offshore equipment.

The supercritical fluid arrives onshore at about 1650 psig and is let down in pressure to about 1100 psig in the first stage turbo-expander, which generates power to operate the second stage residue gas compressor. The expansion process also chills the feed gas to a lower temperature. The refrigeration content of this stream is utilized in the deethanizer reflux condenser for separation purposes.

The heated streams arriving from offshore are partially condensed and separately fed to the demethanizer. To improve NGL recovery levels, vapor is split into two portions; one portion is routed to the reflux condenser and the other portion is let down in pressure in the second stage turbo-expander, which chills the gas, supplying the cooling duty for fractionation in the demethanizer.

The process can also operate in an ethane recovery or propane recovery mode by varying the refrigeration usage from LNG. Very high recovery can be achieved with this process utilizing the free refrigeration from LNG.

1.6 LNG environmental aspects

LNG production has the potential to reduce or eliminate the need for the wasteful process of flaring (burning off) of associated natural gas in oil fields and the resulting environmental impacts. Import LNG can be vaporized to produce natural gas, which can be used in the high efficiency combined cycle power plants to replace coal. Alternatively, LNG can be used as a transportation fuel, replacing import oil, and is a cleaner fuel. In effect, LNG production has recovered a low carbon fuel to displace the high carbon oil and coal, and contribute significantly in reducing the environmental impacts. Despite this positive outcome, LNG plant construction and plant operation may produce pollutions and effluent, which can be an environmental concern if not properly addressed in the project execution, plant design, and operation.

LNG plants have vapor emissions from different sources (i.e., process vents and driver exhausts), liquid effluents from sumps and drains and cooling medium return, and solid waste from spent molecular sieve and mercury removal catalyst. Other major environmental contributions are construction impacts, emissions from sour gases, and stack emissions from fired equipment. Each of these potential contaminants might require different mitigation methods in different jurisdictions around the globe.

A typical LNG liquefaction terminal exporting 4.5 million tonnes of LNG can be expected to produce in the order of 1.2 million tonnes equivalent carbon dioxide of direct emissions. About 10% of the fuel value in LNG is consumed in the production of LNG. If fuel gas is used to regasify LNG using submerged combustion vaporizers, the greenhouse gas emissions are increased by an equivalent of combusting about 1.5% of the import LNG, or about 0.2 million tonnes of carbon dioxide emissions.

The designs of the LNG facilities are no different than other hydrocarbon processing plants. Emissions must comply with local and federal environmental regulations. Environmental impact assessment must be completed; emissions must be estimated, starting from the well head to the consumers. In the United States, the Federal Energy Regulatory Commission (FERC) and US Coast Guard must be involved in the permitting processes. The environmental impact statement is part of the overall safety and security assessment.

1.6.1 Emission sources

From an emissions viewpoint, an LNG plant is relatively clean when compared to refineries or petrochemical plants. Some of the major sources of emissions from the different units within the LNG supply chain can be summarized as follows.

1.6.1.1 Feed gas conditioning

- Condensate Stabilization Unit – waste water and hydrocarbons
- Glycol Regeneration – chemical waste, hydrocarbons, and waste water
- Acid Gas Removal Unit – mercury, solvent waste, and waste water
- Dehydration – mercury, spent mole sieve waste, and waste water
- NGL Recovery Unit – not significant

- Sulfur Recovery, Tail Gas Unit, and Incineration – CO_2, SOx, H_2S, and NOx emissions, and spent catalysts
- Refrigeration Units – CO_2, SOx, H_2S, NOx emissions, and particulate matters from gas turbine drivers
- Power Generation – CO_2, SOx, H_2S, NOx emissions, and particulate matters from gas turbine drivers, waste water, chemical waste.

1.6.1.2 Liquefaction

- CO_2, SOx, H_2S, NOx emissions, and particulate matters from the exhaust stacks from gas turbine drivers, a major contributor of CO_2 emissions
- Increase cooling water and ambient air temperatures.

1.6.1.3 Transport

- CO_2, SOx, NOx emissions, and particulate matters from LNG carriers propulsion engine drives
- Boil-off gas emissions.

1.6.1.4 Regasification

- CO_2, SOx, and NOx emissions from submerged combustion vaporizers
- Cold seawater return
- Fog generation if ambient air vaporizers are used for LNG heating.

The distribution of CO_2 emissions for the LNG supply chain, from the well head production, gas processing, and natural gas liquefaction; LNG shipping to the receiving terminals; and regasification is shown in Figure 1-58. As can be seen, the majority of the CO_2 emissions is from the gas processing and liquefaction step, following by LNG shipping and the LNG regasification.

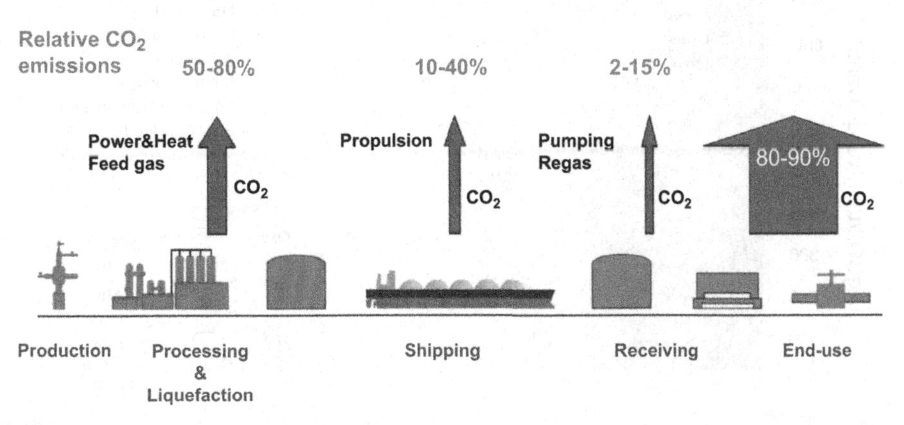

FIGURE 1-58

Typical emission sources in the LNG supply chain.

(Neeraas and Marak, 2011)

Although these emissions can be minimized through new technologies and innovation, emissions from combustion of natural gas contributes to 80 to 90% of the emissions. CO_2 emissions from the end users can only be reduced through energy conservation.

1.6.2 Near zero CO_2 emission

Figure 1-59 shows the potential improvement in terms of kg of CO_2 per ton of LNG delivered for the various technologies. The base case technologies for a typical LNG supply chain consist of the following:

- The liquefaction plant is located in a warm climate area processing feed gas containing 10 mole % CO_2. CO_2 from the gas treating unit is vented to the atmosphere.
- The liquefaction drivers are based on the simple cycle Frame 7 gas turbine.
- There is no waste heat recovery.
- The air cooler is used for the refrigeration units.
- Ship propulsion uses steam turbine drivers. It is assumed 5000 km LNG transportation distance using a conventional LNG carrier.
- The regasification terminal uses fuel gas for heating the LNG.

The improvement of natural gas delivered to the consumer results in lower CO_2 emissions by employing updated technologies, such as CO_2 reinjection for sequestration, more efficient mixed refrigeration liquefaction cycle, waste heat recovery, cold seawater for cooling, more efficient aero-derivative gas turbines, and combined cycle power plant. The use of ORV instead of SCV can save 15% of the natural gas fuel. If LNG cold is used for air separation, the energy efficiency will be further improved (see Chapter 10).

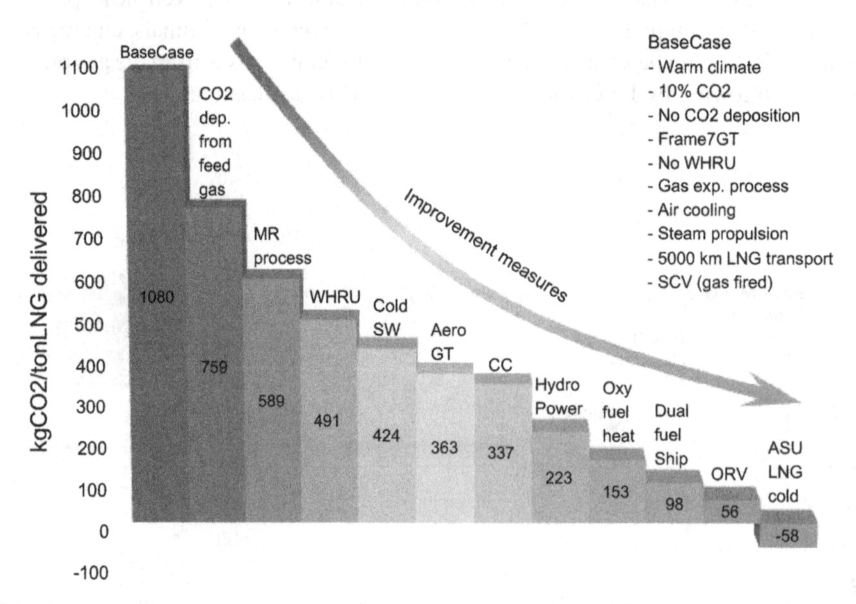

FIGURE 1-59

CO_2 reduction stairs.

(Neeraas and Marak, 2011)

In summary, CO_2 emissions from the LNG supply chain can be significantly reduced with the best available technologies as follows:

- Improved plant efficiency on liquefaction cycles and equipment
- CO_2 reinjection from the amine unit for sequestration
- More efficient power plant for ship propulsion
- Utilization of LNG cold in the receiving terminals.

If the distance for transportation is short and there is renewable energy such as hydroelectric power or solar available, CO_2 emissions can be further reduced. In short, a zero emission LNG supply chain is possible, as shown in Figure 1-60.

1.6.3 Environmental impacts

The environmental impacts associated with LNG facilities are project specific. On the liquefaction side, it depends on the feed gas compositions and acid gas contents. High acid gas content feed gas would generate higher sulfur oxides and carbon dioxide emissions. On the regasification side, the location of the terminal determines the source of heating used for LNG vaporization. A cold climate regasification terminal would require submerged combustion vaporizers, which would generate more emissions than a warm climate location where seawater is available. However, there are impacts on marine life from the cold water discharge from an LNG plant. There are other impacts and risks that must be assessed for the LNG facilities in determining its overall environmental impact. Some of the environmental impacts, potential hazards, and design considerations to reduce the impacts are discussed in the following section.

1.6.3.1 Site selection

LNG plants, regasification plants, and associated terminal facilities should be located at sites that minimize the effects of natural hazards, and measures should be adopted to provide for the safe operation of LNG vessels at the terminal. LNG plants, regasification plants, and terminal facilities

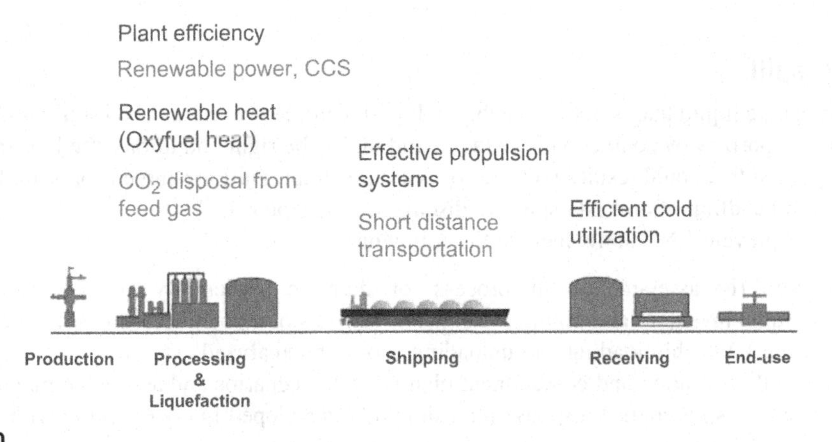

FIGURE 1-60

Zero emissions in LNG supply chain.

(Neeraas and Marak, 2011)

should be located, designed, constructed, and operated to minimize the risk to the project from natural hazards such as earthquakes, tidal waves, floods, and fires from surrounding areas. Exclusion zones should be established to minimize the potential hazards of the project to immediate areas from the results of accidents. Site selection should minimize the project's impact on sensitive ecosystems such as sea life, corals, fisheries, and resources that support local communities.

A site should be selected that can comply with National Fire Protection Association LNG Standards or equivalent standards. Terminal operating procedures, security measures, and emergency response plants must comply with the best industry practice and international guidelines, such as the World Bank Group (1999) Operational Policies and guidelines set forth in the Pollution Prevention and Abatement Handbook.

The construction activities also have significant impact on sea life and the surroundings. Construction; dredging; construction of piers, wharves, breakwaters, and other water-side structures; and soil erosion may lead to long-term impacts on sea life and shoreline habitats. Construction debris, soil sediment, and wastewater discharges may deteriorate water quality in the nearby communities.

1.6.4 LNG handling hazard and risks

Storage, transfer, and transport of LNG may have leaks and accidental vapor releases. Misoperation during storage, unloading, and loading of LNG may pose a fire hazard. To manage such risks, LNG storage tanks, piping systems, and instrumentation must be designed to comply with international safety standards to prevent fires and explosion hazards. The design must handle the different operating modes, from holding operation to LNG unloading operation, from startup to shutdown, as well as natural and climate hazards. The hazard risk can be minimized by design provisions such as storage tank overfill protection, secondary storage containment, and boil-off gas management control. Periodic inspection and maintenance programs must be established.

Loading and unloading of LNG cargoes must be carried out by licensed and well-trained personnel. The procedure should include all phases of the LNG delivery, loading and unloading the LNG, connect and disconnect of the unloading arms or hoses, and strict enforcement of a no-smoking policy for any personnel.

1.6.5 LNG spill

LNG is a cryogenic liquid that is not flammable in liquid form. However, an LNG spill produces boil-off gas as it is vaporized by contact with surfaces, and under the right conditions, the boil-off gas can form a vapor cloud that could result in jet fires or pool fires if an ignition source is present. Details on the hazards and handling of an LNG spill are discussed in Chapter 9.

Measures to prevent LNG spills include the following:

- Conduct a spill risk assessment of the process flow diagram, particularly on vessels with a significant liquid inventory and piping system. The impacts of a potential spill during operations such as during LNG ship loading and unloading should be analyzed.
- Develop a spill prevention and containment plan for spill scenarios and assess the magnitude of vapor releases. A spill control response plan should be developed in coordination with local regulatory agencies.
- Facilities should be equipped with hydrocarbon detection systems to monitor gas leakages.

- An ESD system should be configured to initiate automatic shutdown of the LNG transfer system in case of emergency.
- For unloading/loading of LNG, proper procedure should be stressed to safely purge and drain the liquid inventory prior to disconnect.
- Onshore LNG storage tanks should be designed with secondary containment, such as the use of a full containment tank.
- Impound areas should be designed to contain a credible spill resulting from breakage of outlet piping from vessels and equipment.
- Impound areas should be ventilated to minimize gas accumulation.
- The facility drainage system should be designed to contain the spill and to reduce fire and explosion risk.

1.6.6 Process heating and cooling

In the design of the natural gas liquefaction cycle, air coolers should be used as much as possible instead of cooling water in order to minimize water consumption of the facility.

In LNG regasification terminals where seawater is used for LNG heating, cold water should be discharged to the ocean body in a location that will allow maximum mixing and cooling of the thermal plume. The cold water discharge must be within 3°C of surrounding water at the edge of the mixing zone or within 100 meters of the discharge point, as noted in the International Finance Corporation (IFC) guideline.

If biocides or chemicals are used, chemical additives must be selected considering dose concentration, toxicity, biodegradability, bioavailability, and bioaccumulation potential. The residual effects of these chemicals on marine life must be carefully evaluated.

1.6.7 Waste water streams

Nonhazardous and hazardous wastes include general wastes such as waste oils, hydraulic fluids, waste chemicals, filter elements, solvent wastes, spent molecular sieves, and spent catalysts and chemicals. These chemicals and solid waste must be properly disposed of or recycled. Efforts should be made to recycle and reclaim the waste materials where possible. If recycling is not practical, wastes must be disposed of in an environmentally safe manner using the best technologies and practices, which must also comply with local laws and regulations.

1.6.7.1 Air emissions

Air emissions in LNG facilities are mainly from the exhaust stacks from the gas turbines in refrigeration units, utility steam boilers, and gas turbines in power plants. Maximum emissions from the equipment must be specified in the duty specifications of the rotating equipment.

In the regasification terminals, the selection of LNG heaters must consider the difference in emissions and flexibility among a variety of heat exchange equipment; for example, submerged combustion vaporizers (SCV), open rack vaporizers (ORV), shell and tube vaporizers, and ambient air vaporizers. If waste heat is available nearby, such as hot water discharge from a power plant, they should be considered as a heat source for LNG regasification to reduce the overall environmental impact.

1.6.7.2 Venting and flaring

Flaring or venting is necessary to avoid build-up of excess gas in the system and to safely dispose of the excess gas in the event of an emergency, such as system shutdown, or other plant upset conditions. Every effort should be made to avoid venting and flaring during normal operation by recovering the excess gas as fuel gas.

1.6.7.3 Boil-off gas management

In the liquefaction plant, boil-off from storage tanks is compressed to the gas turbines inlet, to be used as fuel gas or reliquefied in the process. In the regasification terminal, BOG from storage tanks is compressed by a BOG compressor and recondensed in a BOG recondenser using a portion of the cold LNG sendout. No BOG venting is allowed during normal operation.

1.6.7.4 Fugitive emissions

Fugitive emissions generally originate from vents, leakage from instruments, valve packing, seals, and pressure relief valves. Equipment and instruments should be designed to meet the emission requirements according to local industrial standards.

1.6.7.5 Noises

The main noise generation sources in LNG facilities are pumps, compressors, generators, drivers, fin fan air coolers, LNG vaporizers, and heaters. Installation of noise-dampening devices should be used where necessary.

1.6.7.6 LNG transport

Emissions from tugs and LNG carriers, especially where the jetty is close to local communities, are a significant emission source that may impact the surrounding air quality. Air emissions from LNG vessels during its period at port must be evaluated for compliance with the local and state regulatory requirements.

1.7 Costs and economic implications of LNG facilities

1.7.1 LNG overall project costs

LNG projects are inherently capital-intensive, with the liquefaction plant accounting for up to 50% of total project cost. Table 1-7 summarizes the cost components of delivered LNG in 2003. As can be seen, the average cost for gas production from the reservoir to the LNG plant (including gas processing and associated pipelines) will vary between 0.5 and 1.0 $/MMBtu for most large conventional natural gas fields. The contribution of the liquefaction plant cost to the cost of delivery of LNG ranges from 1.5 to 2.0 $/MMBtu. However, from 2003 to 2012 capital costs in the oil and gas sector have more than doubled due to cost inflation. Table 1-8 shows the approximate percentage cost distribution for the LNG supply chain, which remains representative in 2012.

The total cost to bring LNG to market in 2003 was in the range of around 2.80 to 4.60 $/MMBtu. In 2012 a more realistic range of costs to deliver LNG along its full supply chain from gas field to customer was in the range of $6/MMBtu to $9/MMBtu. This sets the base price under which LNG would not be produced and exported. What will ultimately set the price for LNG above the base price are supply and demand and the cost of competing fuels.

Table 1-7 Typical Cost Allocation of the LNG Project

Components	Low-End Cost ($/MMBtu)	High-End Cost ($/MMBtu)
Gas production	0.5	1.0
Base load liquefaction	1.5	2.0
LNG transportation	0.5	1.2
Receiving terminal	0.3	0.4
Total	2.8	4.6

(Coyle et al., 2003)

Table 1-8 Approximate Cost Distribution of the LNG Plant

Percent of Total Cost	Gas Treatment	Liquefaction/ Refrigeration/ N2 Removal	Fractionation	Utilities and Offsites	LNG Storage and Loading	Total
Equipment	4	14	1	10	1	30
Bulk material	3	7	1	5	4	20
Construction	4	6	2	8	15	35
Miscellaneous	1	5	1	4	4	15
Total	12	32	5	27	24	100

(Kotzot, 2003)

Accurate data on LNG plant costs are difficult to pinpoint since costs vary widely depending on location and whether a project is greenfield (i.e., built in a new location) or an expansion of an existing plant. LNG transportation costs vary based on the tanker's operating and amortization costs, the size of the tanker, tanker availability, and the distance transported.

A typical contribution of the transportation cost to the cost of delivered LNG was about 0.5$/MMBtu to 1.2 $/MMBtu in 2003, but by 2012 this had risen to a range of about $1.5 $/MMBtu to 3.2 $/MMBtu, reflecting in part longer supply chains and higher demand for LNG carriers.

The costs of building regasification or receiving terminals show wide variation and are very site-specific. The typical cost of the LNG terminal ranged from 0.3$/MMBtu to 0.4 $/MMBtu in 2003, and the costs more than doubled by 2012. The allocation of different units in an LNG regasification terminal is shown in Table 1-9.

The most expensive items in an LNG receiving terminal are the LNG storage tanks, which can account for about one-third to one-half of the entire facility cost, depending on the type, local planning code regulatory requirements, and the number and size of tanks.

1.7.1.1 Liquefaction plant cost

The engineering, procurement, and construction (EPC) costs of liquefaction projects are typically quoted in dollars per metric ton of annual plant capacity ($/TPA) rather than in $/MMBtu. Phalen and

Table 1-9 Approximate Cost Allocation in an LNG Receiving Terminal

Unit	%
Jetting	11
LNG storage	45
Sendout	24
Utilities	16
General facilities	4

(Tarlowski et al., 2005)

Scotti (2008) provided unit cost data for liquefaction plants built between 1964 and 2005, which showed, in money of the day, costs rising from some $60/TPA in the late 1960s to a peak of some $400/TPA in the late 1980s (NWS plant train 1 and 2, Australia) and then falling to about $200/TPA by 2003. Oman and Trinidad (Jamieson et al., 1998) greenfield liquefaction plants were among the lowest in the late 1990s in terms of $/TPA (Yost and DiNapoli, 2003). Using data from Poten & Partners, Phalen and Scotti (2008) indicated how capital costs for liquefaction plants started to increase from their 2003 low point, with some plants in the 2005 to 2007 period exceeding $300/TPA in 2005 dollars between 2005 and 2008. Rising cost trends were also noted during that period by many EPC contractors (e.g., Humphrey, 2008).

Poten & Partners (Hartnell, 2009), using additional data, extrapolated the liquefaction plant cost trend to 2012 suggesting that greenfield liquefaction plant costs were likely to rise steeply to some $1000/TPA in 2009 dollars in 2012. That prediction did indeed materialize. At the beginning of 2012, based upon information available to the authors new liquefaction plants in the 4.0 to 4.5 MTPA capacity range, under construction and approaching completion (e.g., Angola LNG and Pluto Australia) were expected to be delivered in the range of $1100/TPA and $1600/TPA. Historically only the Snohvit LNG plant built by Statoil and partners in Norway (on-stream in 2007) and the Sakhalin II plant built by Shell and partners in Russia (on-stream in 2009) have exceeded $1200/TPA in 2009 dollars. Both projects were completed with significant budget overruns.

EPC contracts awarded for gas liquefaction plants in the past eight years were sanctioned with project budgets ranging from some $400/TPA to $1,200/TPA including projects from Equatorial Guinea, Yemen, Peru, Angola, Nigeria, Algeria, and Australia. In this period unprecedented materials price escalations, extreme market fluctuations, and competition for resources and skills led to the noted trend of increasing unit capital costs. Peru LNG (Bruce and Lopez-Piñon, 2009) and Equatorial Guinea LNG managed to be delivered at the low end of the unit capital cost trend, demonstrating that lump sum contracting and focused contracting strategies can be effective even during inflationary periods. Cost trends over the past decade suggest that greenfield liquefaction plants in the past decade cost between about $25/TPA and $40/TPA more than expansion projects (i.e., adding additional trains to an existing site). Adding an additional train to an existing liquefaction site is clearly cheaper and less risky than developing a greenfield site.

The full costs of building a new liquefaction plant (i.e., including gas field development, feed gas pipelines, and port facilities) in 2012 dollars are clearly much greater than the cost of the liquefaction train EPC costs described earlier. Examples from company reports of full unit cost ranges for plants recently built, under construction, or in planning are:

- 1,500 to 2,000 $/TPA: Sabine Pass (US), East Africa, Kitimat (Canada)
- Over 2,000 to 2,500 $/TPA: PNG LNG, Queensland CSG-LNG projects (most of these CSG-LNG projects breached their capital budgets in 2012 / 2013)
- Over 2,500 to 3,000 $/TPA: Australia NWS projects (Gorgon, Browse), Tangguh Train 3 (Indonesia)
- Over 3,000 $/TPA: Australia NWS projects (Pluto, Wheatstone, Prelude, and Ichthys), Abadi (Indonesia), Shtokman and Yamal (Russia Arctic Coast).

Clearly, the problem with LNG supply is that the facility to produce LNG (i.e., the liquefaction plant) requires a lot of time and capital to build. In addition, a liquefaction plant is generally not built until there are enough proven reserves of natural gas established to feed into it. In general, for a base load liquefaction plant to be economical, it is necessary to have a large gas production stream (>500–600 MMscf/d) available for 20 to 30 years; this translates to a need for recoverable gas reserves on the order of 5 to 10 tcf to supply these long-haul LNG projects (Kellas, 2003). Typically, for one million tons per year of LNG (47 Bcf per year) produced by a liquefaction plant for a 20-year period, about 1.5 tcf of natural gas reserves are required (Sen, 2002).

The large capital cost involved with each element of the LNG value chain imposes substantial financial demands on LNG businesses. Consequently this has limited participation in the industry primarily to large, well-financed corporations with experience in LNG handling. One way to minimize the substantial risks has been to obtain long-term supply contracts (20–25 years in duration), with a "take or pay" clause that obligates buyers to pay for gas at a certain price, even if the customer is unable to take delivery of the LNG.

1.7.1.2 Cost reduction

Significant reductions in LNG supply costs were experienced between 1999 and 2004, which promoted the involvement of smaller oil and gas companies to enter into operatorship of liquefaction projects (e.g., Marathon in Equatorial Guinea; Hunt in Peru). Note that LNG costs vary considerably in practice, largely as a function of capacity, particularly the number and size of individual trains in liquefaction plants and shipping distance.

Those cost reductions came largely from improved efficiency through design innovations, economies of scale through larger train sizes, and competition among manufacturers. Despite the fact that cost inflation has taken its toll on LNG supply chain costs as mentioned earlier, smaller companies are eager to join the liquefaction sector (e.g., Apache in one of the liquefaction projects proposed for Kitimat on Canada's Pacific coast).

Further advances in LNG technology can be expected in liquefaction, regasification and shipping, which should lead to lower overall project costs in the future. Shipping cost reductions have been achieved in the past decade by increasing carrier size (e.g., Qatar's Q-flex and Q-max vessels), where larger LNG carriers equipped with BOG reliquefaction capabilities are able to reduce unit transportation costs on long LNG supply routes. However, the main obstacle to very large LNG carriers across the industry is the capability of existing ports and jetties to receive the LNG tankers, as they were designed for smaller vessels.

Reducing storage tank costs, from a materials and construction perspective, has the potential to reduce the costs of LNG receiving terminals. Today's onshore storage tanks are based upon the self-supporting tank technology, and the membrane technology can bring significant cost reductions.

The perception of steadily falling costs for LNG projects has been dashed in the last several years. The main issues that have caused rising costs in the LNG industry are low availability of EPC contractors as result of extraordinary high levels of ongoing petroleum projects worldwide, high raw material prices as result of a surge in demand for raw materials, lack of skilled and experienced workforce in the LNG industry, and the project's financial risks (Hashimoto, 2011). Increasing LNG project costs are adding uncertainty to already complex LNG value-chain projects. Uncertainties in costs, coupled with the ever-present uncertainties of gas price and changes in regulation in supply and receiving countries, increase the investment risk and can delay investment decisions. Additional delays in the time to first gas for an LNG project can affect the value of the

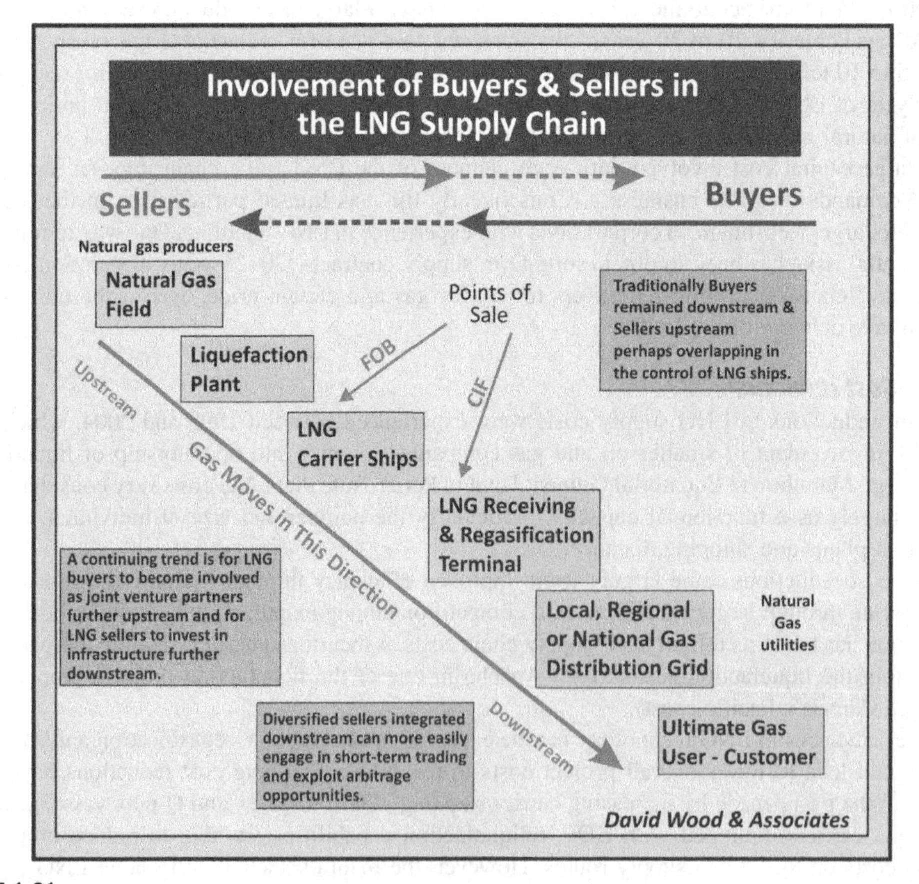

FIGURE 1-61

Expanding roles for traditional buyers and sellers along the components of the LNG supply chain.

(Wood, 2005a; Wood and Mokhatab, 2006a)

project negatively and may result in missing a competitive gas supply or demand window (Resley and Reinsvold, 2008).

1.8 LNG contracts and project development

As the LNG industry grows and diversifies more sophisticated supply chain structures and contractual interactions have developed in conjunction with the more flexible, short-term LNG trading arrangements that seem destined to expand in coming years. The multiple components of the LNG supply chain (Figure 1-61) mean that robust contractual relationships between the parties operating and controlling infrastructure along that supply chain are essential for the trade to operate effectively.

Cross-involvement of participants (Figure 1-61) from their traditional positions in the LNG supply chain became common during the 1990s in conjunction with new liquefaction projects (e.g., Qatar, Oman, Trinidad). This trend expanded significantly in the past decade and seems likely to continue to do so (e.g., many Asian utilities and state-owned entities have taken positions in upstream liquefaction projects around the world, while many major oil and gas producing companies have taken positions in regasification, shipping, and gas distribution networks).

1.8.1 Evolution away from integrated LNG supply chains

Some 20 years ago the traditional model for all LNG supply chains saw integrated upstream groups (gas production plus liquefaction plus shipping), consisting of major international oil and gas companies (IOCs) and state-owned national oil and gas companies (NOCs), selling LNG to integrated downstream groups, consisting of creditworthy state- controlled gas and/or electricity utilities (SGUs) primarily in Asia and Western Europe (Figure 1-62). Moreover, the LNG sales contract was long-term (20 years or more), with CIF or DES (i.e., INCOTERMS abbreviations for cost insurance and freight or delivered ex-ship) delivery terms, involving rigid take-or-pay terms with prices indexed to oil price (i.e., Asian buyers) and/or liquid petroleum products and competing (i.e., European buyers), but including a floor price to protect investors in liquefaction plant construction from price collapse.

Sales contracts with at least two LNG buyer consortia were involved in providing diversified off-take security for each liquefaction plant. Such contractual arrangements suited both buyers (i.e., long-term security of supply at prices linked to main competing fuels) and sellers (i.e., guaranteed sales at or above a minimum price to underpin investment and guarantee a long-term, low-risk healthy return on investment). Contracts were relatively simple and projects relatively easy to finance and insure with limited credit risks for lenders, because of the take-or-pay terms, the floor price, and creditworthy buyers enjoying the benefits of monopoly gas markets.

During the 1990s partially nonintegrated LNG supply chains emerged with buyers involved in separate arms-length consortia operating the shipping and purchasing the LNG on an FOB (i.e., INCOTERMS abbreviations for free on board) basis at the liquefaction plant port (Figure 1-63). This provided long-term buyers, particularly Japan and South Korea, more flexibility in managing their LNG supplies. It also offered those buyer nations and their utilities opportunities to share more directly in the profitability of LNG shipping, expand their domination of LNG ship building, and increase their influence over long-term shipping charters associated with the major LNG supply chains. Equity interests in the arms-length shipping company usually involved participants from both upstream and downstream consortia.

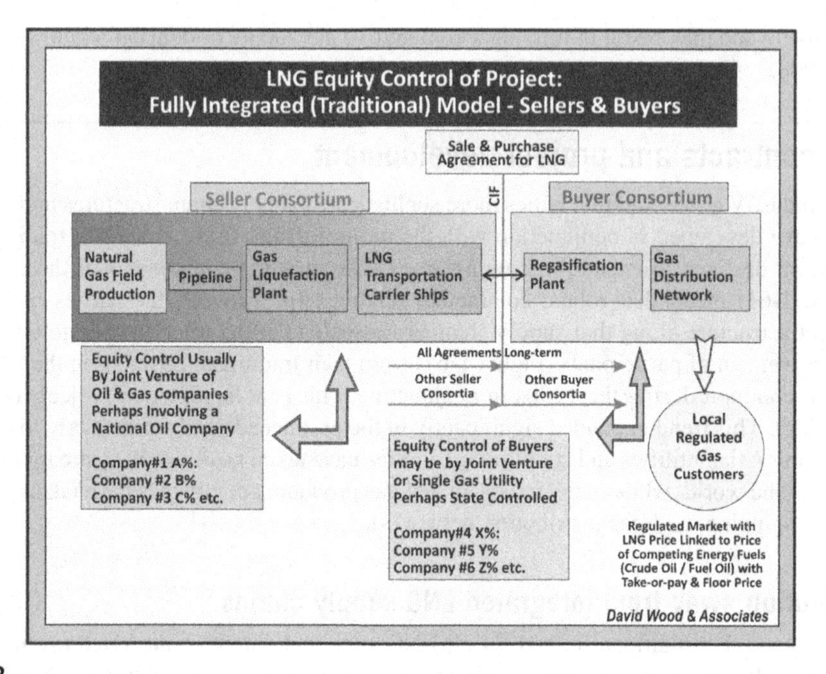

FIGURE 1-62

The two fully integrated upstream and downstream segments to a traditional LNG supply chain have a clear single contractual interface where the ownership of LNG changes hands. The sale and purchase agreements (SPA) governing the terms of that change of ownership is of paramount importance to LNG buyers and sellers.

(Wood, 2005a)

The LNG markets gradually evolved to include substantial amounts of FOB sales contracts as well as the more traditional CIF contracts. Sales terms otherwise remained similar with long-term take-or-pay arrangements. In Japan the Japanese Crude Cocktail (JCC) became the oil price benchmark for LNG (Figure 1-64). In some contracts floor prices were replaced by moderated crude pricing equations that softened LNG price increases in high oil price environments and LNG price decreases in low oil price environments, providing a more stable LNG contract pricing mechanism (e.g., the so-called S-curve pricing formulas secured by Japanese buyers in the 2000 to 2003 period). LNG delivered prices remained higher in Asia than in Europe due to more competitive gas supply and prices in Europe.

Tightness of LNG supplies to Asian markets at several periods since 2005 has resulted in an increase in short-term LNG cargo trades and redirections. This, and high oil prices, have provided LNG suppliers the upper hand in the Asian market LNG price negotiations in recent years. This is quite different from the prevailing market situation in 2001 to 2003 when buyers had the advantage and it was possible to negotiate low prices with very limited price escalation (e.g., supply to the first two receiving terminals in China, involving supply from NWS Australia and Tangguh, Indonesia; initial supply to India from Qatar). Prevailing market conditions often determine what pricing formulas can be agreed upon at any given time. In contrast to the oil-indexed LNG pricing in Asia, in the liberalized gas markets of the United States and United Kingdom, LNG buyers struggle to guarantee delivery of fixed volumes to specific customers on a long-term basis. Shorter-term contracts with flexible volumes at prices indexed to the regional gas

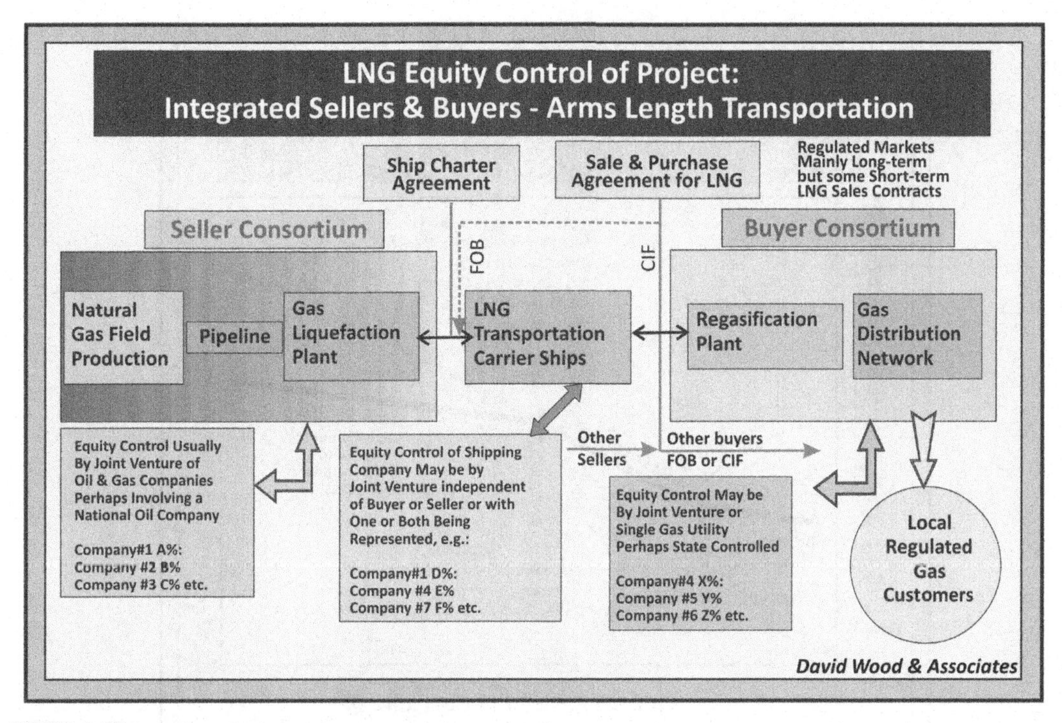

FIGURE 1-63

Two fully integrated upstream and downstream segments of the LNG supply chain separated by arms-length transportation involve more than one contractual interface and often also an overlap of equity interests.

(Wood, 2005a)

benchmark prices (e.g., Henry Hub in the United States; NBP in the United Kingdom), not oil prices, and with no specified floor price are favored in those markets (Wood and Mokhatab, 2006a; Wood et al., 2008b). However, these fail to provide long-term offtake security for LNG suppliers as illustrated by the impact of sustained low Henry Hub gas prices on LNG imports to the United States since 2009.

Since the 1990s Japanese and South Korean LNG buyers have sought minority nonoperated equity interests in the development of upstream components of the LNG supply chain (i.e., liquefaction plants and gas fields). This has enhanced their security of LNG supply and also provided them with valuable knowledge of the complex technical and operational issues associated with LNG supply. IOCs also realized that there was potential to extract value along the supply chain by taking equity positions in LNG shipping and regasification to tie in with their gas trading businesses in emerging deregulated gas markets of Europe. Contractual initiatives and risk taking by IOCs will also be involved in FLNG developments (Wood and Mokhatab, 2006b).

1.8.2 Evolution of less-integrated LNG supply chains

In certain LNG supply chains the components have become even more fragmented since the late 1990s. Liberalization, and in some cases full deregulation, of the downstream gas market sector,

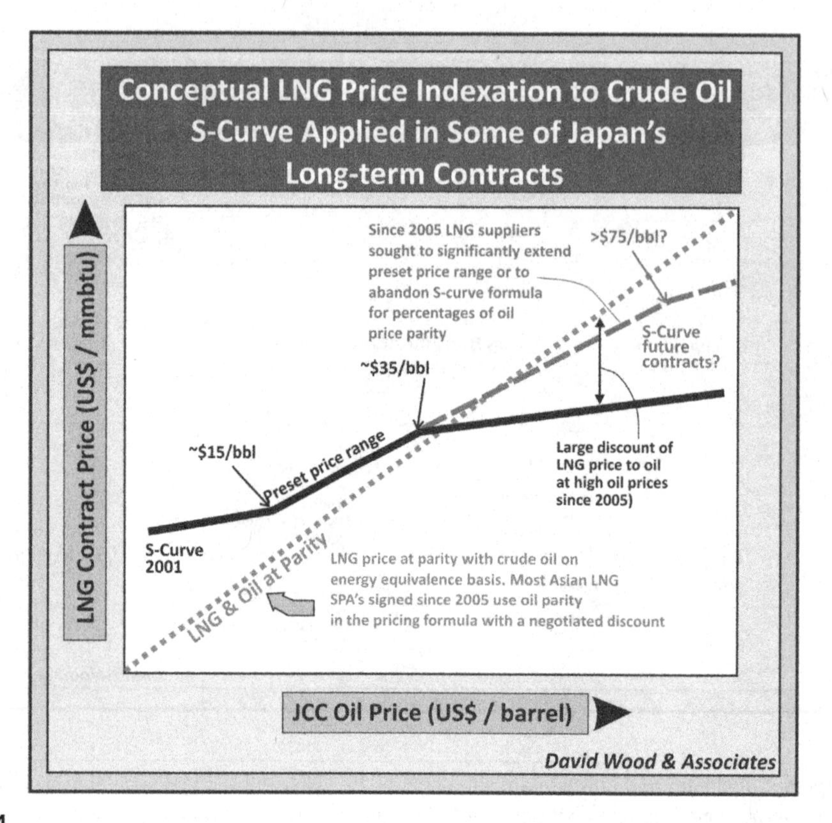

FIGURE 1-64

Changing Asian gas market conditions have led to changing pricing and price indexation terms since 2001 in Japan and other Asian LNG buyers.

(Wood, 2007)

short-term contracts, swap sales, and introduction of more flexibility in the destination clause in many of the more recent sales contracts have introduced much more complexity to the LNG supply chains and the contracts governing them. The building of many new receiving terminals worldwide, commencing in Europe, has also resulted in both long-term and short-term LNG sales becoming more diversified. IOCs have exploited their competitive advantage acting as aggregators along the supply chains (Figure 1-65) and developing a merchant model. By owning shipping capacity that is uncontracted to specific LNG supply chains and with access to LNG receiving terminals in many countries IOC have become effective LNG aggregators of being able to secure and deploy cargoes at short notice to exploit opportunities to supply LNG on short-term SPAs to different markets at different times.

The merchant model also involves the aggregators purchasing some LNG on an uncontracted basis without a specified destination, with the aggregators taking the market risk from the LNG suppliers (Figures 1-65 and 1-66). In certain upstream markets gas fields located in different licenses held by different joint venture groups have combined to fund the building of tolling liquefaction plants. In such cases the liquefaction plant and upstream gas fields supplying the liquefaction plants are not integrated

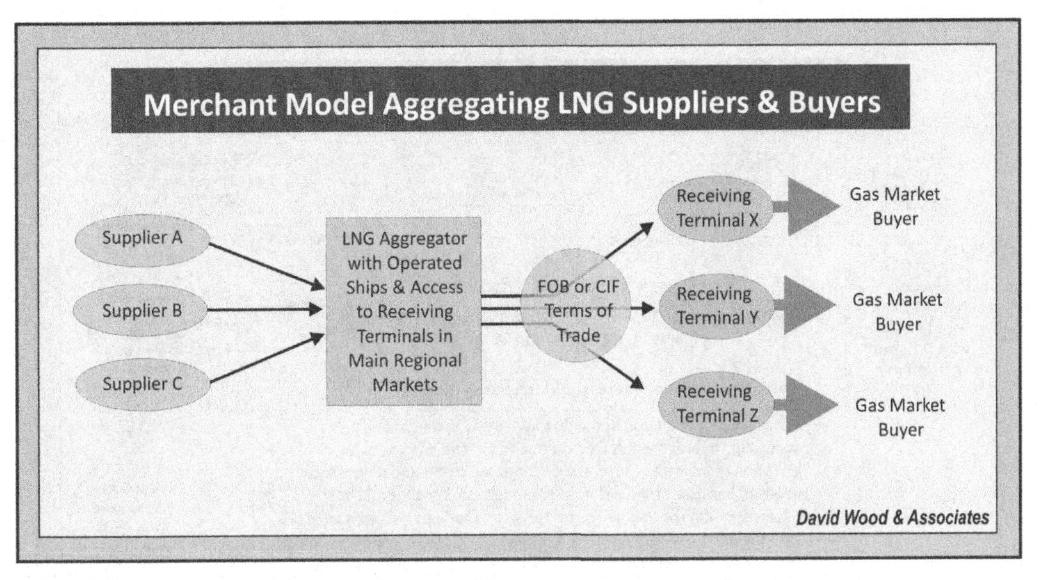

FIGURE 1-65

Merchant model with large IOCs acting as LNG aggregators to supply LNG around the world on a short-notice, short-term basis.

(Wood, 2005a)

in either ownership or operatorship (e.g., Liquefaction projects in Trinidad and Egypt). In such arrangements it is now possible to have several upstream components to the LNG supply chain:

- Gas fields involving several equity groupings (often including the NGC) subject to different types of fiscal arrangements (e.g., negotiated production sharing agreements and tax–royalty licenses). Because governments take the major share of revenues under production sharing agreements (commonly 70–90%) they are often keen to see LNG revenues flow all the way back to the gas field licensees rather than revert to the liquefaction plant that may benefit from tax holidays or uplifts that would limit the government take.
- Feed gas pipelines to liquefaction plants involving distinct equity holdings, tariff structures, and taxation terms.
- One or more liquefaction plants with several trains, each with distinct equity holdings established on a fixed rate of return construction basis, perhaps reverting to state ownership once payback is achieved, and charging each gas supplier a negotiated tariff for liquefying gas and loading LNG for export.

Similarly it is possible to have several components in the downstream LNG supply chain if open-access rules (i.e., often termed third-party access) are applied to the import and regasification terminal. Several different companies could contract portions of the capacity available in an LNG receiving terminal from its owners for specified periods at market rates. This would enable each of these capacity holders the ability to source LNG from different supply chains and deliver regasified gas to different buyers through capacity purchased in the transmission system. The import and

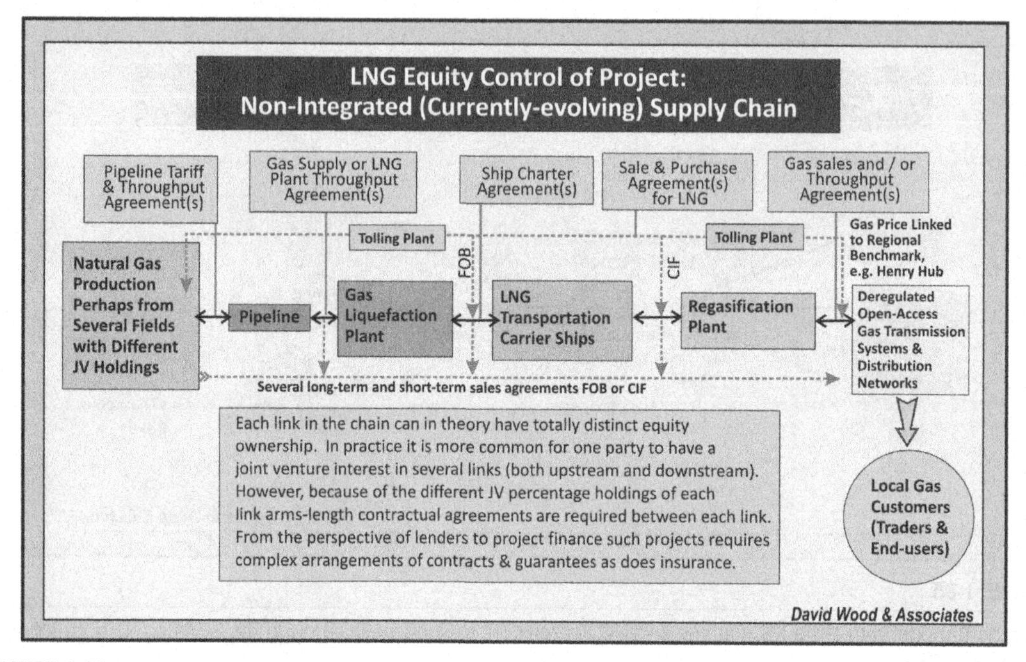

FIGURE 1-66

Fully nonintegrated LNG supply chain taking a merchant LNG model to its logical extreme. However, it is important to note that this model does not currently reflect reality in most LNG supply chains.

(Wood, 2005a)

regasification terminal also then operates essentially as a tolling plant charging each capacity holder a processing fee for receiving and regasifying its LNG (e.g., Grain LNG terminal in the United Kingdom with National Grid as operator and more than five LNG suppliers contracting long-term capacity in the plant through periodic auctions).

1.8.3 Issue of open-access to regasification terminals

Selling LNG into the deregulated and liberalized markets of North America and Western Europe has led to somewhat more restrictive contractual arrangements evolving at the downstream end of the LNG supply chain (Figure 1-67). Rather than move progressively toward this less-integrated supply model many LNG suppliers have secured exemptions from open-access rules providing them with competitive advantage in return for committing the capital investment to build or expand regasification terminals. Some issues also revolve around open-access rules for the gas pipeline transmission systems and distribution networks, coupled with many exemptions to such open-access rules negotiated for new LNG regasification terminals.

In December 2002 the US Federal Energy Regulatory Commission (FERC) ruled that the proposed LNG import terminal at Hackberry, Louisiana could be built without complying with the open-access requirements that previously had been strictly applied to all parts of the gas transmission chain as part of ensuring open competition in the deregulated gas market. This "Hackberry Decision," as it is

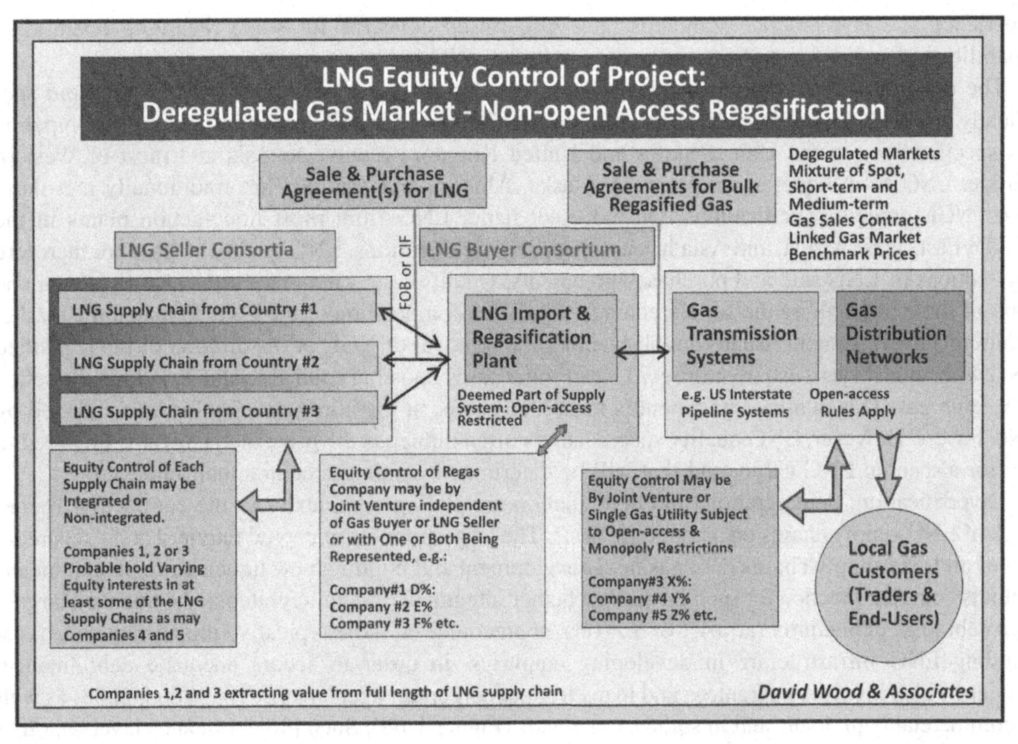

FIGURE 1-67

Nonintegrated LNG supply chain, but with restricted access to many new-build regasification terminals leading to severe access restrictions to deregulated markets for those without equity holdings in import and regasification infrastructure.

(Wood, 2005a)

known, encouraged proposals to build several new receiving terminals in the United States without developers being forced to offer capacity to third parties at "market" rates. Such open-access exemptions have proved critical in securing equity investment to build terminals in the United States and Europe by providing the shareholders in the regasification projects guarantees that they will maintain control over a large percentage or all of the LNG capacity of the plant for a substantial period of time, enabling them to realistically recover and benefit from their investment of risk capital.

1.8.4 Project finance of LNG facilities

Several recently constructed LNG supply chains have led the diversification of LNG supplier and buyer countries away from those traditional LNG buyers with high credit ratings (e.g., Argentina, Chile, China, India, Mexico, and Turkey). These LNG buyers pose financing and insurance challenges and concern for sellers over the long-term fiscal and contractual stability. At the upstream end countries such as Nigeria, Egypt, Yemen, Equatorial Guinea, Peru, Russia, Trinidad, and Angola pose challenges for financing, insurance, security of supply, and fiscal stability. Concerns over lack of experience and best-practice standards have also raised concerns for some regarding operational reliability and safety for new-entrant countries to the LNG industry.

The ongoing diversification of the LNG markets has raised issues about LNG quality and specifically about NGL content with respect to lower calorific value and Wobbe Index limits on pipeline gas specifications in the United States and United Kingdom relative to Asia and most of Western Europe. LNG from liquefaction plants in Alaska, Algeria, and Trinidad has traditionally met these lower NGL content specifications. On the other hand, LNG from most liquefaction plants in the Middle East, West Africa, and Asia has not met such specifications. LNG quality clauses are therefore key sections of LNG sale and purchase agreements. Quality issues may require investment from one party at some point along the supply chain to adjust the composition of the LNG to match the market requirement. Adjustment usually involves either the removal of NGL or the dilution of the regasified LNG with an inert gas (usually nitrogen), or in some cases boosting calorific value by adding liquefied petroleum gas. This quality adjustment is most likely to be at the import terminals and conducted by LNG buyers. However, LNG quality specifications often influence the price that a buyer is prepared to pay for a specific LNG cargo, and that will be determined within the contractual framework.

Diversification, deintegration, and deregulation are adding complexity to the contractual framework of LNG supply chains on an ongoing basis. The large amounts of capital required at the upstream end of all LNG supply chains (i.e., gas field development and building new liquefaction plants) means that it is common practice for such projects, whether integrated or nonintegrated, to be financed largely with debt (e.g., debt:equity ratio 4:1 or 3:1) on a nonrecourse or, more typically, limited-recourse basis. Building LNG infrastructure in developing countries, in order to secure adequate debt finance, frequently also involves guarantees and loans from export credit agencies (ECAs; Wood, 2008) as well as commercial bank loans and in some cases bonds (Figure 1-68). Such project finance layers another level of complexity into the contractual framework and introduces more rigorous requirements for the contracts to dovetail, guarantees, insurance, and such to provide security to lenders. The production sharing agreement or license terms under which the Government approves the plant remains the ultimate contractual driver.

Because so much debt finance and high upstream taxation rates are involved in such projects the rules allowing deduction of interest paid on project debt from taxation liabilities can make significant impact on the overall profitability of a liquefaction project. The government approval of the credit agreements is therefore often an integral part of the project approval process.

The LNG industry has evolved and diversified quite dramatically on a global scale in recent years. Increasingly more countries supply and buy LNG for both peak-load and base load gas supplies. This has resulted in more complex contractual relationships and structures of some modern LNG supply chains. Less integration exists along several recently developed LNG supply chains. IOCs, NOCs, and utilities now participate at several points along a typical supply chain in order to extract more value, spread risk, and establish more security of offtake and /or supply. Although long-term supply contracts still underpin the supply chains, short-term trading opportunities continue to increase. Contractual complexity, increased diversification, and deintegration result in greater risks and opportunities (Wood, 2005b).

The Asian market for LNG continues to be the most important LNG market worldwide with significant demand growth forecast in China, Japan, Korea, and Taiwan over coming years (Wood and Mokhatab, 2011). The development of a Singapore LNG gas hub (due onstream in 2013) is also likely to change the dynamics of Asian short-term LNG market. Gas liquefaction projects linked to deep-water and/or distal offshore field developments and unconventional gas (e.g., coal bed methane in

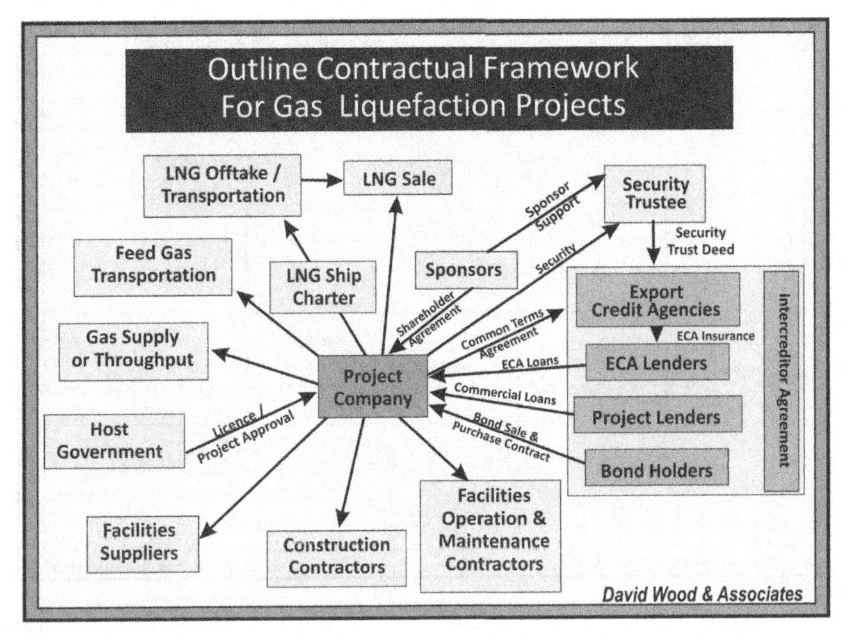

FIGURE 1-68

Contractual framework, loan, and security agreements typically involved in securing approval and debt finance to build a new liquefaction plant.

(Wood, 2005a)

Queensland) in a number of Southeast Asia countries and Australia seem best placed geographically to meet a significant portion of the demand growth of this market over the next decade. However, competition is emerging to supply this market from Russia, the Middle East, East Africa, and North America (e.g., Kitimat and other liquefaction projects in Western Canada). Contractual and pricing terms that for several years have favored LNG suppliers are likely to come under pressure as more competition and diversity of short-term supply develops.

1.9 LNG trade

International LNG trade continues to experience significant growth and diversification and has done so for the past decades. Such growth will require continued huge capital investments along the whole LNG supply chain and continue to be accompanied by an expanding spectrum of risks and opportunities for the industry.

1.9.1 Global LNG market growth

Figure 1-69 illustrates the growth in LNG trade from 1990 to 2011, extrapolated from 2012 to 2020. The trade has evolved from 55 MTPA (equivalent to 74 bcm of natural gas) in 1990 to 220 MTPA

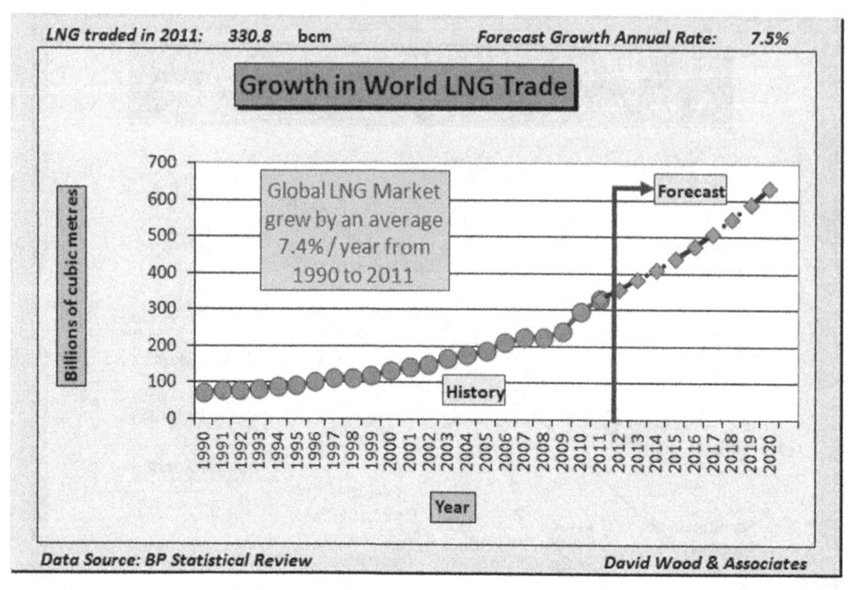

FIGURE 1-69

Evolution of the LNG trade 1990 to 2011 (historical data from BP Statistical Reviews) with a forecast to 2020 (updated with 2011 actual data from original analysis of Wood, 2004, 2012).

(298 bcm) in 2010 representing a compound annual growth rate (cagr) of 7.2%. A review of projects in 2012 in engineering, construction, and advanced stages of planning suggests that LNG production could reach some 320 MTPA (~430 bcm) in 2015 and some 450 MTPA (~610 bcm) by 2020 assuming a cagr for the period of 7.5% and that global LNG demand keeps pace with new installed capacity. An independent review of projects in 2008 in engineering, construction, and planning suggested that production in 2015 could be 360 MTPA, assuming LNG trade is equivalent to 80% of new installed capacity (Collins, 2008).

However, forecasts for future LNG trade are difficult to predict due to volatile regional LNG market conditions, competition from other sources of natural gas, emergence of new technologies, late development and start-up of some planned projects, and so on. In addition, global gas demand is inextricably linked to global economic growth and persistent economic downturns can slow demand and reduce utilization of installed supply chain capacities. However, because LNG offers a global commodity that can meet the growth in demand for natural gas forecast in many countries around the world many are confident that the LNG sector will continue to sustain growth at high average annual rates. The combination of secure energy supplies, higher natural gas prices, lower LNG production costs, rising gas import demand, with increasing demands for clean energy in the developed and developing world, and the desire of gas producers to monetize their gas reserves, are setting the stage for increased LNG trade in the years ahead.

Over the past few years many of the large oil and gas companies already involved in developing and operating longstanding LNG supply chains have taken large investment decisions to expand these facilities and have developed new projects for importing, shipping, and exporting LNG. New entrants, some much smaller companies with limited previous experience of the international gas trade, have also become aware of the benefits of LNG trade and are developing new projects, some in politically challenging countries. In such cases the new-entrant operators are relying heavily on more experienced LNG engineering, construction companies, service providers, and critical-path item suppliers to ensure the safety, engineering integrity, and ultimate efficient performance of these installations. A trend toward increased competitiveness has been an increase in the economies of scale; from expansion of existing LNG receiving facilities to construction of larger liquefaction trains and ships. This trend has also led to the development of new technologies such as floating regasification, gas ports, and LNG FPSOs, which have emerged and continue to evolve and take some important niche market positions around the world.

Higher materials and service costs and skilled manpower shortages in the LNG industry, due to the high demand of upstream and downstream oil, gas, and petrochemical plants since 2004, however have reversed the trend of lower unit costs for LNG facilities developed from 1990 to 2005. However, because of higher gas demand and persisting high gas prices in Europe and Asia, higher unit costs have slowed but far from extinguished the growth of LNG industry capacity in the period from 2006 to 2012. Contracts being signed for future supplies suggest sustained growth of LNG capacity in the medium and long-term, particularly in Asia and Europe.

The impact of shale gas in North America has depressed natural gas prices there since 2008 to the extent that imported LNG is only competitive on a small scale during seasonal peaks in demand. Indeed projects to build liquefaction plants to export LNG from Western Canada and the United States are now at an advanced stage of planning. Rapid deployment of shale gas technologies to other continents outside North America seems unlikely, due to the proprietary technologies and associated technical risks, in addition to the large number of wells required and regulators' reservations about the environmental sustainability of large-volume hydraulic fracture stimulation on water supplies (e.g., an outright ban on hydraulic fracturing introduced in France and several other countries tightening regulations). Nevertheless shale gas projects are seen as likely to impact some markets (e.g., China, India, South Africa, parts of Europe, and South America) without displacing significantly the continued growth in LNG imports. Contrary to this view, it should be appreciated that some analysts are forecasting a more significant impact on LNG trade in coming years from unconventional gas.

Over the past few years, greater competition, economies of scale, and market liquidity have increased the importance of controlling LNG infrastructure to improve margins at every point across the value chain. Companies, such as BG, BP, GdFSuez, Petronas, QP, Shell, Statoil, and Total recognized that access to markets through trading LNG, and their ability to control infrastructure in each part of the LNG value chain, would assist them in monetizing their gas reserves and provide them with the flexibility to switch cargo destinations to access the best LNG netbacks. Such LNG strategies have enabled these large companies to rapidly exploit evolving LNG trading opportunities in gas markets around the world, while their competitors have lacked the capability to do so. Other key players in LNG supply are following their lead (e.g., ExxonMobil, Gazprom, Chevron, ENI, Repsol), but some of them have a significant gap to breach or are regretting that they did not chose to do so much earlier.

1.9.2 LNG import markets

Table 1-10 and Figures 1-70 and 1-71 compare the status of the LNG imports in various countries and regional markets in 1996 and in 2011. These two snapshots in time highlight where the key developments in demand have occurred and just how rapidly the industry has grown in some countries.

Table 1-10 LNG Imports by Country in 1996 and 2011

LNG Imports (bcm) by Country	Year	
	1996	2011
United States	1.2	10.01
Canada		3.30
Argentina		4.38
Brazil		1.05
Chile		3.86
Dominican Republic		0.91
Mexico		4.05
Puerto Rico		0.74
Belgium	4.0	6.57
France	7.8	14.57
Greece		1.29
Italy		8.75
Netherlands		0.78
Portugal		3.01
Spain	6.9	24.16
Turkey	2.3	6.23
United Kingdom		25.31
Kuwait		3.18
United Arab Emirates		1.43
China		16.62
India		17.10
Japan	63.8	106.95
South Korea	13.0	49.31
Taiwan	3.4	16.31
Thailand		0.98
Global	**102.4**	**330.9**

Data comes from BP statistical reviews and the color code separates Atlantic, Middle East, and Pacific market regions (updated with 2011 actual data from original analysis of Wood, 2004, 2012).

Table 1-10 shows how the trade has diversified and grown almost threefold during the period. In Europe, Spain and France established significant expansions of trade and the UK LNG demand evolved to become the highest in Europe in just a few recent years. In Asia, South Korea's LNG imports have grown more than threefold, whereas Japan has posted a respectable 68% growth during that period. It is through the emergence of China and India during the period that is signaling where much of the future growth is expected to come from. US LNG imports rose steadily to 21.82 bcm in 2007, but contracted to 10.01 bcm in 2011 due to shale gas competition and price collapse. Nonetheless in 2010 the United States still imported 10 times more LNG than it did in 1996. The LNG industry is not dead in North America, but is likely to play a more complex import–export role in the future.

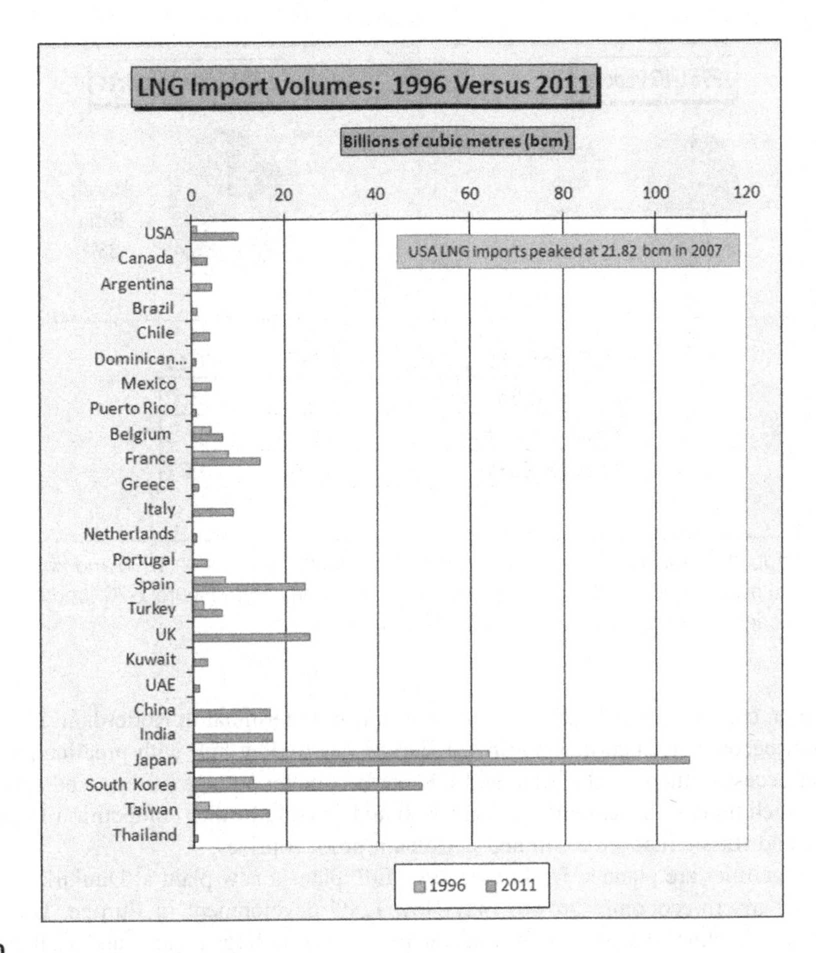

FIGURE 1-70

Graphical comparison of LNG imports by country in 1996 and 2011. Data comes from BP statistical reviews; the color code separates Atlantic, Middle East, and Pacific market regions (updated with 2011 actual data from original analysis of Wood, 2004, 2012).

Table 1-10 displays the scale of expansion and diversification of the LNG industry during the 1996 to 2011 period. In addition to the large market shares dominated by the countries mentioned earlier, the emergence of a diverse Latin American import market is revealed, something that was not envisaged until just a few years ago. Indeed, Argentina, Chile, and Brazil are all developing LNG infrastructure to facilitate access to secure energy supplies and promote economic growth. The LNG industry has managed to circumvent access to the politically and geographically stranded gas resources of Bolivia.

The emergence of LNG markets in some of the European countries with low gas demand (e.g., Portugal and Greece) and Italy, together with expansions in the Turkish market (Figure 1-70) are likely to be followed over the next decade by new market entrants from Eastern Europe (e.g., Poland, Baltic States, and some Balkan nations). In addition the Netherlands, the only net gas exporter in the

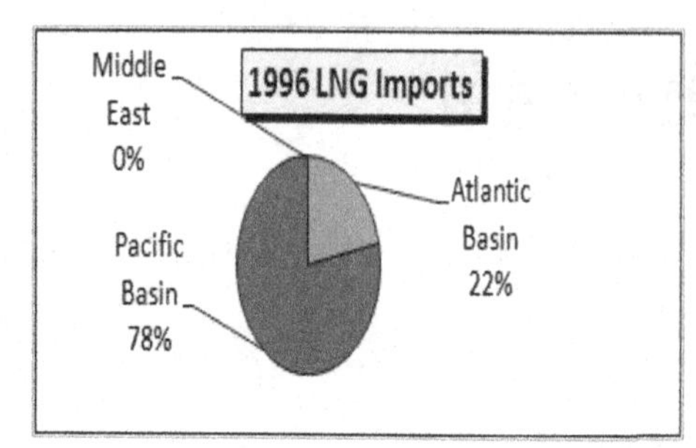

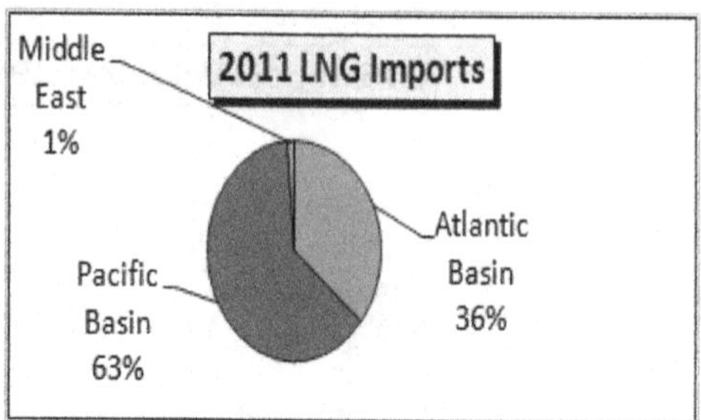

LNG Imports (bcm)	1996	2011
Atlantic Basin	22.2	119.0
Pacific Basin	80.2	207.3
Middle East	0.0	4.61

FIGURE 1-71

Pie charts and supporting data table in bcm comparing LNG imports by region in 1996 and 2011. Data comes from BP statistical reviews and country combination color codes come from Figure 1-70 (updated with 2011 actual data from original analysis of Wood, 2004, 2012).

European Union, commissioned in 2011 its first regasification terminal in Rotterdam. It is likely that Rotterdam will become an important northwest Europe gas trading hub with pipeline gas and LNG competing for access to the large German and UK markets under a range of term and spot contracts. Indexation of such trade to benchmark gas prices should provide further competition to pipeline gas from Norway and Russia indexed to oil and petroleum product prices.

New LNG facilities are planned for France (e.g., EdF plans a new plant at Dunkirk sanctioned in 2011). Stagnant macro-economic growth may slow LNG development in Europe. However, with North Sea gas in decline and EU environmental policies curtailing nuclear and coal power plant developments, even if energy demand remains flat in the region, the quantity of gas imports seems destined to increase significantly. The European drive to substitute fossil fuels and nuclear by intermittent renewable energies (mainly solar and wind) also requires back-up supplies, and natural gas provides the most efficient option to provide such back-up supplies.

The emergence of small LNG import markets in Kuwait and Dubai is testament to the versatility and flexibility of the industry, with short-term cargoes coming not from neighboring Qatar but from around the world (Wong and Fesharaki, 2010). More short-term spot markets such as these are destined to develop over the coming decade and will play an important role in market liquidity, diversity, and competition for spot LNG cargoes.

Figure 1-71 illustrates how the Atlantic markets for LNG have grown to take a larger share of the trade in 2011 than they did in 1996. The emergence of the small but significant niche Middle East LNG market is also illustrated. Whereas these regional markets were completely distinct and not really competing for LNG supply in the 1990s, this has all changed over the interim period, particularly with rapid growth of Middle Eastern suppliers that can serve both Atlantic and Pacific markets. The usefulness of the classification of LNG trading regions as "Atlantic" versus "Pacific" markets is likely to disappear over the coming decade as more and more LNG travels along diverse and lengthy supply chains around the world for long-term and short-term contracts.

Table 1-11 lists the 26 countries that imported LNG in 2011 and a long list of countries that are considering or in some cases undertaking the construction of new regasification facilities. In fact the list

Table 1-11 Countries Importing LNG in 2012 or Planning to Import in the Future

Countries Importing LNG	
Situation in August 2012	
Current (26)	**Planned/Possible**
Argentina	Albania
Belgium	Aruba
Brazil	Bahamas
Canada	Bahrain
Chile	Bangladesh
China	Bulgaria
Dominican Republic	Croatia
Dubai	Cuba
France	Dominican Republic
Greece	El Salvador
India	Finland
Italy	Fujairah (UAE)
Japan	Germany
Kuwait	Indonesia
Mexico	Ireland
Netherlands	Jamaica
Portugal	Jordan
Puerto Rico	Kenya
South Korea	Lebanon
Spain	Lithuania
Sweden	Malaysia
Taiwan	New Zealand
Thailand	Pakistan
Turkey	Philippines
United Kingdom	Poland
United States	Singapore
	South Africa
	Ukraine
	Uruguay
	Vietnam

(Updated from original analysis of Wood, 2004, 2012)

of countries planning to develop LNG import infrastructure continues to get longer year by year as the industry further diversifies. Clearly it will take probably the best part of the next decade before many of the countries in the right-hand column of Table 1-11 actually establish operational LNG import supply chains. The Netherlands made this step in 2011 with the commissioning of its Gate LNG terminal in Rotterdam, opening up that location as a potentially important northern European gas trading hub.

There is a niche market for isolated island nations with limited overall energy demand, but limited to expensive distillate and fuel oil power generation facilities in the absence of LNG. These markets may import only two or three cargoes per year and in some cases their demand will be seasonal. Such markets individually are of little consequence to the overall LNG trade. However, collectively several island markets can be attractive, especially if they are located on or close to the main base load supply chain routes. In such cases it is possible to deliver partial spot cargoes at minimum incremental cost, or even backhaul cargoes with vessels that would otherwise be sailing back to their main loading ports essentially with just a small heel cargo to maintain the vessel's tanks at cryogenic temperatures.

1.9.3 LNG export markets

Figure 1-72 summarizes LNG exports by country in 2011. The four largest LNG producers—Qatar, Malaysia, Indonesia, and Australia, in that order by volume—accounted for 58% of all LNG exports in

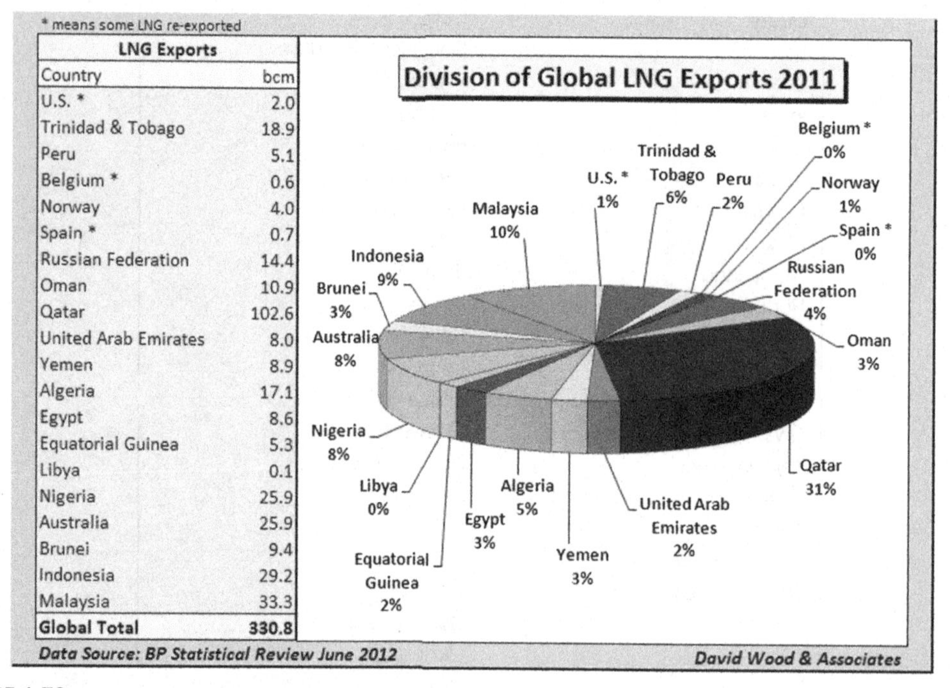

* means some LNG re-exported

LNG Exports

Country	bcm
U.S. *	2.0
Trinidad & Tobago	18.9
Peru	5.1
Belgium *	0.6
Norway	4.0
Spain *	0.7
Russian Federation	14.4
Oman	10.9
Qatar	102.6
United Arab Emirates	8.0
Yemen	8.9
Algeria	17.1
Egypt	8.6
Equatorial Guinea	5.3
Libya	0.1
Nigeria	25.9
Australia	25.9
Brunei	9.4
Indonesia	29.2
Malaysia	33.3
Global Total	**330.8**

Data Source: BP Statistical Review June 2012

FIGURE 1-72

LNG exports by country 2011. Data from BP Statistical Review June 2012 (updated with 2011 actual data from original analysis of Wood, 2004, 2012).

2010. In 2003 the top four exporting countries (Indonesia, Algeria, Malaysia, and Qatar, in that order) accounted for 63% of all LNG exports. This highlights the diversification of the LNG export markets and that some countries are increasing their share of it through massive capital investments (i.e., Qatar and Australia), while others are losing ground through much slower investment and expansion strategies, aging infrastructure, and depleting reserves supplying the older plants (i.e., Algeria, Indonesia, and UAE).

Hovering just below the top four exporting nations in 2011 were Nigeria and Trinidad; both expanded their capacity rapidly between 2000 and 2005 but subsequently their growth has been curtailed. In the case of Nigeria there are plenty of reserves but the problems include political unrest in the Niger Delta areas where the gas is located, unstable fiscal terms (i.e., the Nigerian government is seeking to increase its fiscal take and pay little for gas consumed domestically within Nigeria), and procurement constraints that inhibit import of specialist technology. This situation has resulted in several potential investors delaying commitments for several years on new plants. In the case of Trinidad the problem is both limited proven gas reserves inhibiting rapid expansion by building new liquefaction trains and the rapid and significant fall in demand for LNG in the United States, its main and geographically closest customer.

Countries to watch for future potential LNG export growth, not among those already mentioned, include Russia (expensive, remote, but large arctic gas resources), Libya (depends upon pace of regeneration), and Egypt (depends upon pricing policy and fiscal incentives to investors).

Table 1-12 lists the 19 countries that are exporting LNG in 2012 and 14 countries that are considering or in some cases undertaking the construction of new liquefaction facilities. Angola was the most recent country to commission a new facility in 2012. Papua New Guinea has one project under construction led by Exxon Mobil. The United States has several Gulf of Mexico projects looking to liquefy and export shale gas as LNG from terminals built solely as receiving terminals. The Sabine Pass terminal made its final investment decision to build a liquefaction plant in August 2012 for commissioning in 2015/2016; others await the outcome of a US government review to approve further LNG export projects. Three liquefaction export projects planned to be located at Kitimat on the West Coast of Canada, one led by Shell with Asian gas buyers and another being planned by Apache. EOG and EnCana are also planning to export shale gas as LNG to Asian markets. The Kitimat projects involve a feed gas pipeline across the Rocky Mountains and in 2012 delays in regulatory and first nation approvals for that line are holding up investment decisions on these projects. Alaska is also now looking at possible LNG export projects from Valdez linked to a gas pipeline from the North Slope fields. High costs and fiscal uncertainties mean that LNG exports from Alaska are unlikely before 2020.

Other interesting potential LNG export projects at the planning stage in 2012 include the following. (1) Shell's associated gas gathering project in Southern Iraq, finally signed at the end of 2011 after long negotiations, is also likely to involve a gas liquefaction plant, but this could not be operational before 2016. (2) A series of large deepwater gas discoveries offshore Mozambique by Anadarko, ENI and their partners and Tanzania by BG, Statoil, and their partners hold the necessary gas reserves to justify a gas liquefaction export project from those countries. Pre-FEED[5] studies commenced in mid-2011, but such plants are unlikely to be operational, if sanctioned, until 2016 at the earliest.

In conclusion the evolution of global and regional LNG trade over the past 20 years has been a story of rapid growth, diversification, and increased flexibility in LNG cargo movements. Asia continues to

[5]FEED stands for Front End Engineering Design.

Table 1-12 Countries Exporting LNG in 2012 or Planning to Export in the Future

Countries Producing and Exporting LNG	
Situation in August 2012	
Current (19)	**Planned/Possible**
Angola	Azerbaijan/Georgia
Abu Dhabi	Bolivia
Algeria	Brazil
Australia	Cameroon (GdFSuez)
Brunei	Canada (British Colombia)
Egypt	Cyprus
Equatorial Guinea	Iran
Indonesia	Iraq
Libya	Mozambique
Malaysia	Papua New Guinea
Nigeria	Tanzania
Norway	United States (Alaska, Gulf of Mexico, Oregon)
Oman	Venezuela
Peru	
Russia	
Qatar	
Trinidad and Tobago	
United States (Alaska)	
Yemen	

Due to start-up delays the Angola LNG plant did not finally export its first LNG cargo until 2013 (updated from original analysis of Wood, 2004, 2012)

dominate global LNG trade, but the European LNG market has evolved significantly in the past decade and seems destined for sustained growth and diversification over the next decade or so. The promising expansion of LNG imports to the United States up to 2007 proved to be a false dawn due to indigenous shale gas production causing gas price collapse. However, North America is likely to play a more complex role in global LNG trade with some terminals both importing and exporting LNG depending upon market conditions. It remains to be seen how such terminals will compete with already established supply chains to Europe and Asia and how it will influence gas prices and price indexation in those markets.

1.9.4 LNG prices diverge from the three regional markets

Asia continues to dominate global LNG trade (see Figures 1-70 and 1-71), but the European LNG market has evolved significantly in the past decade and seems destined for sustained growth and diversification over the next decade or so. The promising expansion of LNG imports to the United States up to 2007 proved to be a false dawn due to indigenous shale gas production causing gas price collapse. These three main markets have quite different LNG pricing mechanisms and Figure 1-73

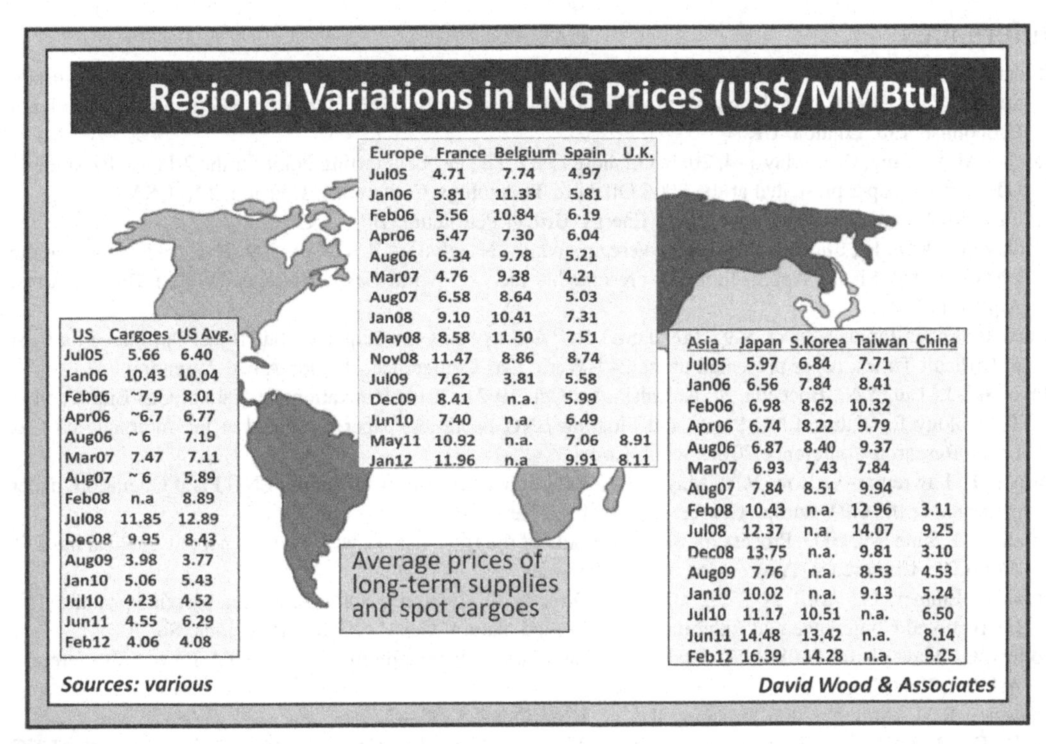

US	Cargoes	US Avg.
Jul05	5.66	6.40
Jan06	10.43	10.04
Feb06	~7.9	8.01
Apr06	~6.7	6.77
Aug06	~6	7.19
Mar07	7.47	7.11
Aug07	~6	5.89
Feb08	n.a	8.89
Jul08	11.85	12.89
Dec08	9.95	8.43
Aug09	3.98	3.77
Jan10	5.06	5.43
Jul10	4.23	4.52
Jun11	4.55	6.29
Feb12	4.06	4.08

Europe	France	Belgium	Spain	U.K.
Jul05	4.71	7.74	4.97	
Jan06	5.81	11.33	5.81	
Feb06	5.56	10.84	6.19	
Apr06	5.47	7.30	6.04	
Aug06	6.34	9.78	5.21	
Mar07	4.76	9.38	4.21	
Aug07	6.58	8.64	5.03	
Jan08	9.10	10.41	7.31	
May08	8.58	11.50	7.51	
Nov08	11.47	8.86	8.74	
Jul09	7.62	3.81	5.58	
Dec09	8.41	n.a.	5.99	
Jun10	7.40	n.a.	6.49	
May11	10.92	n.a.	7.06	8.91
Jan12	11.96	n.a	9.91	8.11

Asia	Japan	S.Korea	Taiwan	China
Jul05	5.97	6.84	7.71	
Jan06	6.56	7.84	8.41	
Feb06	6.98	8.62	10.32	
Apr06	6.74	8.22	9.79	
Aug06	6.87	8.40	9.37	
Mar07	6.93	7.43	7.84	
Aug07	7.84	8.51	9.94	
Feb08	10.43	n.a.	12.96	3.11
Jul08	12.37	n.a.	14.07	9.25
Dec08	13.75	n.a.	9.81	3.10
Aug09	7.76	n.a.	8.53	4.53
Jan10	10.02	n.a.	9.13	5.24
Jul10	11.17	10.51	n.a.	6.50
Jun11	14.48	13.42	n.a.	8.14
Feb12	16.39	14.28	n.a.	9.25

FIGURE 1-73

Regional variations in LNG prices among importers in Asia, Europe, and the United States (updated from original analysis of Wood et al., 2008b).

highlights how gas prices have diverged from those three markets over the past few years. In 2012 the differences between LNG import prices in Asia, Europe, and the United States are the greatest they have ever been. It is these price differences that are driving the enthusiasm for new-build LNG export projects in the United States. It seems unlikely though that such price differences will last long.

North America is likely to play a more complex role in global LNG trade after 2015 with some terminals both importing and exporting LNG depending upon market conditions. It remains to be seen how such terminals will compete with already established supply chains to Europe and Asia and how it will influence gas prices and price indexation in those markets. Many gas consumers in those regions are expecting, over the course of the next decade, continued diversification of LNG supply from new sources (e.g., North America, East Africa, and East Mediterranean) to improve security of supply and bring down the delivered price of LNG into their respective markets.

References

Avidan, A., Richardson, F., Anderson, K., Woodard, B., 2001. LNG Plant Scaleup Could Cut Costs Further. Section 7.1: Design and Technology, 128–132, Fundamentals of the Global LNG Industry. Petroleum Economists Ltd, London, UK.

Barclay, M.A., Yang, C.C., May 1–4, 2006. Offshore LNG: The Perfect Starting Point for the 2-Phase Expander?. OTC 18012, paper presented at the 2006 Offshore Technology Conference, Houston, TX, USA.

BP, June 2012. Statistical Review of World Energy. British Petroleum (BP), UK.

Bradley, A., Duan, H., Elion, W., van Soest-Vercammen, E., Nagelvoort, R.K., Oct. 5–9, 2009. Innovations in the LNG Industry: Shell's Approach, paper presented at the 24th World Gas Conference (WGC), Buenos Aires, Argentina.

Bruce, B., Lopez-Piñon, C., Oct. 5–9, 2009. Peru LNG: A Grassroots Gas Liquefaction Project Optimized for Cost in Difficult Times, paper presented at the 24th World Gas Conference, Buenos Aires, Argentina.

Bukowski, J., Liu, Y.N., Boccella, S., Kowalski, L., Oct. 19-21, 2011. Innovations in Natural Gas Liquefaction Technology for Future LNG Plants and Floating LNG Facilities, paper presented at the International Gas Union Research Conference 2011, Seoul, Korea.

Carnell, P., Lawrence, A., Row, V.A., May 14-16, 2008. Introducing Flexibility into LNG Plant Operation, paper presented at the GPA Europe Conference, Ashford, Kent, UK.

Chretien, D., June 5-9, 2006. Process for the Adjustment of the HHV in the LNG Plants, paper presented at the 23rd World Gas Conference, Amsterdam, The Netherlands.

Chrz, V., Emmer, C., April 24–27, 2007. LNG Directly to Customer Stations. Poster presented at the 15th International Conference and Exhibition on Liquefied Natural Gas (LNG 15), Barcelona, Spain.

Collins, C., May 14–16, 2008. LNG Update – What's New?, paper presented at the GPA Europe Conference, Ashford, Kent, UK.

Conachey, R.M., Autumn 2006. Breaking the Ice. LNG Industry, 38–42.

Coyle, D., De La Vega, F.F., Durr, C., April 24–27, 2007. Natural Gas Specification Challenges in the LNG Industry, paper presented at the 15th International Conference and Exhibition on Liquefied Natural Gas (LNG 15), Barcelona, Spain.

Coyle, D., Durr, C., De La Vega, F.F., Hill, D.K., Collins, C., March 13–15, 1995. LNG Plant Design in the 1990s, paper presented at the 74th GPA Annual Convention, San Antonio, TX, USA.

Coyle, D., Durr, C., Shah, P., Oct. 5–8, 2003. LNG: A Proven Stranded Gas Monetization Option. SPE 84251, paper presented at the 2003 SPE Annual Technical Conference and Exhibition, Denver, CO, USA.

Coyle, D., Patel, V., Feb.. Processes and Pump Services in the LNG Industry, paper presented at the 22nd International Pump Users Symposium, Houston, TX, USA.

Cuellar, K.T., Hudson, H.M., Wilkinson, J.D., March 11–15, 2007. Economical Options for Recovering NGL/LPG at LNG Receiving Terminals, paper presented at the 86th GPA Annual Convention, San Antonio, TX, USA.

Dendy, T., Nanda, R., April 6–10, 2008. Utilization of Atmospheric Heat Exchangers in LNG Vaporization Processes: A Comparison of Systems and Methods, paper presented at the 2008 AIChE Spring Meeting, New Orleans, LA, USA.

DNV, March 2011. Managing Risk—Floating Liquefied Gas Terminal. Offshore Technical Guidance OTG-02, Det Norske Veritas (DNV), Høvik, Norway.

Durr, C., Coyle, D., Hill, D., Smith, S., March 14–17, 2005. LNG Technology for The Commercially Minded, paper presented at the GasTech 2005 Conference and Exhibition, Bilbao, Spain.

Durr, C., Coyle, D., Patel, H., March 11–13, 2002. LNG Import Terminals, paper presented at the 81st Annual GPA Convention, Dallas, TX, USA.

Economides, M.J., Mokhatab, S., 2007. Compressed Natural Gas: Another Solution to Monetize Stranded Gas. Hydrocarbon Processing 86 (1), 59–64.

Eriksen, R., Brandstorp, J.M., Cramer, E., Oct. 13–16, 2002. Evaluating The Viability of Offshore LNG Production and Storage, paper presented at the GasTech 2002 Conference and Exhibition, Doha, Qatar.

Faber, F., Resweber, L.R., Jones, P.S., Bliault, A.E.J., May 6–9, 2002. Floating LNG Solutions from the Drawing Board to Reality, paper presented at the Offshore Technology Conference, Houston, TX, USA.

Festen, L., Leo, J., Vis, R., Sept. 2009. CB&I Lummus and Partners to Turn LNG FPSO Concept into a Reality. LNG Journal, 43–46.

Finn, A.J., Oct. 2007. Nitrogen Rejection Strategies. Hydrocarbon Engineering 49–52.

Finn, A.J., March 9–11, 2009. Floating LNG Plants – Scaleup of Familiar Technology, paper presented at the 88th Annual GPA Convention, San Antonio, TX, USA.

Finn, A.J., Johnson, G.L., Tomlinson, T.R., March 13–15, 2000. LNG Technology for Offshore and Mid-scale Plants, paper presented at the 79th Annual GPA Convention, Atlanta, GA, USA.

Foss, M.M., Delano, F., Gülen, G., Makaryan, R., Oct. 2003. LNG Safety and Security. Research Report, Commercial Frameworks for LNG in North America Consortium, Center for Energy Economics (CEE) as the Institute for Energy. Law & Enterprise, University of Houston' Law Center, Houston, TX, USA.

Garcel, J.C., May 14–16, 2008. Liquefaction of Non-Conventional Gases, paper presented at the GPA Europe Conference, Ashford, Kent, UK.

Hartnell, G., Sept. 23–25, 2009. Fundamentals of Global LNG – a Commercial Perspective, paper presented at the 26th Annual GPA Europe Conference, Venice, Italy.

Hashimoto, H., Feb. 21–23, 2011. Evolving Roles of LNG and Asian Economics in the Global Natural Gas Markets, paper presented at the 2011 Pacific Energy Summit, Jakarta, Indonesia.

Humphrey, G., Dec. 9–12, 2008. The Rising Costs of LNG Projects, Why? paper presented at the CWC 9th Annual World LNG Summit, Barcelona, Spain.

Huang, S., Hartono, J., Shah, P., April 24–27, 2007. BOG Recovery from Long Jetties During LNG Ship Loading, paper presented at the 15th International Conference and Exhibition on Liquefied Natural Gas (LNG 15), Barcelona, Spain.

Huang, S., Kotzot, H., de la Vega, F.F., Durr, C., March 13–16, 2005. Selecting The Optimal Scheme for N2 Injection in an LNG Terminal, paper presented at the 84th Annual GPA Convention, San Antonio, TX, USA.

Hudson, H.M., Wilkinson, J.D., Cuellar, K.T., Pierce, M.C., March 9–12, 2003. Integrated Liquids Recovery Technology Improves LNG Production Efficiency, paper presented at the 82nd GPA Annual Convention, San Antonio, TX, USA.

Hunter, P., Avidan, A., Duty, J.M., Eaton, A., Hernandez, R., Risley, A., April 25–29, 2004. Lowering LNG Unit Costs Through Large and Efficient LNG Liquefaction Trains—What is the Optimal Train Size? paper presented at the 2004 AIChE Spring Meeting, New Orleans, LA, USA.

ILEX Energy Consulting, Nov. 2003. Importing Gas into the UK-Gas Quality Issues. A Report to Department of Trade and Industry, Ofgem and the Health and Safety Executive. ILEX Energy Consulting Limited, Oxford, UK.

Jamieson, D., Johnson, P., Redding, P., May 4–7, 1998. Targeting and Achieving Lower Cost Liquefaction Plants, paper presented at the 12th International Conference and Exhibition on Liquefied Natural Gas (LNG 12), Perth, Australia.

Janseens, P., Oct. 16, 2012. Floating LNG. Presented at the 2012 SNAME UK Collegium 2012 Conference, London, UK.

Jung, M.J., Cho, J.H., Ryu, W., June 1–5, 2003. LNG Terminal Design Feedback from Operator's Practical Improvements, paper presented at the 22nd World Gas Congress, Tokyo, Japan.

Kellas, G., April 5–8, 2003. Comparison of LNG Frameworks and Fiscal Systems. SPE 82023, paper presented at the SPE Hydrocarbon Economics and Evaluation Symposium, Dallas, TX, USA.

Kerbers, I., Hartnell, G., Oct. 5-9, 2009. A Breakthrough For Floating LNG? paper presented at the 24th World Gas Conference, Buenos Aires, Argentina.

Kim, H., Lee, J.H., March 14–17, 2005. Design and Construction of LNG Regasification Vessels, paper presented at Gastech 2005 Conference and Exhibition, Bilboa, Spain.

Kitzel, B., Spring 2008. Choosing The Right Insulation. LNG Industry 114–117.

Kotzot, H.J., March 9–12, 2003. Overview of the LNG Industry, paper presented at the 82nd Annual GPA Convention, San Antonio, TX, USA.

Kulish, K.M., Shah, K., Wanvik, L., March 13–16, 2005. Offshore and Onshore LNG Terminals—Design and Economic Considerations, paper presented at the 84th Annual GPA Convention, San Antonio, TX, USA.

Lee, J.H., Janssens, P., Cook, J., May 2–5, 2005. LNG Regasification Vessel—The First Offshore LNG Facility. OTC 17161, paper presented at the Offshore Technology Conference (OTC), Houston, TX, USA.

Lee, S.K., March 10–13, 2008. Combining Ice Class Rules with Direct Calculations for Design of Arctic LNG Vessel Propulsion, paper presented at the Gastech 2008 Conference and Exhibition, Bangkok, Thailand.

Lee, Y.M., Cho, T.-I., Kwon, O.-Y., May 5–8, 2008. Trends and Technologies in LNG Carriers and Offshore LNG Facilities. OTC 19339, paper presented at the Offshore Technology Conference (OTC), Houston, TX, USA.

Lemmers, S.P.B., 2009. Designing the LNG Terminal of The Future. Hydrocarbon Processing 88 (9), 105–111.

Liu, Y.N., Edwards, T., Gehringer, J., Lucas, C., May 25–28, 1992. Design Considerations of Larger LNG Plants, paper presented at the 10th International Conference and Exhibition on Liquefied Natural Gas (LNG 10), Kuala Lumpur, Malaysia.

Mak, J.Y., Sept. 28–29, 2005. Technical Considerations in the Design and Optimization of LNG Terminals, paper presented at INFONEX LNG Conference, Vancouver, BC, Canada.

Mak, J.Y., May 14–16, 2008. LNG Wobbe Index Control, paper presented at the GPA Europe Conference, Ashford, Kent, UK.

Mak, J.Y., February 7, 2012. Configurations and Methods for Offshore LNG Regasification and BTU Control. US Patent No. 8,110,023.

Mak, J.Y., Amott, N., Graham, C., Patel, D., Sept 15, 2013. Vaporizer Selection Based on Site Ambient Temperatures, GPA Europe, Edinburg, UK.

Malvos, H., May 24–25, 2012. Technical Development: New Technologies for a Changing LNG Environment, paper presented at the GIE Annual conference, Krakow, Poland.

McCulley, R., July 6, 2011. Prelude in the supermajor key of FLNG. OE Digital.

McCulley, R., June 01, 2012. FMC Answers Prelude Call to Arms. OE Digital & AtComedia LLC, Houston, TX, USA.

McGuire, J.J., White, B., 2000. Liquefied Gas Handling Principles on Ships and in Terminals, third ed. Witherby Publishers, London, UK.

Mokhatab, S., Economides, M.J., Wood, D.A., 2006. Natural Gas and LNG Trade—A Global Perspective. Hydrocarbon Processing 85 (7), 39–45.

Mokhatab, S., Purewal, S., 2006. Is LNG a Competitive Source of Natural Gas? Petroleum Science and Technology 24 (2), 243–245.

Mokhatab, S., Wood, D.A., 2007a. Why Consider Exploiting Stranded Gas? Petroleum Science and Technology 25 (3), 411–413.

Mokhatab, S., Wood, D.A., 2007b. Breaking the Offshore LNG Stalemate. World Oil 228 (4), 139–148.

Mokhatab, S., Finn, A.J., Shah, K., 2008. Offshore LNG Industry Developments. Petroleum Technology Quarterly 13 (5), 105–109.

Mokhatab, S., 2010. Selecting Technologies for Onshore LNG Production. Petroleum Technology Quarterly 15 (3), 115–119.

Natural Gas Technology. DNV Report 2000–3526. Rev. 1, Feb. 1, 2001.

Neeraas, B.O., Marak, K.A., Nov. 2–3, 2011. Energy Efficiency and CO2 Emissions in LNG Chains, presented at the 2nd Trondheim Gas Technology Conference, Trondheim, Norway.

OCT System LNG Offshore Loading Technology. Pusnes News 1 (1–4), 2012.

OTA, Sept. 28, 1977. Transportation of Liquefied Natural Gas, Report prepared by the Office of Technology Assessment (OTA). Congress of the U.S. Washington, DC, USA.

Phalen, T., Scotti, J., May 5–8, 2008. Update on LNG Facility Construction, OTC 19306, paper presented at the Offshore Technology Conference (OTC), Houston, TX, USA.

Poldervaart, L., Oomen, H., May 1–4, 2006. Offshore LNG Transfer: A Worldwide Review of Offloading Availability, paper presented at the 2006 Offshore Technology Conference, Houston, TX, USA.

Powell, J., Thomas, G., March 10–13, 2008. Construction of All-Concrete LNG Tanks, paper presented at the Gastech 2008 Conference and Exhibition, Bangkok, Thailand.

Price, B.C., McCartney, D., March 13–16, 2005. LNG Terminals-Concepts for Increased Efficiency and Economic Enhancement, paper presented at the 84th Annual GPA Convention, San Antonio, TX, USA.

Resley, D.D., Reinsvold, C., Oct. 20–22, 2008. Reducing Time to First Gas: Lessons Learned in Expediting and Informing International Investment Chain Decisions. SPE 115251, paper presented at the 2008 SPE Asia Pacific Oil and Gas Conference and Exhibition, Perth, Australia.

Scherz, D.B., Jan. 30–31, 2006. Arctic LNG: Keys to Development, paper presented at 6th Annual LNG Economics and Technology Conference, Houston, TX, USA.

Sen, C.T., 2002. Trends and Developments in the LNG Industry", Report of the Potential Gas Committee, 89–98. Colorado School of Mines, Golden, CO, USA.

Sheffield, J.A., Mayer, M., May 16–18, 2001. The Challenges of Floating LNG Facilities, paper presented at the GPA Europe Spring Conference, Norwich, UK.

Skovholt, O., March 10–13, 2008. CryoTank—A Concrete/Concrete LNG Tank, paper presented at the Gastech 2008 Conference and Exhibition, Bangkok, Thailand.

Smaal, A., June 1–5, 2003. Liquefaction Plants: Development of Technology and Innovation, paper presented at the 22nd World Gas Conference, Tokyo, Japan.

Tarlowski, J., Sheffield, J., Durr, C., Coyle, D., Patel, H., 2005. LNG Import Terminals—Recent Developments. www.cheresources.com/.

Teramoto, K., June 1–5, 2003. LNG Reception Terminals: Operational Aspects of the Regasification System, including LNG Cold Utilization, paper presented at the 22nd World Gas Conference, Tokyo, Japan.

Thackeray, F., Leckie, G., 2002. Stranded Gas: A Vital Resource. Petroleum Economist 69 (5), 10.

Tustin, R., April 27–29, 2005. Recent Developments in LNG and Ice-Class Tanker Design and the Potential Application to Future Arctic LNG Ships, paper presented at the 1st Arctic Shipping Conference, Helsinki, Finland.

Tustin, R., April 2006. From Russia with LNG. Ice Focus, Lloyd's Register.

K. Gerdsmeyer, "Design concepts for LNG FSRU/FPSO with IMO type C cargo tanks", TGE Marine Gas Engineering Presentation.

U.S DOE/FE-0489, Aug. 2005. Liquefied Natural Gas: Understanding the Basic Facts. Report by US Department of Energy (DOE). Washington, DC, USA.

Vovard, S., Bladenet, C., Cook, C.G., May 25–27, 2011. Nitrogen Removal on LNG Plant—Select the Optimum Scheme, paper presented at the GPA Europe Spring Conference, Copenhagen, Denmark.

White, J., McArdle, H., Oct. 5–9, 2009. Floating LNG: A Review of the Forces Driving the Development of FLNG, Challenges to be Overcome, Project Structures and Risk Allocation in a Viable FLNG Project, paper presented at the 24th World Gas Conference, Buenos Aires, Argentina.

Wong, S.C., Fesharaki, S., Oct-Dec, 2010. Global LNG Supply and Demand—Review and Outlook. Hydrocarbon Asia 6–17.

Wood, D.A., Feb.2004. LNG's Share of Global Market to Soar. Petroleum Review, 38–39.

Wood, D.A., Jan. 24, 2005a. LNG Risk Profile-1. Where We Are: Relationships, Contracts Evolve Along Supply Chain. Oil & Gas Journal 54–59.

Wood, D.A., Feb. 14, 2005b. LNG Risk Profile-2. Where We Are Going: SWOT Analysis Aids Risk Assessment. Oil & Gas Journal 54–58.

Wood, D.A., Dec. 2007. Japan's LNG Prices: Trending Upwards. Energy Tribune 29–31.

Wood, D.A., Feb. 25, 2008. Lenders Likely to Tighten LNG Project Financing. Oil & Gas Journal 20–24.

Wood, D.A., May 1, 2009. Floating Gas Liquefaction; Competing Technologies Make Progress. Harts E&P.

Wood, D.A., Nov. 2012. A Review and Outlook for the Global LNG Trade. Journal of Natural Gas Science & Engineering 9, 16–27.

Wood, D.A., Mokhatab, S., Sept. 2006a. LNG Trading Strategies Require Diverse Approaches from Buyers and Sellers. LNG Journal, 12–15.

Wood, D.A., Mokhatab, S., Nov. /Dec., 2006. LNG Liquefaction Ventures Offshore Require Contract Initiatives to Progress. LNG Journal, 16–21.

Wood, D.A., Mokhatab, S., 2007a. Global LNG: Profitable Year So Far, But Cost and Demand Challenges Confront Suppliers. World Oil 228 (10), 129–132.

Wood, D.A., Mokhatab, S., Feb. 2007b. Natural Gas Interchangeability in Focus as Sources of LNG Widen. LNG Journal 14–18.

Wood, D.A., Mohhatab, S., 2008a. Gas Monetization Technologies Remain Tantalizingly on the Brink. World Oil 229 (1), 103–108.

Wood, D.A., Mokhatab, S., 2008b. Commercial Breakthroughs in LNG Technology. World Oil 229 (10), 135–140.

Wood, D.A., Mokhatab, S., 2009. What Are the Opportunities to Construct Liquefaction Facilities at the Arctic Circle? Hydrocarbon Processing 88 (3), 55–58.

Wood, D.A., Mokhatab, S., 2011. Successful Realization of Gas Liquefaction Projects in Southeast Asia. World Oil 232 (1), 79–80.

Wood, D.A., Mokhatab, S., Economides, M.J., Nov. 11–14, 2007. Offshore Natural Gas Liquefaction Process Selection and Development Issues. SPE-109522, paper presented at the 2007 SPE Annual Technical Conference and Exhibition, Anaheim, CA, USA.

Wood, D.A., Mokhatab, S., Economides, M.J., March 2–5, 2008a. Technology Options for Securing Markets for Remote Gas, paper presented at the 87th Annual GPA Convention, Grapevine, TX, USA.

Wood, D.A., Mokhatab, S., Economides, M.J., March 2–5, 2008b. Diverse LNG Pricing Arrangements in Major Gas Consuming Markets, paper presented at the 87th Annual GPA Convention, Grapevine, TX, USA.

World Bank Group, 1999. Pollution Prevention and Abatement Handbook 1998: Toward Cleaner Production. World Bank Publications, The World Bank, Washington, DC, USA.

Yang, C.C., Kaplan, A., Huang, Z., March-April 3–30, 2003. Cost-effective Design Reduces C_2 and C_3 at LNG Receiving Terminals, paper presented at the 2003 AIChE Spring National Meeting, New Orleans, LA, USA.

Yost, C., DiNapoli, R., 2003. Benchmarking Study Compares LNG Plant Costs. Oil & Gas Journal 101 (15), 56–59.

CHAPTER

2 Gas Conditioning and NGL Recovery Technologies

2.1 Introduction

The acid gas contents in most oil and gas production fields have been increasing gradually as CO_2 is injected into the reservoirs for sequestration, and nitrogen is injected to stimulate oil production for enhanced oil recovery. The treating and conditioning processes are becoming more challenging, particularly for conditioning gas to the high purity that is required for liquefied natural gas (LNG) projects.

This chapter discusses the different technologies in treating and conditioning the sour feed gas to the natural gas liquefaction plant, and the technology in the removal of contaminants to meet environmental and emissions regulations and LNG feed gas specification. NGL recovery is necessary to produce a residual gas with heating value or Wobbe Index that meets the sales gas specification while reaping the economic benefits of the liquid products. Nitrogen rejection technologies are required to maintain the feed gas compositions under control and minimize the liquefaction horsepower within the acceptable value, which are addressed in this chapter.

2.2 LNG production plants

The feed gas from the well heads typically contains various contaminants and acid gases that cannot be sent directly to the liquefaction unit. It must pass through different processing units for removal of the undesirable components. The processing units of a typical LNG production plant are shown in Figure 2-1.

These units work together to deliver an LNG product that would meet the specifications given in Table 2-1. As can be seen, there are two LNG product specifications: one for the LNG from the liquefaction plant and the other one for the LNG unloaded at customer port. The difference is due to the boil-off of the volatile components from heat leaks during storage, ship loading and unloading, and ship voyage.

Note that the actual LNG plant configuration may vary, depending on the feed gas compositions, product slates and specifications, environmental regulations, and emissions limits. The plant complexities and process unit setup are also functions of the plant size, the acid gas contents, sulfur content, and the contaminant levels, which must be studied in depth to arrive at the viable plant configuration.

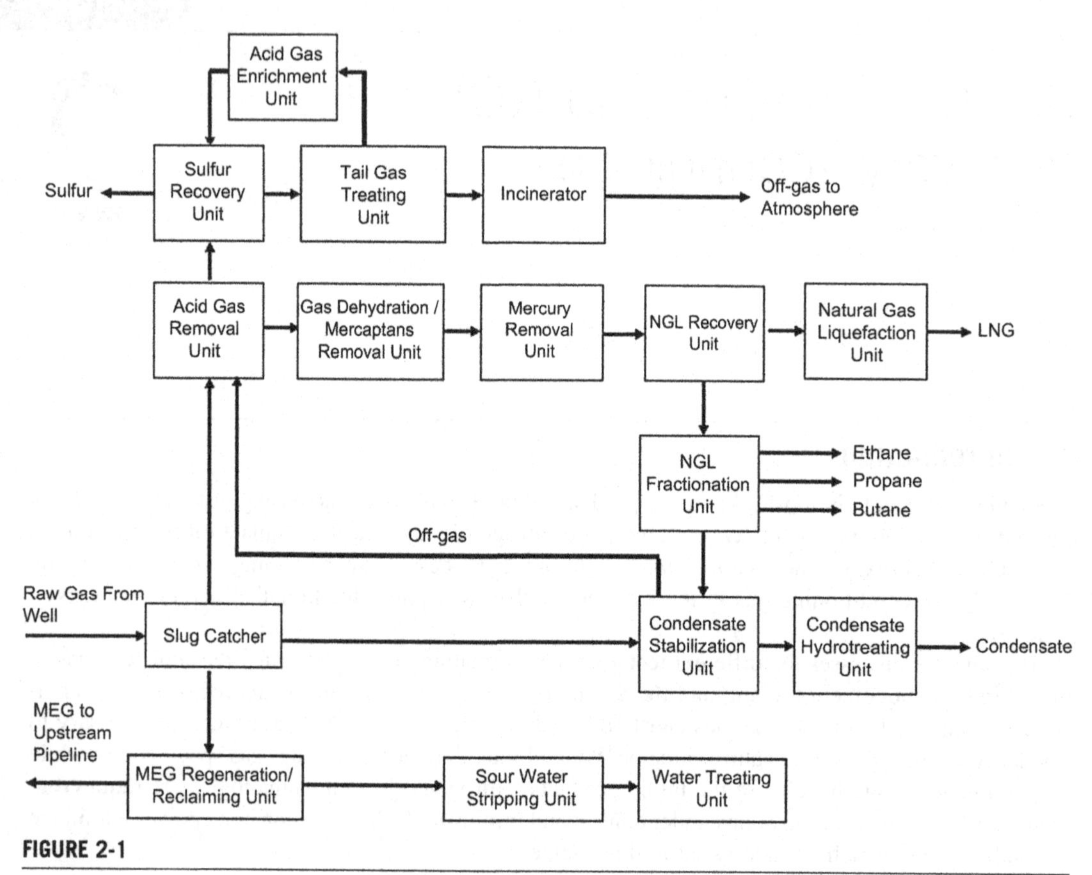

FIGURE 2-1

Typical LNG production plant.

Table 2-1 Typical LNG Product Specifications

	Rundown to Storage	Unloaded at Customer Port	Units
Properties			
Higher Heating Value	42–44	42–45	MJ/Sm^3 (15°C)
Wobbe Index	51–53	51–54	MJ/Sm^3 (15°C)
Compositions			
Methane	84–96	84–96	Mole %
Max C4+	2.4	2.5	Mole %
Max C5+	0.1	0.1	Mole %
N_2	1.4	1.0	Mole %

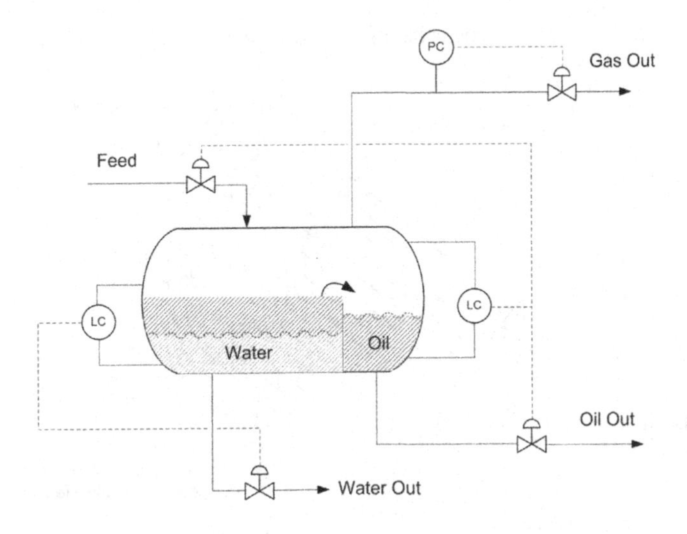

FIGURE 2-2

Typical vessel-type slug catcher.

2.2.1 Inlet separation facility

Field production upon arrival at the LNG production plant will be processed in a slug catcher, which catches the largest liquid slugs expected from the upstream operation[1] and then allows them to slowly drain to the downstream processing equipment to prevent overloading the system.

The slug catcher design is either a vessel-type or finger-type. A vessel-type slug catcher is essentially a knockout vessel (Figure 2-2). The advantages of the vessel-type are that they require significantly less installation area, are simple in design, and are easy to maintain. The traditional finger-type slug catcher consists of multiple long pieces of pipes ("fingers"), which together provide the surge volume (Figure 2-3). The finger-type design is generally less expensive than the vessel-type in high pressure operation. The disadvantage is the large footprint requirement. It is limited to land base facilities where there is no space constraint.

The vessel-type slug catchers typically are used only if the incoming liquid volumes are fairly stable and relatively small. When the incoming liquid flow is uncertain, such as in long pipelines, the potentially large liquid volumes would require the use of the finger-type slug catcher (Shell, 1998).

The slug catcher serves as a three-phase separator where the gas, hydrocarbon liquids (condensate), and aqueous phase are separated. The flash gas from the slug catcher is further separated in a downstream separator to remove any liquid entrainment, prior to entering the gas treatment section. The condensate is processed in the condensate stabilization unit to reduce the vapor pressure to allow storage in atmospheric storage tanks (Figure 2-1). If the condensate contains mercaptans and other sulfur components, it must be hydrotreated to meet the total sulfur specification, typically, 50 ppmw in order to meet the export requirement.

[1]Some slugs will grow as they travel down the pipeline, while others are dampened and disappear before reaching the oil and gas separation facility.

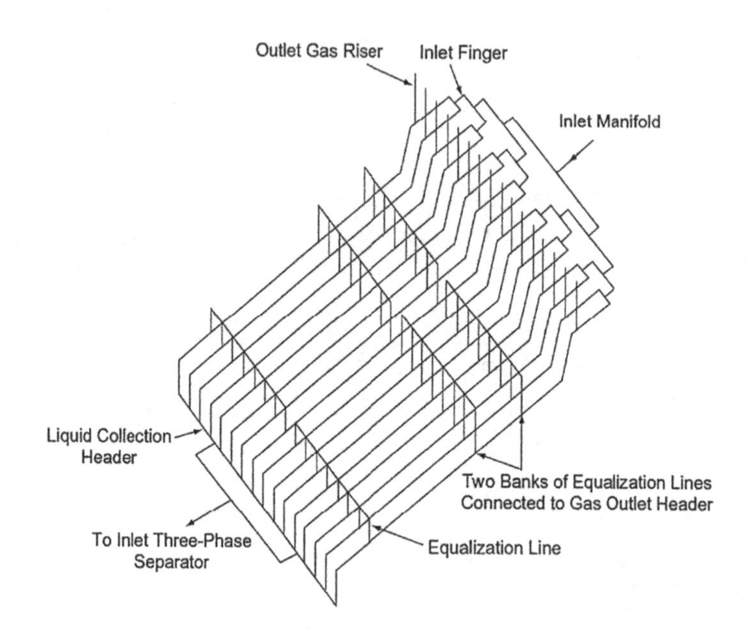

FIGURE 2-3

Typical finger-type slug catcher.

Monoethylene glycol (MEG), which is required for hydrate prevention, is present in the aqueous phase. The MEG is often contaminated with the salts contained in the formation waters. Since salt is nonvolatile, it will remain in the lean glycol during regeneration, which can cause serious corrosion and fouling problems with equipment and pipelines (Son and Wallace, 2000). There are MEG reclamation packages currently available to remove these salts and other contaminants to maintain the required purity. The processed MEG is collected in a lean MEG tank for reinjection to the production field.

Sour water separated from the slug catcher and other sources typically is treated in a sour water stripping unit for removal of the acid gas and ammonia contents. The stripped water can be recycled to the process units or be further processed in a wastewater treatment system prior to disposal.

2.2.1.1 Condensate stabilization unit

The condensate, a hydrocarbon liquid, separated from the slug catchers contains the dissolved light hydrocarbons and H_2S, which must be removed from the liquid to meet the export condensate specifications. A condensate stabilization unit typically is designed to produce a condensate with 4 ppm H_2S and RVP (Reid Vapor Pressure) specification of 8 to 12 psi.

A typical condensate stabilization unit is shown in Figure 2-4. The hydrocarbon liquids are let down to an intermediate pressure in the medium pressure (MP) flash drum. The flash gas is compressed back to the feed gas section, and the flashed liquid is further reduced in pressure prior to entering the condensate stabilizer column. The stabilizer is a fractionator that is heated with steam. The overhead vapor from the stabilizer is compressed and recycled back to the feed section. The combined gas stream is sent to the acid gas removal unit (AGRU). The condensate is heat exchanged with the

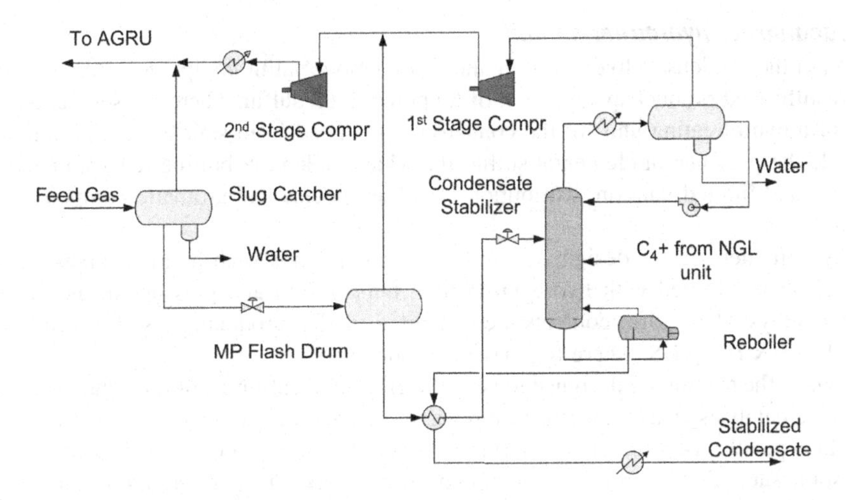

FIGURE 2-4

Typical condensate stabilization.

stabilizer feed, then cooled and exported as the stabilized condensate. If significant amounts of mercaptans are present in the feed gas, the condensate would need to be further treated, typically by hydrotreating, to meet the product sulfur specification.

The condensate stabilization unit can also be designed to produce an NGL product, as shown in Figure 2-5. In this configuration, a feed liquid stripper (FLS) is added before the stabilizer, which removes the C_2 and lighter components. The stripper overhead is recycled back to the feed section, and the bottom is fractionated in the stabilizer into a LPG overhead and C_5+ condensate bottom products.

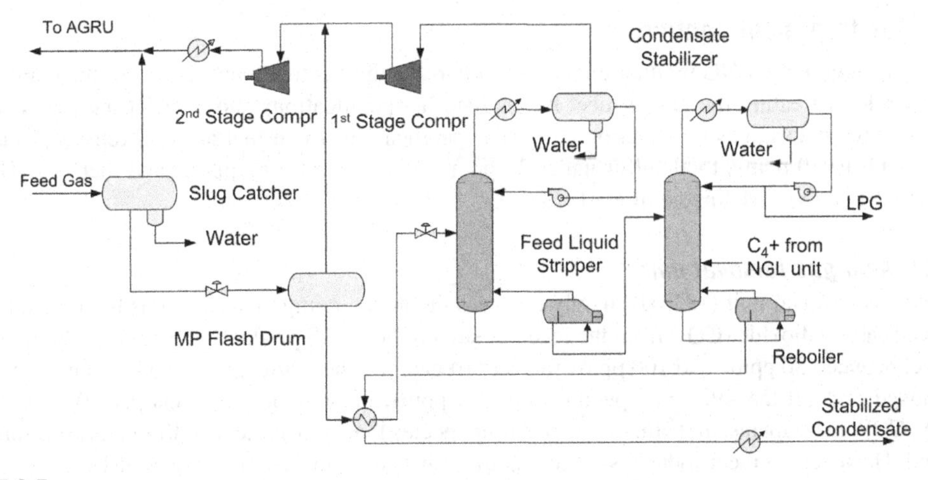

FIGURE 2-5

Typical condensate stripper with LPG production.

2.2.1.2 Condensate hydrotreating unit

The function of the condensate hydrotreating unit is to remove sulfur compounds from the condensate to meet a desulfurized product specification of 50 ppmw total sulfur. There are several challenges in the design of a hydrotreating unit for the condensate produced from sour gas (Schulte et al., 2009). These include the presence of elemental sulfur, the wide condensate boiling ranges, uncertainties on compositions, and limited data on distribution of sulfur species and aromatic compound and metal contaminants.

In the hydrotreater reactor design, various sulfur compounds such as mercaptans, sulfides, and elemental sulfur are reacted with hydrogen at high temperature and pressure in the presence of a hydrotreater catalyst. The sulfur contents are converted to H_2S producing a sulfur-free hydrocarbon ($R\text{-}S\text{-}H + H_2 \rightarrow R\text{-}H + H_2S$, where R = Hydrocarbon chain).

Conditions in the reactor are designed to keep the elemental sulfur in solution and avoid deposition of solids on the catalysts. This typically requires a lower operating temperature and higher operating pressure than typically used in naphtha service. Although the purpose of the hydrotreater is desulfurization, some saturation of aromatic compounds also occurs. The configuration of the hydrotreater unit is conventional and similar to refinery application, and primarily consists of the following sections:

- High pressure reactor loop: Sulfur compounds and elemental sulfur in the condensate are converted to H_2S. The reactor effluent is cooled, and product liquids are separated from the recycled gas in a product separator.
- Low pressure stripping section: The condensate product from the separator is stripped to remove H_2S and H_2.
- Makeup hydrogen compression: Hydrogen from the hydrogen plant is compressed to the reactor loop pressure.
- Hydrogen plant: Supplies high purity makeup hydrogen to the high pressure reactor loop.

2.2.2 Gas treatment section

While each facet of the LNG production plant is important, the gas treatment section of the plant plays a critical role in treating the gas to meet its final sulfur specifications and to meet the purity levels required by the natural gas liquefaction unit. The specifications to be met are H_2S removal to under 4 ppmv, CO_2 to 50 ppmv, total sulfur under 30 ppmv as S, water to 0.1 ppmv, and mercury (Hg) to levels of 0.01 $\mu g/Nm^3$ (Klinkenbijl et al., 2005).

2.2.2.1 Acid gas removal unit

The acid gas removal unit (AGRU) mainly removes the acidic components such as hydrogen sulfide (H_2S) and carbon dioxide (CO_2) from the feed gas stream. For LNG production, CO_2 must be removed to a level between 50 ppmv and 100 ppmv to avoid freezing in the cryogenic exchanger, and H_2S must be removed to meet the sales gas specification of 4 ppmv, or one-quarter grains per 100 scf. Additionally, COS, mercaptans, and other organic sulfur species that contribute to sulfur emissions must be removed. However, to meet today's stringent sulfur emission requirements, the AGRU alone cannot meet the required specifications, and treated gas from these units must be polished with other processes such as molecular sieves, which are specifically designed for removal of the other sulfur components.

A number of processes are available to remove H_2S and CO_2 from natural gas (Stone et al., 1996; Clarke and Sibal, 1998). Some have found wide acceptance in the LNG segment of the industry while others are currently being considered. The selection of an acid-gas removal process can have a significant impact on project economics, especially when the feed gas contains a high percentage of acid gas. Carbon dioxide and hydrogen sulfide in the feed gas will also significantly impact the thermal efficiency of the LNG liquefaction process due to the fact that AGRU is an energy intensive process.

The main considerations for treating a high CO_2 content feed gas are energy consumption and the extent of hydrocarbon coabsorption. A high energy consumption process such as amine will lower the thermal efficiency of the liquefaction plant and generate more greenhouse gases, whereas a physical solvent process will lower the energy consumption but will coabsorb more hydrocarbons. Coabsorption of hydrocarbons has two negative effects. First, it reduces the heating value of the product gas, and second, the higher hydrocarbon contents in the sulfur plant (SRU) feed will cause equipment fouling problems and increase emissions.

There are three commonly used solvent absorption processes for acid gas removal in base load LNG plants: chemical absorption, physical absorption, and the mixed solvents processes (Klinkenbijl et al., 1999). The other processes have limited applications. Membrane separation is only suitable for bulk acid gas removal, and the other processing methods, such as cryogenic fractionation and fixed beds adsorption, are not cost competitive.

Chemical solvent processes

Chemical absorption processes, which chemically absorb the H_2S, CO_2 and to some extent COS, will not remove mercaptans down to low levels due to the low solubility of these components. The advantage of a chemical solvent process such as amine is that the solubility of aromatics and heavy hydrocarbons in the aqueous solvent is low, hence lower hydrocarbon losses. The disadvantage is their high energy consumption in amine regeneration heat duty and cooling duty.

Common examples of amine processes are aqueous solutions of alkanol amines such as monoethanolamine (MEA), diglycolamine (DGA), diethanolamine (DEA), diisopropanolamine (DIPA), and methyldiethanolamine (MDEA). With the exception of MDEA, amines are generally not selective and will remove both CO_2 and H_2S from the gas. Amine can also be formulated by solvent suppliers to increase their selectivity and/or absorption capacity (Hubbard, 2009). Typically, the MDEA selectivity toward H_2S is highest at low operating pressures such as in the tail gas unit, but its selectivity is significantly reduced at high pressure.

When used in treating sour gases to meet the tight CO_2 specification for LNG plant, the activity of CO_2 absorption is too slow with pure MDEA, which must be enhanced with a promoter. The most widely used promoted MDEA process is the aMDEA process, which was originally developed by BASF. The aMDEA process uses piperazine as an activator in MDEA for CO_2 absorption. Since the patent on the use of piperazine with MDEA has expired, the solvent can now be available from several amine suppliers such as Dow, Huntsman, and INEOS. The process can also be licensed from technology licensors such as BASF, UOP, Shell, Lurgi, and Prosernat.

A typical amine process is shown in Figure 2-6. The feed gas is scrubbed in an amine contactor that consists of an amine absorption section and a water wash section. The amine absorption section removes the acid gases from the sour feed gas by contacting with a lean amine. The treated gas is washed in the water wash section to recover amine from the treated gas. The water wash section reduces amine makeup requirement and minimizes fouling in the molecular sieve unit.

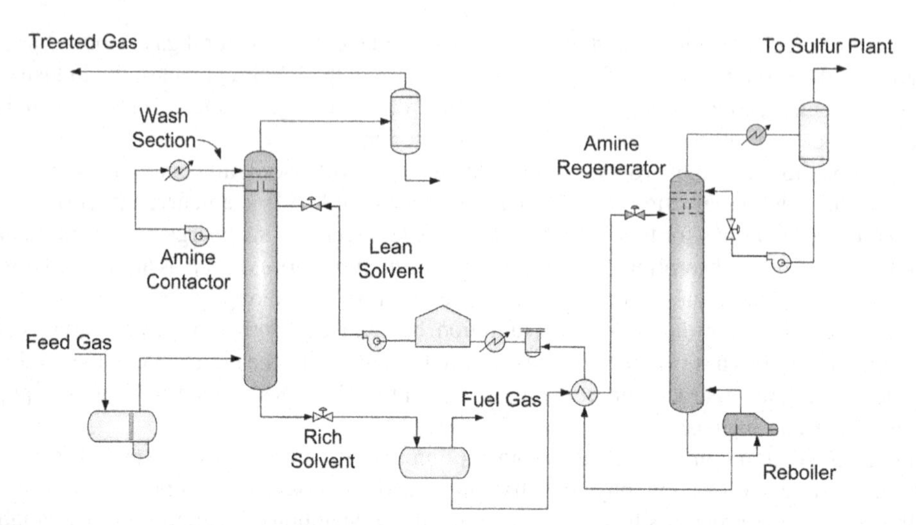

FIGURE 2-6

Typical amine process flow.

The rich amine from the amine absorber is flashed in a rich amine flash drum, producing a flash gas that can be used for fuel after being treated with amine. Typically, a lean rich exchanger is used to reduce the regeneration reboiler duty. The amine is regenerated using steam or another heating medium. The lean amine is cooled, pumped, and recycled back to the amine absorber.

The common problem in operating an amine unit is foaming in the amine contactor. Sour gas from the high pressure separator is at its hydrocarbon dewpoint, such that the lean amine temperature must be controlled at some margin above the sour gas temperature, typically at 10°F to prevent condensation and subsequent foaming in the absorber. In hot desert areas, where cooling water is not available, process cooling must be by air coolers. In most areas, it is difficult to cool the process gas to below 150°F. In the amine absorber, removal of the acid gases would increase the gas dewpoint temperature, which means that the lean amine temperature would need to be further increased. A high lean amine temperature would lower the equilibrium loading of the rich amine, increasing the amine flow rate, making treating difficult. To allow the absorber to operate in hot climate areas, feed gas chilling is required, which can be done with the use of propane refrigeration. The feed gas can be chilled to remove the hydrocarbons, reducing the hydrocarbon dewpoint, which would allow the amine contactor to operate at a lower temperature, hence reducing the solvent circulation.

Physical solvent processes

Physical absorption processes use a solvent that physically absorbs CO_2, H_2S, and organic sulfur components (COS, CS_2, and mercaptans). Physical solvents can be applied advantageously when the partial pressure of the acid-gas components in the feed gas is high, typically greater than 50 psi. Note that the physical solvent acid gas holding capacity increases proportionally with the acid gas partial pressure, according to Henry's law, and can be competitive to chemical solvent processes because of the higher loading and less heating duty. However, physical solvents are not as aggressive as chemical solvents in deep acid gas removal, and may require additional processing steps. Depending on the acid gas contents, a hybrid treating system, such as a physical solvent unit coupled with a sulfur scavenger, may be a better choice than a single amine system (Mak et al., 2012).

There are several proven physical solvent processes such as Selexol® (licensed by UOP), Fluor® Solvent (licensed by Fluor), or Purisol® (by Lurgi). The main advantages of physical solvent processes are that the solvent regeneration can be partially achieved by flashing of the solvent to lower pressures, which significantly reduces the heating requirement for regeneration. In some processes, such as the Fluor Solvent process (Figure 2-7), no heating is required as the solvent is regenerated by vacuum flashing or by stripping with treated gas or inert gases. When used in treating high pressure, high acid gas content gases, greenhouse gas emissions from physical solvent units are significantly lower than the amine units.

The main disadvantage of the physical solvent unit is the coabsorption of hydrocarbons, which results in Btu losses in the CO_2 effluent. Energy savings could be offset by the losses in product gas. Unlike an amine unit where the process design is quite straightforward, physical solvent unit designs are more complex. Most of the physical solvent processes are licensed processes and the process configurations and operating conditions vary depending on the licensed solvent used. Typically, physical solvent units require additional equipment, such as recycle compressor, refrigeration, stripper, and flash drums. Gas recycling reduces the hydrocarbon losses, which, however, also increase the

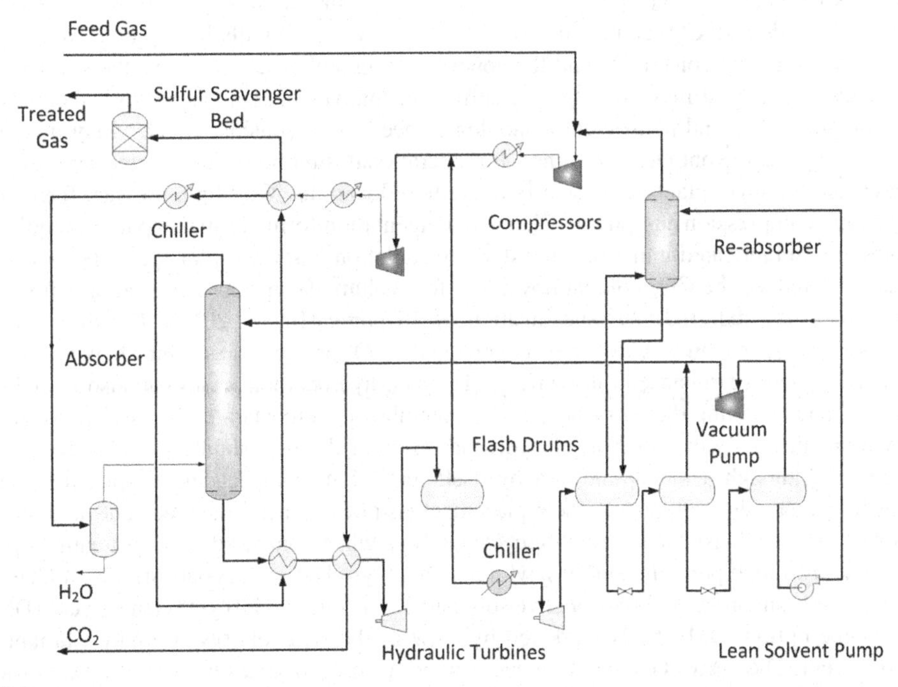

FIGURE 2-7

Fluor physical solvent process.

(Mak et al., 2012)

solvent circulation. If treated gas is used for stripping, the stripper overhead gas can be recycled back to the absorber, which also increases the solvent circulation.

Therefore, the design of a physical solvent unit is more involved and must be carefully optimized in order to realize the advantages of the physical properties of the solvent.

Mixed solvents processes

Mixed solvents processes use a mixture of a chemical and a physical solvent. They are used to treat high acid gas content gases while meeting the deep removal of the chemical solvents. To some extent, these favorable characteristics make them a good choice for many natural gas treating applications. The Shell Sulfinol process is the one of the proven mixed solvent processes.

Process selection considerations

The selection of the acid gas removal technology in an LNG plant is dependent on the feed gas composition and conditions, the product gas specifications, and the availability of utilities. For example, if waste heat is available, then amine units may be selected since they can be used as a heat sink for heat rejection. However, when a heat source is not readily available, a physical solvent may be a better choice. But other criteria such as hydrocarbon coabsorption, removal of aromatics, and operating flexibility should be considered in conjunction with the capital and operating costs.

Feed gases with low acid gas contents are easy to treat, but these fields are becoming scarce. Most of today's gas fields are sour gas fields. These fields commonly contain 10 to over 20 mole % acid gases. Sour gas fields with CO_2 content of over 50 mole % will eventually be explored. With the ever-increasing environmental constraints and the concerns on greenhouse emissions, the straightforward sweetening unit in the past may no longer be sufficient. Innovative acid gas removal methods, more complex configurations, and nonsolvent technologies need to be considered. This may include bulk removal of CO_2 via membranes. Designed for operational simplicity, membrane systems are an excellent choice for small plants, particularly in offshore locations. With high pressure feed gas, they may not require compression equipment. Chemical reagent monitoring or makeup is not required. The membranes are prefabricated units, which reduces installation costs and plot space. However, membranes can only reduce the CO_2 content down to a few volume % and further treating with amine is required to meet the required CO_2 specification of 50 ppmv (Bauer, 2011). The disadvantage of membrane separator is the high hydrocarbon content in the CO_2 permeate gas, which may not be a loss if the CO_2 is reinjected for enhanced oil recovery. However, hydrocarbon losses can also be reduced by multiple units and recycling the permeate gas. The membrane elements are also susceptible to degradation by aromatics and condensation of hydrocarbons, and the unit should be protected by a pretreatment unit. Although a membrane unit by itself may seem compact and simple, the additional pretreatment, compression equipment, the replacement cost of the membrane elements and the hydrocarbon losses must be considered in the life cycle analysis when compared to other treating options.

When treating a high pressure sour gas, with a high CO_2 to H_2S ratio, coabsorption of CO_2 cannot be avoided. In this situation, an H_2S selective solvent such as dimethylpolyethylene glycol (DMEPG) can be a viable option. DMEPG is marketed by Dow as the Selexol solvent or by Clariant as the Genosorb solvent. The DMEPG solvent process can be licensed from UOP as the Selexol process or from Fluor as the EconoSolv process. The process consists of an H_2S removal section and a CO_2 removal section (see Figure 2-8). In the H_2S removal section, the H_2S content is removed from the feed gas using a CO_2 saturated lean solvent from the CO_2 section. In order to concentrate the H_2S content in

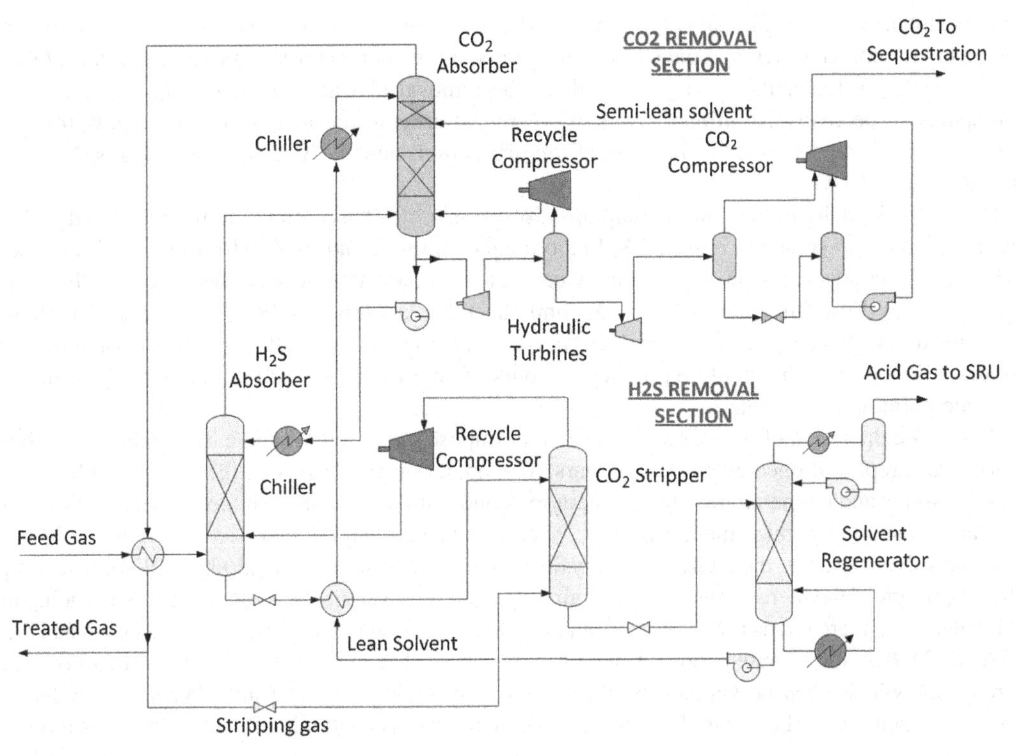

FIGURE 2-8

Two-stage physical solvent selective process.

(Mak et al., 2010)

the acid gas, the CO_2 content in the solvent is stripped using a slip stream of the treated gas. The stripper overhead gas is recycled back to the H_2S absorber to reduce hydrocarbon losses. This operation effectively reduces the CO_2 content in the rich solvent, hence increasing the H_2S content in the acid gas to the Sulfur Recovery Unit. The rich solvent is regenerated with steam in the solvent regenerator to produce a lean solvent that is used in the CO_2 section. The CO_2 content in the sulfur-free gas from the H_2S section is removed in the CO_2 absorber. In the CO_2 section, the CO_2 rich solvent is regenerated by flash regeneration by reducing pressure at various levels. The high pressure CO_2 can be fed directly to the CO_2 compressor, which significantly reduces the compression horsepower for CO_2 sequestration. This configuration can be used to capture over 95% of the carbon content in gasification application (Mak et al., 2010).

2.2.2.2 Dehydration and mercaptans removal unit

The second step in the treating process is a Molecular Sieve Unit (MSU), which is designed to meet the product specifications on water and mercaptans (RSH) content. The MSU can be designed to remove water to 0.1 ppmv and most mercaptans to 2 to 3 ppmv. When molecular sieves are used for mercaptans removal, water must be removed first before the mercaptans removal bed. Removal of water and mercaptans on molecular sieves can be installed in a single vessel, where the first layers of molecular sieves remove water, and the subsequent layers of molecular sieves remove mercaptans. Often, 4A, 5A, and 13X molecular sieves are used for the removal of water, light mercaptans, and heavy mercaptans, respectively. Additional layers of specific adsorbents may also be used to remove traces of mercury and to protect the molecular sieves from plant upsets and unexpected contaminants (Northrop and Sundaram, 2008).

Design of the dehydration and mercaptan-adsorption unit is based on the number of fixed beds in a parallel lineup, as shown in Figure 2-9. In a typical operation, four molecular sieves beds are used, with two of the adsorbers on drying the sweet gas from AGRU, one adsorber is being thermally regenerated by desorbing the compounds, and the other adsorber is being cooled before being returning to the next cycle. Each molecular sieve adsorber is typically in adsorption mode for 6 hours, followed by 3 hours of heating and 3 hours of cooling. Every 12 hours, the cycle returns to the same point and is repeated.

The cyclic operation of the molecular sieve unit imposes stress on the sieve materials and the beds are typically replaced once every 3 to 5 years. The operation conditions of the cycles, such as flow rates, pressures and temperatures, and regeneration conditions should be evaluated in a dynamic model that can be used to optimize the dehydration operation and extend the bed life (Mock et al., 2008).

If the solvent used in the AGRU is a physical solvent or mixed chemical/physical solvent, a significant portion of the mercaptans may have already been removed in the AGRU, thereby reducing the molecular sieve mercaptans removal requirement. On the other hand, chemical solvents such as DEA, DGA, or MDEA (accelerated) do not absorb any appreciable amount of heavy hydrocarbons and mercaptans, which must be handled by the downstream molecular sieve unit. When the molecular sieves are regenerated, the wet molecular sieve regeneration gas containing the mercaptans is returned

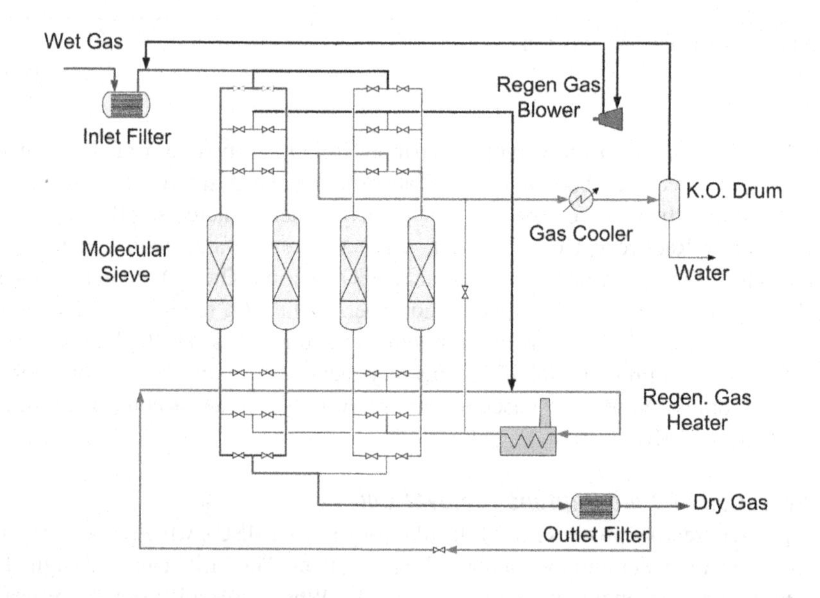

FIGURE 2-9

Typical molecular sieve unit.

to the AGRU inlet. The mercaptans concentration in the gas entering the AGRU will build up until the amount removed in the AGRU equals the incoming mercaptans.

The split of the mercaptans removal between the ARGU and the molecular sieves should consider factors such as operating flexibility on different gas compositions and flow rates, environmental and emissions, and operating and capital costs (Klinkenbijl et al., 1999; Grootjans, 2005).

When using the molecular sieves for both dehydration and mercaptans removal, the amount of mercaptans must be purged from the system by one of several means, such as:

- Use the wet regeneration gas as fuel gas. This is unlikely as the regeneration gas typically far exceeds the plant's fuel gas requirement. Also the sulfur oxides from incineration of the mercaptans may exceed the plant's sulfur emissions limit.
- Add a physical solvent AGRU to treat the regeneration gas.

The regeneration gas from the AGRU is processed in the Sulfur Recovery Unit (SRU). If an amine unit is used as the AGRU, the majority of mercaptans will come from the molecular sieve beds during regeneration. The mercaptans release will vary during the regeneration cycle and will peak at some point. This has an impact on the SRU as the SRU catalyst is designed for a certain mercaptans level and is very sensitive to mercaptans load changes (Carlsson et al., 2007; Bradley et al., 2009). To avoid fouling of the SRU catalysts from excessive mercaptans loads, some leveling facilities should be provided to even out the mercaptans flow to the SRU.

The branched-type mercaptans that slip through the AGRU usually are not caught by the molecular sieves and are left in feed gas to the downstream unit. Since these branched type mercaptans have volatility close to pentane, they are removed together with the NGL stream and will concentrate in the pentane plus condensate. To meet the sulfur specification in the condensate product, further processing such as hydrotreating is required.

2.2.2.3 Mercury removal unit

Almost every LNG plant will have a mercury removal unit installed. This is because the consequence for the LNG plant from mercury attack is severe and because it is difficult to predict the mercury contents from production reservoirs. Low levels of mercury can result in severe corrosion of the brazed-aluminum heat exchangers used in cryogenic systems, and potentially can pose environmental and safety hazards. The presence of mercury in the feed stocks to petrochemical plants will also cause poisoning of precious metal catalysts (Carnell and Row, 2007). For this reason, the LNG plant is designed with a conservative design that requires mercury removal to levels below 0.01 $\mu g/Nm^3$.

Most of the current methods for removing mercury from natural gas and hydrocarbon liquids use fixed beds of mercury removal materials. The fluid flows through the fixed bed in which mercury reacts with the reactive reagent in the mercury removal vessel producing a mercury-free product (Kidnay and Parrish, 2006).

There are two types of mercury removal materials: nonregenerative mercury sorbents, and regenerative mercury adsorbents.

Nonregenerative mercury sorbents

In the nonregenerative mercury removal process, the mercury reacts with the sulfur to form a stable compound on the sorbent surface. A number of different mercury removal sorbents are available with various tolerances to operating temperature, liquid hydrocarbons, and water (Markovs and Clark,

2005). The use of sulfur-impregnated activated carbon is a proven commercial process for removing the mercury.[2] However, there are drawbacks to this method of mercury removal, where sulfur-impregnated carbon can only be used with dry gas since it has a high surface area and small pore size. This also restricts the access of mercury to the sulfur sites and increases the length of the reaction zone. Also, sulfur can be lost by sublimation and dissolution in hydrocarbon liquids. This again reduces mercury removal capacity. Furthermore, it is often difficult to dispose of the spent mercury-laden carbon material (Abbott and Oppenshaw, 2002).

Recognition of these problems has led to examining technologies other than sulfided carbon, where a range of nonregenerable absorbents utilizing transition metal oxides and sulfides instead of carbon have been developed to improve on existing mercury removal technologies, in which the discharge absorbents can be safely handled. In these systems, the reactive metal is incorporated in an inorganic support and the absorbent is supplied with reactive sulfide component by either ex-situ or in-situ sulfiding (Carnell and Row, 2007). Johnson Matthey has taken this concept further and supplied an established range of absorbents marketed under the PURASPEC™ brand. The PURASPEC™ materials are a mixture of copper sulfide/copper carbonate, zinc sulfide/zinc carbonate, and aluminum oxide, which can operate in a wet gas environment (Row and Humphrys, 2011).

Note that the nonregenerative methods appear to be simple since no regeneration equipment and special valving is required. However, disposal of the used sorbent can be a problem since the sorbent not only picks up the mercury, but it will often contain other hazardous material such as benzene and other hydrocarbons (Markovs and Clark, 2005).

Regenerative mercury adsorbents

The regenerative mercury removal process utilizes silver on a molecular sieve (such as UOP's HgSIV sieve) to chemisorb elemental mercury. The mercury saturated bed is then regenerated by hot regeneration gas. The regeneration gas is typically heated to 550°F. The mercury can later be recovered in the condensed water. In this method of mercury removal, since the mercury does not accumulate on the adsorbent, it would avoid the disposal problems with spent adsorbent. However, there will be trace of mercury left in the regeneration gas, which can be removed with a nonregenerative mercury bed (Markovs and Clark, 2005).

Process selection considerations

There are four possible options for mercury removal.

- Option 1: Installing nonregenerative mercury removal sorbents at the plant inlet. This option removes all the mercury and ensures no mercury contamination in the rest of the plant (Edmonds et al., 1996). However, the large volume of feed gas and acid gases require a large removal system, which may be challenging in design and operation.
- Option 2: Installing a nonregenerative mercury removal sorbent downstream of the acid gas removal unit, just before the molecular sieve unit. This option reduces the size of the beds to some extent, but it poses the risks of mercury contamination in the AGRU solvent system.
- Option 3: Add a silver-impregnated mercury sieve section to the molecular sieve beds. Although this option can remove water, mercaptans, and mercury at the same time, and avoid the need of a

[2] Sulfur is the active ingredient, securely fixing mercury as sulfide in the microporous structure of the carbon.

separate mercury bed, it presents problems with a high mercury content in the regeneration water that would pose operating hazards unless treated by another mercury removal step (Hudson, 2010).

- Option 4: Installing a nonregenerative mercury removal bed or a silver-impregnated molecular sieve bed after the molecular sieve unit. While this option yields good mercury removal performance, as the feed gas is dried and clean, it cannot avoid the mercury contamination problem in the AGRU and molecular sieves upstream and may pose operating hazards (Eckersley, 2010).

Life cycle costs, adsorbent disposal methods, mercury levels, environmental limits, operating hazards, and plant operator procedure must be evaluated in the selection of a suitable mercury removal system. The optimum mercury removal method can also be a combination of nonregenerative and regenerative mercury removal system (Markovs and Clark, 2005).

2.2.2.4 Sulfur recovery unit

In a typical gas treating unit, acid gas from the regenerator contains significant amounts of H_2S that must be further processed and cannot be vented or flared. In the past, the acid gas could be reinjected for sequestration, but recent research indicated that the sulfur compounds would have long term negative impacts on the reservoirs and formation. Today, acid gas is generally processed in a sulfur recovery unit that is closely coupled with a tail gas unit. With this combination, the sulfur recovery system can meet the 99.9% sulfur removal target, which is needed to meet today's emission requirements.

There are many sulfur technologies that are available with different performance in terms of operation and results. The selection of the sulfur technology is mainly depending on the amount of H_2S, CO_2, and other contaminants in the feed gas. As a rule of thumb, liquid redox technology is suitable for small sulfur plants (below 20 tons per day), and for the larger units, such as those used for LNG production, the Claus sulfur technology is the most common (Hubbard, 2009).

Claus technology

The common method for converting H_2S into elemental sulfur in a gas plant is the Claus-based process, which is available from several sulfur plant licensors. The Claus process is basically a combustion unit, and to support the sulfur conversion reaction, the acid gas must contain sufficient H_2S to support the heat of combustion. Typically, the H_2S content in the acid gas must be greater than 40 mole %. If the feed gas contains insufficient H_2S, additional processing steps are required, which may require supplemental preheating, oxygen enrichment or acid gas enrichment (Mokhatab and Poe, 2012). In addition, if the acid gas contains other contaminants such as ammonia, BTEX, and mercaptans, a higher combustion temperature is necessary, which may require even higher H_2S content gases.

In a conventional Claus process (see Figure 2-10), the reaction is carried out in two stages. The first stage is the thermal section where air is used to oxidize about one-third of the H_2S content in the sour gas to SO_2. This reaction is highly exothermic and typically about 60 to 70% of the H_2S in the sour gas is converted to sulfur. In the thermal stage, the hot gases are cooled to 600°F to 800°F and the waste heat is used to generate high pressure steam. During this process, the S_2 sulfur species content are converted to other sulfur species, primarily S_6 and S_8. The gases are finally cooled, to 340°F to 375°F, in a sulfur condenser by generating low pressure steam.

The second stage is a catalytic stage. The residual H_2S is converted in three stage reactors where sulfur is converted by reaction of the residual H_2S with SO_2 at lower temperatures, typically 400°F to

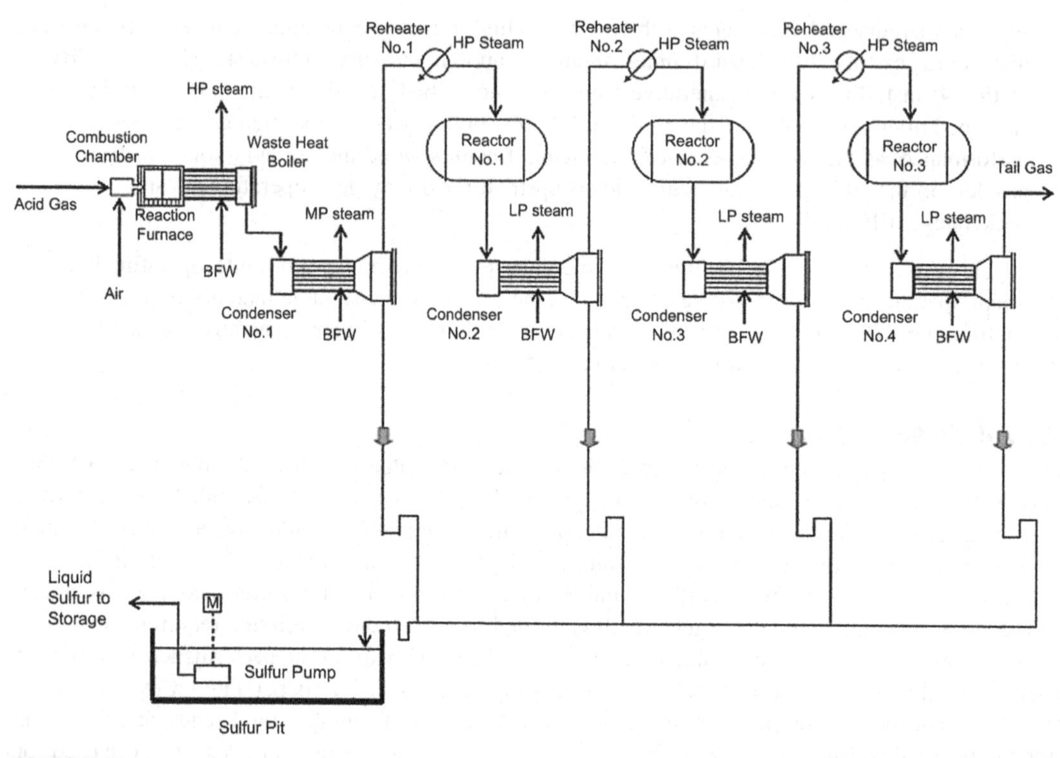

FIGURE 2-10

Typical three-stage Claus sulfur recovery unit.

650°F. The gas must be preheated first to avoid sulfur deposition on the reactors. Sulfur liquid is condensed from the condenser and is routed to the sulfur pit for degassing.

Sulfur recovery efficiencies for a two-stage catalytic process are about 90 to 96%, and for a three-stage process, the efficiencies can be increased to about 95 to 98%. With an additional selective oxidation stage, sulfur conversion can be further increased to 99%, such as the SUPERCLAUS® process, licensed from Jacob Comprimo.

Sulfur degassing

Liquid sulfur produced from the condensers typically contains 200 to 350 ppmw H_2S, partially dissolved and partly present in the form of polysulfides. If liquid sulfur is not degassed, H_2S will be released in the storage tanks, which would create a toxicity hazard and noxious odor problems. Sulfur can be degassed by oxidation and stripping to meet the 10 ppmw H_2S specification, using the patented D'GAASS process from Goar, Allison & Associates, licensed from Fluor.

2.2.2.5 Tail gas treating unit (TGTU)

To meet today's stringent sulfur emissions requirements, use of the Claus process or the SUPERCLAUS® process alone is not enough, as they are limited by chemical equilibrium. To comply with today's emissions requirement, the residual sulfur in the effluent from the Claus unit must be further removed by amine treatment in a tail gas treating unit.

Before the tail gas can be processed in the gas treating unit, the SO_2 content must be converted to H_2S, which can then be absorbed by a H_2S selective amine. The amine absorbs the H_2S content, which is recycled back to the front section of the Claus unit. This recycle process was first developed by Shell as the SCOT unit (Shell Claus Off-gas Treating), which became a standard unit to meet today's emissions requirements. With the use of a selective amine, sulfur recovery of over 99.9% can be met (Harvey and Verloop, 1976). This recycle design can also be integrated with an acid gas enrichment unit as discussed in the following section.

The tail gas unit designs that are offered by several licensors typically consist of two sections, the hydrogenation section and the tail gas treating section (Figure 2-11). In the hydrogenation, SO_2 is converted to H_2S by the hydrogenation catalyst in the presence of hydrogen. The conversion of SO_2 must be complete as it will react with amine in the treating unit resulting in amine degradation. The hydrogenated tail gas is then cooled by water in a quench tower. Excess water condensate is removed from the tower and the overhead tail gas is sent to an amine unit.

The tail gas to the amine unit consists of mainly CO_2 and a very low level of H_2S. The tail gas treating unit is designed to selectively absorb H_2S while rejecting the CO_2 content before it can be sent to a thermal oxidizer or incineration. The H_2S selective amine can be a formulated MDEA for sulfur removal or sterically hindered amines. The formulated MDEA is available from several amine suppliers, and the sterically hindered amine, such as FLEXSORB®, can be licensed from ExxonMobil.

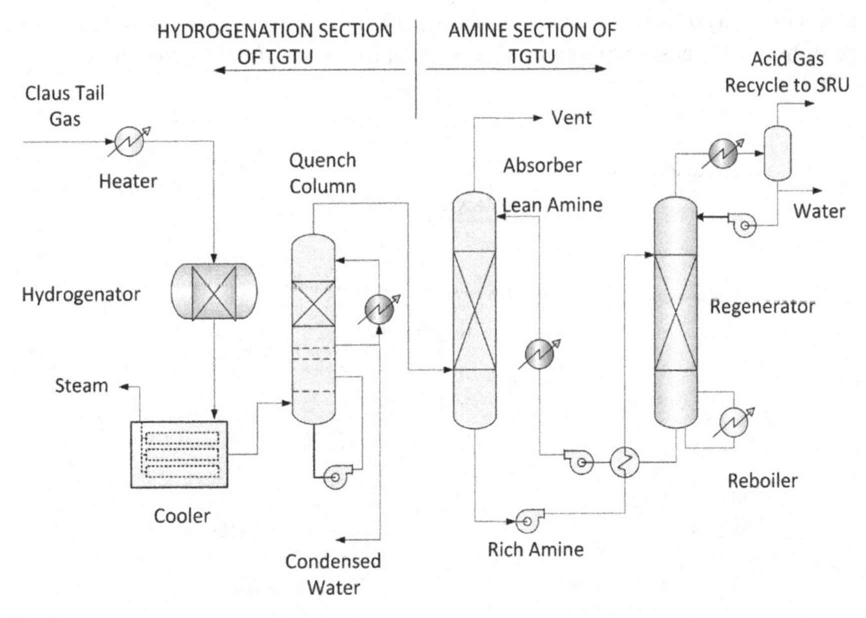

FIGURE 2-11

Typical tail gas treating and hydrogenation unit.

It should be noted that H_2S absorption equilibrium is favored by low amine temperatures, which may not be achievable with air cooling alone, especially in hot climate areas. Typically a chilled water system or refrigeration is used in the tail gas treating unit.

2.2.2.6 *Acid gas enrichment unit*

In a conventional Claus sulfur recovery unit, the acid gas typically contains over 40 mole % H_2S, which would provide sufficient heating value to support the heat of reaction. When a sour gas with high CO_2 content is treated, the acid gas produced from the treating unit will contain a lower H_2S content. To operate the Claus unit, the H_2S content in the acid gas must be concentrated by rejecting its CO_2 content using an acid gas enrichment unit (AGEU).

Similar to the tail gas treating unit, the acid gas enrichment unit uses an H_2S selective solvent such as the formulated MDEA or the sterically hindered amines. There are several patented configurations that can be used to treat lean acid gases. These processes can be used to enrich an acid gas with 10 mole % H_2S to as high as 75% (Wong, 2007). In addition to rejection of CO_2, the enrichment unit can also be used to reject hydrocarbons, mercaptans, and other contaminants that are known to be problems with Claus unit operation. These processes are based on recycling a portion of the acid gas from the regenerator back to the absorber, as shown in Figure 2-12.

2.2.2.7 *Sulfur scavenger unit*

With the concerns for global warming, the application of SRU and TGTU may not be sufficient. If less than 1 ppm sulfur specification is required, a sulfur scavenger fixed bed can be used. One of the common sulfur scavenger processes is the PURASPEC process, which can be licensed from Johnson Matthey. PURASPEC has been used as a polishing unit downstream of a gas treating unit. When used in conjunction with a physical solvent unit such as the FLUOR Solvent process, it can be used to treat a wide range of high CO_2 gases without the use of heat, hence minimizing greenhouse gas emissions (Mak et al., 2012).

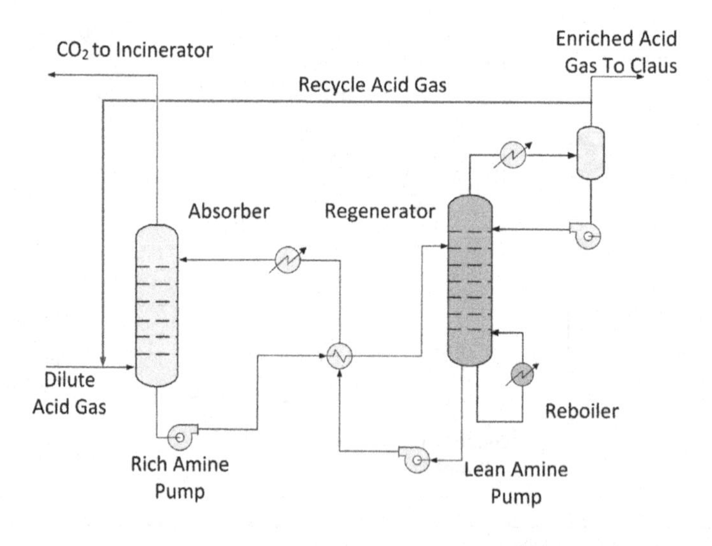

FIGURE 2-12

Typical acid gas enrichment process.

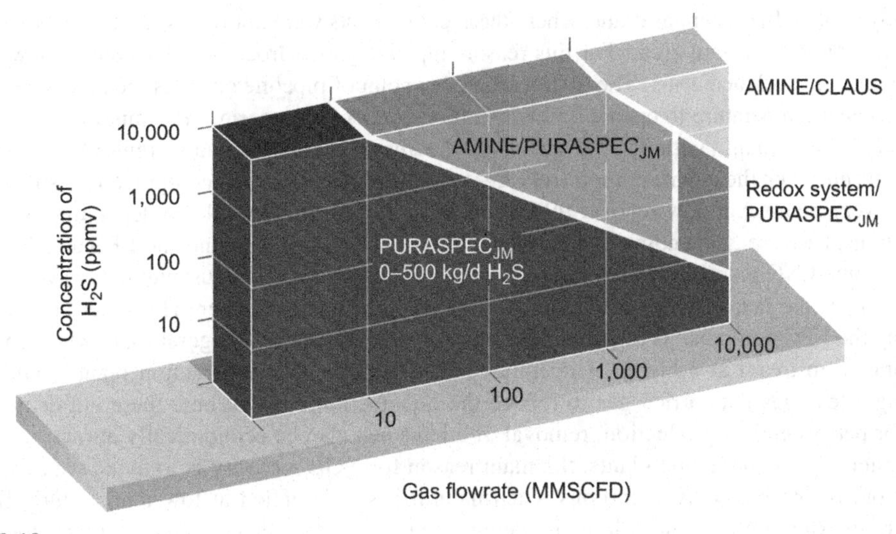

FIGURE 2-13

Recommended ranges of applications for PURASPEC.

(Mak et al., 2012; Row, 2012)

Compared to other sulfur options, a sulfur scavenger is expensive. However, it can produce a treated gas with an H_2S content that cannot be economically achieved by other methods. When used by itself alone, the scavenger bed is usually economical for low sulfur feed gases, typically below 500 kg/day. A process selection chart is shown in Figure 2-13. For higher sulfur throughput, the sulfur scavenger system can best be used as a polishing unit downstream of a gas treating unit in combination with a Claus unit or Redox unit.

2.2.2.8 Integrated gas treating and sulfur recovery technologies

In the past, when treating sour gas with lower CO_2, mercaptans and other contaminants, the gas treating process, dehydration, and sulfur technology can be independently selected based on the owner's preference and there are less technological issues. However, as the feed gas is getting more sour and contains a higher CO_2 and mercaptans content, the process selection must be carefully evaluated to ensure the selected technologies are compatible. Today, many licensors and contractors are offering their own suite of integrated technologies for the acid gas removal, dehydration, mercaptans removal as well as sulfur recovery, tail gas treating, and acid gas enhancement. Such integration by a single licensor can avoid the coordination difficulties and can ensure the overall plant performance, as described by Mokhatab and Meyer (2009) and Mokhatab and Poe (2012).

2.2.3 NGL recovery unit

In most gas processing plants, the C_5+ gasoline components are separated in the condensate stabilization unit and are sold as a high value liquid product. The residual gas after the condensate unit cannot be sent directly to the gas pipeline as it still contains significant amounts of the hydrocarbon tails and LPG components. These heavy components will result in hydrocarbon condensation in the pipeline when sufficiently cooled. In earlier gas plants where these components were not removed, there were frequent reports on explosion in rural areas. For this reason, pipeline gas hydrocarbon dewpoint is now strictly enforced to avoid such accidents. The hydrocarbon dewpoint of pipeline gas must be at least below the coldest ambient temperature to ensure no hydrocarbon condensation during transmission.

In today's gas plant, with the advances of the turbo expander plants, removal of the LPG components and even the ethane components are very efficient and can be economically justified. Most NGL recovery plants can now recover 99% propane and close to 90% ethane depending on the gas compositions. However, depending on the price differential between methane and LPG and the market demands, some LNG plants may prefer to recover the minimum amounts, just enough to avoid waxing and free-out of the heavy hydrocarbons in the cryogenic main exchanger. The other aspect is that liquefying the LPG components together with methane requires more refrigeration power than if they are separated upstream at a higher temperature. For this reason, a liquefaction plant would prefer liquefying a lean gas than a rich gas to reduce the liquefaction costs. Where there are demands for ethane for petrochemical production, removal of ethane can also be economically attractive.

In earlier LNG liquefaction plants, the main reason for NGL recovery is to avoid wax formation and free-out of the heavy hydrocarbons when natural gas is liquefied at low temperature. In most design, the use of a simple scrub column is adequate to meet the benzene (C_6) requirements for export. Although propane and butane pose no freezing problem, they are removed together with the C_6+ hydrocarbons and can later be separated and sold as liquid products.

NGL recovery processes can be classified by the lean oil process, the Joule-Thomson process, the refrigeration process, and the turbo-expander processes. The selection of an NGL recovery process depends on the gas compositions, the gas heating content targets and the opportunities on exporting the LPG or ethane product. The following section describes the different methods that can be used for removal and recovery of these components.

2.2.3.1 Lean oil absorption

The lean oil absorption process was developed in the 1910s and was common in earlier recovery plants. The process can typically recover about 40% of the ethane, 90% of the propane, and 95% of the pentane and heavier hydrocarbons.

In the lean oil absorption process, the ethane and heavier components in the feed gas is scrubbed and absorbed by a lean oil (molecular weight typically about 100 to 150) in an absorber column. The lean oil can be supplied at ambient temperature or refrigeration temperature. The rich oil exiting the bottom of the absorber is sent to a rich oil depropanizer, which separates the propane and lighter components as an overhead gas. The column bottom is fractionated in a still, where the NGL's (C_4+) are recovered as an overhead product and the lean oil is recycled back to the absorber. A typical flow sheet is shown in Figure 2-14.

The lean oil process requires a significant amount of heating and cooling. For base-load LNG plant applications, especially when high NGL recoveries are required, lean oil plants are not cost competitive when compared to turbo-expander processes.

2.2.3.2 Dewpointing by joule-thomson cooling

A dewpointing unit is a cost-effective method to remove water and C_5+ hydrocarbon contents for pipeline gas transmission. Dewpointing is achieved by chilling the gas and separating the heavy hydrocarbons.

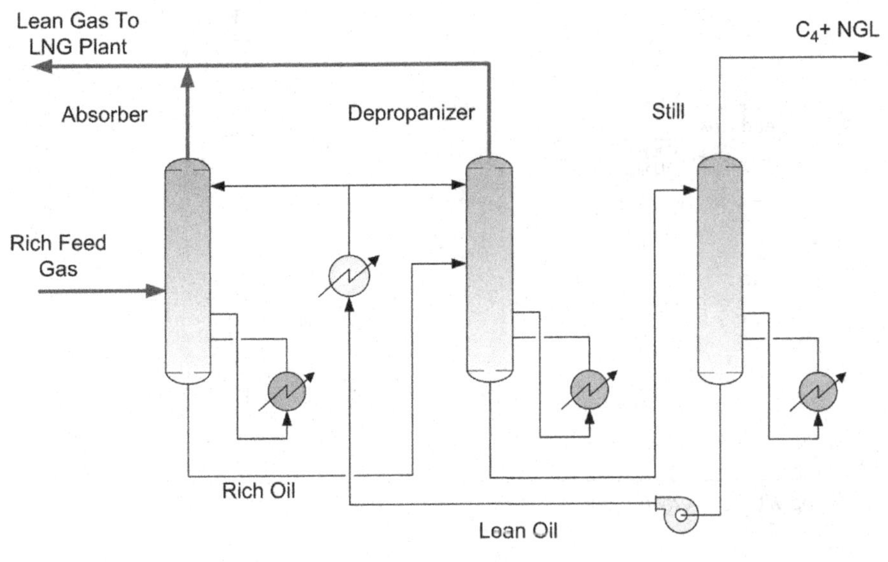

FIGURE 2-14

Typical lean oil absorption process.

For high pressure feed gas, chilling can be achieved with the use of a Joule-Thomson (JT) valve. The flow schematic of a dewpointing unit is shown in Figure 2-15. The feed gas is first dried by the molecular sieve unit, and is cooled in a gas-gas exchanger, typically in a plate fan fin exchanger, using the cold gas and liquid from the low temperature separator (LTS). The cold gas is then let down in

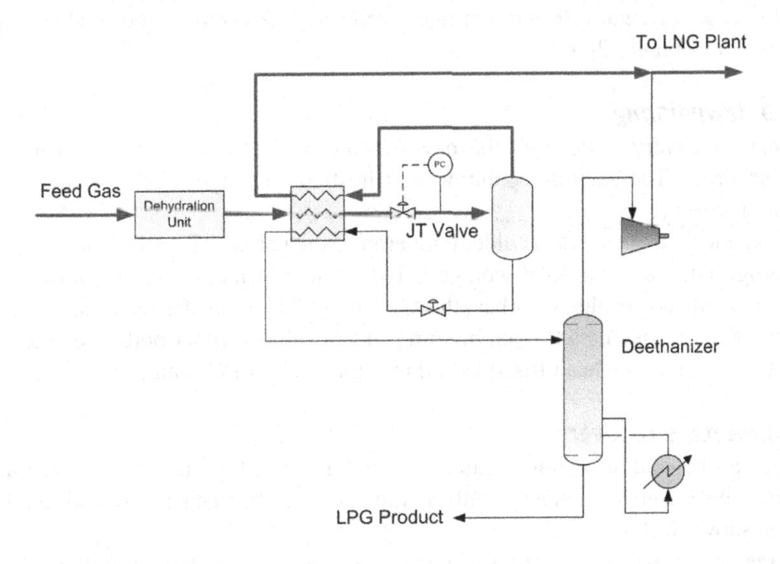

FIGURE 2-15

Typical dewpointing unit using JT cooling unit.

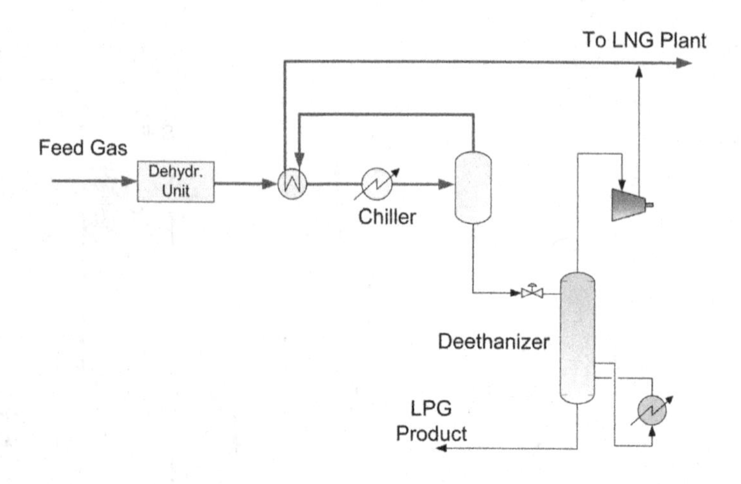

FIGURE 2-16

Typical dewpointing with refrigeration unit.

pressure in the JT valve and is routed to the LTS. The C_3+ liquid is separated and is fed to a deethanizer that fractionates the light components from the C_3+ heavier liquid. The deethanizer is heated with steam. The overhead vapor is compressed to combine with the LTS overhead gas and fed to the LNG plant. A typical dewpointing unit can achieve 30 to 50% propane recovery.

2.2.3.3 Dewpointing by refrigeration cooling

If the feed gas pressure is low, there is not sufficient pressure differential to generate cooling by the JT valve. In this case, external refrigeration using propane refrigerant is required. The process flow schematic is shown in Figure 2-16.

2.2.3.4 Deep dewpointing

If higher propane recovery is desired, the dewpointing unit can be modified to include a turbo-expander and absorber. The deep dewpointing unit is shown in Figure 2-17, which can achieve 70 to 80% propane recovery.

The turbo-expander can generate a colder temperature than either JT cooling or refrigeration. In the deep dewpointing unit, the expander discharge is fed to an absorber, which is refluxed by the ethane rich deethanizer overhead. In the absorber, the C_3+ components in the feed gas are condensed by the ethane rich reflux, producing a C_3 rich bottom product. The absorber bottom is fractionated in the deethanizer into a C_2 rich overhead (used as reflux) and a C_3+ LPG bottom product.

2.2.3.5 High propane recovery

Where propane can be sold at premium value over natural gas, high recovery is desirable. Propane recovery of over 99% can be achieved with a refluxed absorber and a refluxed deethanizer. This configuration is shown in Figure 2-18.

Similar to the deep dewpointing unit, refrigeration is generated by the turbo-expander. Feed gas is cooled by the cold absorber overhead vapor. The refrigerant contents in the separator liquid and the

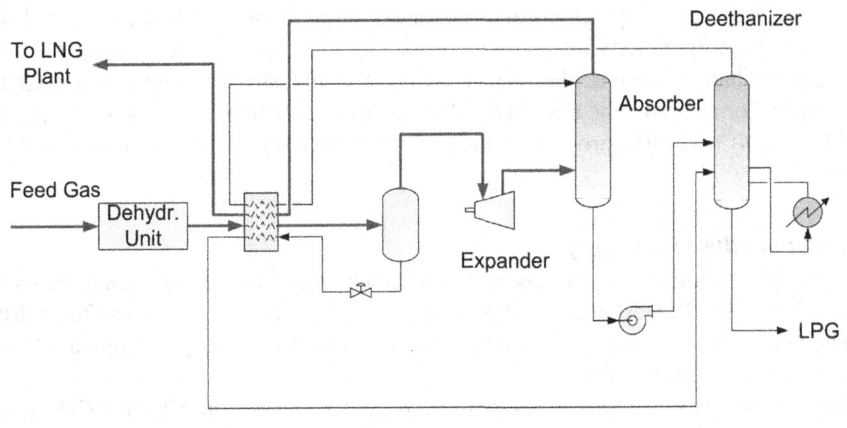

FIGURE 2-17

Typical deep dewpointing unit.

absorber bottoms are also recovered in chilling the feed gas. Side reboilers on the deethanizer can also be used to reduce the refrigeration duty. The heat exchange among these streams occurs in a plate and fin aluminum exchanger, which is designed with a close temperature approach.

The absorber bottom is sent to the deethanizer using a pump or by hydraulic head. The deethanizer operates at about the same pressure as the absorber. To improve propane recovery the deethanizer overhead is condensed with refrigeration, producing a C_2 rich liquid that is refluxed to both the deethanizer and the absorber.

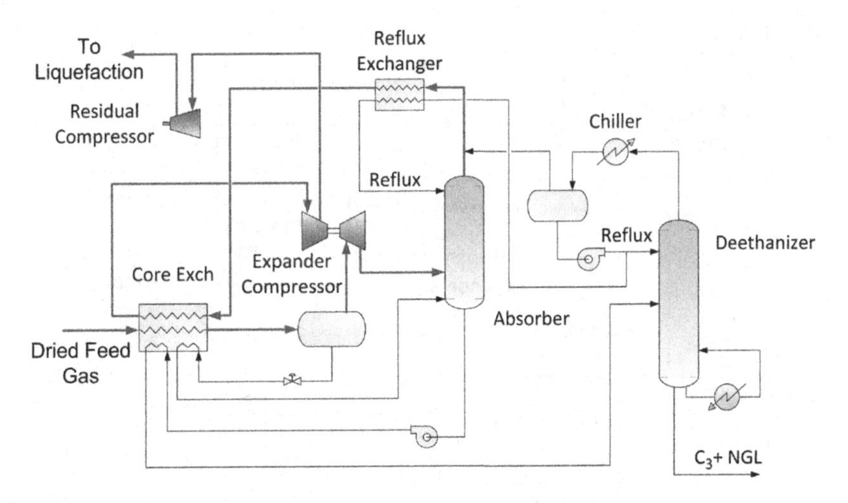

FIGURE 2-18

High propane recovery process.

(Mak et al., 2003)

Depending on feed gas compositions and pressures, the optimum configuration can differ. There are many patented configurations for NGL recovery, and they are different in the respect of flow sequences, heat exchanger design, the use of different reflux streams, and column configuration (Mak et al., 2003). For example, the absorber and the deethanizer can be combined as a single column, as patented by Ortloff's SCORE process, or the use of multiple reflux streams as patented by Fluor's TRAP process.

2.2.3.6 Medium ethane recovery

Where ethane can be exported to a petrochemical feedstock, high purity ethane can be recovered using with the Gas Subcooled Process (GSP). Typically, the GSP process can achieve 70 to 80% ethane recovery. The GSP process was originally developed by Ortloff over 20 years ago and is a common process for the gas processing industry.

In LNG production, since CO_2 has been removed in AGRU to meet the 50 ppmv CO_2 specification, CO_2 freezing is not a concern in NGL recovery. Because of the absence of CO_2, the demethanizer column can operate at a very cold temperature without CO_2 freezing problems.

As shown in Figure 2-19, the GSP process uses a portion of the feed gas to reflux the demethanizer. The feed gas from the separator is split into two portions; about 60% is let down in the expander to generate cooling, while other portion is chilled and subcooled by the demethanizer overhead to produce a liquid reflux to the demethanizer.

Typically, the demethanizer operates at about 450 psig to 500 psig. Higher pressure would minimize recompression, however the operating pressure must stay below the critical pressure for separation efficiency. The relative volatility between methane and ethane gets closer at high pressure, which is the reason for the fractionation difficulty. The ethane product must contain no more than 1 to 2 volume % of methane, which is difficult to achieve at high pressure.

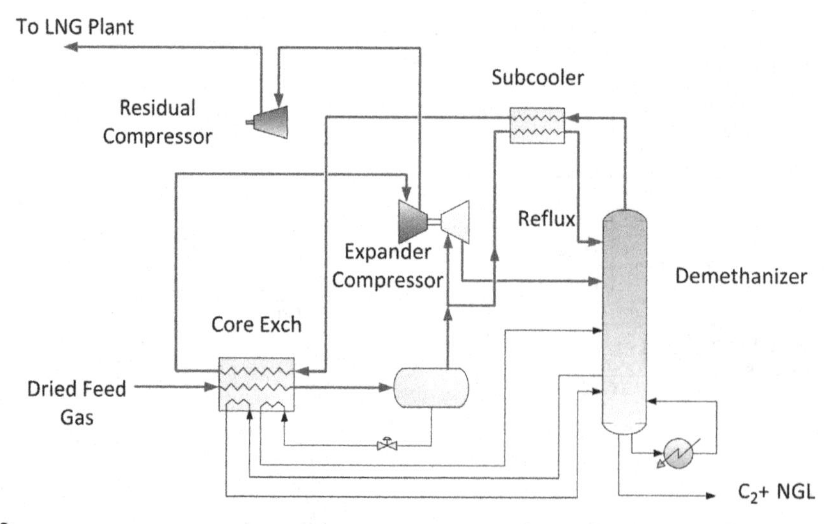

FIGURE 2-19

Typical ethane recovery—GSP process.

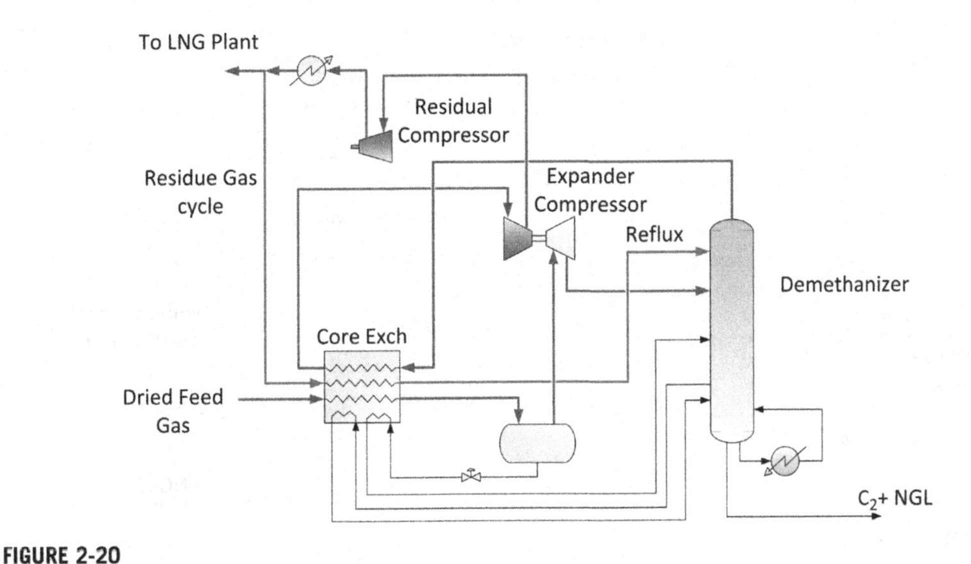

FIGURE 2-20

Typical high ethane recovery—residue gas recycle.

2.2.3.7 High ethane recovery

When ethane recovery of over 90% is required, additional lean reflux must be added by recycling the lean gas from the residue gas compressor as shown in Figure 2-20. To generate additional cooling, propane refrigeration can be added and the expander can operate at a higher expansion ratio by lowering the demethanizer pressure. The extent of refrigeration requirements depends on the richness of the feed gas, the feed gas pressure, and the desirable level of ethane recovery. Typically, over 95% ethane recovery is difficult to meet without spending excessive energy for refrigeration.

Similar to the propane recovery configurations, there are variations in the process configurations based on this concept. Some of the configurations are patented technologies that specify different lean refluxes and heat integration methods.

2.2.3.8 High pressure absorption

In LNG liquefaction, a high feed pressure is desirable because it reduces the refrigeration cycle compression requirements and increases the liquefaction efficiency. However, when NGL recovery is required, the demethanizer pressure must be at 500 psig or lower for fractionation efficiency reasons. To solve this problem, Fluor has developed the patented TCHAP (Twin Column High Absorption process) concept that uses a two-column approach with the first column (the absorber) operating at a high pressure, typically over 600 psig, and the second column (the demethanizer) operates at a lower pressure, at 450 to 500 psig. The innovation is that the high pressure absorber bottom is let down in pressure, and the refrigeration content in the letdown liquid is used to provide cooling to the reflux condenser in the demethanizer column. The overhead vapor from the demethanizer is compressed with a recycle gas compressor, cooled, and used as a reflux to the high pressure absorber. The final fractionation of methane from the ethane product takes place in the second column operating at a lower pressure. The TCHAP process is shown in Figure 2-21, which can achieve recovery of 70% ethane

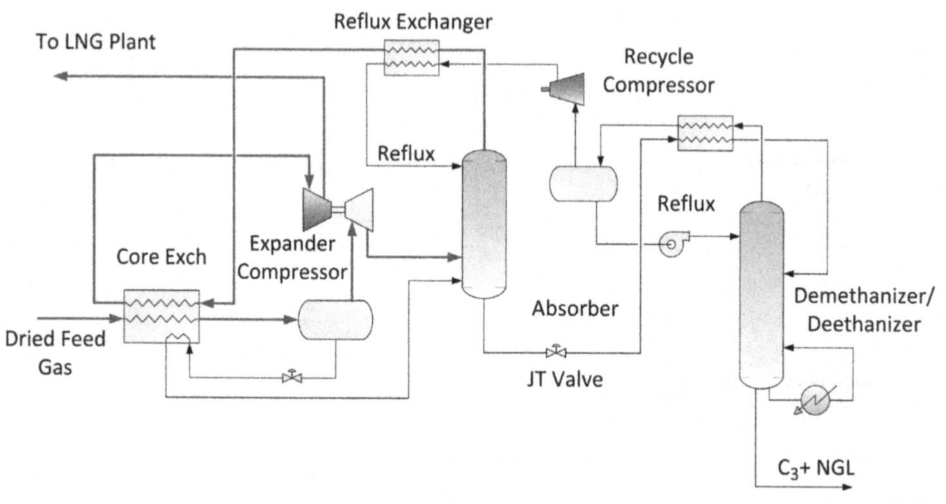

FIGURE 2-21

High pressure absorption process.

(Mak, 2005)

and 99% propane. The high pressure demethanizer overhead vapor is compressed by the expander compressor to typically about 900 psig feeding the LNG plant. There are several variations of this patented process concept. The process can also be used for propane recovery.

2.2.4 NGL fractionation unit

Once natural gas liquids (NGLs) have been removed from the natural gas stream, they must be fractionated into their base components, which can be sold as high purity products. Fractionation of the NGLs may take place in the gas plant but may also be performed downstream, usually in a regional NGL fractionation center. A typical process flow schematic is shown in Figure 2-22. NGLs are fractionated by heating the mixed NGL stream and processing them through a series of distillation towers. Fractionation takes advantage of the differing boiling points of the various NGL components. As the NGL stream is heated, the lightest (lowest boiling point) NGL component boils off first and separates. The overhead vapor is condensed and a portion is used as reflux and the remaining portion is routed to product storage. The heavier liquid mixture at the bottom of the first tower is routed to the second tower where the process is repeated and a different NGL component is separated as product. This process is repeated until the NGLs have been separated into their individual components.

2.2.5 Natural gas liquefaction plant

Natural gas liquefaction requires removal of sensible heat and latent heat from the natural gas at high pressure over a wide temperature range. A number of liquefaction processes have been developed in the past decades with the differences mainly on the choice of refrigerants, refrigeration driver match,

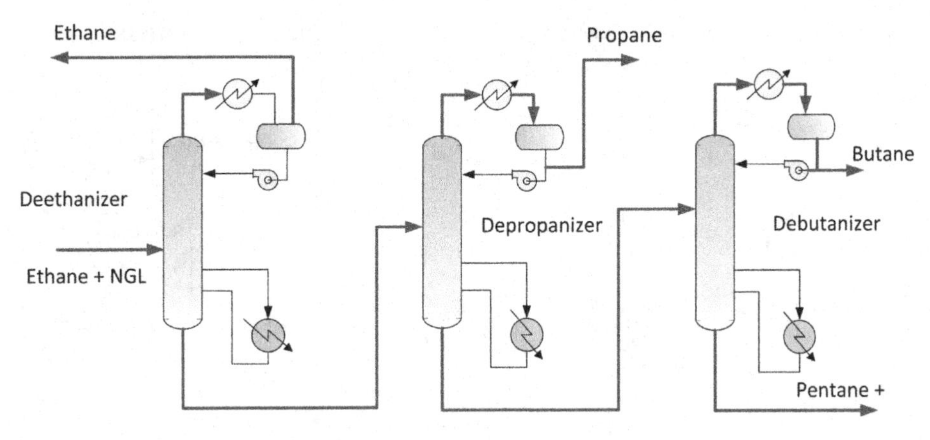

FIGURE 2-22

Typical NGL fractionation process.

and refrigeration configurations. Besides capital and operating costs, the primary objective of these technological innovations is to increase the throughput of a single LNG train. The different liquefaction processes are described in Chapter 3.

2.2.6 Integrating NGL recovery and natural gas liquefaction plant

NGL recovery plants can be operated with or without an LNG plant downstream. There are advantages of the standalone NGL recovery plants. They can be operated on NGL recovery, producing sales gas to the pipelines and sending only the required volume to the LNG plants. On the other hand, the liquefaction plant can be designed to have the minimum gas conditioning equipment that allows LNG production without the upstream NGL recovery plants. This independent operation increases the reliability and availability of both facilities but would require additional costs. The process block flow diagram of a nonintegrated NGL recovery plant and an LNG liquefaction plant is shown in Figure 2-23. The recovery of NGL can maximize revenues from the LNG plants. The extracted LPG propane can be sold at a premium price over natural gas or they can be blended back to enrich the heating value of LNG for export to Asian markets. Lean LNG is required for export to the United States or Western European markets because of their heating value specifications (see Chapter 1). Such a blending operation provides the flexibility of exporting LNG to meet the heating value requirements by different markets.

When compared to the standalone process, an integrated NGL/LNG process can eliminate duplication of the heat exchangers and reduce the pressure drop between the two units. Traditionally, an LNG plant is integrated with a scrub column for removal of the heavy hydrocarbons to avoid freeze-up in the cryogenic exchanger. A typical flow schematic is shown in Figure 2-24. The scrub column normally operates at the feed gas pressure, typically at 700 psig. Because of the high pressure, the scrub column can recover only about 50 to 70% of the propane while producing a bottom product of the heavier components containing significant amounts of ethane and lighter components. Consequently, a deethanizer is required in additional to the traditional depropanizer and debutanizer in an NGL recovery plant. The deethanizer overhead containing the methane and ethane components

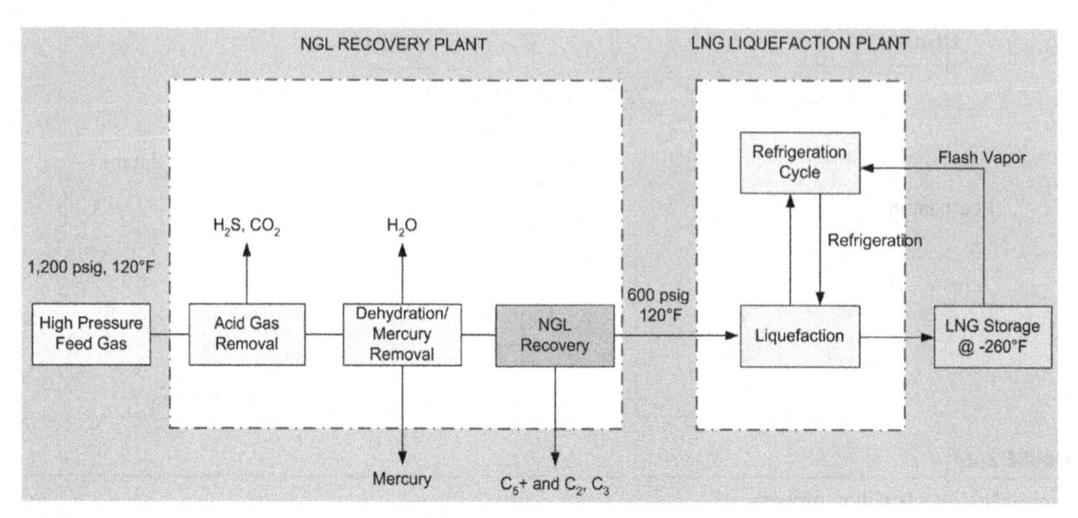

FIGURE 2-23

Block flow diagram of standalone NGL and LNG plants.

(Mak et al., 2007)

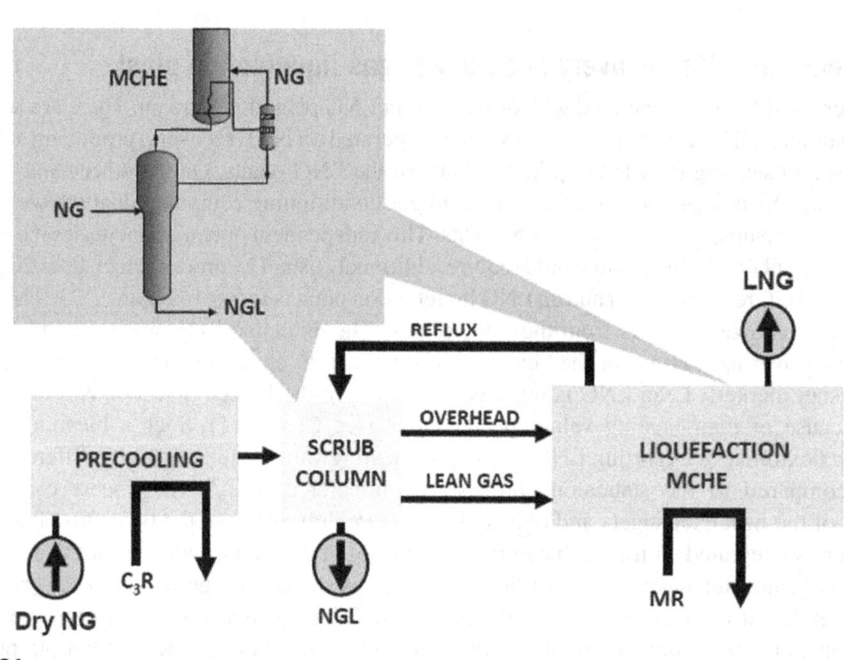

FIGURE 2-24

Integrated LNG Plant—Scrub Column.

(Hagyard et al., 2004)

typically are used in the fuel gas system. The scrub column uses propane refrigeration to generate reflux at about –20°F, producing a lean gas that is sent to the main cryogenic heat exchanger (MCHE).

The scrub column is simple and easy to operate, however, component separation is not very sharp and LPG recovery is limited due to the high operating pressure and relatively high reflux temperature. The advantage of a scrub column is that it can produce a high pressure residue gas to reduce liquefaction plant horsepower. On the other hand, the rough separation may be problematic with rich feed gas containing high levels of benzene and aromatics and heavier hydrocarbons. If they cannot be completely removed to the ppm level, they may cause waxing and freeze-out in the cryogenic exchanger. For this reason, high propane recovery processes (see Sections 2.2.3.5 and 2.2.3.7) are necessary for LNG production to ensure complete removal of these heavier components, especially when processing rich gases.

There are many advanced NGL recovery processes that can be integrated with the liquefaction plant. However, these high propane recovery turbo-expander processes typically use a demethanizer operating at 450 psig to 500 psig for good separation. Although these processes can recover 99% propane, they also require additional recompression to meet the LNG plant pressure requirement.

If feed gas inlet pressure is high, say over 1200 psig, the turbo-expander will need to operate at a higher expansion ratio, which would also means that excessive refrigeration is generated that may not be fully utilized in the NGL recovery process. For this reason, a more efficient patented process has been developed by Fluor that uses a high pressure absorber operating at 600 psig or higher followed by

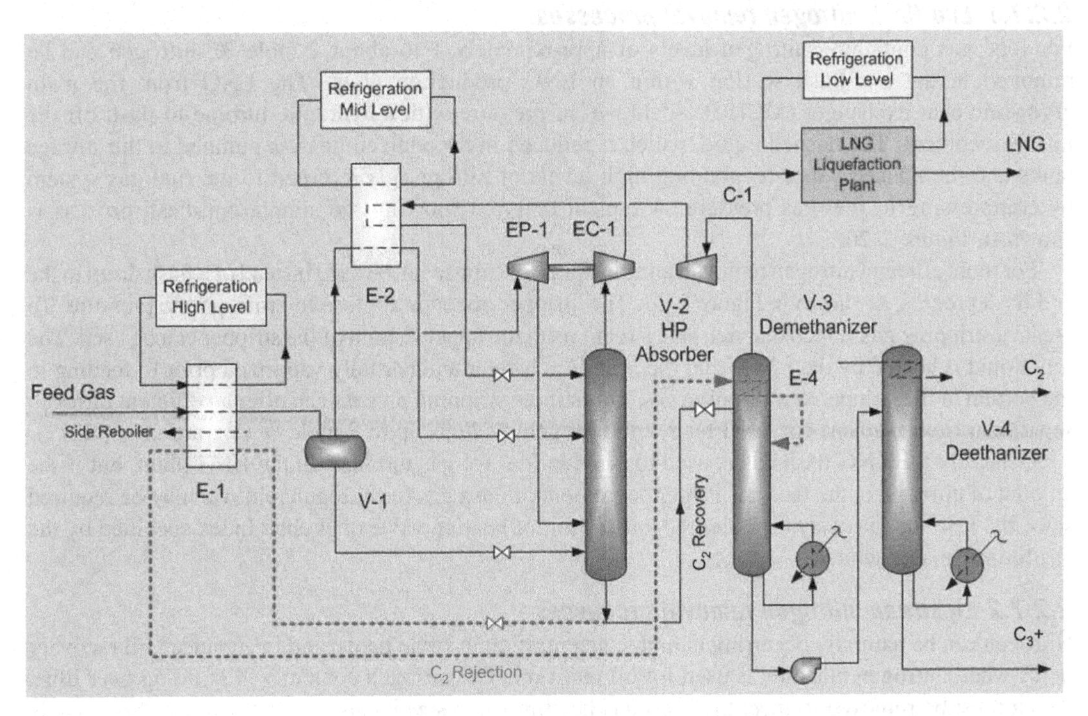

FIGURE 2-25

Fluor TCHAP NGL/LNG integration.

(Mak et al., 2007)

a low pressure deethanizer, with a recycle gas compressor integrated to the process. The Fluor TCHAP process, shown in Figure 2-25, can be used in the integrated plant and can achieve 99% propane recovery while producing a high pressure residue gas to the LNG plant. Details of the process are described in Section 2.2.3.8. The Fluor TCHAP integrated process can save about 9% of the energy consumption in LNG production.

2.2.7 Nitrogen removal unit

Nitrogen content in natural gas varies depending on characteristics of the gas fields. When present in high concentrations (greater than around 5 mole %), it must be removed from the LNG product to levels of below 1 % mole in order to improve its calorific value and simplify problems associated with the management of boil-off gas during storage and transportation. The presence of greater than about 1 mole % nitrogen in LNG may lead to auto-stratification and rollover in storage tanks, which presents a significant safety concern (Johnson et al., 2011). The high percentage of nitrogen content also impacts the liquefaction process itself by reducing liquefaction efficiency (additional refrigeration requirements per unit of LNG produced, due to the need to condense nitrogen in feed gas). There is therefore a need for efficient technique for the removal of nitrogen from LNG, even for relatively low nitrogen levels.

2.2.7.1 End flash nitrogen removal processes

For feed gas containing nitrogen levels of approximately 1 to about 2 mole %, nitrogen can be removed in the end-flash section within an LNG production plant. The LNG from the main cryogenic heat exchanger (MCHE) is letdown in pressure with a hydraulic turbine to flash off the nitrogen content. The flashed liquid, which is reduced in nitrogen content, is pumped to the storage tanks and the flashed vapor (containing high levels of nitrogen) is returned to the fuel gas system by compressing to fuel gas pressure. A typical diagram showing the simple end-flash process is shown in Figure 2-26.

For more efficient nitrogen removal, an atmospheric stripper can be used instead of a flash drum in the end flash process, as shown in Figure 2-27. The stripper operates at close to atmospheric pressure. To create a stripping gas, a methane rich side stream from the upper section of the stripper can be used. The tray liquid is heated by the LNG from the Main Exchanger and partially vaporized prior to feeding to the bottom of the stripper as a stripping gas. This simple stripping process can offer an efficient nitrogen separation from methane for LNG feed with nitrogen contents up to 5 mole % (Vovard et al., 2011).

Generally the LNG flash gas is used to provide fuel for gas turbines on the LNG plant, but if the amount of nitrogen in the feed gas is high, treatment of flash gas for nitrogen removal may be required since the lean fuel gas may not meet the lower limit of heating value or Wobbe Index specified by the gas turbine manufacturer.[3]

2.2.7.2 Upstream nitrogen removal processes

Nitrogen can be naturally occurring in high concentration in some fields, and in enhanced oil recovery fields where nitrogen injection is used for oil recovery. The nitrogen content will build up over time, which must be removed to meet the sales gas heating value specification.

[3]If the amount of nitrogen in the fuel gas is high, the burners in gas turbines may not be able meet the emissions requirement.

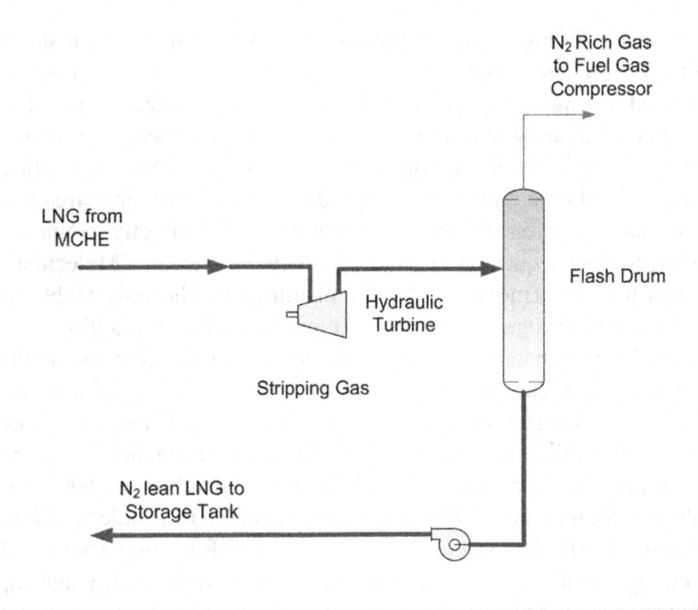

FIGURE 2-26

Typical simple end-flash process for nitrogen removal.

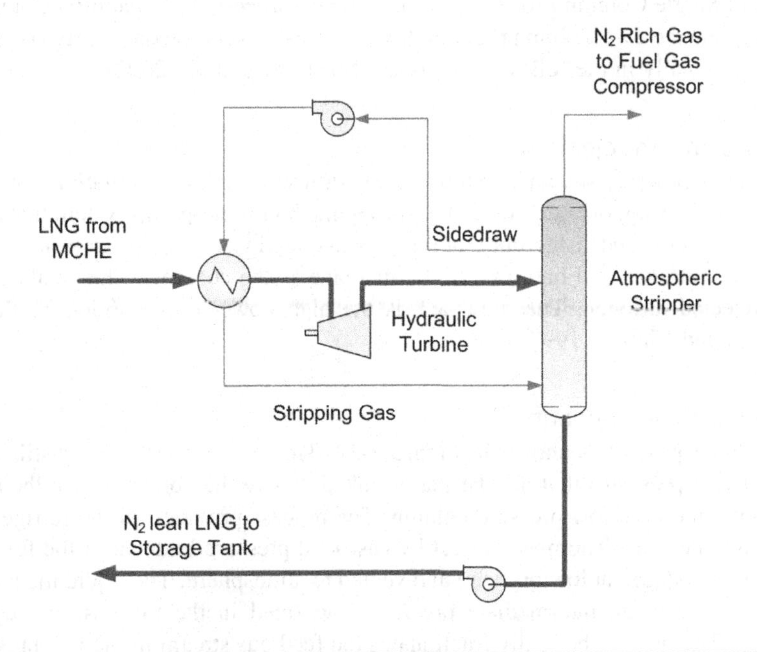

FIGURE 2-27

Typical end-flash process utilizing nitrogen stripper.

When nitrogen is present in high concentrations (greater than 5 mole %), it should be removed in the front section of the LNG plant to minimize liquefaction requirements. There are several processing methods that can be used to remove nitrogen before it enters the liquefaction unit. There are also other options within the LNG technology to remove nitrogen in the liquefaction process.

Membranes and molecular sieve technology can remove nitrogen but the application is generally limited for small units. Membrane units typically produce a waste nitrogen stream with a fairly high hydrocarbon content that is a revenue loss and cannot be vented directly to the atmosphere. Waste nitrogen must be reinjected for sequestration or disposed by other means. Molecular sieve technology is uneconomical when used to remove high levels of nitrogen. The only viable nitrogen rejection technology is with the use of cryogenic separation (Finn, 2007; Garcel, 2008).

Several cryogenic schemes are known to reject nitrogen from variable-content nitrogen natural gas. Each type of these processes can produce a nitrogen waste stream suitable to meet the EOR requirements. The rejected nitrogen stream usually contains a small quantity of hydrocarbon (predominantly methane). If the nitrogen is reinjected, the hydrocarbon contained in the nitrogen is not lost but becomes deferred revenue. If nitrogen is vented, the hydrocarbon content of the nitrogen vent stream must meet the environmental regulation, typically set at 0.5 to 1 mole % (Millward et al., 2005; Wilkinson and Johnson, 2010). The selection of the Nitrogen Rejection Unit (NRU) process must therefore consider the optimum hydrocarbon recoveries, as well as the utility and capital tradeoffs. If the unit is installed in an existing facility, the design selection must be suitable to be integrated with the existing equipment.

Similar to the LNG plant feed gas, the high boiling materials such as aromatics and CO_2 content in the feed gas must be removed to a very low level. There are four generic cryogenic processes for nitrogen removal: Single Column process, Double Column process, Preseparation Column (or Three Column) process, and the Two Column process. These processes vary in complexity and efficiency and are discussed individually in the following sections (MacKenzie et al., 2002).

Single column nitrogen rejection

The Single Column process, shown in Figure 2-28, utilizes a single distillation column typically operating at 300 to 400 psig, operated by a closed-loop methane heat-pump system that provides both the reboiler duty and the condensing duty. In this process, feed gas is cooled in heat exchanger HE-1 using the overhead nitrogen and bottom reboiler methane as the coolants. This method can produce high-pressure rejected nitrogen. The drawback is the high power consumption by the heat pump compressor (Kohl and Nielsen, 1997).

Double column nitrogen rejection

The Double Column process is shown in Figure 2-29. This process uses two distillation columns operating at different pressures that are thermally linked, where the condenser for the high pressure column is used to reboil the low pressure column. The process provides all the refrigeration for the separation through the Joule-Thompson effect by cascaded pressure letdown of the feed.

The nitrogen is produced at low pressure and vented to atmosphere. The liquid methane product is cryogenically pumped to an intermediate pressure, vaporized in the process, and compressed to pipeline pressure. The process basically fractionates the feed gas stream in the low pressure column, which operates at the cold portion of the NRU, typically at –250°F to –310°F, which is prone to CO_2

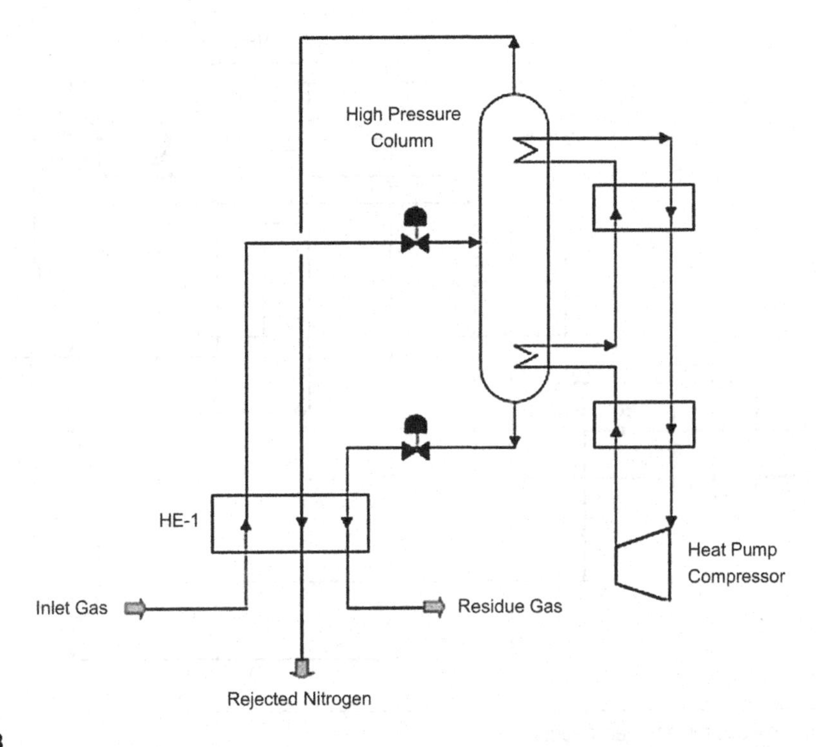

FIGURE 2-28

Single column process for nitrogen rejection.

(Modified after Kohl and Nielsen, 1997)

freezing. The CO_2 content in the feed gas must therefore be removed to a very low level to avoid freezing in the tower.

In this process, the feed gas is cooled and partially condensed in heat exchanger HE-1 using the nitrogen vent stream and methane product and fed to a high pressure rectification column, which produces a high purity nitrogen liquid off the top. The bottom product, a mixture of methane and nitrogen, is subcooled in heat exchanger HE-2 and then fed to the low pressure column, which completes the separation.

The overhead product from the high pressure column, relatively pure nitrogen, is condensed in HE-4, which also serves as a reboiler for the low pressure column. A portion of this liquid nitrogen is used as reflux for the high pressure column while the remainder is subcooled in heat exchanger HE-3 and used as reflux for the low pressure column. The methane product from the bottom of low pressure column is pumped, and then heated to ambient temperature in heat exchangers HE-2 and HE-1 (Kohl and Nielsen, 1997).

Three column nitrogen rejection

The Three Column (Preseparation Column) process, as shown in Figure 2-30, is a variation of the Double Column process where the process is made up of the high pressure column (prefractionator),

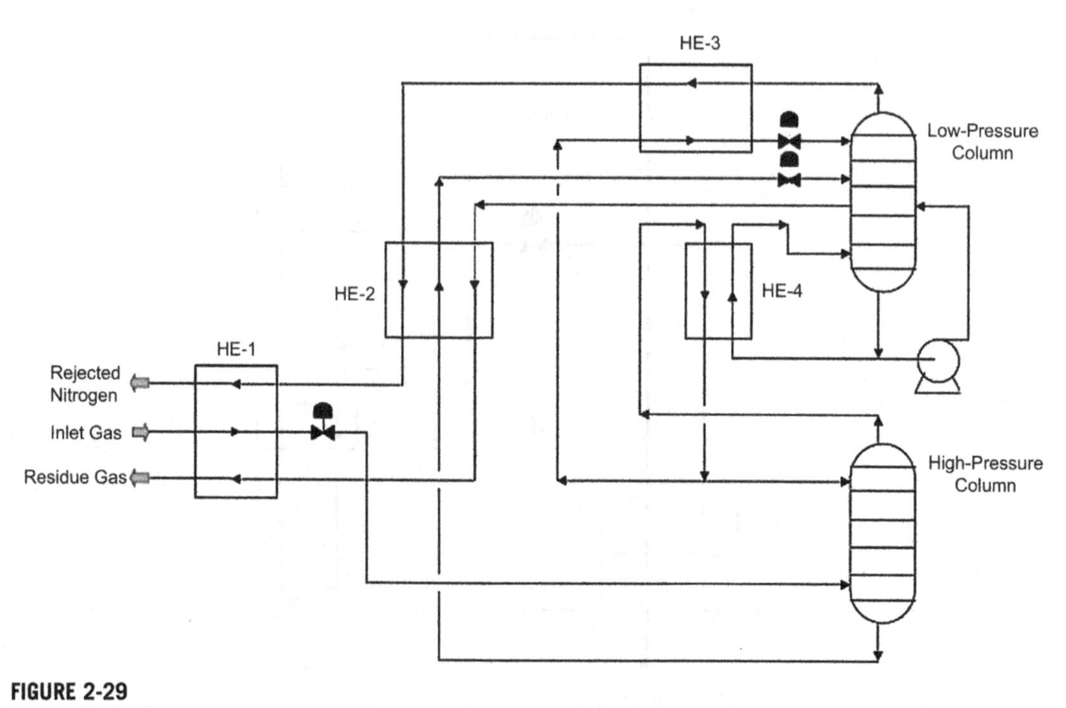

FIGURE 2-29

Double column process for nitrogen rejection.

(Modified after Kohl and Nielsen, 1997)

intermediate pressure column, and the low pressure column. The prefractionator can remove the bulk of the methane and CO_2 content as a bottoms product, thereby concentrating the nitrogen content in the overhead. With a lower CO_2 content to the cold section, the process is more CO_2 tolerant and can operate with a higher CO_2 content feed gas with CO_2 content up to 1.5 mole %.

The prefractionator column operates at a higher pressure with temperature ranging between –150°F and –180°F, which would avoid any CO_2 freezing problem. Removing the bulk of the CO_2 from the prefractionator column greatly reduces the CO_2 content in the cold end of the NRU, increasing the process tolerance to CO_2. This was a very important consideration in the design selection since the streams coming from the NGL plants are already dry and treated for CO_2. Additional CO_2 removal would add to the capital and operating cost of the project.

Additionally, the heavy hydrocarbons in the feed are recovered in the residue stream from the bottoms of the prefractionator column. This increases the hydrocarbon recoveries and revenues from the NRU.

Two-column nitrogen rejection

The Two Column Process, as shown in Figure 2-31, is similar to the Three Column process, without the intermediate column. The process comprises a high pressure prefractionator and a low pressure column. Similar to the Three Column design, the prefractionator reduces the CO_2 content in the feed gas to the low pressure column and is therefore more CO_2 tolerant.

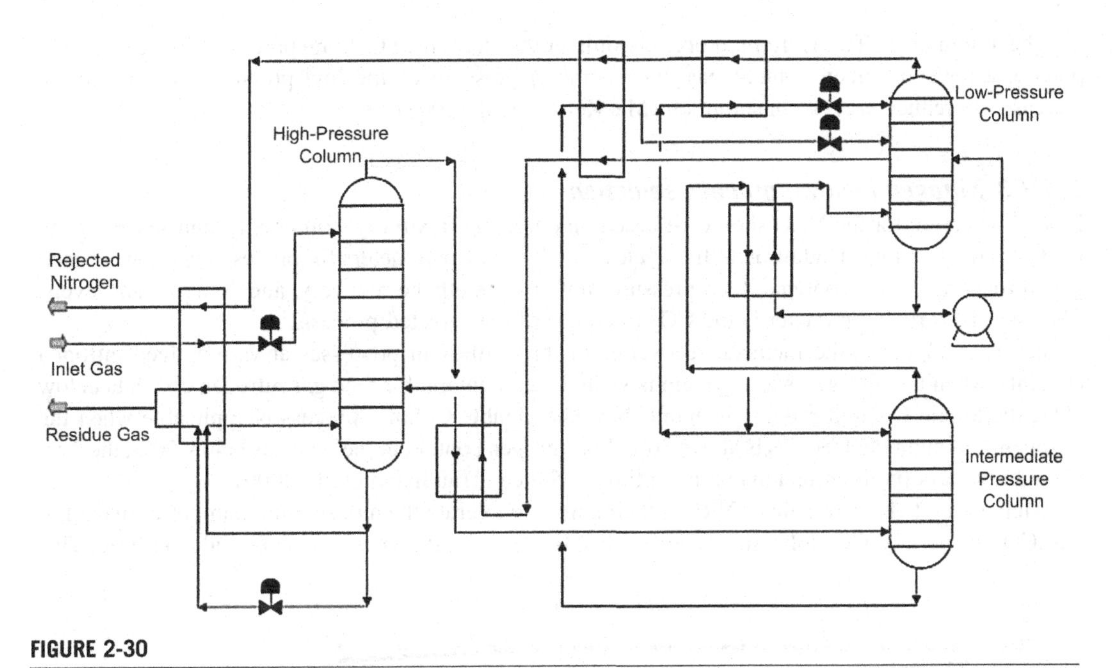

FIGURE 2-30

Three column process for nitrogen rejection.

(MacKenzie et al., 2002)

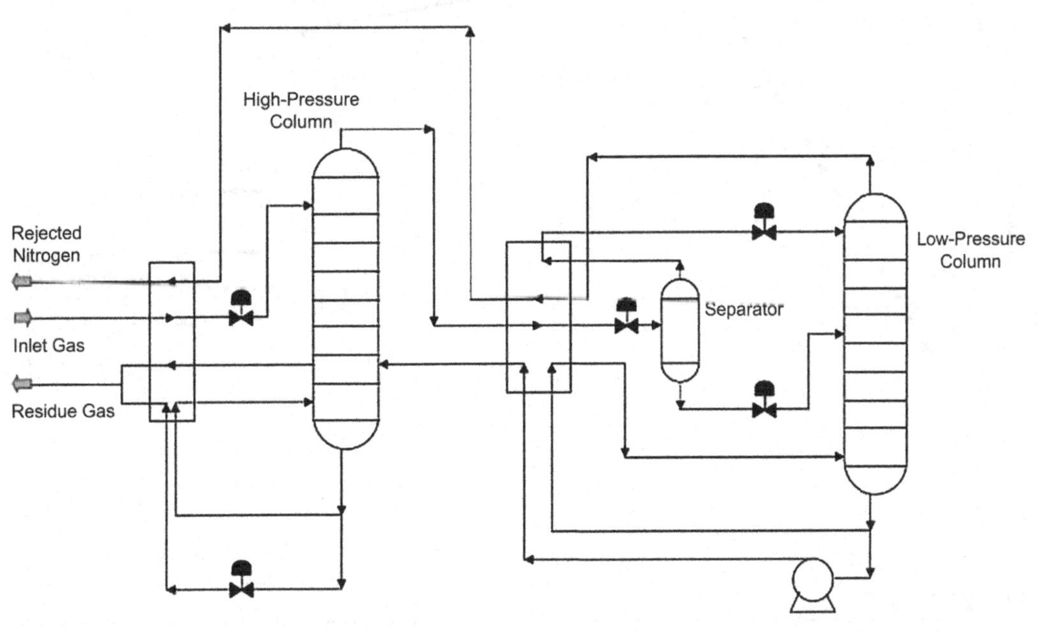

FIGURE 2-31

Two column process for nitrogen rejection.

(MacKenzie et al., 2002)

The design of the Two Column process is simpler than the Three Column process. If it were used to process a lower nitrogen content gas, the operating pressure of the low pressure column can be increased to reduce energy consumption (MacKenzie et al., 2002).

2.2.7.3 Nitrogen removal process selection

Process selection for the NRU should be based on operating flexibility, complexity, and sensitivity to feed gas compositions in addition to life cycle costs. The key parameters for process selection are feed gas nitrogen and CO_2 contents, feed pressure, flow rate, methane recovery, and contaminant levels. The more important parameter is the CO_2 tolerance of the selected process.

Figure 2-32 shows the methane recoveries of three different processes at various feed nitrogen contents, when the nitrogen waste stream is vented. In summary, for feed gas nitrogen contents below 30%, the Single Column process is applicable. The Double Column process is applicable when the nitrogen content in the feed gas is above 30%. For nitrogen content in the feed gas below 50%, the Two Column process (prefractionator) is the optimum choice (Trautmann et al., 2000).

An important characteristic of NRU is their ability to tolerate the most common impurity in natural gas, CO_2. Generally, CO_2 tolerance is determined by the coldest spot where CO_2 tends to freeze. This

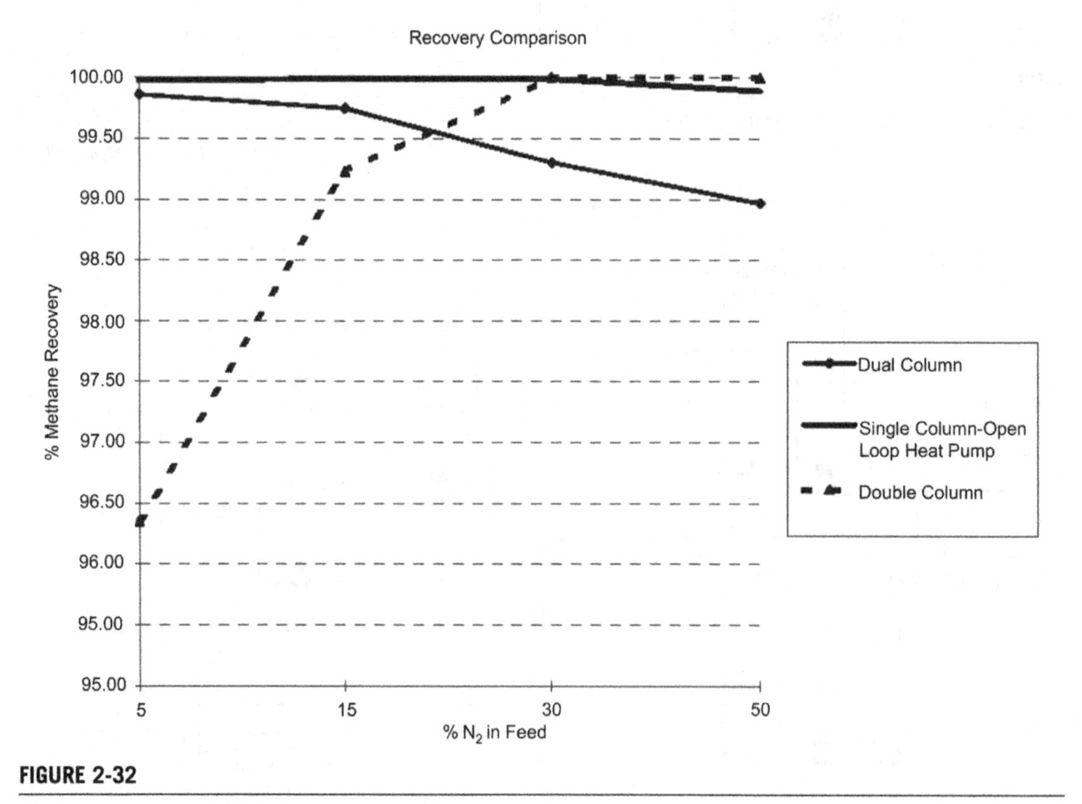

FIGURE 2-32

Methane recovery of NRU processes in nitrogen venting mode.

(Trautmann et al., 2000)

is usually a function of the column pressure and nitrogen content in the feed gas. The higher the column pressure the warmer the temperature and the higher the CO_2 tolerance.

Startup, transient, and upset conditions also need to be evaluated in determining if CO_2 freezing may pose an operating problem. The CO_2 tolerance of the process is probably more important in the selection process. A process that has very little CO_2 tolerance may require a costly deep CO_2 removal system like a molecular sieve whereas a more CO_2 tolerant process may require only an amine system. A more CO_2 tolerant process is also more reliable, since it can handle CO_2 removal upsets and avoid shutdown required for derimming due to a CO_2 freezing problem.

2.2.7.4 Nitrogen removal process integration

As mentioned before, the specific liquefaction power consumption (kW/ton) increases with the nitrogen content in the feed gas. Therefore, there is an economic justification to remove the nitrogen from the feed gas before liquefaction.

When the nitrogen removal is performed within the liquefaction section, it avoids the NRU product compression system (with refrigeration provided by a liquefaction unit refrigeration system) as well as losses associated with reheating and cooling feed gas for nitrogen rejection. However in this scheme, a high level of integration with liquefaction system adds to process complexity and operation risk as neither the NRU nor liquefaction system is conventional.

When nitrogen rejection is to be installed in conjunction with AGRU and NGL recovery, the opportunity exists to integrate both facilities by eliminating repeated heat exchange equipment and recompression. For example, the selection of the NGL unit outlet pressure can be set to match the best efficiency point of the NRU columns, and the rejected nitrogen from the NRU can be reused in stripping in the AGRU section. Such a simple integration concept can be incorporated into the design to achieve higher energy efficiency and reduce equipment counts while maintaining the operability of the overall process design (Trautmann et al., 2000).

References

Abbott, J., Oppenshaw, P., March 11–13, 2002. Mercury Removal Technology and Its Applications, paper presented at the 81st Annual GPA Convention, Dallas, TX, USA.

Bauer, H.C., Oct. 19–21, 2011. Modular Design of a Baseload LNG Plant, paper presented at the 2011 IGU Research Conference (IGRC), Seoul, Korea.

Bradley, A., De Oude, M., van der Zwet, G., Aug. 23–27, 2009. Sulfinol-X Process-Efficiently Achieving Complex Contaminant Removal Objectives, paper presented at the 8th World Congress of Chemical Engineering (WCCE8), Montreal, Canada.

Carlsson, A.F., Last, T., Smit, C.J., March 11–14, 2007. Design and Operation of Sour Gas Treating Plants for H2S, CO2, COS and Mercaptans, paper presented at the 86th Annual GPA Convention, San Antonio, TX, USA.

Carnell, P.J.H., Row, V.A., April 24–27, 2007. A Rethink of the Mercury Removal Problem for LNG Plants, poster presented at the 15th International Conference & Exhibition on Liquefied Natural Gas (LNG 15), Barcelona, Spain.

Clarke, D.S., Sibal, P.W., March 16–18, 1998. Gas Treating Alternatives for LNG Plants, paper presented at the 77th Annual GPA Convention, Dallas, TX, USA.

Dalrymple, D., Deberry, D., Thompson, P., March 2–5, 1997. CrystaSulf TM Process for Desulfurizing Natural Gas and Other Streams, paper presented at the Laurence Reid Gas Conditioning Conference, Norman, OK, USA.

Eckersley, N., 2010. Advanced Mercury Removal Technologies. Hydrocarbon Processing 89 (1), 29–35.
Edmonds, B., Moorwood, R.A.S., Szcepanski, R., March 1996. Mercury Partitioning in Natural Gases and Condensates, paper presented at the GPA Europe Conference, London, UK.
Finn, A.J., Oct. 2007. Nitrogen Rejection Strategies. Hydrocarbon Engineering 49–52.
Fong, H.L., et al., 1987. Shell Redox Desulfurization Process Stresses Versatility. Oil & Gas Journal, 54–62.
Gall, A.L., Gadelle, D., 2003. Technical and Commercial Evaluation of Processes for Claus Tail Gas Treatment, paper presented at the GPA Europe Technical Meeting, Paris, France.
Garcel, J.C., May 14–16, 2008. Liquefaction of Non-Conventional Gases, paper presented at the GPAE Europe Conference, Ashford, Kent, UK.
Goar, B.G., Fenderson, S., 1996. Fundamentals of Sulfur Recovery by the Claus Process, paper presented at the Laurance Reid Gas Conditioning Conference, Norman, OK, USA.
GPSA Engineering Databook. twelveth ed., 2004. Gas Processors Suppliers Association (GPSA), Tulsa, OK, USA.
Grootjans, H., Nov. 21–23, 2005. Sustained Development in the Pearl GTL Treating Design, paper presented at the 2005 International Petroleum Technology Conference, Doha, Qatar.
Hagyard, P., Paradowski, H., Gadelle, D., Morin, P., Garcel, J.C., March 21–24, 2004. Simultaneous Production of LNG and NGL, paper presented at the 14th International Conference & Exhibition on Liquefied Natural Gas (LNG 14), Doha, Qatar.
Hardison, L.C., 1984. Catalytic Gas-Sweetening Process Selectively Converts H2S to Sulfur, Treats Acid Gas. Oil & Gas Journal, 60–62.
Harvey, C.G., Verloop, J., 1976. Experience Confirms Adaptability of the SCOT Process, paper presented at the 2nd International Conference of the European Federation of Chemical Engineers, University of Salford, Greater Manchester, UK.
Heisel, M.P., Marold, F.J., 1992. How New Tail Gas Treater Increases Claus Unit Throughput. Hydrocarbon Processing 71 (3), 83–85.
Holloway, C.S., 1996. GRI Testing of ARI-LO-CAT II for the Direct Treatment of High-Pressure Natural Gas at NGPL's Kermit, Texas Site. Final Report GRI-96/0007. Gas Research Institute (GRI), Chicago, IL, USA.
Hubbard, R., Aug. 2009. The Role of Gas Processing in the Natural-Gas Value Chain. Journal of Petroleum Technology 65–71.
Hudson, C., April 18-21, 2010. "Implications of Mercury Removal Bed Material Changeout: Brownfield Versus Greenfield, poster presented at the 16th International Conference & Exhibition on Liquefied Natural Gas (LNG 16), Oran, Algeria.
Johnson, G.L., Finn, A.J., Tomlinson, T.R., April 2011. Process and Apparatus for Separation of Nitrogen from LNG. UK Patent 2462555.
Kidnay, A.J., Parrish, W.R., 2006. Fundamentals of Natural Gas Processing, first ed. CRC Press, Boca Raton, FL, USA.
Klinkenbijl, J.M., Dillon, M.L., Heyman, E.C., March 1–3, 1999. Gas Pre-Treatment and Their Impact on Liquefaction Processes, paper presented at the 78th Annual GPA Convention, Nashville, TN, USA.
Klinkenbijl, J., Grootjans, H., Rajani, J., March 14–17, 2005. Best Practice for Deep Treating Sour Natural Gases (to LNG and GTL), paper presented at GasTech 2005 Conference & Exhibition, Bilbao, Spain.
Kohl, A., Nielsen, R., 1997. Gas Purification, fifth ed. Gulf Publishing Company, Houston, TX, USA.
Lagas, J.A., Borsboom, J., Heijkoop, G., 1989. Claus Process Gets Extra Boost. Hydrocarbon Processing 68 (4), 40.
Mak, J.Y., Jan. 4, 2005. High Propane Recovery Process and Configurations. U.S. Patent No. 6,837,070.
Mak, J.Y., Deng, E., Nielsen, R.B., Aug. 5, 2003. Methods and Apparatus for High Propane Recovery. U.S. Patent No. 6,601,406.
Mak, J.Y., Nielsen, R.B., Chow, T.K., Sept/Oct. 2010. Zero Claus Plant SOx Emissions. Sulphur 331.

Mak, J.Y., Nielsen, R.B., Graham, C., Feb. 22–23, 2007. A New Integrated NGL Recovery/LNG Liquefaction Process, paper presented at the GPA Europe Conference, Paris, France.

Mak, J.Y., Row, A.R., Varnado, C., April 15–18, 2012. Production of Pipeline Gas from a Raw Gas with a High and Variable Acid Gas Content, paper presented at the 91st Annual GPA Convention, New Orleans, LA, USA.

Markovs, J., Clark, K., March 13–16, 2005. Optimized Mercury Removal in Gas Plants, paper presented at the 84th Annual GPA Convention, San Antonio, TX, USA.

McIntush, K.E., Petrinec, B.J., 1995. GRI Testing of SulFerox for the Direct Treatment of High-Pressure Natural Gas at NGPL's Kermit, Texas Site. Final Report GRI-94/0432. Gas Research Institute (GRI), Chicago, IL, USA.

McIntush, K.E., Seeger, D.M., Rueter, C.O., DeBerry, K.E., 2001. Comparison of Technologies for Removing Sulfur from High-Pressure Sour Natural Gas with Sulfur Throughputs Between 0.1 and 30 Long Tons/Day, paper presented at the 80th Annual GPA Convention, San Antonio, TX, USA.

McKenzie, D., Cheta, I., Burns, D., Nov. 2002. Removing Nitrogen. Hydrocarbon Engineering 7, 57–63.

Millward, R.J., Finn, A.J., Kennett, A.J., Sept. 21–23, 2005. Pakistan Nitrogen Removal Plant Increases Gas Quality, paper presented at the GPA Europe Annual Conference, Warsaw, Poland.

Mock, J.M., Hahn, P., Ramani, R., Messersmith, D., Feb. 24–27, 2008. Experiences in the Operation of Dehydration and Mercury Removal Systems in LNG Trains, paper presented at the 58th Annual Laurance Reid Gas Conditioning Conference (LRGCC), Norman, OK, USA.

Mokhatab, S., Meyer, P., May 13–15, 2009. Selecting Best Technology Lineup for Designing Gas Processing Units, paper presented at the GPA Europe Sour Gas Processing Conference, Sitges, Spain.

Mokhatab, S., Poe, W.A., 2012. Handbook of Natural Gas Transmission & Processing, second ed. Gulf Professional Publishing, Burlington, MA, USA.

Northrop, S., Sundaram, N., August 2008. Modified Cycles, Adsorbents Improve Gas Treatment, Increase Mole-Sieve Life. Oil & Gas Journal 106, 29.

Row, V.A., Oct. 9, 2012. Desulphurization and Mercury Removal from Natural Gases, presented at the GasTech Centre of Technical Excellence, London, UK.

Row, V.A., Humphrys, M., May 25–27, 2011. The Impact of Mercury on Gas Processing Plant Assets and Its Removal, paper presented at the GPA Europe Spring Conference, Copenhagen, Denmark.

Rueter, C.O., Winter 2002. CrystaSulf Process Fills Mid-Size Niche for Sulfur Recovery in Multiple Applications. GasTIPS, 7–12.

Shell, Shell DEP 31.40.10.12–Gen, 1998. Design Manual of Multiple-Pipe Slug Catchers. Design and Engineering Practice (DEP) Publications, Shell Global Solutions International B.V, The Hague, The Netherlands.

Schulte, D., Graham, C., Nielsen, D., Almuhairi, A.H., Kassamali, N., March . The Shah Gas Development (SGD) Project – A New Benchmark, paper presented at the 5th International Sour Oil & Gas Advanced Technology Conference, Abu Dhabi, UAE.

Son, K.V., Wallace, C., Nov. 7–9, 2000. Reclamation/Regeneration of Glycols Used for Hydrate Inhibition, paper presented at the 12th Annual Deep Offshore Technology Conference, New Orleans, LA, USA.

Stone, J.B., Jones, G.N., Denton, R.D., March 11–13, 1996. Selection of an Acid-Gas Removal Process for an LNG Plant, paper presented at the 75th Annual GPA Convention, Denver, CO, USA.

Trautmann, S., Davis, R., Harris, C., Ayala, L., May 10–12, 2000. Cryogenic Technology for Nitrogen Rejection from Variable Content Natural Gas, paper presented at the XIV Contencion Internacional de Gas, Caracas, Venezuela.

Vovard, S., Bladenet, C., Cook, C.G., May 25–27, 2011. Nitrogen Removal on LNG Plant – Select the Optimum Scheme, paper presented at the GPA Europe Spring Conference, Copenhagen, Denmark.

Wong, V., Mak, J.Y., Chow, T., March/April 2007. Fluor Technology for Lean Acid Gas Treatment. Sulphur Magazine 309, 39–42.

Wilkinson, D., Johnson, G., April 28, 2010. Nitrogen Rejection Technology for Abu Dhabi, paper presented at 18th Annual GPA Gulf-Chapter Conference, Muscat, Oman.

Yates, D., Oct. 13–16, 2002. Thermal Efficiency-Design, Lifecycle, and Environmental Considerations in LNG Plant Design, paper presented at the GasTech 2002 Conference & Exhibition, Doha, Qatar.

CHAPTER

Natural Gas Liquefaction

3

3.1 Introduction

The natural gas liquefaction process is the key component in LNG plants in terms of cost, complexity, and operational importance. A good understanding of design and operational requirements and efficiencies of natural gas liquefaction systems is essential for the success of the gas liquefaction plant. There are several licensed natural gas liquefaction processes available with varying degrees of complexities and experience. The thermal efficiency and capital cost for the various licensed processes are competitive and the differences are typically small with respect to thermodynamics and cost. The real keys in developing a successful liquefaction plant are equipment selection and its configurations to meet a plant's capacity goals.

This chapter provides a critical overview of the process technology options available for the liquefaction of natural gas, and discusses the factors that should be considered by project developers in order to select the most appropriate process for their situation.

3.2 Natural gas liquefaction technology

3.2.1 Liquefaction background

Liquefaction technologies are based on refrigeration cycles, which take warm, pretreated feed gas and cools and condenses it to cryogenic temperatures into a liquid product. The refrigerant may be part of the natural gas feed (open-cycle process) or a separate fluid continuously recirculated through the liquefier or heat exchanger (closed-cycle process). To achieve the extremely cold or cryogenic temperatures required to produce LNG, work must be put into the refrigerant cycle(s) through a refrigerant compressor(s), and heat must be rejected from the cycle(s) through air or water coolers. A number of natural gas liquefaction processes have been developed over the last five decades based upon this fundamental principle.

Besides seeking to reduce unit investment and operating costs, the primary objectives of these technological innovations are to increase the volume of LNG production gas and optimize the efficiency of the refrigeration process employed. In theory, the most thermodynamically efficient liquefaction process is one with a refrigerant or a mixed refrigerant system that can duplicate the shape of the natural gas cooling curve at operating pressure (see Figure 3-1). Observing the cooling curve of a typical natural gas liquefaction process, three zones can be noted in the process of the gas being liquefied. A precooling zone, followed by a liquefaction zone, and completed by a subcooling

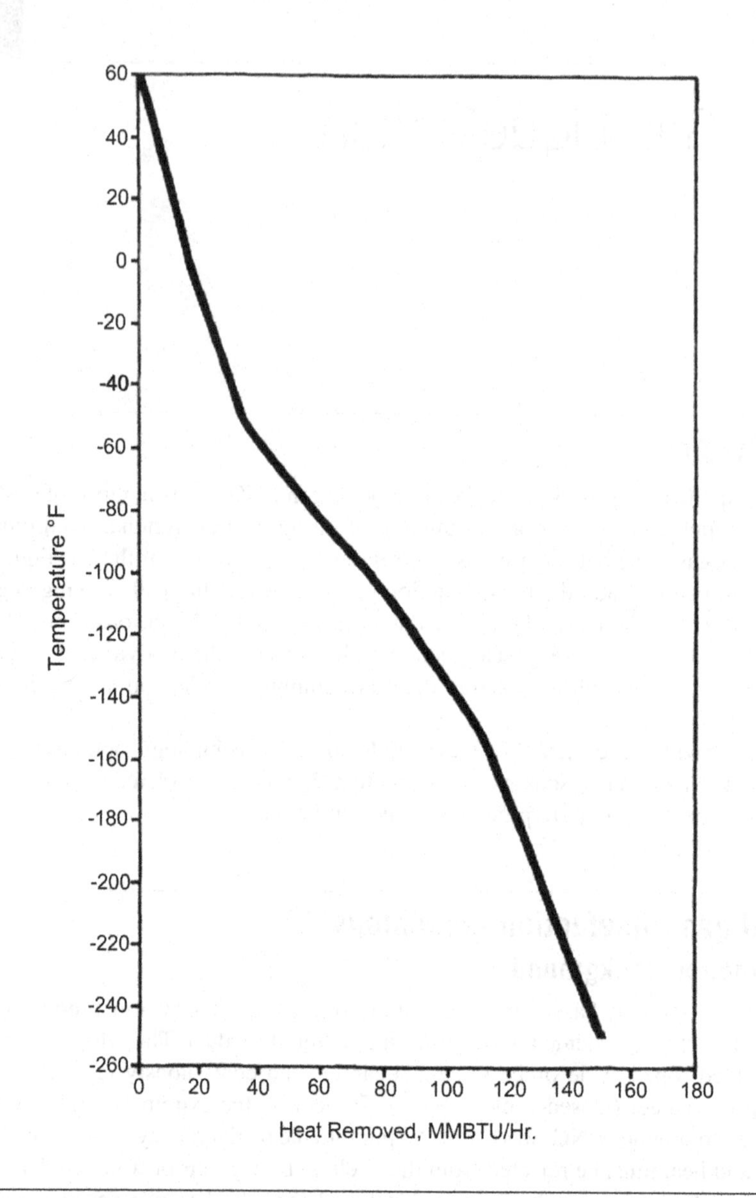

FIGURE 3-1

Typical natural gas cooling curve.

(Younger, 2004)

zone. All of these zones are characterized by having different curve slopes, or specific heats, along the process.

The duty curve associated with a specific gas liquefaction process can be used as a design optimization tool (Ransbarger, 2007). In such an approach the key liquefaction equipment components are designed to match as closely as possible the cooling curve of the gas being liquefied at the different

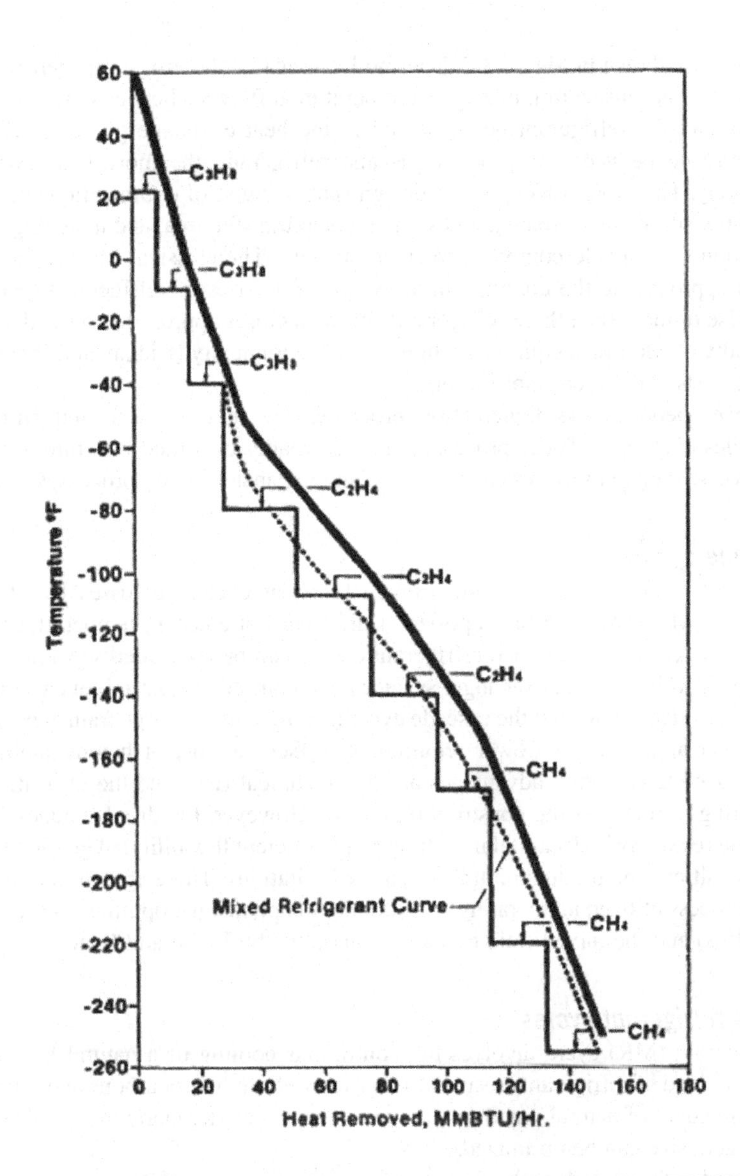

FIGURE 3-2

Cooling curve for natural gas and the corresponding warming curves for the propane precooled/mixed refrigerant (C3-MR) and cascade cycles.

(Younger, 2004)

zones/stages of the liquefaction process in order to achieve high refrigeration efficiency and to reduce energy consumption. Figure 3-2 illustrates the cooling curves for a natural gas system, and the heating curves for a propane-precooled/mixed (C3-MR) refrigerant system and a three-refrigerant classical cascade refrigerant system. Another mixed refrigerant in some plant designs is substituted for the propane in the process shown in Figure 3-2. Thermodynamically, the mixed refrigerant comes closest to a reversible process because it minimizes the temperature difference between the two fluids (i.e., the gas being cooled and the refrigerant being heated in the heat exchange process). The smaller the temperature difference between the process gas and refrigerant, the more heat exchange area is required for the duty. Therefore, LNG process design is an exercise of optimization among refrigerant selection and composition, heat exchanger design and heat transfer area, and matching the refrigerant power consumption to available compressor/driver capacity. The classical cascade liquefaction process attempts to approximate the cooling curve by use of a series of refrigerants (usually three) in separate loops. Use of more than three refrigerants allows a closer approximation to the cooling curve but with the penalty of additional equipment, higher cycle complexity (Kidnay and Parrish, 2006), and higher operating costs and larger plant footprint.

The differences between gas liquefaction processes are mainly associated with the type of refrigeration cycles employed. These processes can be broadly classified into three groups: cascade liquefaction processes, mixed refrigerant processes, and expansion-based processes.

3.2.1.1 Cascade cycle

The classical cascade process reduces the irreversible heat exchange losses by utilizing several refrigeration cycles whose refrigerants vaporize at different but constant temperatures. The cascade cycle is flexible in operation, since each refrigerant circuit can be controlled separately. The cascade cycle has a comparatively low heat exchanger surface area requirement per unit of capacity (Finn et al., 1999). Economics of scale show that the cascade cycle is most suited to large train capacities, since the low heat exchanger area and low power requirement offset the cost of having multiple machines (Jamieson et al., 1998). The other advantages are low technical risks and the utilization of standard equipment resulting in reducing the construction period. However, the disadvantages of the cascade technology are the relatively high capital investments, insufficient flexibility/adaptation to variations in natural gas composition, and production train capacity limitations. There are two modifications of the classic cascade process utilized in operating gas liquefaction plants: the optimized cascade (developed by Conoco-Phillips) and the mixed-fluid cascade (elaborated by Linde and Statoil).

3.2.1.2 Mixed refrigerant cycles

The mixed refrigerant (MR) cycle involves the continuous cooling of a natural gas stream using a carefully selected blend of refrigerants (usually a mixture of light hydrocarbons and nitrogen) that can mimic the cooling curve of natural gas from ambient to cryogenic temperatures, so that energy usage and heat exchangers size can be optimized.

When compared to the cascade technology process, the advantages of the mixed refrigerant process are better proximity to operational temperature of heat exchangers; lower number of compressors and heat exchanger services; and its ability to adjust refrigerant compositions to accommodate the changes in gas composition, feed gas throughput, and plant operating pressure. On the other hand, a single MR cycle typically has a lower thermal efficiency than the more complex cascade cycle because a single mixed refrigerant composition just cannot optimally match the wide range of cooling temperatures in the liquefaction of natural gas. The refrigeration process also takes longer to start up and line up because of the need for precise blending of the refrigerant mix. This is a significant consideration in an environment where frequent start-up and shutdown are to be expected that would require frequent adjustment of the refrigerant composition. The MR concept has been applied in several liquefaction plant designs.

3.2.1.3 Single mixed refrigerant cycle

The single mixed refrigerant (SMR) process involves an inverse or reverse Rankine cycle, in which the gas is chilled and liquefied in a single heat exchanger. The Rankine cycle generally refers to a cycle that converts heat into work, using a working fluid such as steam or hydrocarbons. The reverse Rankine cycle uses power to generate cooling by rejection of heat, and in the case of LNG plant, propane or a mixed refrigerant can be used as the working fluid.

The mixed refrigerant is a mixture of several compounds (mainly hydrocarbon with low boiling points and nitrogen), and the optimum composition is determined by the feed composition, feed pressure, liquefaction plant pressure, and ambient temperature. The refrigeration process follows a reverse Rankine cycle in the following stages: compression–cooling–condensation–expansion–evaporation. Cooling of the refrigerant occurs at ambient temperature whereas evaporation of the refrigerant occurs at a low temperature that is used for liquefaction. Due to its relatively low thermal efficiency, the single mixed refrigerant cycle is mainly suited to mid-sized and small-scale plants where low cost and simplicity are the deciding factor in the plant economics.

3.2.1.4 Dual mixed refrigerant cycle

The dual mixed refrigerant (DMR) process achieves the liquefaction of the feed gas using two independent MR cycles. In the first cycle the natural gas is precooled by a heavier mixed refrigerant. Subsequently the cooled natural gas is condensed in a second heat exchanger by a lighter mixed refrigerant. The heat exchangers are commonly half the height and size of the heat exchangers used in a single mixed refrigerant (SMR) process due to the split of the cooling duty into two cycles.

Numerous versions of the dual-stage cooling cycles with one or both cycles involving mixed refrigerants have been developed. The propane precooled mixed refrigerant process (C3-MR) is the most widely used (i.e., the first cycle employing a single propane refrigerant and the second cycle a mixed refrigerant). The C3-MR cycle uses a closed loop propane refrigeration circuit to precool the natural gas stream. The balance of the liquefaction process uses a mixed refrigerant (MR). The precooled process results in a more efficient plant design and uses less power when compared to SMR systems. The downside of this modification is the higher process complexity and the higher processing equipment counts.

3.2.1.5 Gas expander cycles

The expansion-based gas liquefaction processes utilize turbo-expanders to produce the refrigerant for liquefaction. The turbo-expander refrigeration cycle works by compressing and expanding a fluid to generate refrigeration. The expansion cycle has been greatly improved due to the advances in high efficiency turbo-expanders (typically over 85%).

The process can be configured as single, dual, or multiple turbo-expander designs that can be driven by electric motors or gas engines. The heat curves of expander liquefaction processes have a relatively large temperature gap between refrigerant and cooling gas, typically at the warm end of the natural gas cooling curve. The refrigerant, either nitrogen or methane, is a light volatile component that is a better refrigerant for low temperature cooling rather than the high temperature range in the beginning of the gas cooling, especially when the feed gas contains significant amounts of C_3+ components.

Table 3-1 Comparison of Efficiencies for Different Refrigeration Cycles

Refrigeration Cycle	Approximate Power Consumption Relative to Cascade Cycle
Cascade	1.00
Single stage mixed refrigerant	1.25
MR cycle with propane precooling	1.15
Multistage MR	1.05
Single expander	2.00
Single expander cycle with propane precooling	1.70
Double expander	1.70

(Finn et al., 1999)

The refrigerants remain in the gaseous state throughout the expansion cycles. Being a single component, there is no need to adjust the composition, thus simplifying the process operation. Also, because the heat exchangers operate with relatively wide temperature approaches, they are less sensitive to changes in feed gas compositions. For these reasons precise temperature control is not as critical as it is for mixed-refrigerant cycles. This cycle is considered to be more stable over a range of liquefaction conditions (Finn et al., 1999). Nevertheless, the expander cycle is less efficient when compared to the cascade and mixed refrigerant cycles (see Table 3-1), which makes it suitable for small LNG plants (such as BOG liquefaction) and less suitable for large-scale base load plants.

Nitrogen expander cycle efficiency can be improved to some extent with additional equipment, such as multiple expanders and the use of propane precooling but are seldom justified for small plants (e.g., Finn et al., 1999). However, because there is no hydrocarbon liquid inventory, the design is inherently safe. Being a gas phase cycle, the system performance is not much impacted by ship motion and is more suitable for ship-based floating liquefaction plants.

3.2.2 Liquefaction process selection criteria

LNG process selection has often been highly influenced by the specific power consumption, (i.e., kW/ton of LNG). This is certainly an important parameter, since refrigerant compressors are the single largest cost and energy consumption components in a gas liquefaction train. However, in order to compare different natural gas liquefaction systems, it is necessary to compare (1) the compressor power required, (2) the heat exchanger surface area requirement, and (3) and the temperature approaches between the heating and cooling curves in the main cryogenic heat exchanger.

The cooling performance curve is a benchmark that is commonly used in LNG technology comparison. However, it can be misleading if energy performance of a liquefaction system is the only consideration without taking life-cycle costs into account. A detailed knowledge of the design and operation, and the capital and operating cost of the liquefaction process is necessary to make a meaningful comparison.

The first few natural gas liquefaction plants and a few current plants were based on the classical cascade processes operating with pure components such as methane, ethylene, and propane. The majority of existing base-load natural gas liquefaction plants operate with at least one mixed refrigerant cycle. However, other processes such as the cascade system operating with mixed refrigerants and dual mixed refrigerant processes have been built in recent years (e.g., Sakhalin in Russia and Snohvit in Norway). Expander-based cycles that operate with nitrogen as the working fluid are used in a number of small-scale gas liquefaction plants, peak shaving plants, and the BOG recovery units, such as the BOG reliquefaction systems in Q-Max LNG ships.

In the following section, the most commonly licensed processes utilizing one and/or combinations of the previously mentioned technologies are discussed in more detail.

3.2.3 Onshore natural gas liquefaction processes

The most common onshore gas liquefaction technologies are described next. These processes have been successfully applied to land base liquefaction plants. The descriptions do not attempt to disclose the full details of the licensor processes, but rather the basic design concepts and design considerations. The liquefaction processes are continuously evolving, and there are many versions of the same licensed processes that take advantage of current equipment advancement. Involving mainly gas turbine drivers and heat exchanger designs, in most cases the new versions are design enhancements built upon the original design concepts.

3.2.3.1 APCI propane precooled mixed refrigerant process

The Propane Precooled Mixed Refrigerant (C3-MR) process, developed by Air Products & Chemicals Inc. (APCI), is the most widely used liquefaction process to date. This process has dominated the base load LNG technology since the late 1970s with about 75% of the natural gas liquefaction market.

The C3-MR process, as shown in Figure 3-3, is composed of a multistage propane precooling system followed by liquefaction using a mixed refrigerant system (made up of nitrogen, methane, ethane, and propane). The natural gas feed is initially cooled by a separate propane chiller package to an intermediate temperature, approximately –35°C (–31°F). The natural gas is then liquefied and subcooled in the main cryogenic heat exchanger (MCHE), composed of a large number of small-diameter spiral-wound tube bundles (i.e., spiral-wound or spool-wound heat exchanger). The mixed refrigerant is partially condensed by the propane chiller before entering the heat exchanger. The separate liquid and vapor streams are then chilled further before being flashed across Joule-Thomson valves, which provide the cooling for the final gas liquefaction.

The heart of the APCI C3-MR process is its proprietary spiral wound heat exchangers (SWHE). Large capacity trains over 5 MTPA can be designed using the split MR compressor/driver arrangement, where the available power of each gas turbine driver and its helper motor/turbine is fully utilized for LNG production with a minimum number of refrigerant compressor casings.

A modification by APCI made in the past decade has increased the size of LNG plants (over 6 MTPA; e.g., several liquefaction trains commissioned in Qatar 2008 to 2011 are over 7.8 MTPA). The AP-X™ process adds a third refrigerant cycle (nitrogen expander) to provide the LNG subcooling duties outside the main cryogenic heat exchanger -MCHE (Roberts et al., 2002). With the nitrogen cycle, the size of the main exchanger can be kept the same, with the subcooling duty shared by the

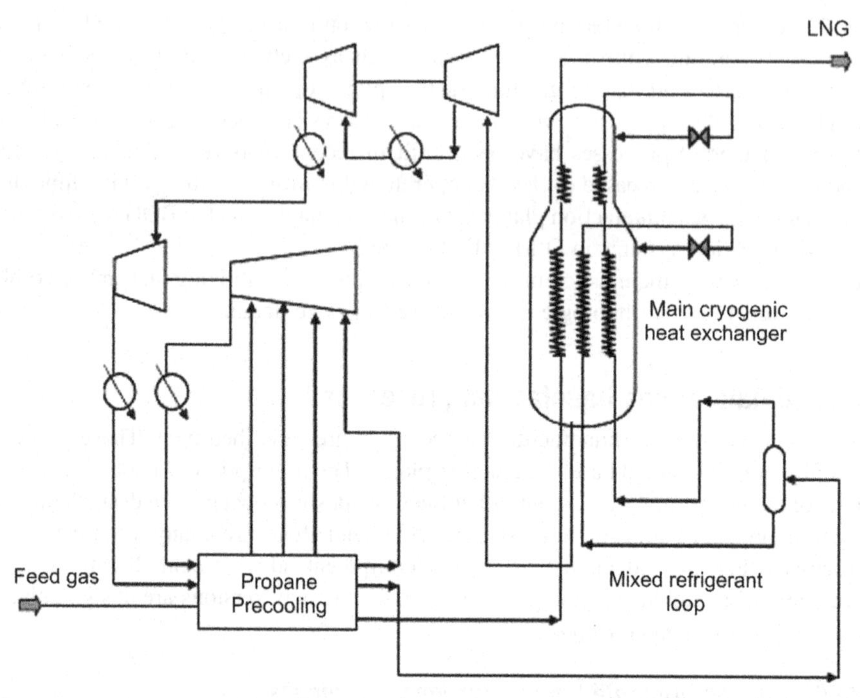

FIGURE 3-3

Typical APCI propane precooled mixed refrigerant process.

(Bronfenbrenner, 1996)

nitrogen cycle. This design approach makes liquefaction of 10 MTPA possible without development of a larger main exchanger. More details on the AP-X™ process are discussed in Chapter 10.

3.2.3.2 Phillips optimized cascade LNG process

Phillips Petroleum Company developed the original cascade LNG process in the 1960s. This process was first used in 1969 at the ConocoPhillips' Kenai LNG plant in Alaska. Figure 3-4 provides a design schematic of a typical Phillips Optimized Cascade Liquefaction Process (POCLP). This process uses propane and ethylene systems and a multiple stage methane refrigeration system to balance refrigeration loads.

In this process, feed gas is routed successively through each stage of propane and ethylene chillers. Air or cooling water removes the compression heat and condenses propane, while propane removes heat and condenses ethylene. Heavier components (NGL) typically are removed from the feed gas after one or more stages of chilling and the resulting methane-rich feed is routed through the methane refrigeration system. If methane refrigerant contains nitrogen, a slip stream is drawn off to be used as fuel to prevent build-up of inert. LNG from the last stage flash drum is sent to the LNG tanks by the LNG transfer pumps where it is stored at approximately 70 mbar above atmospheric pressure and at −161°C.

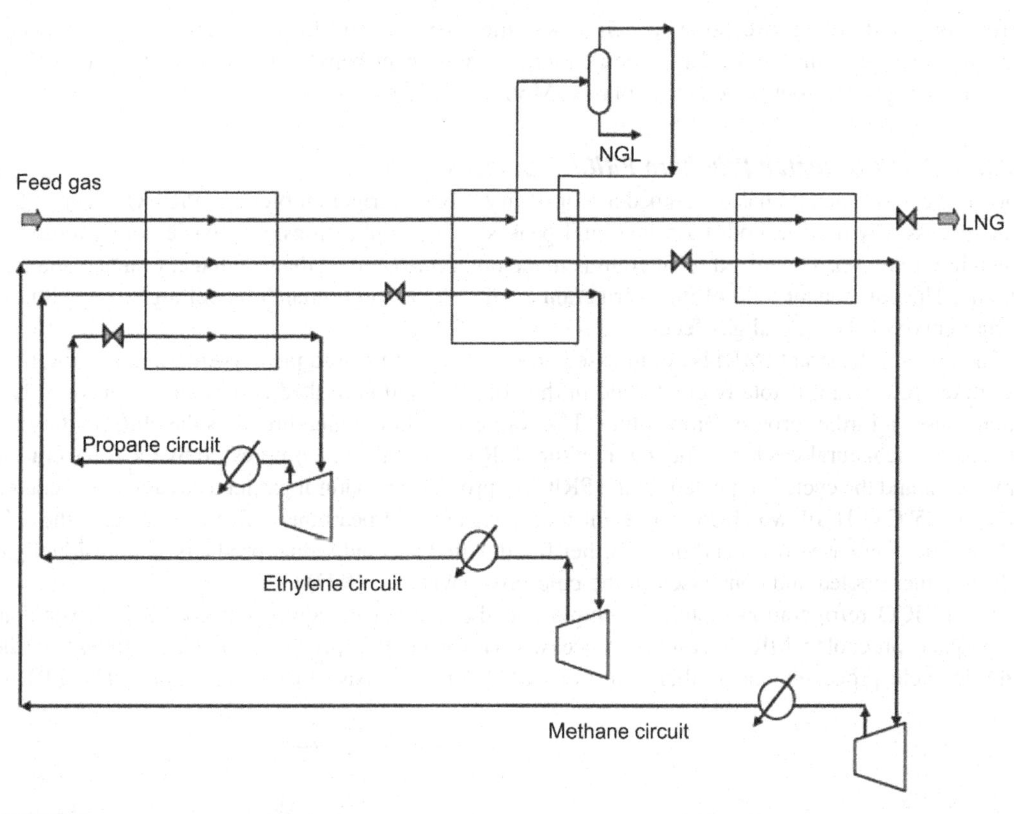

FIGURE 3-4

Phillips optimized cascade process.

(Houser and Krusen, 1996)

Each refrigeration circuit uses two 50% compressors with common process equipment. Brazed aluminum heat exchangers and core-in-kettle heat exchangers can be used for the refrigeration circuits. These exchangers are open arts designs that can be supplied by several manufacturers. The heat exchanger designs are less complex than the proprietary spiral-wound heat exchanges used in the C3-MR process.

Train sizes of liquefaction plants built to date with POCLP technology are less than 5.0 MTPA but the process is claimed to be capable of larger train sizes. The POCLP is claimed to be able to provide designs with high thermal efficiency (exceeding 93%). Design thermal efficiencies for the optimized cascade facilities commissioned at Darwin Australia in 2006 were reported to range from 90 to 93% and could reach 94% in some cases (Ransbarger, 2007). Turndown rates to 10% are also claimed (Houser et al., 1997).

Phillips and Bechtel have entered into a global collaboration providing design, construction, commissioning, and startup for the POCLP plants. The most recent plant of this design to come onstream is the Soyo plant operated by Chevron in Angola (2012). The Phillips/Bechtel alliance provides a single contractor working with a technology licensor to be responsible for the process performance and project execution, which lowers the operation and financial risk on large projects. The avoidance of joint venture EPC contractors has significant benefits to the financial institutions when financing grass-root liquefaction plants (Mølnvik, 2003).

3.2.3.3 Black & Veatch Pritchard PRICO® process

Black & Veatch Pritchard has developed a proprietary mixed refrigerant process, PRICO®, which has been successfully used in both base load and peak shaving applications. The basic configuration is shown in Figure 3-5. The mixed refrigerant is made up of nitrogen, methane, ethane, propane and iso-pentane. The component ratio of this refrigerant is chosen to closely match its boiling curve with the cooling curve of the natural gas feed.

The mixed refrigerant (MR) is compressed and partially condensed prior to entering the cold box. The mixed refrigerant is totally condensed in the cold box and is flashed across an expansion valve, which causes a further drop in temperature. This vaporizing liquid then provides the chilling duty for condensing the natural gas feed. The low-pressure MR vapor is then recompressed by the refrigeration compressor and the cycle is repeated. In the PRICO® process, the natural gas feed stream is first cooled to about −35°C (−31°F), which condenses most of the heavy hydrocarbons to avoid freezing in the cold section. The condensed hydrocarbon is further fractionated into individual products. The residual gas is then further cooled and condensed in the cold box (Swenson, 1977).

The PRICO refrigeration system is simple and the number of equipment counts is lower than the propane precooled MR or cascade processes. However, this process is not as efficient as the multiple-cycle processes and is therefore not suited for large base load LNG plants. The PRICO

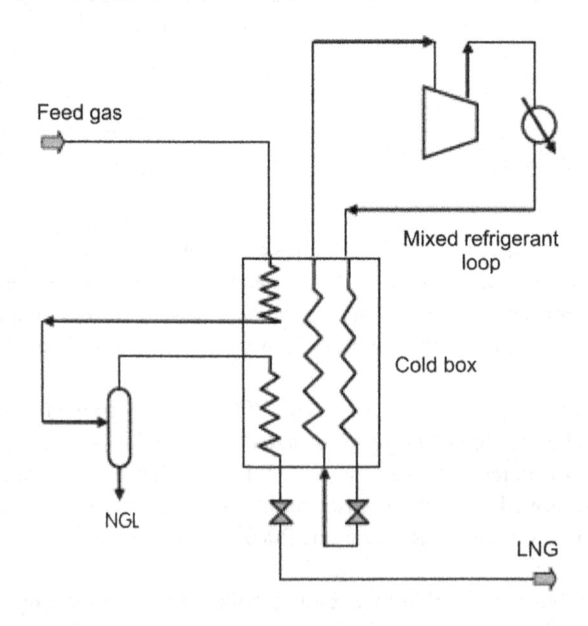

FIGURE 3-5

Black & Veatch Pritchard PRICO process.

(Swenson, 1977)

process is mainly reserved for smaller-scale or peak shaving LNG plant applications. The Black & Veatch PRICO liquefaction process was recently used in China for peak shaving and vehicle fuel applications because of its simple operation, low capital and operating costs, and its flexibility in handling a wide range of feed gas compositions.

3.2.3.4 Statoil/Linde mixed fluid cascade process

The Statoil/Linde LNG technology alliance was established to develop alternative LNG base load plants particularly suitable for harsher environments. It was pioneered in the Snohvit plant deployed on Melkoya Island offshore Hammerfest in the Northern North Sea (i.e., latitude 71° North) of Norway. This plant remains Europe's only base load export gas liquefaction plant.

The technology incorporates the Mixed Fluid Cascade (MFC) process. The MFC process is a classic cascade process with the difference that the mixed components refrigerant cycles replace single component refrigerant cycles, thereby potentially improving the thermodynamic efficiency and operational flexibility.

As shown in Figure 3-6, the MFC process takes purified natural gas and precools, liquefies, and subcools it by means of three separate mixed refrigerant cycles. The precooling cycle involves two plate fin heat exchangers (PFHEs), whereas the liquefaction and subcooling cycles utilize two

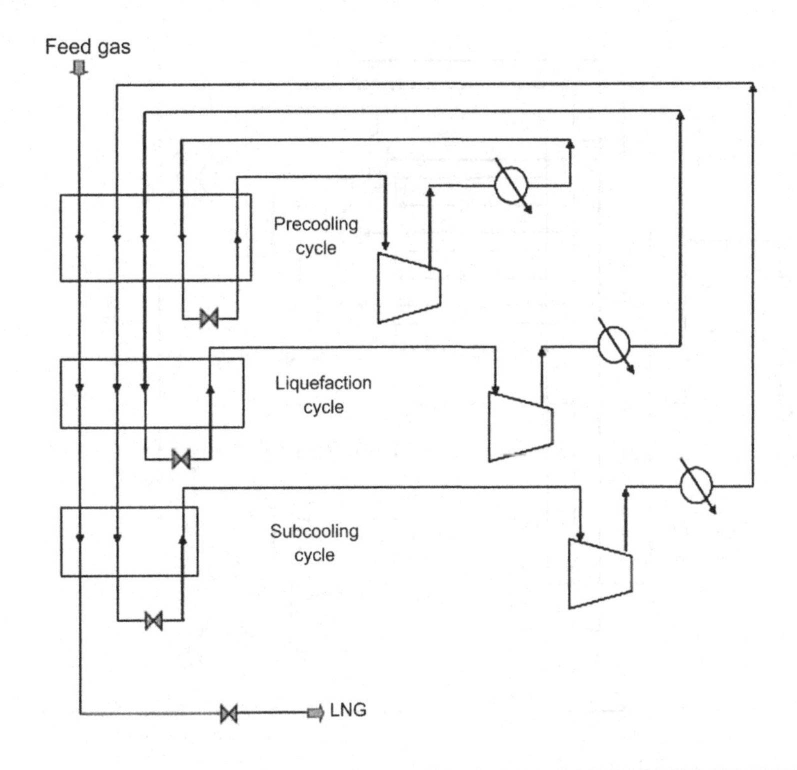

FIGURE 3-6

Statoil/Linde mixed fluid cascade (MFC) process.

(Heiersted et al., 2001)

spiral-wound heat exchangers (SWHEs; Bach, 2000). The SWHE could also be used for the precooling stage. The refrigerants are made up of components selected from methane, ethane, propane, and nitrogen. The three refrigerant compression systems can have separate drivers or be integrated to have just two strings of compression. Frame 6 and Frame 7 gas turbine drivers are used in the large liquefaction trains (> 4 MTPA).

To date, only one gas liquefaction plant has been built using the MFC process. The Snohvit plant was built by Aker in a joint venture with Linde and has a capacity of 4.3 MTPA. However, since commissioning in 2007 (more than one year behind schedule and significantly over budget) the plant has experienced poor operating performance. In particular, reportedly heat exchangers are not performing and require replacement.

3.2.3.5 IFP/Axens Liquefin™ process

IFP and Axens proposed the Liquefin™ liquefaction process, aimed at achieving high capacities with simple equipment configurations and standard compressors (Martin et al., 2003; Fisher and Boutelant, 2002). It is a two-mixed refrigerant process as illustrated in Figure 3-7.

Feed gas enters the liquefaction train's precooling section where it is cooled between –50°C and –80°C (–60°F and –110°F) using a mixed refrigerant in a bank of brazed aluminum plate-fin heat

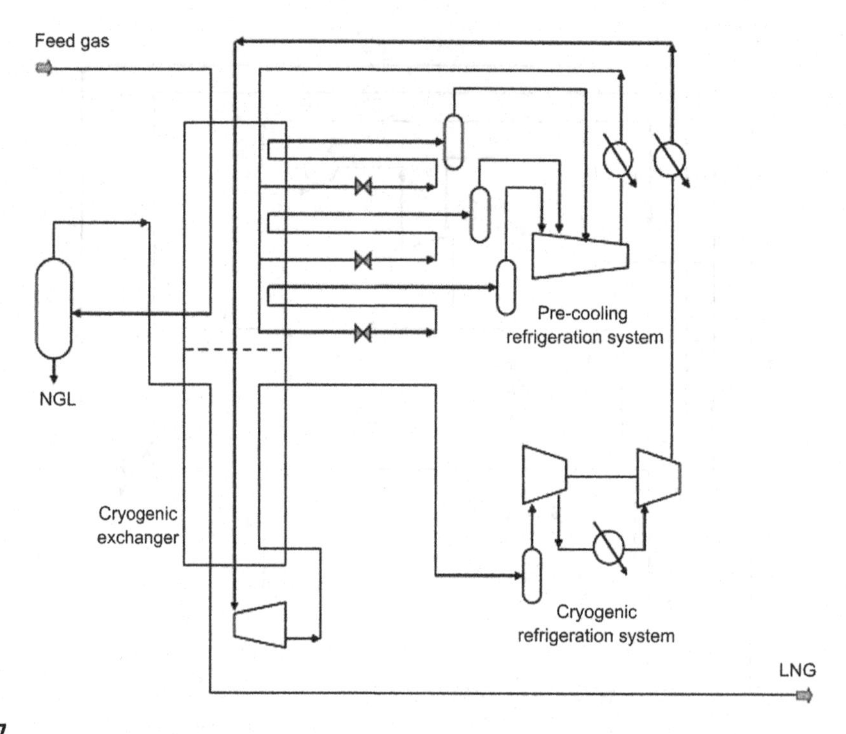

FIGURE 3-7

IFP/Axens Liquefin™ process.

(Fisher and Boutelant, 2002)

exchangers (PFHEs). The cooled feed gas stream is separated for NGL removal. The gas returns to the heat exchanger and enters the cryogenic section where it is liquefied with a second mixed refrigerant.

The first mixed refrigerant is used at three different pressure levels to precool the process gas and the second mixed refrigerant is used to liquefy and subcool the process gas. The mixed refrigerant gas entering the precooling section is completely condensed by the time it leaves the cryogenic section. After leaving the cryogenic section, the refrigerant is expanded and reenters the cryogenic section where the process gas and refrigerant are condensed.

This process involves shifting a good part of the condensation duty from the cryogenic section to the prerefrigeration cycle. Theoretically, this shifting of duty provides opportunity for a more even distribution of the duty, and more optimum heat exchanger designs (Fisher and Boutelant, 2002). This process was initially developed to obtain a 50–50 sharing of power between the liquefaction refrigerant cycle and the precooling refrigerant cycle (Burin de Roziers and Fischer, 1999). The main advantage claimed is in the use of a single refrigerant composition and a simplified PFHE design (Paradowski and Hagyard, 2000).

The Liquefin process providers claimed a total cost savings per ton LNG of some 20% compared to the C3-MR process (Knott, 2001). The cost savings were claimed from (1) higher plant capacity, (2) lower heat exchanger costs, (3) use of plate-fin heat exchangers, (4) compact plot area, and (5) the multisource of all equipment, especially heat exchangers (Mølnvik, 2003). However, the industry continues to be skeptical about the commercial benefits of the Liquefin technology, because no base load plants have been built yet.

3.2.3.6 Shell dual mixed refrigerant process

Shell has developed a dual mixed refrigerant (DMR) process for natural gas liquefaction, as shown in Figure 3-8, with two separate mixed refrigerant cooling cycles, one for precooling of the gas to approximately –50°C and one for final cooling and liquefaction of the gas. This concept allows the designer to choose the load on each cycle. The technology was deployed for the first time at base load scale at the Sakhalin Liquefaction plant in eastern Russia (i.e., two 4.8 MTPA trains using SWHEs and air cooling enhanced by the cold climate), which was commissioned in early 2009.

DMR process configuration is similar to the propane precooled mixed refrigerant (C3-MR) process, but with the precooling conducted by a mixed refrigerant (made up mainly of ethane and propane) rather than pure propane. The DMR process can be highly efficient in cold climates since the precooling mixed refrigerant can be formulated to avoid temperature and handling limitations associated with propane. Another main difference is that the precooling is carried out in SWHEs rather than core-in-kettle or plate-fin heat exchangers. The cooling duty for liquefaction of the natural gas is provided by a second mixed refrigerant cycle. The refrigerant of this cycle consists of a mixture of nitrogen, methane, ethane, and propane. Mixed refrigerant vapor leaving the main cryogenic heat exchanger is compressed in an axial compressor followed by a two-stage centrifugal compressor. Intercooling and initial desuperheating is achieved by air cooling. Further desuperheating and partial condensation is achieved by the precooling mixed refrigerant cycle. The mixed refrigerant vapor and liquid are separated and further cooled in the main cryogenic heat exchanger, except for a small slipstream of vapor mixed refrigerant, which is routed to the end flash exchanger (Dam and Ho, 2001).

A further development of the DMR process is the electrically driven Parallel Mixed Refrigerant PMR design (Heyman et al., 2002) using a parallel line-up of electrically driven refrigerant

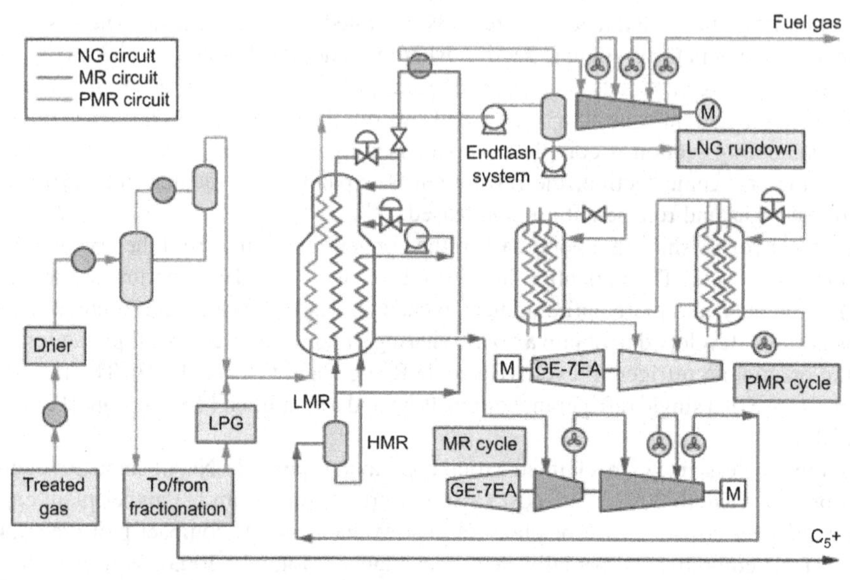

FIGURE 3-8

Schematic overview of the Shell DMR cycle.

(Dam and Ho, 2001)

compressors around a common set of cryogenic spool wound exchangers. Electric motors of 65 MW have already been constructed for LNG service. Motors up to 80 MW are considered feasible (Kleiner and Kauffman, 2005). The current electrically driven DMR design is particularly attractive in the 5 to 10 MTPA capacity range (Del Vecchio and De Jong, 2004). Electrically driven LNG trains can compete with mechanically driven trains because the increase in availability compensates for the increase in cost. Other benefits of the electric option are the variable size and speed of the driver, the higher vendor base, and the potential to make a step change reduction in overall plant carbon dioxide (CO_2) emissions by using the highly efficient combined cycle electric power generation plants (Van de Graaf and Pek, 2005). See Chapter 10 for more details on electric motor drives and combined power cycle power plants.

Shell has also developed technology to further push the capacity of the propane cycle by employing double-casing instead of single-casing equipment. This is a reliable method to bring the propane-MR process closer to a capacity of 5 MTPA. With a single precooling cycle and two parallel mixed refrigerant cycles, the capacity can also be boosted up to 8 MMTPA with three General Electric frame seven (GE-F7) compressors in a tropical climate. The process can use either C3 or an MR in precooling. Proven refrigerant cycles can be used and the design can currently be applied without step changes in technology. The capacity can be increased further with different (larger) drivers. Another possibility for the propane-MR process is to transfer power from the propane cycle to the mixed refrigerant cycle, a concept developed by APCI. The closer coupling between the two cycles by mechanical interlinking of compressors is an operational challenge.

Table 3-2 Onshore Liquefaction Cycles Evaluation

Cycles	Cascade/C3-MRC/ Dual Cycle	MRC	Expander
Efficiency	High	Moderate/High	Low
Complexity	High	Moderate	Low
Heat exchanger area	Low	High	Low
Flexibility	High	Moderate	High

(Finn et al., 2000)

3.2.3.7 Process selection

A qualitative comparison of the different liquefaction cycle is summarized in Table 3-2. In general, expander plants are favored on the small peak-shave facilities, and the mixed refrigerant cycle (MRC) plants are more attractive for mid-scale LNG plants. For base load LNG plants, propane precooled MRC, cascade, or dual cycle is favored due to their higher efficiency (Finn et al., 2000).

It should be noted that liquefaction process selection is critical to the LNG economics, so it is a key activity that starts at an early stage of an LNG project and should be addressed at the feasibility study and pre-FEED stages. Sufficient process and utility details must be developed to define the capital and operating costs for each licensor. Quotations from the various licensors and main equipment vendors must be obtained for the different processes.

In selecting the most appropriate technology, both technical and economic considerations must be addressed. Technical considerations include licensor experience, plant reliability and availability, energy efficiency, and environmental impacts. Economic evaluation must address the life cycle costs, including units upstream and downstream of the LNG plant; and utility and offsite cost. In most cases, process licensors have to adjust and optimize their designs to meet the owners' preferences, economic criteria, site conditions, and product specifications. Most often, several sets of feed gas compositions must be considered to make sure that the selected process would meet any future requirements.

Thermal efficiency

Thermal efficiency, or train efficiency, is expressed as the ratio of the total higher heating value (HHV) of the liquefied product to the total higher heating value of the feed gas. When evaluating thermal efficiency of a particular LNG liquefaction technology, all the energy being consumed in the process must be considered. Thermal efficiency is a common benchmark used to compare competing processes for new projects. The value of thermal efficiency can have more weighting when the feed gas to the LNG plant is relatively expensive or supply is limited.

Thermal efficiency depends on numerous factors, such as gas composition, inlet gas pressure and temperature, site temperature and pressure, and the other more obscure factors such as location of the loading area relative to the liquefaction process. Thermal efficiency calculated for the LNG loaded to the ship will be less than the thermal efficiency calculated based on the LNG run down from the liquefaction train, as the former will include boil-off losses in LNG storage and energy consumed in loading the LNG to the ship.

Thermal efficiency is a trade-off between capital and life cycle costs. The gas turbine driver selection, waste heat recovery, boil-off gas recovery, end-flash design, utility, and offsite system must be included in calculating the liquefaction thermal efficiency (Yates, 2002).

Equipment selection

The economics of the liquefaction processes, more than any other units in the LNG supply chain, are dependent on the equipment selection. The efficiencies and costs of the major equipment are some of the important parameters in the selection process. The other more important factors are proven equipment with successful track records, and the confidence of the plant owners in the technology. Equipment must be supplied by reliable suppliers with good financial backgrounds.

Despite the emphasize on proven equipment, the LNG would accept step changes on equipment improvement, such as size increase and efficiency improvement, as long as they are based on familiar technology and equipment.

The advancement in technology is discussed in Chapter 10. This section provides general background information on equipment that is commonly used in the liquefaction processes.

Compressors. The rotating equipment selection is affected by the characteristics of the process, such as gas composition, molecular weight, pressure and temperature, compression ratio, and flow rate of the refrigerant. Typically, the economical choice of a refrigeration compressor in base load plants is the centrifugal compressor because it can be designed with high volumetric flow. Axial compressors are more suited to higher flow rates but at lower pressure ratios than that required by the MR compressors. However, if the refrigeration process is configured to take advantage of the high efficiency of the axial compressor, it can be used as the first stage of compression, but the design has to be optimized together with the rest of the compression train. As for the small to mid-range LNG plants, a wide range of compressors can be considered.

Drivers. Many earlier base load liquefaction plants used steam turbine drivers since they were a proven technology at the time. But as gas turbine size increased and efficiency improved, gas turbines have replaced the steam turbine driver. Operation of a gas turbine is much simpler than a steam power plant. Operating a steam boiler plant is significantly more complex, which is not suitable in remote locations with limited technical supports or in desert locations where water supply is scarce. For these reasons, gas turbines are the dominating drivers for today's liquefaction plant design.

Gas turbines are more compact than steam systems. The gas turbine delivery schedule and installation time are also shorter than for a steam turbine system (Finn et al., 1999). The gas turbine types can be broadly categorized as heavy-duty industrial types (Frame turbines) and industrial aero-derivatives, characterized by their light weight and higher efficiency (Avidan et al., 2001). The performance and design of some of the common gas turbines for LNG plants are shown in Table 3-3.

Up to now, almost all the refrigerant compressors in liquefaction plants are driven by frame-type gas turbines. The industrial type turbines are heavier in weight but more robust and able to operate under harsh conditions. These turbines have been workhorses for the gas processing industries. The GE Frame 5 turbine has established itself as the prime driver for most compression units and is well accepted by the industry. The Frame 5 output has been slowly increased throughout the years. As shown in Table 3-3, the power output of Frame 5D can reach 32.6 MW. The bigger frames are now available with Frame 9E reaching 123 MW power. With the larger frame turbine, the cost of power ($/kW) is reduced. To take advantage of the lower cost, the refrigeration cycles must be reconfigured to utilize the larger turbine power. The objective is to design the liquefaction process to utilize the maximum power output from these large drivers.

For the small frame turbines, the power output can also be increased with the use of electric helper motors (Kleiner et al., 2003). This is necessary during summer operation at high ambient temperatures, when the turbine power output will drop and LNG production will suffer. This deficiency can be

Table 3-3 Gas Turbine Performance Comparison

Frame Turbines					
Model	**GE 5C**	**GE 5D**	**GE 6**	**GE 7**	**GE 9E**
ISO power, MW	28.3	32.6	43.5	85.4	123.4
Relative cost	1.0	1.2	1.3	2.7	3.0
Relative specific cost	1.0	1.04	0.85	0.89	0.89
Number of shafts	2	2	1	1	1
Efficiency %	29.3	30.3	33.2	32.7	33.8

Aero-derivatives				
Model	**GE LM2500+**	**GE LM6000**	**RR Cobera 6761**	**RR Trent**
ISO power, MW	30	44.6	33.4	52.55
Relative cost	1.2	1.7	1.2	1.9
Relative specific cost	1.13	1.08	1.02	1.02
Number of shafts	2	3	2	3
Efficiency %	40.3	42.6	40.5	42.5

(after Avidan et al., 2001)

compensated by inlet air chilling or with the use of a helper motor. See Chapter 10 for more details on turbine inlet air chilling.

Aero-derivative gas turbines As shown in Table 3-3, aero-derivative gas turbines can achieve about 25% higher thermal efficiencies than industrial gas turbines. Aero-derivative turbines operate with a higher compression ratio and a combustion higher inlet temperature. However, the maintenance costs are typically higher than frame turbines. The turbine availability is also lower due to the less robust design compared to the frame type turbines. However, the change-out of the turbine part is easier because of the compact and lightweight design.

Aero-derivative gas turbines were successfully installed and operated in the optimized cascade liquefaction trains of the Darwin LNG facility (onstream 2006; Meher-Homji et al., 2008). The positive feedback from plant operation was the ability to swap out a gas turbine generation set in less than three days. This compares to the typical 14-plus days turnaround for a major overhaul on an industrial gas turbine. The short turnaround time has improved the overall plant availability of an aero-derivative turbine, despite the more frequent outages (Yates, 2002).

Note that the original designs using the Phillips Optimized Cascade Liquefaction Process (POCLP) were based on the Frame 5 series gas turbines. The LNG output capacities using the Frame 5 turbines ranges from 3.3 to 3.7 MTPA (e.g., Atlantic LNG facility, Trinidad Trains 1 to 3; Idku LNG Egypt; Darwin LNG Australia and Bioko LNG, Equatorial Guinea).

Large gas turbine drivers To take advantage of the larger gas turbine drivers, the liquefaction train design is configured to match the available driver capacity, and the lower cost of power $/kW (see Table 3-3).

For larger LNG trains, Frame 6, 7EA or Frame 9E industrial gas turbines are a better fit (Avidan et al., 2003).

The Soyo LNG Plant in Angola (5.2 MTPA single train design, onstream 2012) utilizes a combination of Frame 6 and Frame 7 compressor drivers in a two-train-in-one configuration.

A Frame 9 gas turbine with 130 MW ISO rating provides 50% more power than a Frame 7. The six large (7.8 MTPA single train capacities) trains in Qatar (2007 to 2011) use three Frame 9 gas turbines for each train. Operating at slower speeds (3000 rpm versus 3600 rpm) further increases the power output from the Frame 9 turbines.

Driver selection criteria With direct gas turbine drive, the match of the drivers' design to the refrigeration compressors is crucial in the liquefaction process selection. If the compressor driver train is not limiting, then the other equipment in the liquefaction train needs to be evaluated. Most importantly, the licensor must be confident in the plant performance when extending their technology on building larger heat exchanger equipment; that is, the main cryogenic exchanger.

If the large equipment size is not viable or economical, then parallel trains can be considered. The parallel trains would undoubtedly drive up the capital costs, but the additional equipment may also improve the plant availability and flexibility that may be favorable in some cases.

In most cases, the selection criteria is the match of the gas turbine sizes to the liquefaction designs that would result in higher efficiency and lower capital cost. However, with the availability of large electric motors, the liquefaction process is no longer limited by the mechanical drivers. This would open up more opportunities for innovation and efficiency improvement.

See Chapter 10 for discussion of the advances in combined cycle and motor drives.

Heat exchangers. The main heat exchangers used in gas liquefaction process trains are: plate-fin heat exchangers (PFHEs) and spiral-wound heat exchangers (SWHEs). These types of exchangers have large internal surface areas consisting of a large number of heat exchanger cores or circuits. These designs can achieve a close temperature approach between the refrigerant and the natural gas in the liquefaction cycle.

However, with today's technology, there is a limit on the heat exchanger size. A single PFHE can typically be designed to liquefy about 1.5 MTPA of LNG. Higher capacity would require multiple exchanger cores grouped together to provide the heat exchange areas. On the other hand, a single SWHE can typically be designed for 4 MTPA of LNG (Finn et al., 1999).

PFHEs, often referred to as brazed aluminum heat exchangers (BAHXs), consist of aluminum fins. PFHEs can be configured with different flow configurations: cross-current, parallel, and multipass directions. This type of heat exchanger is also very common in the NGL recovery unit to achieve high recovery. The PFHEs are manufactured by several heat exchanger manufacturers, and are readily available. The main advantages of the PFHE are their compactness, low equipment weight, small footprints, and typically lower capital cost. They can be configured to accommodate different processes with multiple exchangers. Since they are constructed of aluminum, they are also more vulnerable to mechanical damages or damage from thermal shocks. The process must be operated to ensure a temperature difference among the different passes of no more than 50°F to minimize thermal stress.

SWHEs, or spiral (or coiled) wound heat exchangers, are designed with a greater internal heat transfer area and therefore can operate with a larger temperature gradient. In addition to the larger surface area, the other advantage of the SWHE is its higher proven tolerance to thermal shocks. However, because of the fixed heat exchange configuration, the exchanger design has a limited flexibility to handle different feed gas compositions. Compared to PFHEs, this type of exchanger is higher in capital cost, size, and weight. The spiral wound exchanger suppliers are also limited, making them less cost competitive (Martin et al., 2004).

In general, the SWHE are used for large MR liquefaction plants, whereas the PFHEs are used in expander and cascade liquefaction plants. However, there are no strict rules in the selection process. The heat exchange equipment selected should be evaluated on a case by case basis, considering the latest advances in technologies. For example, the Statoil/Linde Mixed Fluid Cascade (MFC) process uses both technologies; a PFHE for the first cascade stage and SWHE for the two latter stages. The propane precooled mixed refrigerant (C3-MR) process also uses a combination of PFHEs for pre-cooling and SWHEs for liquefaction.

Heating and cooling

Heating. The main steam consumers in LNG plants are amine reboilers in the amine unit, dehydration regeneration gas heater, tail gas unit reboiler, and reboilers in NGL fractionation. When processing high content sour feed gases, the heating duties for regenerating the amine can be significant. The heating requirement can be supplied by a heat transfer fluid or steam using waste heat from the gas turbines.

If excessive steam is still available after being supplied to the process users, it can be used to produce power using back pressure turbines or condensing steam turbines. The additional power produced can be used to drive the process compressors or one of the refrigeration compressors.

However, with the use of the more efficient gas turbine drivers, the amount of waste heat is expected to be reduced. Worst still, if a combined cycle is used to produce power using gas turbine exhaust heat, the amount of heat available for process heating will be further curtailed. Process heating will then be supplied by steam boiler or by duct firing in the gas turbine exhaust. In this case, the gain in power generation efficiency will be offset by the fuel required for process heating.

To arrive at an optimum process configuration, the plant steam balance and the power block steam and power configuration must be evaluated together with the refrigeration cycles and the plant steam consumption.

Cooling. The main coolant consumers are compressor interstage and discharge coolers in refrigeration units. Process units also require cooling, such as the lean amine cooler, the molecular sieve regeneration gas cooler, sulfur plant product cooler, reflux condensers in the stabilization unit, and the NGL fractionation unit.

There are basically two types of cooling medium: air and water. Therefore, selection of the type of cooling medium is site dependent.

In desert areas where water supply is not available, air cooling is the only solution. Since most of the base load LNG plants are located in desert areas, air cooling is the primary choice. Because of the hot desert ambient, the process may require further cooling below the ambient temperature using refrigeration provided by a chilled water system. For example, gas precooling before the molecular sieve unit, which is necessary to reduce the water content before the molecular sieve bed, would significantly reduce the size of the molecular sieve beds. Another example is gas turbine inlet air chilling using refrigeration or evaporative cooling to maximize gas turbine power output. See Chapter 10 for more details on air chilling.

For plant sites located close to the ocean, seawater can be considered. Seawater supply temperature is typically lower than air temperature, which can potentially reduce the refrigeration compressor horsepower and increase the liquefaction efficiency. However, the higher efficiency must be evaluated against the higher seawater system costs, such as water inlet and outfall structure and piping, seawater pumps, seawater treatment requirement, and seawater environmental impact.

Comparison of power consumption for different technologies

Along the entire LNG supply chain, the liquefaction process plants represent the largest capital expenditure component. For this reason, technology providers and operating companies strive to improve the liquefaction technologies that would result in cost savings in production through economy of scale and process innovations. Besides the liquefaction process, improvements in other processing units that would increase the plant operability and reliability are just as important (Bronfenbrenner et al., 2009).

ExxonMobil is in the process of developing a dual mixed refrigerant (DMR) process, utilizing BAHX exchangers and large gas-turbine-driven centrifugal compressors. The objective is to improve scalability and expandability for large-scale gas liquefaction project developments using more generic equipment, while promoting competition among different suppliers, keeping production cost down. The main aim is reducing the specific power requirement and lowering capital cost over traditional technologies.

The specific power requirement and the unit cost for some of the latest technologies including the ExxonMobil DMR-BAHX design can be compared in Figures 3-9 and 3-10 (Stone et al., 2010). In these figures, SMR stands for Single Mixed Refrigerant; C3MR stands for Propane Precooled Mixed Refrigerant; C3MRN2 stands for Propane Precooled Mixed Refrigerant plus Nitrogen expander cycle; Cascade uses pure components, such as methane, ethylene and propane; DMR-SWHE stands for Dual Mixed Refrigerant with single pressure levels using SWHEs; DMR-BAHX stands for Dual Mixed Refrigerant with multiple pressure levels using BAHXs; and TMR stands for Triple Mixed Refrigerant.

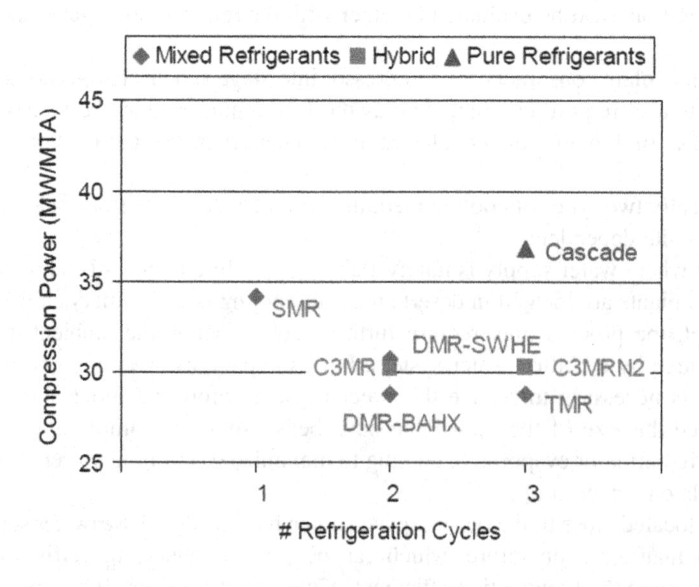

FIGURE 3-9

Comparison of specific power of different liquefaction technologies.

(Stone et al., 2010)

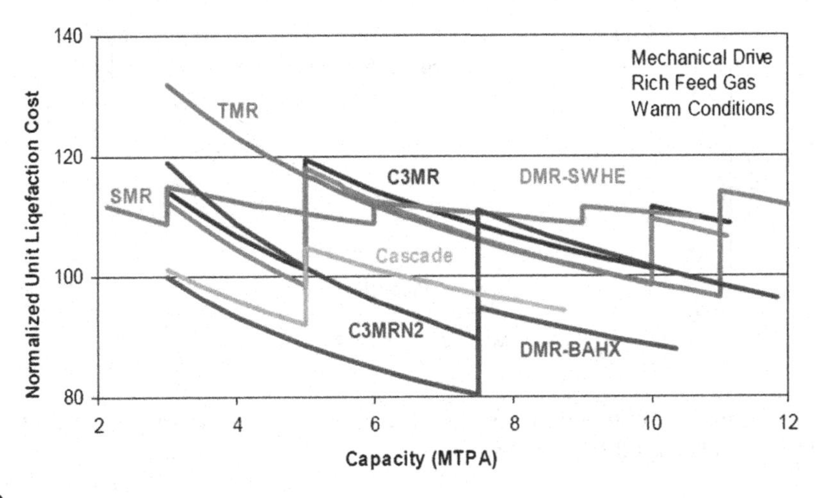

FIGURE 3-10

Unit cost comparison of different liquefaction technologies.

(Stone et al., 2010)

3.2.4 Liquefaction in cold climates

Exploiting natural gas using gas liquefaction in high latitudes is proven to be commercially viable at some locations (e.g., Alaska, Snohvit, and Sakhalin). The process design of the liquefaction plants in arctic climates presents some unique challenges (Wood and Mokhatab, 2009).

Cold ambient temperature increases the operating efficiencies and reduces the energy consumption in the cryogenic facilities, independent of the liquefaction technology. When compared to hot climate areas, cold climate or arctic areas refrigeration requirements are lower, mainly due to the reduction in refrigeration in the precooling section. The temperature fluctuation affects the precooling section the most since this is where most of the heat is rejected to ambient air.

Arctic locations have a wide seasonal variation in ambient temperature, resulting in significant seasonal variations in output (Figure 3-11), which would require process adjustments throughout the year to maximize the plant throughput. The cold ambient also increases the power output from the gas turbines as the air gets denser, increasing the mass flow to the turbines. This presents a challenge to the owner on whether to invest additional process equipment to take advantage of the higher horsepower output in the winter, while the equipment is underutilized in the summer (Omari et al., 2001; Spilsbury et al., 2006; Martinez et al., 2007).

For the extreme weather conditions in the Sakhalin liquefaction plant, ambient temperature can vary from 26°C in the summer to −18°C in the winter. By adopting the dual mixed refrigerant process and using mixed refrigerants (propane and ethane) for precooling, the ratio of these components can be adjusted to match the seasonal changes in ambient temperature, to minimize energy consumption. Waste heat must be recovered from the gas turbine exhaust to provide heating for heat tracing on equipment to avoid water from freezing in the gas processing units. Operating a steam plant in cold climate areas requires more operator attention.

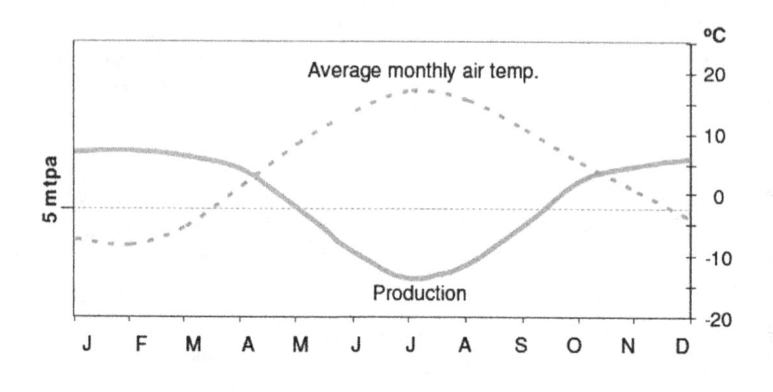

FIGURE 3-11

Typical seasonal temperature profile, and the corresponding variation in theoretical maximum plant output for a 5 MTPA liquefaction unit utilizing C3-MR process.

(Josten and Kennedy, 2008)

If liquefaction plants are to be operated at varying throughput based on ambient temperatures, LNG shipping logistics must also be evaluated to cope with such variations. However, this may not always be possible. For instance, colder weather conditions may sometimes lead to shipping delays at a time when the unit is capable of maximum output. For this reason, liquefaction plants are usually not designed to utilize the full capacity of the drivers during winter; the plant capacity must be optimized by simulation model based on logistics and climatic factors, which set a limit or "cap" on peak production.

3.2.4.1 Cooling medium selection

Cooling medium selection (direct air versus direct seawater) has significant cost and design impacts on LNG plants. Air temperatures vary more widely than seawater temperatures. The colder air in winter has two impacts on refrigeration system performance. Cold inlet air to gas turbines increases gas turbine power output. Colder air also lowers compressor interstage and discharge temperature, resulting in a lower compression requirement. In particular, in the precooling section, process cooling can be partially achieved by air, which significantly reduces the refrigeration requirement. Conversely, during summer, power output from a gas turbine would be reduced and compression cooling is less effective, leading to reduction in production. Unfortunately, the loss in production also coincides with typical plant turnaround in the summer, which further worsens the plant output. Production may be somewhat compensated by the increase in production in winter.

Seawater temperature generally stays fairly constant throughout the year, and the average seawater temperature in high latitudes is typically about 2 to 3°C higher than the air temperature. While operating in winter, the colder seawater may increase the production somewhat, but the increase is limited by the freezing seawater. In most cases, the capital costs for a seawater cooling system cannot justify the marginal increase in production in winter months. Because of this, coupled with the environmental sensitivity on warm seawater discharge, air cooling is generally the likely choice in arctic locations, except where space is severely limited (Josten and Kennedy, 2008).

3.2.4.2 Liquefaction process selection in cold climate

The most widely suitable liquefaction processes are divided between those that are precooled with propane and those with a mixed refrigerant. Because the heat sink in arctic LNG plants is available at such a low temperature, reducing the precooling temperature will allow for a better power balance and better machinery selection between precooling and liquefaction duty. This can be accomplished by replacing the propane refrigerant with a lower boiling point gas (e.g., ethane or ethylene) or a multicomponent mixed refrigerant. Although the mixed-refrigerant system is more complex than a single-component refrigerant system, it provides additional flexibility since the composition of both mixed refrigerants can now be adjusted within certain limits to match the wide seasonal variation of heat-sink temperatures. Thus in winter the lower ambient air can be used to condense a lighter refrigerant at a lower temperature. As the refrigerant is condensing at lower temperature, this can be performed at a lower pressure, so that the compressor can move out along its curve and process a greater refrigerant flow at lower compression ratio, all within a given shaft power. Therefore, overall refrigeration duty can be increased to take advantage of the winter conditions (Spilsbury et al., 2006; Josten and Kennedy, 2008; Wood and Mokhatab, 2009).

3.2.4.3 Cold climate design issues

As a further challenge to the LNG plant designers and project managers, freezing temperatures, snowfall, and high winds reduce on-site productivity, complicate the transport of personnel and equipment, and extend the construction schedule. Snow and ice loadings have a significant impact on building and structure designs (Josten and Kennedy, 2008).

If installed outdoors, winterization of the gas liquefaction plants is necessary to prevent fluid freezing, liquid drop-out, and wax and hydrate formation (Martinez et al. 2007). Rotating equipment such as pumps, power generators, and refrigerant gas turbine and compressor units must be housed in heated and ventilated modular buildings.

The construction of the equipment modules is typically completed in modular yards. The completed modules can be shipped to the site for installation. There are special design considerations on modular design, particularly on the piping and electrical layout and configuration, which must be designed to permit easy installation and adequate egress and maintenance (Wood and Mokhatab, 2009).

3.3 Offshore natural gas liquefaction

There are many smaller gas fields in remote offshore locations that can benefit from the floating liquefaction (FLNG) technologies. Ship-to-ship LNG transfer and marine vessel LNG storage tank technologies have now evolved to the stage where offshore liquefaction plants with capacities in excess of 4 MTPA are economically viable.

Despite the high cost of the marine vessel, bringing gas to an onshore LNG liquefaction facility can be just as costly. Onshore installation requires extensive infrastructure, including subsea pipeline, gas separation equipment, storage tanks, dredging for jetties, and other port facility. Construction of such facilities will require extensive time. In some instances, FLNG solution offers the only viable solution that allows completion of the project in a short schedule. Quality of the FLNG design can also be better managed and controlled.

LNG floating production, storage, and offloading (LNG FPSO) has been the focus of research and development since the 1980s, but in 2008 it took a step toward deployment with the commitment by

FLEX LNG Ltd to contract Samsung to construct vessels for service offshore with provisional locations in Papua New Guinea. In 2011, FLNG took another big step forward when Shell announced its decision to build a large FLNG vessel to develop the Prelude gas field offshore Northwest Shelf (NWS) Australia for deployment in 2016/2017.

Assessment of liquefaction technology to offshore FLNG is no different than onshore installation, and the life cycle cost analysis must be evaluated with the same process as a land-based plant, with safety being of paramount importance. However, for offshore floating liquefaction, the criteria for technology selection differ substantially from onshore facilities (Wood et al., 2007; Finn, 2009a; Wood, 2009). While thermodynamic efficiency and train capacity are important criteria for land-based plants, the cost of an offshore facility relies on small footprint, low equipment count and weight, and compact modules. While cost is an important criterion, the offshore design must provide easy access to maintenance, must be easy to start up and shut down, and must be reliable. Being an under–ocean environment, the motion of the vessel must be considered in equipment design, such as in fractionator design where motion can impact flow distribution on trays that can lower fractionation efficiency. The following section reviews the different liquefaction cycles in respect to the offshore criteria.

3.3.1 Refrigeration cycles

A key decision in the development of FLNG projects is the selection of the liquefaction process cycle that best meets the project objectives. Three generic types of refrigeration cycle have been considered for offshore natural gas liquefaction: cascade, mixed refrigerant, and expander. Previous sections have discussed the performance and requirements of these technologies in land-based environments. The following section reviews the characteristics of each process with respect to offshore functionality and explains why the cascade technologies have been largely discounted for FLNG deployment. Figure 3-12 highlights the trade-off between process efficiency and plant complexity for expander and mixed refrigerant technologies. The selection process is more difficult, may not be cost dependent, and may depend on the availability of technical staff to operate a complex facility.

3.3.1.1 Cascade and optimized cascade

The cascade cycle specific power is low but the process is relatively complex and involves a large number of equipment items. Each of the three refrigerants requires its own storage. The refrigerants would need to be fractionated from the feed gas, and the purity must be monitored and controlled.

For offshore installation, the cascade cycle has the disadvantage despite its high efficiency. It has three refrigeration circuits driven by individual compressors, and each refrigerant has its own liquid storage. The large number of equipment counts and the large plot requirement increase the cost of the FLNG where space is a premium.

3.3.1.2 Propane precooled mixed refrigerant

Compared to the cascade processes, the propane precooled mixed refrigerant (C3-MR) process by APCI has less equipment counts but is still quite complex to operate in an offshore environment (Bukowski et al., 2011). The use of propane refrigerant in the precooling loop requires propane storage, which is a safety hazard in the close environment of an offshore facility. The liquid inventory in kettle-type heat exchangers is a potential hazard. The heat exchange equipment can be impacted by

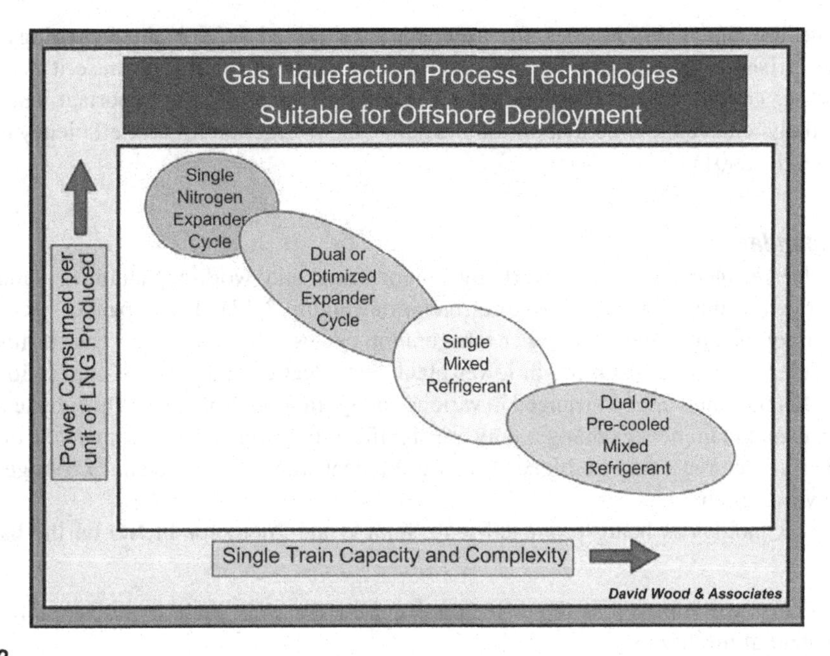

FIGURE 3-12

Qualitative comparison of the efficiency and complexity of gas liquefaction technologies for FLNG.

(Wood, 2009)

the vessel motion and the sloshing of liquid. A different type of heat exchanger may need to be used that are more resistant to motion to ensure good distribution and effective heat transfer.

3.3.1.3 Mixed refrigerant

Mixed refrigerant technology has been assessed for offshore liquefaction based on both single mixed refrigerant (SMR) and dual mixed refrigerant (DMR) cycles (Johnsen and Christiansen, 1998; Naklie, 1998; Price and Mortko, 1998; Bliault, 2001; Sheffield and Mayer, 2001).

For smaller-scale facilities, the simplicity of the SMR technology has some attractions for offshore deployment. However, the SMR process requires a single large liquefaction coil-wound heat exchanger (CWHE) as opposed to the DMR cycle because it must handle all the liquefaction duty including precooling in one exchange. This exchanger is more difficult to install in FLNG for structural and process reasons. At capacities greater than 2 MTPA, two CWHEs would likely be required (Bukowski et al., 2011). A basic SMR process has a lower efficiency than a C3MR or DMR cycle. The process can be somewhat improved with an additional compression and heat exchanger, but at the expense of equipment count and complexity.

The DMR cycle is more efficient, but more complex. It has less hydrocarbon inventory compared to the SMR cycle and the C3MR cycle. The relief rates are lower than other processes in the event of compressor trip and overpressure. However, all mixed refrigerant cycles require storage of hydrocarbons and fractionation facilities for refrigerant makeup. The safety concerns of hydrocarbon liquid storage are the major hurdles for acceptance of mixed refrigerant technology offshore. Operating an offshore facility is more complex than an onshore facility because it is exposed to extreme weather conditions and other hazards. Thermal efficiency is less important, and operators will be less likely to have the time to monitor and adjust the refrigerant mix for efficiency gain (Finn, 2009b; Pekdemir, 2001).

3.3.1.4 Expander

Turbo-expander refrigeration cycles work by compressing and work-expanding a suitable fluid, typically nitrogen or methane, to generate refrigeration (Figure 3-13). The expander cycle is simple and has fewer equipment counts than other refrigeration cycles. This equipment can be modularized, requiring smaller plot space, resulting in lower plant cost. Heat exchangers use conventional brazed aluminum exchangers and can be arranged in various configurations. Note that although the refrigerant circulation rate and main heat exchanger duty are significantly lower in expander cycles, the required heat transfer surface area could be higher because the heat transfer coefficient of nitrogen is much lower than hydrocarbons.

Expander technology is being promoted (e.g., Finn et al., 2000) for FLNG on the basis that it demonstrates:

- Inherent safety with nitrogen as the refrigerant, and there are no liquid hydrocarbon refrigerants and no potential fire hazards
- Insensitivity to vessel motion as the refrigerant is in gas phase, and there is no concern on refrigerant distribution in heat exchangers
- Flexibility to changes in feed gas conditions and ease of operation
- Rapid startup and shutdown
- Less equipment counts, smaller plant footprint, and relatively low topside weight.

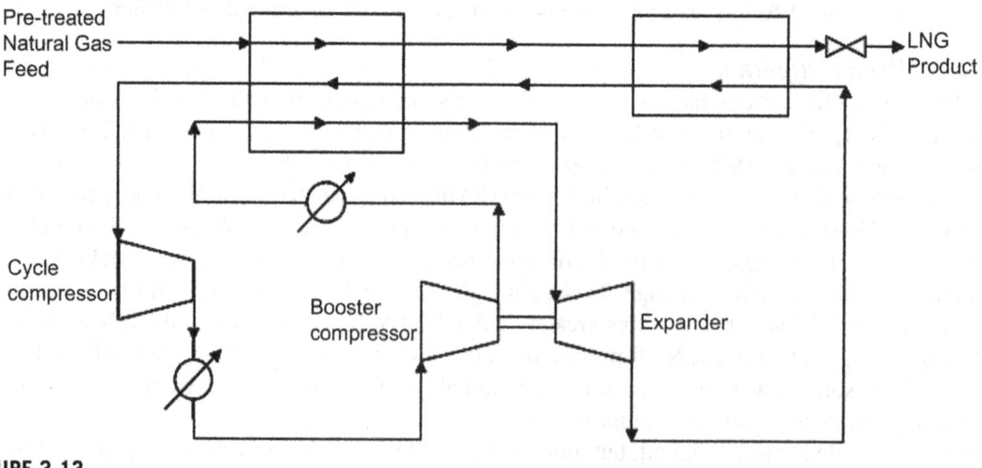

FIGURE 3-13

Turbo-expander cycle for gas liquefaction.

(Finn et al., 2000)

The trade-off is its lower thermal efficiency compared to other technologies. The expansion cycle can be operated with a closed loop nitrogen cycle or open loop feed gas cycle.

N_2 expansion cycles are proven at small LNG plants, in peak-shaving plants, and in reliquefaction of boil-off gas in recent LNG carriers. The main advantage of nitrogen as the cycle fluid is that it is inherently safe. Other more efficient cycles require hydrocarbon liquid storage, which is potentially a fire hazard (Finn et al. 2000). In addition, the nitrogen cycle system is compact as it does not require any refrigerant storage and makeup. System turndown changes can be accommodated with a nitrogen gas surge drum. There is no mal-distribution in the heat exchangers because it is in gas phase. Exchanger and column performance where distribution is critical is not affected by motion in the offshore vessel.

The nitrogen expander design is flexible to changes in feed gas composition and feed gas conditions. There is a side benefit of the less efficient cycle. The temperature approaches on heat exchangers are significantly larger than the mixed refrigeration cycles, and hence adjusting the process parameters to accommodate any feed gas composition changes is simple. With this flexibility, the FLNG can be used to process other gas fields when the gas resource is depleted, extending the usable life of the liquefaction plant.

Compared to the mixed refrigerant cycle, there is no composition adjustment required in handling different gases, because it is a pure component system. Startup and shutdown is quick. Venting in an emergency situation is not a safety concern. No flaring is necessary. However, the refrigerant flow is significantly higher than the mixed refrigeration cycle, since it depends on sensible heat of the gas phase for cooling, and single train capacity is typically limited to 1 to 2 MTPA.

There are several configurations that strive to improve the efficiency of the basic expander technology. The expander cycle efficiency can be increased with the use of a propane precooled cycle. Propane is a more efficient cycle as it can better match the higher end of the LNG cooling curve than the nitrogen cycle. However, the use of propane for refrigerant defeats the advantage of the safety claims of the nitrogen cycle, and is not favored in an offshore environment. Alternatively, a second colder turbo-expander (Figure 3-14) can be used to close up the temperature difference in the cold end of the nitrogen/ LNG heat curve (Finn et al., 2000). However, this configuration would require more heat exchanger surfaces and the capital cost is higher (Finn, 2002).

The dual turbo-expander process has been proposed for several projects. In the late 1980s, Costain advocated a dual turbo-expander flowsheet based on nitrogen refrigerant for the Pandora field, offshore Papua New Guinea, in conjunction with Three Quays Marine Services Ltd. BHP Ltd and Linde AG also proposed the use of a dual nitrogen expander cycle to develop the Bayu-Undan field in the Timor Sea (Cottrill, 1997). The use of dual turbo-expander cycle is commonly accepted for offshore deployment.

Some expander cycles also use feed gas as the refrigerant; for example, Mustang Engineering's OCX-2® process (Walther et al., 2008) and Kryopak's EXP® cycle. The gas expansion cycle uses the feed gas as the refrigerant and can be operated in an open cycle configuration, eliminating the need for refrigerant supply, further simplifying the expander process. The feed gas expander cycle has the same attributes as the nitrogen cycle with the exception that special attention is required for operating the expander compressor as any relief must be vented to a safe location to avoid fire hazard.

3.3.1.5 Process selection

To select the most appropriate offshore natural gas liquefaction cycle, we must compare the options with the major criteria that influence commercial acceptance of floating LNG production. Table 3-4 summarizes the evaluation of liquefaction cycles for offshore use. Although the dual nitrogen expander

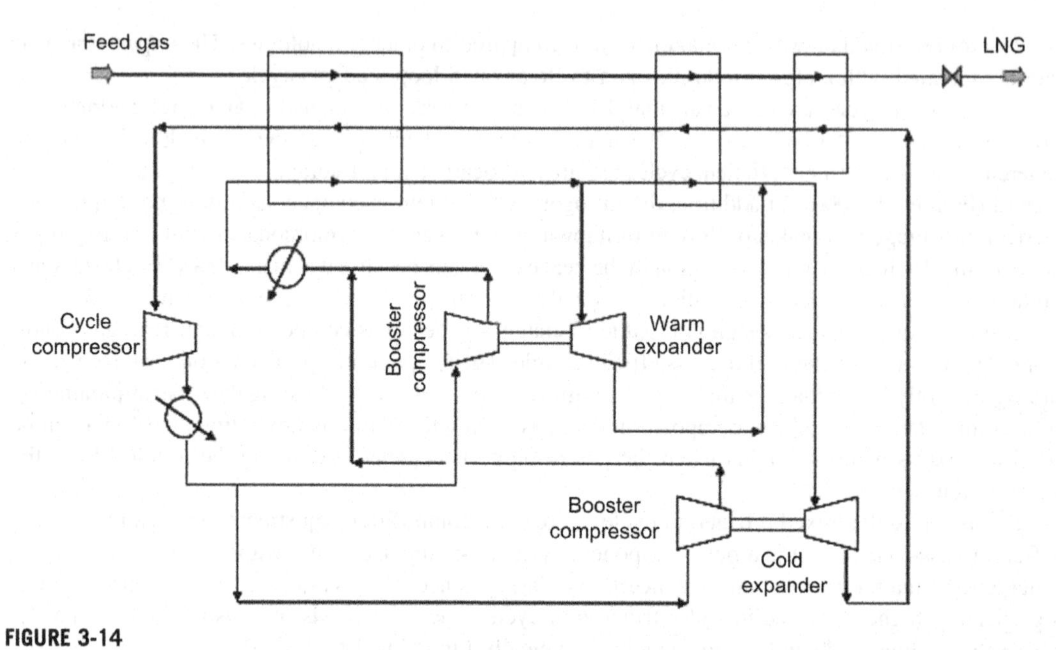

FIGURE 3-14

Dual turbo-expander cycle for gas liquefaction.

(Finn et al., 2000)

Table 3-4 Evaluation of Offshore Natural Gas Liquefaction Cycles

	Cascade	SMR	C3-MR	DMR	N2 Expander
Thermal Efficiency	High	Medium	High	High	Low
Equipment Count	High	Low	Medium	Medium	Medium
Hydrocarbon Refrigerant Storage	Large	Medium	Large	Medium	None
Capital Investment	Medium	Low	Medium	Medium	High
Offshore Suitability	Medium	High	Medium	High	High
Compactness	Low	Medium	Low	Medium	High
Motion Impacts	Medium	Medium	High	Medium	Low

(Chiu and Quillen, 2008)

cycle requires more power than the more complex and thermodynamically efficient cycles,[1] the simplicity of the process and the other critical factors listed make it a strong candidate for many floating liquefaction projects.

[1]Dual nitrogen expander processes for gas liquefaction are expected to achieve unit power requirements of between about 15 and 20 kW*day/ton. Propane precooled dual expander processes may approach the efficiencies (about 13.5 kW*day/ton) of state-of-the-art C3MR cycles (Barclay and Denton, 2005).

In addition to the various items outlined earlier, there are a number of factors such as machinery configuration and types of drivers, heat exchanger type and area, and utility and offsite systems that also must be addressed to determine the optimal FLNG process technology.

3.3.1.6 Choice of compressor driver

Onshore liquefaction plants typically use refrigerant compressors driven by industrial heavy-duty gas turbines, but these are difficult to deploy on a FLNG vessel. Aero-derivative gas turbines are likely to be used in FLNG vessels because of their:

- High power-to-weight ratio resulting in smaller footprints and about half the weight of a standard industrial compressor of comparable power output
- Proven availability and reliability
- Modular engine sections that are easy to service and maintain
- High thermal efficiency (i.e., greater than 40%) compared to industrial units that results in lower fuel consumption and less carbon emissions.

The LNG industry has been using the industrial gas turbines for refrigeration compressor drivers for the past decades. The adoption of the aero gas turbine for the gas processing industry has been slow. However, with the many advantages of the aero turbines and the potential capital cost and energy savings, the use of the aero gas turbines is being accepted for process drivers, particularly in smaller LNG plants and in offshore applications. For example, the Darwin LNG plant (Australia) uses GE LM2500+ aero-derivative gas turbines as the refrigerant compressor drivers and has been in successful operation since 2006 (Rockwell, 2008).

3.3.1.7 Liquefaction heat exchangers

Aluminum plate-fin heat exchangers, which are lightweight and compact, are ideal for floating liquefaction plants. The plate-fin heat exchanger consists of several heat exchanger blocks or cores that are installed inside an insulated cold box. While the exchangers are designed for close temperature approaches, excursion to high temperature differential must be avoided as it would impose excessive thermal stress, which may occur during startup or off-design conditions. There are no special mechanical design or exchanger support issues as long as the design temperature not exceeding 150°F. The challenge is the piping integrity in the transition from stainless steel piping to aluminum flanges of the exchangers which can now be resolved with cryogenic aluminum to stainless steel transitional joints. The system must also be designed considering the stress imposed by the vessel movement (Finn, 2009a).

3.3.1.8 Utility systems

Process cooling systems in FLNG typically use seawater, because air coolers would take up extensive deck space. Seawater cooling can be in either an open or closed loop. In an open loop system, seawater is drawn in, filtered, treated to avoid fouling, pumped through the heat exchangers in the refrigeration systems, and then discharged back to the sea. The heat exchangers are shell and tube exchangers with the exchanger tubes constructed of seawater corrosion-resistant material such as titanium. In the closed loop system, inhibited water is pumped to the heat exchangers and heat is rejected to seawater using a plate and frame exchanger. The closed loop system can be constructed of the less expensive carbon steel material. The indirect heat exchange system in the closed loop circuit increases the coolant supply temperature, which may have a minor impact on the thermal efficiency of the refrigeration cycle.

Waste heat from gas turbine exhaust is recovered to provide process heating in the gas condition units. For operation simplicity, heat transfer fluid such as Dowtherm or Therminol is preferred to steam, as it avoids the operation complexity and the weight of a steam generation plant. For higher efficiency, the steam generation can produce high pressure steam to drive one of the refrigeration compressors. However, in cold climate operation, a steam plant is difficult to operate and is generally avoided.

3.3.2 Small to mid-scale liquefaction processes

The cost index of LNG liquefaction plants was decreasing because of economy of scale up until 2005, but since then, the cost index has been increasing significantly, especially the Australian plants due to high labor cost and escalation of materials. This trend increases the risks on the return on capital, dampening the venture capital's appetite on commitment to large-scale projects. However, a number of medium and smaller sized gas reservoirs, especially less than 5 TCF, are undeveloped. The number of these reserves is over 1,000. These reservoirs are held not only by major oil companies but also by small oil companies or national oil companies. If a good solution for monetizing these reservoirs, such as stranded-gas or associated-gas, is found, projects will be active for such medium-sized gas development.

There are many applications for the small-scale LNG plants: peaking shaving plants, satellite LNG plants for gas transport where pipeline is nonexistent, on-board ship liquefaction, coal bed methane recovery, and bio-gas, landfill gas and today's shale gas liquefaction. For example in Queensland Australia, four mid-scale LNG projects are being developed to liquefy CBM gas on a base load scale project targeted for the Asian markets.

Small to mid-scale liquefaction plants, typically up to 1 MTPA capacity, generally focus upon standardizing compact, preassembled modular designs (Matsui and Yatsuhashi, 2010) that can be deployed quickly minimizing design and construction costs. These plants typically are designed with the following objectives:

- Process simplicity
- Safety
- Easy operation
- Low Cost
- Plant reliability.

There are several small and medium-scale gas liquefaction plants in operation around the world. Plants of this scale are typically sized to process between about 2.5 MMscfd and 100 MMscfd, which is about 10% of the capacity of a base load liquefaction plant. Existing LNG production plants in the small-scale area employ a variety of liquefaction technologies. In general, the small to mid-scale liquefaction plant use one of two processes; that is, single mixed refrigerant or gas expansion-based technologies. The characteristics of the representative processes are discussed next.

3.3.2.1 PRICO process

The PRICO (Poly Refrigerated Integrated Cycle Operation) process, as discussed in Section 3.2.3.3, uses a Black & Veatch patented, single-mixed refrigerant loop for natural gas liquefaction. This process has made Black & Veatch an industry leader for both onshore and offshore small to mid-scale liquefied natural gas applications (Price et al., 2008). As a reference, 25% of the peak-shaving US plants use this process, and, as already mentioned, several liquefaction plants using this technology have been built in China in recent years.

3.3.2.2 APCI single mixed refrigerant process

JGC (Japan) have worked with APCI for possible Asian deployments for a liquefaction plant with 0.5 to 1.0 MTPA capacity (Matsui and Yatsuhashi, 2010). The generic compact design offered by JGC/APCI involves feed gas inflow at 60-bar pressure and 30°C temperature and includes a plant configuration with the following process components:

- Gas receiving and acid gas removal unit (up to 6% CO_2)
- Dehydration unit
- Liquefaction units (scrubber and liquefaction; each unit of 0.5 MTPA capacity)
- Fractionation unit (for refrigerant make-up)
- Fuel/hot oil unit
- Centrifugal refrigerant compressor
- Gas turbine Frame 5D compressor driver
- Air cooling
- Hot oil process heating
- No end-flash or expander units involved.

The specific power requirements for this design was quoted as between 385 kWh/ton of LNG (holding mode) and 398 kWh/ton of LNG (loading mode) and claimed to be lower than alternative processes for comparable plant capacities (Matsui and Yatsuhashi, 2010). The plant as described fit into an area of 218 m × 136 m. Preliminary cost estimates for installation of process units only quoted by Matsui and Yatsuhashi (2010) varied from $400/tpa (Indonesia location) to $600/tpa (East Coast of Australia location).

3.3.2.3 Linde multistage mixed refrigerant process

The Linde Multistage Mixed Refrigerant (LiMuM®) process consists of a Spiral Wound Heat Exchanger (SWHE) and one three-stage single mixed refrigerant loop for the precooling, liquefaction, and subcooling of the natural gas. This process can also be scaled up if required (e.g., up to 2.5 MTPA) by adopting a two-in-one configuration for the refrigerant compressors (barrel type). An operating plant using this process is the Shan Shan LNG plant (China), with a capacity of 0.43 MTPA (Berger et al., 2003).

3.3.2.4 Kryopak precooled mixed refrigerant process

The Precooled Mixed Refrigerant (PCMR®) is a process offered by Kryopak, which consists of a precooling stage (ammonia or propane cycle) followed by a single mixed refrigerant cycle, where the mixed refrigerant is a mixture of nitrogen, methane, ethane, propane, and butanes. The heat exchangers are of the BAHX type. This process is used in some small-scale liquefaction plants with a capacity less than 0.1 MTPA. For example the Karratha LNG project in Western Australia produces some 200 tons of LNG/day (i.e., approximately 0.07 MTPA). The LNG produced at the Karratha plant is then delivered via "road-train" (i.e., very large road vehicles) to small communities in Northwestern Australia up to distances of 2000 km from the liquefaction plant.

3.3.2.5 Optimized single mixed refrigerant process

The Optimized Single Mixed Refrigerant (OSMR™) process, offered by LNG Limited, is a single mixed refrigerant process complemented with a standard package ammonia absorption process.

Within the SMR cycle, the main compressor comprises a single stage unit and the cold box optimizes the passes of the streams (3 main streams plus 2 minor streams). The utilization of an ammonia process allows an improvement of the efficiency of the process and an increase of the LNG output compared to traditional SMR processes. It also allows a reduction in the cold box size. Plans to utilize this technology on larger-scale coal bed methane to LNG are being advanced by LNG Limited for the Gladstone LNG Project—Fisherman's Landing, in Australia. The first phase of that project as planned involves a plant capacity of 1.5 MTPA with a possible scale-up to 3 MTPA in a second phase. Several features of the OSMR® LNG process, including the use of a combined heat and power plant recovering waste heat from the gas turbines, allows the Fisherman's Landing LNG project as designed to offer reduced CO_2 emissions (i.e., with a confirmed design rate of 0.21 tCO2e/tLNG) relative to the other CBM to LNG projects under construction at Gladstone.

3.3.2.6 Hamworthy closed nitrogen expansion cycle

This system is designed and deployed mainly on LNG ships to reliquefy boil-off gas during their voyages. It is used on several membrane tankers serving Qatar LNG supply chains. The Moss RS™ concept is based on a closed nitrogen expansion cycle extracting heat from the boil-off gas. The BOG is cooled and condensed to LNG in a cryogenic heat exchanger (cold box) involving a nitrogen-compander with coolers (one expander). Noncondensable items, mainly nitrogen, are removed in a separator vessel. From the separator, the LNG is returned to the ship's cargo tanks by the differential pressure in the system. Hamworthy's Mark III design adds a compressor in order to reduce power consumption and increase liquefaction capacity of BOG.

3.3.2.7 Mustang OCX-2 process

The OCX-2 (Open Cycle Expander Refrigeration, second generation), offered by Mustang, is an expansion-based liquefaction technology that uses the inlet gas as a refrigerant in an open refrigerant cycle with turbo-expanders. OCX-2 is claimed to achieve 89 to 92% thermal efficiency and is targeted for mid-scale (0.5 to 2 MTPA) applications (Walther et al., 2008).

3.3.2.8 CB&I Lummus Niche LNG process

The CB&I Lummus patented Niche LNG process (targeting small to mid-scale LNG applications up to 2 MTPA) is a dual expansion methane/nitrogen technology in which the methane cycle provides cooling at moderate and warm levels while the nitrogen cycle provides refrigeration at the lowest temperature level. There are no operating references for this process, but it is being actively promoted for FLNG (Nayak, 2009).

Note, for small to medium-size natural gas liquefaction capacities, expander processes have attracted growing interest over the last decade. One of the main reasons for this is the applicability for floating LNG production, because of the simplicity, safety, and the operational advantages of a gaseous refrigerant. However, the selection of the most appropriate small scale liquefaction (SSL) technology is very case-specific. The most relevant parameters to be considered in the selection will depend on the particular conditions of a project (including feed gas composition and feed gas pressure), where every project has its own priorities and the selection criteria may change when the bases of design of the project change. Likewise, the weighting factors assigned to the main parameters in the selection process will vary in each project.

3.3.2.9 Pressurized LNG concepts

Pressurized LNG (PLNG) containment was described by Fairchild et al. (2005) and patented by ExxonMobil. Such storage conditions allow LNG to be contained at warmer temperatures at a higher pressure than conventional. The higher temperature reduces the refrigeration requirement significantly, leading to possible cost savings. The key component of the ExxonMobil's technology is the containment system. The PLNG pressure vessels are made of a newly developed high strength, low temperature (HSLT) steel along with a novel welding technique and welding consumable. In order to confirm the design and fabrication methods for these PLNG containers, ExxonMobil fabricated and hydro-tested PLNG prototype container (i.e., 5 m diameter by 18 m tall) using the new HSLT steel and welding method.

The potential savings of PLNG are, however, somewhat offset by the higher cost of the vessels, and the heavier vessels. To commercialize the PLNG technology, the design of the PLNG ship must be fully developed to assess the economic viability of this concept.

More recently Daewoo Shipbuilding & Marine Engineering (DSME) has developed and patented their "Cluster LNG" concept (Lee et al., 2011) as an integrated liquefaction, transportation, and regasification supply chain of pressurized LNG. LNG is maintained at a temperature of –110°C and at a pressure of 20 bars along the supply chain. DSME claim substantially reduced energy requirements and lower capital and operating costs for the supply chain as a whole, but note that the costs of the transport and regasification are expected to be higher when compared to conventional systems. This is in part because per unit volume the PLNG at Cluster LNG conditions contains about 20% less gas than conventional LNG.

DSME also claim that their Cluster LNG design can cope with lower quality gas (i.e., higher CO_2, water and sulfur), significantly reducing the requirement for pretreatment of feed gas in many cases. The Cluster LNG system claims that it can accept approximately 1% of CO_2 for the liquefaction plant and related systems (Lee et al., 2012).

The PLNG concept has yet to be used as part of a commercial development but is likely to be further developed over the coming years and deployed as part of a small-scale supply chain development (e.g., island-to-island gas movements within Indonesia focused on developing small-scale reserves).

References

Anderson, T.N., Ehrhardt, M.E., Foglesong, R.E., Bolton, T., Jones, D., Richardson, A., Jan. 10–12, 2009. Shipboard Reliquefaction for Large LNG Carriers, paper presented at the 1st Annual Gas Processing Symposium, Doha, Qatar.

Avidan, A., Richardson, F., Anderson, K., Woodard, B., 2001. LNG Plant Scaleup Could Cut Costs Further. Section 7.1, 128–132. Fundamentals of the Global LNG Industry. Petroleum Economist, London, UK.

Avidan, A., Varnell, W., Martinez, B., Aug.18, 2003. Study Evaluates Design Considerations of Larger, More Efficient Liquefaction Plants. Oil & Gas Journal 101, 32.

Bach, W.A., Sept. 14–15, 2000. Developments in the Mixed Fluid Cascade Process (MFCP) for LNG Baseload Plants, paper presented at the World LNG Conference, London, England.

Barclay, M., Denton, N., Oct. 2005. Selecting Offshore LNG Processes. LNG Journal 34–36.

Berger, E., Xiang, D., Jin, G.Q., Meffert, A., Atzinger, L., June 1-5, 2003. LNG Baseload Plant in Xinjiang, China – Commercialisation of Remote Gas Resources for an Eco-responsible Future, paper presented at the 22nd World Gas Conference, Tokyo, Japan.

Bliault, A., May 2001. Shell Floating LNG Plant, Technology Ready for Project Development. Offshore 61 (5), 102–104.

Bronfenbrenner, J.C., Nov./Dec. 1996. The Air Products Propane Precooled/Mixed Refrigerant LNG Process. LNG Journal 25–27.

Bronfenbrenner, J.C., Pillarella, M., Solomon, J., Summer 2009. Selecting a Suitable Process Technology for the Liquefaction of Natural Gas. LNG Industry 17–25.

Bukowski, J., Liu, Y.N., Boccella, S., Kowalski, L., Oct. 19–21, 2011. Innovations in Natural Gas Liquefaction Technology for Future LNG Plants and Floating LNG Facilities, paper presented at the International Gas Union Research Conference 2011, Seoul, Korea.

Buijs, K., Pek, B., Nagelvoort, R., Nov. 21–23, 2005. Shell's LNG Technology for 7–10 MTPA LNG Trains. IPTC 10681, paper presented at the International Petroleum Technology Conference (IPTC), Doha, Qatar.

Burin de Roziers, Th., Fischer, B., Feb. 19, 1999. New Trends in LNG Process Design, paper presented at the GPA Europe Meeting, London, England.

Chiu, C.-H., Quillen, L.D., March 10–13, 2008. A New Frontier- Offshore Natural Gas Liquefaction, paper presented at the Gastech 2008 Conference and Exhibition, Bangkok, Thailand.

Choi, M.S., Sept. 19–22, 2010. LNG for Petroleum Engineers. SPE 133722, paper presented at the 2010 SPE Annual Technical Conference and Exhibition, Florence, Italy.

Cottrill, A., April 1997. An Offshore LNG First for Undan-Bayu. Offshore Engineer 21.

Dam, W., Ho, S.-M., March 12–14, 2001. Engineering Design Challenges for the Sakhalin LNG Project, paper presented at the 80th Annual GPA Convention, San Antonio, TX, USA.

Del Vecchio, J., De Jong, E., 2004. The Mariscal Sucre LNG Project – A Downstream Perspective, paper presented at the 4th Topical Conference on Natural Gas Utilization, AIChE Spring National Meeting, New Orleans, LA, USA.

Fairchild, D.P., Smith, P.P., Biery, N.E., Farah, A.M., Lillig, D.B., Jackson, T.S., Sisak, W.J., June 19–24, 2005. Pressurized LNG: Prototype Container Fabrication, paper presented at the 15th International Offshore and Polar Engineering Conference, Seoul, Korea.

Fallaize, R.A., Phillips, R.S., Feb. 21, 2002. Next Generation LNG – eDrive, paper presented at the GPA Europe Technical Meeting, London, England.

Finn, A.J., 2002. New FPSO Design Produces LNG from Offshore Sources. Oil & Gas Journal 100 (34), 56–62.

Finn, A.J., March 9–11, 2009a. Floating LNG Plants – Scaleup of Familiar Technology, paper presented at the 88th Annual GPA Convention, San Antonio, TX, USA.

Finn, A.J., 2009b. Are Floating LNG Facilities Viable Options? Hydrocarbon Processing 88 (7), 31–38.

Finn, A.J., Johnson, G.L., Tomlinson, T.R., 1999. Developments in Natural Gas Liquefaction. Hydrocarbon Processing 78 (4), 47–59.

Finn, A.J., Johnson, G.L., Tomlinson, T.R., March 13-15, 2000. LNG Technology for Offshore and Mid-scale Plants. Paper presented at the 79th Annual GPA Convention, Atlanta, GA, USA.

Fisher, B., Boutelant, P., Feb.21, 2002. A New LNG Process is Now Available, paper presented at the GPA Europe Technical Meeting, London, England.

Garcia-Cuerva, E.D., Sobrino, F.S., Oct. 5–9, 2009. A New Business Approach to Conventional Small Scale LNG, paper presented at the 24th World Gas Conference, Buenos Aires, Argentina.

Heiersted, R.S., Jensen, R.E., Pettersen, R.H., Lillesund, S., May 14–17, 2001. Capacity and Technology for the Snohvit LNG Plant, paper presented at the 13th International Conference and Exhibition on Liquefied Natural Gas (LNG 13), Seoul, Korea.

Heyman, E., Bliault, A., Pek, B., May 27–30, 2002. The LNG GameChanger™ Technology, paper presented at the Asia Congress, Gasex, Brunei.

Houser, C.G., Krusen, L.C., Dec. 3–6, 1996. Phillips Optimized Cascade LNG Process, paper presented at Gastech 96 Conference and Exhibition, Vienna, Austria.

Houser, C.G., Yao, J., Andress, D.L., Low, W.R., Sept.23, 1997. Efficiency Improvement of Open-Cycle Cascaded Refrigeration Process. U.S Patent 5,669,234.

Hudson, H.M., Wilkinson, J.D., Cuellar, K.T., Pierce, M.C., March 9–12, 2003. Integrated Liquids Recovery Technology Improves LNG Production Efficiency, paper presented at the 82nd GPA Annual Convention, San Antonio, TX, USA.

Jamieson, D., Johnson, P., Redding, P., May 4-7, 1998. Targeting and Achieving Lower Cost Liquefaction Plants, paper presented at the 12th International Conference and Exhibition on Liquefied Natural Gas (LNG 12), Perth, Australia.

Johnsen, R.J., Christiansen, P., Nov. LNG Production on Floating Platforms, paper presented at the GasTech 98 Conference and Exhibition, Dubai.

Josten, M., Kennedy, J., May 14–16, 2008. Liquefaction in a Cold Climate, paper presented at the GPA Europe LNG Conference, Ashford, Kent, UK.

Kidnay, A.J., Parrish, W., 2006. Fundamentals of Natural Gas Processing. CRC Press, Taylor & Francis Group, Boca Raton, FL, USA.

Kleiner, F., Kauffman, S., March 14-17, 2005. All Electric Driven Refrigeration Compressors in LNG Plants Offer Advantages, paper presented at the Gastech 2005 Conference and Exhibition, Bilbao, Spain.

Kleiner, F., Rausch, S., Knabe, J., Jan. 2003. Increase Power and Efficiency of LNG Refrigeration Compressor Drivers. Hydrocarbon Processing 1, 67–69.

Knott, T., Dec. 2001. Cool Future for Gas. Frontiers 10–16.

Lee, J.H., Lee, J.Y., Yoo, S.J., May 2–5, 2011. Development of Innovative LNG Production, Transportation and Regasification System. OTC 21958, paper presented at the Offshore Technology Conference (OTC), Houston, TX, USA.

Lee, J.H., Jung, J., Kim, K., April 30–May 03, 2012. Development of CO_2 Tolerant LNG Production System. OTC 23261, paper presented at the Offshore Technology Conference (OTC), Houston, TX, USA.

Martin, P., Pigourier, J., Boutelant, P., June 1-5, 2003. Liquefin™: An Innovative Process to Reduce LNG Costs, paper presented at the 22nd World Gas Conference, Tokyo, Japan.

Martin, P., Pigourier, J., Fisher, B., 2004. LNG Process Selection, No Easy Task. Hydrocarbon Engineering 9 (5), 75–79.

Martinez, B., Meher-Homji, C.B., Paschal, J., Eaton, A., March 14–17, 2005. All Electric Motor Drives for LNG Plants, paper presented at the Gastech 2005 Conference and Exhibition, Bilbao, Spain.

Martinez, B., Huang, S., McMullen, C., Shah, P., March 11–14, 2007. Meeting Challenges of Large LNG Projects in Arctic Regions, paper presented at the 86th Annual GPA Convention, San Antonio, TX, USA.

Matsui, S., Yatsuhashi, Y., April 18–21, 2010. Mid Small Scale LNG Plant—Compact, Pre-Assembled and Standardized Design, poster presented at the 16th International Conference and Exhibition on Liquefied Natural Gas (LNG 16), Oran, Algeria.

Meher-Homji, C.B., Messersmith, D., Hattenbach, T., Rockwell, J., Weyermann, H., Masani, K., June 9–13, 2008. Aeroderivative Gas Turbines for LNG Liquefaction Plants – Part 1: The Importance of Thermal Efficiency and Part 2: World's First Application and Operating Experience, paper presented at the ASME Turbo Expo 2008 Congress, Berlin, Germany.

Mokhatab, S., Economides, M.J., 2006a. Process Selection Is Critical to Onshore LNG Economics. World Oil 227 (2), 95–99.

Mokhatab, S., Economides, M.J., Sept. 24–27, 2006b. Onshore LNG Production Process Selection. SPE 102160, paper presented at the 2006 SPE Annual Technical Conference and Exhibition, San Antonio, TX, USA.

Mølnvik, M.J., Aug. 29, 2003. LNG Technologies – State of the Art, paper presented at the Statoil – NTNU Global Watch Seminar: Gas Technology, Trondheim, Norway.
Naklie, M., May 4–7, 1998. Mobil's Floating LNG Plant, paper presented at the 12th International Conference and Exhibition on Liquefied Natural Gas (LNG 12), Perth, Australia.
Nayak, K., Feb. 18–20, 2009. Design of an LNG FPSO, paper presented at the GPA Europe Offshore Processing Technical Meeting, London, UK.
Omori, H., Konishi, H., Ray, S.A., de la Vega, F.F., Durr, C.A., May 14-17, 2001. A New Tool – Efficient and Accurate for LNG Plant Design and Debottlenecking, paper presented at the 13th International Conference and Exhibition on Liquefied Natural Gas (LNG 13), Seoul, Korea.
Paradowski, H., Hagyard, P., Sept. 27–29, 2000. An LNG Train Capacity of 1 BSCFD is a Realistic Objective, paper presented at the GPA Europe Annual Conference, Barcelona, Spain.
Pekdemir, T., July 2001. Private Communication. Department of Mechanical and Chemical Engineering, Herriot-Watt University, Edinburgh, UK.
Pérez, S., Díez, R., Oct. 5–9, 2009. Opportunities of Monetizing Natural Gas Reserves Using Small- to Medium Scale LNG technologies, paper presented at the 24th World Gas Conference, Buenos Aires, Argentina.
Price, B.C., Mortko, R.A., Nov. 29–Dec. 2, 1998. Development of Mid-Scale and Floating LNG Facilities, paper presented at the GasTech 98 Conference and Exhibition, Dubai.
Price, B.C., Winkler, R., Hoffart, S., March 13–15, 2000. Developments in the Design of Compact LNG Facilities, paper presented at the 79th GPA Annual Convention, Atlanta, GA, USA.
Price, B.C., Fossella, J., Hoffart, S., July, 2008. Development of Mid-Scale LNG Projects from a Contractor's Perspective. Hydrocarbon Processing, 59–64.
Ransbarger, W., Spring 2007. A Fresh Look at LNG Process Efficiency. LNG Industry 73–80.
Roberts, M.J., Liu, Y.N., Petrowski, J.M., Bronfenbrenner, J.C., Oct. 13–16, 2002. Large Capacity LNG Process – the AP-X Cycle, paper presented at the Gastech 2002 Conference and Exhibition, Doha, Qatar.
Rockwell, J., Spring 2008. 40 Years of Innovation. LNG Industry 34.
Sheffield, J.A., Mayer, M., May 16–18, 2001. The Challenges of Floating LNG Facilities, paper presented at the GPA Europe Spring Meeting, Norwich, UK.
Shell, Aug. 2001. Shell Pushes World's First LNG Floater in Timor Sea. International Gas Report.
Shukri, T., 2004. LNG Technology Selection. Hydrocarbon Engineering 9 (2), 71–74.
Spilsbury, C., Liu, Y.N., Petrowski, J., Kennington, W., Nov. 28–29, 2006. Evolution of Liquefaction Technology for Today's LNG Business, paper presented at the 7th Journées Scientifiques et Techniques of SONATRACH (JST 7), Oran, Algeria.
Stone, J.B., Rymer, D.L., Nelson, E.D., Denton, R.D., April 18–21, 2010. LNG Process Selection Considerations for Future Developments, poster presented at the 16th International Conference and Exhibition on Liquefied Natural Gas (LNG 16), Oran, Algeria.
Swenson, L.K., July 5, 1977. Single Mixed Refrigerant Closed Loop Process for Liquefying Natural Gas. U.S Patent 4,033,735.
Van de Graaf, J.M., Pek, B., Oct.2005. Large-capacity LNG Trains, The Shell Parallel Mixed Refrigerant Process. LNG Review, 41–45.
Walther, S., Franklin, D., Ross, P., Hubbard, B., Feb.2008. Liquefaction Solutions for Challenge of New Offshore FPSO Developments. LNG Journal, 31–34.
Wood, D., 2009. Floating Gas Liquefaction; Competing Technologies Make Progress. Harts E&P 82 (5), 83–85.
Wood, D., Mokhatab, S., Economides, M.J., Nov. 11–14, 2007. Offshore Natural Gas Liquefaction Process Selection and Development Issues. SPE-109522, paper presented at the 2007 SPE Annual Technical Conference and Exhibition, Anaheim, CA, USA.

Wood, D., Mokhatab, S., 2009. What Are the Opportunities to Construct Liquefaction Facilities at the Arctic Circle? Hydrocarbon Processing 88 (3), 55–58.

Yang, C.C., Kaplan, A., Huang, Z., March . Cost-effective Design Reduces C_2 and C_3 at LNG Receiving Terminals, paper presented at the 2003 AIChE Spring National Meeting, New Orleans, LA, USA.

Yates, D., Oct. 13–16, 2002. Thermal Efficiency-Design, Lifecycle, and Environmental Considerations in LNG Plant Design, paper presented at the Gastech 2002 Conference and Exhibition, Doha, Qatar.

Younger, A.H., April 2004. Natural Gas Processing Principles and Technology. University of Calgary, Calgary, AB, Canada.

CHAPTER 4

Energy and Exergy Analyses of Natural Gas Liquefaction Cycles

4.1 Introduction

One of the most important sections in the LNG supply chain is the energy consumption in natural gas liquefaction utilizing the energy-intensive refrigeration cycles. Therefore, selection and development of efficient refrigeration cycles to liquefy natural gas offer potential energy savings and cost benefits. The cycle selection can best be evaluated by energy and exergy analyses.

The first law of thermodynamics is the theoretical basis of energy analysis. It is simply an energy conservation principle, asserting that energy is a thermodynamic property and that during a reaction, energy can change from one form to another but total amount of energy remains constant. An energy analysis of an energy-conversion system is essentially an accounting of the energies entering and exiting. The exiting energy can be broken down into products and wastes (for refrigeration cycle, they are typically refrigeration effect and heat and/or mechanical losses). Efficiencies often are evaluated as ratios of product energy to the energy quantities in the feed streams, and are often used to assess and compare various systems. However, energy analysis is often insufficient to evaluate system performance in that it cannot quantify the performance of a system compared to an ideal cycle (Carnot or reverse Carnot cycle). Further, energy analysis only takes into account the quantity of the thermodynamic losses that occur within a system, rather than the quality of the energy (potential to produce work).

Exergy analysis permits many of the shortcomings of energy analysis to be overcome. Exergy analysis is based on the second law of thermodynamics, and is useful in identifying the causes, locations, and magnitudes of process inefficiencies. Exergy (or availability) is a thermodynamic property that represents the maximum work that can be done with a given fluid stream in a reversible process.[1] Thus, the exergy associated with an energy quantity is a quantitative assessment of its usefulness or quality.

As an energy-conversion system, the real liquefaction/refrigeration cycle includes various irreversible processes. It is of great importance to analyze work losses and thermodynamic efficiency at low temperatures to avoid excessive power consumption. Therefore, the energy and exergy analyses

[1]Exergy is defined only in terms of enthalpy and entropy of the system that is brought to equilibrium with its surroundings in a hypothetical reversible process. The change of this property states the maximum amount of useful work that can be produced (or the minimum work that needs to be supplied) as a system undergoes a process change between the specified initial and final states.

provide a very important criterion to evaluate the thermodynamic performance of a liquefaction/refrigeration system. By analyzing the irreversibility of the system, it can be known how the actual cycle deviates from the ideal cycle. Exergy analysis is usually aimed at determining the maximum performance of the system and identifying the locations of exergy destruction and to show the direction for potential improvements. An important object of exergy analysis for systems that consume work such as liquefaction of gases is finding the minimum work required for a certain desired result and comparing it to the actual work consumption. The ratio of these two quantities is often considered the exergy efficiency of such a liquefaction process. In recent decades, exergy analysis has been accepted as an alternative tool to traditional energy analyses for evaluation of cryogenic cycle optimization.[2]

An exergy analysis of a complex system can be performed by analyzing the components of the system separately. The general principles and methodologies of exergy analysis have been published in various textbooks and are not repeated in this chapter. In this chapter, theoretical and practical aspects of energy and exergy analyses that are most relevant to refrigeration and natural gas liquefaction cycles are described. The first section reviews fundamental principles of refrigeration and liquefaction cycles. The second section summarizes the category of various refrigerants and cryogenic fluids. The third section reviews relevant theories of energy and exergy modeling and analysis. The last section provides the case analysis of some natural gas liquefaction cycles using the methods and theories introduced in the previous sections.

4.2 Refrigeration/liquefaction cycle principles

4.2.1 Constant-temperature refrigeration cycle

A refrigerator is a device used to transfer heat from a low- to a high-temperature medium. They are cyclic devices. Figure 4-1a shows the schematic of a vapor-compression refrigeration cycle (the most common type). The working fluid (called refrigerant) absorbs heat (Q_L) from a low-temperature medium (at T_L) in the evaporator. Power (W) is added in a compressor to compress the refrigerant to the condensing pressure. The high-temperature refrigerant cools into the liquid phase by rejecting the heat (Q_H) to a high-temperature medium (at T_H) in the condenser. The refrigerant in the liquid phase enters the expansion valve and is expanded to give a low temperature and pressure two-phase mixture at the evaporator inlet. The cycle is demonstrated in a simplified form in Figure 4-1b.

The Carnot cycle is a theoretical model that is useful for understanding a refrigeration cycle. As known from thermodynamics, the Carnot cycle is a model cycle for a heat engine where the addition of heat energy to the engine produces work. Conventionally, the Carnot refrigeration cycle is known as the reversed Carnot cycle (Figure 4-2). The maximum theoretical performance can be calculated, establishing criteria against which real refrigeration cycles can be compared.

The following processes take place in the Carnot refrigeration cycle as shown on a temperature-entropy (T-s) diagram in Figure 4-3.

1. ($1 \rightarrow 2$) is the ideal compression at constant entropy, and work input is added. The temperature of the refrigerant increases.
2. ($2 \rightarrow 3$) is the rejection of heat to the surrounding at a temperature T_H.

[2]In general, cycle work consumption and/or exergy efficiency are chosen as the objective of optimization.

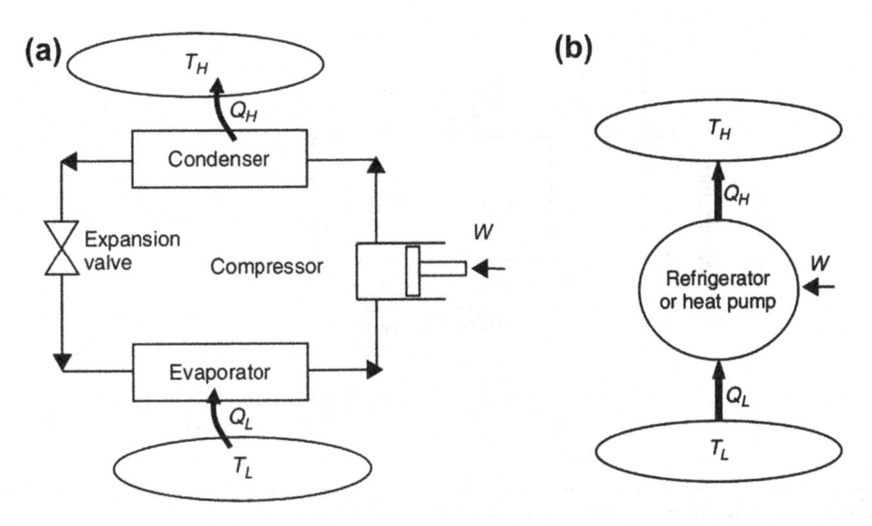

FIGURE 4-1

(a) The vapor compression refrigeration cycle. (b) Simplified schematic of refrigeration cycle.

3. (3→4) is the ideal expansion at constant entropy. The temperature of the refrigerant decreases.
4. (4→1) is the absorption of heat from the heat source at a constant evaporation temperature T_L.

The heat transfer between the refrigerator and the heat source/sink is assumed to occur at a zero temperature difference in all reversible refrigerators. The condensing and evaporating temperatures of the refrigerant are therefore the same as that of the ambient (T_H) during the heat rejection process and that of the load (T_L) during the heat absorption process, respectively.

The refrigeration effect Q_L is represented as shown here:

$$Q_L = T_L(S_1 - S_4) \tag{4-1}$$

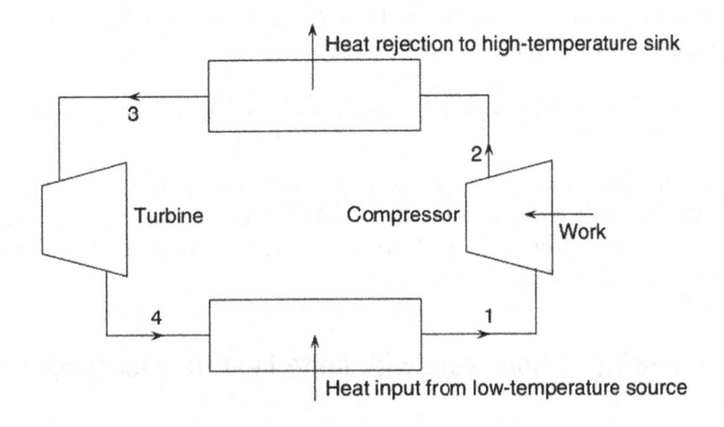

FIGURE 4-2

The reversed Carnot refrigeration cycle.

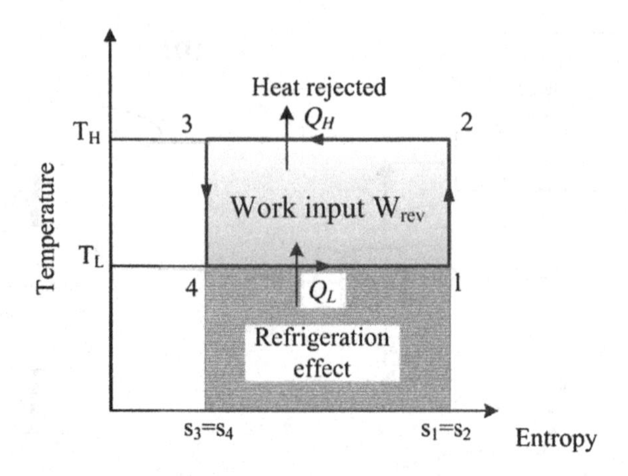

FIGURE 4-3

Temperature-entropy diagram of the reversed Carnot refrigeration cycle.

Based on the first law of thermodynamics, the theoretical work input (e.g., compressor work) for the reverse cycle (W_{rev}) is represented as the area within the cycle line 1-2-3-4-1, as follows:

$$W_{rev} = (T_H - T_L)(S_1 - S_4) \tag{4-2}$$

The coefficient of performance (*COP*) of any refrigerator is defined as the ratio of heat absorbed at a low temperature to compressor work input:

$$COP = \frac{Q_L}{W_{rev}} = \frac{T_L}{T_H - T_L} \tag{4-3}$$

This relation indicates that *COP* for the reverse cycle depends only on the source/sink temperatures (T_H and T_L), not on the thermal-physical properties of the working fluids. In order to increase the cycle *COP*, T_H should be decreased but T_L should be increased as much as possible. Besides, variation of heat sink temperature T_L has more influence on the cycle *COP* than variation of heat source temperature T_H. Therefore, the derivative of Equation 4-4 should satisfy with the following inequality equation:

$$\left| \left(\frac{\partial COP}{\partial T_L} \right)_{T_H} \right| > \left| \left(\frac{\partial COP}{\partial T_H} \right)_{T_L} \right| \tag{4-4}$$

The Ericsson and Stirling cycles are principally of theoretical interest as examples of cycles that exhibit the same thermal efficiency as the reverse Carnot cycle. As shown in Figure 4-4, they respectively have two constant-pressure and constant-volume processes, replacing the constant-entropy processes in the reverse Carnot cycle.

4.2.2 Irreversible refrigeration cycle with finite heat transfer temperature differences

The analysis in Section 4.2.1 assumes zero temperature difference between the working fluid (refrigerant) and ambient/load at heat rejection and absorption processes, respectively. However, such zero

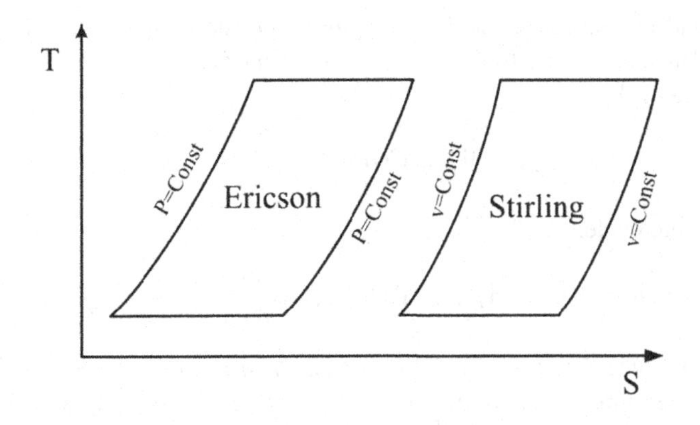

FIGURE 4-4

Temperature-entropy (T-s) diagrams of Ericson and Stirling cycles.

heat transfer temperature difference requires an infinitely large area of the heat exchanger, which is impossible for practical applications. In reality, in the heat rejection process the temperature of the working fluid (T_H') is higher than the ambient temperature (T_H), and in the heat absorption process the working fluid temperature (T_L') is lower than the load temperature (T_L).When T_H and T_L are fixed, the reverse refrigeration cycle with finite heat transfer temperature difference is denoted as 1′-2′-3′-4′-1′ in Figure 4-5.

Since heat transfer through a finite temperature difference is an irreversible process, this cycle with two constant-entropy and two heat transfer processes is a reverse Carnot cycle with external

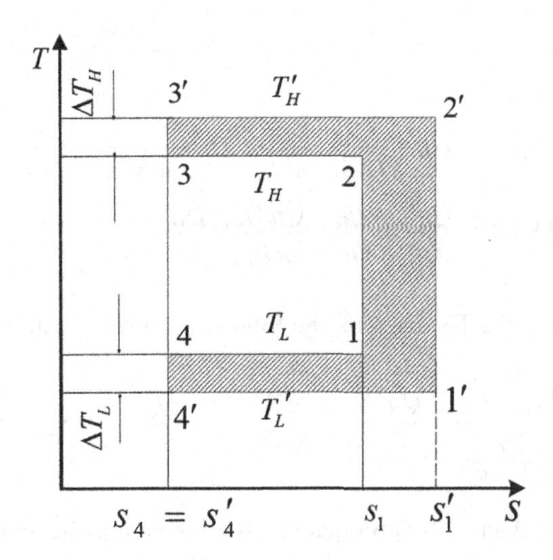

FIGURE 4-5

The reversed Carnot cycle with finite heat transfer temperature difference.

irreversibility. Q_H' and Q_L' represent the heat rejection and the cooling capacity in the irreversible cycle, respectively. Thus, according to the first law of thermodynamics:
The internal reversible cycle:

$$Q_H = Q_L + W_{rev}, \; COP_{rev} = \frac{Q_L}{W_{rev}} = \frac{T_L}{T_H - T_L}$$

The external irreversible cycle:

$$Q_H' = Q_L' + W_{irrev}, \quad COP_{irrev} = \frac{Q_L'}{W_{irrev}} = \frac{T_L'}{T_H' - T_L'}$$

Due to finite heat transfer temperature differences $\Delta T_H = T_H' - T_H$ and $\Delta T_L = T_L - T_L'$, the range of the cycle temperature is increased; that is, $(T_H' - T_L') > (T_H - T_L)$ and $T_L' < T_L$. Compared with the reversible Carnot cycle 1-2-3-4-1, the coefficient of performance for the cycle with heat transfer temperature difference is lower than that with zero temperature difference; that is, $COP_{irrev} < COP_{rev}$. If two cycles produce the same refrigeration capacity $Q_L = Q_L'$, the additional work required by the irreversible cycle (1′-2′-3′-4′-1′) can be expressed as

$$\Delta W = W_{irrev} - W_{rev} = Q_H' - Q_H \tag{4-5}$$

This indicates that the value of the additional work input (ΔW) is equal to the difference of heat rejection to the ambient in the irreversible cycle and in the reversible cycle. In Figure 4-5, ΔW can also be represented as the shadow area difference within two cycle lines. Entropy generation in this irreversible cycle is due to heat transfer at $\Delta T_H > 0$ and $\Delta T_L > 0$. Thus, the total entropy generation can be expressed as

$$\Delta S_{irrev} = Q_H'\left(\frac{1}{T_H} - \frac{1}{T_H'}\right) + Q_L'\left(\frac{1}{T_L'} - \frac{1}{T_L}\right) \tag{4-6}$$

Since in the reversible cycle the entropy generation $\Delta S_{rev} = -\frac{Q_H}{T_H} + \frac{Q_L}{T_L} = 0$, Equation 4-6 can be simplified to

$$\Delta S_{irrev} = Q_H'\left(\frac{1}{T_H} - \frac{1}{T_H'}\right) + Q_H\frac{T_L}{T_H}\left(\frac{1}{T_L'} - \frac{1}{T_L}\right) \tag{4-7}$$

$$\Delta S_{irrev} = \frac{Q_H'}{T_H} - \frac{Q_H}{T_H} + \frac{Q_H T_L}{T_H T_L'} - \frac{Q_H'}{T_H'} \tag{4-8}$$

Using the relationships in the T-s diagram, the following equation can be derived:

$$\frac{Q_H T_L}{T_H T_L'} = \frac{Q_H \frac{Q_L}{S_1 - S_4}}{T_H \frac{Q_L'}{S_1' - S_4'}} = \frac{Q_H(S_1' - S_4')}{T_H(S_1 - S_4)} = \frac{Q_H}{T_H(S_2 - S_3)}(S_2' - S_3') = \frac{Q_H'}{T_H'} \tag{4-9}$$

Substituting Equations 4-5 and 4-9 into Equation 4-8, the entropy generation can be expressed as

$$\Delta S_{irrev} = \frac{Q_H'}{T_H} - \frac{Q_H}{T_H} = \frac{1}{T_H}(Q_H' - Q_H) = \frac{1}{T_H}\Delta W \tag{4-10}$$

Thus,

$$\Delta W = T_H \Delta S_{irrev} \tag{4-11}$$

Equation 4-11 demonstrates that to obtain the same amount of cooling capacity, additional work input for the irreversible refrigeration cycle is equal to the ambient temperature (T_H) multiplied by the entropy generation in the system (ΔS_{irrev}).

4.2.3 Refrigeration cycle between two varying temperatures

In a real refrigeration cycle, the temperatures of the low temperature medium (heat source) and the high temperature medium (heat sink) are usually varying. In this case, we can assume that the heat source is cooled down from T_a to T_d and the heat sink is heated up from T_c to T_b, as shown in Figure 4-6a. Line *a-d* and *c-b* represent thermodynamic processes of the heat source and the heat sink, respectively; area within the line *a′-a-d-d′-a′* represents the cooling capacity Q_L. Under conditions just given, in order to consume the minimum input work, cycle *a-b-c-d-a* is assumed to be a reversible refrigeration cycle. Heat transfer temperature differences between the refrigerant and the heat source/sink are assumed to be infinitely small. Therefore, thermal state lines of the refrigerant *d-a* and *b-c* represent the heat absorption process from the low temperature medium (heat source) and the heat rejection process to the high temperature medium (heat sink), respectively; arrows indicate the direction of the reversible refrigeration cycle. Process *a-b* and *c-d* are the reversible adiabatic compression and expansion. It can be seen that the cycle *a-b-c-d-a* is a reversible cycle without work loss due to the entropy generation. Therefore, the work input for this cycle (W_0) is minimum, and $COP = \frac{Q_L}{W_0}$.

For the reversible Carnot cycle with varying temperatures, its *COP* can be presented based on average equivalent temperatures. As shown in Figure 4-6b, we can use a series of adiabatic lines to divide the Carnot cycle *a-b-c-d-a* into infinite numbers of element Carnot cycles. For each element

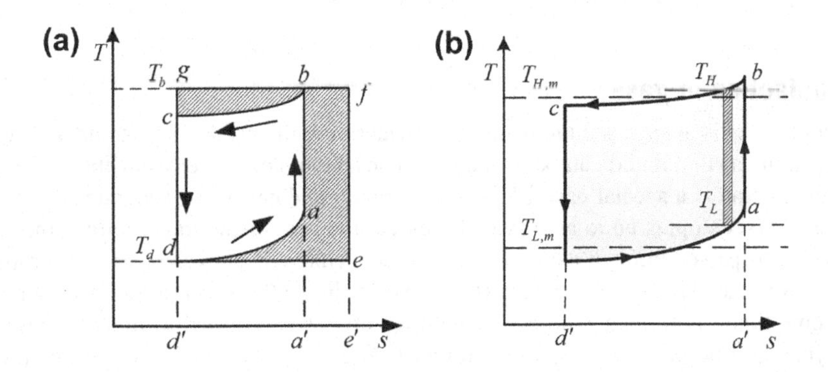

FIGURE 4-6

(a) The reversed Carnot cycle between heat source and heat sink with varying temperatures. (b) The reversed Carnot cycle based on average equivalent temperatures.

cycle, heat source temperature $T_{H,i}$ and heat sink temperature $T_{L,i}$ can be assumed to be constant. Thus, its COP_i is

$$COP_i = \frac{dq_L}{dq_H - dq_L} = \frac{T_{L,i}}{T_{H,i} - T_{L,i}} \tag{4-12}$$

For the entire Carnot cycle, COP can be expressed as

$$COP = \frac{Q_L}{Q_H - Q_L} = \frac{\int_d^a dQ_L}{\int_c^b dQ_H - \int_d^a dQ_L} = \frac{\int_d^a T_L ds}{\int_c^b T_H ds - \int_d^a T_L ds} \tag{4-13}$$

The heat exchange can be expressed as

$$Q_L = \int_d^a T_L ds = T_{L,m}(s_a - s_d) \tag{4-14a}$$

$$Q_H = \int_d^a T_H ds = T_{H,m}(s_a - s_d) \tag{4-14b}$$

where $T_{L,m}$ is defined as the average equivalent temperature of working fluid absorbing heat from the heat source, which indicates that the amount of heat absorption at the constant $T_{L,m}$ is equal to that at varied temperatures in process *d-a*. Similarly, $T_{H,m}$ is the average equivalent heat rejection temperature.

Substituting Equations 4-14a and 4-14b into Equation 4-13, COP for the reversible Carnot cycle at varied temperatures can be written as

$$COP = \frac{T_{L,m}}{T_{H,m} - T_{L,m}} \tag{4-15}$$

4.2.4 Liquefaction of gases

Liquefaction of gases is always accomplished by refrigerating the gas to some temperature below its critical temperature so that liquid can be formed at some suitable pressure below the critical pressure. Thus, gas liquefaction is a special case of gas refrigeration and cannot be separated from it. In both cases, the gas is first compressed to an elevated pressure in an ambient temperature compressor. This high pressure gas is passed through a countercurrent recuperative heat exchanger to a throttling valve or expansion engine, as shown in Figure 4-7. Before expanding to the lower pressure, cooling may take place, and some liquid is formed. The cool, low pressure gas is returned to the compressor inlet to repeat the cycle. The purpose of the countercurrent heat exchanger is to warm the low-pressure gas prior to recompression and simultaneously to cool the high-pressure gas to the lowest temperature possible prior to expansion. Both refrigerators and liquefiers operate on this basic principle (Flynn, 2004).

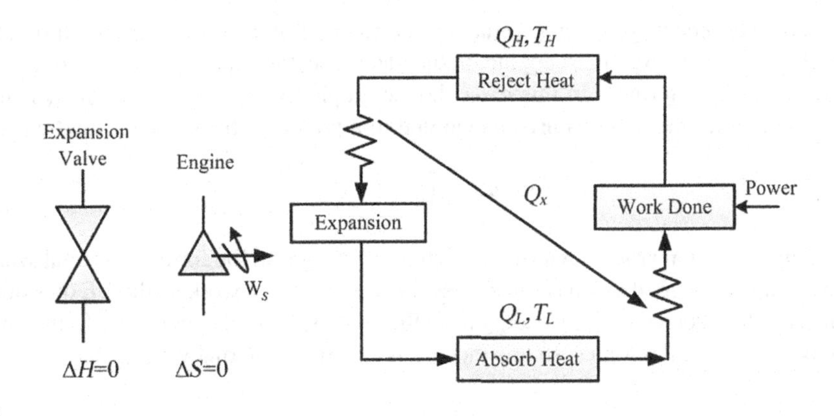

FIGURE 4-7

Closed cycle cryogenic refrigerator.

(Flynn, 2004)

There is nonetheless an important distinction between refrigerators and liquefiers. In a continuous refrigeration process, there is no accumulation of refrigerant in any part of the system. This contrasts with a gas liquefying system, where liquid accumulates and is withdrawn. Thus, in a liquefying system, the total mass of the gas that is warmed in the counter-current heat exchanger is less than that of the gas to be cooled by the amount liquefied, creating an imbalance of flow in the heat exchanger. In a refrigerator, the warm and cool gas flows are equal in the heat exchanger. This results in what is usually referred to as a "balanced flow condition" in a refrigerator heat exchanger. The thermodynamic principles of refrigeration and liquefaction are identical. However, the analysis and design of the two systems are quite different because of the condition of balanced flow in the refrigerator and unbalanced flow in liquefier systems.

The prerequisite refrigeration for gas liquefaction is accomplished in a thermodynamic process when the process gas absorbs heat at temperatures below that of the environment. As mentioned, a process for producing refrigeration at liquefied gas temperatures always involves some equipment at ambient temperature in which the gas is compressed and heat is rejected to a coolant. During the ambient temperature compression and cooling process, the enthalpy and entropy, but not usually the temperature of the gas, are decreased.

The reduction in temperature of the gas usually is accomplished by recuperative heat exchange between the cooling and warming gas streams and further by an expansion of the high-pressure stream. One method of producing low temperatures is the isenthalpic expansion through a throttling device. The effect has come to be known as the Joule-Thomson effect, and the Joule-Thomson coefficient (μ_{J-T}) of a gas is defined as follows:

$$\mu_{J-T} = \left(\frac{\partial T}{\partial P}\right)_H \tag{4-16}$$

The Joule-Thomson coefficient is a property of each specific gas, a function of temperature and pressure, and may be positive, negative, or zero.

Another method of reducing gas temperatures is the adiabatic expansion of the gas through a work-producing device such as an expansion engine. In the ideal case, the expansion would be reversible and adiabatic and therefore isentropic. In this case, the isentropic expansion coefficient μ_s is defined as follows, which expresses the temperature change due to a pressure change at constant entropy.

$$\mu_S = \left(\frac{\partial T}{\partial P}\right)_S \tag{4-17}$$

The isentropic expansion process removes energy from the gas in the form of external work, so this method of low-temperature production is sometimes called the external work method. Expansion through an expansion valve does not remove any energy from the gas but moves the molecules farther apart under the influence of intermolecular forces, so this method is called the internal work method.

4.2.5 Ideal Linde-Hampson liquefiers

Figure 4-8 shows an ideal Linde-Hampson liquefier, originally invented by Carl von Linde and William Hampson independently in 1895 to liquefy air, and the corresponding temperature-entropy (T-s) diagram. In an ideal liquefaction system, the gas to be liquefied is compressed in a process of multistage adiabatic compression and multistage isobaric cooling. Therefore, this process is regarded as isothermal compression at ambient temperature (1-2). The high-pressure gas is cooled in the heat exchanger (2-3) and then expanded isenthalpic in a throttling device (3-4). The liquid (6) and gaseous phases (5) are separated in the phase separator, and the unliquefied gas is used to cool the high-pressure warm stream in the heat exchanger. The temperature of the low-pressure return gas at the exit of the heat exchanger depends on the effectiveness of the heat exchanger used. In the case of an ideal heat exchanger with a heat exchanger effectiveness of 100%, $T_7 = T_1$.

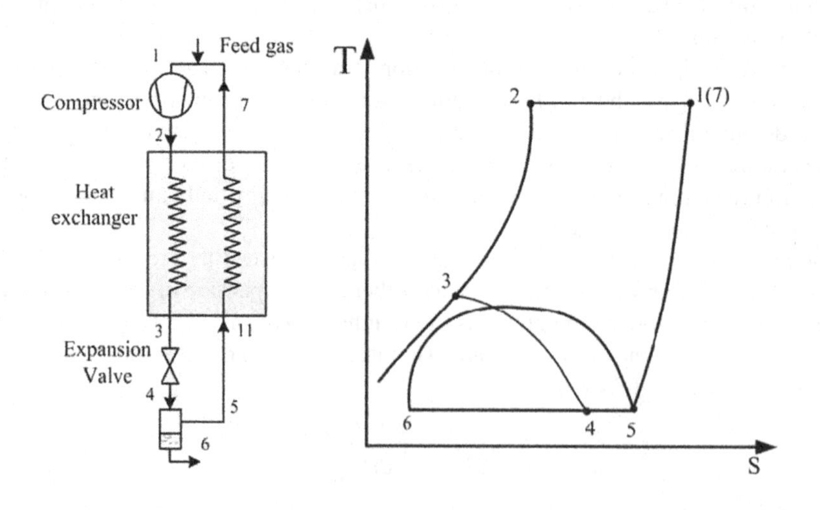

FIGURE 4-8

Ideal Linde-Hampson liquefaction process.

In order to quantify the fraction of the gas that gets liquefied (the liquid yield (y) in an ideal Linde-Hampson process), y kg gas is liquefied from 1 kg feed gas after compression, cooling, and expansion. Taking a control volume that includes the heat exchanger, the expansion valve, and the phase separator, the energy conservation gives

$$1 \cdot h_2 = yh_6 + (1 - y)h_1 \tag{4-18}$$

Therefore, the liquid yield y can be expressed as

$$y = \frac{h_1 - h_2}{h_1 - h_6} \tag{4-19}$$

where h_1, h_2, and h_6 represent enthalpy at corresponding state points of 1, 2, and 6.

4.2.6 Ideal liquefiers with expanders

Figure 4-9 shows an ideal expander liquefier and the corresponding temperature-entropy diagram. A large part of the high pressure gas is diverted to an expander and undergoes a large temperature drop. This cold gas is used to cool and condense the high-pressure gas in the second heat exchanger (HX-2). The refrigeration obtained in the expander also helps in precooling the high-pressure gas in the first heat exchanger (HX-1) before it enters the expander and the second heat exchanger (HX-2). Similarly, the expression for liquid yield (y) in this liquefier is obtained by an energy balance over a control volume that excludes the compressor as follows:

$$y = \frac{h_1 - h_2}{h_1 - h_7} + \frac{m(h_9 - h_{10})}{h_1 - h_7} \tag{4-20}$$

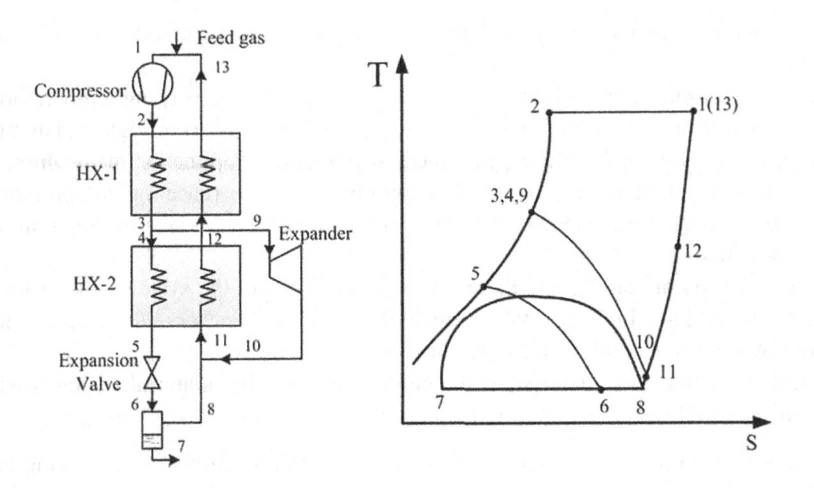

FIGURE 4-9

Ideal expander liquefaction process.

where h_1, h_2, h_7, h_9, and h_{10} represent enthalpy at corresponding state points of 1, 2, 7, 9, and 10; m is the fraction of gas passing through the expander.

4.3 Refrigerant selections

Applications requiring refrigeration are extremely diverse and broad-based. They span a broad range of temperatures ranging from near absolute temperature (liquid helium) of 4 K to near ambient temperature (human comfort) of 300 K. Apart from the refrigeration temperature, the refrigeration capacity requirement also plays a significant role in influencing the refrigeration method and refrigerants that can be employed. Practices in the industry also show that for any given application, defined by its refrigeration temperature and capacity requirements, a number of different refrigeration methods and corresponding refrigerants have been implemented. In general, refrigerants are classified into two categories. The first group includes refrigerants working at near-ambient temperatures. They are used in refrigerators operating as a closed loop. There is no accumulation or withdrawal of product in a refrigerator. Typical refrigerants are inorganic compounds, halocarbon compounds, and hydrocarbons. The common feature of this type of refrigerants is that they have relatively high critical temperatures and can be liquefied near ambient or medium cold temperatures. The other type is called cryogenic fluids—cryogenic gases including methane, air, oxygen, nitrogen, and so on. They have very low boiling temperatures, usually below 120 K. Cryogenic gases are used in open liquefiers where liquid products are removed and equivalent make-up stream must be added.

4.3.1 Basic requirements on refrigerants

The basic requirements on the common refrigerants used for room or low temperature refrigeration are the following:

1. Critical temperatures are not too low such that refrigerants can be liquefied at ambient or medium low temperatures.
2. It should have appropriate saturation pressures at the range of the working temperatures of the refrigerator. In general, the evaporating pressure is preferred close to or higher than atmosphere pressure, preventing the air infiltrating into the low pressure components; meanwhile, the condensing pressure should not be too high, otherwise it may increase the compressor work.
3. The volumetric refrigerating capacity is high, which will reduce the size of the compressor and refrigerant flow rate.
4. Viscosity and density are small, and therefore pressure drops in the system will be low.
5. Refrigerants should have high thermal conductivity, which can increase the effectiveness of the evaporator and condenser and decrease the heat exchanger area.
6. With respect to chemical characteristics, the nontoxicity, suitable material of construction, and compatibility with lubricant are preferred.

Since the cryogenic fluids are mainly used for cryogenic applications at a working temperature below 120 K, in order to reach such extremely low working temperatures the working fluids should have a very low normal boiling temperature (below 120 K) and triple point. Any fluids with such characteristics potentially could be used as cryogenic refrigerants.

4.3.2 Type of refrigerants for refrigeration and liquefaction

The typical refrigerants used for refrigeration and liquefaction cycles are categorized, based on their own chemical compositions, as the following:

- **Halocarbon compounds.** The halocarbon compounds group includes refrigerants that contain one or more of the three halogens chlorine, fluorine, and bromine.
- **Inorganic compounds.** Many of the early refrigerants were inorganic compounds and some have maintained their prominence to this day, such as ammonia and carbon dioxide.
- **Hydrocarbons.** Many hydrocarbons are suitable as refrigerants, especially for service in the petroleum and petrochemical industry.
- **Cryogenic gases.** Many gases and their mixtures are used as cryogenic fluids such as nitrogen, oxygen, air, methane, helium, etc.

4.3.3 Type of refrigerant mixtures

Refrigerant mixtures can be broadly classified into two groups based on the temperature change during the phase change process (evaporation and condensation): (1) zeotropic mixtures and (2) azeotropic mixtures. A mixture of chemicals is zeotropic if the compositions of the vapor and the liquid phases at the vapor-liquid equilibrium state are never the same. Dew point and bubble point curves do not touch each other over the entire composition range with the exception of the pure components. An azeotropic mixture of two substances is one that cannot be separated into its components by distillation. An azeotrope evaporates and condensates as a single substance with properties that are different from those of either constituent. Typical advantages of using zeotropic mixtures for refrigeration include (1) reducing compression ratio; (2) increasing the capacity of the refrigerator; (3) reaching nonconstant temperature refrigerating (refrigerant temperature decreases at the condensation and increases at the evaporation), consequently reducing the compressor work and increasing COP. With respect to azeotropic, its benefit is that at the same operating conditions, evaporating temperature is lowered compared to using either constituent. Thus, the refrigerating capacity will be increased and compressor discharge temperature will be reduced.

Figure 4-10 shows the relationship between the dew and bubble point temperatures of a typical zeotropic mixture of nitrogen and methane at a pressure of 0.1 MPa. Consider four different states, a (107.5 K), b (103.87 K), c (81.49 K), and d (77.5 K) at a nitrogen mole fraction of 0.5. The mixture is in a superheated vapor state at a, saturated vapor state at b, saturated liquid state at c, and subcooled liquid state at d. The temperature of the mixture at saturated vapor state b and at saturated liquid state c is called the dew and bubble point temperature, respectively. The line passing through the dew points is called the dew line, and that through the bubble points is called the bubble line. The equilibrium composition of vapor and liquid will be different in the two-phase region. For example, the mole fraction of vapor in equilibrium with liquid at state c will be greater than 0.5 (state f), whereas the mole fraction of liquid in equilibrium with vapor at state b would be less than 0.5 (state e).

Figure 4-11 shows the typical variation of the dew and bubble point temperature of an azeotropic mixture of refrigerants R22 ($CHClF_2$) and R115 ($CClF_2CF_3$). The temperature glide becomes zero when the mole fraction of R22 in the mixture is 0.488. The mole fraction of the vapor and liquid phases is also the same at this mole fraction. Azeotropic mixtures are widely used for constant-temperature refrigeration. However, they are not suitable for the process described in this monograph (except as a precooling refrigerant in some cases).

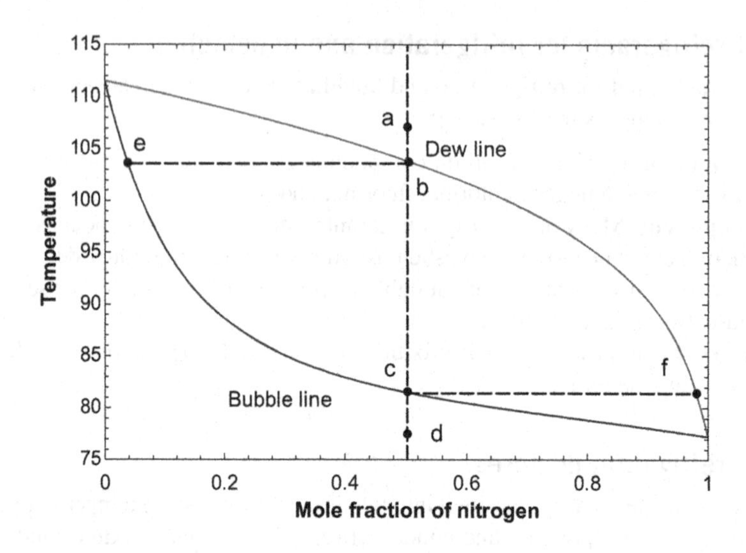

FIGURE 4-10

Zeotropic mixture of nitrogen and methane at 0.1 MPa.

4.3.4 Choice of refrigerant mixture

The energy and exergy efficiency of any mixed refrigerant cycle (MRC) depends on the mixture's constituents and their concentration. The exergy efficiency of MRC refrigerators will be high when a second liquid phase occurs in the evaporator. Liquid-liquid immiscibility is observed at low temperatures in multicomponent mixtures of nitrogen-hydrocarbon, fluorocarbon-hydrocarbon,

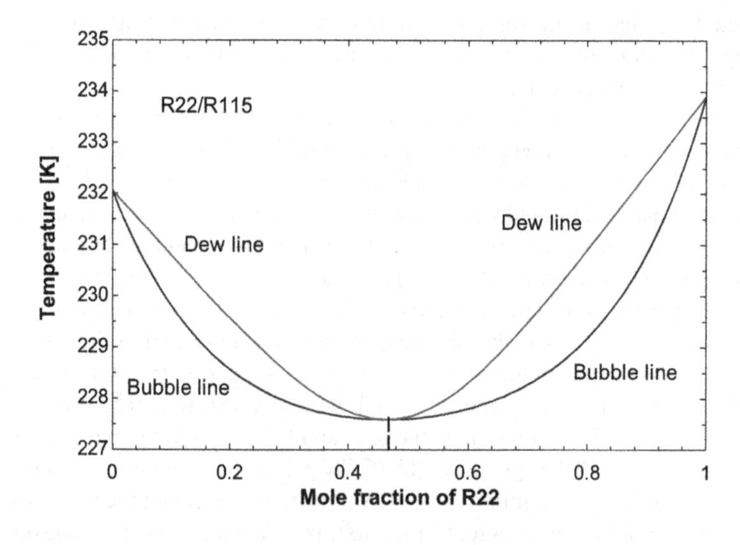

FIGURE 4-11

Mixture of R22 and R115 at 0.1 MPa exhibiting an azeotropic behavior.

fluorocarbon-hydrochlorofluorocarbon and fluorocarbon-hydrofluorocarbon refrigerants. This immiscibility can be exploited to obtain a near-constant temperature evaporation with binary or multicomponent mixtures. The method for determining the most basic components of a nitrogen-hydrocarbon mixture was first given by Alfeev et al. (1973) in their patent. These principles later were extended to other fluid mixtures by Venkatarathnam (2008). The proposed guidelines for choosing the components of a mixture are as follows:

1. Choose a first fluid whose boiling point temperature at 1.5 bar is less than the desired refrigerating temperature. For example, nitrogen can be used for temperatures between 80 K and 105 K, tetrafluoromethane (Refrigerant R14) between 150 K and 180 K.
2. Choose a second fluid whose boiling point is about 30 to 60 K above that of the basic fluid and that does not exhibit liquid-liquid immiscibility at low temperature with the primary fluid. For example, we can choose methane with argon and nitrogen, trifluoromethane (R23) with tetrafluoromethane (R14), etc.
3. Choose a third fluid that exhibits a liquid-liquid immiscibility at low temperature with the first fluid and whose boiling point is at least 30 K above that of the second fluid, for instance ethane, ethylene, etc., which exhibit liquid-liquid immiscibility at low temperatures with nitrogen. Ethylene also exhibits liquid-liquid immiscibility at low temperatures with argon. Propane, butanes, and chlorodifluoromethne (R22) exhibit liquid-liquid immiscibility with R14.
4. Choose a fourth and an optional fifth fluid that exhibit liquid-liquid immiscibility at low temperatures with the first fluid.

4.4 Fundamentals of energy and exergy analysis

This section introduces the thermodynamic fundamentals for energy and exergy analysis of typical components and overall cryogenic cycles.

4.4.1 First law and second law of thermodynamics

The first law of thermodynamics can be defined as the law of conservation of energy, and it states that energy can be neither created nor destroyed. It can be expressed for a general system since the net change in the total energy of a system during a process is equal to the difference between the total energy entering and leaving the system. Let us consider a control volume (CV) involving a steady-flow process. Mass is entering and leaving the system and there are heat and work interactions with the surroundings as shown in Figure 4-12.

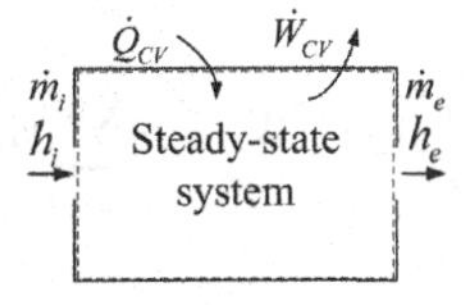

FIGURE 4-12

A general steady-flow control volume with mass, heat, and work interactions.

During a steady-flow process, the total energy content of the control volume remains constant, and thus the total energy change of the system is zero. If the change in kinetic and potential energies are ignored, then the first law of thermodynamics can be expressed as

$$\dot{Q}_{CV} - \dot{W}_{CV} + \sum_i \dot{m}_i h_i - \sum_e \dot{m}_e h_e = 0 \tag{4-21}$$

where i denotes inlets and e denotes exits, $\dot{Q}_{CV}$ and $\dot{W}_{CV}$ denote rate of energy transfer to the control volume as heat and rate of work done by the control volume, $\dot{m}$ and h represent fluid mass flow rate across the boundary of the control volume and the corresponding specific enthalpy associated with the fluid flow.

However, the energy conservation idea alone is inadequate for depicting some important aspects of resource utilization. In principle, work can be developed as the systems are allowed to come into equilibrium. When one of the two systems is a suitably idealized system referred to as an exergy reference environment or an environment, and the other is some system of interest, exergy is the maximum theoretical work obtainable as they interact to equilibrium. When mass flow is across the boundary of a control volume, there is an exergy transfer accompanying mass flow. Additionally, there is an exergy transfer accompanying flow work. The specific flow exergy accounts for both of these, and is given by

$$e_f = h - h_0 - T_0(s - s_0) + \frac{V^2}{2} + gz \tag{4-22}$$

where h and s denote specific enthalpy and specific entropy, respectively; $\frac{V^2}{2}$ and gz represent the specific kinetic and potential energy, respectively; T_0 represents the temperature at the environment; and h_0 and s_0 represent the respective values of these properties when evaluated at the environment. Similarly, the exergy rate balance for a control volume in steady-state flow can be expressed as

$$\sum_j \left(1 - \frac{T_0}{T_j}\right)\dot{Q}_j - \dot{W}_{CV} + \sum_i \dot{m}_i e_{fi} - \sum_e \dot{m}_e e_{fe} - \dot{E}_d = 0 \tag{4-23a}$$

In Equation 4-23a, the term $\dot{Q}_j$ represents the time rate of heat transfer at the location on the boundary where the instantaneous temperature is T_j. The accompanying exergy transfer rate is given by$(1 - T_0/T_j)\dot{Q}_j$. The term $\dot{W}_{CV}$ represents the time rate of energy transfer rate by work. The terms $\dot{m}_i e_{fi}$ and $\dot{m}_e e_{fe}$ account for the time rate of exergy transfer accompanying mass flow and flow work at inlet i and exit e, respectively. Finally, the term $\dot{E}_d$ accounts for the time rate of exergy destruction due to irreversibilities within the control volume. This equation indicates that the rate at which exergy is transferred into the control volume must exceed the rate at which exergy is transferred out, the difference being the rate at which exergy is destroyed within the control volume due to irreversibilities. The exergy loss $\Delta\dot{E}(= \dot{E}_d)$ can be expressed as:

$$\Delta\dot{E} = \sum_j \left(1 - \frac{T_0}{T_j}\right)\dot{Q}_j - \dot{W}_{CV} + \sum_i \dot{m}_i e_{fi} - \sum_e \dot{m}_e e_{fe} \tag{4-23b}$$

4.4.2 Exergy analysis of different components of a cryogenic liquefaction system

Typical components in a cryogenic liquefaction system include multistream heat exchangers, compressors, condensers, or aftercoolers exchanging heat with ambient or water, mixers, and expansion devices, as shown in Figure 4-13.

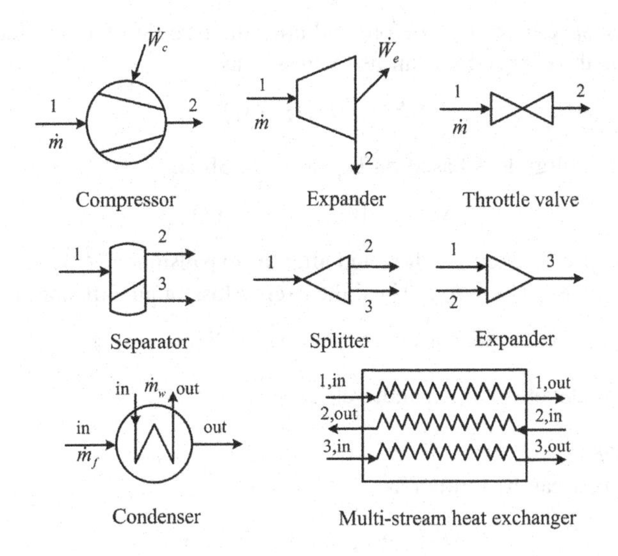

FIGURE 4-13

Different components in cryogenic systems.

Ignoring the kinetic and potential energy of the working fluid, the flow exergy e in a steady state flow can be expressed as

$$e = h - h_0 - T_0(s - s_0) \tag{4-24}$$

where h and s represent the specific enthalpy and entropy, respectively; h_0 and s_0 represent the respective values of these properties when evaluated at the environment; and T_0 is the environment temperature.

4.4.2.1 Compressor

According to Equation 4-23b, the exergy loss in a compressor can be written as

$$\Delta\dot{E} = -\dot{W}_c + \dot{m}(e_1 - e_2) \tag{4-25}$$

where e_1 and e_2 are the specific exergy before and after the compression; $\dot{m}$ is the working fluid mass flow rate; $\dot{W}_c$ is the compressor work, and $-\dot{W}_c = \frac{1}{\eta_c}(h_2 - h_1)\dot{m}$. The exergy loss can be expressed as

$$\Delta\dot{E} = \dot{m}((h_2 - h_1)/(1/\eta_c - 1) + T_0(s_2 - s_1)) \tag{4-26}$$

In Equation 4-26, η_c denotes the compressor efficiency.

4.4.2.2 Expander and throttle valve

The exergy loss in a throttle valve can be written as

$$\Delta\dot{E} = \dot{m}e_1 - \dot{m}e_2 \tag{4-27}$$

where e_1 and e_2 are the specific exergy before and after the throttle process. Because of isenthalpic throttling, $h_1 = h_2$, thus the exergy loss can be expressed as

$$\Delta\dot{E} = \dot{m}T_0(s_2 - s_1) \tag{4-28}$$

For an expander, the exergy loss based on Equation 2-23b is

$$\Delta\dot{E} = -\dot{W}_e + \dot{m}(e_1 - e_2) \tag{4-29}$$

where e_1 and e_2 are the specific exergy before and after the expansion; and $\dot{W}_e$ is the output work from the expander, and $-\dot{W}_e = \dot{m}\eta_e(h_2 - h_1)$. Thus, the exergy loss in an expander can be expressed as

$$\Delta\dot{E} = \dot{m}((h_1 - h_2)/(1 - \eta_{ex}) - T_0(s_1 - s_2)) \tag{4-30}$$

In Equation 4-30, η_{ex} denotes the expander efficiency.

4.4.2.3 Stream mixer

The exergy loss in a mixer can be written as

$$\Delta\dot{E} = \dot{m}_1e_1 + \dot{m}_2e_2 - \dot{m}_3e_3 \tag{4-31}$$

where e_1 and e_2 are the specific exergy of two steams before the mixer; e_3 is the specific exergy of mixed stream. Due to energy and mass conservations, $\dot{m}_3h_3 = \dot{m}_1h_1 + \dot{m}_2h_2$ and $\dot{m}_3 = \dot{m}_1 + \dot{m}_2$, the exergy loss can be expressed as

$$\Delta\dot{E} = \dot{m}_1T_0(s_3 - s_1) + \dot{m}_2T_0(s_3 - s_2) \tag{4-32}$$

4.4.2.4 Phase separator or stream splitter

The exergy loss in a phase separator or stream splitter can be written as

$$\Delta\dot{E} = \dot{m}_1e_1 - (\dot{m}_2e_2 + \dot{m}_3e_3) \tag{4-33}$$

where e_1, e_2, and e_3 are the specific exergy of main stream before the separator or splitter and two afterward streams. Similarly, the exergy loss can be expressed as

$$\Delta\dot{E} = T_0(\dot{m}_2s_2 + \dot{m}_3s_3 - \dot{m}_1s_1) \tag{4-34}$$

4.4.2.5 Multistream heat exchanger

The exergy loss in a multistream heat exchanger can be written as

$$\Delta\dot{E} = \sum_i \dot{m}_{i,in}e_{i,in} - \sum_i \dot{m}_{i,out}e_{i,out} \tag{4-35}$$

Based on the energy conservation equation $\sum_i \dot{m}_{i,in}h_{i,in} = \sum_i \dot{m}_{i,out}h_{i,out}$, the exergy loss can be expressed as

$$\Delta\dot{E} = \sum_i \dot{E}_{i,in} - \sum_i \dot{E}_{i,out} = T_0\left(\sum_i \dot{m}_{i,in}s_{i,out} - \sum_i \dot{m}_{i,out}s_{i,in}\right) \tag{4-36}$$

4.4.2.6 Water condenser or aftercooler

The exergy loss can be written as

$$\Delta \dot{E} = \dot{m}_f e_{f,in} + \dot{m}_w e_{w,in} - \left(\dot{m}_f e_{f,out} + \dot{m}_w e_{w,out}\right) \tag{4-37}$$

Substituting the energy conservation equation, the exergy loss can be expressed as

$$\Delta \dot{E} = \dot{m}_f T_0 \left(s_{f,out} - s_{f,in}\right) + T_0 \dot{m}_f c_{p,w} \ln \frac{T_{w,out}}{T_{w,in}} \tag{4-38}$$

In Equation 4-38, $s_{f,in}$ and $s_{f,out}$ are the specific entropy of working fluid at the inlet and outlet of the heat exchanger; $c_{p,w}$is the water specific heat (assumed to be constant); and $T_{w,in}$ and $T_{w,out}$ are water inlet and outlet temperatures, respectively.

4.4.3 Overall energy and exergy efficiency of a cryogenic liquefaction system

The exergy efficiency is introduced and used to assess the effectiveness of energy resource utilization in a thermal system. The exergy efficiency of any refrigeration or cryogenic liquefaction system is defined as follows:

$$\eta_{ex} = \frac{\textit{minimum power required by a reversible system}}{\textit{actual power supplied}} \tag{4-39}$$

or

$$\eta_{ex} = 1 - \frac{\sum \textit{exergy loss in each component}}{\textit{actual power supplied}} \tag{4-40}$$

For the processes or equipment (control volume) where there is no work transfer involved, the concept of exergy efficiency can also be determined. Instead, the actual power supplied is replaced by exergy expenditure as follows:

$$\eta_{ex} = 1 - \frac{\sum \textit{exergy loss in each component}}{\textit{exergy expenditure}} \tag{4-41}$$

or

$$\eta_{ex} = 1 - \frac{\sum \dot{m}_{in} e_{in} - \sum \dot{m}_{out} e_{out} + \sum \dot{Q}_i \left(1 - T_0/T_i\right)}{\textit{exergy expenditure}} \tag{4-42}$$

where $\dot{Q}_i$ and T_i denote the obtained refrigerating effect and the corresponding refrigerating temperature. T_0 denotes the temperature at the environment.

The exergy expenditure depends on the type of system or process. When a system receives heat and produces work as in a heat engine, the exergy expenditure of the system is $\dot{Q}_i(1 - T_0/T_i)$. When a system receives work and absorbs heat as in a refrigerator, the exergy expenditure of the system is $\dot{W}$.

4.4.4 Energy and exergy analysis of an ideal Linde-Hampson liquefaction cycle

A methodology for the first- and second-law-based performance analyses of the simple ideal Linde-Hampson cycle was reported by Kanoglu et al. (2008), and they investigated the effects of gas inlet and liquefaction temperatures on various cycle performance parameters. The ideal Linde-Hampson cycle and its temperature-entropy diagram are shown in Figure 4-8. Makeup gas is mixed with the uncondensed portion of the gas from the previous cycle, and the mixture at state 1 is compressed by an isothermal compressor to state 2. The temperature is kept constant by rejecting compression heat to a coolant. The high-pressure gas is further cooled in a regenerative counterflow heat exchanger by the uncondensed portion of gas from the previous cycle to state 3, and throttled to state 4, which is a saturated liquid–vapor mixture state. The liquid (state 6) is collected as the desired product, and the vapor (state 5) is routed through the heat exchanger to cool the high-pressure gas approaching the throttling valve. Finally, the gas is mixed with fresh makeup gas, and the cycle is repeated.

The refrigeration effect for this cycle may be defined as the heat removed from the makeup gas in order to turn it into a liquid at state 6. The fraction of liquefied gas (i.e., liquid yield) can be written as

$$y = \frac{h_1 - h_2}{h_1 - h_f} \tag{4-43}$$

where h_f is the specific enthalpy of saturated liquid that is withdrawn. From an energy balance on the cycle, the refrigeration effect per unit mass of the gas in the cycle may be expressed as

$$q_L = h_1 - h_2 \tag{4-44}$$

An energy balance on the heat exchanger gives

$$h_2 - h_3 = x(h_1 - h_5) \tag{4-45}$$

where x is the quality of the mixture at state 4. Then the fraction of the gas that is liquefied may also be related as

$$y = 1 - x \tag{4-46}$$

An energy balance on the compressor gives the work of compression per unit mass of the gas in the cycle as

$$w_{actual} = h_2 - h_1 - T_1(s_2 - s_1) \tag{4-47}$$

The coefficient of performance (COP) of this ideal cycle is then given by

$$COP = \frac{q_L}{w_{actual}} = \frac{h_1 - h_2}{h_2 - h_1 - T_1(s_2 - s_1)} \tag{4-48}$$

Engineers usually are interested in comparing the actual work used to obtain a unit mass of liquefied gas and the minimum work requirement to obtain the same output. Such a comparison may be performed using the second law. For instance, the minimum work input requirement (reversible work) and the actual work for a given set of processes may be related to each other by

$$w_{actual} = w_{rev} + T_0 s_{gen} = w_{rev} + ex_{dest} \tag{4-49}$$

where T_0 is the environment temperature; s_{gen} is the specific total entropy generation; and ex_{dest} is the specific total exergy destruction during the processes. The reversible work for the simple Linde–Hampson cycle may be expressed by the stream exergy difference of states 1 and 6 as

$$w_{rev} = y(ex_6 - ex_1) = y(h_6 - h_1 - T_0(s_6 - s_1)) \tag{4-50}$$

where state 1 has the properties of the makeup gas, which is essentially at the environment. This expression gives the minimum work requirement for a unit mass of the gas. The exergy efficiency can be defined as the reversible work input divided by the actual work input:

$$\eta_{ex} = \frac{w_{rev}}{w_{actual}} = \frac{y(h_6 - h_1 - T_0(s_6 - s_1))}{h_2 - h_1 - T_1(s_2 - s_1)} \tag{4-51}$$

Here we present a numerical example for the ideal Linde-Hampson cycle. Major assumptions are (1) a reversible and isothermal compressor; (2) an ideal heat exchanger (effectiveness = 1 and no pressure drop); (3) an ideal phase separator (no pressure drop, complete separation of phases); (4) the isenthalpic expansion process in the expansion valve; and (5) no heat leak to the cycle. The gas is methane, at 25°C and 1 atm at the compressor inlet and at 20 MPa at the compressor outlet. With these assumptions and specifications, the various properties at the different states of the cycle and the performance parameters just discussed are shown in Table 4-1.

As part of the analysis, the effects of liquefaction temperature and gas inlet temperature on various energy- and exergy-based performance parameters are investigated considering the methane as the gas being liquefied. The results are given in Figures 4-14 to Figure 4-17. They show that as the liquefaction temperature increases and the inlet gas temperature decreases the liquefied mass fraction, the actual COP, and the exergy efficiency increases while the actual work consumptions decreases.

4.4.5 Exergy losses in a nonideal Linde-Hampson liquefaction cycle

Consider the exergy losses in a nonideal Linde-Hampson methane liquefier. It can be assumed that the isothermal efficiency of the compressor is 60%, and the effectiveness of the heat exchanger is 90%. For the sake of simplicity, the pressure drop in the heat exchanger and phase separator is assumed to be zero. The compressor work is utilized in the liquefaction of methane and in overcoming the exergy

Table 4-1 Properties and Performance Parameters of the Cycle

State Point	T [°C]	P [MPa]	h [kJ kg^{-1}]	s [kJ kg^{-1}K^{-1}]
1	25	0.1	−0.97	0.004431
2	25	20	−181.9	−3.202
3	−55.62	20	−501.9	−4.458
4	−161.6	0.1	−501.9	−3.009
5	−161.6	0.1	−400.4	−2.098
6	−161.6	0.1	−911.5	−6.682
q_L [kJ kg^{-1}]	**w_{actual} [kJ kg^{-1}]**	**COP**	**COP$_{rev}$**	**η_{ex}**
180.9	775.1	0.2334	0.8407	27.76%

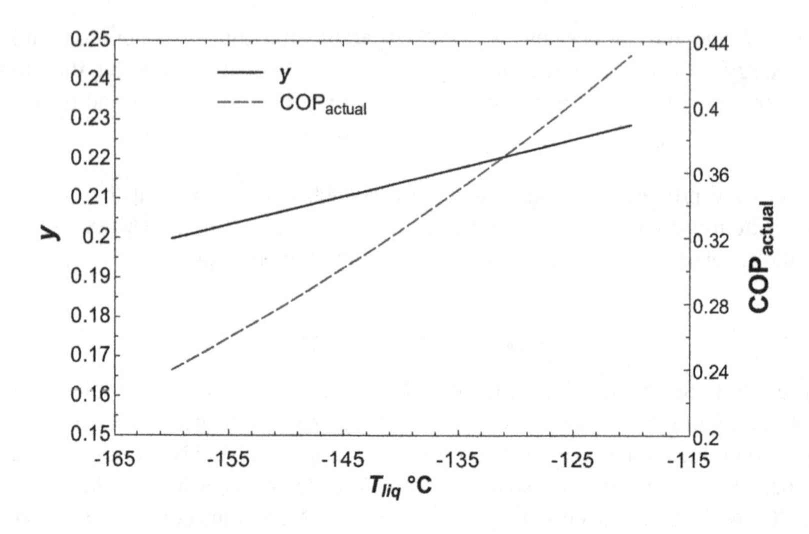

FIGURE 4-14

The liquefied mass fraction *y* and the actual COP versus liquefaction temperature.

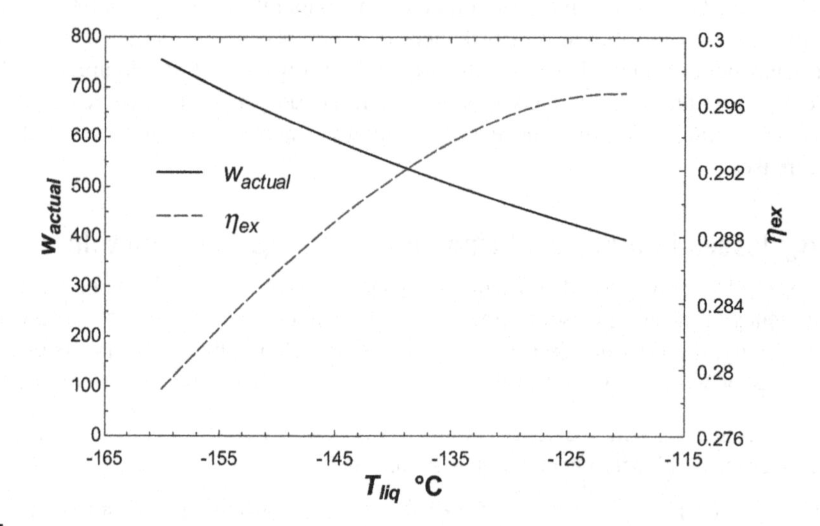

FIGURE 4-15

The work input and the exergy efficiency versus liquefaction temperature.

losses in the compressor, heat exchanger, and valve. As shown in Figure 4-18, most of the exergy loss occurs in the compressor and then in the expansion valve even when the heat exchanger effectiveness is 90%. Compared to the ideal liquefier as discussed previously, the exergy efficiency is decreased from 27.67 to 11.31% due to the irreversibility in the nonideal heat exchanger and compressor. Figure 4-19 shows the utilization of the input exergy in this nonideal cryogenic when high (P_2) and low (P_1) pressures are 25 MPa and 0.1 MPa, respectively.

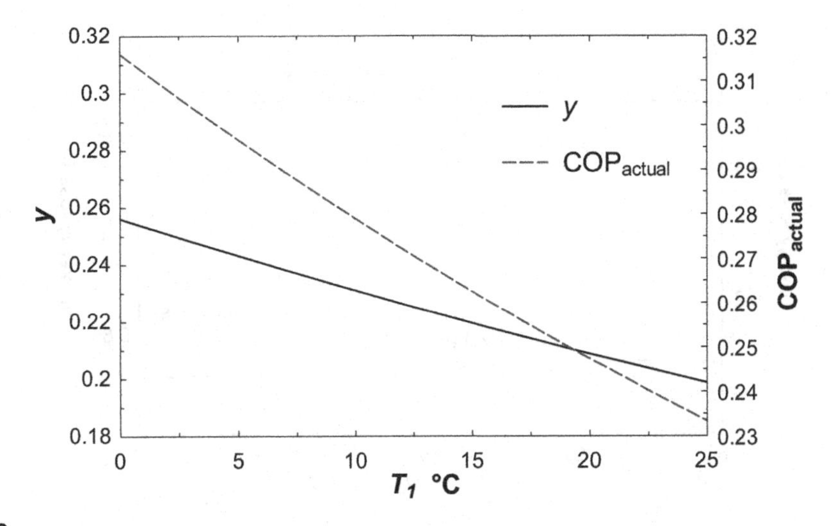

FIGURE 4-16

The liquefied mass fraction *y* and the actual COP versus gas inlet temperature.

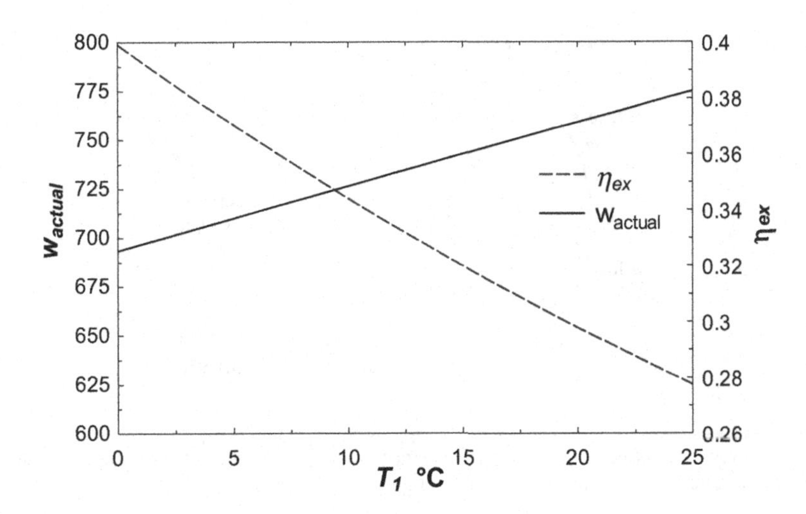

FIGURE 4-17

The work input and the exergy efficiency versus gas inlet temperature.

4.4.5 Energy and exergy analyses of a nonideal Kapitza liquefaction system

A Kapitza liquefaction cycle has been shown schematically and on a temperature-entropy diagram in Figure 4-9, in order to describe energy and exergy analyses of liquefaction cycles. The fundamental difference between it and a Linde-Hampson liquefaction cycle is in the process used for expansion of the fluid from high to low pressure. Makeup gas is mixed with the uncondensed portion of the gas from the previous cycle, and the mixture at state 1 is compressed by an isothermal compressor to state 2. The

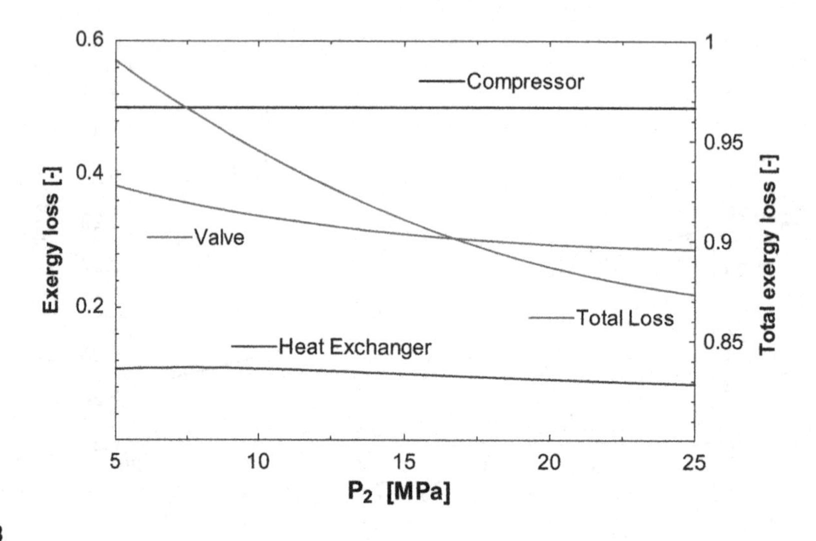

FIGURE 4-18

Variation of exergy loss in a nonideal Linde-Hampson methane liquefier versus compressor outlet pressure.

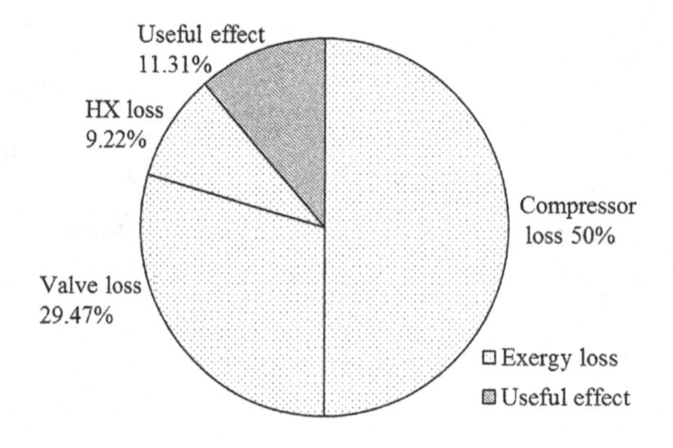

FIGURE 4-19

Utilization of input exergy in a nonideal Linde-Hampson methane liquefier at heat exchanger effectiveness 90%, compressor efficiency 50%, $P_2 = 25$MPa, $P_1 = 0.1$ MPa.

temperature is kept constant by rejecting compression heat to a coolant. The high-pressure gas is further cooled in a first regenerative counterflow heat exchanger by the uncondensed portion of gas from the previous cycle to state 3, and then the main stream splits into two streams. One portion (9) passing through the expander undergoes a large temperature drop (10). This cold gas is used to cool and condense the high-pressure fluid in a second heat exchanger. The refrigeration obtained in the expander also helps in precooling the high-pressure methane in the first heat exchanger (HX-1) before it enters the expander and the second heat exchanger (HX-2). The high-pressure fluid is in a subcooled liquid state (5) at the entry of the throttle valve when the operating pressure (P_2) is lower than the critical pressure (P_c). The temperature of the high-pressure gas at the entry of the throttle valve is much lower than that in a Linde-Hampson system.

The expression for liquid yield (y) in a Kapitza liquefier is obtained by an energy balance over a control volume that excludes the compressor as follows:

$$y = \frac{h_{13} - h_2}{h_{13} - h_f} + \frac{m(h_9 - h_{10})}{h_{13} - h_f} \tag{4-52}$$

The first term on the right-hand side of the equation is the liquid yield in a simple Linde-Hampson liquefier. The second term represents the additional yield due to refrigeration effects from the expander. Thus, the refrigeration effect per unit mass of the gas in the cycle may be expressed as

$$q_L = h_{13} - h_2 + m(h_9 - h_{10}) \tag{4-53}$$

where m represents the fraction of gas passing through the expander. The power extracted from the turbine $[m(h_9 - h_{10})]$ can be used to reduce the compressor power in a Kapitza liquefier. The actual compressor input required in a Kapitza cycle can be written as

$$w_{actual} = h_2 - h_1 - T_1(s_2 - s_1) - m(h_9 - h_{10}) \tag{4-54}$$

Similarly, the reversible work for Kapitza cycle may be expressed by the stream exergy difference of states 1 and 7 as

$$w_{rev} = y(ex_7 - ex_1) = y(h_7 - h_1 - T_0(s_7 - s_1)) \tag{4-55}$$

Thus, the energy and exergy efficiency of a Kapitza liquefier can be expressed as follows:

$$COP = \frac{q_L}{w_{actual}} = \frac{h_{13} - h_2 + m(h_9 - h_{10})}{h_2 - h_1 - T_1(s_2 - s_1) - m(h_9 - h_{10})} \tag{4-56}$$

$$\eta_{ex} = \frac{w_{rev}}{w_{actual}} = \frac{y(h_7 - h_1 - T_0(s_7 - s_1))}{h_2 - h_1 - T_1(s_2 - s_1) - m(h_9 - h_{10})} \tag{4-57}$$

Consider the exergy losses in a nonideal Kapitza methane liquefier. It can be assumed that the isothermal efficiency of the compressor is 50%; the turbine adiabatic efficiency is 80%; the minimum temperature approach is 10°C in the first heat exchanger and 20°C in the second heat exchanger. The operating pressure is assumed to be 4 MPa. Table 4-2 shows the thermodynamic parameters at different streams of the Kapitza methane liquefier.

Figure 4-20 shows the utilization of input exergy in a Kapitza methane liquefier. It can be seen that the exergy loss in the throttling valve is only 0.97% in this case, compared to about 29% in Linde-Hampson liquefiers. The smaller exergy loss in the throttling valve is due to the expansion of a subcooled liquid with a temperature change of about 20°C, compared to the expansion of a superheated vapor with a temperature change of about 110°C. The exergy loss in the two heat exchangers and that in the expander are 11.18% and 6.94%, respectively. The exergy loss in the second heat exchanger is larger than that in the first heat exchanger, because the finite temperature difference in the HX2 is larger and also the temperature level is lower. The useful effect (exergy efficiency) of the Kapitza liquefier is 22.54%, much higher than 11.31% in the Linde-Hampson liquefier mainly due to the reduction of exergy loss in the throttling valve and compressor work input.

Table 4-2 Thermodynamic Parameters of the Different Streams of a Kapitza Methane Liquefier

Stream	T [°C]	P [MPa]	*h* [kJ kg^{-1}]	*s* [kJ kg^{-1} K^{-1}]	*e* [kJ kg^{-1}]	*m* [kg kg^{-1} gas]
1	25	0.1	−0.9762	0.004431	0	1
2	25	4	−40.01	−2.004	559.7	1
3	−33.64	4	−190.8	−2.568	577.1	1
4	−33.64	4	−190.8	−2.568	577.1	0.2344
5	−141.6	4	−835.6	−6.134	995.7	0.2344
6	−161.6	0.1	−835.6	−6.001	955.9	0.2344
7	−161.6	0.1	−911.5	−6.682	1083	0.1996
8	−161.6	0.1	−400.4	−2.098	227.5	0.0348
9	−33.64	4	−190.8	−2.568	577.1	0.7656
10	−161.6	0.1	−400.4	−2.098	227.5	0.7656
11	−161.6	0.1	−400.4	−2.098	227.5	0.9652
12	−72.84	0.1	−211.6	−0.85	44.16	0.9652
13	15	0.1	−23.18	−0.07132	0.3803	0.9652

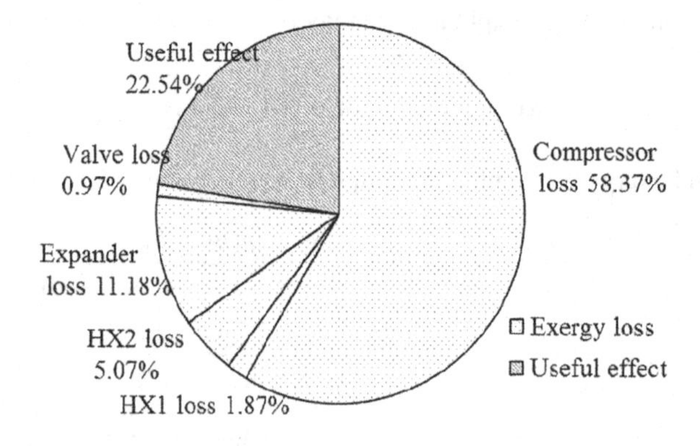

FIGURE 4-20

Utilization of input exergy in a nonideal Kapitza methane liquefier at compressor efficiency 50%, $P_2 = 4$MPa, and $P_1 = 0.1$ MPa.

4.4.6 Pinch analysis

Figure 4-21(a-c) shows typical temperature profiles of the hot and cold fluid streams in a counterflow heat exchanger with constant heat capacities but different relative values: (a) the same heat capacity of the hot and cold fluid streams; (b) larger heat capacity of the cold streams than that of hot streams; (c) larger heat capacity of the hot streams than that of cold streams. Consequently, the minimum temperature approach between the streams occurs at the warm or cold end of the heat exchanger in Figures 4-21(b) and 4-21(c). In some cases, the specific heat of the fluids varies along the length of the heat exchanger, for instance zeotropic mixtures undergoing phase change or real fluids at temperatures close to the critical point when the pressure is greater than the critical pressure and critical temperature for all fluids. Thus, the minimum temperature approach between the hot and cold fluid streams of the heat exchanger may

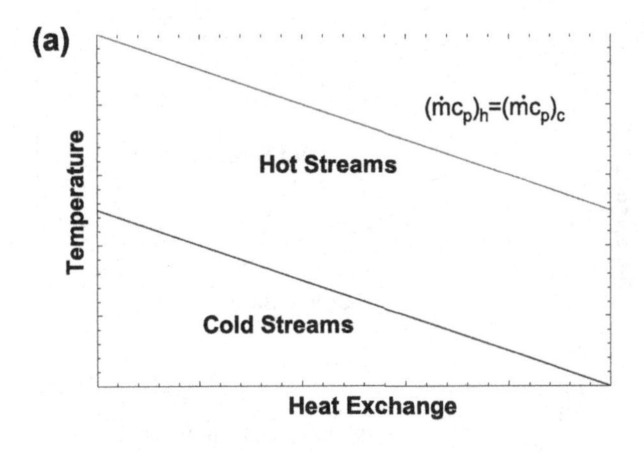

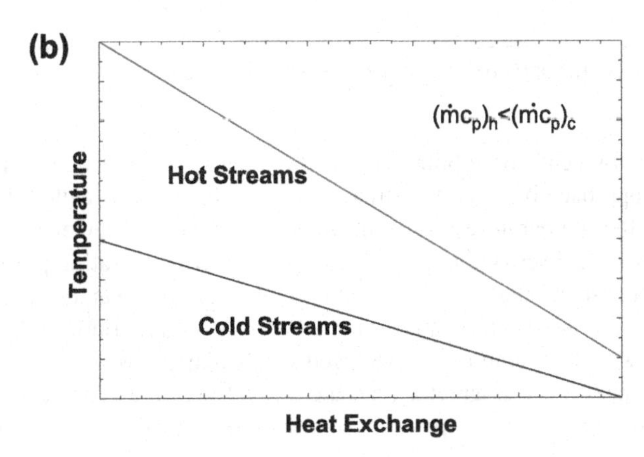

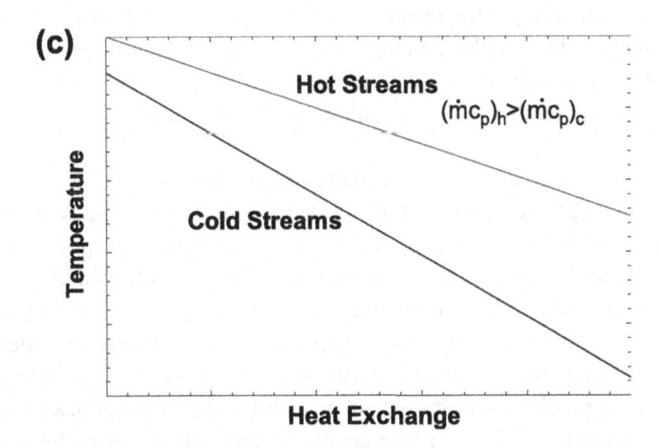

FIGURE 4-21

Different types of temperature profiles that can exist in heat exchangers operating with single phase fluids. The subscripts h and c refer to the hot and cold streams, respectively.

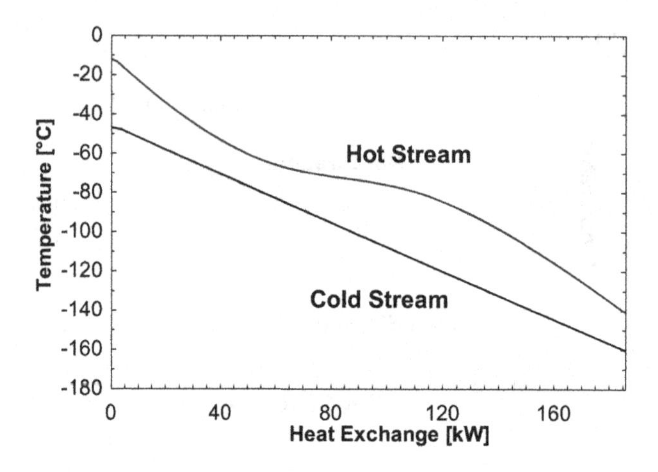

FIGURE 4-22

Variation of the temperature of hot and cold streams in the second heat exchanger of a Kapitza methane liquefier with heat transferred.

occur either between the two ends or at both warm and cold ends of the heat exchangers. The location where the temperature approach between the streams is at its minimum is called the pinch point.

Figure 4-22 shows the temperature profile in the second heat exchanger (HX-2) in the Kapitza methane liquefaction process discussed in the previous section with operating pressures of 6/0.1 MPa. Pinch point occurs because of the large variation of the specific heat of methane at 6 MPa at temperatures close to the critical temperature of methane. A large variation of the specific heat at constant pressure (c_p) with temperatures is observed in all real fluids at temperatures close to the critical point when the pressure P is greater than the critical pressure. The specific heat at constant pressure (c_p) tends to infinity at critical pressure and critical temperature for all fluids, as shown in Figure 4-23.

Pinch analysis is a methodology for minimizing energy consumption of chemical processes by calculating thermodynamically feasible energy targets (or minimum energy consumption) and achieving them by optimizing heat recovery systems, energy supply methods, and process operating conditions. It is also known as process integration, heat integration, energy integration, or pinch technology.

The process data is represented as a set of energy flows, or streams, as a function of heat load against temperature. These data are combined for all the streams in the plant to give composite curves, one for all hot streams (releasing heat) and one for all cold streams (requiring heat). The point of closest approach between the hot and cold composite curves is the pinch point (or just pinch) with a hot stream pinch temperature and a cold stream pinch temperature. This is where the design is most constrained. Hence, by finding this point and starting the design there, the energy targets can be achieved using heat exchangers to recover heat between hot and cold streams in two separate systems, one for temperatures above pinch temperatures and one for temperatures below pinch temperatures. In practice, during the pinch analysis of an existing design, often cross-pinch exchanges of heat are found between a hot stream with its temperature above the pinch and a cold stream below the pinch. Removal of those exchangers by alternative matching makes the process reach its energy target.

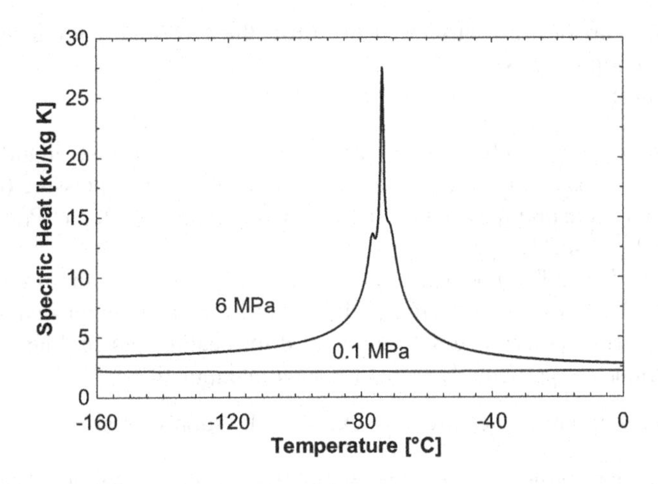

FIGURE 4-23

Variation of specific heat of methane with temperature and pressure.

However, the major limitation with the heat pinch methodology is that only temperature is used as a quality parameter of a stream, thus neglecting pressure and composition. The advantage of exergy analysis is the inherent capability of including all stream properties (temperature, pressure, and composition). In a cryogenic liquefaction system (such as the Kapitza cycle discussed earlier), the temperature is closed related to both pressure and power, since compression and expansion (valve or turbine expander) will change the boiling and condensation temperature in order to be able to transfer heat from a cold source to a hot sink. The required energy (shaft work) and resulting refrigeration duty are determined by the hot and cold temperature levels, which dictates the required pressure increase or decrease and thereby the need for shaft work.

Aspelund et al. (2007) developed the Extended Pinch Analysis and Design (ExPAnD) methodology with particular focus on subambient processes to design a system of heat exchangers, expanders, and compressors in such a way that the irreversibilities are minimized for given a set of process streams with a supply state (temperature, pressure, and the resulting phase) and a target state, as well as utilized for external heating and cooling. They generalized the following conclusions based on general theoretical analysis:

1. Pressure-based exergy can effectively be transformed to cooling duty and work and may give a significant reduction in the required utility cooling for subambient processes.
2. Expanding a cold stream at high temperature will generate more work and cooling duty than at a lower temperature.
3. By heating the cold stream prior to expansion, more cooling duty will be produced and less irreversibility may occur in the heat exchanger, due to reduced driving forces.
4. Additional expansion steps and heat exchange passes may reduce the irreversibilities and create more cold duty and work.
5. If too much cold duty is produced, the heating requirement will increase and the driving forces will be reduced, however some additional work can be generated.

A set of detailed heuristics are introduced to utilize the ExPAnD methodology to design and optimize the subambient processes.
Some general heuristics:

1. Available pressure ($P_s > P_t$) can be utilized through expansion to reduce cold utility requirements with power generation as an important by-product. Contrary, lack of pressure ($P_s < P_t$) requires power; however, this situation may reduce hot utility requirements (P_s and P_t denote the supply pressure and target pressure, respectively).
2. Temperature gap ($\Delta T > \Delta T_{min}$) between the hot and cold composite curves results in unnecessary irreversibilities. The pressure of the streams may be manipulated to decrease the irreversibilities, generate power, and reduce the need for heating and cooling utilities. Streams with phase transitions are particularly suited for such manipulation.

Heuristics for streams with target pressure different from the supply pressure:

1. Compression of a vapor or dense phase stream requires power and will add heat to the system. Hence, from a Pinch analysis point of view, compression should preferably be done above the Pinch point.
2. Expansion of a vapor or dense phase stream in an expander will provide cooling to the system, and at the same time generate power. Hence, expansion should preferably be done below Pinch. In subambient processes, a stream with a supply pressure higher than the target pressure should always be expanded in an expander (not a valve) if the stream is located below the pinch point.
3. If expansion of a vapor or dense phase stream above Pinch is required, a valve should be used to minimize the increase in utility consumption, unless the main purpose of the expansion is to produce work.

Heuristics for streams with target pressure equal to the supply pressure:

1. A hot gas or dense phase fluid that is compressed above the Pinch point, cooled to near Pinch point temperature, and then expanded will decrease the need for both cold and hot utilities. Additional work is required, however.

Heuristics for liquid streams and streams with phase change:

1. A fluid with $P_s < P_t$ should be compressed in a liquid phase if possible to save compressor work.
2. In a liquid stream with $P_s = P_t$, a phase transition is necessary for the composite curves to be manipulated, since the effect of expansion/compression in the liquid phase alone is marginal.
3. If a cold liquid stream to be vaporized does not create a Pinch point, it should be pumped to avoid vaporization at a constant temperature, reduce the total cooling duty, and increase the pressure-based exergy. Work and cooling duty should be recovered by expansion of the fluid in the vapor phase at a later stage (higher temperature).

The ExPAnD methodology combines Pinch analysis and exergy analysis. It may be a very useful design tool for some cryogenic applications in which some of the heat exchangers are of the multi-stream type, enabling several streams with different entering and leaving temperatures to be joined in one heat exchanger, and expansion and compression work are involved to manipulate the temperatures of the hot and cold streams.

4.5 Energy and exergy analyses of natural gas liquefaction cycles

In this section, the energy and exergy analyses of three typical natural gas liquefaction cycles are discussed.

4.5.1 Propane-precooling mixed refrigerant cycle (C3/MRC)

Gu et al. (2003) reported a thermodynamic modeling and parametric analysis of a propane-precooling mixed refrigerant cycle (C3/MRC) based on the first- and second-law analysis. The schematic drawings of the mixed refrigerant cycle and propane-precooling cycle are shown in Figure 4-24(a) and (b). The pressurized natural gas is precooled in the propane refrigeration unit, and then cooled down to the subcooled liquid state through three multistream heat exchangers. Finally, it expanded to the ambient pressure at the throttle valve. Refrigerant mixture first is compressed to a high pressure of about 2.5 MPa in the two-stage compressor, and then is passed through a water cooler. It comes out at a temperature of about 305 K, and then is further cooled down in the propane refrigeration unit to about 238 K. Afterward, it comes through three multistream heat exchangers to provide the cooling capacity for liquefying supplied natural gas. In a propane-precooling refrigeration cycle, the propane experiences three-stage expansions, operating at three discrete evaporation temperature levels.

In the parametric effects on the cycle energy and exergy analysis, the following parameters and assumptions are given:

1. The process streams (e.g., natural gas) have an initial temperature of 298 K and pressure of 5 MPa with a flow rate of 1mol/s and with a composition of nitrogen (0.7%), methane (82%), ethane (11.2%), propane (4%), i-butane (1.2%), and n-butane (0.9%).
2. Liquefied natural gas is stored at the temperature of about 117.2 K and the pressure of about 0.15 MPa.
3. At the warm end of the first heat exchanger (HX-1), the refrigerant mixture has the composition of nitrogen (5%), methane (44%), ethane (34%), and propane (17%).
4. The heat transfer temperature difference at each end of the heat exchangers are assumed to be about 3 K.
5. The compressor efficiency is 0.75.
6. The high pressure of the propane-precooling cycle is about 1.3 MPa.

Table 4-3 summarizes the parametric effects on C3/MRC performances, and based on this case analysis some conclusions can be made as follows:

1. Natural gas feed pressure and the propane precooling temperature have relatively significant influences on the refrigerant mixture flow rate ($q_{n,r}$), total work input (W_{tot}), total refrigeration capacity (Q_{cmr}) and natural gas liquefaction (Q_{cng}) by MRC unit, and natural gas liquefaction by propane-precooling unit (Q_{cng}). Especially, increase of Q_{cng} is up to 16.35%.
2. Natural gas precooling temperature T_2 by the propane-precooling unit has the most significant effect on the required cooling capacity (Q_{png}); similarly for the effect of the suction temperature in MRC unit (T_9).
3. Compared to the high pressure (P_{13}), variation of the refrigerant mixture low pressure (P_9) affects total work input more significantly.

(a)

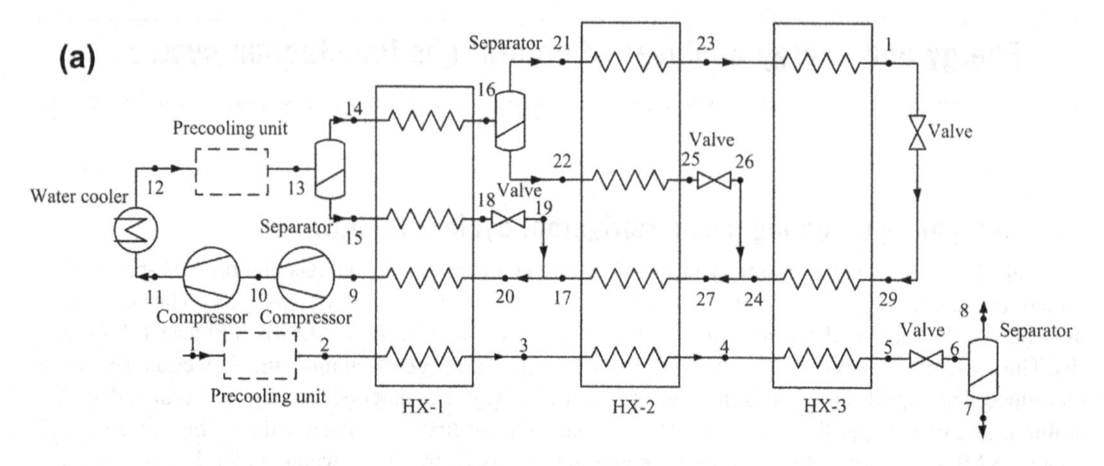

(b)

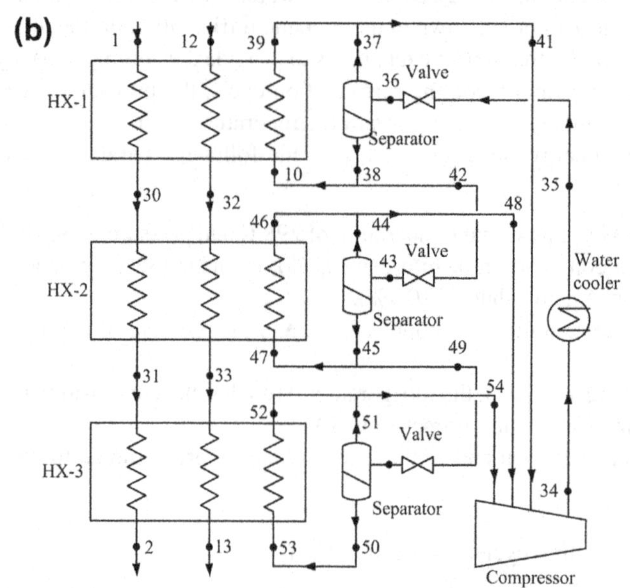

FIGURE 4-24

Propane-precooling mixed refrigerant cycle for natural gas liquefaction.

(Gu et al., 2003)

4.5.1.2 Exergy analysis

Based on the earlier parametric analysis, operating parameters of the C3/MRC cycle are determined, and therefore the corresponding analysis of exergy loss on each component of the entire cycle is performed. Figure 4-25(a) and (b) show the relative exergy loss in the mixed refrigerant cycle (MRC) and in the propane (C3)-precooling cycle, respectively. It can be seen that for both cycles, the maximum exergy loss occurs in the compressor—60.88% for the MRC and 45.14% for the C3-precooling cycle. In

Table 4-3 Summary of Parametric Effects on C3/MRC Performances

Parameter Variation	P_{ng}(+31.58%)	T_2(+4.23%)	T_9(+2.79%)	P_9(+25%)	P_{13}(+5.04%)
$q_{n,r}$ (%)	−6.83	+6.01	−3.59	+0.48	−2.16
W_{tot}(%)	−5.10	+3.86	−2.00	−5.34	−0.42
Q_{cmr}(%)	−6.11	+5.41	−1.52	+0.71	−2.20
Q_{cng}(%)	−6.88	+6.07	0	0	0
Q_p(%)	−3.26	+1.53	−2.94	+0.40	−1.00
Q_{png}(%)	+16.35	−25.69	−9.19	0	0

(Gu et al. 2003)

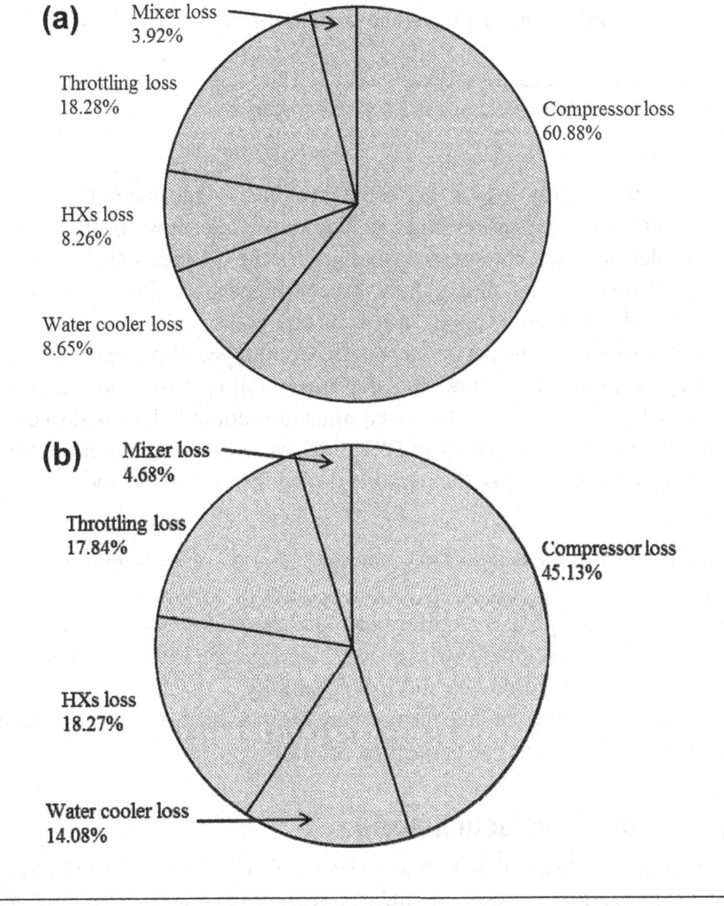

FIGURE 4-25

(a) Exergy loss distribution in mixed refrigerant unit. (b) Exergy loss distribution in propane-precooling unit.

(Gu et al., 2003)

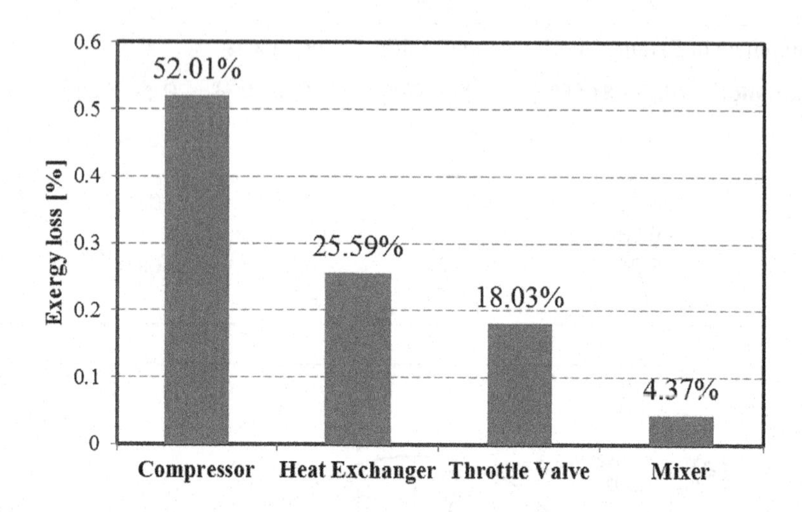

FIGURE 4-26

Exergy loss distribution in the C3/MRC natural gas liquefaction system.

(Gu et al., 2003)

the MRC, the second highest exergy loss occurs in the throttling valves due to the significant amount of irreversibility in the isenthalpic expansion process. The results are those in the multistream heat exchangers and water coolers after the compressor. Similarly, in the C3-precooling cycle the exergy loss in the throttling valve is also significant, although the second highest loss occurs in the multistream heat exchangers. For both units, the exergy loss in stream mixers is the lowest.

If comparing the exergy loss in the perspective of different types of process equipment in the entire natural liquefaction cycle, Figure 4-26 shows that the irreversibility in the compressors accounts for up to 52% loss; the exergy loss in the heat exchange equipment is about half of that in compressors, 25%; throttling loss in the expansion valves are about 18%; the loss in all the stream mixers is only about 4%. According to the exergy analysis results, the possible directions to improve the cycle efficiency can be proposed as follows:

1. Choose the appropriate compressor suction temperature and compression ratio in the cycle design.
2. Improve the compressor design and therefore increase the compressor efficiency.
3. Utilize the heat transfer enhancement structure and increase the heat exchanger size to decrease the heat transfer temperature difference and irrepressibility.
4. Appropriately design the cycle to reduce the pressure drop in the throttling process or increase the subcooling of the process natural gas before the throttling process.

4.5.2 N_2-CH_4 expander liquefaction cycle

Pu et al. (2007) performed a thermodynamic analysis of N_2-CH_4 expander natural gas liquefaction cycles. As a typical liquefaction process, the nitrogen expansion liquefaction cycle is a good option for small-scale liquefaction applications. The typical advantages and benefits of this process are its simplicity when compared with other liquefaction processes; adaptability to varied contents of feedstock by changing the composition of N_2-CH_4 refrigerant mixture; certain requirement of refrigerant inventory and phase separators are eliminated since nitrogen and methane refrigerants are always in gas phase.

Two different cycles are considered in this case. The first process is single expander with a propane-precooling unit, as shown in Figure 4-27(a). Adopting a propane precooling unit, which is a more efficient refrigeration method in the high temperature range, can consequently reduce the power consumption of the overall liquefaction cycle. The pressurized natural gas is cooled down to the subcooled liquid state through four multistream heat exchangers, and in the first heat exchanger

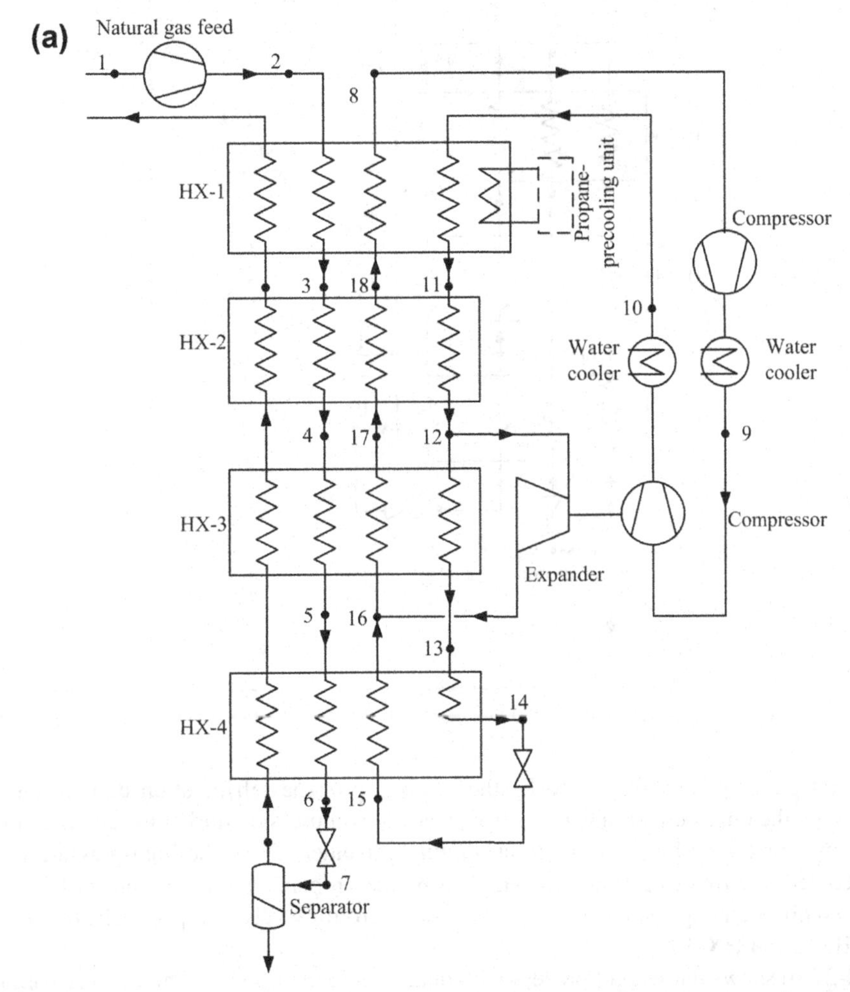

FIGURE 4-27

N_2-CH_4 expander natural gas liquefaction.

(Pu et al., 2007)

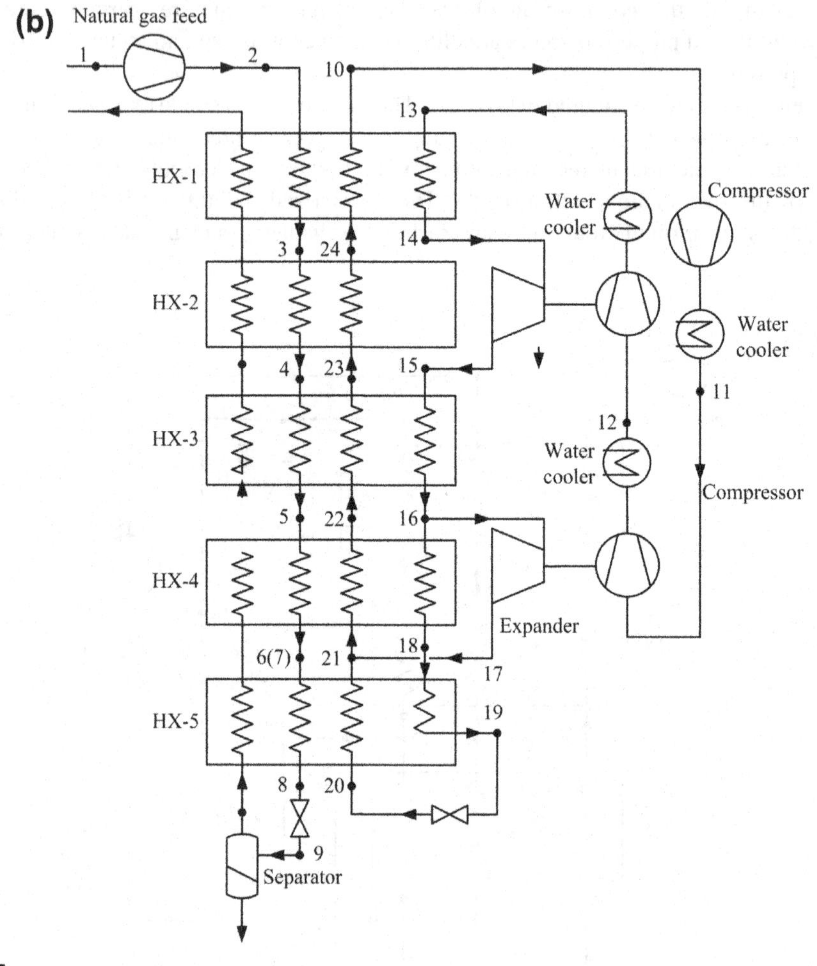

FIGURE 4-27

(*continued*).

(HX-1) natural gas is mainly precooled by the external propane refrigeration unit. In the N_2-CH_4 expansion cycle, the mixture first undergoes two stages of compression during the compressor and the booster compressor driven by the refrigerant mixture expander. Then, the high-pressure refrigerant mixture is cooled first by using water coolers, and by the propane precooling unit in HX-1. Then it undergoes two different expansions in a expander and a throttling valve, respectively, to produce cold energy for HX-3 and HX-4.

Figure 4-27(b) shows the second process with dual expanders in series. The first stage expander is used to produce cold energy at a high temperature range and thus to replace the propane precooling unit employed in the first process. By this means, no additional mechanical refrigeration is used, simplifying the process accordingly. But the associated penalty is to compress the N_2-CH_4 mixture refrigerant to a higher pressure to produce greater cold energy, which in turn may increase the total power consumption. To make a fair comparison of two processes, the cold energy produced by the first stage expander is maintained the same as that produced by the propane refrigeration unit in the first process, which can determine the expansion ratio of the first stage expander. In this study, the following parameters and assumptions are made:

1. The feed natural gas has a temperature of 303 K and a pressure of about 0.2 MPa with a mass flow rate of 2.08 kg/s. It is liquefied and stored at the pressure about 0.12 MPa.
2. The minimum heat transfer temperature difference at each end of the heat exchangers is assumed to be about 3 K, and compressor and expander efficiencies are 0.7 and 0.78, respectively.

4.5.2.1 Effect of methane concentration in N_2-CH_4 refrigerant mixture

In both processes, the N_2-CH_4 mixture is used as the working refrigerant to provide external cold energy for natural gas liquefaction. Methane and nitrogen are the high and low boiling point components, respectively. The methane mole fraction of the refrigerant mixture has two effects on the liquefaction cycle power consumption. The compressor work consumption per unit flow rate of the refrigerant mixture increases with the methane mole fraction. On the other hand, methane has a higher specific heat capacity than the nitrogen, thus the cold energy provided by the expander is greater per unit flow rate of the mixture with a higher methane concentration, which in turn will reduce the flow rate of the refrigerant mixture and possibly decrease the compressor power consumption. Figure 4-28 shows the effect of methane mole concentration in the refrigerant mixture on the total power consumption in the single expander cycle with a propane precooling unit at the same natural gas liquefaction rate. The power consumption reaches the minimum when the methane fraction is about 50 to 55% for this case. In addition, Table 4-4 shows refrigerant temperatures before and after the throttling valve at varied methane mole fractions when the thermodynamic state after the throttling valve is fixed (P = 0.55 MPa, T = 105.15 K). It is clear that as the methane concentration increases, the refrigerant

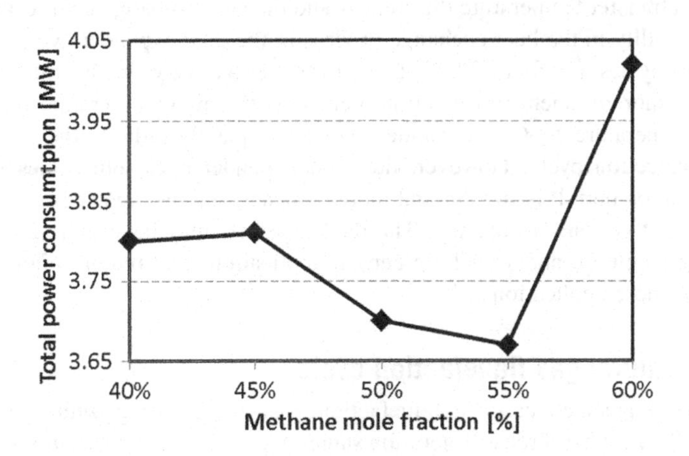

FIGURE 4-28

Effect of methane mole fraction in N_2-CH_4 mixture on the total power consumption of the liquefaction process.

(Pu et al., 2007)

Table 4-4 Temperatures of the Refrigerant Mixture before and after the Throttling Valve

Methane Mole Fraction [%]	Temperature Before Throttling Valve T_{14} [K]	Temperature After Throttling Valve T_{15} [K]
40	132	105.15
45	125.78	105.15
50	118.79	105.15
55	111.46	105.15
60	103.83	105.15

(Pu et al., 2007)

mixture must be precooled to a lower temperature before the throttling valve, and especially at 60% the temperature begins to increase after the throttling. This indicates that the cold energy at a low temperature decreases when the mole fraction of the methane, high boiling point component, increases in the mixture. Based on this result, the mole fraction of the methane in the refrigerant mixture is chosen as 50% in the following exergy analysis for both liquefaction processes.

4.5.2.2 Exergy analysis

Figure 4-29 shows the relative exergy loss in each component for two different liquefaction processes. It can be seen that for both processes the compressor unit has the maximum exergy loss, up to about 42%. Then, the results are the exergy losses in expanders, cycle heat exchangers, and throttling valves. In the first process, about 7.6% exergy loss occurs in the additional propane refrigeration unit. Thus, it has only about 29.8% exergy loss in the single-stage expander, about 10% lower than in the dual expanders in the second process. Table 4-5 compares the power consumption in two liquefaction processes. Results reveal that the single expander cycle adopting a propane precooling unit saves about 5.5% power consumption compared to the dual expander cycle. This is mainly due to the fact that the relatively large heat transfer temperature differences and the heat exchange load result in a significant amount of irrepressibility in the heat exchange process in the dual expander cycle. Thus, the exergy efficiencies for two cycles are about 32.9% and 31.0%, respectively. It illustrates that the propane precooling unit is a more efficient refrigeration method in the high temperature range than the corresponding high temperature N_2-CH_4 expander, and consequently reduces the power consumption of the overall liquefaction cycle. However, the dual expander cycle eliminates the conventional mechanical refrigeration unit. It is simpler and more compact, and thus requires less installation area, has greater safety, and is easier to operate. The dual expander may have more advantages than the propane precooling single expander cycle in certain applications that require high compactness and mobility such as offshore applications.

4.5.3 Cascade natural gas liquefaction cycle

Figure 4-30 shows a patented cascade liquefaction process operating with refrigerant mixtures (Venkatarathnam, 2008). It has three refrigeration stages with different refrigerants for each stage: one for desuperheating the natural gas feed, the second for condensation, and the third for subcooling. Consider the conventional cascade liquefaction process shown in Figure 4-30. The large number of phase separators and heat exchangers that need to be used makes the system quite complex.

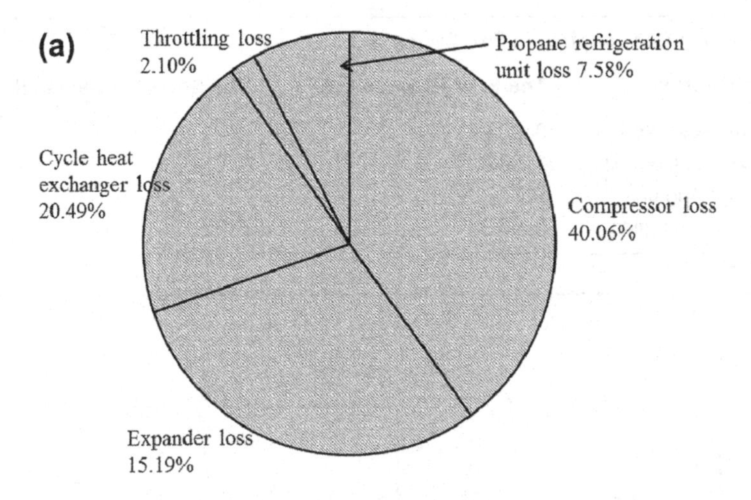

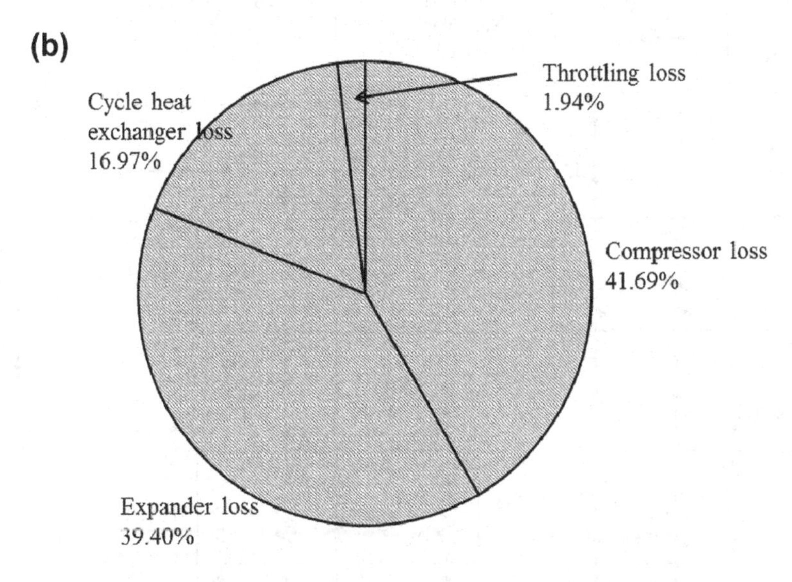

FIGURE 4-29

Exergy loss distribution in N_2-CH_4 expander natural gas liquefaction cycle. (a) Single expander cycle with propane precolling unit, (b) Duel expander cycle.

(Pu et al., 2007)

Additionally, the main disadvantage with the conventional cascade process operating with pure fluid (single-component) refrigerants is that the refrigeration is provided at constant temperature at discrete temperature levels. On the other hand, mixed refrigerant processes provide refrigeration over a range of temperatures.

Table 4-6 shows the temperature, pressure, vapor fraction, and flow rate of different streams of the process. The precooling refrigerant (stream 16) is completely condensed in the condenser. All the

Table 4-5 Power Consumption in Two Cycles

Main Components	The First Process [kW]	The Second Process [kW]
N_2-CH_4 compressor unit	3698.2	4624.5
Feed gas compressor unit	1508.6	1508.6
Propane precooling unit	589.5	–
Total power consumption	5796.3	6133.1

(Pu et al., 2007)

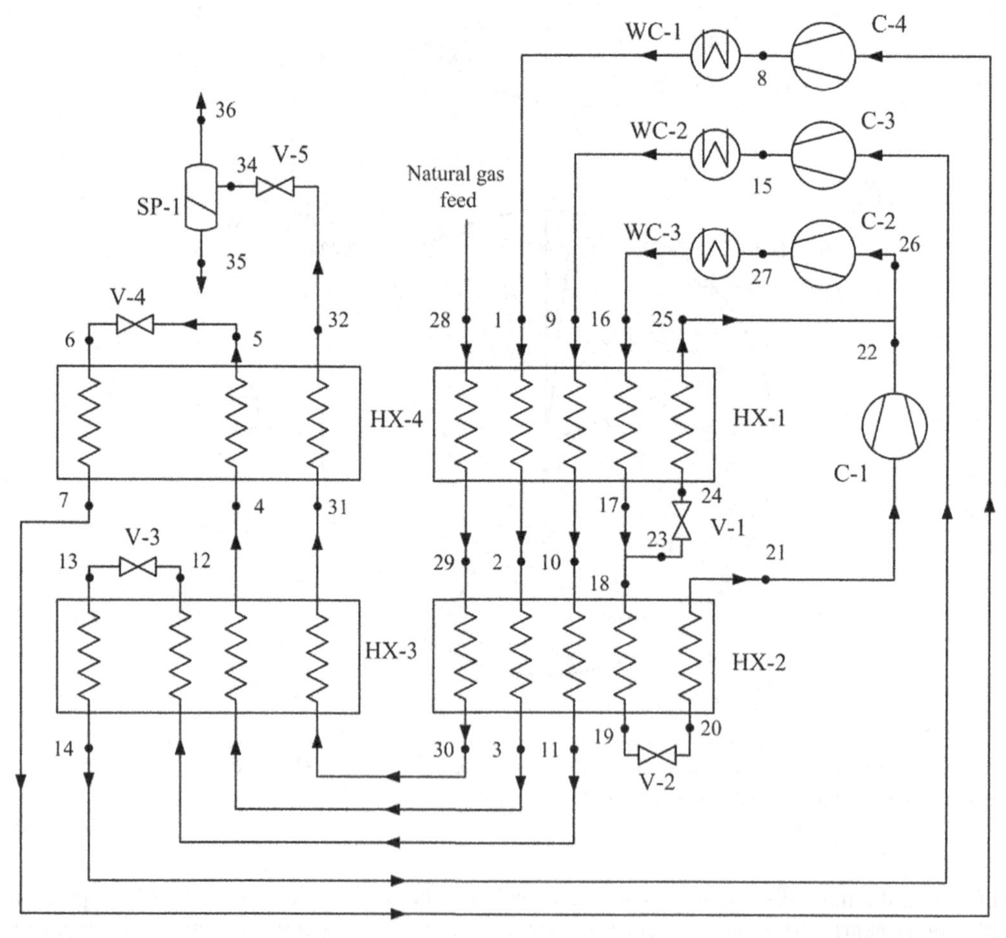

FIGURE 4-30

Cascade refrigeration cycle operating with refrigerant mixtures. [C – Compressor; HX – Heat exchanger; SP – Separator; V – Valve; WC – Water cooler]

(Stockmann et al., 2001)

Table 4-6 Thermodynamic States of Different Streams in the Cascade Refrigeration Process

Stream	1	2	3	4	5	6	7	8	9
Temperature [K]	310	276.2	247.9	186.7	113.9	106.6	181.1	310	310
Pressure [MPa]	3.39	3.39	3.39	3.39	3.39	0.35	0.35	3.39	2.79
Vapor fraction [-]	1	1	1	0.297	0	0.109	1	1	1
Flow rate [mol/s]	0.721	0.721	0.721	0.721	0.721	0.721	0.721	0.721	1.023
Stream	**10**	**11**	**12**	**13**	**14**	**15**	**16**	**17**	**18**
Temperature [K]	276.2	247.7	191.5	180.9	242.7	326.1	310	282	282
Pressure [MPa]	2.79	2.79	2.79	0.31	0.31	2.79	1.69	1.69	1.69
Vapor fraction [-]	0.395	0	0	0.098	1	1	0	0	0
Flow rate [mol/s]	1.023	1.023	1.023	1.023	1.023	1.023	1.369	1.369	0.513
Stream	**19**	**20**	**21**	**22**	**23**	**24**	**25**	**26**	**27**
Temperature [K]	251	243.8	275.9	311.1	307.5	282	272.8	305.2	307.5
Pressure [MPa]	1.69	0.3	0.3	0.67	0.67	1.69	0.67	0.67	0.67
Vapor fraction [-]	0	0.054	1	1	1	0	0.077	1	1
Flow rate [mol/s]	0.513	0.513	0.513	0.513	1.369	0.855	0.855	0.855	1.369
Stream	**28**	**29**	**30**	**31**	**32**				
Temperature [K]	300	276.2	247.9	186.8	113				
Pressure [MPa]	6.5	6.5	6.5	6.5	6.5				
Vapor fraction [-]	1	1	1	1	1				
Flow rate [mol/s]	1	1	1	1	1				

(Venkatarathnam, 2008)

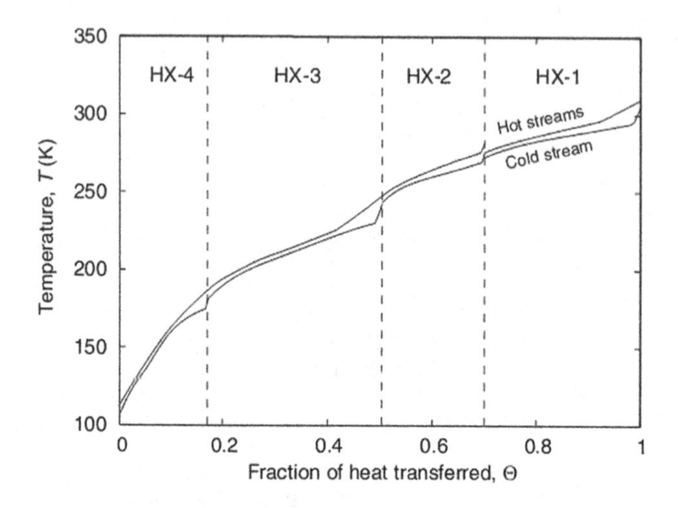

FIGURE 4-31

Temperature profiles in the cascade refrigeration cycle operating with mixed refrigerants.

(Venkatarathnam, 2008)

other streams entering the first heat exchanger (HX-1) at the warm end are in a superheated state. The high-pressure condensation refrigerant (stream 9) is condensed partially in the first heat exchanger and leaves the second heat exchanger (HX-2) as stream 11 in a subcooled state. On the other hand, the subcooling refrigerant (stream 1) leaves the third heat exchanger (HX-3) as stream 4 in a partially condensed state. All the low-pressure refrigerant streams enter the different compressors in a superheated condition. Figure 4-31 shows the temperature of the hot and cold fluid streams in the four heat exchangers. It can be seen that the temperature approach between the streams is small and nearly uniform throughout the length, except at temperatures close to the dew point temperature of the low-pressure refrigerant (cold) stream. This results in a relatively small log mean temperature difference. The small temperature difference between the cold and hot streams, particularly at low temperatures, significantly reduces the irrepressibility in the heat transfer process and results in a high exergy efficiency of the overall natural gas liquefaction process.

Figure 4-32 shows the exergy loss distribution in the overall process. The maximum exergy loss occurs in the compression units about 37.5%, and it is lower than that in mixed refrigerant cycle (about 52%). This is mainly because in the cascade cycle, each stream of refrigerant mixtures working at different temperature levels is compressed into different operating pressures. For instance, maximum pressures for the subcooling, condensation, and precooling refrigerants are 3.39 MPa, 2.79 MPa, and 1.69 MPa, correspondingly. Thus, the relative compressor work and irreversibility that occurs will be expected to be lower than that in the mixture refrigerant cycle where all the refrigerant mixtures are compressed to the same maximum pressure. Besides, the cascade cycle also has a relatively low cycle HX exergy loss compared to the N_2-CH_4 expander cycle. As discussed earlier, this is mainly due to the relatively small and uniform temperature approach between hot and cold streams in the cycle HXs.

It is evident that for all the natural gas liquefaction cycles the combined losses for the compression system, including aftercoolers, are the most significant contributor to the work loss.

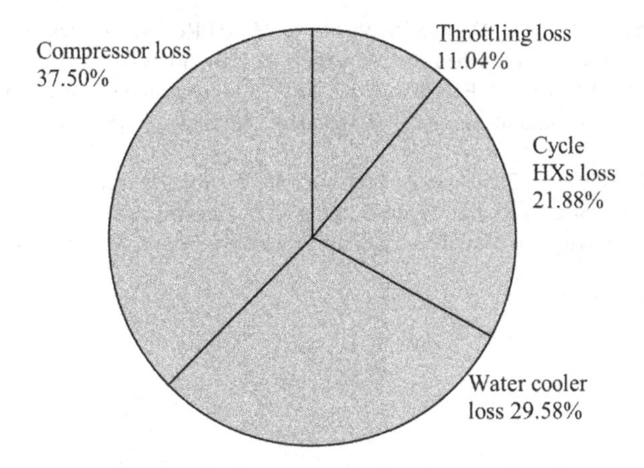

FIGURE 4-32

Exergy loss distribution in the cascade refrigeration cycle.

(Venkatarathnam, 2008)

The work or exergy loss in this component is a direct result of the efficiency of the compression system and thus there is great potential to increase the cycle exergy efficiency by improving the compressor efficiency. In addition, for practical application, adopting a multistage compression system to split the compression system into multiple stages with intercooling will also help to reduce the compression work. In the case of the single expander liquefaction cycle, expander loss takes up to about 15.9% of the total exergy loss. Any increase in the isentropic efficiency of the expander would reduce the required refrigerant flow rate since more cooling effect per flow rate will be generated in the expansion process, and also more useful work would be extracted to drive the generator or the compressor. Both effects would result in a power saving. For the heat exchangers system, using the mixed refrigerants will cause the lower exergy loss because the mixture composition can be optimized to minimize the heat transfer temperature difference between the hot and cold streams. Another option to enhance the cycle efficiency is to add a precooling refrigeration system since the same amount of refrigerating capacity will consume more power when generated at a lower temperature. But precooled liquefaction processes are preferable for large liquefaction systems.

References

Alfeev, V., Brodyanskii, V., Yagodin, V., Nikolsky, V., Ivantosv, A., 1973. Refrigerant for a Cryogenic Throttling Unit. UK Patent 1,336,892.

Aspelund, A., Berstad, D., Gundersen, T., 2007. An Extended Pinch Analysis and Design Procedure Utilizing Pressure Based Exergy for Sub-ambient Cooling. Applied Thermal Engineering 27 (16), 2633–2649.

Flynn, T., 2004. Cryogenic Engineering, second ed. CRC Press, Boca Raton, FL, USA.

Gu, A., Lu, X., Wang, R., 2003. Liquefied Natural Gas Technology. China Machine Press, Beijing, China.

Kanoglu, M., Diner, I., Rosen, M., 2008. Performance Analysis of Gas Liquefaction Cycles. International Journal of Energy Research 32 (1), 35–43.

Lee, G.C., Smith, R., Zhu, X.X., 2002. Optimal Synthesis of Mixed-Refrigerant Systems for Low-Temperature Processes. Industrial & Engineering Chemistry Research 41 (20), 5016–5028.

Pu, L., Sun, S.X., Yan, Z.L., Tuo, H.F., Zhao, M., 2007. Calculation and Thermodynamic Analysis on Liquefaction Processes of Natural Gas with Expanders. Journal of Xi'an Jiaotong University 41 (9), 1116–1119.

Stockmann, R., Forg, W., Bolt, M., Steibauer, M., Pfeiffer, C., Paurola, P., Fredheim, A., Sorensen, O., 2001. Method for Liquefying a Stream Rich in Hydrocarbons. U.S. Patent 6,253,574.

Venkatarathnam, G., 2008. Cryogenic Mixed Refrigerant Processes, first ed. Springer, New York, NY, USA.

CHAPTER 5

Natural Gas Liquefaction Cycle Enhancements and Optimization

5.1 Introduction

LNG plants are massive energy consumers. This is due to the fact that the energy required for liquefying one kilogram of natural gas is around 1,188 kJ (Finn et al., 1999). However, this liquefaction energy varies depending on the liquefaction cycle and site conditions. Practically about 8% of the feed gas to the LNG plants (Patel, 2005) is consumed for the liquefaction. Most of the LNG plant energy consumption occurs in the compressor drivers where fuel energy (usually natural gas) is converted to mechanical work (or electricity in case of electrically driven compressors). Due to the energy consumption scale of the LNG plants, any enhancement to the energy efficiency of LNG plants will result in a significant reduction in gas consumption and consequently CO_2 emission.

There are two ways to increase the energy efficiency of natural gas liquefaction cycles: liquefaction cycle enhancement and driver cycle enhancement. Liquefaction cycle enhancements reduce the compressor power and consequently the compressor driver's fuel consumption. Driver cycle enhancement reduces the amount of fuel consumption to generate a specific amount of power. Typical natural gas liquefaction cycles utilize either pure refrigerant in cascade cycles, expansion-based cycles, or mixed refrigerant cycles. Pure refrigerant cycles have a constant evaporating temperature that is a function of the saturation pressure. Mixed refrigerant cycles, on the other hand, do not maintain a constant evaporating temperature at a given pressure. Their evaporating temperature range, called temperature glide, is a function of their pressure and composition. Refrigerant mixture of hydrocarbons and nitrogen is chosen so that it has an evaporation curve that matches the cooling curve of the natural gas with the minimum temperature difference. Small temperature difference reduces entropy generation and, thus, improves thermodynamic efficiency and reduces power consumption (Townsend and Linnhoff, 1983).

Due to the high complexity of natural gas liquefaction cycles design, their refrigerant mass fractions selection has been done by trial and error and guided only by heuristics (Lee et al., 2002). Mass fractions of refrigerant mixtures for certain cooling processes were patented (Alffev et al., 1972; Boyarsky et al., 1995). Typical refrigerant mixtures are a mixture of three or more of hydrocarbons (like methane or ethane) and nitrogen. Finding optimal mass fractions of constituents of refrigerant mixtures, which have the closest match between the cooling and heating curves, would require optimizing several variables (e.g., component mass fractions and pressure levels). Such complex design tasks would be impossible to handle without using optimization techniques. Optimization techniques could be employed to ensure optimal designs of liquefaction cycles and refrigerants mass fractions if robust models were applied (Arora, 2004).

The objective of this chapter is to reduce the energy consumption of the natural gas liquefaction cycle through process optimization and cycle enhancement. Optimization is briefly introduced and applied to natural gas liquefaction and driver cycles. Several energy enhancement options and waste heat utilization options are discussed and applied on the natural gas liquefaction cycle.

5.2 Natural gas liquefaction cycle enhancement types

There are two types of natural gas liquefaction cycle enhancements described in this chapter as shown in Figure 5-1. The first type is modifying the liquefaction cycle by changing the cycle configuration by adding a new process/component or replacing the current process/component with the new process/component. This type of enhancement will be discussed in Section 5.3 for both liquefaction cycle and driver cycle enhancements. Enhancement of the liquefaction cycle by using expanders and enhancement of driver cycles by using absorption chillers are used as examples for this enhancement type. The second type of enhancement is optimization of the current cycle processes/components. Here optimization refers to the procedure of setting process design variables in such a way that all the design criteria are met and the energy consumption is minimized. An optimal natural gas liquefaction cycle should use lower energy consumption than an unoptimized natural gas liquefaction cycle. A brief overview of the optimization methods with an example explaining Genetic Algorithm (GA) are presented in Section 5.4. In Sections 5.5 and 5.6, the application of optimization in design enhancements of the liquefaction cycles and compressor driver cycles are discussed. The challenges in optimizing mobile natural gas liquefaction cycles are described in Section 5.7.

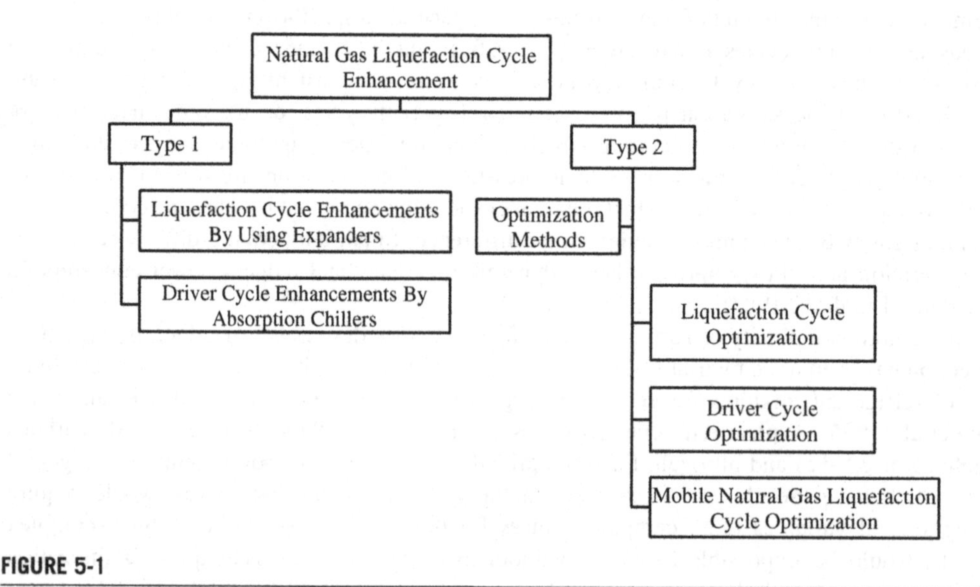

FIGURE 5-1

Two types of natural gas liquefaction cycle enhancements.

5.3 Energy consumption enhancement options by recovering process losses and waste heat

The first step in enhancing natural gas liquefaction cycles is allocating the location where unnecessary entropy is generated. In the liquefaction cycle, the entropy is generated in heat exchangers, expansion valves, and compressors. In the driver cycles, the entropy is generated in the compressor, combustion chamber, and turbine. Moreover, the exhaust of the gas turbine generates significant amounts of entropy due to its temperature difference with the ambient temperature (Borgnakke and Sonntag, 2008). Here in this section, as a case of enhancing the liquefaction cycle, the effect of reducing the entropy generation by replacing expansion valves with expanders is considered. For the driver cycle enhancement case, the effect of employing absorption chillers powered by the gas turbine waste heat is considered.

5.3.1 Expanders

The liquefaction cycle efficiency could be improved by replacing expansion valves with expanders. Here the expanders refer to any mechanical device that converts the pressure exergy (maximum recoverable work by reducing the refrigerant pressure) of a fluid to work by reducing its pressure. The reason for enhancement is that the fluid undergoes isenthalpic expansion process through the expansion valve while it ideally undergoes isentropic expansion process through the expander. The entropy increases during the isenthalpic expansion process is expressed in Equation 5-1 based on the Maxwell relation.

$$dh = Tds + vdP \tag{5-1}$$

where h, T, s, v, and P are specific enthalpy, temperature, specific entropy, specific volume, and pressure. This means that the enthalpy of the fluid after expansion process will be higher in the case of isenthalpic expansion process than that of isentropic expansion process. Hence in the case of expanding the refrigerant using expansion valves, the vapor quality of the refrigerant will be higher than that using the expander. This will result in lower cooling capacity per unit mass of the refrigerant. Meanwhile compression work per unit mass of the refrigerant stays constant for both cases. Therefore, replacing expansion valves with expanders will increase the liquefaction cycle energy efficiency.

There are three types of expanders: liquid turbines, two-phase expanders, and gas expanders. Liquid turbines or hydraulic turbines are a well-established technology. They are available with efficiencies over 90% (Gordon, 2001). In order to apply them to a refrigeration cycle, the refrigerant should be subcooled before entering the turbine. This should be done to prevent evaporation of the refrigerant inside of the liquid turbine. Two-phase expanders are under development with current efficiencies in the vicinity of 80% (Renaudin, 1995; Kanoglu, 2001) and can easily replace expansion valves used in vapor compression cycles. For expanding gases, gas expanders could be used instead of expansion valves. Gas expanders or gas turbines are a readily available technology and typically exist with efficiencies greater than 80% (Ordonez, 2000). Mortazavi et al. (2011) studied the effect of replacing expansion valves with expanders on the performance of propane precooled mixed component refrigerant (MCR) liquefaction cycle. The schematic diagram of this

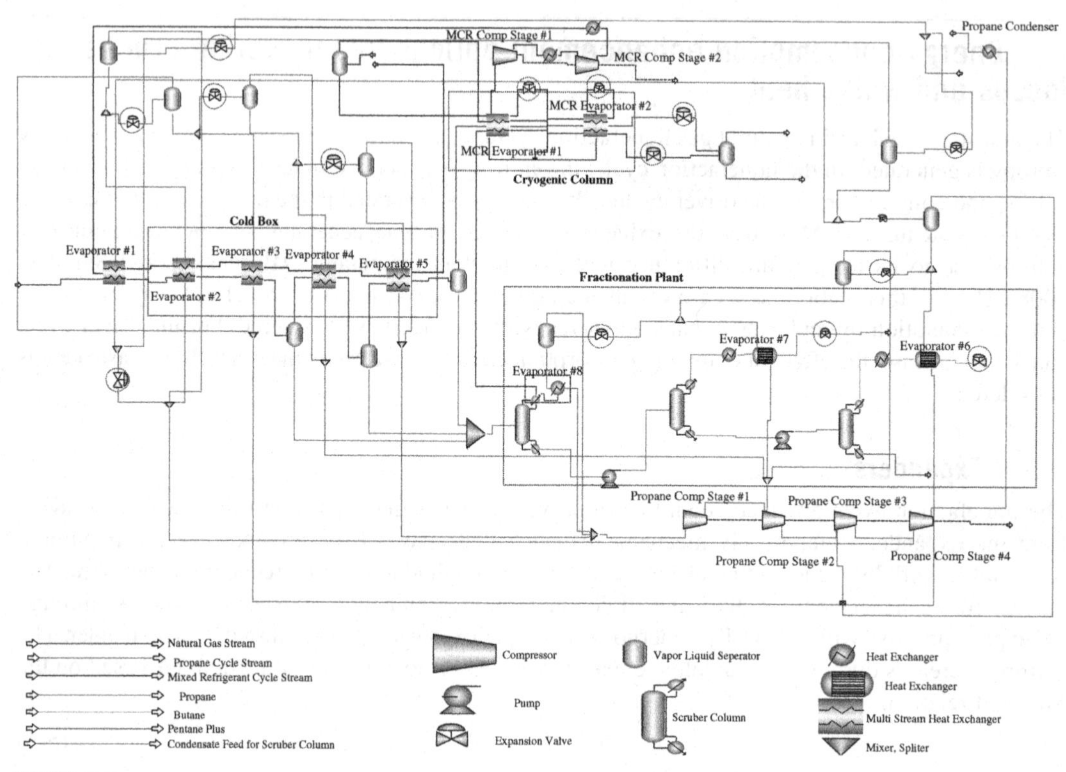

FIGURE 5-2

Schematic diagram of the propane precooled MCR liquefaction cycle with the points that have a potential of recovering expansion losses shown by dashed circles.

(Mortazavi et al., 2011)

cycle is shown in Figure 5-2. In the figure, the expansion processes that potentially can be recovered are marked with blue dashed circles.

The effect of replacing expansion valves used in the MCR and propane cycles and the LNG expansion process with expanders was investigated by Mortazavi et al. (2011). In their study, for replacing expansion valves used in the MCR cycle and the LNG expansion process, only two-phase expanders were considered. However, in the propane cycle both two-phase expanders and liquid turbines used for expansion of the liquid propane and the gas expander used for expansion of gaseous propane were considered for replacing the expansion valves. In their study the following four different enhancement options were modeled:

- **Option 1.** Enhancing the liquefaction cycle by replacing the LNG expansion valves by two-phase expanders. In Figure 5-2, the LNG expansion valves are located after the cryogenic column.
- **Option 2.** Improving the liquefaction cycle by replacing the expansion valves of the MCR cycle with two-phase expanders in addition to the enhancement of Option 1.

- **Option 3.** Improving the plant by replacing all the expansion valves with two-phase expanders for expanding liquid, and with gas expanders for expanding gas.
- **Option 4.** Enhancing the cycle by replacing the propane cycle expansion valves with liquid turbines for expansion of liquid propane and replacing the rest of the expansion valves with two-phase expanders and gas expanders.

It should be noted that when replacing the propane cycle expansion valves with liquid turbines the refrigerant (propane) should be subcooled before the expansion process to minimize the evaporation of the refrigerant inside of the turbine. Therefore, a slight change should be made to the propane cycle to subcool the refrigerant to near the saturation temperature corresponding to the expander outlet pressure. In the study, the isotropic efficiency of the gas expander, liquid turbines, and two-phase expanders were assumed to be 0.86, 0.85, and 0.85, respectively (see Mortazavi et al., 2010). The modeling results of these enhancement options are shown in Table 5-1.

As can be seen from Table 5-1, the liquefaction cycle enhanced by Option 4 is the most efficient cycle among the cycles investigated. Based on their results, its total power consumption, flash gases after the LNG expander, and energy consumed per unit mass of LNG are lower than those of the base cycle approximately by 2.43%, 96.09% and 3.68%, respectively. Moreover the expanders were expected to recover about 3.47% of total consumed power. Based on their modeling, the LNG production was also higher than that of the APCI base cycle by 1.24% from the same amount of feed gas. They claimed, if the recovered power is deducted from the total power consumption, the energy consumed per unit mass of LNG could be reduced by 7.07%. Since replacing expansion valves with

Table 5-1 Modeling Results of APCI Enhanced Cycles

Cycle Option	Base Cycle	Option 1	Option 2	Option 3	Option 4
Propane cycle compressor power [MW]	43.651	43.651	42.953	42.774	42.143
MCR cycle compressor power [MW]	66.534	66.534	65.375	65.375	65.368
Propane cycle cooling capacity [MW]	115.469	115.469	113.939	113.937	113.962
MCR cycle cooling capacity [MW]	67.635	67.635	67.634	67.635	67.631
LNG vapor fraction after the expander	0.0142	0.0006	0.0006	0.0006	0.0006
Total power consumption [MW]	110.185	110.185	108.324	108.148	107.506
Flash gases after LNG expander [kg/s]	1.28	0.05	0.05	0.05	0.05
Energy consumption per unit mass of LNG [MJ/kg]	1.115	1.101	1.083	1.081	1.074
Recovered power from expanders [MW]	—	0.648	2.528	3.296	3.821
LNG production [kg/s]	98.83	100.06	100.06	100.06	100.06

(Mortazavi et al., 2010)

expanders mainly utilizes the pressure exergy of the refrigerant, which is not significant, the efficiency enhancements were small. Although the enhancements were not considerable from a percentage point of view, if it is seen from an energy perspective it will lead to considerable savings and recovered power due to the scale of the plant.

5.3.2 Absorption chiller

One of the avoidable entropy generation sources in the natural gas liquefaction plant is waste heat produced from driver cycles. A waste heat source is defined as any hot exhaust gases or other streams that are cooled by rejecting heat to the ambient. The amount of avoidable entropy is proportional to the temperature difference between these streams and ambient temperature. The main sources of waste heat typically available at LNG plants are:

- Gas turbine exhaust gases
- Flared gases
- Boiler exhaust gases.

Based on the waste heat temperature, the amount of waste heat, and utility requirement, such waste heat could be utilized for water desalination, air conditioning, refrigeration, gas turbine inlet cooling, heating, steam generation, and power generation. As an example case, the use of waste heat in enhancing the energy efficiency of a propane precooled MCR natural gas liquefaction cycle is considered. This study is taken from Mortazavi et al. (2010). In their study the waste heat from the gas turbine that drove the propane and MCR compressors was utilized to generate cooling from absorption chillers. Absorption chillers are the machines that convert heat to the refrigeration effect. There are a number of absorption chiller technologies with different applications. One of the measures used to evaluate absorption chiller performance is coefficient of performance (COP). COP is defined as the amount of cooling provided by the absorption chiller divided by the amount of energy consumed by the absorption chiller. Most absorption chillers use either water/lithium-bromide or ammonia/water as their working fluid pair. In working fluid nomenclature the first part represents the refrigerant and the second part represents the absorbent fluid. Water/lithium-bromide absorption chillers tend to have higher COPs compared to ammonia/water chillers. Meanwhile, the water/lithium-bromide absorption chillers evaporator cannot be below about 2°C due to water freezing; moreover there is crystallization issue if they are operated with high absorber temperature. On the other hand, ammonia/water absorption chillers evaporator temperature can be lower than –30°C. Conventional absorption chillers can be single or double-effect. Single-effect absorption chillers are less complex by having less number of components and requiring lower temperature waste heat; however their COPs are lower than those of double-effect chillers. For more information regarding the absorption chillers please refer to Herold et al. (1996).

To quantify the amount of available waste heat from the gas turbine, gas turbine exhaust temperature and mass flow rate should be assessed. In the Mortazavi et al. (2010) study such information was derived by modeling the gas turbine. The comparison of the gas turbine simulation results with vendors' data at ISO condition, which is 15°C and 1 atm inlet pressure, is shown in Table 5-2.

In the study a double-effect design was selected based on the gas turbine exhaust temperature and the minimum temperature that the exhaust can be cooled. It should be noted that lowering the exhaust temperature below 180°C increases the chance of corrosion due to condensation. They

Table 5-2 Comparison of Gas Turbine Modeling Results with Vendors' Data at ISO Condition

Parameter	ISO Rated Power [MW]	Efficiency [%]	Exhaust Temperature [°C]
Actual gas turbine	130.100	34.6	540.0
ASPEN model	130.103	35.0	540.4
Discrepancy	+0.003 (0.0%)	+0.4 (1.16%)	+0.4

(Mortazavi et al., 2010)

selected water/lithium-bromide as a refrigerant/absorbent pair due to its superior COP compared to ammonia/water. In their investigation they assumed that the plant is located on the Persian Gulf coast with 45°C ambient temperature and 35°C seawater temperature. Based on these ambient conditions they modeled a double-effect water/lithium-bromide chiller for two different evaporating temperatures corresponding to the two highest temperature stages of the propane cycle. Their modeling results are shown in Table 5-3. The COP values were used to estimate the amount of waste heat that is needed to produce the amount of cooling required for different enhancement options.

For a fixed production capacity, the energy consumption of a liquefaction cycle could be reduced by either reducing the compressor power demand or increasing the gas turbine power generation efficiency. The absorption chiller's cooling can be used to reduce the compressor's power demand. This reduction can be done by lowering the propane cycle condenser temperature, replacing the propane evaporators by waste heat powered absorption chillers, and/or intercooling the compressor of the MCR cycle. Moreover, the absorption chiller's cooling can be used to increase gas turbine power generation efficiency by cooling the inlet air of the gas turbine.

Mortazavi et al. (2010) developed an integrated gas turbine and liquefaction cycle model to investigate the available amount of waste heat for different enhancement options. In their integrated model, their gas turbine model was scaled to satisfy the liquefaction cycle power demand at 45°C ambient temperature under its full load condition. Scaling the gas turbine is not an unreasonable assumption due to the fact that some gas turbine manufactures scale their gas turbine design to meet for different demands. Eight enhancement options of waste heat utilization using absorption chillers were considered, as summarized in Table 5-4.

To account for part load effects they considered two cases for each enhancement option. The first case, which was referred to as an unscaled case, was based on the assumption that for each option the gas turbine would be the same size as that of the base cycle (cycle without enhancement). This

Table 5-3 Modeling Results of Absorption Cycle

Evaporating Temperature	COP
9°C	1.284
22°C	1.489

(Mortazavi et al., 2010)

Table 5-4 Different Waste Heat Utilization Options

Option	Description
1	Replacing 22°C propane cycle evaporators with absorption chillers
2	Replacing 22°C propane cycle evaporators and cooling the inlet of gas turbine with absorption chillers
3	Replacing 22°C and 9°C propane cycle evaporators with absorption chillers
4	Replacing 22°C and 9°C propane cycle evaporators and cooling the inlet of gas turbine with absorption chillers
5	Replacing 22°C and 9°C evaporators and cooling the condenser of propane cycle at 27°C with absorption chillers
6	Replacing 22°C and 9°C evaporators and cooling the condenser of propane at 27°C cycle and turbine inlet with absorption chillers
7	Replacing 22°C and 9°C evaporators and cooling the condenser of propane cycle at 14°C with absorption chillers
8	Replacing 22°C and 9°C evaporators and cooling the condenser of propane at 14°C cycle and intercooling the compressor of mixed refrigerant cycle with absorption chillers)

(Mortazavi et al., 2010)

assumption allowed them to observe some part load degradation effects. The second case, referred to as a scaled case, was based on the assumption that for each option a gas turbine is sized to provide the maximum liquefaction cycle power plant demand at its full load. In their modeling they assumed the gas turbine exhaust gas could be cooled down to 200°C. Their simulated enhancement option results are shown in Table 5-5.

As Table 5-5 shows, the options are placed based on their energy efficiency enhancement order, where Option 8 has the highest amount of enhancement. Based on their results, by implementing Option 8 the compressor power demand and hence gas turbine fuel consumption could be reduced by 21.3%. Within all the options the scaled gas turbine cases have lower fuel consumption than those of the unscaled cases. This difference indicates that the efficiency of the scaled case is higher than the unscaled case for the same option. This difference is due to the fact that in the unscaled case the gas turbine is operated at part load running condition, whereas in the scaled case the gas turbine is operated in full load. Meanwhile the gas turbine firing temperature in part load conditions are lower than the full load condition, which leads to lower efficiency of the gas turbine. Based on Table 5-5, the better the option, the more amount of waste heat is required to operate the absorption chillers, which means less entropy generation.

5.4 Brief introduction to optimization

Design of a system involves selecting variables that meet system requirements that are defined beforehand by a customer. For example, a heat exchanger designer needs to know the required heating load and other product requirements specified by the customer. Then the designer would select design variables such as the heat exchanger material, dimensions, and so on that meet the heat exchanger requirements. Several heat exchanger designs could be developed that meet the same heat load requirements. However, the heat exchanger's performance and cost depend on the selected variables.

Table 5-5 Enhancement Results of Different Waste Heat Utilization Options

Gas Turbine Sizing			Scaled Turbine Size Case			Unscaled Turbine Size Case		
Option	**Compressor Power [MW]**	**Power Reduction [MW]**	**Required Amount of Waste Heat [MW]**	**Fraction of Available Amount of Waste Heat [%]**	**Fuel Consumption [MW] (% saving)**	**Required Amount of Waste Heat [MW]**	**Fraction of Available Amount of Waste Heat [%]**	**Fuel Consumption [MW] (% saving)**
Base cycle	110.185	—	—	—	329.448	—	—	329.448
Option 1	107.510	2.675 (2.43%)	8.865	5.948	321.444 (2.43)	8.865	5.958	322.754 (2.03)
Option 2	107.510	2.675 (2.43%)	12.572	8.985	314.002 (4.69)	12.913	9.341	318.175 (3.42)
Option 3	100.334	9.851 (8.94%)	34.998	25.162	299.999 (8.94)	34.998	25.311	304.859 (7.46)
Option 4	100.334	9.851 (8.94%)	41.579	33.537	287.624 (12.70)	43.112	36.306	296.482 (10.01)
Option 5	94.043	16.142 (14.65%)	98.306	75.408	281.186 (14.65)	98.306	76.133	289.482 (12.21)
Option 6	94.043	16.142 (14.65%)	104.474	89.899	269.598 (18.17)	106.420	96.977	281.006 (14.70)
Option 7	88.420	21.765 (19.75%)	105.553	86.112	264.378 (19.75)	105.553	87.227	275.340 (16.42)
Option 8	86.696	23.489 (21.32%)	116.391	96.844	259.217 (21.32)	116.391	98.195	271.086 (17.72)

(Mortazavi et al., 2010)

The optimal design represents the design that has the most desired objectives, such as minimum cost or highest performance.

In the conventional design approach, the process of selecting design variables is based on a designer's intuition and experience. A designer could achieve optimal design with the conventional design approach when designing simple products, which have only a limited number of design variables. On the other hand, it is difficult and could be impossible to achieve optimal design with the conventional design approach when designing complex systems with many variables.

The process of finding the optimal design is called optimization. Optimization uses mathematical techniques to achieve the optimum design. In the optimization formulation, there is an objective function whose value should be optimized. Moreover, there might be some constraints that should be satisfied by the optimum design. In an optimization procedure, any point that satisfies the constraints is called a feasible point. The variables that are varied during an optimization procedure are called design variables. The variables that are assumed to be fixed during an optimization procedure are called design parameters. It should be noted that an optimization problem can have multiple objective functions. This kind of optimization is called a multiobjective optimization.

There are two types of optimization problems, deterministic (conventional) and stochastic (Lee, 2001). In the deterministic optimization it is assumed that there is no uncertainty in the design variables and parameters. However, in the stochastic optimization problems the uncertainties in the design variables and parameters are considered. One class of stochastic optimization problems is robust optimization. In robust optimization the solutions should be feasible for all variations of the uncertain variables and parameters. Mathematically the deterministic optimization problems are stated as follows:

$$\min_{\mathbf{x}} f(\mathbf{x}, \mathbf{p}) \tag{5-2}$$

Subject to:

$$g_i(\mathbf{x}, \mathbf{p}) \leq 0 \quad i = 1, \ldots, I$$
$$h_j(\mathbf{x}, \mathbf{p}) = 0 \quad j = 1, \ldots, J$$

where f, g_i, and h_j are called objective function, inequality constraint, and equality constraint, respectively. $\mathbf{x}$ is vector of design variables, and $\mathbf{p}$ is vector of design parameters. In deterministic optimization, design parameters are fixed but design variables are varied. An overview of conventional optimization methods is presented in Section 5.4.1. Section 5.4.4 provides a brief overview of robust optimization.

5.4.1 Conventional optimization methods

Generally optimization methods can be classified into general methods and methods tailored for a specific class of problems. Specific methods such as linear programming and quadratic programming are more efficient than the general methods in solving the problems because they are tailored for it. However, they are not applicable to general problems. General methods can be divided to local optimization methods and global optimization methods. Except for specific problems, local optimization methods only provide results that are locally optimal. However, their computational cost is lower than those of global search methods. Newton method and sequential quadratic programming are examples of local optimization methods. Global optimization methods are heuristic-based methods.

This means that there is no guarantee for their result to be globally optimal. Genetic algorithm (GA) and simulated annealing are examples of methods that do not have any restriction in the type of functions that are used in stating the objective and constraint functions. The GA optimization method will be explained briefly in the next section because it was used in the LNG plant refrigerant and driver optimization sections. Interested readers are referred to optimization books such as Bazaraa et al. (1993), Arora (2004), or Gen and Cheng (2007).

5.4.2 Genetic algorithm

Genetic algorithm (GA) is a class of heuristic optimization methods. GA mimics the process of natural evolution by modifying a population of individual solutions. Design points, x's, are represented by chromosomes. The method randomly selects individuals from the current population to be parents and uses them to produce the children for the next generation. Over consecutive generations, the population approaches an optimal solution because "good" parents produce "good" children. The "bad" points are eliminated from the generation. GA can be applied to solve a variety of optimization problems in many applications that are not suited for conventional optimization methods, including problems in which the objective function is discontinuous, nondifferentiable, or nonlinear. GA has the potential to reach the global optimal solution if it does not stick at a local optimal solution. Since GA is a probabilistic approach, different solutions could be generated by different runs. Therefore, multiple runs are required to verify the optimal solution. A GA flowchart is shown in Figure 5-3. The steps involved in using GA are described in the following sections, using a simple refrigerant composition optimization problem detailed next.

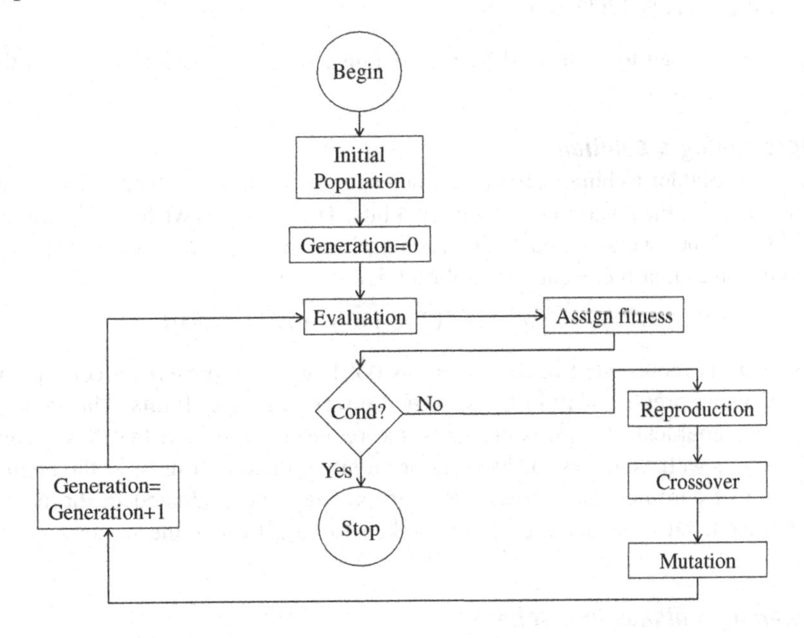

FIGURE 5-3

Flowchart showing working principle of GA.

(Deb, 2001)

5.4.3 Example of GA

Finding the optimal refrigerant composition is difficult without optimization. Suppose we would like to find the optimal refrigerant composition of mixture for a vapor compression cycle that has a given cooling load. The optimal refrigerant would require less compression power for the same cooling provided. The formulation of the optimization problem is as follows:

Objective function:

Minimize compressor power $(x_1, x_2, x_3, x_5, x_6)$

where

x_1, x_2, x_3, x_4 are mass fractions of constituents
x_5 total refrigerant mass flow rate
x_6 refrigerant condensing pressure

Subjected to:

Evaporator cooling capacity = Baseline value
Evaporator cooling temperature = Baseline value

$$x_4 = 1 - (x_1 + x_2 + x_3)$$

$$0 \le x_1, x_2, x_3, x_4 \le 1$$

$$80\%\ x_{5,\text{Baseline}} \le x_5 \le 120\%\ x_{5,\text{Baseline}}$$

$$80\%\ x_{6,\text{Baseline}} \le x_6 \le 120\%\ x_{6,\text{Baseline}}$$

x_5 and x_6 were assumed to be limited by a 20% range from the baseline values so that we have a limited design space.

5.4.3.1 Representing a solution

GA represents the variables as binary strings. In the example, we have five variables because x_4 is not an independent variable. Each variable is coded in 4 bits. The more bits we have, the more precise the variables would be. Since we assumed 4 bits, each variable can have 2^4 possible values. One 20-bit string (or chromosome) that represents one solution is as follows:

$$x_1,\ x_2,\ x_3,\ x_5,\ x_6 = 0110\ 1100\ 0101\ 1110\ 0001$$

For example, x_6 is represented in this string as 0001, which represents a certain value for the refrigerant condensing pressure within the specified lower and upper limits. The same principle is true for the other variables. Thus, this chromosome represents a design that has a corresponding compressor power as well as values for meeting or violating the constraints. In the beginning of the GA, a population of chromosomes is randomly generated to be evaluated in the next steps. The number of the generated chromosomes can be taken to be 20 times the number of the decision variables.

5.4.3.2 Assigning a fitness to a solution

Once the initial population was generated, a fitness function is assigned to evaluate the "goodness" or "badness" of a solution or a chromosome. In a minimization problem, the good solution has small fitness function value and the bad solution has high fitness function value. The generated solutions are evaluated based on the objective function value and constraints violation. In our example, the objective function is the minimization of the compressor power and the constraints are the evaporator capacity and temperature. Following fitness assignment, the GA checks for termination conditions and, if not met, a new population is generated based on three genetic operators, which are the reproduction, crossover, and mutation operators.

5.4.3.3 Reproduction operator

The reproduction operator duplicates "good" solutions and eliminates "bad" solutions in the population. Different methods exist in the reproduction operator such as tournament selection or ranking selection (Deb, 2001). Tournament selection method, for example, compares two solutions and the best one survives in the mating pool. Ranking selection method lists the population based on individual solutions' fitness values. As the GA runs, only "good" solutions survive in the population.

5.4.3.4 Crossover operator

The reproduction operator does not generate a new solution. It just duplicates "good" solutions and eliminates "bad" solutions. On the other hand, the crossover operator is used to generate new solutions. It selects two chromosomes, called parents (e.g., A and B in Figure 5-4), from the mating pool and exchanges bits at certain points to generate a new solution, which is called offspring (e.g., A1 and B1 in Figure 5-4). The "good" parents hopefully will produce "good" offspring.

5.4.3.5 Mutation operator

The mutation operator has the same objective as the crossover operator—to generate new solutions and possibly jump out from a local optimum. It works by exchanging a string in the mating pool from 0 to 1 or vice versa as shown in Figure 5-5.

5.4.3.6 Stopping criteria

Different stopping criteria can be implemented in GA. Common stopping criteria are reaching the maximum number of generations or the change in the best solution in successive iteration is less than the prespecified value (convergence tolerance).

5.4.4 Robust optimization methods

One of the main challenges in engineering optimization is that the deterministic optimization algorithms do not account for the uncertainty involved in the design process. Neglecting the uncertainties (in design parameters and variable) could result in designs that are infeasible or have a poor performance. The uncertainty in design variables might be reducible by using more precise manufacturing

Solution A: 1110 1101 0011 1011 1111 → Solution A1: 1110 1101 0011 1010 0001
Solution B: 0110 1100 0101 1110 0001 → Solution B1: 0110 1100 0101 1111 1111

FIGURE 5-4

Crossover operator. (Solution A and B are called parent solution while A1 and B1 are called offspring.)

Solution A1: 1110 1101 0011 1110 001 → Solution A2: 1111 1101 0011 1110 0001

FIGURE 5-5

Mutation operator. (Solution A1 is old solution and Solution A2 is the new solution.)

techniques, which would increase production cost. However, there are uncontrollable parameters such as weather conditions, energy price, and market demand, whose uncertainties are not reducible. Therefore, deterministic optimization might not be the best optimization technique for many of the real world engineering problems. The goal of robust optimization algorithms is finding an optimal solution that is insensitive to design variables and parameter uncertainty. This design is called a robust design. There are two types of robustness: objective robustness and feasibility robustness (Li et al., 2006). A design $\mathbf{x}$ is said to be objectively robust if the objective variation remains in the specific range for all realizations of the design variables and parameters in their uncertainty range. Here the robustness formulation is developed for problems in which the uncertainty is in the form of a symmetric interval with the nominal point as its center. Mathematically objective robustness can be stated as

$$\max_{\Delta\mathbf{x},\Delta\mathbf{p}} \frac{|f(\mathbf{x}+\Delta\mathbf{x},\mathbf{p}+\Delta\mathbf{p})-f(\mathbf{x},\mathbf{p})|}{|f(\mathbf{x},\mathbf{p})|} \leq \Delta f^{*} \tag{5-3}$$

$$-\Delta\tilde{\mathbf{x}} \leq \Delta\mathbf{x} \leq \Delta\tilde{\mathbf{x}}$$

$$-\Delta\tilde{\mathbf{p}} \leq \Delta\mathbf{p} \leq \Delta\tilde{\mathbf{p}}$$

where $\Delta f^{*}$is the acceptable variation of the objective function. $\Delta\tilde{\mathbf{x}}$ and $\Delta\tilde{\mathbf{p}}$ are the design variables and parameters uncertainty range and they are nonnegative vectors.

For the problem where the objective robustness is based on the degradation of the objective function Equation 5-3 becomes (Mortazavi et al., 2012):

$$\max_{\Delta\mathbf{x},\Delta\mathbf{p}} \frac{f(\mathbf{x}+\Delta\mathbf{x},\mathbf{p}+\Delta\mathbf{p})-f(\mathbf{x},\mathbf{p})}{|f(\mathbf{x},\mathbf{p})|} \leq \Delta f^{*} \tag{5-4}$$

In Equation 5-4 it is permissible if the variation of the uncertain design variables and parameters reduce the objective function value. However, if it is increasing the objective function value it should be a less than acceptable deviation of the objective function.

A design $\mathbf{x}$ is defined to be robust feasible if it stays feasible for all the realizations of the design variables and parameters in their uncertainty range. Mathematically it is shown as

$$\max_{\Delta\mathbf{x},\Delta\mathbf{p}} g_i(\mathbf{x}+\Delta\mathbf{x},\mathbf{p}+\Delta\mathbf{p}) \leq 0 \quad i=1,\ldots,I \tag{5-5}$$

$$-\Delta\tilde{\mathbf{x}} \leq \Delta\mathbf{x} \leq \Delta\tilde{\mathbf{x}}$$

$$-\Delta\tilde{\mathbf{p}} \leq \Delta\mathbf{p} \leq \Delta\tilde{\mathbf{p}}$$

Note that it is not possible to have a robust design for equality constraints.

The process of finding the most desired robust design is called robust optimization. Mathematically it is stated as

$$\min_{\mathbf{x}} f(\mathbf{x}, \mathbf{p}) \tag{5-6}$$

subject to:

$$\underset{\Delta\mathbf{x},\Delta\mathbf{p}}{\text{Max}}\ g_i(\mathbf{x} + \Delta\mathbf{x}, \mathbf{p} + \Delta\mathbf{p}) \leq 0 \quad i = 1, \ldots, I$$

$$-\Delta\tilde{\mathbf{x}} \leq \Delta\mathbf{x} \leq \Delta\tilde{\mathbf{x}}$$

$$-\Delta\tilde{\mathbf{p}} \leq \Delta\mathbf{p} \leq \Delta\tilde{\mathbf{p}}$$

Note that the equality constraints do not exist in robust optimization. This is due to the fact that it is not possible to maintain an equality constraint when there is a variation (due to uncertainty) in design variables and parameters.

Up to now all the natural gas liquefaction cycle optimizations were conducted without implementing the robust optimization techniques. Therefore their results might become infeasible in the real world operating conditions. In the real world application, designers are using safety factors to account for the uncertainties. Although this approach might lead to a robust design it usually does not lead to an optimum robust solution. It is proposed to use robust optimization techniques to find the realistic enhancement design option for the liquefaction cycle. This will increase the applicability of these enhancement options to the real-world LNG plants.

5.5 Liquefaction cycle optimization

As explained earlier, the GA has the capability of optimizing a natural gas liquefaction cycle. Since among all available cycles, the APCI liquefaction cycle, developed by Air Products and Chemicals Inc., is the most predominant cycle in the LNG industry (Paradowski et al., 2004), it was selected to demonstrate the power of optimization using GA. The optimization approach followed here would be similar when it is applied to different liquefaction cycles optimization. More details on the liquefaction cycle optimization problem and a comparison between different optimization results obtained from different optimization methods can be found in Alabdulkarem et al. (2011).

5.5.1 Model development

The model of the APCI liquefaction cycle can be found in Mortazavi et al. (2010). The model was based on ASPEN Plus software v7.1 (2010), which is a robust steady-state process simulation software. ASPEN Plus has a built-in optimization tool. However, it does not have GA methods available and we were not able to link it with MatLab (2009), which has GA as well as different optimization tools. On the other hand, HYSYS v7.1 (2010), a steady state and transient process simulation software, does not have a GA either but we were able to link it with MatLab and utilize its optimization capabilities. Therefore, a model was developed using HYSYS for the APCI liquefaction cycle with the same inputs and assumptions used in the ASPEN Plus model. The inputs and assumptions used in the two models are essentially identical.

Table 5-6 Gas Composition after Gas Sweetening

Component	Mole Fraction (%)
Nitrogen	0.100
Carbon Dioxide	0.005
Methane	85.995
Ethane	7.500
Propane	3.500
i-Butane	1.000
n-Butane	1.000
i-Pentane	0.300
n-Pentane	0.200
Hexane Plus	0.400
Total	100

(Mortazavi et al., 2010)

Gas sweetening units were not modeled for simplification purposes and the feed gas was assumed to be after the sweetening units with the composition shown in Table 5-6. The two sections of the spiral-wound heat exchanger (SPWH) were modeled using the segmented UA method. Using the segmented UA method means applying the UA method on several intervals inside the heat exchanger so that the cooling curve inside the SWHX is calculated more accurately. The first section of the SPWH has 53 segments and the second section has 38 segments. Simulation results from models in HYSYS and ASPEN Plus are compared in Table 5-7. It can be seen that there is a good agreement between the two software models. The reason for the discrepancies in some parameters is due to the fact that the binary coefficients of the equation of state used in HYSYS and ASPEN Plus are slightly

Table 5-7 Simulation Results of APCI Base Cycle Model in ASPEN Plus and HYSYS

Model	ASPEN Plus	HYSYS	Discrepancy (%)
Propane compressor power (MW)	43.651	43.698	−0.1
Mixed refrigerant compressor power (MW)	66.534	66.48	0.08
Propane cycle cooling capacity (MW)	115.469	115.733	−0.22
Mixed refrigerant cycle cooling capacity (MW)	67.635	67.508	0.18
Propane cycle coefficient of performance	2.645	2.648	−0.13
LNG vapor fraction after expansion valve (%)	1.4	1.43	−2.14
LNG production (kg/s)	98.83	98.89	−0.06
LPG (propane, butane, pentane, and heavier hydrocarbons) (kg/s)	11	10.97	0.27
Flash gas flow rate after LNG expansion valve (kg/s)	1.28	1.24	3.13

(Alabdulkarem et al., 2011)

different from each other. The schematic of the APCI liquefaction cycle modeled in HYSYS is shown in Figure 5-6. There is a break point in the process flow sheet, which is required for HYSYS and ASPEN Plus model convergence. However, inside the MatLab code the two sides of the break point are set to be identical to each other.

5.5.2 Optimization approach

Since there are many variables involved in designing the APCI liquefaction cycle, the optimization problem is computationally very expensive. Thus, the optimization was carried out in two stages as shown in Figure 5-7. First, the MCR cycle was optimized, and then the propane cycle was optimized. The only effect of separating the two cycles is that the propane cycle precools the MCR cycle. This precooling load changes with refrigerant mixture composition, temperature, pressure, and mass flow rate. This effect was considered as optimization constraints. MatLab has a powerful optimization tool. It has different optimization methods such as gradient-based and GA. To verify the results of the optimization, different optimization methods (e.g., gradient-based or pattern search; Arora, 2004) from MatLab as well as optimization tools from the HYSYS and ASPEN were used. Results show that the optimum results (i.e., minimum power consumption) were found with the GA method. For instance, the pattern search optimization method resulted in a 7.16% saving in MCR power consumption whereas the GA resulted in a 13.28% saving in MCR power consumption for an identical optimization problem.

Different optimization methods lead to different results (e.g., different refrigerant composition mixtures). A comparison of different optimization methods to find refrigerant composition mixtures was conducted by Alabdulkarem et al. (2011). The number of design variables for the MCR cycle is eight, compared with 14 for the propane cycle. Each cycle takes between 16 and 24 hours to solve with a desktop computer having an Intel Core 2 Duo processor (2.83 GHz) CPU with 2 GB of RAM. Typical GA tuning parameters used in the optimization are listed in Table 5-8.

The objective function of the MCR cycle optimization and propane cycle optimization is to reduce the total power consumption of the APCI liquefaction cycle. The power consumption comes from the compressors and seawater pumps, as calculated by HYSYS software. The model in HYSYS is treated as a black box in this optimization study. The variables of the MCR optimization are listed in Table 5-9. The range of the optimization variables is taken to be $\pm 20\%$ of the baseline values. The MCR optimization's seven constraints with limits taken from the baseline model are:

1. Temperature of the LNG $\leq -160°C$
2. LNG mass flow rate $= 98.89$ kg/s
3. Precooling load of the propane cycle ≤ 88.22 MW
4. Two compressors with inlet vapor quality ≥ 0.99
5. Two heat exchangers with pinch temperatures ≥ 3 °K.

The propane cycle optimization variables are listed in Table 5-10. The range of the optimization variables is also taken to be $\pm 20\%$ of the baseline values. Changing an intermediate compressor outlet pressure is equivalent to changing a corresponding evaporation temperature because each heat exchanger/evaporator is connected to a compressor outlet. The propane cycle optimization's 17 constraints with limits taken from the baseline model are:

1. Three LPG produced vapor quality ≤ 0.01
2. Propane fuel production $= 2.95$ kg/s

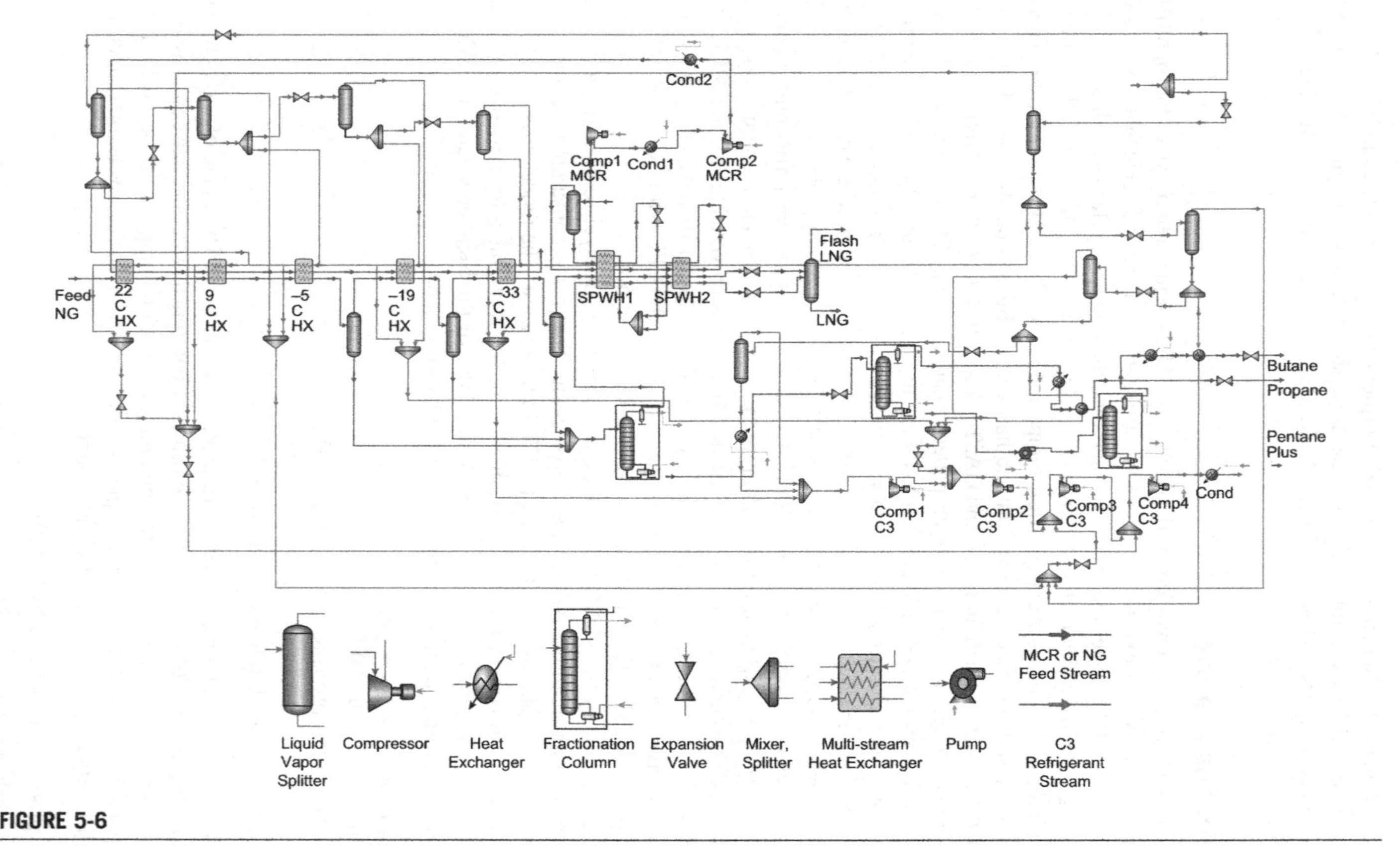

FIGURE 5-6

APCI liquefaction cycle's HYSYS model.

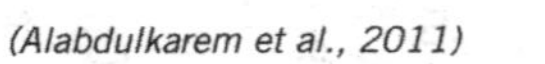

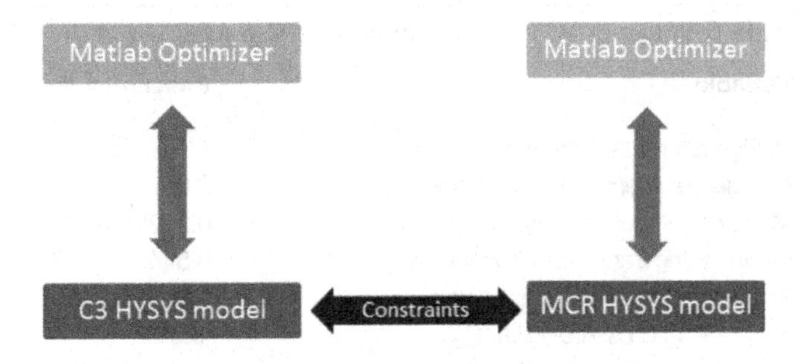

FIGURE 5-7

APCI liquefaction cycle optimization approach.

(Alabdulkarem et al., 2011)

3. Pentane plus fuel production = 5.1 kg/s
4. Butane fuel production = 2.92 kg/s
5. Condensing load of the fractionation unit = 5.07 MW
6. Four compressors with inlet vapor quality $\geq$ 0.99
7. Six heat exchangers with pinch temperatures $\geq$ 3 °K.

5.5.3 Optimization results

The optimization results of the APCI natural gas liquefaction cycle are described in Section 5.5.3.1 for the MCR cycle optimization and Section 5.5.3.2 for the propane cycle optimization.

5.5.3.1 MCR cycle optimization

The optimized MCR cycle has a power consumption of 63.63 MW, which is 4.48% less than the baseline power consumption. The optimized MCR cycle has lower refrigerant mass flow rate and lower

Table 5-8 Typical GA Tuning Parameters

Tuning Parameters	Value
Population size	20 × number of design variables
Reproduction count	50% of the population size
Maximum number of generations	100
Crossover fraction	0.8
Selection method	Tournament
Tournament size	8
Fitness scaling method	Top
Number of crossover points	1
Mutation method	Adaptive feasible

(Alabdulkarem et al., 2011)

Table 5-9 List of MCR Cycle Optimization Variables

Variable	Baseline Value
Refrigerant mass flow rate (kg/s),$\dot{m}$	270
Nitrogen refrigerant mass fraction, x_{N2}	0.0971
Methane refrigerant mass fraction, x_{C1}	0.2225
Ethane refrigerant mass fraction, x_{C2}	0.5445
Propane refrigerant mass fraction, x_{C3}	0.1359
Evaporating pressure (kPa), P_{Ex}	420
Intercooling pressure (kPa), P_i	2,300
Condensing pressure (kPa), P_H	4,000

(Alabdulkarem et al., 2011)

Table 5-10 List of Propane Cycle Optimization Variables

Variable	Baseline Value
Refrigerant mass flow rate, $\dot{m}$ (kg/s), x_1	447
Condensing temperature, T_c (°C), x_2	43
Compressor 1 outlet pressure (kPa), x_3	253
Compressor 2 outlet pressure (kPa), x_4	406
Compressor 3 outlet pressure (kPa), x_5	618
Compressor 4 outlet pressure (kPa), x_6	1,540
22°C Heat exchanger pressure (kPa), x_7	882
Refrigerant mass splitter to the −19°C HX split ratio, x_8	0.465
Refrigerant mass splitter to the −5°C HX split ratio, x_9	0.758
Refrigerant mass splitter to the 9°C HX split ratio, x_{10}	0.143
Refrigerant mass splitter to the 16°C HX split ratio, x_{11}	0.886
Refrigerant mass splitter to the 22°C HX split ratio, x_{12}	0.705
Refrigerant mass splitter to liquefy the propane produced, x_{13}	0.909
Refrigerant mass splitter to liquefy the butane produced, x_{14}	0.938

(Alabdulkarem et al., 2011)

overall compression ratio than those of the baseline cycle as shown in Table 5-11. Nitrogen and propane mass fraction increased while the methane and ethane decreased. Since nitrogen has the lowest boiling temperature among the constituents, it lowers the lowest refrigeration temperature. On the other hand, propane has the highest boiling temperature, which increases the refrigeration capacity of the refrigerant. A plot of the cooling curves for the optimized and baseline MCR cycles is shown in Figure 5-8. The SWHX has two sections with log mean temperature difference (LMTD) for the optimized cycle of 5.24°C and 4.91°C, whereas the LMTDs of the baseline cycle are 7.12°C and 5.17°C. Figure 5-8 and the LMTD values show that the cold curve in the optimized cycle is closer to

Table 5-11 MCR Cycle Optimization Results

Cycle	Variables								Objective Function
	$\dot{m}$ (kg/s)	x_{N2}	x_{C1}	x_{C2}	x_{C3}	P_i (kPa)	P_H (kPa)	P_{Ex} (kPa)	Min. Power Cons. (MW)
Baseline	270	0.0971	0.2225	0.5445	0.1359	2300	4000	420	66.62
Optimized	267	0.1027	0.218	0.5306	0.1487	2346	4137	451	63.63

(Variable definitions are given in Table 5-10)
(Alabdulkarem et al., 2011)

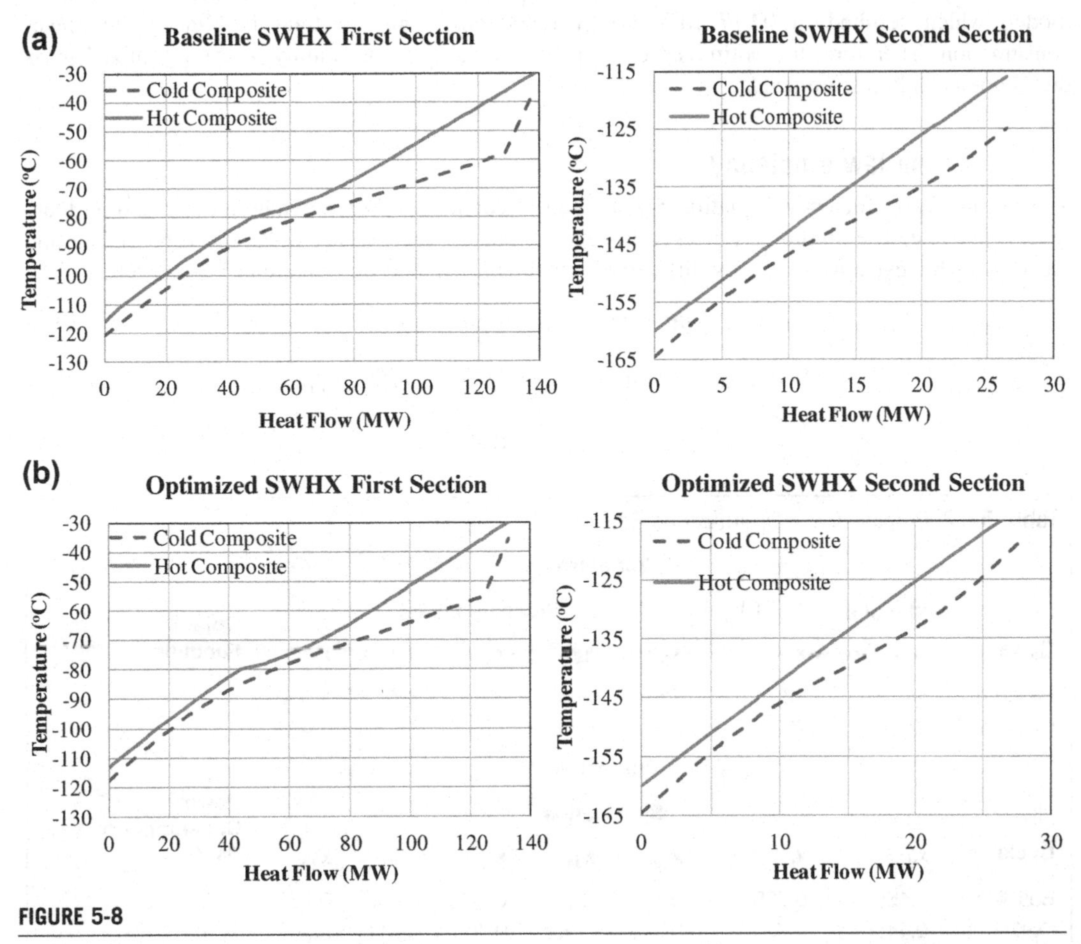

FIGURE 5-8

Cooling curves in SWHX of the (a) baseline and (b) optimized MCR cycles. The cooling curves are closer to the heating curves in the optimized cycle than the baseline cycle at equivalent pinch temperature of 3°K.

(Alabdulkarem et al., 2011)

the hot curve than the baseline cycle, which means more efficient heat transfer or less entropy generation in the heat exchanger.

5.5.3.2 Propane cycle optimization

The optimized propane cycle has a power consumption of 37.15 MW, which is 15.98% less than the baseline power consumption. The optimized propane cycle has a higher refrigerant mass flow rate, maximum subcooling, and slightly lower overall compression ratio than those of the baseline cycle as shown in Table 5-12. Split ratios, x_8 and x_9, are reduced so that an optimized amount of refrigerant is provided to low temperature heat exchangers at –19°C and –33°C, respectively, which has low expansion pressure that requires more compression power. The other split ratios, x_{11}, x_{13}, and x_{14}, were adjusted by the optimizer to meet the change in the total mass flow rate. In order to know how much power savings were obtained due to lower condensing temperature, 40°C was applied on the baseline model, which resulted in 0.817 MW power reductions from the total baseline cycle power consumption. Therefore, the optimized cycle power savings is due mainly to the optimized mass distributions and pressure levels.

5.5.4 Second law efficiency

The Second Law efficiency, Equation 5-7, is used to compare a thermodynamic cycle and an ideal reversible cycle. The minimum power to liquefy the natural gas (NG) to LNG and LPG is from the ideal reversible cycle using exergy difference, which is calculated by Equations 5-8 and 5-9.

$$\eta_{II} = \dot{W}_{Minimum}/\dot{W}_{Calculated} \tag{5-7}$$

$$\dot{W}_{Minimum} = (\dot{m}\ ex)_{NG} - (\dot{m}\ ex)_{LNG} - (\dot{m}\ ex)_{LPG} \tag{5-8}$$

$$ex = h - T_o s \tag{5-9}$$

Table 5-12 Propane Cycle Optimization Results

	Variables							Objective Function
	$\dot{m}$ (kg/s)	T_c (°C)	Pressures (kPa)					
Cycle	x_1	x_2	x_3	x_4	x_5	x_6	x_7	
Baseline	447	43	253	406	618	882	1,540	
Optimized	465	40	240	406	758	847	1,433	
	Variables							Power Consumption (MW)
	Split Ratios							
Cycle	x_8	x_9	x_{10}	x_{11}	x_{12}	x_{13}	x_{14}	
Baseline	0.465	0.758	0.143	0.886	0.705	0.909	0.938	44.22
Optimized	0.449	0.7	0.143	0.905	0.705	0.911	0.902	37.15

(Variable definitions are given in Table 5-10)
(Alabdulkarem et al., 2011)

where $\dot{W}$ is the power, $\dot{m}$ is the mass flow rate, ex is the exergy, h is the enthalpy, T_o is the ambient temperature, and s is the entropy.

The minimum reversible compressors power required for LNG and LPG production was calculated from Equation 5-8 to be 50.36 MW. The total power consumption of the baseline cycle, as well as that of the optimized cycle, was calculated by the model in HYSYS to be 110.84 MW and 100.78 MW, respectively. This resulted in a Second Law efficiency of 45.43% for the baseline cycle and 49.97% for the optimized cycle. The baseline natural gas liquefaction cycles model consumes 5.66 kWh energy per kmol of LNG produced whereas the optimized plant consumes 5.14 kWh energy per kmol of LNG produced.

5.5.5 Effect of pinch temperature

APCI liquefaction cycle uses SWHX, which is a proprietary heat exchanger developed by Linde Inc. This expensive heat exchanger is used in the cryogenic column and it has a low pinch temperature with a range as low as 3 °K (Hasan et al., 2007). The other type of less expensive heat exchanger that is used for liquefying natural gas is a plate-fin heat exchanger (Shukri, 2004). The effect of pinch temperature on the liquefaction cycle power consumption was investigated as shown in Table 5-13. The optimizer, coupled by HYSYS, was run with four pinch temperatures: 0.01, 1, 3, and 5°K. Different pinch temperatures represent the performance of different heat exchangers. Lowest pinch temperature (0.01°K) represents extremely large, efficient, and high UA value heat exchangers as an ideal case. As shown in Figure 5-9, the savings in power consumption increases with the decrease in the pinch temperature. These savings can be translated to operating cost and then compared with initial cost for the economic evaluation of different s selections. Equation 5-10 fits the resulting optimized MCR power for different pinch temperatures.

$$P_{MCR} = 1.538\ TP^2 - 3.316\ TP + 59.453 \quad (5\text{-}10)$$

where P_{MCR} is optimized MCR power consumption, MW; and TP is pinch temperature, °K.

5.6 Driver cycle optimization

The three components that have the most impact on the efficiency of a natural gas liquefaction cycle are heat exchangers, compressors, and compressor drivers. Most of the liquefaction plant energy consumption is due to the compressor drivers. In most cases the chemical energy of the fuel is converted to mechanical work to drive the compressors directly or to drive the electric motor driving the compressors. Gas turbines and steam cycles are conventionally used as drivers. The new plants are using mostly gas turbines as their drivers. However, the efficiencies of the gas turbines are below 50%. Gas turbine efficiency value degrades if the gas turbine is operated below its nominal capacity. The efficiency of the combined cycles can reach 60%.

In selecting the best option the economic aspects should also be considered. In optimization of the driver the main objectives are minimizing the annual energy consumption of the drivers or maximizing the annual production or profit. The typical constraints that should be met are the compressor power and the availability and reliability of the plant. Either the conventional optimization or stochastic optimization methods can be used for optimizing the driver cycles. However, if the conventional

Table 5-13 Liquefaction Cycle Optimization Results with Different Pinch Temperatures

Cycle	Variables								Objective Function	LMTDs
	$\dot{m}$ (kg/s)	x_{N2}	x_{C1}	x_{C2}	x_{C3}	P_i (kPa)	P_H (kPa)	P_{Ex} (kPa)	MCR Power Cons. (MW)	(°K)
Baseline	270	0.0971	0.2225	0.5445	0.1359	2,300	4,000	420	66.62	7.12/5.17
Optimized, TP ≥ 0.01 °K	268	0.1072	0.2101	0.5308	0.1519	2,700	3,880	506	57.77 (13.28% savings)	0.74/3.90
Optimized, TP ≥ 1 °K	268	0.0978	0.2186	0.5376	0.146	2,870	4,003	501	58.63 (11.91% savings)	2.86/3.06
Optimized, TP ≥ 3 °K	267	0.1027	0.2180	0.5306	0.1487	2,346	4,137	451	63.63 (4.48% savings)	5.24/4.91
Optimized, TP ≥ 5 °K	267	0.1175	0.2132	0.5287	0.1407	2,137	4,458	399	70.69 (6.61% more power consumption)	7.9/7.39

(Alabdulkarem et al., 2011)

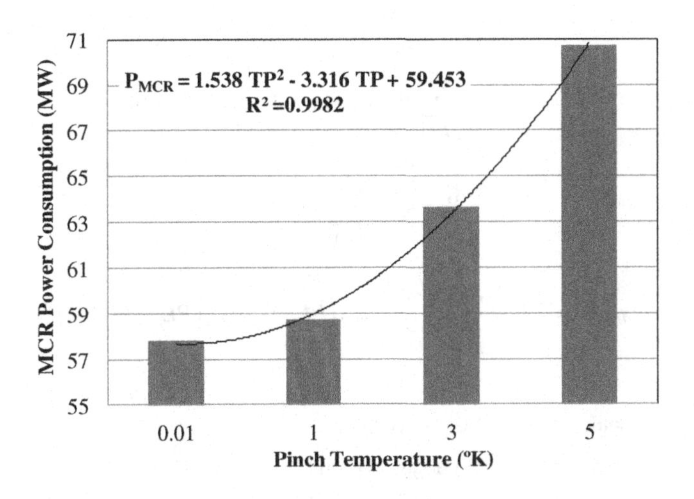

FIGURE 5-9

Optimized MCR power consumption at different heat exchanger pinch temperatures.

(Alabdulkarem et al., 2011)

optimization is used, the nominal point should account for the typical worst scenario. Moreover, to cover the annual operation of the plant it is suggested dividing the year into several periods (bean) and setting the parameters as typical worst scenario for that period.

As a demonstration, a simple optimization driver problem is formulated. In this example the task is maximizing the profit for the plant where the compressors are electrically driven and the electricity is generated on site. There are "k" electrical power generator options. The goal is to select the best combination of these generators, their monthly operation schedule, and LNG monthly volumetric production.

$$\max_{v_{m,LNG}, n_j, PL} Profit = \sum_{m=1}^{12} \left(C_{m,LNG} v_{m,LNG} - C_{m,NG} \left(v_{m,driver} + v_{m,LNG} \right) - C_{m,operation} \right) \qquad (5\text{-}11)$$

Subject to:

$$g_1 : P_{LNG,m}\left(v_{m,LNG}\right) - P_{GT,m} \leq 0 \; for \; m\text{: } 1..12 \qquad (5\text{-}12)$$

$$g_2 : \sum_{j=1}^{k} n_j G_{j,m,P} G_{j,Ava} - d_m\left(v_{m,LNG}\right) P_{LNG,m}\left(v_{m,LNG}\right) \leq 0 \; for \; m\text{: } 1..12 \qquad (5\text{-}13)$$

$$g_3 : P_{LNG,m}\left(v_{m,LNG}\right) = \sum_{j=1}^{k} \sum_{i=1}^{n_j} G_{j,m,p} PL_{j,i} \qquad (5\text{-}14)$$

$$0 \leq PL \leq 100 \qquad (5\text{-}15)$$

$$v_{m,max,LNG} \geq v_{m,LNG} \geq v_{m,min,LNG} \; for \; m\text{: } 1..12 \qquad (5\text{-}16)$$

$$C_{m,\ operation} = \sum_{j=1}^{k} n_j G_{j,m,cost} + C_{m,\ other}\left(v_{m,LNG}\right) \tag{5-17}$$

$$P_{GT,m} = \sum_{j=1}^{k} n_j G_{j,m,P} \tag{5-18}$$

$$v_{m,driver} = 2.6e9 \sum_{j=1}^{k} \sum_{i=1}^{n_j} G_{j,m,p}\left(PL_{j,i}\right) / \left(\eta_{j,m}\left(PL_{j,i}\right) LHV_{NG}\right) \tag{5-19}$$

where

$C_{m,LNG}$:	Volumetric price of LNG in month *m*
$C_{m,NG}$:	Volumetric price of natural gas in month *m*
$C_{m,\ operation}$:	Adjusted operational cost (including the capital cost) except fuel cost of drivers
$C_{m,\ other}$	Adjusted operational cost of the liquefaction plant excluding the drivers
d_m:	Total operation time in month *m* in hours
$G_{j,Ava}$:	Adjusted availability of generator driver type *j* in hours per month
$G_{j,m,cost}$:	Adjusted operational cost of generator driver *j* in month *m*
$G_{j,m,P}$:	Nominal capacity of generator driver *j* in month *m*
n_j:	Number of power generator type *j*
$P_{GT,m}$:	Maximum power generation capacity in month *m*
$P_{LNG,m}$:	Liquefaction power demand in month *m*
PL	Percentage of part load (ratio of driver output to the nominal output at that condition)
$v_{m,LNG}$:	LNG production in month *m*
$v_{m,driver}$:	Total natural gas consumption of generator drivers in month *m*
$v_{m,min,LNG}$:	Minimum LNG production capacity in month *m*
$v_{m,max,LNG}$:	Maximum LNG production capacity in month *m*
$\eta_{j,m}$:	Efficiency of the generator driver type *j* at PL% part load condition

The goal in this optimization problem is finding the optimum values of $v_{m,LNG}$, n_j and PL. The objective function (Equation 5-11) calculates the annual plant profit. It is based on the monthly profit of the natural gas liquefaction cycle. The monthly profit is the value of the produced LNG minus the adjusted operational, raw material, and fuel cost. Constraint g_1 (Equation 5-12) ensures that the maximum power generation capacity in month *m* is higher than the power demand in that month. The power demand in month *m* is assumed to be a function of LNG production in that month. Constraint g_2 (Equation 5-13) ensures that there will be enough generation capacity in month *m*. In constraint g_2, $G_{j,Ava}$ is the average period that driver *j* is available for power generation. In the parameter $G_{j,Ava}$ both average maintenance time and average down time of driver *j* are accounted.

Down time refers to the periods that the driver is not operational due to a failure. In constraint g_3 (Equation 5-14), which is an equality constraint, the power demand in month m is set to the summation of generated power by the drivers in that month. Equation 5-14 contributes in setting the operation strategy for the driver (i.e., how much should each driver cycle generates in month m). Equation 5-17 calculates the $C_{m,operation}$, which is used in the objective function. Equation 5-18 calculates the maximum power generation capacity in month m, which is the summation of the nominal capacity of the drivers in the design condition of that month. Equation 5-19 calculates the fuel consumption by the drivers. It should be noted that $PL_{j,i} = 0$ means the ith driver of type j is not operating. As a further elaboration consider the following example.

5.6.1 Driver cycle optimization example

In this test example the goal is selecting a generator driver cycle for an electrically driven natural gas liquefaction cycle. For simplicity it is assumed that the values are constant throughout the year and the $C_{m,other}$ value does not vary significantly with LNG production. Table 5-14 shows the problem assumptions. Table 5-15 shows different driver cycle options and their specifications. The results are shown in Table 5-16. It should be noted that the amount of production had the highest influence. However the driver selection did not have a significant influence. For the case of maximum production the profit difference between the optimum driver cycle configuration and the worst feasible driver cycle configuration was less than 2%. This is true for the real world since the gas price is cheap and the cost of driver cycle is small in comparison to the natural gas liquefaction cycle. Therefore in many natural gas liquefaction cycles the goal is maximizing the reliability of the power generation not the power generation energy efficiency.

5.7 Mobile LNG plants optimal design challenges

One of the barriers in the development of small remote gas fields is transportation of natural gas from these reservoirs to a specific market since it is not economical to transport natural gas for long distances by the pipeline. On the other hand, it is not economical to build a stationary LNG plant for a small natural gas reservoir. One solution to this problem might be the development of mobile LNG plants.

Table 5-14 Problem Assumptions

Parameter	Value	Parameter	Value
$v_{m,min,LNG}$ [cf]	10.0e9	$C_{m,NG}$ [$/cf]	0.114
$v_{m,max,LNG}$ [cf]	20.0e9	$C_{m,LNG}$ [$/cf]	0.15
$C_{m,other}$ [$]	1.0e8	$P_{LNG,m}$[MW]	$40 + 110v_{m,LNG}/20e9$
d_m[hr]	$720v_{m,LNG}/20e9$	LHV_{NG} (MJ/cf)	0.7

Table 5-15 Driver Cycle Specifications

Driver	$G_{j,m,P}$ [MW]	$G_{j,m,cost}$[$]	$\eta_{j,m}$	$G_{j,Ava}$ [hr]
Small heavy duty gas turbine	50	30000	30PL	690
Medium heavy duty gas turbine	70	40000	32PL	690
Large heavy duty gas turbine	120	70000	34PL	690
Aeroderivative gas turbine	50	60000	$45(1-(1-PL)^2)$	700
Combined cycle type 1	100	80000	$52(1-(1-PL)^2)$	680
Combined cycle type 2	150	120000	$60(1-(1-PL)^2)$	680

Table 5-16 Results of the Driver Cycle Optimization Example

Production						
Profit [$]	71.88e6		$v_{m,LNG}$ [cf]	20e9		
Driver Configuration						
	Small Heavy Duty Gas Turbine	Medium Heavy Duty Gas Turbine	Large Heavy Duty Gas Turbine	Aeroderivative Gas Turbine	Combined Cycle Type 1	Combined Cycle Type 2
n	1	0	1	0	0	0
PL	0.833	0	1	0	0	0

There are several uncertainties involved in designing mobile LNG plant, including the natural gas composition (feed gas composition). It should be noted that for a mobile LNG plant the design should be insensitive to the natural gas composition of the gas field. Moreover, the mobile LNG plant should be energy efficient. The enhancement options of Section 5.3 could be implemented in the design of liquefaction cycles for mobile LNG plants. However, the big challenge is the development of a refrigerant mixture that is both efficient and insensitive to the natural gas composition. Here, the efficient refrigerant mixture refers to a refrigerant mixture composition that leads to minimum amount of energy consumed per unit mass of LNG produced. To develop this refrigerant mixture, optimization techniques should be employed. However, conventional optimization techniques cannot handle problems that involve uncertainty. Robust optimization techniques would be the most suitable choice based on the design goal, which is operability of the mobile LNG plant for different natural gas compositions. The result from robust optimization would be a refrigerant mixture, which is insensitive to the feed gas composition.

References

Alabdulkarem, A., Mortazavi, A., Hwang, Y., Radermacher, R., Rogers, P., 2011. Optimization of Propane Pre-cooled Mixed Refrigerant LNG Plant. Applied Thermal Engineering 31, 6–7, 1091-1098.

Alffev, V.N., Yagodin, V.M., Nikol'skii, V.A., 1972. Cooling Articles to Cryogenic Temperatures. Patent No. 362978, S.U.

Arora, J.S., 2004. Introduction to Optimum Design, second ed. Elsevier Academic Press, San Diego, CA, USA.

Aspen Technology, Inc. Aspen and Hysys Software. www.aspentech.com/.

Bazaraa, M., Sherali, H., Shetty, C., 1993. Nonlinear Programming: Theory and Algorithms, second ed. John Wiley & Sons Inc., New York, NY, USA.

Borgnakke, C., Sonntag, R.E., 2008. Fundamentals of Thermodynamics, seventh ed. John Wiley & Sons Inc., New York, NY, USA.

Boyarsky, M., Yudin, B., Mogorychny, V.I., Klusmier, L., 1995. Cryogenic Mixed Gas Refrigerant for Operation Within Temperature Ranges of 80 °K–100 °K. U.S. Patent No. 5441658.

Deb, K., 2001. Multi-objective Optimization Using Evolutionary Algorithms, first ed. John Wiley & Sons Inc., Chichester, UK.

Finn, A., Johnson, G., Tomlinson, T., 1999. Developments in Natural Gas Liquefaction. Hydrocarbon Processing 78 (4), 47–59.

Gordon, J.L., 2001. Hydraulic Turbine Efficiency. Canadian Journal of Civil Engineering 28 (2), 238–253.

Hasan, M.M.F., Karimi, I.A., Alfadala, H., Grootjans, H., May 27–30, 2007. Modeling and Simulation of Main Cryogenic Heat Exchanger in a Base-load Liquefied Natural Gas Plant, paper presented at the 17th European Symposium on Computer Aided Process Engineering, Bucharest, Romania.

Herold, K., Radermacher, R., Klein, S.A., 1996. Absorption Chillers and Heat Pumps. CRC Press, Boca Raton, FL, USA.

Kanoglu, M., 2001. Cryogenic Turbine Efficiencies. Exergy International Journal 1 (3), 202–208.

Lee, G.C., Smith, R., Zhu, X.X., 2002. Optimal Synthesis of Mixed-Refrigerant Systems for Low-Temperature Processes. Ind. Eng. Chem. Res. 41 (20), 5016–5028.

Lee, K., 2001. Robust Optimization Considering Tolerances of Design Variables. Computers & Structures 79 (1), 77–86.

Li, M., Azarm, S., Boyars, A., 2006. A New Deterministic Approach Using Sensitivity Region Measures for Multi-Objective Robust and Feasibility Robust Design Optimization. ASME Journal of Mechanical Design 128 (4), 874–883.

MathWorks. MatLab software. www.mathworks.com.

Mortazavi, A., Azarm, S., Gabriel, S., 2012. Adaptive Gradient Assisted Robust Design Optimization under Interval Uncertainty. Engineering Optimization.

Mortazavi, A., Somer, C., Alabdulkarem, A., Hwang, Y., Radermacher, R., 2010. Enhancement of APCI Cycle Efficiency with Absorption Chillers. Energy 35 (9), 3877–3882.

Mortazavi, A., Somers, C., Hwang, Y., Radermacher, R., Al-Hashimi, S., Rodgers, P., 2012. Performance Enhancement of Propane Pre-cooled Mixed Refrigerant LNG Plant. Applied Energy 93, 125–131.

Ordonez, C.A., 2000. Liquid Nitrogen Fueled, Closed Brayton Cycle Cryogenic Heat Engine. Energy Conversion & Management 41 (4), 331–341.

Paradowski, H., Bamba, M.O., Bladanet, C., March 21–24, 2004. Propane Precooling Cycles for Increased LNG Train Capacity, paper presented at the 14th International Conference and Exhibition on Liquefied Natural Gas (LNG 14), Doha, Qatar.

Patel, B., July 10–14, 2005. Gas Monetization: A Techno-Economic Comparison of Gas-To-Liquid and LNG. paper presented at the 7th World Congress of Chemical Engineering, Glasgow, Scotland.

Renaudin, G., July 3–6, 1995. Improvement of Natural Gas Liquefaction Processes by Using Liquid Turbines. paper presented at the 11th International Conference on Liquefied Natural Gas (LNG 11), Birmingham, UK.

Shukri, T., Feb. 2004. LNG Technology Selection. Hydrocarbon Engineering 9 (2), 71–76.

Townsend, D.W., Linnhoff, B., 1983. Heat and Power Networks in Process Design. AIChE Journal 29 (5), 742–771.

CHAPTER

Process Control and Automation of LNG Plants and Import Terminals

6

6.1 Introduction

The Liquefied Natural Gas (LNG) marketplace is inherently complex and is becoming quite competitive. Automation strategies have become a key element for aligning business needs with operating practices to increase competitiveness. There are various components of a best in class automation system. From the field sensors to boardroom information, choices for the controls and automation impact the effectiveness and profitability of an LNG operation. A well-designed, maintained, and leveraged control system will reduce start-up time, maintain maximum operating profit, avoid forced shutdowns, keep operating and maintenance costs as low as possible, assist in the management of environmental compliance, and support plant safety and security needs without adversely affecting construction costs. Though automation represents only 2 to 6% of the typical project spending, its impact on the operations over the plant life could be around 15 to 30%. The main objectives of the LNG plant automation are the safe, reliable operation and optimization of both the process operations and the plant asset utilization.

The history of plant automation is one of rapid technological progress. Early days of automation were characterized by manual operation of the valves by operators in the field. This was followed by pneumatic control systems, whereby the signals were transmitted pneumatically via air pressure and calculations were performed by mechanical computation devices. This enabled the use of automatic control using algorithms like proportional, integral, and derivative (PID) control. The sensor technology also started to improve accordingly. The pneumatic control systems were gradually replaced by electronic systems.

The introduction of distributed control system (DCS) in the 1970s enabled several new innovations in the plant automation field. The control became centralized with operator displays, alarm panels, and field stations. Another important development was the rise of microcomputers, which resulted in the widespread acceptance of computer-based control. This also facilitated the development of several control algorithms and plant optimization technologies that have since matured. The future of automation in the LNG plant is geared toward intelligent field devices and further innovations in asset management and asset optimization. The concept of Collaborative Process Automation, which involves delivering information at the right time to the right people, is also expected to become widespread. Advanced process control (APC) and other optimization techniques are expected to become more standard tools in LNG plant operations.

The key aspects related to the automation and controls of LNG plant and import terminals are described in this chapter. The chapter gives an overview of main automation functionalities in the LNG

Handbook of Liquefied Natural Gas.

plant and their applications. The control of individual units in the LNG plant is discussed, where the main control loops in each unit are described. The chapter includes a discussion on APC, which is being deployed widely in LNG plants. A brief overview of the control and automation of LNG import terminals is also included.

6.2 Objectives of LNG plant automation

The key objectives of LNG plant automation are the following:

- **Safety.** Safety is the first and foremost objective in the LNG plant automation. Design and functionalities of key areas of the automation systems are directly are related to this. The first step in achieving this goal is to maintain the important process variables in the safe operating limit. This is accomplished by the base and advanced regulatory loops in the plant. The alarms and trips in the DCS are designed to warn the operators or take action about any impending violation of the safe operating limit. The SIS system is designed to ensure that the plant is safely shut down in case any significant violation is detected.
- **Stable operation.** The smooth and stable operation of the LNG plant is another important objective of the automation system. This is achieved by the base regulatory controls in the plant. Hence, the design and tuning of these control loops are critical for the stable operation. Apart from these, advanced regulatory controls and advanced process controls can be utilized to enhance the operational performance of the plant.
- **Operations optimization.** The optimal operation of the LNG plant that considers the various market factors and operating conditions is economically beneficial. This is best achieved by using APC/Real Time Optimization.
- **Asset performance optimization.** The proper management of plant assets is important in the overall optimization of plant life cycle. This includes both asset management and maintenance management.

6.3 LNG plant process control/automation development and functionalities

The selection of the process control and automation platform is a strategically important decision that impacts the long-term performance of the facility (Maffeo, 2005). The process control and automation system development for LNG plants during construction normally is done by the EPC contractor. In some cases, a Main Automation Contractor (MAC) is identified, who acts as a single point responsible party for all automation related aspects of the project (Phople, 2009; and Honeywell White Paper). This is one approach for achieving sustained benefits across all three (construction, startup, and operational) phases of LNG asset life.

6.3.1 Distributed control system

Distributed control systems (DCSs) were introduced in the early 1980s to avoid the proprietary PLC and other systems. Initially, the control was performed at the local level. With the appearance of microcomputers, distributed control became possible in the plants.

The backbone of a DCS is a data highway, used to communicate between various elements. There could be several data highways depending on the scope. It also acts as the main data link between the main control computer and other parts of the network. In the common DCS architecture, micro-computers attached to the process are used to perform lower level calculations like unit conversion. The results from this are then passed on to higher level computers that are responsible for more complex calculations like control algorithms.

The following are the main elements of the typical DCS system (Figure 6-1):

- **Local control unit.** These units are capable of handling a number of individual PID units.
- **Data acquisition unit.** Both discrete and analog I/O can be handled with these units.
- **Batch sequencing unit.**
- **Local display.**
- **Bulk memory unit.** The various historical data are archived and stored in these units.
- **General purpose computer.** These are used to perform more complex calculations like optimization, advanced control, etc.
- **Central operator display.** This typically consists of consoles for operator communication with the system.
- **Data highway.** This is the main digital data transmission link connecting the various components of the system.
- **Local area network.**

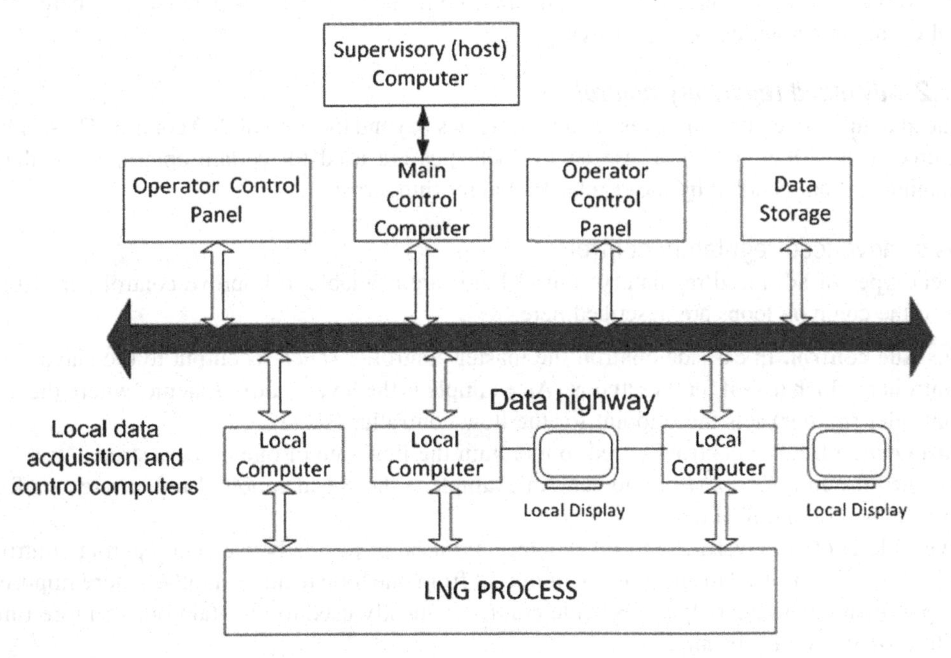

FIGURE 6-1

Components of a Distributed Control System (DCS).

The key role of the DCS system in the day-to-day plant operation is to maintain the various process variables for smooth and stable operation. This is accomplished using the base and advanced regulatory control schemes in the DCS.

6.3.1.1 Base regulatory control

The base regulatory control layer is the main control layer in the plant, which ensures that the operating variables are maintained at desired values. These are normally PID control loops that control the pressure, temperature, flows, etc., in the plant. These loops are executed with a frequency of one second or less to ensure fast response. The proper design of this control layer is the key to stable and optimal operation of the facility.

PID control

The PID control is the main control technology used in the plant. The PID control algorithm is a single-loop control algorithm that utilizes proportional, integral, and derivative action to maintain the variable at a setpoint. One or more of these actions can be selectively utilized, depending on the variables that are being controlled.

There are various PID equations that are available in commercial DCS systems, which are improvements over the base PID equation. The main control tuning parameters for the PID control are the proportional gain, integral time, and the derivative time. These are selected to achieve desired closed loop performance like minimum overshoot, fast settling, and so on. Various tuning methodologies developed over the years are available for PID controllers (Astrom and Hagglund, 1995).

An overview of the various PID loops that are used in the LNG plant is given later, where process control of the individual units is discussed.

6.3.1.2 Advanced regulatory control

Advanced regulatory control involves control strategies beyond the normal PID control. These are still configured in the DCS system. These control schemes are used for certain control loops that are challenging and not handled by the regular PID algorithm alone.

Types of advanced regulatory controls

Different types of advanced regulatory control loops are available to improve control performance. Some of the common loops are described here.

- **Cascade control.** In cascade control, the master controller sends its output to the slave controller, which uses it for its setpoint. An example is the level control scheme, where the level controller (master) sets the setpoint for the flow controller (slave).
- **Ratio control.** Ratio control is used to maintain the flow rate of one process stream at a specified proportion to that of another. An example is the specification of reflux rate as a fixed ratio to the column feed rate.
- **Override control.** Override control schemes are used to decide between competing control objectives. This is used to take control of output from one loop to allow another, more important loop to manipulate the output. Override control is mostly used to maintain one or more limits with a single control handle.
- **Feed-forward control.** This is used to handle disturbances that affect the process in an adverse manner, but may not be compensated by the control handles in a timely manner.

In feed-forward control, a model that predicts the effect of a disturbance on the controlled variable is utilized to determine the appropriate control actions in advance to offset the effect of the disturbance. Ratio control can be considered as a very basic feed-forward loop.

Other advanced regulatory control schemes are available in the commercial DCS systems. With proper implementation and tuning, advanced regulatory control schemes can improve the overall controllability of the LNG plant significantly (Shinskey, 1996).

6.3.1.3 Loop management

Loop management functionality offers comprehensive online control loop and valve diagnostics and reporting, expert loop tuning, and continuous remote monitoring to help ensure maximum control loop performance over time. A brief description of the capabilities that are usually present in the loop management system scope is given here.

- **Controller monitoring.** Here, the performances of the various PID loops in the plant are monitored on a continuous basis. Typically, it is found that some of the control loops are run in manual, which is not desirable. This issue can be addressed using diagnosis of the controller performance problem using statistical techniques. The problem could be controller tuning, sticky valves, actuator problems, and the like. Once the problem is diagnosed, an appropriate solution can be implemented. All the loops can be monitored using key performance indicator (KPI) values developed for the best performance enhancement.
- **Controller tuning.** Different tools to perform optimal controller tuning are available in the plant DCS. These tools can be used to identify the tuning parameters for the various PID controllers in open-loop or closed-loop mode. Various tuning criteria can be used to optimize the tuning of controllers for either setpoint tracking or disturbance rejection.
- **Remote monitoring.** Remote monitoring is a valuable tool to identify and troubleshoot any field problems remotely. With the advent of Ethernet-based and Internet-enabled systems, it is possible to perform remote monitoring at any location with a smartphone or other device.
- **Valve diagnostics.** DCS systems with the latest technology provide enhanced functionality to utilize field device data for enhanced monitoring and diagnosis. One of the valuable applications of this is the valve diagnostics. This includes identifying any sticky valves, valves with excessive wear or other issues, and subsequent troubleshooting of the problem. This can significantly contribute toward enhancing the overall plant control performance.

6.3.2 Intelligent field devices

An automation strategy begins with the selection of field devices such as transmitters, valves, and rotating equipment monitors. Intelligent field devices can lead to reduced capital costs while allowing a digital automation network to receive the maximum amount of information on which to act. Digital systems run throughout a facility carrying important process and equipment information from the devices to enable process control, asset optimization, and safety functionality. These digital plant networks create a wealth of data enabling unprecedented information about overall plant performance.

The increased use of intelligent field devices has resulted in safer, more reliable, and more productive processes with the added benefits of lower costs and faster implementation. Intelligent field devices with digital capabilities that include predictive diagnostics for monitoring equipment and process health as well as open-standard communications offer the LNG process system designer many benefits. These include precise measurement and control to reduce process variability and the ability to use multivariable field devices to reduce the number of products required. The higher level of information that is available enables more informed decision-making from the control room to the executive suite.

Open communications standards including Fieldbus and other digital protocols enable timely transmission of data generated by field devices. Built-in diagnostics provide the capability to monitor asset health and predict problems before they develop into faults that could interrupt production or result in a safety or security situation. Asset management software can deliver these predictive diagnostics from the most critical production assets in the facility, including control valves, transmitters, and motor-pumps.

Fieldbus protocols are an all-digital, two-way communications technology for plant instrumentation. Automation architectures based on Fieldbus improve operations by providing access to data not previously available while reducing maintenance costs through expanded access to diagnostics and other digital device information. Engineering drawings are simplified by Fieldbus technology, and wiring and connection costs are decreased. Fieldbus technology also permits control in field devices, providing greater engineering flexibility and reliability options in contrast with traditional control systems. Fieldbus protocols were developed specifically to meet the needs of the process industry. They have become worldwide standards accessible to multiple suppliers and open to innovation.

The diagnostic capabilities of digital intelligent field devices enable users to manage the health of the devices themselves as well as the process to which they are connected. Fieldbus devices report a good, bad, or uncertain status, and asset management software enables individual devices to be interrogated to determine their status. Screen graphics allow operations and maintenance personnel to "look inside" these assets from a centralized control room, making it easier to pinpoint the location and nature of a problem and dramatically reduce the troubleshooting effort required.

Online, real-time diagnostics running in intelligent field devices provide the information needed to streamline maintenance activities. The diagnostics run continuously in the devices and generate alerts that enable operations and maintenance personnel to identify issues and take actions that prevent abnormal situations. This enables the scarce maintenance resources, typical of most production environments, to focus on assets that require immediate attention.

By using diagnostic and asset management software to identify deteriorating equipment or process conditions and potential problems, maintenance engineers at an LNG plant can predict how long it will be before equipment will fail or a potential safety or security issue will develop. Remedial action can be scheduled on the basis of actual conditions rather than a predetermined schedule, thereby safely maintaining production at optimal levels. By resolving problems before they interfere with production, the unscheduled breakdowns and work stoppages that are typical of a reactive maintenance regime can be avoided.

Leading automation suppliers offer a complete range of intelligent devices and process automation solutions that extend the digital architecture across the complete LNG supply chain, including control valves and regulators, transmitters and analyzers, motor-pumps, metering skids, gauging, and ultrasonics. Typical integrated solutions include best-in-class turbine/compressor antisurge protection, LNG storage tank management, terminal management and custody transfer, gas processing and transmission, safety instrumented systems, and marine and offshore applications.

6.3.2.1 Fieldbus in LNG plants

A fieldbus is an industrial network communication system for real-time distributed control. It connects field instruments and measurement with control devices via serial, two-way communication systems. It is increasingly used in the area of process automation. Fieldbus encourages the move away from switches to transmitters, thus enhancing performance and flexibility (Figure 6-2). The potential paybacks include reduced life cycle costs and improved access to information.

Fieldbus is also an enabling technology for knowledge-based decision-making to improve plant operations. It enables real time diagnostics and preincident information. The information and knowledge gained can be used to enable real-time information management applications and predictive maintenance functions that improve the overall plant performance.

A prominent fieldbus technology that is widely used is Foundation fieldbus. It is an all-digital, serial, two-way communication system that serves as the base level network in a plant. It is the local area network for the instruments used in process automation with built-in capacity to distribute control applications across the network. LNG plants can employ Foundation fieldbus to interconnect the field equipment, such as transmitters, sensors, and valves, on a single network (Mokhatab, 2007). In this technology, true distributed control is achieved by enabling control capability at the device level across the plant (Lancaster and Parkes, 2002).

LNG plants usually are located in remote environments, which make fieldbus an attractive technology to consider. The capability for remote diagnostics, predictive maintenance, Internet/web-based

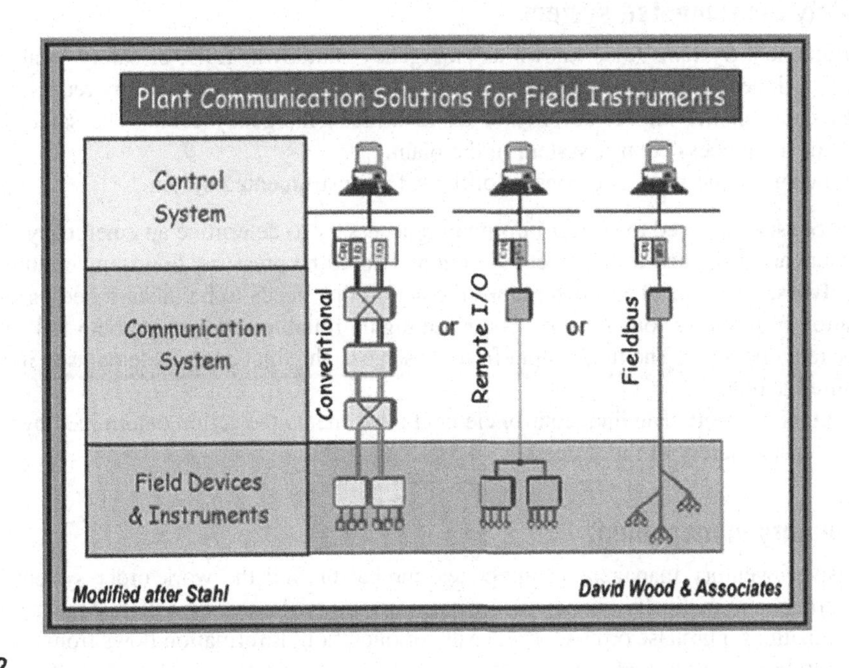

FIGURE 6-2

Fieldbus configuration compared to the traditional device communication.

(Wood and Mokhatab, 2007)

connectivity, and remote reporting has the potential to minimize downtime and drive down the life cycle project costs substantially.

6.3.3 Data historian and alarm management

Plant historians allow real-time event management, retrieval, and archiving of process data to analyze and generate reports. These historians provide information for performance management and tracking. By employing interactive graphics, dashboards can be generated to display results similar to information delivery in cars and aircraft. The data from the historian can be used for process improvements any time in the plant life cycle.

Alarm management is designed to dramatically minimize nuisance alarms. Alarms indicate abnormal conditions in the process. However, there may be times when an operator cannot do anything to correct the abnormal condition. If this happens frequently, the alarms can become a nuisance, or even worse, the operator can become desensitized to the alarm, which is a potentially dangerous situation.

There are certain key areas in alarm management that can be very beneficial to the plant operations. First is the management of alarm rate and the prevention of alarm flood that is potentially detrimental to operator performance. The priority distribution of the alarms can be optimized to better deliver the critical alarms. The analysis of alarm type can help to determine and set the priority and limit the alarm flood.

6.3.4 Safety instrumented system

Safety Instrumented Systems (also known as emergency shutdown, ESD, or safety shutdown, SSD systems) are designed to protect plant personnel, equipment, and environment by reducing the likelihood (frequency) or the impact severity of an identified emergency event. The safety system is independent of the process control system in the plant.

The following are the main components of the safety instrumented system:

- **Sensors.** Sensors are used to collect information necessary to determine an emergency situation. The sensors are dedicated to the safety system and could be pressure, flow, temperature, etc.
- **Logic solvers.** This part of SIS determines the action that needs to be taken based on the information from the various sensors. These are highly reliable and provide both fail-safe and fault-tolerant operation. The final output from these go to the final control elements to implement the required action.
- **Final control element.** The final control element implements the action determined by the logic solvers to ensure safety of the process.

6.3.5 Inventory management

Comprehensive inventory management integrates purchasing and the work order system. Demand from work orders automatically creates pick lists in inventory. Inventory replenishment may create purchase requisitions, purchase orders, or requests for quotation. Information flows from one function to another seamlessly and in real time.

Inventory management addresses the main challenges of maintenance repair and operations (MRO), enabling the control of a large number of unique and low-unit-value items. The system automates the reorder process by recognizing calculated safety stock levels, replenishment lead times, and sophisticated "available-to-promise" logic based on expected receipts (open purchase orders). The module also provides the ability to uniquely identify and track repairable items and critical parts through serialization.

The main objective of the procurement function in a maintenance and materials management environment is to minimize the cost of buying high volumes of MRO inventory items, and to ensure that parts are available when needed. Purchasing receives demand from work orders and inventory, maintaining tight links to those systems for visibility into the progress of a purchase order.

6.3.6 Asset management system

Asset performance management (Figure 6-3) with LNG-specific performance services increases the visibility between plant assets and the enterprise, aligning their operation and maintenance with key business metrics. Asset management solutions are ideal for implementing and supporting a total productive maintenance program. A total productive maintenance program impacts the corporate culture by forming a partnership with engineering, maintenance, and operations to focus on improving equipment effectiveness and product quality while reducing waste.

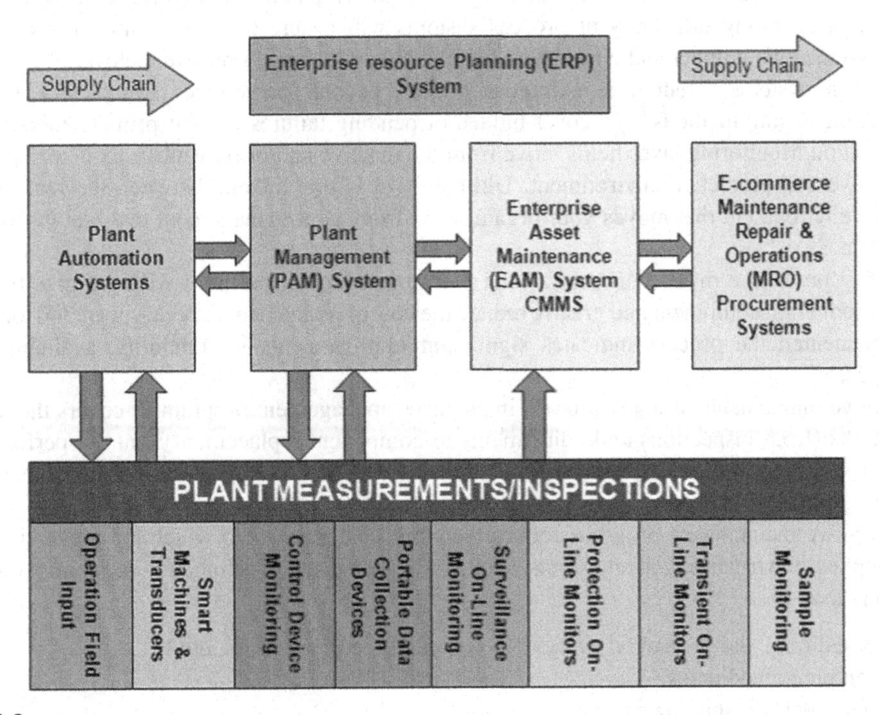

FIGURE 6-3

Plant asset management system and its relationship with other plant systems.

(Hills, 2001)

Total productive maintenance begins with a comprehensive program that includes preventive maintenance, planned and scheduled maintenance, predictive maintenance, materials management, training programs, and an enterprise asset management system. In addition, it involves a more quality-focused employee empowerment approach. Reliability centered maintenance also can be embraced in a total productive maintenance approach.

The collection of accurate maintenance and materials management activity data and the transformation of that data into information through relevant compilation and presentation are critical for an effective total productive maintenance program. With insight, direction, and involvement, this information provides the knowledge to guide the total productive maintenance teams in their tactical planning and decision-making processes.

Maintenance software acts as the central collection point for all asset information (cost, performance, history).

Predictive maintenance is a valuable addition to a comprehensive, total plant maintenance program. Where preventive maintenance incorporates routine machinery servicing, predictive maintenance schedules specific tasks, as they are actually required by the equipment.

Recent surveys report that one-third—33 cents out of every dollar—of all maintenance costs is wasted due to unnecessary or improperly implemented maintenance. The major cause is a lack of factual data that quantifies the actual need for repair or maintenance of plant equipment and systems. The premise of predictive maintenance is that regular monitoring of the actual mechanical condition of machines and operating efficiency of process systems will ensure the maximum interval between repairs; minimize the number and cost of unscheduled downtime, and improve the overall availability of production systems. Predictive maintenance goes beyond preventive maintenance by using nondestructive testing methods to uncover hidden or pending failures in their primary mode.

A Condition Monitoring layer helps move from the reactive and/or preventive mode of operations to a proactive and predictive environment. Ultimately, it is this linkage between the real-time and operational environment that moves your organization from asset management to Asset Performance Management.

Including predictive maintenance in a total plant management program will optimize the availability of production equipment and greatly reduce the cost of maintenance. A survey of 500 plants that have implemented the process indicates significant improvements in reliability, availability, and operating costs.

Preventive maintenance is a key process in any asset management program. It covers the range of periodic tasks (from inspections and adjustments to component replacement) that are performed on assets at a specific date, on an elapsed time, or preferably, a usage basis in order to keep assets functioning properly.

A preventive maintenance program defines the work plan, entities to which the preventive maintenance applies, and triggering mechanisms. Multiple triggers may be defined for each entity on a task, and triggers may be:

- Time-based (e.g., daily, weekly, every 45 days, yearly, every 2 years, etc.)
- Based on predefined dates
- Based on a defined shutdown period
- Based on other tasks being due (within a specified timeframe)
- Based on user-defined statistics (e.g., meter reading, throughput, temperature, pressure, etc.)

- Future statistic-based trigger dates calculated based on specified average values for the entity or actual values over a specified period of time
- Future triggers based on work task creation or work task completion
- On demand.

Reliability centered maintenance (RCM) is a process that determines what must be done to ensure that any plant asset continues to function in the desired manner within its present operating context. There are numerous variations and derivatives of the classic RCM process in use today, most of which are aimed at facilitating the failure modes and affects analysis and developing the appropriate plan of action. These methodologies are often supported by tools such as:

- RCM analysis software
- RCM forms generators
- RCM spreadsheets.

6.4 Process control of key units in LNG plants

6.4.1 Inlet facilities

Details of the inlet facility and feed gas treatment operation are described in Chapter 7. The following describes the special considerations that are necessary for stable plant operation.

In the inlet facility, the level control loop is important as the liquid flows from the slug catcher to the condensate stabilizer system. Attempts to control the level too tightly can introduce unnecessary disturbances in the condensate stabilizer. Appropriate controller tuning strategies like averaging level control tuning can be used to address this issue (Driedger, 2000).

The main control of the stabilizer is to produce a bottom product with acceptable RVP. An RVP analyzer or GC can be used to measure the RVP in real time. This measurement can be performed on a daily basis. An alternative is to use an RVP soft sensor with a pressure compensated temperature (PCT) as the input. The condensate stabilizer bottom temperature is normally used for controlling the RVP.

The feed from the feed drum needs to be maintained as relatively stable. Typically this drum has a relatively large surge volume, and coupled with the tuning of the level controller of this drum, has to accommodate rapid feed rate changes. The stabilizer level is controlled using a feed forward controller cascading to the flow control from the bottom products. The stabilizer overhead pressure control scheme typically is controlled using the capacity controller of the overhead compressors. Alternatively, the column pressure can be controlled by recycling the compressor discharge vapor.

The overall capacity of the LNG plant can be controlled in the back-end LNG product stream, (as is done in the APCI C3MR process), or in the front-end feed gas stream. A stable feed gas pressure is critical in the LNG plant operation, which is typically achieved with a feed pressure control valve. The pressure downstream of the valve is controlled by a flow control valve to maintain the process required by the LNG process and to avoid overpressuring. This can be achieved by an override control system that will reduce feed rate when the maximum operating pressure is reached.

An important consideration in the design of an LNG plant feed system is monitoring the feed gas temperature, since this is needed to prevent hydrate formation. Hydrate formation typically occurs across control valves when the inlet pressure is high and temperature is low. In this case, provision, such as a feed heater, needs to be used for controlling the temperature above the hydrate formation temperature. In most cases, methanol injection points must be provided on the control valve and heat exchanger in case the feed gas temperature unexpectedly drops below the hydrate temperatures.

The main equipments in a typical LNG feed treatment system are the amine system, molecular sieve dehydrators, and mercury removal system. The molecular sieve dehydrators normally are operated in batch mode, with a fixed number in dehydration mode and others in regeneration mode. The sequencing of the operation is automated and the switching is done automatically with a pre-determined period for operation during absorption, heating, cooling, and repressurizing. Moisture analyzers with high accuracy to the ppb levels are provided at the outlet of each dehydrator and the combined stream to ensure the gas is completely dried. The regeneration gas flow and temperature has to be controlled to ensure proper operation of this system.

6.4.1.1 Amine treating system

The amine treating system may use DGA or formulated MDEA to meet the acid gas content required for LNG production. Excessive CO_2 will result in freezing in the downstream natural gas liquid (NGL) unit or LNG liquefaction unit, which must be avoided. The design of the amine processes are described in Chapter 2 and the typical operation is described in Chapter 7.

To treat the sour gas, lean amine is supplied to the amine contactor at a fixed flow rate. The rate is controlled by a flow controller that is set at a rate high enough to remove the required acid gas content from the feed gas. The composition of the feed gas and treated gas are monitored and the lean amine rate can be set appropriately.

To properly operate the amine unit, the lean loading and the rich loading must be monitored using the procedure recommended by the amine licensor. Rich loading is determined by the amount of amine used. Typically rich loading is determined by the approach to equilibrium (amine to acid gases) which is set at 80% for design cases. This would allow sufficient operating margins to avoid a breakthrough of acid gases. Rich loading in most operation is limited to 0.4 to 0.5 depending on the acid gas content in the feed gas.

The lean amine loadings are different for different types of amine. The lean loading can be controlled by adjusting the steam flow to the reboiler in the regenerator. Typically, the optimum stripping ratio in the regenerator overhead is about 1 to 1.3 (mole steam per mole of acid gas). Alternatively, steam supply can be controlled using amine circulation as the criteria. Typically, in treating a lean gas, such as the tail gas treating unit, the steam requirement can be set based on amine circulation, typically at 1 to 1.3 lb of steam per gallon of circulation.

The amine strength is maintained by the addition of amine or demineralized water. As CO_2 tends to react irreversibly with amine to form heat stable salt, amine must be reclaimed on a continuous or periodic basis. The strength of the amine can be maintained by purging the amine system while supplying fresh amine. If DGA is used as the amine solvent, amine can be reclaimed thermally using an amine reclaimer.

6.4.1.2 Molecular sieve dehydration system

The molecular sieve dehydration process is described in Chapter 2 and the typical operation is described in Chapter 7.

The objective of the molecular sieve dehydrators is to produce a treated gas with low water content suitable for LNG production. Accurate moisture analyzer must be used on each molecular sieve bed and on the common discharge line to monitor the bed performance.

The main control in the molecular sieve operation is the switching valve operation. Depending on the number of beds in operation, the duration of the bed operation under different conditions (absorption, regeneration, heating, and cooling) should be adjusted and optimized as needed to produce a dry gas.

The temperature profile in the molecular sieve beds must be monitored to ensure that the beds are fully regenerated before they are returned to the absorption cycle. The regeneration heater is controlled by fuel gas to maintain the heater outlet at the optimum regeneration temperature.

The water condensate may contain mercury and other hydrocarbons and should be analyzed and disposed of properly.

6.4.2 NGL recovery unit

Most of today's LNG plants are designed to recover propane and heavier components, which is necessary to meet the heating value specification of the export LNG. The various NGL recovery processes are described in Chapter 2 and the typical operation is described in Chapter 7.

The control philosophy of an NGL recovery is to meet the NGL recovery target and the LPG product specification. Typically, 99% propane recovery can be achieved and the LPG product should not contain more than 1 mole % ethane. A typical control scheme for an NGL recovery unit is shown in Figure 6-4.

The NGL recovery process requires refrigeration for condensation and for refluxing in the fractionation column. The refrigeration requirement is provided by turbo-expansion of at least a part of the feed gas, and externally supplied by propane refrigeration. In some feed cases, propane refrigeration is not necessary if the feed gas is lean and the feed pressure is high.

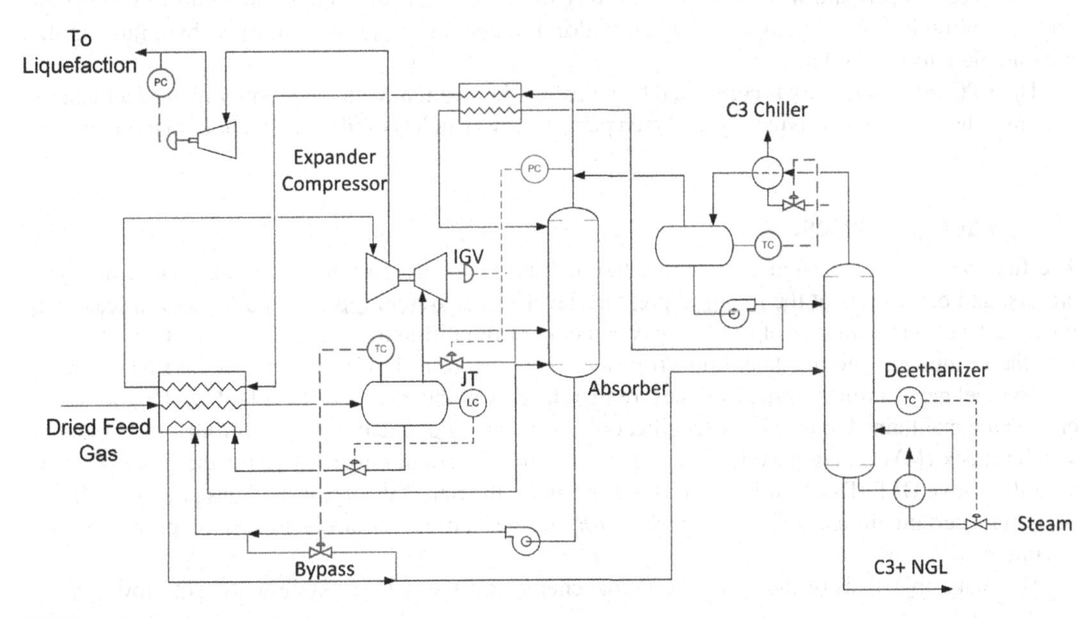

FIGURE 6-4

Control scheme for NGL recovery.

The column pressure must be maintained stable during normal operation. However, the column pressure can be increased when a high throughput is required. Higher column pressure reduces the pressure ratio across the expander, and hence less refrigeration is generated, which ultimately results in lower recovery.

There are two control devices that can be used for column pressure control: the inlet guide vane (IGV) valve of the turbo-expander, and the JT bypass valve. Normally, the absorber pressure is controlled by closing or opening of the IGV valve. Only when the IGV valve is fully open and the column pressure is still dropping, the JT valve will open to maintain the pressure. This typically occurs during plant startup or shutdown operation.

The residue gas pressure is controlled by varying the speed of the residue gas compressor as necessary to meet the LNG liquefaction plant inlet requirement.

Another critical operating parameter is the temperature of the expander suction drum temperature. This temperature determines the composition and amount of NGL condensed in the drum. Lower temperature will reduce the flow rate to the expander while increasing the liquid from the suction drum that is eventually fed to the column. Conversely, higher temperature will increase the flow to the expander and reduce the liquid to the column. Changing this temperature will impact the control of the expander and the vapor traffics in the columns.

One of the control options on controlling the expander suction drum temperature is to bypass the absorber bottom. The absorber bottom is used to provide cooling to the feed gas, and bypassing this stream will reduce the cooling effect on the feed gas, increasing the chilled temperature. The operation will reduce the feed temperature to the deethanizer, which will indirectly reduce the propane chiller duty.

Another measure that needs to be monitored is the deethanizer overhead temperature. Decreasing the overhead temperature will increase the reflux to the deethanizer, which will result in better separation and higher NGL recovery. The reflux drum temperature can be controlled by adjusting the propane flow to the chiller.

The LPG product quality is controlled by adjusting the steam flow to the reboiler in the deethanizer column. The C_3+ liquid is typically analyzed periodically, which is used to set the bottom temperature.

6.4.3 Fuel gas system

The fuel gas system is designed to supply the fuel gas to the gas turbines, fired heaters, steam generators, and other users at the required pressure levels. This system typically consists of at least two pressure levels. The sources of the fuel gas can come from different sources. It can be from the feed gas, the residue gas, storage tank boil-off gas, or other off-gases in various parts of an LNG plant.

The fuel gas control system has to ensure that the dew point criteria and the fuel gas heating value criteria are maintained to be within specifications. Depending on the type of gas turbine, the Modified Wobbe Index (MWI) may have to be maintained within a certain range. Also, for the more efficient aero-derivative DLE (Dry Low Emissions) gas turbines, the rate of change of fuel gas properties has to be within certain limits. This is important for the gas turbine controls to ensure proper burner operation.

The main objectives of the process control schemes for the fuel gas system are as follows:

- Maintain adequate fuel gas supply. This requires adequate backup supply of fuel gas sources in case of a surge in fuel gas consumption during transient conditions.

- Maintain the temperature and Wobbe Index. It is critical to prevent any condensation in the system. The temperature has to be maintained at a superheat state, avoiding the condensing temperature. Also, the fuel gas Wobbe Index has to be within a certain range for the gas turbine to operate properly. This may require blending with lean fuel gas or with nitrogen.
- Mitigate rate of change of fuel properties. This could be important to ensure proper burner operations during transient conditions. Sudden changes in the fuel compositions may result in gas turbines misfire, resulting in a shutdown of the gas turbine.

Pressure controllers could be cascaded to flow controllers, which are tuned for fast response. The pressure controls could also be set up with staggered setpoints or split range control schemes to manage the various makeup sources for the fuel gas. One of the challenging issues for control is the transition from one fuel source to other. The regular PID control scheme may not react fast enough to prevent a drop in the header pressure. Feed-forward schemes may be required to avoid an unacceptable drop in header pressure. Appropriate advanced regulatory control schemes to operate under this scenario can be developed and tested using dynamic simulation.

Fuel gas temperature can be controlled by heating using a steam or electric heater. A temperature controller is normally used, with the setpoint set with a good margin above the dew point. If frequent changes in the fuel gas composition are expected, the dew point calculation can be automated based on the fuel gas composition.

One way to mitigate problems during transition is to ensure that the proper instrumentation and control schemes are set up to handle it. Normally, a Wobbe meter would be needed to measure the properties of fuel gas for the burner control scheme. The positioning of the Wobbe meter can be tailored to ensure that the deviation is within acceptable limits. Dynamic simulation can be used to determine the appropriate location of the instrumentation and control logic that is effective under all transitions.

6.4.4 Liquefaction system

The liquefaction system is the heart of the LNG plant, because it cools and condenses the pre-treated natural gas to produce LNG. This system is characterized by extensive heat integration utilized to enhance process efficiency. The process temperatures at various points need to be maintained to ensure proper temperature approaches for heat exchangers. Another important requirement here is the reliable operation of the turbo-machinery used to compress the refrigerants in the system. The specific aspects of the liquefaction process control scheme depend on the process technology that is used. There are several liquefaction process technologies available on the market. These include the APCI mixed refrigerant process (C3MR), cascade refrigeration process, the Shell DMR process, among others. The discussion will be targeted mainly at the former three processes, since these are used in the majority of the operating LNG plants and LNG plants under construction.

The main objectives and challenges in the control of liquefaction section of the LNG plant are:

- Ensuring that condensing and subcooling are maintained for the LNG stream. This is achieved by controlling the LNG temperature, directly or indirectly.

- Ensuring the reliable operation of the LNG plant with changing ambient conditions and feed composition. The specifications for LNG and NGL products have to be met under various feed gas compositions.
- Proper operation of the turbo-machinery strings, which is critical for the reliability of plant operation. This includes surge protection for the refrigeration compressors.
- Maximizing the LNG production or maximizing efficiency under fixed LNG production.

The selection of appropriate controlled variables and manipulated variables is important to ensure adequate operability of the LNG plant. Methodologies to perform this selection have been reported by Jensen and Skogestad (2006) and Singh et al. (2008). This is based on optimizing certain objective functions and evaluating the degrees of freedom in the LNG process. This methodology was utilized to develop a regulatory control scheme for the LNG plant.

6.4.4.1 Control of refrigeration processes

Since the LNG processes utilize principles of refrigeration systems in their design and operation, the control philosophies from traditional refrigeration processes are applicable. A typical single-stage refrigeration process normally consists of a chiller, compressor, condenser, and surge tank (Dossat and Horan, 2001). The chiller cools the hot stream and evaporates the refrigerant in the process. The liquid from the accumulator goes through a JT valve to the chiller. A schematic of the typical refrigeration process with the control schemes is given in Figure 6-5. The main control loop here is the temperature of the cooled stream, which is maintained by controlling the pressure in the chiller. This can be achieved by manipulating the compressor speed by implementing a cascade loop.

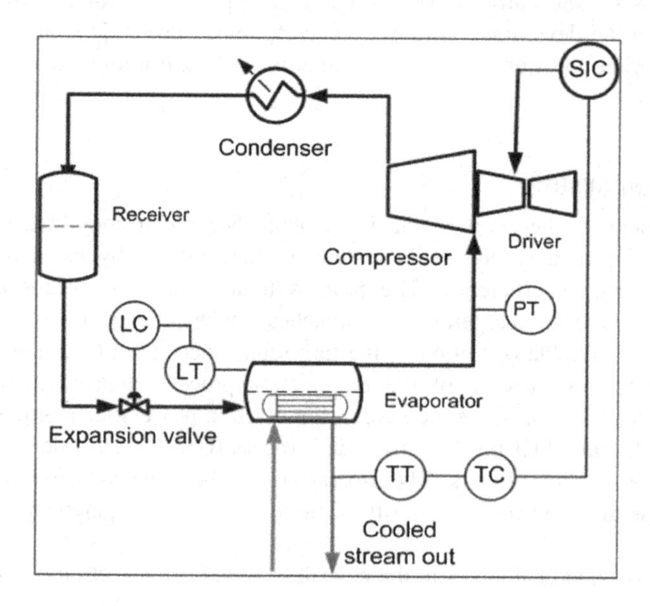

FIGURE 6-5

Control scheme for a typical single-stage refrigeration process.

Either the temperature itself or the chiller pressure can be the process variable for the master loop. The rest of the process essentially floats on the chiller duty. The chiller level is normally maintained using the JT valve. Multistage refrigeration processes are used to enhance the efficiency of the process. The basic control philosophy in this case is still the same as for a single-stage process. A systematic procedure to analyze the available degrees of freedom in refrigeration systems is given in Jensen and Skogestad (2009).

6.4.4.2 Control of liquefaction unit

The main control objective in the liquefaction section is to ensure that the condensing of natural gas to form LNG and subcooling of LNG is achieved under all conditions. The other important consideration is to maintain the desired plant capacity. These two objectives should be achieved while rejecting disturbances like ambient temperature and feed composition. The capacity control can be achieved by feed rate control or back-end LNG flow control. The cascade refrigeration process normally uses the first, whereas the APCI C3MR process utilizes the second. Instead of feed flow control in the front end, pressure control in the front end can be utilized as an alternative. This scheme can be designed to take as much feed gas as needed to produce LNG at a particular temperature (Mandler et al., 1998).

It is important to maintain the LNG temperature to ensure condensing of natural gas and any amount of subcooling that is required. If the LNG is too warm, the boil-off gas vapor generation will exceed the capacity of the BOG system, resulting in flaring. For cascade refrigeration processes, the LNG temperature is maintained by maintaining the pressure in the chillers, as in a typical refrigeration process. This is controlled by cascade control with the refrigeration compressor speeds (Valappil et al., 2007). The condensing pressure in the back end is also controlled in this case.

For APCI C3MR processes, if independent control of production and temperature is not desired, the LNG temperature is controlled by varying the production rate. The refrigeration capacity is adjusted by the compressor speeds and JT valve positions (warm and cold). The amount of refrigeration is determined by these handles. This will also set the production rate to maintain the LNG temperature. The compressor pressure ratio is also controlled using these variables. The ratio of liquid mixed refrigerant (LMR) to vapor mixed refrigerant flow rate is an important variable in the C3MR process and is controlled using the JT valve positions.

For independent control of LNG temperature, temperature can be controlled by manipulating the amount of refrigeration in the system (Mandler et al., 1998) as shown in Figure 6-6. This can easily be achieved by manipulating the MR compressor speed. Other parameters like Warm JT valve can also be used as a handle to accomplish this. Increasing the compressor speed would lower the LNG temperature for a given production rate. The LNG production can then be set at the desired value that is independent of the temperature. The cold JT valve can be used to maintain the ratio of liquid MR to vapor MR (Figure 6-6).

6.4.4.3 Compressor and gas turbine control

The refrigeration compressors are critical equipment in any LNG plant and their proper operation is important for the overall reliability of the plant. The control schemes used for these compressors depend on the driver that is selected for them. A vast majority of the drivers used in the LNG process are gas turbines, including frame-type turbines and aero-derivative turbines (steam turbines and

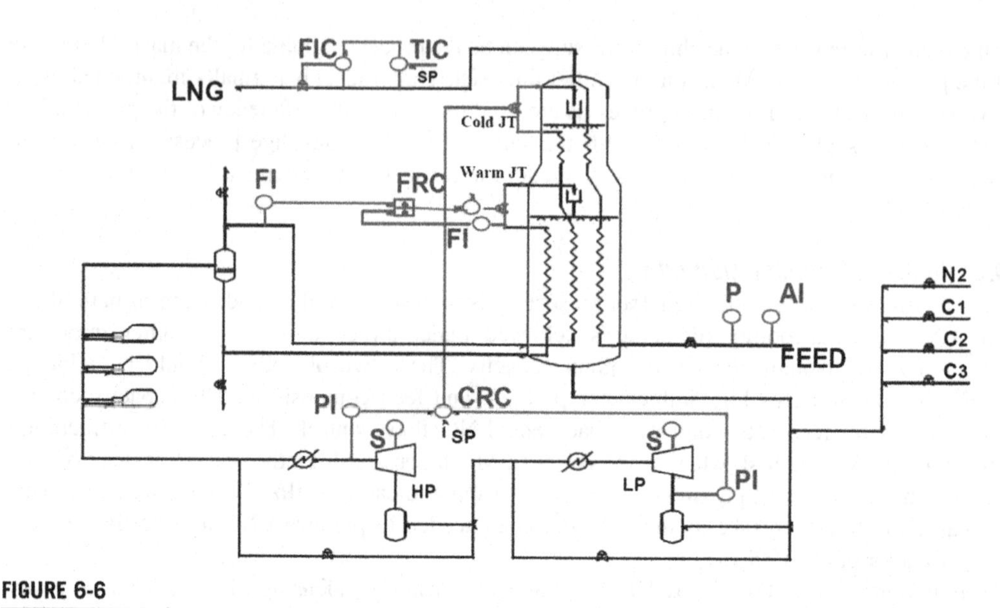

FIGURE 6-6

Traditional regulatory control scheme for a C3MR unit.

(Mandler et al., 1998)

electric motors are used as drivers, but rarely). The capacity control of these machines is accomplished using compressor speed as a variable. The compressor speed can be used as an independent variable and set by the operator. Alternatively, the compressor suction pressure is cascaded to the speed control to maintain the capacity. This way, the machine speed, and hence the load, change depending on the refrigeration load.

There are different options as to the implementation of this capacity control function for refrigeration compressors. This could be implemented in the plant DCS system or can be included in the compressor control package from different vendors. The second one is the preferred option (as opposed to DCS-based capacity control). This is due to the fact that the capacity control interacts with the antisurge control, and this has to be accounted for via a loop decoupler for improved control performance.

There are several important capabilities that are required in a well-designed compressor-driver control system. Some of the key ones are:

- **Capacity control.** A well-designed capacity control system is important to ensure the smooth operation of the LNG process. The selection of the controlled variable is important here.
- **Antisurge control.** An effective antisurge control system is critical for the protection of refrigeration compressors and to prevent damage due to surge.
- **Load balancing.** Load balancing is important to enhance the operation of compressors in parallel. This helps to ensure that both machines operate with close to equal loads, thus preventing the possibility of one machine operating in recycle (unloaded) with the parallel one fully loaded.

- **Loop-decoupling.** The interaction between antisurge and the capacity control can be significant. This is due to the fact that opening the antisurge valves affects the pressure and changing speed varies the antisurge control setpoint (flow requirements). Hence, loop-decoupling is important to ensure the interactions between these do not result in persistent oscillations.
- **Override control.** The override control of suction or discharge pressure should be considered to prevent undesirable operating conditions. These control overrides normally utilize the antisurge valve openings to prevent any abnormal suction or discharge pressures.
- **Load-limiting control.** This is relevant for the cases where the driver power becomes a limiting factor. In these cases, the load on the machines has to be maintained below the upper limit to prevent driver overloading (which could result in a turbine underspeed trip). This can be accomplished using load limiting controls that use handles like suction throttling valves or other variables.

The various features of the compressor-driver control system are shown in Figure 6-7. These capabilities may be provided by a control system vendor specialized in control of compressor systems. These could be used to separate the system from the plant DCS system. Certain DCS systems also provide functionality for implementing these functionalities. The final decision has to be made considering a number of project-specific factors.

6.4.4.4 Antisurge control systems

The antisurge control system is critical in ensuring the safe operation of the refrigeration compressors in the LNG plant. The refrigeration compressor stages usually run at high power levels, which can

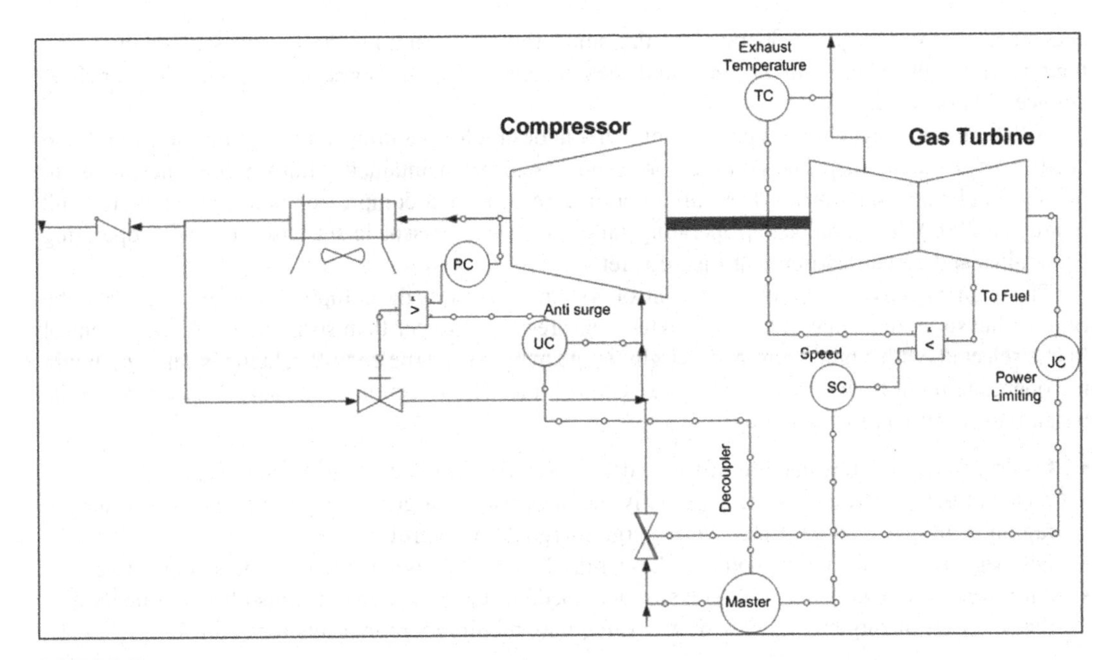

FIGURE 6-7

Typical compressor-gas turbine string control scheme.

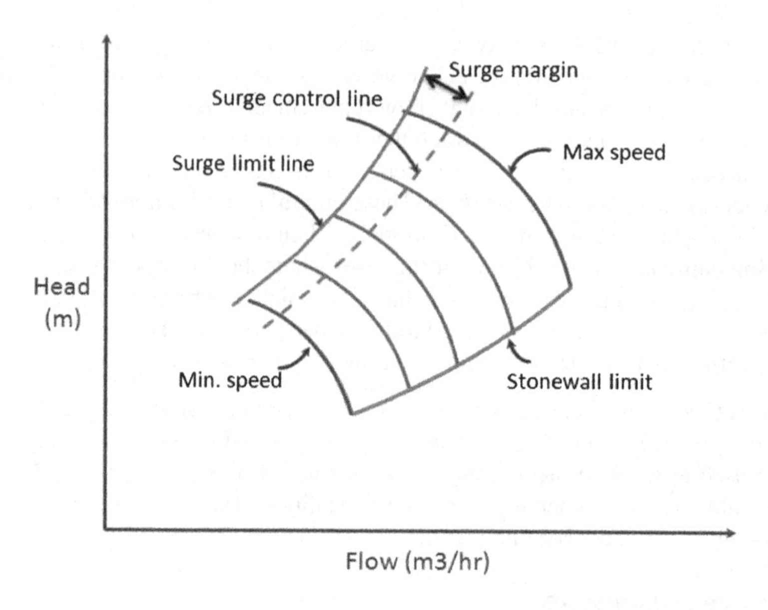

FIGURE 6-8

The compressor map with the acceptable operating region.

encounter severe damage if operated in the surge region. Operating the compressor in the surge region can result in intermittent and rapid flow reversals, which can severely affect the impellers (Boyce, 2003).

Although surge prevention is paramount, other undesirable operating conditions like stonewall also need to be avoided. Also, the compressor speed has to be maintained within acceptable ranges to prevent unplanned shutdowns. The safe operating region for a compressor is shown in Figure 6-8 (Narayan, 2005). The acceptable operating range for a compressor is set based on these operating constraints and certain chosen margins of safety.

The main objective of the antisurge control system is to ensure the compressor stages operate to the right of the surge limit line in the map (which ensures flow greater than surge flow). A surge control line is selected, with a predetermined margin (surge margin) and the control scheme is set to maintain the operating point to the right of this surge control line. The selection of the surge margin has to be based on several factors, including:

- Control functionalities that are utilized (derivative response, recycle trip response, etc.).
- Shape of the compressor curve, especially the head rise to surge. Having flat curves in the surge region would necessitate using higher surge margins for control.
- Discharge and suction piping volumes. Larger discharge volumes may require higher surge margins.
- Antisurge valve characteristics. The size and speed of the antisurge valve also has an impact on the selection of surge margin. A slower valve may require more margins for control.

The surge control margin will have to be developed empirically based on actual field testing. Reliable initial estimates can be obtained using dynamic simulation analysis.

The control functionalities used for the antisurge control are important in deciding the performance of the surge protection system. The following are some of the capabilities that are desirable in the antisurge control systems:

- **Normal Proportional-Integral (PI) control.** The base antisurge control is usually a proportional and integral-based control response. This looks at the distance from the surge control line (SCL) and tries to maintain the operating point to the right of SCL.
- **Derivative response.** The derivative response evaluates at the speed at which the operating point moves toward the surge control line. Based on this, the antisurge response is varied to enhance the control performance.
- **Recycle trip response.** The recycle trip response is used once the operating point moves to the left of the recycle trip control line (closer to surge limit line). The antisurge controller output is proportionally opened in this case to bring the operating point back to the right of SCL.
- **Surge counter reset.** The surge counter is used to evaluate the number of times the surge happens. This is used to reset the surge margin and move the surge control line to the right to prevent further surge events.

6.4.4.5 Gas turbine control

The control of gas turbine drivers is handled by the gas turbine vendor package and not in the scope of plant DCS. It is still important to understand the various control features for effective LNG plant design and operation. The main features in the gas turbine controls are listed here (Boyce, 2011).

- **Speed governor.** The gas turbine speed control is the most important loop because this sets the refrigeration compressor speeds and hence the capacity. This could be an independent variable that is set by the operator (alternatively APC system) or could be cascaded to the suction pressure for capacity control. The speed governor normally manipulates the fuel gas flow to the turbine, which delivers varying shaft power. This is a PID controller with all three modes active and tuned for fast response.
- **Exhaust temperature control.** The exhaust temperature control is normally an override loop that will act on the fuel gas if the limit is reached. This loop should not be active normally since this is a physical limit.
- **Acceleration control.** Acceleration control is used to ensure that the gas turbine rate of change of speed does not vary beyond safe limits. This is normally applicable in transients like startup to limit thermal stresses due to overacceleration. During normal operation, this may become active to prevent rapid overspeeding in the event of a sudden load reduction.
- **Other.** Other control functionalities are also present in a gas turbine control package, which are important from a mechanical and machine integrity perspective.

6.4.5 Storage and loading

The storage and loading section of the LNG plant consists mainly of LNG storage tanks, the boil-off gas compressors, the LNG loading pumps, and the loading system.

The main objective of the process control system here is to maintain the LNG storage tank pressure to prevent any flaring due to overpressuring of the tank. The vapors from the tank are fed to the BOG compressor, which then compress it and sends it back into the liquefaction process or uses it as fuel gas. The tank pressure has to be controlled by manipulating the flow of BOG vapors. This can be accomplished in two ways. If the BOG compressor is variable speed or variable IGV machine, speed or IGV control could be used to control the pressure. On the other hand, if it is a fixed speed machine, discharge valve throttling is used to control the pressure.

The control of tank pressure is also impacted by the liquefaction system performance, especially the LNG temperature. If the LNG product is too warm, excess vapor generation will overload the BOG system. Hence, specifying and maintaining the correct setpoint LNG temperature control is important for efficient plant operation.

6.5 Advanced process control and optimization of LNG plants

APC has established itself as a widely used technology in process and other industries. Currently, there are hundreds of industrial units that benefit from this technology. The main reason for its popularity is the ability to run plants at optimal conditions while handling various operational constraints and disturbances. Model predictive control (MPC) is the most popular advanced control technique. MPC consists of a model capable of predicting the behavior of key variables in the plant. Adjustments are made to the corresponding manipulated variables to keep the control variables tracking close to their desired target values. These target values are obtained from an optimizer that aims to increase the profit function. A typical MPC application resides on top of the regulatory control layer and sends the setpoints to the PID controllers in the plant.

There are many benefits to be derived from application of APC to an LNG process (Poe and Mokhatab, 2007). LNG plants are characterized by operational challenges including varying ambient conditions, heat integration, and product quality constraints. For LNG plant owners, the main benefit from APC is product maximization, which increases the operating profit. Other benefits like increased process efficiency, increased NGL production, and reduced operator intervention are also valuable. A properly designed APC application can achieve these objectives with reasonable capital investment.

APC has been applied extensively in refining and petrochemical industries in the early days of this technology. There was not much interest in applying advanced control to LNG units then. This has changed in recent years, and different applications of APC to LNG processes are being reported. Recently, Shell has reported applications of APC to the LNG processes based on the Shell DMR process (den Bakker, 2006b). Application of APC to a LNG train in Qatar was reported by Aspen Technology (El Hadi et al., 2007). More interest in the application of APC to LNG plants is expected in the future. This is due to changing market conditions, whereby operating plants will be able to sell excess LNG production in the spot market. This provides significant economic incentives in maximizing the existing plant potential. This also raises the issue of tailoring the LNG quality to the appropriate market and changing the product mix based on economics. APC is a proven tool to achieve all these objectives.

6.5.1 Model predictive control technology

Model Predictive Control (MPC) is a class of methods that uses an explicit model for predicting future response of the plant and for control. The variables in MPC are divided into manipulated variables (MV), which are changed to maintain the controlled variables (CV) within constraints. This is accomplished in the presence of disturbance variables (DV) that upset the process. The plant model that is used for MPC is an empirical model derived from performing step tests in the plant or a dynamic model. This empirical model is developed from the step test data and model identification tools. At each time step (every minute or so) MPC solves an optimization problem to calculate the best steady state operating point for the plant. This optimization is normally formulated as a linear or quadratic program.

Once the optimization is solved and the optimal steady state operating point is determined, the best method to reach the target is calculated. This calculation is done based on the empirical model and the current operating point. The first manipulated variable move of this trajectory is implemented and the whole calculation is repeated in the next time step with the new measurements (Blevins et al., 2003).

6.5.2 Advanced process control implementation steps

The main steps in the implementation of APC are detailed here.

1. **Economic benefit analysis.** The first step in the implementation of APC is the evaluation of the economic benefits that can be achieved from APC. This is done using statistical methods based on the plant operating data. In this case, the performance of the plant is analyzed in terms of the variability in the key variables, normally measured in terms of standard deviation. The benefit of implementing advanced control is to reduce the variability in these key constraints. This allows the operation of the plant closer to the constraint, thus improving economic gain (Figure 6-9).
2. **APC Controller design.** In this step, a design of the controller is developed. This is based on the operational expertise of the plant personnel and operators. The main objective here is to identify the various capacity limiting constraints and the economic benefits from optimization. Also, the main controlled variables, manipulated variables, and the disturbance variables are identified at this stage. The need for any inferential variables is also determined here.
3. **Plant pretest.** This step involves a thorough review of the existing plant regulatory controllers and analyzers. The APC sends the setpoints to the base regulatory controls. The tuning of these controllers has to be verified and improved. Also, the reliability of any analyzers used in the plant will be established in this stage.
4. **Plant step test.** Here, the main objective is to develop empirical models for the APC controller. This can be done in the plant or a dynamic simulation model (Valappil et al., 2006). The models developed can be updated using closed loop identification at a later stage during commissioning.

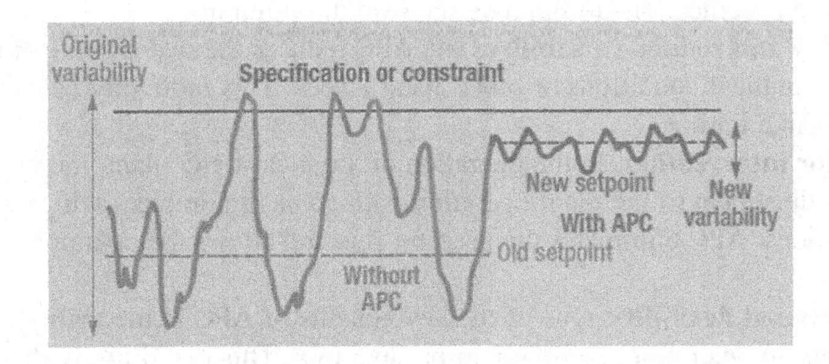

FIGURE 6-9

The benefits gained from APC by reducing the variability in key constraints and moving closer to the optimum.

(Poe and Mokhatab, 2007)

5. **APC controller development.** In this step, the data collected from the plant step test is used to identify empirical models for the controller. Also, the controller is configured with the initial tuning parameters and limits for field implementation.
6. **APC commissioning.** In this step, the developed APC controller is implemented in the plant. The controller is tested in open and closed loop modes. The performance of the APC under various disturbances and other operational scenarios are verified.
7. **Operator training and post implementation review.** The operators have to be trained to understand the plant operation with APC. This is done in this stage of the project. Also, the actual benefits derived from APC are estimated in this step using the plant data after implementation.

6.5.3 Benefits and challenges in applying advanced process control in LNG plants

There are several key benefits to be derived from application of APC to LNG plants as detailed here.

- **Maximize plant throughput.** The main benefit of APC is to maximize the LNG production by operating at the capacity limiting constraints. These could be hydraulic, equipment, or other constraints. These constraints vary with changing conditions like ambient temperature, feed composition, and so on.
- **Enhance plant efficiency.** There are two different objectives to be considered in this case. First is to minimize the plant fuel consumption to increase the net LNG production with a fixed plant feed rate. The second is to minimize the energy consumption for the operation of energy-intensive unit operations like separation columns. This can be achieved by operating the column under optimal conditions.
- **Improve yield of valuable products.** A typical LNG plant has products like NGL, propane, or butane, in addition to LNG. The benefit of applying APC here is to increase the production of more valuable product by operating at the best constraints. This is accomplished while honoring the product quality.
- **Maintain product quality.** It is important to meet the product specification for LNG and NGL products to minimize product waste. For LNG, the HHV and Wobbe Index are important parameters. For NGL, the RVP has to be maintained. The APC can move the operation closer to the limits of these specifications to improve the overall optimum.
- **Improve stability and reduce variability.** The APC reduces the overall variability in operation due to changing ambient conditions or other disturbances. This indirectly helps to minimize abnormal events like flaring.
- **Reduce operator intervention.** In the operation of a typical LNG plant, the operator has to make frequent adjustment to the plant operation with changing ambient temperature or other operating conditions. APC eliminates the need for this and allows the operator to focus on other important tasks.
- **Enhance operational flexibility.** One of the key benefits of APC is the ability to incorporate market conditions to determine the optimum product mix. This could imply that under certain market conditions, maximizing the production of NGL would be more valuable than LNG or vice versa. APC objective functions can incorporate this to reach an overall economic optimum.

There are certain features of LNG plant operation that are challenging and will have to be considered in the APC controller design. These include the following:

- Ambient temperature variation, both typical day–night cycle and sudden changes due to extreme weather conditions. This affects the gas turbine operation and hence the LNG process and will have to be handled by the APC.
- Feed gas composition changes, with changes from lean feed gas to rich and vice versa. The APC should be able to change the various operating conditions, including separation column variables to handle transitions likes this.
- Strict limitations on temperature differentials in plate and frame exchangers, which needs to be maintained during dynamic transitions.
- Ensuring proper CO_2 and H_2O removal for feed to liquefaction units to prevent freezing in downstream equipment.
- Variations in feed gas supply temperature and pressure, which acts as disturbances to the APC.
- Ability to accommodate ship-loading, which results in extra boil-off-gas return to the liquefaction section of the plant.
- Other unmeasured disturbances that can impact the LNG process.

6.5.4 Advanced process control for individual units in LNG plants

6.5.4.1 Upstream and inlet facilities

The upstream and inlet facilities in the LNG plant include the slug catcher, the inlet stabilizer, and associated equipment. This does not include the offshore equipment, which is usually implemented in a separate controller. The slug catcher does not have significant APC benefits, other than to control the level and to use the buffer capacity efficiently.

The main area of application of APC in this area is to the stabilizer columns. The inlet stabilizer is a key unit that removes the NGL from the inlet stream. The main control objective here is to maintain the C_5+ concentration in the overhead and the RVP of the bottom product. The APC can be beneficial in the control of RVP, which can be pushed to its high limit, thus increasing the condensate production.

Another application of APC in this unit is to reduce the energy consumption. It is beneficial to operate this column at a low pressure, thus reducing the energy consumption. This pressure is constrained by the column flooding, control valve openings, or the capacity of the overhead compression system. Further economic benefits can be derived from increasing the stabilizer throughput. This is especially beneficial in the case where stabilizer capacity is the limiting factor in increasing the plant production (Hodges, 2001).

6.5.4.2 Acid gas removal unit

The objective of an acid gas removal unit is to remove the CO_2 and H_2S in the feed gas to a very low concentration before sending the gas to liquefaction. This is achieved using absorption with MDEA or other solvents. The solvent is regenerated in the regenerator column and recycled back to the absorber. The main benefit to be derived from applying APC to the acid gas removal unit is the improvement in control of CO_2 concentration in the column overhead. This is achieved by means of reduced variability via model-based control. The reduction in variability in CO_2 concentration makes it possible to minimize the energy consumption in the column reboiler. Normally, steam or hot oil is used for reboiler energy supply. Other benefits include the reduction in amine recirculation rate, which can reduce the solvent losses. The main manipulated variables that are used to achieve these objectives include the lean amine flow rate, the reflux ratio setpoint, and the regenerator bottom temperature.

The benefits to be gained from applying APC to the acid gas removal unit may not be significant compared to the other units like liquefaction. For plants with a sulfur recovery unit, increasing the sulfur recovery using APC could yield significant economic benefits (El-Hadi et al., 2007). A comprehensive benefit analysis would be needed to evaluate this and to make a decision to include this unit in APC.

The other main components of the feed treatment system in LNG plants are the molecular sieve units for dehydration (to ppm level) and mercury removal beds. There are no significant economic benefits to be gained from applying APC to these units. There could be cases where the water or mercury composition in the feed gas at the outlet of these can be a capacity limiting constraint. The composition could be then included as a constraint in the APC.

6.5.4.3 Liquefaction system

The liquefaction system is the main unit in the LNG plant and is also the unit where significant benefits can be derived from APC. The design and specific variables used for APC in a liquefaction system are dependent on the liquefaction technology that is used. The application of APC has been reported for the APCI C3MR process (El-Hadi et al., 2007; Poe and Mokhatab, 2007), Conoco Phillips Optimized cascade LNG process (Valappil et al., 2007), and Shell DMR process (den Bakker, 2006a).

The main APC objectives in the liquefaction system are to increase capacity and optimize the overall plant efficiency. All LNG processes are affected by the ambient temperature, and hence the need to push the plant capacity to the maximum at any ambient is applicable to various LNG technologies. The plant efficiency is enhanced by operating the plant at the optimal conditions that minimize the fuel usage for the drivers.

The capacity limiting constraints could be different for different LNG processes. For the plants based on the same LNG process technology, individual plants have unique constraints in addition to the generic constraints. The main constraints applicable to all processes are the limitations in the refrigeration compressors and drivers (gas turbines or other), including turbine power or compressor conditions (flow limitations, speeds, discharge pressure, and temperature). Further, all LNG plants are characterized by limits on heat exchanger temperature differentials, which should be included as constraints in the APC design.

For the C3MR process, the MR compressor/driver and propane compressor/driver variables are key constraints. Also, LNG temperature is one of the key controlled variables. These are maintained by mainly varying compressor speeds or MR compositions and flow rates (den Bakker, 2006b). The design of APC for this process is discussed in more detail later.

For the cascade refrigeration process, optimal distribution of load between various refrigeration cycles can be maintained using APC. This load balancing between the various refrigeration systems is very important to maximize the capacity under all conditions. The key variables that are used to achieve this are the suction pressures for the refrigeration compressors (Valappil et al., 2007). In addition, other load-balancing handles are normally provided, which can be used to optimize the plant capacity.

The other main consideration in this system is maintaining the product quality for LNG (HHV). This is discussed further in the next section. Also, the BOG compressors need to be able to handle the BOG return from the LNG tank. This means that the LNG product sent to the storage tank has to be at an appropriate temperature and the vapor return at an appropriate pressure, such that the BOG vapor return compressors are able to handle the return vapors from the tank.

6.5.4.4 Scrub column and fractionation units

There are significant economic benefits to be gained from using APC in the NGL and fractionation section of the LNG plant. This is the area of the plant that dictates the LNG properties and the split between various products. The main objectives of the APC controller in this unit are:

- **Determine the right operating LNG/NGL split.** The APC can incorporate the market values of LNG and NGL to determine the optimum allocation of one versus the other.
- **Control the product qualities.** The APC can be valuable in maintaining the product qualities in the NGL and fractionation systems and hence the final LNG product quality.
- **Minimize the energy consumption.** Further operational optimization can be achieved by minimizing the energy consumption in the columns in this section.

The scrub column (Heavies removal column) is first used to remove the heavier components in the feed stream, before it is sent to liquefaction. There are two main objectives of APC in this column.

- Maintain the overhead composition of the C_5+ below limit. This is important to ensure that heavier components do not cause freezing in the liquefaction.
- Depending on the market conditions, the remaining objective of the scrub column could be different.
  - If the NGL is a more valuable component, then the bottoms from the scrub columns can be maximized with the overhead composition maintained such that the LNG product meet its HHV spec.
  - If it is more important to maximize LNG, the HHV of the LNG can be pushed to the maximum allowable limit by sending heavier components to the overhead of the column.

The LNG plant fractionation section may or may not be present depending on the facility and plant design. A typical fractionation system consists of deethanizer, depropanizer, and debutanizer. The following key objectives are to be considered for APC on this unit.

- Maintain the operating temperature and pressure values for each column.
- Maximize the recovery of propane and butane in the overhead of the columns.
- Maintain the RVP spec for the final bottom product.
- Minimize the energy consumption to optimize efficiency.

APC has been applied extensively to fractionation columns, and the design of the controller for these columns is well established.

The main controlled variables in the NGL recovery section are the product qualities; hence, reliable measurement for these is important. It is preferred to have online analyzers to measure these in real time. Even if these properties are measured, the measurement may be very infrequent. It is beneficial to develop and use property inferentials (soft sensors) for use in APC. These inferentials are used to determine a product quality variable as a function of other measured variables. These can also be used along with the analyzers to provide more frequent updates for control.

If the plant has a DOMGAS and LPG unit, further economic benefits can be gained from application of APC to this section. The main benefit in this unit is ensuring the quality and increasing the recovery of the LPG, which is a sellable product. Additional optimization can be achieved by coordinating APC here and in the fractionation section APC, thus maximizing the overall facility LPG recovery (Poe and Mokhatab, 2007).

6.6 Process control and automation in LNG import terminals

There are several challenges in designing the automation and control layer in LNG terminals. These include tough environmental regulations, volatile market conditions, and operational challenges. The main components of the LNG import terminal automation system include the process control system (PCS), the safety system, the fire and gas system and the emergency shutdown system (ESD).

6.6.1 Basic process control systems

The process control in LNG terminals can be divided into three distinct areas of operation:

- LNG ship unloading
- LNG tank operation
- LNG send-out and vaporization.

The LNG ships unload their cargoes to the LNG storage tanks using the ship pumps. The unloading rate is controlled by the ship. The operation of LNG ship pumps, unloading arms, BOG compressors, and vapor return blowers are coordinated during unloading. There is a significant increase in the vapor return during ship loading. The operator will have to set the number of BOG compressors based on the ship unloading rate to maintain the system pressure. Normally, a Jetty unloading and mooring PLC will monitor the unloading line and automatically stop unloading when unsafe conditions are detected.

The LNG storage tank pressure normally is controlled to prevent it from going too high or low. When the tank pressure goes high, the BOG compressors will be manipulated to increase the flow and bring the pressure down. When there is no ship loading, the tank pressure can go lower than the desired values. This is prevented by introducing vacuum breaker gas on pressure control. In addition to this, the instrumentation with alarms is provided to monitor levels, temperatures, and pressures in the LNG storage tanks.

During ship unloading, the BOG can be returned to the ship using return blowers. The pressure of the vapor return will be maintained by pressure control at the blower outlet. The return vapor will also have to be desuperheated on temperature control to maintain appropriate temperature. The BOG gases, which are compressed, are normally used as a fuel gas supply for the submerged combustible vaporizers (SCV). The remaining BOG can be condensed and recovered in the BOG recondensors. The fuel flow to SCV will vary based on the demand and is set to maintain the pressure. The pressure control at the discharge of the BOG compressors open the valve to the recondensor based on the total vapor flow.

The LNG is sent to the vaporizers using the send-out pumps on flow control. The temperature at the outlet of the vaporizers is monitored to ensure a sufficiently hot stream. This product natural gas is then delivered to the pipeline. A pressure controller is used to maintain the back pressure in this line.

6.6.2 Terminal information management system

The main role of the Terminal Information Management System (TIMS) is to ensure operational safety and efficiency and to provide integration between the corporate management systems and terminal process control system (Barash and Rangnow, 2007). This is especially important for LNG receiving terminals because they are a crucial link between the gas field and the consumer. Also, there is a tight linkage between terminal scheduling and operating orders, with cyclical switching between loading and unloading. This coordination is more important with the changing market conditions, where the LNG terminals are serving more customers. The application of an operating information system (OIS) for the Incheon LNG terminal was discussed by Lee and Yang (2007). The OIS in this case included several advanced functions including online monitoring and reporting, data reconciliation, PID auto-tuning, and operator support.

The main functionalities of the typical LNG TIMS, as shown in Figure 6-10, are:

- **Supply chain management.** The objective of SCM is to ensure proper delivery of timely information for supply chain optimization. Several functions like cargo tracking, energy balance, and terminal scheduling are included in this.
- **Operations management.** Assists the terminal operations by providing the right information and the operating parameters.
- **Asset management.** Assists the maintenance personnel with maintenance data and schedules.
- **Compliance management.** Helps to ensure that the regulatory reporting requirements are met by facilitating reporting.
- **Business management.** Ensures alignment of business processes to meet stakeholder requirements.

6.6.3 Optimization in LNG receiving terminal operations

Several options are available to enhance the reliability and efficiency of the LNG terminal operation (Teramine and Sakamoto, 2001). These include the traditional methods like process control enhancements and advanced process control. Some of the other possible options in this regard are:

- Basic process control and advanced control. This entails integration of various control systems so that they can be monitored and operated from common consoles.
- Operator training with simulators. The improvements in basic control systems mean that there are less frequent upsets. Hence, the operators have to be trained to handle emergency situations using operator training simulators.
- Facility condition monitoring. Enables early detection of fault signs and response to the same. This increases the reliability of gas production and reduces maintenance costs. For example, this allows detection of excessive vibration from cryogenic power generation and LGN pumps before any failure, thus enabling early measures to correct the problems.
- Tank operation planning system. Used to determine a tank operation plan that efficiently balances the costs including LPG for calorific adjustment and electrical power, while honoring the operating constraints. This system takes tank level, tank calories, cargo schedule, production schedule, etc., as its inputs and outputs the optimal LNG discharging and receiving plans.
- Productivity management system. Utilizes the operating and productivity data for four main types of equipment in the LNG terminal: LNG pumps, BOG compressors, seawater pumps, and LNG

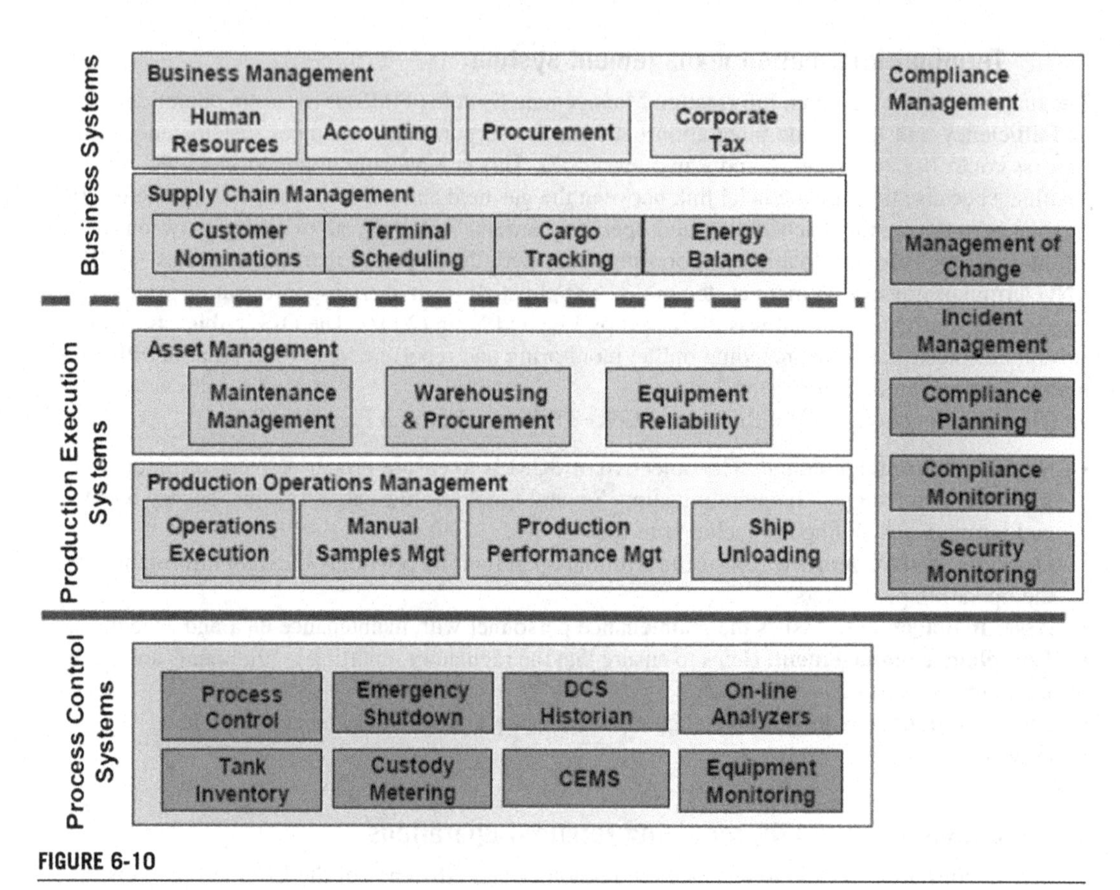

FIGURE 6-10

The main functionalities of a TIMS used in LNG terminal operations.

(Baresh and Rangnow, 2007)

vaporizer units. This is then used to provide operation support that improves the productivity of the entire terminal.

A very important economic objective is the efficiency of operation of the LNG import terminal. This is achieved by reducing the power consumption for various equipment like BOG compressors, LNG pumps, and seawater pumps (if used). Alternatively, efficient operation of SCV can increase the overall LNG terminal operational efficiency.

6.7 Case study 1: advanced process control for APCI C3MR LNG process

The evaluation, design, and implementation of an APC controller for the APCI C3MR process is discussed in this case study. The APCI C3MR process is the most common LNG process used in the industry. Detailed descriptions of this process are available in the literature. The case study considers the liquefaction and the heavies removal sections of the plant. The application of APC to condensate stabilizer system was presented in Mokhatab and Poe (2012).

6.7.1 Estimation of benefits from APC implementation

A feasibility study is first conducted to identify the economic benefits from APC and the potential implementation issues. The main benefits identified included maximizing the LNG and LPG production and enhancing the overall plant efficiency. Other intangible benefits like stabilizing plant operations are also considered valuable.

The scrubber column APC objectives could be different, depending on the relative LNG and LPG prices. If the LPG is more valuable, the objective is to maximize the LPG production. Otherwise, it is more beneficial maximize the C_3+ content in LNG (with due consideration of the LNG HHV spec). For the plant under consideration, LPG maximization was beneficial. Hence, the main objectives of APC for the scrubber unit include the following:

- Keep C_5+ in the overhead below the specified limit to prevent freezing.
- Keep the overhead stream HHV within range.
- Maximize the C_3 in scrubber bottoms.

The main objective of the APC in the deethanizer is to ensure that the overhead stream meets the specified limits. This stream is send to liquefaction or fuel gas for further processing.

The liquefaction APC is where the majority of the economic benefits are realized. The main objective here is to maximize the LNG production. The LNG HHV has to be maintained within the specification range. Also, the LNG has to meet the required specifications; including the one that the amount of C_1 should be higher than the specification value. If the production is constrained by the market, the efficiency has to be maximized to realize the benefits. The efficiency for the C3MR process is specified by Specific Brake Horse Power (SBHP). Alternate measures of efficiency including the ratio of the amount of fuel gas to the amount of LNG can be utilized for optimization purposes.

The APC benefit estimation is based on reducing the variability in plant operation, hence pushing the key variables to the constraints (Figure 6-11). The standard deviation of the main controlled variables before APC implementation is determined from the historical plant data. The reduction in standard deviation by means of APC is known from other similar implementations. This difference in

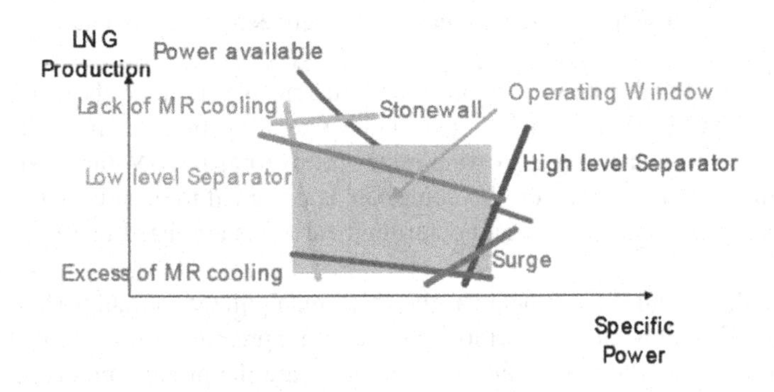

FIGURE 6-11

Operating window for a typical MR LNG process.

(Sturm et al., 2009)

standard deviation can be translated to an economic benefit and can be determined (Figure 6-9). For calculating the benefits from LNG production maximization, the percent increase in the LNG is determined based on the reduction in standard deviation. For example, for a plant with 10MTPA LNG production rate and 1% increase in LNG production, the benefits can be calculated as $EconomicBenefits = 10 \times .001 \times (LNG_P - GIP_P)$, where LNG_P is the price of LNG and GIP_P is the price of gas in place (GIP).

The other main economic benefit for this case is in maximizing the NGL production by three means:

- Reducing C_3+ content in LNG up to specification
- Increasing C_2 content in C_3 product up to specification
- Pushing the amount of condensate in LPG C_4.

The benefits from these actions can be calculated by determining the extra NGL that is produced. For this, the variability in the LNG composition (HHV) is first determined from operating data. The reduction in variability is assumed, thus making it possible to push more C_3 in the NGL products. The economic benefits can be determined from this and the NGL price. In a similar manner, the benefits from maximizing the C_2 components in a C_3+ product can be estimated. These calculations need to account for the need for additional feed gas (hence added cost) due to reduction in LNG production/increase in NGL.

The APC benefits from increasing plant efficiency can also be calculated. This involves looking at the current operational efficiency and estimating a typical improvement with APC. This can then be translated to a decrease in fuel consumption, which increases the net LNG output.

6.7.2 Design and implementation of LNG advanced process control

The objectives of the APC implementation for the entire LNG plant can be accomplished in different ways depending on the user preferences. One option is to have one APC controller for the liquefaction, scrubber, and deethanizer sections. This controller will be large in scope. An alternative is to have different controllers for the liquefaction and the column sections. This approach is utilized here. These controllers could still pass data between them to enable the true optimum operating point.

The design of the APC for scrub column is undertaken to maximize the condensate extraction and to control the LNG HHV in a tight range while operating the unit in a steady manner. The objective function is formulated so as to minimize the overhead HHV (and hence maximize the condensate extraction). For deethanizer, the controller is designed to ensure more C_2 components in the bottom flow. The main manipulated and controlled variables are listed in Table 6-1 and shown in Figure 6-12.

For the scrub column and the deethanizer, the main manipulated variables (MV) are the reboiler steam flow and the reflux flow rate. In addition, the operating pressure is used as an MV to add further flexibility for optimization. The main controlled variables are the product properties and the column differential pressure (to prevent column flooding). The operating temperatures are also maintained within constraints. The HHV of LNG is also included to control this in a tight range, so as to minimize giveaway of heavier components.

Table 6-1 Variables in the Advanced Control Design for Scrubber and Deethanizer

Manipulated Variables	
1	Scrubber reflux rate
2	Scrubber reboiler steam flow
3	Scrubber inlet feed temperature
4	Scrubber operating pressure
5	Deethanizer reboiler steam flow
6	Deethanizer reflux flow
Disturbance Variables	
1	Natural gas feed flow rate
Controlled Variables	
1	Scrubber column top temperature
2	HHV of scrubber overhead
3	HHV of LNG product
4	Scrub column bottom temperature
5	Scrub column differential pressure
6	C_3 in deethanizer overhead
7	C_2 in deethanizer bottoms
8	Deethanizer column pressure
9	Deethanizer column top temperature
10	Deethanizer column bottom temperature
11	Deethanizer reflux ratio
12	RVP of NGL product
13	Other plant-specific constraints

Since the properties are important variables in the controller, inferential properties for the column overhead and LNG HHV are developed. The HHV of the scrub column overhead is a function of column top PCT (Pressure Compensated Temperature). The LNG HHV can be calculated as a function of the scrub column overhead HHV and reinjected C_2, C_3, and C_4. These inferential property estimators are used along with the analyzers (bias update) to obtain frequent, reliable measurements for APC. Similarly, inferential measurements are developed and used for deethanizer top and bottom quality variables.

The design of the liquefaction section APC is developed to achieve the key economic objectives. The manipulated and controlled variables selected for this purpose are listed in

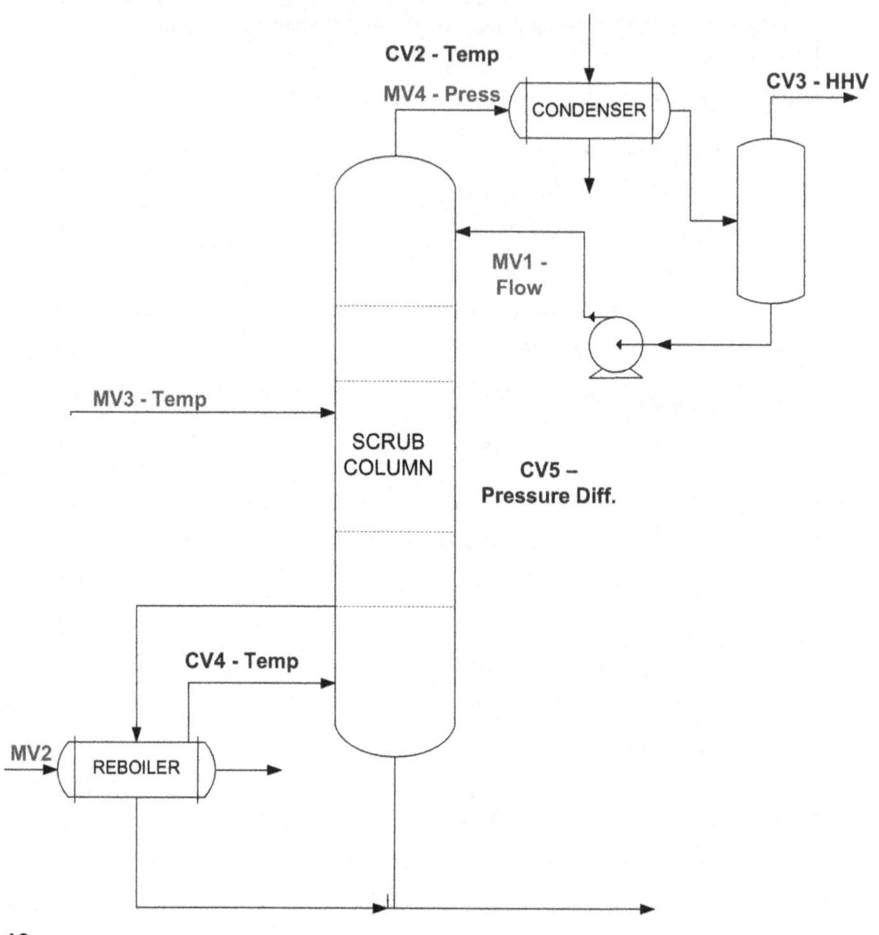

FIGURE 6-12

The scrub column variables used in APC (MV and CV denotes the manipulated and controlled variable, respectively).

Table 6-2 and shown in Figure 6-13. The main economic objective here is to maximize the LNG production by pushing against the capacity limiting constraints. The constraints could be compressor power/speed, exchanger temperature approaches, or the flash drum pressures. In addition, load shifting between C_3 and MR compressors need to be performed with varying ambient conditions to maximize capacity. If the LNG plant capacity is limited by the market conditions, the efficiency needs to be maximized. These objectives are formulated in the APC controller, which is then used to calculate the steady state targets for the manipulated and control variables.

In the APC design, the compressor speeds, light mixed refrigerant flow (LMR) and heavy mixed refrigerant flow (HMR) are main manipulated variables for the liquefaction section. The makeup flows

Table 6-2 Variables in the Advanced Control Design for APCI C3MR Process

Manipulated Variables	
1	Plant feed rate
2	Propane compressor speed
3	MR compressor speed
4	LMR flow
5	HMR flow
6	MR C_3 makeup
7	MR C_2 makeup
8	MR C_1 makeup
9	MR N_2 makeup
10	LNG flash vessel pressure
Disturbance Variables	
1	Natural gas feed pressure
2	Natural gas feed temperature
3	Cooling medium supply temperature
Controlled Variables	
1	Production rate
2	LNG temperature
3	HMR flow
4	HMR/LMR ratio
5	HMR/NG ratio
6	Valve positions, as needed
7	Propane compressor suction pressure
8	MR compressor suction pressure
9	Propane compressor exhaust temperature/Power
10	MR compressor pressure ratio
11	MR compressor exhaust temperature/Power
12	Warm temperature approach
13	Cold temperature approach
14	Exchanger warm bundle pressure drop
15	Other plant specific constraints

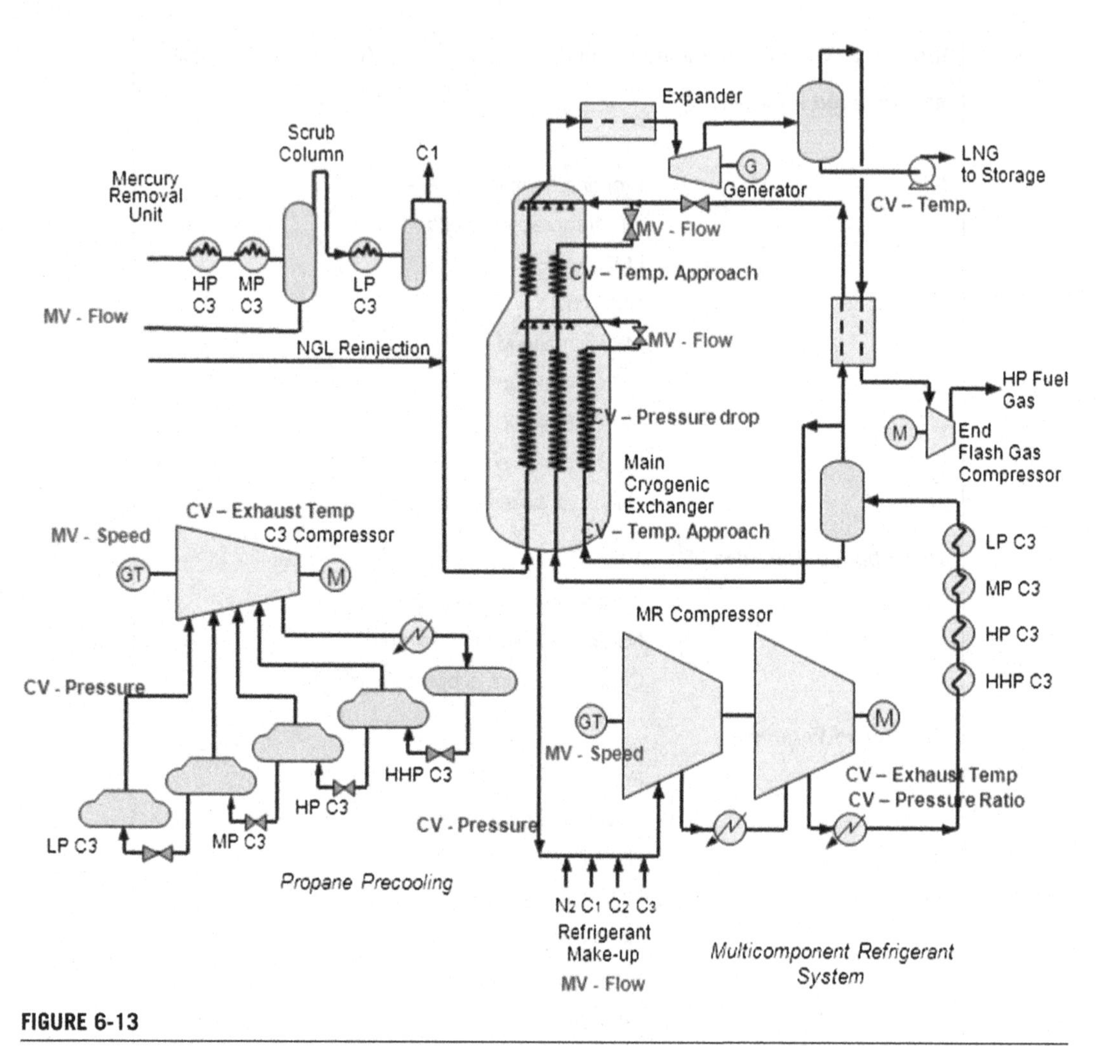

FIGURE 6-13

The main manipulated and control variables for the liquefaction area of APCI C3MR process.

for the mixed refrigerants (MR) are used as manipulated variables as they add further flexibility for control. The main disturbance variable is the cooling medium temperature, either ambient air or seawater temperature.

The manipulated variables are used to control the LNG temperature and HMR/LMR ratio. In addition, the compressor constraints like pressure ratio and driver constraints like power are included. The valve outputs are also included in the controller to ensure that the valves do not become saturated (mainly LNG off-take valve, HMR flow valve, and LMR flow valve). The key approach temperatures for the MCHE are controlled variables, since these are important for equipment protection (Figure 6-13).

References

Astrom, K.J., Hagglund, T., 1995. PID Controllers: Theory, Design and Tuning. ISA Press, Instrumentation, Automations, and Systems (ISA) Society, Research Triangle, Park, NC, USA.

Barash, M., Rangnow, D., April 24–27, 2007. Designing and Implementing Information Management Systems for LNG Receiving Terminals, poster presented at the 15th International Conference & Exhibition on Liquefied Natural Gas (LNG 15), Barcelona, Spain.

Blevins, T.L., McMillan, G.K., Wojsznis, W.K., Brown, M.W., 2003. Instrumentation, Automations, and Systems (ISA) Society, Research Triangle, Park, NC, USA. Advanced Control Unleashed. ISA Press.

Boyce, M.P., 2003. Centrifugal Compressors: A Basic Guide, first ed. PennWell Book Publishing Company, Tulsa, OK, USA.

Boyce, M.P., 2011. Gas Turbine Engineering Handbook, fourth ed. Butterworth-Heinemann, Boston, MA, USA.

Bramoullé, Y., Morin, P., Capelle, J., Oct. 11, 2004. Different Market Quality Specs Challenge LNG Producers. Oil & Gas Journal 48–55.

Buckley, P.S., Luyben, W., Shunta, J., 1985. Design of Distillation Column Control Systems, first ed. Butterworth-Heinemann, Boston, MA, USA.

Coyle, D., de la Vega, F., Durr, C., April 24-27, 2007. Natural Gas Specification Challenges in the LNG Industry, paper presented at the 15th International Conference and Exhibition on Liquefied Natural Gas (LNG-15), Barcelona, Spain.

den Bakker, K., April 23–27, 2006a. Unlocking the Potential of Advanced Process Control in LNG Plants, paper presented at the 6th Topical Conference on Natural Gas Utilization, AIChE Spring Meeting, Orlando, FL, USA.

den Bakker, K., June 5–9, 2006b. A Step Change in LNG Operations Through Advanced Process Control, paper presented at the 23rd World Gas Conference. Amsterdam, The Netherlands.

Dossat, R.J., Horan, T., 2001. Principles of Refrigeration, fifth ed. Prentice Hall, Englewood Cliffs, NJ, USA.

Driedger, W.C., 2000. Controlling Vessels and Tanks. Hydrocarbon Processing 79 (3), 101–115.

El-Hadi, B., Abdelkader, B., April 24-27, 2007. Advanced Process Control Qatar Gas - One year experience, paper presented at the 15th International Conference on Liquefied Natural Gas (LNG-15). Barcelona, Spain.

Hills, P.W., 2001. Plant Asset Management. CM2001. Oxford, UK.

Honeywell Whitepaper, "An Integrated Approach to Accelerating and Optimizing LNG Construction and Operations", Honeywell web site.

Hodges, D., May 14–17, 2001. Advanced Process Control at North West Shelf LNG Plant, paper presented at the 13th International Conference and Exhibition on Liquefied Natural Gas. Seoul, South Korea.

Jensen, J.B., Skogestad, S., April 2–5, 2006. Optimal Operation of a Simple LNG Process, paper presented at the ADCHEM 2006 Conference. Gramado, Brazil.

Jensen, J.B., Skogestad, S., 2009. Steady State Operational Degrees of Freedom with Application to Refrigeration Cycles. Ind. Eng. Chem. Res. 48, 6652–6659.

Lancaster, D.S., Parkes, B., Nov./Dec. 2002. Fieldbus Technology for the LNG Process – A Viable Option? LNG Journal 11–13.

Lee, S., Yang, Y., 2007. Operating Information System for LNG Facilities. Ind. Eng. Chem. Res. 46, 6540–6545.

Maffeo, R., 2005. Automation Strategy – A Key to LNG Project Success. LNG Review 36–38.

Mandler, J.A., Brochu, P.A., Fotopoulos, J., Kalra, L., May 4–7, 1998. New Control Strategies for the LNG Process, paper presented at the 12th International Conference and Exhibition on Liquefied Natural Gas (LNG 12), Perth, Australia.

Mokhatab, S., 2007. Foundation Fieldbus Is Revolutionizing LNG Facilities. Hydrocarbon Processing 86 (3), 13.

Mokhatab, S., Poe, W.A., 2012. Handbook of Natural Gas Transmission and Processing, second ed. Gulf Professional Publishing, Burlington, MA, USA.

Nagelvoort, R.K., June 5–9, 2006. LNG Quality – Report of Study Group D1. presented at the 23rd World Gas Conference, Amsterdam, The Netherlands.

Narayan, K., April 10–14, 2005. Controlling Complex Turbomachines in LNG Plants, paper presented at the 5th Topical Conference on Natural Gas Utilization, AIChE Spring Meeting, Atlanta, GA, USA.

Phople, P., Jan–Mar 2009. Main Automation Contractor (MAC) Approach for Achieving Operational Excellence. Hydrocarbon Asia 44–51.

Poe, W.A., Mokhatab, S., 2007. Model Predictive Control for Liquefied Natural Gas Processing Plants. Hydrocarbon Processing 86 (6), 85–90.

Shinskey, G., 1996. Process Control Systems: Application, Design and Tuning, fourth ed. McGraw-Hill Publishing Company, New York, NY, USA.

Singh, A., Hovd, M., Kariwala, V., July 6–11, 2008. Controlled Variable Selection for Liquefied Natural Gas Plant, paper presented at the 17th IFAC World Congress, Seoul, South Korea.

Sturm, W., Parra-Calvache, M., Chantant, F., Opstal, J., Jan. 10–12, 2009. Unlocking the Potential of Modern Control and Optimization Strategies in LNG Production, paper presented at the 1st Annual Gas Processing Symposium, Doha, Qatar.

Tailor, A., Jamaludin, S., 2012. Innovative APC Boosts LNG Train Production. Hydrocarbon Processing 91 (1), 49–53.

Teramine, T., Sakamoto, Y., May 14–17, 2001. New Optimization Scheme of the LNG Receiving Terminal Considering the Terminal Wide and Inter-terminal Efficiency and Reliability, paper presented at the 13th International Conference and Exhibition on Liquefied Natural Gas (LNG 13), Seoul, South Korea.

Valappil, J., Messersmith, D., Wale, S., Mehrotra, V., June 2006. Lifecycle Dynamic Simulation in the Design and Testing of Advanced Process Control, paper presented at the IEEE Advanced Control Applications in Industry (APC 2006) Conference, Vancouver, BC, Canada.

Valappil, J., Wale, S., Mehrotra, V., Ramani, R., Gandhi, S., April 22–27, 2007. Improving the Profitability of LNG Plants Using Advanced Process Control, paper presented at the 7th Topical Conference on Natural Gas Utilization, 2007 AIChE Spring Meeting, Houston, TX, USA.

Wood, D., Mokhatab, S., 2007. Fieldbus Technology in LNG and GTL Plants. Q3 Issue. Petroleum Technology Quarterly 12 (4), 91–95.

CHAPTER

7 LNG Plant and Regasification Terminal Operations

7.1 Introduction

An LNG production plant is designed to meet the production target with specifications that meet the contractual agreement (Table 2-1) while satisfying the emission and environmental regulations. The LNG liquefaction plant is a complex process, and for this reason, it is important to understand the design limitations and the process interaction among the different units for plant operation. The focus of this chapter is to discuss the process parameters and typical pitfalls that operators may encounter in a day-to-day operation. Note that only the major process units that are critical to the LNG plant and regasification operation are addressed in the following sections. The generic units such as NGL fractionation units, condensate hydrotreating unit, refrigeration unit, and nitrogen rejection unit are left to the reader. This chapter also briefly addresses the general guidelines on the startup and shutdown of the LNG plant and the regasification terminal. In addition to this chapter, we recommend you read Chapter 2 for technology selection, Chapter 6 for the process control and automation methods, and Chapter 8 on dynamic simulation to understand the interaction among the different units.

7.2 LNG plant normal operation

The success of a plant operation is judged by its reliability, availability, and safety records. A high availability factor enables the plant to meet the contractual obligation to deliver the product on schedule. As such, the plant must be designed to operate under various feed gas conditions and the equipment must be designed with flexibility in mind to maintain production during equipment failure or unit turnaround. They must also be able to meet the plant turn-down or ramp-up requirements.

Most often, multiple units are necessary and spare capacities and design margins must be included such that when one unit is down for maintenance, the spare capacities can be applied to make up the losses. For equipment that is susceptible to mechanical fatigue and fouling such as pumps and compressors, plate and frame heat exchangers, and filters, they are often spared. Major equipment must be inspected and maintained regularly to meet industrial and safety standards.

A typical LNG production plant is shown in Figure 2-1 in Chapter 2. The following discusses the various process units that must be successfully operated to maintain LNG production.

Handbook of Liquefied Natural Gas.

7.2.1 Slug catchers

Slug catcher is a critical unit that safeguards the LNG production plant against surge in flow and pressure swing from upstream oil and gas production wells. The separator serves as a three-phase separator where the gas, hydrocarbon liquids (condensate), and aqueous phase are separated.

Daily monitoring of the pipeline operation, such as pressure drop and temperatures, are necessary to ensure a stable feed to the plant. When pressure drop in the pipeline system becomes excessive, a pipeline pigging operation should be started to remove the contaminant, wax, and corrosion debris from the pipeline.

The inlet gas compositions should be analyzed daily, especially the acid gas contents and contaminant levels such as mercury and mercaptans. In most cases, these contaminants are very dilute in the gas phase, which may not be in the detectable levels. They are concentrated in the liquid phase and can typically be detected in the condensate from the slug catcher. The amount of contaminants in the vapor phase can be calculated using the liquid composition distribution and the frequency of pigging.

7.2.2 Condensate stabilization unit

The function of the condensate stabilization unit is to process the condensate from the slug catchers and to remove the light components to meet the export condensate specification, which is typically set at 4 ppm H_2S and a RVP of 8 to 12 psi. A typical condensate stabilization unit is shown in Figure 2-4.

The stabilizer column operates as a stripper column. The pressure of the column is controlled by removal of the light components using a two-stage compression system. The H_2S and light hydrocarbons are recycled back to the feed gas section and then to the acid gas removal unit. The column bottom product specification is controlled by heat input (steam) to the reboiler. The condensate product is cooled with a heat exchanger and product cooler and is sent to the condensate storage tanks under level control of the stabilizer column.

Although condensate can be considered a side product from the LNG production plant, the liquid production adds significant value to the facility. If the condensate contains other sulfur components such as mercaptans, it is worthwhile to be further treated to meet the sulfur specification of the condensate product.

With respect to the contaminants in the liquid hydrocarbon stream, the operation of the stabilizer shall take the following effects into consideration (Bras et al., 2007):

- **Sulfur components.** It is desirable to strip off as much H_2S with the overhead gas as possible to meet the condensate specification. Preferably, the condensate should contain mercaptans only. This can be achieved by overstripping. However, this may result in sending heavier hydrocarbons to the stabilizer overhead stream and introduces a risk of hydrocarbon condensation in the AGRU with potentially foaming operational problems as a consequence.
- **Water.** To meet the H_2S specification in the bottom product, the required bottom temperature is higher than the bubble point of water at the prevailing operating pressure, and consequently water will not be able to leave with the stabilized condensate. Water entering the stabilizer must leave the system as an overhead liquid or vapor. Water must be removed from draw trays above the feed tray or from the reflux drum. A stabilizer column will be required at the appropriate position in the stabilizer. The drawn off fluid is routed to a separator vessel, where the excess water is removed and the hydrocarbons are returned to the stabilizer column.

- **Salts.** Any water entering the stabilizer may contain dissolved salts. These may precipitate in the stabilizer or on the reboiler tubes. At sufficiently high temperature these salts will hydrolyze to form hydrochloric acid, which may cause material failure.
- **LPG.** If the condensate contains excessive LPG, the propane content may create a recycle within the unit and eventually overload the compression system. In this case the use of a feed liquid stripper installed prior to the stabilizer and the use of a refluxed stabilizer to produce an LPG stream may be necessary, as shown in Figure 2-5.
- **Fouling from degradation.** If corrosion inhibitors are injected upstream for corrosion protection, they will end up in the condensate. Check if they are thermally stable at the column bottom temperatures. If these components degrade into heavier products, fouling of the column reboiler may result.

7.2.3 Acid gas removal unit

The function of the acid gas removal unit (AGRU) is removal of the acid gases to meet the specifications for an LNG liquefaction plant, typically, 50 ppmv of CO_2 to avoid freezing in the main exchanger and 4 ppmv H_2S to meet the LNG sales specification. A typical amine process is shown in Figure 2-6.

The common problems with the AGRU operation are foaming and flooding problems in the absorber or stripper. Amine solution foaming is a persistent and troubling operational problem encountered in natural gas production. Solution foaming contributes to excessive solution losses, amine carryover, and reduction in treating capacity, and may result in unstable operations and off-specification product.

Failure of the AGRU to meet treated gas specifications may be contributed by one of many factors. To determine the root causes of the problem, the operator should check the followings:

- Are the contaminant levels in the feed gas high? The gas feed flowing to the AGRU may contain chemicals, such as MEG, corrosion, and hydrate inhibitors. Although the concentration in the gas may be very low, the contaminants will be absorbed by the solvent in the AGRU and build up in the circulation system. To avoid the reduction in AGRU capacity, the contaminants can be removed by a water wash column before the amine absorber. The spent water from the water wash, containing the glycol washed from the feed gas, can be recovered in the glycol regenerator.
- Is the solvent circulation rate adequate? Sample the rich amine solution loading. If greater than the design loading limit (which is corrosive), increase the solvent circulation rate. Alternatively, if the circulation rate is already at a maximum, but with a lower than design loading increase the amine strength to the maximum limit.
- Is the lean amine solution loading satisfactory? Sample the lean solution and compare to a typical operation and the process design basis. If the lean loading is abnormal, check for the following:
  - Check for leaks in the lean/rich exchanger. Check the amine loading between and after the lean/rich exchanger; if there is a significant change, it may be due to exchanger leakage.
  - Check the regenerator operations. If the regenerator overhead temperature is low ($< 210°F$ [98.9 °C]), increase the steam flow to the reboiler. If there is no change in the lean amine

loading with the proper overhead temperature, the cause may be due to mechanical damage in the regenerator.
  - Check the regenerator for foaming. Check for high differential pressure or erratic liquid levels. Anti-foam may be used to regain column stability until the root source of the foaming can be determined.
- Is the lean amine solution temperature too high? Check the absorber feed and solvent temperatures and compare to the process design. For most gas treating applications, as a practical maximum, the lean amine temperature should not exceed 135°F to 140°F (57.5°C to 60°C). If the solution temperature is high, check the lean amine cooler and check if the lean amine cooler bypass line is opened.
- Is the absorber operating condition different than design? Check the inlet process conditions and compare to the design basis and historical data to see if there has been a change in the operating conditions.
- Is the absorber installed properly? If the absorber has multiple feed locations, check that the top feed location is in service. A column scan of the absorber may be used to determine poor distribution in the column, which may be due to damaged trays or plugging.
- Is the filtration system operating properly? Iron sulfide is the common particulate found in most amine solutions. Additionally, particulates can enter the amine system with the feed gas or makeup water. These may include rust particles, dirt, pipe scale, and salts. Removal of particulate matters in the amine circuit can best be accomplished by continuous filtration of a slip stream of the amine solution, typically 10% of the amine circulation. More efficient inlet gas separation and filtration system has been proven successful in reducing foaming problems. Additionally, particulate filters should be installed downstream of the carbon filter to prevent fouling from carryover of carbon fines. Carbon filtration can be used to remove surface-active contaminants such as oil and heavy hydrocarbons that contribute to foaming.
- Is the Heat Stable Salt (HSS) level acceptable? HSS precursors found in natural gas streams include well/pipeline treating chemicals and oxygen from vapor recovery systems. A few of the more common HSS anions in natural gas applications are acetate, formate, sulfate, oxalate, and thiosulfate. In some instances, HSS can be attributed to the quality of makeup water. Chlorides and sulfate are common contaminants found in amine plants that use poor quality process water. If oxygen is present, the amine can also degrade to form corrosive by-products such as bicine, glycine, and other amino acids.
  There are several methods that can be used to control the level of HSS. The obvious method is bleeding a portion of the HSS contaminated amine solution from the amine system, replacing it with fresh solvent. While this operation appears simple, disposal of the contaminated amine to a wastewater system to form corrosive by-products such as is prohibited due to environmental regulations, and is not a permanent solution. Other methods of removal of the HSS include electro-dialysis reclamation, distillation reclamation, and ion exchange reclamation, which can be provided by the amine licensors or the amine suppliers.

7.2.4 Sulfur recovery unit

The modified Claus process is the most common sulfur recovery process used in the LNG industry because of its simplicity and large sulfur capacity. However, there is a limitation of the modified Claus process. The acid gas stream must be relatively rich in H_2S and therefore in most applications, when processing dilute acid gases with contaminants such as ammonia, mercaptans, and heavy hydrocarbons, special design and operation considerations must be made to ensure complete destruction of these contaminants that is necessary to meet the sulfur emission requirements.

In the Claus process (see Figure 2-12), the overall reaction is carried out in two stages. The first stage is the reaction furnace section where air is added to oxidize only one-third of the incoming H_2S to SO_2. The unconverted H_2S then reacts with the produced SO_2 to produce elemental sulfur vapor in the later stages.

The reaction furnace is the critical piece of equipment in the Claus process. Due to the complexity of the reactions, the design and operation are more dependent on operating experience than theoretical calculations. The minimum temperature for effective operation of the reaction furnace on "clean" acid gas must be controlled at above 925°C (1700°F). Below this temperature, the flame stability is poor and adverse side reactions occur, making operation difficult and unreliable. For units processing ammonia-laden sour water stripper off-gas, or feeds containing heavy hydrocarbons, BTX, and cyanide, the minimum furnace operating temperature required must be increased above 925°C (1700°F) to ensure destruction of these components. Ammonia, in particular, requires a high combustion temperature, and plants with inadequate temperature are prone to shutdown from plugging due to deposition of ammonia salts. The consequence of not achieving the required destruction temperature is detrimental to the sulfur plant operation. Coke formation from cracking and incomplete reaction of heavy hydrocarbons can cause irreversible deactivation of the downstream Claus converter catalyst.

For sulfur recovery unit in an LNG plant, a straight-through process is the most common, due to its simplicity and lower costs. There are other techniques to raise the furnace temperature for ammonia destruction, such as supplemental fuel gas firing, preheating acid gas, and/or combustion air and oxygen enrichment. However, these techniques introduce a number of negative factors. Notably, there will be a decrease in overall sulfur recovery efficiency due to the formation of undesirable by-products such as COS and CS_2. Oxygen enrichment requires an air separation that cannot be justified for grass root design. The equipment count is higher and the equipment sizes are larger, which increases the capital and operating cost.

In grass root design, a straight-through process is preferred, which requires the H_2S concentration in the acid gas to be greater than 50%. Figure 7-1 provides an operating guideline of the theoretical, adiabatic flame temperature, as a function of the H_2S content in the acid gas feed. The two horizontal dashed lines provide the operator with the temperature targets, depending on the nature of the hydrocarbons contained in the acid gas feed.

When the acid gas is too dilute in H_2S, a selective absorption technique can be used to enrich the acid gas prior to entering the Claus unit. By selectively absorbing H_2S from the acid gas and then stripping the rich solvent, two gas streams are produced. The gas passing through the absorber is primarily CO_2. The CO_2 stream can be sold as a product for enhanced oil recovery. The gas stream leaving the regenerator is acid gas enriched in H_2S that can be processed in a straight-through Claus plant. The process design and configuration is described in Chapter 2.

For the catalytic stages, some technology licensors, such as Fluor, employ titanium-oxide (TiO_2) catalyst in one or more of the Claus converters, which greatly improves the ability to hydrolyze COS and CS_2 (Flood et al., 2011). Traditional activated alumina Claus catalyst gives high conversion of H_2S, but only about 65% hydrolysis of COS and about 30% hydrolysis of CS_2 in the first Claus

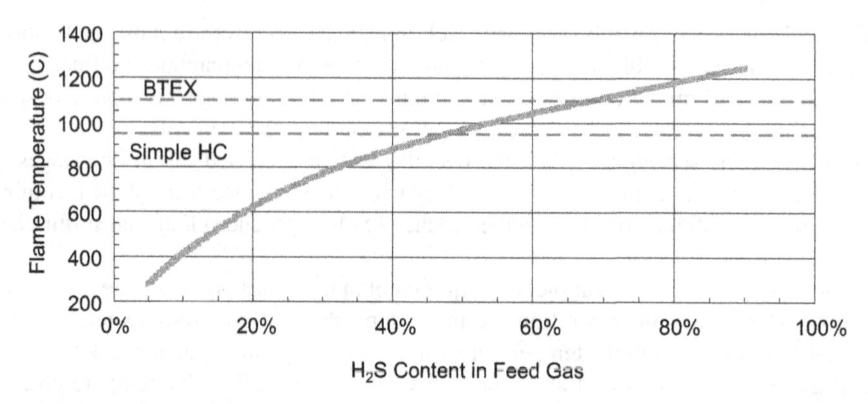

FIGURE 7-1

Flame temperature vs. H_2S content.

(Flood et al., 2011)

Converter. Conversions are reduced in subsequent Claus stages. For the second stage, activated alumina catalyst gives about 20% hydrolysis of COS and only about 5% hydrolysis of CS_2. In contrast, even at end-of-run, TiO_2 catalyst is capable of over 90% hydrolysis of COS and CS_2 in the first Claus converter when operated at proper bed temperature. Therefore the proper operation of the TiO_2 catalyst enhances the overall sulfur recovery efficiency necessary to meet the environmental requirements.

7.2.5 Tail gas treating unit

The tail gas unit process consists of two sections, the hydrogenation section and the tail gas treating section (Figure 2-11). The hydrogenation section consists of a hydrogenation reactor, which converts the SO_2 content to H_2S in the presence of a catalyst and hydrogen. The hydrogenated product is cooled in a quench column by circulating with process water. The water formed in the quenching process is removed under a level control from the quench tower. The sour water is sent to the sour water stripper and then to the waste water treater prior to disposal.

The tail gas treating unit is designed to reject CO_2 from the dilute acid gas using an H_2S selective solvent such as MDEA. The CO_2 reject stream containing ppm levels of H_2S can be sent to a thermal oxidizer for disposal. The operation of the tail gas treating unit is similar to the AGRU discussed in the previous section. The solvent quality and purity must be maintained by filtration and controlling the HSS level in the solvent. The operation of the absorber and regenerator must be closely monitored. Typically the absorber is designed with several feed trays for the lean solvent; the lower feed tray is preferred as it will minimize CO_2 absorption as long as H_2S in the absorber overhead meets the emission limits. Typically, a lean solvent loading can be achieved with steam heating equivalent to 1.0 to 1.2 lb steam per gallon of amine circulation.

7.2.6 Molecular sieve unit

The dehydration unit is designed to meet the product specifications on water and mercaptans (RSH) content. The Molecular Sieve Unit (MSU) is designed to remove water to 0.1 ppmv and most mercaptans to 2 to 3 ppmv. In a natural gas plant the molecular sieve dryer is considered a reliable continuous process, characterized by a high degree of complexity and automation, and low manpower requirements. The molecular sieve unit can typically operate without interruption for 3 to 5 years with the same molecular sieves.

The molecular sieve unit feed gas is typically precooled to 40°C or lower and condensate removed prior to entering the unit. In hot climate operation, propane refrigeration can be used to remove the bulk of its water content, which would reduce the duty of the dehydration unit.

The most significant factor affecting the performance of the molecular sieve unit is the regeneration conditions. If the unit has been producing satisfactory dewpoints but suddenly goes bad then the following items should be checked (Mock et al., 2008).

- **Mercaptans buildup.** When the molecular sieve unit is designed for dehydration and mercaptan removal, a mass balance of the different mercaptans for the regeneration gas is required. This is necessary when the AGRU is an amine base that is not capable of removal of mercaptans. See Chapter 2 for detailed discussion. The branched mercaptans that slip through the AGRU are usually not caught by the molecular sieves and are left in feed gas and fed to the NGL recovery unit. Since the lightest branched mercaptans have volatility similar to C_5^+, they will end up in the C_5^+ fraction and sent to the condensate stabilization unit. In order to perform a mass balance of the mercaptans, it is not only required to know the mercaptan concentration in the feed gas, but also the breakdown into individual mercaptans (Bras et al., 2007).
- **Regeneration gas temperature.** Check that the thermocouples are reading correctly. A low temperature shifts the equilibrium upward causing a higher than design dewpoint. Also the lower temperature results in a lower gas velocity and a slower desorption rate.
- **Flow rate.** Check the calibration of the flow meter. Mole sieve beds typically are designed with a regeneration flow close to laminar flow. A small change in feed condition can force the flow regime into the laminar flow regime, which is undesirable from the standpoint of heat and mass transfer operation.
- **Flow channeling.** Good distribution must be maintained through all of the bed to insure that there are no pockets or poorly regenerated adsorbent. Check the pressure drop during regeneration. If channeling is confirmed due to fouling then the bed should be replaced.
- **Oxygen.** Occasionally oxygen can get into the regeneration gas from the gas gathering system through valve seals and such. If the oxygen concentration is above 20 ppmv then there is a chance of side reactions when passing through the regeneration heater forming CO_2 and water. The iron oxide coating on the furnace tubes can act as a catalyst for these oxidation reactions. The end result is that the regeneration gas is wet, which will result in off-spec gas.
- **Inlet separator.** The frequently underdesigned component in the mole sieve system is the inlet separator. If liquid droplets enter the beds, they can weaken the mole sieves and will cause erosion damage. On the extreme side, a massive failure can cause large slugs of liquids to hit the beds, which can crush the adsorbent.
- **Bed fouling.** The most common problem is fouling in the adsorber beds. If the inlet separator is not performing, liquid carryover will wet the bed rendering the mass transfer zone ineffective, which will result in an early water breakthrough. Sometimes, what is thought to be a dew point problem is actually a breakthrough problem. Amines, glycols, compressor oils, corrosion inhibitors, and heavy hydrocarbons will destroy the crystalline structure of the molecular sieves.

Some of these liquids will decompose at temperatures between 200 and 350°F resulting in coking, loss of capacity, reduction of mass transfer efficiency, and early breakthrough.

- **Dehydration bed failure.** This can be due to improper support screen installation resulting in buckling and loss of the mole sieve from the dehydration bed. A more sturdy support using JOHNSONSCREENS™ can provide more mechanical support strength (Mock et al., 2008).
- **High pressure drop in bed.** Typical molecular sieve pressure drop across the bed is approximately 0.5 bar. Higher pressure drop may indicate bed support failure, maldistribution due to clogging with aging mole sieve materials. This will cause loss of mole sieve from the dehydration bed, which subsequently ends up in the mercury removal bed resulting in high pressure drop.
- **Moisture analyzer probes locations.** The online moisture analyzers typically get fouled due to contact with contaminants and are prone to failure. Locating the sampling point close to the bottom of the mole sieve bed is helpful for troubleshooting and for breakthrough tests. A sampling point a few feet downstream of each dehydration bed provides a more reliable measurement of moisture contents. A sample point in the main line downstream is another useful check (Mock et al., 2008).
- **Even distribution across the bed.** Proper design of the inlet distributors is important to achieve uniform adsorption and efficient utilization of the mole sieve beds. The use of a distributor device on top of the mole sieve beds helps the mixing and distribution, avoiding the potential for bed channeling.
- **Feed bypass.** Check obvious sources such as leaking valves using ultrasonic leak detectors. Also look for bypasses through instrument lines or a pinhole leak in a diaphragm. Even the smallest leak can result in off spec dewpoint.
- **Degradation of molecular sieve.** The excessive thermal stresses of each regeneration cycle can eventually break some of the crystal bonds of the sieves. Hydrocarbons tend to decompose or crack forming coke in the macro pore structure, slowing down the adsorbing rate. An obvious way to increase the adsorbents' life is to run longer cycles; that is, to water breakthrough stage each cycle, thus minimizing the number of regenerations and the amount of hydrocarbons on the bed.
- **Change in inlet conditions.** Check the inlet conditions against the design levels. Often a small change in inlet temperature can increase the water load more than would be suspected; for example a temperature rise in the inlet of 10°F might increase the water content by 40% and result in an adsorbing time decrease of approximately 40%. Also the adsorbent's capacity drops as the temperature increases. A drop in pressure will cause a proportional increase in water content and the increased velocity resulting from the lowered pressure will increase the mass transfer zone length causing an earlier breakthrough. Changes in gas composition should be monitored. If the inlet gas is coming predominantly from lean gas fields and the feed changes to a rich gas feed, more liquids will enter the plant.
- **Retrograde condensation.** The recent trend of operating the mole sieves system at high pressures presents a new problem. Heavy hydrocarbons may drop out due to cooling of the system due to pressure drop. This may result in wetting the adsorbent throughout the bed. Retrograde condensation is very difficult to predict but can be checked with Hysys simulation. Typically, a steadily increasing pressure drop will occur during the adsorbent cycle, which then

drops back to a base level after regeneration. This is normally due to liquid accumulation in the adsorbent bed.

7.2.7 NGL recovery unit

The function of an NGL recovery unit is to remove the C_2^+ or C_3^+ hydrocarbons from the feed gas, producing an NGL product and a lean gas feed to the LNG plant. The operation of a high recovery NGL unit is based on the deep chilling generated by turbo expansion to produce refluxes to the NGL recovery columns. During propane recovery, the column must produce an NGL product that contains maximum 1% volume of ethane and during ethane recovery, the NGL product must contain no more than 0.1% volume of methane.

In order to meet the product specifications, the fractionation column must operate at a pressure well below the critical pressure of the feed stream such that reasonable separation is practical. While fractionation is easier at lower pressures, it will increase the refrigeration driver power consumption for LNG production (Chapter 10). Therefore it is desirable to produce a residue gas at high pressure. One of these NGL processes is the Fluor TCHAP process. This process employs a high pressure absorber and a low pressure deethanizer for NGL production. Refer to Chapter 2 for detailed discussion of this process.

To achieve higher recovery, the NGL columns must be refluxed with a lean NGL. For example for high propane recovery (99%) as shown in Figure 2-21 in Chapter 2, both the absorber and deethanizer are refluxed with a lean ethane NGL stream produced from the deethanizer overhead. In the medium ethane recovery process as shown in Figure 2-22, the demethanizer column is refluxed using the feed gas as reflux. In the high ethane recovery process described in Chapter 2, the demethanizer column is refluxed by recycling a slip stream of the residue gas.

To achieve high recovery while minimizing energy consumption, the reflux flow rates and the temperature profile of the process must be optimized with simulation. This also requires a close temperature approach brazed aluminum exchanger and an efficient expander compressor design. Operation of the NGL unit is very sensitive to changes in feed gas compositions and conditions and the NGL plant operating condition must be adjusted periodically for the best efficiency operation.

Some of the common problems encountered in the operation of an NGL recovery unit are summarized as follows.

- **Feed gas composition is different than design.** When processing a richer feed gas, more refrigeration is required, and the flow to the expander is lower. Conversely, when processing a leaner gas, less condensation will occur, and the vapor flow is higher. The turbo-expander may not have sufficient capacity which may require opening of the expander bypass valve. The tray loading in the upper section of the NGL column would also increase, which may result in column flooding and loss in NGL recovery.
- **Column operation.** One of the common problems in operating an NGL recovery unit is high pressure drop in the columns due to the formation of hydrates. Hydrates are mainly due to unsatisfactory operation of the upstream units; that is, the molecular sieve dehydration unit and the AGRU. This typically occurs during ethane recovery operation when the demethanizer is required to operate at –100°F or lower temperatures. The high pressure drop is caused by equipment blockage by hydrates of CO_2 or water, and will eventually require the unit to be placed on deriming mode. To resolve these problems, the operation of the molecular sieve unit and the AGRU must be investigated as discussed in the previous sections.

- **Plate and fin exchanger.** The plate and fin exchangers in the NGL recovery unit are prone to fouling by hydrate. The inlet nozzles of the exchanger are equipped with methanol injection, which can be used to dissolve the hydrates.
- **Strainers.** Dual parallel strainers and block valves should be installed upstream of the plate and fin heat exchangers. The pressure drop indication should be monitored and the strainer should be cleaned once the pressure drop becomes excessive.
- **Turbo-expander.** The turbo-expander typically is equipped with an inlet guide vane system that allows the expander to operate within a reasonable range, typically 80 to 120% of the design flow. If the inlet flow exceeds the maximum range, the bypass valve can be opened operating on the Joule-Thomson (JT) mode. Typically, the expander control is automatic and does not require operator attention.
- **Level measurements.** A magnetic-type level measurement is commonly used for control purpose or local indication in cryogenic plants. The float must be specified to cover the specific gravity for the expected ranges of liquid compositions. In addition, the lower sensing line to the vessel must be properly insulated to prevent unstable level measurement due to vaporization in the sensing line.
- **Dry-out lines.** Dry-out line should be provided for plant startup. The exchangers, columns, pumps, and piping should be dried as much as possible. The molecular sieve unit can be used to recirculate the hot dried gas for system dry-out. If nitrogen is used for dry-out, the dry-out gas flow can be vented from the equipment to the flare system.
- **Methanol injection.** Residue moisture in piping and low point drains should be flushed out with methanol to avoid hydrate formation during operation. The most likely areas for hydrate formation are around the cold box, and in the demethanizer condenser. All the methanol lines should be hard-piped to the process.
- **Pressure differential transmitters.** Pressure drops across the plate and fin exchanger inlet strainers should be monitored to ensure the performance of the cold box. Pressure drop across the molecular sieve beds and the carbon beds should be provided with high pressure drop alarms. Excessive pressure drop in molecular sieve beds could be due to sieve material degradation contamination, or structural damage.

7.2.8 Liquefaction unit

The LNG production throughput is controlled by the refrigeration from the mixed refrigerant system. The operation of the mixed refrigeration train is based on controlling the MR system refrigeration output by adjusting the vapor inventory and mixed refrigerant composition. The propane system refrigeration is designed to provide cooling to the mixed refrigerant, the feed gas, and the condenser of the scrub column.

LNG production turndown is achieved by venting lighter components at the HP and MR separator. Venting the lighter components reduces the vapor inventory, which, at a constant compression ratio, causes the MR compressor suction pressure and discharge pressures to decrease. With the compressors operating at the same speed and producing the same compression ratio, the suction volumetric flow remains the same but the mass flow reduces. A decrease in mass flow through the system proportionally decreases the available refrigeration. The LNG temperature valve will throttle the LNG flow to maintain the desired LNG temperature from the main exchanger, which is the method used to control LNG production.

The propane compressor is a centrifugal constant speed machine. The discharge pressure is set by the temperature approach to the ambient air temperature in the propane condensers. The suction and side load pressures are functions of the compressor curves and the process cooling load.

During system turndown, when excess power is available from the gas turbine driver of the propane compressor, the starter motor can be operated as a generator to convert the excess power into electricity to be used by other units in the plant.

During hot summer operation, when the gas turbine power degrades, gas turbine inlet cooling should be started to maintain the power output and the liquefaction capacity. In most cases, waste heat or propane refrigeration can be used for cooling (see Chapter 10).

7.2.9 LNG storage tanks

The function of LNG storage tanks is to store the LNG product under a stable pressure. The operating pressure of the storage tank must stay within the design limits of the storage tank. See Table 1-4 in Chapter 1 for typical storage tank operating conditions.

The flashed vapor from tank heat leak and the letdown LNG from the liquefaction plant are compressed and used as gas turbine fuel or recycled back to the liquefaction process. During ship loading the vapor return from the ship is returned to the LNG storage tanks to replenish the volume displaced by the unloaded volume. Under all conditions, the storage tank and the ship pressure must stay below the design limit. Excessive vapors produced during startup or upset are disposed of in the flare system.

In the LNG regasification terminal, tank boil-off is compressed to a higher pressure and condensed in a boil-off gas (BOG) recondenser using a slip stream of the sendout LNG. During ship unloading, the boil-off vapor is returned from the storage tank to the LNG ship to replenish the volume displaced by the unloaded volume. Under all conditions, the storage tank and the ship pressure must stay below the design limits. Excessive vapors produced during startup or upset can be recovered using the vapor condensers or disposed of in the flare system. See Appendix 3 for a detailed discussion of the flare and relief system design.

7.3 General startup sequence

A successful startup requires planning and organization and experienced and knowledgeable staff. A well-executed startup program requires a collaborated effort among contractors, licensors, equipment suppliers, and the plant owner. The startup program typically consists of the following phases.

This startup sequence is generic and can be applied to the liquefaction plant as well as the LNG receiving terminal.

- Preparation and planning
- Precommissioning and operational testing
- Commissioning
- Startup and initial operation
- Performance and acceptance test.

The contractor is responsible for planning and startup of the facility. Only when the owner is satisfied with the plant performance test results, can the contractor then turn over the plant operation to the owner.

7.3.1 Startup preparation

To set up the startup and planning, a startup team must be assembled together. The members can be engineers from the contractor's home office or they can be subcontractors. Most frequently, when the plant is located in a foreign country, it is required to hire locals as much as possible, and therefore a well thought-out training program must be implemented. The success of the plant startup depends on the concerted efforts from the contractors, licensors, equipment vendors, and the plant owner. This team works together throughout the project, from startup and commissioning, to full production. All members of the team must be involved in the development of the startup procedure, including the training requirement.

Typically there is a commissioning manager or a lead commissioning engineer who will hire the commissioning team members and technical consultants. He is also responsible for training and developing safety and risk assessment, commissioning strategy, procedures, and checklist. A detailed plan and budget preparation is also prepared to be approved by the owner.

7.3.1.1 Operator training

Typically the operators are selected from the plant owner's organization or hired from external organizations. The contractor is responsible for setting up the training program for the operator who will carry out plant startup and initial operation before the plant is finally accepted. Engineers who work on the design and are familiar with the plant equipment and layout are assigned for the startup training program. Engineers from the licensor's office should explain their system characteristics in the training class, making sure that their design is performing at the optimum design levels. Vendors of critical equipment, such as gas turbine drivers, compressors, and instrumentation systems should explain the critical design and operating parameters of their machinery.

7.3.2 Startup work plan

The contractor typically follows a systematic procedure whereby their startup engineers review the design drawings line by line to confirm that the plant can be started up as planned. The design is reviewed for potential startup problems such as:

- Is there sufficient capacity for surge?
- Is the system stable and controllable?
- Are there any hydraulic bottlenecks or water hammer problems?
- Are there adequate high point vents and low point drains?
- Are there sufficient dry-out lines and methanol injections?
- Are there sufficient startup lines and bypasses for system fill, cooldown, or defrosting?

A detailed path work plan and procedures and related drawings are prepared. The work path would define the sequence of startup for each unit and system. This should also define the individual activities at each step, operational testing requirement, and cooldown procedure. The technical skills, labor, material, and equipment services are also identified.

Typical milestones for LNG plant startup are as follows:

- Utility systems
- Offsite systems
- Inlet facilities pressurization

- Successful mechanical runs and testing on rotating equipment
- Amine unit startup
- Molecular sieve dryers startup
- Purge and dry-out liquefaction equipment
- Purge and dry-out LNG tanks
- Startup of refrigeration compressors
- Plant cooldown
- NGL production
- LNG production
- Performance test and plant acceptance.

7.3.3 Plant precommissioning

Typical commissioning activities are listed here, roughly in chronological order. In reality, the break between precommissioning and commissioning is not hard and fast.

- As-built P&ID check
- Safety study, prestart-up procedure review
- Utility systems commissioned (e.g., air, steam, nitrogen, fuel gas)
- Power and lighting systems energized
- Rotating equipment rotation check and alignment
- Equipment internal inspection
- Line by line punch-out and punch list
- Installation of temporary strainers/screens
- System pressure testing
- Line flushing
- Chemical cleaning
- Line blowing
- Instrument loop checking
- Install packing, seals, etc.
- Introduce lubricants, etc.
- Load catalysts and chemicals
- Check position of blinds
- Assemble documentation.

The precommissioning stage verifies the integrity of the mechanical equipment by inspection, pressure testing, cleaning and flushing, and machinery checkout. The equipment is checked against specifications based on datasheets, as-built P&IDs, plot plan, layout, and piping isometrics.

For a typical plant checkout, master sets of piping drawings and P&IDs are located in a common area such that designated engineers can walk every line and verify every item as shown on the master drawings. Different color highlighter can be used to identify the system and services. The engineers must check that pipe sizes and ratings are correct, vessel nozzles and valves are correctly installed, vents and drains are in the right place, blind flanges and swing blinds are properly installed. Prepare a punch list for nonconforming items, and identify necessary corrections.

Piping inspection includes checking for piping stress from expansion and contraction due to temperature changes, alignment between flanges on pipe work, and alignment between piping and equipment nozzles. Checking can be by visual inspection and by gauge measurement. Check for thermal growth on piping supports, any surface buckling or crimping, and expansion joints.

Vessel and column inspection includes verifying trays, packing, vessel internals, distributors, demisters, downcomer, vortex breaker, and material of construction. Confirm all column internals have not been damaged during installation.

Pressure test inspection of equipment includes leak tests and verification of the design pressure to meet specifications and applicable codes and standards. Pressure testing must be witnessed by the startup team, who must verify all necessary safety limits.

Electrical system check includes verification of circuit breakers, switches, bus bars, grounding systems of equipment, explosion proof equipment and enclosure, motor control centers, transformers, switch gears, motor starters, and so on.

Equipment cleaning is the last step to ensure no construction debris is left in pipes or vessels. All debris must be removed and flushed out with chloride-free water. Chemical clean may be used if necessary. Before flushing for cleaning, check that startup screens and blinds are properly installed; jumper spool pieces should be added if necessary for startup.

7.3.4 Plant commissioning

Commissioning of the LNG production plant can be started when the equipment and piping system have been checked and verified, and basically consists of the following steps:

- Close-up equipment
- Dry-out
- Heater dry-out
- Rotating equipment run-in
- Purge plant with nitrogen
- Cool-down
- Ready for start-up.

7.4 LNG plant startup

The following is a general guideline for the startup for an LNG production plant. A generic C3MCR liquefaction train process flow diagram is shown in Figure 7-2. Startup of other liquefaction technologies may be slightly different.

7.4.1 LNG plant utility and offsite

Prior to the LNG process plant startup, the utility and offsite systems must be checked to be operational including:

- The flare system
- The fuel gas system

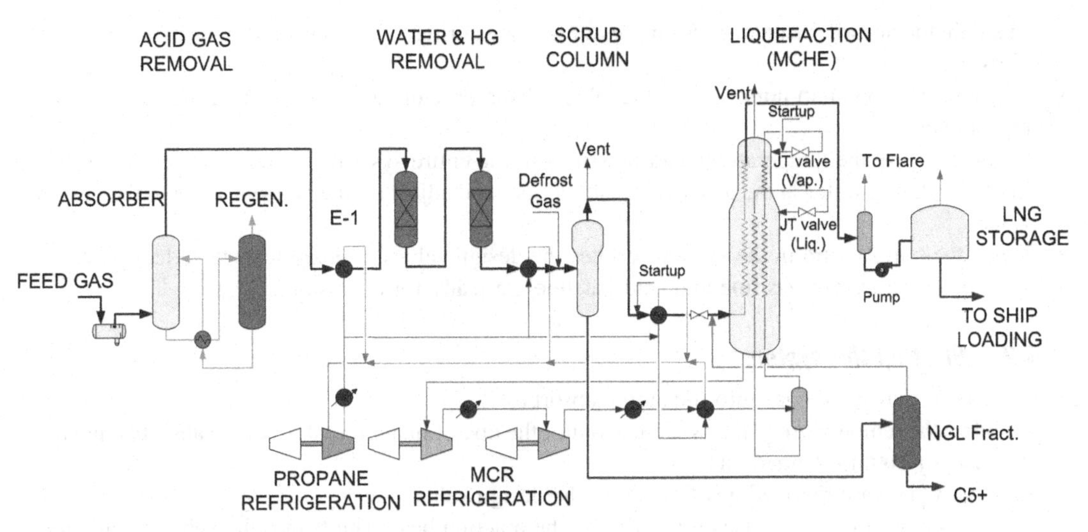

FIGURE 7-2

Typical C3MCR LNG production plant.

- Nitrogen system
- Firefighting system
- Electric power supply
- Cooling water system
- The heat transfer fluid system or waste heat recovery system.

7.4.2 Acid gas removal unit

The startup sequenced for initial operation of the acid gas removal unit is as follows.

7.4.2.1 Heat medium system

- Establish heat medium circulation.
- Blanket the heat medium surge drum with nitrogen.
- Start the heat medium heater using startup fuel gas, gradually heating up the heater to the desirable temperature.
- Charge the heat medium to the amine reboiler.

7.4.2.2 The amine system

- Add the amine solution and start circulation.
- Charge the heat medium gradually to the reboiler, taking caution of the thermal stress.
- Start the lean/rich exchanger and the lean amine cooler and regenerator condenser.
- When steam condensate accumulates in the reflux drum, start the reflux pump. Monitor the reflux flow and avoid sudden change in flowrate to prevent sudden cooling. Failure to do this will damage the regenerator internals.

- Maintain the absorber and flash drum pressures at the normal operating conditions by injection of nitrogen.
- Establish and maintain normal liquid levels in absorber, chimney trays, flash drum, and regenerator.
- Gradually raise the temperatures and pressures in the entire system.
- Start circulating water in the absorber wash section, and adjust amine concentration to the design value.
- When the temperature in the system reaches the design values, prepare to receive feed gas.
- Reconfirm the treated gas line and acid gas line are ready for operation.

7.4.2.3 Start up the system

- Slowly introduce feed gas into the amine absorber.
- Stabilize the temperature in the system. Monitor the operation temperature of the absorber and the regenerator bottom temperature.
- Gradually increase the feed gas flow up to the design value.
- The absorber temperature rises gradually by the reaction heat. The lean/rich exchanger and the regenerator temperatures will also increase.
- After flow rates, pressures, and temperatures in the system stabilize, record the operating conditions.
- Sample and analyze the treated gas product.

7.4.3 Molecular sieve unit

The molecular sieve unit typically consists of four dehydrators, packed with molecular sieves designed to meet the water, mercaptans, and other contaminant specifications for LNG production. The dehydrators follow an adsorption-regeneration cycle. It can be operated on a 3-1 mode (three parallel beds on drying and one on regeneration and cooling) or a 2-2 mode (two parallel beds on drying and one on heating and one on cooling). The switching of beds is accomplished using a programmable logic controller that operates the corresponding valves.

The direction of flow for the dehydration system has been designed so the adsorption, depressurizing, and repressurizing steps are downward flow through the bed, and the heating and cooling steps are upward flow.

For a three-bed parallel operation, the approximate cycle time for adsorption is 18 hours followed by a 6 hour regeneration cycle. The regeneration time typically comprises of a 20 to 30 minute depressurizing step, a 3 hour heating step, a 2 hour cooling step, and a 20 to 30 minute repressurizing step. The cycle times can be adjusted during plant startup according to vendor recommendation.

7.4.3.1 Depressurizing

At the completion of the 18 hour adsorption cycle, the bed must be depressurized to prepare for the regeneration cycle. The depressurization lines have flow restriction orifices to effectively limit the rate of depressurizing. Typically, the depressurizing rate should not exceed 75 psi/minutes to prevent bed movement and fluidization. The gas will flow downward in the bed and should be depressurized in about 20 minutes.

7.4.3.2 Heating

Once the bed is depressurized, the heating step can begin. The residue gas is compressed by the regeneration gas compressor, flows through the bypass valve to the regeneration gas heater, and is heated to about 300°C. The hot gas flows upward through the bed being regenerated and heats up the bed to a temperature at which the adsorptive capacity of the bed is reduced to a very low level. As a result, the water leaving the bed can be removed by the regeneration gas. Typically the heat required to desorb water from the sieve can be as high as 1800 Btu/lb. The wet regeneration gas is then sent to the regeneration gas cooler and water is desorbed from the sieves, plus some hydrocarbons are separated. The wet vapor is recycled back to the process and the condensed water and coadsorbed liquid hydrocarbons are removed from the separator under level control.

7.4.3.3 Cooling

Following the heating step, the bed is cooled with cold compressed regeneration gas to return to the normal adsorption temperature prior to repressurizing. The cool gas flows upward through the bed and is routed to the regeneration gas heater.

7.4.3.4 Repressurizing

After the cooling step, the bed must be pressurized to the feed pressure to prepare the vessel to be brought back online. The repressurization lines for the beds have flow restriction orifices to effectively limit the rate of repressurization to no more than 75 psi/minute.

7.4.4 NGL recovery unit

The following discussion is with reference to a high propane recovery process as shown in Figure 2-21 in Chapter 2. The system consists of a cold box, turbo-expander, refrigeration unit, absorber, and deethanizer.

Startup of the NGL recovery unit starts with nitrogen purging the cold box that houses the brazed aluminum feed gas core exchanger. Nitrogen can be provided with an inert gas generator and is necessary to avoid hydrocarbon accumulation within the box. The core exchanger is typically provided with internal distributors for methanol injection into inlet nozzles. The pressure drop across the exchanger and inlet filters should be monitored. If pressure drop increases above the design value, start methanol injection to remove the hydrate.

When the feed gas is cooled sufficiently, liquid will form and will be collected in the expander suction drum. The liquid level is maintained through a level controller. A high level in the drum could cause liquid carryover into the gas phase. The high liquid level in the drum is monitored to avoid liquid entrained into the expander suction. Methanol injection is started if pressure drop is excessive.

The expander is directly coupled to the residue gas compressor. Therefore, the expansion work transferred from the vapor stream to the expander is used to drive the compressor. The plant can operate without the expander by using the JT bypass valve. The bypass valve is normally closed and will open when the expander outlet line pressure is too low and when the machine is at its maximum allowable speed. Operation of the bypass valve is controlled by the pressure controller on the absorber overhead. The pressure of the absorber is maintained by a split range controller operating on the expander inlet guide vanes and the expander bypass JT valve. The bypass valve can pass the total gas flow in the case where the expander is down.

The expander inlet and JT valve are provided with methanol injection and may be used during startup. The absorber, deethanizer, and condenser system are also provided with methanol injection, which is required during startup.

The absorber is designed to recover most of the propane in the feed gas. The absorber bottom is sent to the deethanizer. The reflux to the absorber is provided by the reflux from the deethanizer after being cooled in the absorber reflux exchanger. The feed streams to the absorber consist of two phase streams, one from the expander and one from the expander suction drum. Liquid formed in the bottom of the absorber is pumped to the deethanizer using a level controller.

The residue gas stream is sent to the cold box to heat exchange with the dehydrated feed gas. The residue gas is then sent to the compressor expander suction drum, and compressed by the compressor and residue gas compressor to the LNG liquefaction plant. The residue gas compressors are controlled by capacity via a master plant capacity controller to meet the sales gas production demand. The compressor is provided with a high residue gas header pressure shutdown protection. The compressor has an antisurge system to protect the compressor from surge. The antisurge system uses suction flow, suction and discharge pressures, and temperature to predict surge and operate the antisurge spill-back valve. The antisurge flow is cooled and recycled back to the suction.

The deethanizer is used to stabilize the recovered condensate and is provided with overhead condenser using propane refrigeration, reboiler, and side reboilers for energy conservation. The reflux pumps draw liquid from the reflux drum and are used to provide reflux for both the absorber and deethanizer. The NGL product from the deethanizer is analyzed using a chromatograph, which is used to adjust the deethanizer bottom temperature and steam flow that is necessary to meet the product specifications.

7.4.5 Liquefaction unit

Start-up of the main exchanger can begin once cold feed gas is available from the scrub column. The main exchanger must be cooled down gradually to the normal operating temperature in two steps. The first step is to precool the main exchanger to –30°C by blending cold feed gas from the scrub column with the defrost gas. The defrost gas heater located downstream of the driers is used to supply defrost gas at 60°C for the precooling operation.

The purpose of the precooldown step is to eliminate large temperature differences by blending cold feed gas from the scrub column and the warm defrost gas. The blended gas is directed into the warm end of the main exchanger. The blended gas flows upward through the shell of the main exchanger, simultaneously cooling the shell and tube circuits. The warm gas is vented at the cold end of the main exchanger.

The MR compressors are charged with feed gas from the scrub column and started up. The exchanger JT valves are opened to start circulation through the MR circuit, which slowly cools the exchangers to the operating temperatures. The blended gas temperature is gradually lowered by reducing the warm defrost gas to cold gas flow ratio until the blended temperature at –30°C. The high pressure MR piping is then cooled by opening the JT valves and vented at the warm end. The MR separator is precooled by circulating the defrost gas through the evaporators that is chilled by propane refrigeration.

Once the system has been cooled down to –20°C, all vent, drain, and defrost connections are closed. The MR compressors are started with the main exchanger isolated. The MR JT valves are opened to continue the cooldown. The cooldown rate is controlled by manually adjusting the JT valves. Typically, the maximum cooldown rate is 28°C/hr with initial cooldown limited to 17°C/hr.

As the exchanger cools down and liquid accumulates in the MR system, cold feed gas is added to the suction of the MR compressor to maintain the suction pressure. Once the cold end of the exchanger is chilled down to the LNG temperatures, the LNG start-up valve is opened to begin LNG production. Nitrogen, ethane, and propane are added to the MR circuit to achieve the design MR composition. LNG production is increased as more refrigeration becomes available.

7.5 LNG plant shutdown

There are emergency situations that will result in shutdown of the process units or the complete facility. Refer to Appendix 3 for the general shutdown causes and sequences for the process pretreatment units. Shutdown of the pretreatment units may also result in shutdown of the liquefaction unit. The following describes the shutdown sequence of the refrigeration systems in the liquefaction unit.

In general, when the emergency is due to process upsets within the process unit, the shutdown period is temporary and the unit can be isolated from the feed and product systems. Heating to the unit will stop, and the unit is placed on hold. Pumps and compressors will be unloaded and placed on recycle or can be stopped if necessary. If extended shutdown is expected, the unit inventory can be depressured to the flare. For maintenance shutdown, the unit operation will stop, and the inventory of the unit can be transferred to offsite storage tanks. The unit can then be purged and repaired.

Shutdown of the liquefaction unit follows a similar approach. If the propane compressor trips, the MR compressor should be placed on total recycle by opening the compressor recycle valves. The main exchanger JT valves should be closed. LNG production flow should be stopped by closing the temperature control valves.

Shutdown of the MR compressor results in unloading the propane compressor, which will be placed on recycle. The feed gas flow to the main exchanger should be stopped by closing the LNG temperature control valve. This will avoid large temperature drops between feed gas and the main exchanger tubes. The main exchanger JT valves should also be closed.

On loss of feed gas flow, the propane refrigeration system will be unloaded. With no feed load on the main exchanger, the warm end temperature of the low pressure mixed refrigerant returning to the compressors will quickly drop in temperature. During this scenario, the main exchanger JT valves should be closed and the MR and propane compressors should be on total recycle.

7.6 Performance and acceptance test

Once the plant is fully operational, performance tests can be started to prove the facility plant can meet the plant capacity and ship transfer rate and sendout rate with quality as designed. During the test runs, values of each independent variable—flow, composition, temperature, pressure—must be recorded. The plant is brought up to those conditions shown on the heat and mass balance table. If the plant feed gas or LNG is different than design, process simulation may be required to establish the optimum operating conditions.

During the test, the plant operating conditions are maintained reasonably constant and there is repeatability of the data. Daily material and energy balances should be collected and analyzed for agreement with the guarantee values. The test results must be confirmed by the plant test engineers and the operators.

When the plant has met the Performance and Acceptance test requirements as set up by the commissioning team, there is usually a formal acceptance process involving signing of acceptance documents. Once the plant is accepted, it is officially considered part of the normal operations and becomes the responsibility of operations and maintenance.

It is also a common practice to prove performance repeatability and plant integrity as part of the performance test, such as shutdown and startup in different times, to the test conditions, to prove plant repeatability. Reinspection of critical process equipment may be necessary to ensure that they have not been damaged during test runs.

Formal acceptance represents formal acknowledgment that the contractor has fulfilled their contractual obligation and the commissioning team has completed their assignment. The capital project is deemed complete and can be transferred to the owner.

7.7 LNG regasification terminal normal operation

The LNG unloading terminal typically is designed to unload LNG from an LNG carrier with an average unloading rate of 12,000 m^3/hr. The total time that the ship stays at port is about 24 hours. Detailed description of the LNG regasification terminal can be found in Chapter 1.

During ship unloading, vapor is returned to the ship to replenish the unloaded volume from the ship in order to avoid vacuum conditions. To meet the hydraulic requirement, the storage tank must operate at a higher pressure, typically at about 8 to 10 KPa higher than the ship, in order to avoid the use of a vapor compressor. Some of the key operating variables to maintain the tank under the design pressure limits are shown in the following:

- **Storage tank.** Typical LNG tank operating conditions are shown in Table 1-4. The normal operating pressure is set at 10 KPag. During ship unloading, the pressure needs to be increased to the maximum pressure of 25 KPag. After LNG unloading is finished, the unloading arms are purged with nitrogen and the LNG inventory in the unloading arms is recovered, either by draining to a drain drum located in the dock area or pressurized to the storage tanks with nitrogen, prior to disconnect from the ship.
- **The storage tanks fill connections.** LNG storage tanks can be filled with top or bottom connections during LNG unloading. Internal piping is configured to permit top and bottom loading. The top loading operation typically is done via a spray device/splash plate to promote flashing and mixing of the unloaded LNG with the LNG inventory in the tank. The bottom loading operation is done using a standpipe that directs the liquid to the bottom of the tank. When unloading a lighter LNG, it should be unloaded into the bottom of the storage tank to enhance mixing. Similarly when unloading a heavier LNG, it should be unloaded from the top.
- **Water hammering.** During unloading, there may be situations that call for the emergency shutdown (ESD) valves to close. Since the storage tanks typically are located at a higher elevation, and these valves are designed for quick closure, there are potential "water hammers" on the system that must be considered in the valve selection and closure time setting.

- **Transfer line cooldown.** A single large unloading line and a small recirculation line usually are installed from the jetty to the storage area. During the holding operation, when the ship is not at port, the unloading line must be kept cold by circulating LNG from the sendout system to the jetty using the small circulation line.
- **Tank cooldown.** Care must be taken to avoid subcooling the vapor space in the storage tank. Overcooling may result in a vacuum inside the tank, causing damage to the tank structure. The tank temperature and pressure must be monitored during the tank cooling operation. Tank cooldown is performed before ship unloading to minimize vapor boil-off during unloading. The tank must be kept below –130°C before LNG can be unloaded.
- **Rollover.** Top filling with LNG of varying density could result, in exceptional circumstances, in stratification of the LNG inside the tank, which may eventually lead to tank rollover. In another instance, when LNG inventory is stored for a long duration, the top layer of liquid is slowly weathered by vaporization in the form of boil-off gas, and its composition becomes richer and heavier than the lower layer. As shown in Figure 1-23, there may be two or more cells of liquid formed with different density, and when the top layer is heavier than the lower layer, and under a stagnant environment, rollover can occur. Rollover is the spontaneous mixing of these two layers, which will result in the release of a significant amount of vapor in a short time, which may result in overpressure and lifting of tank relief valves. More detailed discussions on the rollover phenomenon can be found in Chapter 9.

  Some of the measures to prevent tank rollover are as follows:
  - Store liquids of different density in different storage tanks.
  - Load storage tanks through nozzles or jets to promote mixing.
  - Use filling pipework at an appropriate level in the storage tanks.
  - Monitor the LNG shipment cargo density and temperature before unloading.
  - Monitor liquid density and temperature profile in the tanks.
  - Monitor storage tank boil-off rates.
  - Monitor storage tank pressure.
  - Transfer contents to other tanks.
  - Recirculate the content within the storage tanks using the LP sendout pump.
- **BOG recondenser.** Vapor generation from the LNG tanks vary significantly between the ship unloading and holding operations. Typically, the vapor rate varies from 4 to 40 MMscfd between the holding and ship unloading operation. These vapor flows are compressed by the BOG compressor, and condensed in the BOG recondenser using a slip stream of the cold sendout LNG. The BOG recondenser serves the following functions: Condensing the BOG, maintaining the storage tanks at a low pressure, serving as a mixing drum and suction drum to the HP sendout pump, maintaining a constant suction pressure, and providing sufficient NPSH to operate the HP sendout pump.

  The process design and control system of the BOG recondenser/HP sendout pump system must ensure a stable flow with sufficient subcooling and NPSH for the HP sendout pump operation. Some of the key control parameters are:
  - Liquid flow rate to the packed column is controlled by a flow ratio controller based on the BOG flow rate. The flow ratio is reset by the pressure controller on the recondenser.
  - The main LNG sendout is fed to the lower section of the recondenser under level control on the recondenser. This ensures sufficient liquid is available to the HP sendout pump.

- The liquid level in the recondenser should be kept below the packed section, allowing all the packing to be available for heat transfer.
- The vent from the HP sendout pump is routed to the top of the recondenser, and must be drained back to the pump suction to avoid liquid buildup in the vent header.
- HP sendout pump recirculation flow is returned to the recondenser, which must be designed with surge volume.

7.8 LNG regasification terminal startup

Startup of the LNG regasification terminal is simpler than the liquefaction plant but the basic pre-commissioning and commissioning procedures are quite similar. The units that need to be started up are the LNG unloading system, LNG storage tank, LP pump and recondenser, HP pump, vaporizers, and sales gas piping system. The process flow diagram for a typical LNG regasification terminal is shown in Figure 7-3.

When the unit is fully commissioned, the equipment should be under positive nitrogen pressure. All lines are pressure and leak tested, dried, and free from dirt and debris. The LNG terminal is now ready to be cooled down and started up as described in the following.

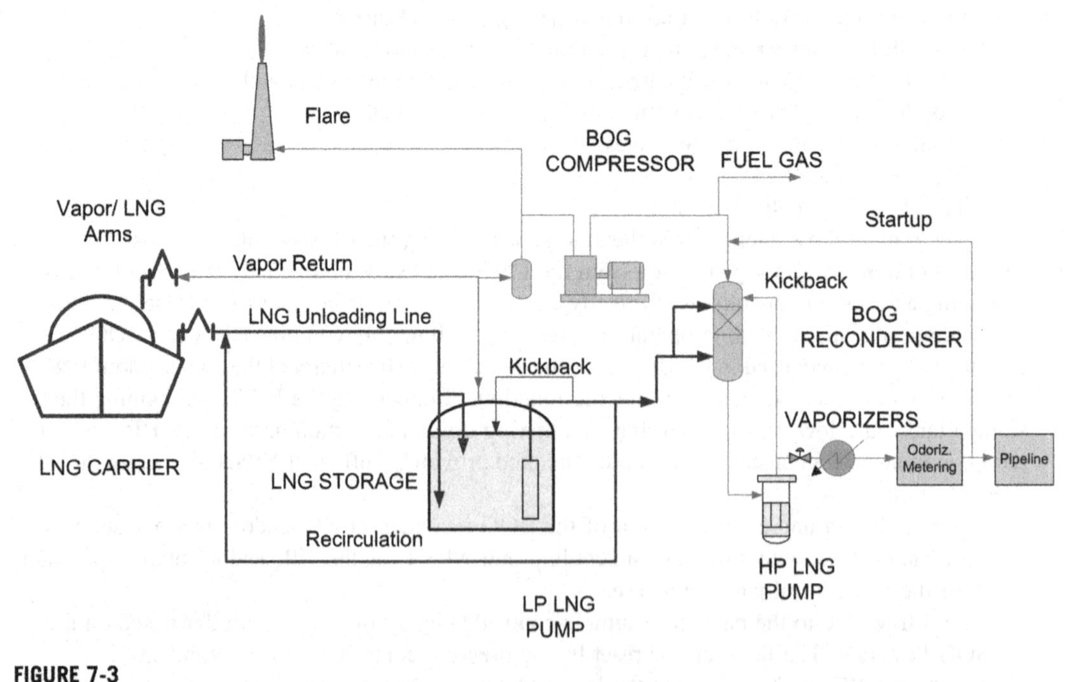

FIGURE 7-3

Typical LNG receiving terminal schematic.

7.8.1 Preparation of flare

To ensure plant safety, the first step in starting up an LNG receiving terminal is purging equipment and piping with nitrogen. All equipment should be purged to the flare to ensure oxygen is removed from the equipment, and a positive nitrogen pressure is maintained to prevent air ingress from the flare. After the flare header is purged, the flare tip can be started. If no fuel gas is available for the firing of the flare pilots, propane from propane bottles can be used.

7.8.2 Cooldown of unloading system

After arrival and mooring of the first LNG carrier, the vapor return arm and the unloading arms can be connected to the LNG carrier. In order to cool down the LNG unloading arms, LNG boil-off gas is routed via the unloading arms through the empty LNG storage tank to the flare. This cold gas from the LNG carrier slowly cools down the unloading arms and the unloading lines. When the system is cooled, one of the LNG carrier's pumps can be started on recirculation while LNG is introduced slowly to the LNG unloading system. The flashed vapor is routed via the LNG storage tank to the flare. No return vapor is available since the BOG compressor is not yet operating. The LNG carrier is equipped with its own vaporizer to produce its own boil-off gas for tank pressure control and fuel gas requirement.

7.8.3 Cooldown of storage tanks

When the unloading lines have been filled, the next step is to cool the LNG storage tank down to the operating temperature. After this cooldown step, the LNG storage tank is ready to receive LNG. The tank is provided with a cooldown spray ring. This ring is located in the upper part of the tank. LNG from the unloading lines is routed to this spray ring from where it is sprayed into the warm tank. The evaporation of the LNG cools down the tank vapor space and ultimately the tank inner container.

The LNG storage tank cooldown is a critical activity. If performed incorrectly it can seriously damage the tank. The cooldown of the LNG storage tank is therefore normally performed under the direct control of the tank vendor, who will advise on permitted cooldown rates. When the tank temperature is low enough, according to the tank vendor, the LNG can be diverted from the spray ring to the main tank fill lines. The flow of LNG from the ship to the LNG storage tank can then be gradually increased up to design rates.

7.8.4 BOG compressor/recondenser

After the nitrogen is sufficiently vented from the LNG storage tank, one BOG compressor can be started with the minimum flow control valve using valve unloaders. After establishing a constant RPM, the compressor is brought into full operation by closing the unloading valves. Boil-off gas from the BOG compressor is directed to the recondenser and then to the flare or fuel gas system while cooling down the recondenser vessel and associated piping.

7.8.5 LP LNG pump

After establishing the liquid level in the LNG storage tanks, the LP LNG system can be started. In the beginning, one LP LNG pump is started with the pump's minimum flow valve open. The LP LNG is slowly directed to the inlet of the vaporizers using startup lines and is recirculated back to the storage tanks. High point vents are slightly opened such that the lines are slowly filled with liquid. When the sendout piping is liquid filled, the LP pump recirculation line can be slowly closed. A slip stream of LNG is routed to the recondenser for cooldown, preparing the recondenser for startup. The HP pump suction line is filled with liquid and the LNG sendout system is ready to be started up. BOG generated from the cooldown operation can be recovered as fuel gas or vented to the flare. When an LNG carrier starts unloading, BOG can also be used as return vapor to the ship.

7.8.6 HP LNG pumps and vaporizers

The next step is starting the gas export. If submerged combustion vaporizers are used, start the burners as required. After establishing the required water-bath temperature, the HP LNG pump is started on minimum flow spill back. Thereafter LNG is introduced slowly to the downstream system by opening the control valves in the vaporizer inlets by activating the downstream pressure controller. Once liquid is routed to the vaporizers, the pump kickback is reduced. When the pressure downstream of the vaporizer is established, gas export can be started. As necessary, one or more LP and HP pumps are started, and other vaporizers are brought on-stream. The LNG terminal is then in the production mode. Flaring will be stopped since the surplus BOG can be condensed in the recondenser.

7.9 LNG regasification terminal shutdown

Normally, in an LNG regasification terminal, equipment maintenance is very minimal, except the high head HP LNG pumps. However, in case of emergency, the facility must be safely shut down to place the LNG system on a standby mode.

During this situation, the ESD valves will operate to isolate the LNG facility from the LNG cargo ship and the natural gas pipeline. The LP and HP LNG pumps will stop with the recirculation valves open to the LNG storage tanks. The inlet valves to the vaporizers will be closed. The heat source to the vaporizers will be stopped. The BOG compressor will stop with the recycle valve open. Excess BOG will be vented to the flare system. The LNG unloading arm's shutdown valves will be closed and the LNG transfer system will be stopped.

References

Bras, E., van der Zwet, G., Klinkenbijl, J., Clinton, P., April 24–27, 2007. Treating Difficult Feed Gases to LNG Plants, paper presented at the 15th International Conference & Exhibition on Liquefied Natural Gas (LNG 15), Barcelona, Spain.

Flood, T.M., Wong, V.W., Chow, T., October 10, 2011. The Optimal Sulfur Recovery Solution for Coal Gasification Plants in Achieving & Maintaining a Green China, paper presented at the 3rd China Energy Chemical International Forum, Urumqi City, Xinjiang, China.

Mock, J.M., Hahn, P., Ramani, R., Messersmith, D., Feb. 24–27, 2008. Experiences in the Operation of Dehydration and Mercury Removal Systems in LNG Trains, paper presented at the 58th Annual Laurance Reid Gas Conditioning Conference (LRGCC), Norman, OK, USA.

CHAPTER

8 Dynamic Simulation and Optimization of LNG Plants and Import Terminals

8.1 Introduction

Dynamic simulation has established itself as a valuable technology in the chemical process industries. Both steady state and dynamic simulation are carried out in the various stages of the process life cycle. These models are useful for a variety of purposes, including but not limited to engineering and process studies, control system studies, and applications in day-to-day operations (Broussard, 2002). Improvements in usability of the dynamic simulation software tools and the innovations in computing technology have led to the further deployment and novel applications of process simulation tools. The operability and profitability of the plant during its life depends on good process and control system design. Dynamic simulation helps to ensure that these aspects are analyzed early in the plant design stage. This helps to eliminate any costly rework that may be needed later.

In the early days of its development, dynamic simulation was used for very specialized applications, mainly due to the lack of user-friendly software applications. These tools required custom development and hence, were not widely deployed. Gradually, the usability of dynamic simulation tools improved over the years. The availability of software packages with built-in unit operation models and physical property packages that can easily be utilized to put together plantwide dynamic models made the application of this technology widely accepted.

In the beginning, the main application of dynamic models was for development of operator training simulators. These models were not necessarily of high fidelity, but were computationally simpler. Since the 1990s, software applications that are easy to use and based on rigorous thermodynamics have entered the market. These made the application of dynamic simulation more widespread. Another very important factor was the increase in computational power that became readily available over the years. With Moore's law becoming a reality, computational capabilities have gone up significantly over time. This made the application of detailed, plantwide high fidelity dynamic models more feasible.

Several benefits can be realized by using the dynamic simulation in the various stages of an LNG plant life cycle. These include both design and operational benefits. On the process design side, it is an important tool for evaluating the antisurge control system for the refrigeration compressors and sizing the antisurge valves. The reliable protection of this key equipment is critical for reliable long-term operation of the LNG plant. Also, dynamic simulation can be pivotal in the sizing of specific key relief valves, and the overall relief system and optimum selection of equipment sizes. LNG plants are also characterized by extensive heat integration, the operational implications of which can be verified by simulation. Further, the effect of external factors like ambient conditions and compositional changes on the future plant operation can be analyzed to further optimize the design. The startup procedures can be developed or verified before any field activities, thus potentially saving valuable startup time.

On the control system side, dynamic simulation can help to ensure that the plant has sufficient design margins to handle the disturbances. Also, dynamic simulation can be used to assess the optimal operation of the LNG plant in the face of changing conditions like ambient temperature. There is a greater emphasis now on considering plant operability in the design stage. Plantwide dynamic models along with operability metrics are valuable tools in attaining this objective. A dynamic simulation model of the LNG plant can also be used for the checkout of the plant distributed control system (DCS) prior to commissioning. This helps to reduce the time needed for commissioning on-site during startup. Also, this simulation model can be leveraged for startup support and for operator training purposes. The direct and indirect benefits of these applications are many, including better plant startup, less plant downtime, and improved safety.

Dynamic simulation has found wide applicability in the LNG industry. The various engineering contractors have used dynamic simulation extensively for improving the plant design (Valappil et al., 2005; Patel et al., 2007; Masuda et al., 2012). Also, process licensors have used these tools to improve and optimize their offerings. Apart from these, the LNG plant owner-operators have also found good use of dynamic simulation, especially in the form of operator training simulators.

The chapter starts with a brief overview of life cycle dynamic simulation of LNG plants. The details of developing LNG plant dynamic models, including the modeling of individual unit operations, is presented. The applications of dynamic simulation in LNG plants, including engineering studies and operator training simulator (OTS) are also presented in this chapter. In addition, the development of dynamic models and their applications in design and operation of LNG import terminals is discussed here.

8.2 Life cycle dynamic simulation of LNG plants and import terminals

Dynamic simulation analysis is used in the various stages of plant life cycle for different applications. The benefits of integrating these modeling activities have been realized over the years (Wedel et al., 1999). This has led to the adoption of life cycle modeling concepts by various process-engineering companies and the vendors of simulation tools (Wedel et al., 1999; Valappil et al., 2005). The dynamic model, evolving with the various stages of a plant life cycle, can be tailored for various applications within the project life cycle (Figure 8-1). A brief description of the main project stages are given next.

1. **Process Studies.** There are many nonproject-specific studies that utilize the benefits of dynamic analysis and investigate the potential for LNG process design improvement. These are aimed at improving the process design rather than a specific plant design. An example is the study of plantwide control structure for the LNG process using a base dynamic model.
2. **FEED.** The Front End Engineering and Design (FEED) stage is characterized mainly by steady state information about the plant. Some of the more detailed information pertinent to dynamic

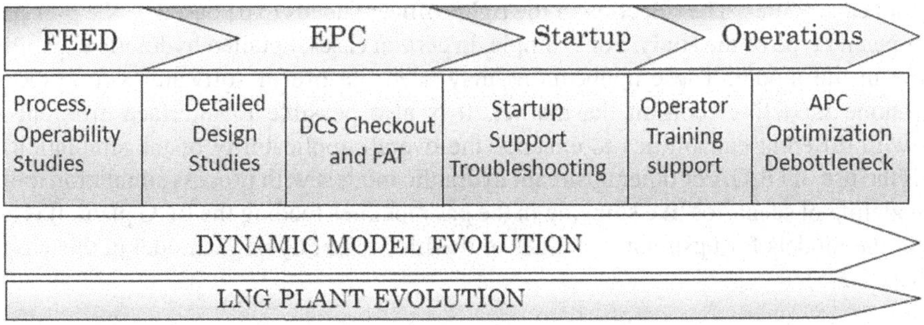

FIGURE 8-1

Applications of dynamic simulation model in the various stages of the LNG project life cycle.

simulation is not available at this stage. For example, equipment and control valve sizes and pressure profiles are not fixed. Dynamic simulation can still be a valuable tool in making qualitative decisions in this stage. The ability to influence the LNG project is highest in the FEED phase, while the cost committed is the lowest. Operability studies that look at the specific plant can be done here. For these studies, a certain level of uncertainty can be accommodated in decision-making.

3. **EPC.** The detailed design of the LNG plant is performed in the Engineering, Procurement, and Construction (EPC) stage. The equipment and line sizes, isometric layout, and other details are available at the end of the EPC stage, as issued for construction. This makes the EPC stage a suitable phase for development of high fidelity dynamic models. Engineering studies and control system checkout studies can be carried out efficiently in this stage. The dynamic model that is plant-specific is best developed at the EPC stage of the project. The engineering studies model may easily be extended to a fully functional virtual plant/OTS at this stage (Mansy et al., 2002).
4. **Startup/Commissioning and Operations.** The startup and commissioning activity involves bringing the LNG plant to normal operation and completing a subsequent performance test of the plant before turnover to the operating firm. This includes a commissioning test for the individual units and a performance test for the overall facility. Dynamic simulation can be valuable in various startup studies, operator training, and additional on-site control system checkout studies in this stage. The model at this stage would have all the details pertinent to the as-constructed plant. The model can then be utilized for troubleshooting and operational support throughout the plant life cycle.

8.3 Dynamic modeling of LNG plants

A dynamic simulation model that is fundamental and based on rigorous thermodynamics is the preferred option for simulation analysis. The scope, level of detail, and fidelity of the model can be tailored to suit the application it is used for and the amount of information available for modeling (Valappil et al., 2004). Several commercially available simulation packages can be utilized for developing dynamic simulation models for LNG plants. Different software packages have different strengths and capabilities. The selection of the right software needs to be based on the user preference and the scope and type of the study. For example, in certain cases, detailed hydraulic capabilities may be required in the model. These applications may need the use of software beyond the standard lumped volume modeling tools in the market. It is also possible to interface different software packages with different capabilities to enhance the overall applicability of the simulation. A good example is the use of OLGA or other upstream hydraulic models with process simulation tools, which enables modeling of scenarios like slugging in the gas pipelines feeding the LNG plant. It is beneficial to integrate the models for upstream pipelines to the LNG plant dynamic model in this case (Szatny and Mullick, 2005).

8.3.1 Differences from steady state modeling

The main advantage of using dynamic simulation is the ability to capture the time-dependent behavior of the system. This functionality comes at the price of increased data and resource requirements. Unlike steady state simulation, dynamic simulation is characterized by the need for accurate pressure profile, equipment volume, and other relevant information. Hence, the amount of information required for a high fidelity dynamic simulation is significantly more compared to steady state models. The internal variable specifications for dynamic simulation are much less. Normally, only the boundary conditions like pressure and temperature need to be specified and almost all the internal variables are calculated. Steady state simulation, especially for conceptual design, allows one to specify many more variables internally.

A clear distinction needs to be made between models and simulation. Models are tools that are used for various purposes, simulation being one of them. There are several other applications for these models (such as optimization, state estimation, etc.). Hence, the dynamic models can be used for steady state simulation in certain special cases, where there is a need to evaluate only a few steady states. This is usually beneficial if a dynamic model is already available and can be leveraged for the intended purpose.

8.3.2 Dynamic modeling of individual equipment in LNG plants

This section describes the main components of the LNG plant dynamic model and the approaches that can be used for modeling this equipment. Different modeling approaches can be used depending on the fidelity desired and the computational resources available. These are explained in the following sections.

8.3.2.1 Piping

The exact modeling approach for the plant piping depends on the purpose of the dynamic model. For detailed engineering studies, accurate volume and pressure drop in the piping becomes important. In these cases, the detailed piping layout drawings or isometric drawings can be used to obtain the required information. For lumped volume simulation applications, the piping data may be condensed to lumped volumes and equivalent lengths or other similar input.

For applications where momentum balance equations in the pipe hydraulics are important, the models need to be developed with more detail. These models tend to be more specialized and limited to smaller sections of the plant. Simulation software packages in the market that are commonly used may not be suitable to handle these applications. An example of the application is analysis of pressure surge in pipelines and heat exchanger tube rupture simulations. Another important consideration in the modeling of piping is the ability to accurately model the choked flow in pipes (Thomas, 1999). For applications that require the solution of momentum balance equations, extensions to the standard simulation tools or specialized hydraulic analysis software tools may be more appropriate.

8.3.2.2 Control and isolation valves

The specification of the control valve and isolation valve data is important for accurately capturing the dynamic responses in the model. The control valves are modeled using the actual valve size, characteristic, and the actuator timings. The valve size can be represented using rated Cv data from the vendor. For the characteristics, actual vendor characteristics need to be used. The speed of the actuator is also an important parameter in the simulation, since this determines the dynamic response of the overall control loop. This could be an input or can be determined from the simulation as a requirement to meet a specified dynamic response.

Further detailed modeling of valves involves using the actual valve sizing equations from the vendor. Some of the simulation software tools allow use of these in addition to the universal gas sizing equation. It is important to use the correct pressure differential ratio factor (X_t) and other vendor-specific parameters for the valve that may impact the sizing calculations in this case.

8.3.2.3 Compressors and pumps

Since LNG plant designs utilize refrigeration processes, compressor performance needs to be accurately captured for the fidelity of the plantwide simulation model. For dynamic modeling of compressors, the following items are important.

- A full performance curve is needed to accurately model compressors. These curves give compressor flow vs. head and flow vs. efficiency relationships at different speeds or Inlet Guide Vane (IGV) positions. The curves would preferably be as-tested curves in the field or final design curves. The full speed range from 0 to maximum operating speed is preferred for modeling. Since the vendor curves normally are limited to a certain set of speeds, extrapolation of the curves using fan laws may become necessary to capture the coast-down and startup behavior at lower speeds.
- Rotational inertia information to capture the effect of transient rotational power during startup, coast-down, and other dynamic events.

To enhance the model fidelity, it is possible to utilize multiple curves for different molecular weights or inlet conditions. Multiple molecular weight curves are useful for cases where compositions are expected to change. This may be more relevant for mixed refrigerant compressors with varying compositions and less for cascade processes with fixed refrigerant compositions. Multiple curves at different suction pressures would allow more accurate modeling for the cases where suction conditions vary significantly. This could be important for some detailed engineering studies.

In dynamic simulation, the operating point in the compressor map is determined based on the overall system hydraulic and other conditions. The variation in speed can be either modeled as an external specification or can be calculated by the compressor model based on the shaft power balance. If the speed variation is to be modeled, the rotational inertia of the compressor needs to be input into the model. The efficiency at the operating point is used to calculate the power consumption (Thomas, 1999; Boyce, 2003).

For accurate modeling of pumps, the pump characteristic curve is needed to model the head-flow relationship under different conditions. Since the pumps used are mostly fixed speed pumps, this usually is specified only at the operating speed.

8.3.2.4 Expanders

The expanders are utilized in the LNG plants to increase the efficiency and recover energy while reducing feed gas pressure. These could be vapor expanders or flashing liquid expanders. The modeling methodology for expanders is very similar to the compressors. Full performance curves (Figure 8-2) for the expanders can be utilized for modeling purposes (Sarah, 2009). These expander curves would provide flow vs. head or flow vs. power relationships (and efficiency) at different expander speeds. This methodology is commonly used in the commercial simulation packages.

If the full curves are not available, the expander can be modeled with the expander flow equation (Thomas, 1999). This is a typical resistance equation, whereby the conductance can be specified based on the flow and pressure drop based on the normal operating conditions. There should be provisions to consider choking in the expander in this case (as the flow would be limited and independent of the downstream pressure). The efficiency can be used along with the isentropic expansion equations to calculate the power generated in the expander.

8.3.2.5 Gas turbine drivers and electric motors

The refrigeration compressors in the LNG plant are mostly driven by gas turbines. In some cases, electric motors or steam turbines may be used as the drivers. The modeling of the gas turbine and the associated dynamics is useful to accurately capture the dynamic behavior of the LNG plant. Normally,

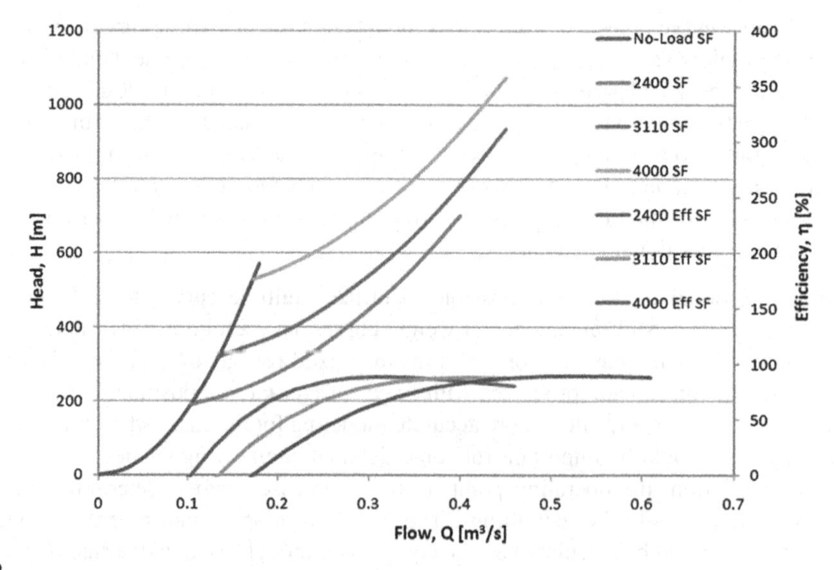

FIGURE 8-2

Typical expander curves that are used for dynamic modeling.

(Sarah, 2009)

the turbine governor speed controller resides in the gas turbine control package. The response of the turbine speed control and hence the power supply has an impact on the overall plant response.

There are two alternate routes to modeling the gas turbines, depending on the application of the model and the fidelity required. The first option is to build a fundamental model based on the combustion of the fuel with air and subsequent expansion. In this case, the air compressor needs to be modeled with actual curves to get the correct representation. Also, the combustion equations need to be modeled rigorously. The flue gas is fed to an expander, which generates the shaft power. The expander curves also have to be accurate to get the right relationship between speed, power, and flue gas exhaust temperature.

The approach to modeling a gas turbine with the combustion and expansion phenomena explained earlier is computationally demanding. This approach can also be challenging due to the lack of accurate data. This type of detailed model is not required for the LNG process related dynamic studies that are commonly performed. An alternative option is to use empirical models for the gas turbines. These models are based on transfer functions that capture the essential dynamics of the system. The various lags and delays in this model are used to represent the dynamics of the fuel gas positioner/valve, fuel system, and the combustor (Rowen, 1983).

An empirical model for the gas turbine reported by Rowen (1983) is shown in Figure 8-3. Here, the speed governor, fuel system, and valve positioner dynamics are represented by first order transfer functions. The shaft power is calculated based on the fuel flow and speed and used in the compressor model to denote shaft power. A balance between power supply and demand is used to compute turbine speed (Rowen, 1983).

The modeling of the control schemes that are internal to the gas turbine can also be important. Apart from speed control, the turbine exhaust temperature control may be a determining factor in the

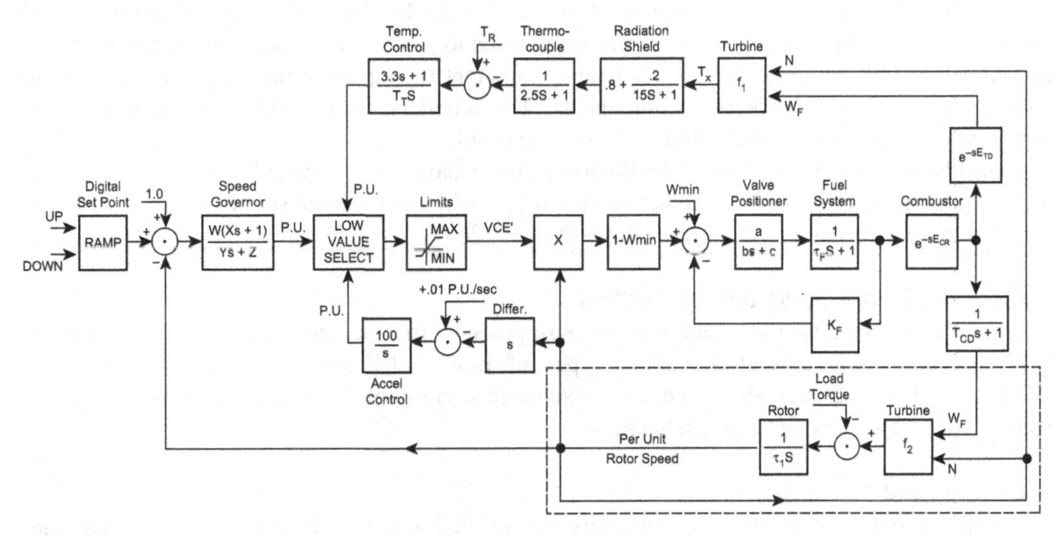

FIGURE 8-3

Dynamic model for the gas turbine utilizing transfer functions.

(Rowen, 1983)

overall system response. This normally acts as an override controller (overriding the speed control), reducing the fuel gas once the turbine reaches the limiting exhaust temperature. In addition, acceleration control (especially relevant during startup) may need to be modeled depending on the purpose of the model. For modeling startup of gas turbines, it is important to accurately capture the available turbine torque at different speeds. This data can be obtained from the vendor and used in the simulation (Heckel and Davis, 1998). Modeling of electric motors is relatively straightforward. The motors are characterized by speed-torque relationships, which are used for modeling purposes (Boyce, 2003).

It is also possible to link the process model with a model of the electrical system (with turbines and synchronous machines) to analyze the interaction between the two (Ismail et al., 2004). In this case, the torque values from the process model are transferred to the electrical model and the frequency values are transferred back to the process model. This is then utilized to verify the operability of the system under turndown conditions and the presence of electrical disturbances.

8.3.2.6 Distillation columns

In LNG plants, the main areas where distillation columns are used are the NGL recovery unit, fractionation section, and nitrogen rejection unit. The columns in these units can be packed bed or trayed columns. The dynamic modeling of the columns is done with the actual trays instead of the theoretical trays. Hence, tray efficiencies need to be specified. This is utilized along with dynamic mass and energy balance equation and vapor-liquid equilibrium relationships. The tray efficiency parameter can be estimated from the design information or actual plant data, if available.

Packed bed columns are relatively straightforward to model with the commercial software tools, as they come with the necessary equations for industry standard packing used. The model for columns need detailed information as to the tray and down-comer details and dimensions and nozzle locations. To accurately capture the hydraulics of the system, static head contributions need to be considered. Another important factor in the modeling of columns is the holdup in the trays. Since the models assume no aeration in the trays, the tray parameters need to be adjusted to accurately capture the tray liquid holdup. The modeling of reboilers is also important for capturing the overall response of the distillation column and associated components. The reboilers need to have the static head and elevations to capture the accurate flow and pressure profile.

Recently, rate-based modeling of distillation columns have been proposed as an alternative to the efficiency-based models. These have not been widely applied to dynamic simulation models, but are expected to be used more in the future (Ramesh et al., 2007).

8.3.2.7 Heat exchangers and air coolers

LNG processes utilize heat integration to increase process efficiency and minimize the capital cost. This heat integration introduces various types of heat exchangers with different amounts of complexity. This includes shell and tube exchangers, air-cooled exchangers, plate and frame exchangers, or spiral wound heat exchangers.

Shell and tube exchanger

The shell and tube heat exchangers can easily be modeled with the normal design data. The main information that is required in this case is the UA,[1] the tube and shell volumes, and the pressure drop

[1]UA for the exchanger is the product of heat transfer coefficient (U) and the heat transfer area (A).

data. For increasing the model details, the exchanger can be divided into additional zones for capturing the accurate temperature changes across the tube length.

Air-cooled exchanger

The air-cooled exchangers can also be modeled in a similar manner as the shell and tube exchangers, with the air flow rate and the UA data. The number of bays and the number of fans per bay are also needed in this case. This approach is an approximation and may not capture the performance under all ambient and operating conditions. For rigorous modeling, more specialized air-cooler modeling tools need to be used.

Plate and frame exchanger and spiral-wound exchanger

The dynamic modeling of plate and frame and spiral wound exchangers is a much more challenging task. These exchangers are mostly multistream exchangers with very close temperature approach. For plate and frame exchangers, the actual geometry and configuration becomes important in developing high fidelity models. These exchangers are characterized by a number of layers, through which the different fluids pass (Figure 8-4). Fins are used to increase the heat transfer area. The layer patterns can be repeating in these exchangers. The fluid volume in the various passes is determined by the geometry of the exchanger.

One way to model the plate and frame exchanger rigorously is to utilize a distributed modeling approach with discretization. This would result in partial differential equations to model the heat transfer along the exchanger length. This involves discretizing the exchanger with distributed model for mass and heat balance. One important parameter here is the heat transfer coefficients, which can be computed in a rigorous manner using empirical correlations (Hammer, 2004; Haarlemmer and Pigourier, 2008). Normally, fixed heat transfer coefficients that are scaled with varying Reynolds numbers are used for computations. This will not capture heat transfer under conditions of phase change in the exchanger. Correlations are available for these exchanger heat transfer coefficients based on geometry. These correlations normally provide the Colburn number (Jh) as a function of operating conditions and can be obtained from the literature or the exchanger vendor. The heat transfer coefficient can then be determined using Equation 8-1 for Nusselt number.

$$Nu = \mathrm{Re}.\mathrm{Pr}^{1/3}.Jh \tag{8-1}$$

where Nu is the Nusselt number, Re is the Reynolds number, Pr is the Prandtl number and Jh is the Colburn number.

The heat transfer between the fluid and the metal can then be calculated using Equation 8-2:

$$Q = \alpha.A.\left(T_{metal} - T_{fluid}\right) \tag{8-2}$$

where α is the film heat transfer coefficient, A is the heat transfer area and T_{metal} and T_{fluid} are the metal and fluid temperatures.

The overall heat exchanger can then be solved using the common wall equation, which assumes one metal at a single temperature. The equation for cell i can be given as:

$$Cp.\frac{\partial T^i_{metal}}{\partial t} - \sum_{j=allfluids} \alpha^i_j A^i_j.\left(T^i_{fluid} - T^i_{metal}\right) = 0 \tag{8-3}$$

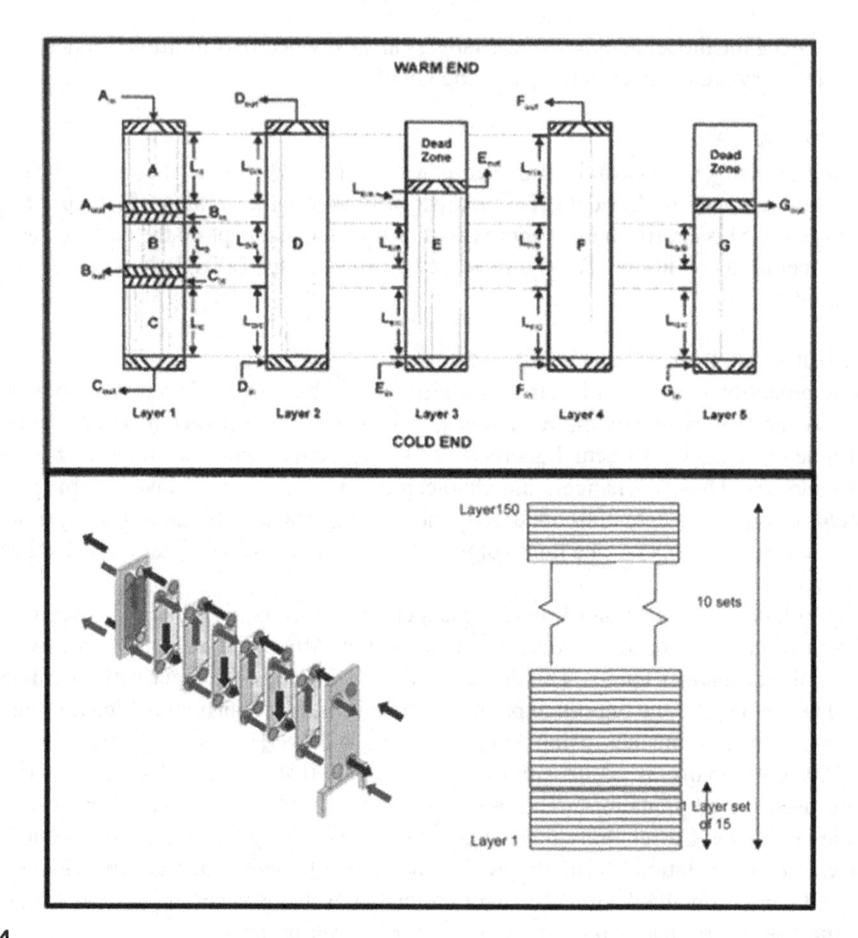

FIGURE 8-4

Plate and frame exchanger layers with multiple streams (top) and lumping of the layers for efficient modeling (bottom).

(Fulton and Collie, 1997)

where C_P is the metal specific heat capacity, T^i_{metal} is the metal temperature for cell i, α^i_j is the film heat transfer coefficient, A^i_i is the heat transfer area for cell i and fluid j, and T^i_{metal} and T^i_{fluid} are the metal and fluid temperatures for cell i.

For the fluid, the equation can be given as (for cell i and fluid j):

$$M^i_j \frac{\partial H^i_j}{\partial t} + F^i_j \frac{\partial H^i_j}{\partial x} - \alpha^i_j A^i_j * \left(T^i_{Fluid} - T^i_{Metal}\right) = 0 \qquad (8\text{-}4)$$

where M^i_j is the linear mass density, F^i_j is the fluid flow rate, and H^i_j is the specific enthalpy for the cell i and fluid j. More information about this modeling approach can be found in the discussion by Haarlemmer and Pigourier (2008).

The modeling of hydraulic phenomenon in the exchanger is another important consideration. This determines the pressure drop in the plate and frame exchanger. The pass hydraulics can be represented in detail by rigorously modeling the pressure drop along the pass. This method involves discretization along the pass length and is computationally very intensive (Haarlemmer and Pigourier, 2008).

The plate and frame modeling approach using discretization is computationally very demanding. This may not be practical for applications where simulation speed is important. An alternate approach is one that approximates the exchanger and splits into a few zones for modeling purposes. As a simplification, the heat transfer coefficients can be fit to match the exchanger performance. The drawback with this approach is that this may not model the heat transfer during the phase change phenomena accurately. For holdup calculations, each pass can be treated as a single holdup. Computational demands are reduced significantly with these approximations.

The dynamic modeling of spiral wound exchangers used in LNG processes is another challenging area (Hammer et al., 2003; Vist et al., 2003). For these exchangers, the use of simplified models may not adequately capture the actual performance under all operating conditions. First principle model for tube bundles that employ fundamental balance equations and thermodynamic calculations is preferred. The modeling of a spiral wound exchanger with an axially distributed model for material flows and an axial with radial distributed model for heat transfer has been reported by Stephenson and Wang (Stephenson and Wang, 2010).

8.3.2.8 Separators

The two-phase and three-phase separation vessels are common equipment in process plants. These are modeled in high fidelity dynamic simulation with volumes, nozzle details, and elevations.

One of the important things to note for the separators is the mixing efficiency in the vessel. The default option in many commercial packages is to use perfect mixing of the incoming stream with the material already in the vessel. This may not be a valid assumption in certain cases, especially where the vapor entering the vessel does not mix with the liquid well. In these cases, mixing efficiency can be used as a parameter to adjust the extent of mixing in these vessels.

8.3.2.9 Process control

The accurate modeling of process control schemes in the LNG plant is very important in dynamic simulation. The responses of these controllers impact the overall dynamic response of the closed loop system and have to be accurately captured. This includes both the process control functionality and the control algorithms.

The base regulatory controllers are easily modeled using the proportional, integral, and derivative (PID) controller functionality. The main issue here is the type of the PID algorithm. The commercially available distributed control systems utilize different PID control algorithm equations. These may not be the same as the ones utilized in the simulation model. If the dynamic model is used for rigorous tuning, this may become a determining factor. The next level of control scheme is the advanced regulatory controls like ratio control, cascade control, and FEED-forward control. These are also modeled easily in the available simulation tools or with custom modeling equations.

Another key consideration in LNG plant process control is the compressor antisurge control and the master performance control systems. These control systems are usually separate from the plant distributed control system (DCS). The high fidelity modeling of these systems can be important for certain applications (Wu et al., 2007). It is possible to replicate the functionality of these control

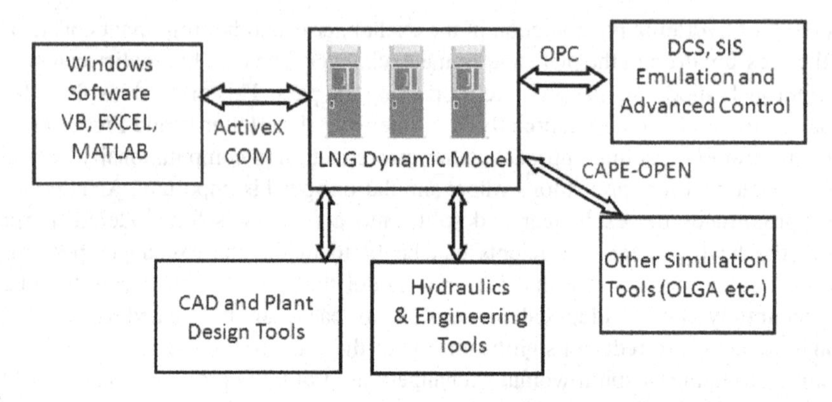

FIGURE 8-5

Dynamic models can be connected to various other tools to enhance the value and application of the models.

schemes in the simulation environment. This requires replicating the various antisurge control functionalities and their interactions. Another option is to use an emulator for the antisurge control system, which is supplied by the vendor. These emulators can be connected to the main process model and run in parallel to capture the exact dynamics of the antisurge control system. Ethernet and Object Linking and Embedding (OLE) for Process Control (OPC) server interfaces are available to provide the data link between the emulator and the plant model (Wu et al., 2007). These standards can also be used to interface the dynamic model with various third-party applications as shown in Figure 8-5.

8.3.3 Fidelity and details in LNG plant dynamic models

The amount of detail and the fidelity of the dynamic models are determined by the objective of the study and the required model speed. The dynamic models can be very detailed, but this would increase computational demands and slow down the simulation, making it less useful. For applications where real-time run is required, this is not acceptable. Hence, a balance needs to be made between the level of details in the model and the simulation speed achieved.

One of the key parameters in the simulation is the integration step size. This parameter is dependent on the dynamics of the system that needs to be captured. For simulations where short time-scale events are important, very small step sizes need to be used. Compressor trip simulations are an example of this. For operator training applications, larger step sizes can be utilized as longer time scale phenomena is investigated.

8.3.3.1 Engineering studies

The dynamic simulation model used for engineering studies has to include sufficient detail for the problem under study. These models are run offline, and hence do not have to necessarily run in real-time. Still, the simulation time and the model complexity must be within reasonable limits. The size and complexity of the model depends on the scope of the study and the process units involved. A lumped parameter modeling approach is normally used for plantwide simulation for engineering studies. The hydraulics and pressure profiles are obtained from the isometric layout information for the plant, if available. This is used to calculate a lumped conductance value and volume for the various pipe elements. For applications where detailed hydraulics with momentum balance capability is required, more specialized models can be developed. Vendor-provided compressor curves are used in modeling the compressors. Actual equipment sizes are used for vessels, heat exchangers, and other equipment. Control valves are modeled with appropriate actuator dynamics, control characteristics, and valve sizes. The process control aspects of the engineering studies model include only those details that are relevant to the study. The engineering studies model is thus highly dependent on the study scope, and should be representative but is not intended to be a replica of the LNG plant.

8.3.3.2 Operator training simulators and other applications

The dynamic model used for LNG plant operator training has to run in real-time or faster to be a realistic tool for training. Hence, certain simplifications become necessary, depending on the model size and complexity. The model that is developed for engineering studies and control system studies can be leveraged as the engine for a startup and operator training tool. The scope of the model needs to be changed in this stage to suit the functionality. For realizing the maximum benefits of using dynamic simulation, a full functionality model (Virtual Plant) has to be developed that can perform startup, shutdown, and other studies (Mansy et al., 2002). The following information is normally added to transform the engineering studies model to a virtual LNG plant.

- **LNG process details.** These details include the startup lines, bypass valves, emergency depressurization (EDP) valves, pressure safety valves (PSVs), and additional equipment that is needed to replicate the plant operation under startup, normal operation, and shutdown.
- **Plant control systems details.** The control functionality for Operator Training Simulator (OTS) has to exactly replicate the actual plant. Coding all the control functionalities within the process model is not practical. Rather, the control system emulations are used to replicate the plant control strategy. These emulations are linked to the dynamic process model using OPC for seamless data transfer.
- **Plant shutdown/interlocks.** The shutdown/interlock logic (SIS) of the LNG plant needs to be implemented for a realistic OTS simulator. These are implemented in separate script files or using the commercially available programmable logic controller (PLC) emulators.

The dynamic model with all the details is the heart of the OTS. The fidelity and robustness of the OTS depends on the scope and accuracy of the dynamic model. Normally, it is ensured that the model closely matches the plant data or design data before it is used for training purposes.

8.4 Applications of dynamic simulation in LNG

As described briefly earlier, dynamic simulation is used for various applications throughout the life cycle of an LNG plant. The key applications of dynamic simulation in LNG plant design and operations are described in the following sections.

8.4.1 LNG process design studies

One of the key applications of dynamic simulation is to identify improvements in the LNG process technology. This is not specific to any particular plant, but the LNG process design in general that is applicable to various plants that use the selected liquefaction technology. Steady state simulation is still the main tool in studying and enhancing the design of the process. Dynamic simulation can be a valuable addition in improving the process technology from the plant operability and reliability viewpoint.

8.4.1.1 LNG driver selection

The selection of appropriate drivers for the refrigeration compressors has a big impact on the overall economics and operability of the LNG plant. There are several factors that go into the selection of the appropriate drivers for the compressors. A number of them are related to the conceptual design, cost, and other related items. The areas where dynamic models can be of assistance here are those related to the operation of the LNG plant.

The first step in utilizing models for driver selection is to identify certain key metrics to quantify and evaluate the operability. These include:

- Time to start up the plant. Some drivers may require depressurization of the casing, which may add additional time.
- Ease of loading and balancing and the time requirement for the same.
- Time to recover from shutdown or time to recover from a compressor trip.
- The ease of load balancing between parallel machines (Patel et al., 2007). For example, single-shaft machines with limited speed ranges may have limited flexibility in load sharing between parallel machines.

Once the metrics are identified, a dynamic simulation model can be utilized to evaluate the various drivers for optimal selection.

8.4.1.2 LNG process sensitivity analysis

The operation of the LNG plant is affected by various parameters including the ambient temperature, feed gas composition, and equipment performance. The equipment design also has a big impact on the actual plant production rate. These sensitivities can be effectively analyzed using dynamic simulation to better understand and manage the risk. A good example is the effect of refrigeration compressor performance curves on the LNG production (Gandhi and de la Vega, 1998). The variations in the compressor head and its impact on the LNG plant production rate can be analyzed to obtain the sensitivity values. This information is valuable for the compressor vendor to optimize the design of the refrigeration compressor. Sensitivity of LNG plant production rate to several key operating and design parameters can also be obtained from dynamic models, as listed (Omori et al., 2001):

- Pressure of natural gas feed
- Molecular weight of natural gas feed
- Surface area of propane condenser
- Surface area of propane subcooler
- Gas turbine power
- Ambient temperature.

This sensitivity information can be useful to optimize the process scheme and evaluate the trade-off between various parameters.

Another important application is the verification of the plant performance with the as-built equipment to verify if the margins that are part of the design will be useful for the owner (Ismail et al., 2004). In this case, the simulation can be used to identify the production above the guaranteed value that can be achieved. This would validate the additional economic benefit gained from the in-built margins in the design. This information can also be valuable in evaluating the ability of the overall plant design to handle the extra throughput from the design margins. Identifying these bottlenecks would enable appropriate debottlenecking to handle the extra LNG plant feed due to the available margins.

8.4.1.3 Other process studies

Other applications of dynamic simulation in the process design area include its use to reduce the weight of offshore LNG installations. In this case, the simulation can be used for material selection and to optimize the size of equipment for an offshore environment (Feng et al., 2009). The material selection is impacted by the maximum pressure, maximum velocity, and forces, which are determined using simulation for startup and other transient operations. Alternatives like Fiber Reinforced Plastics (FRP) can be evaluated as an alternative to metallic pipes. In addition, this tool is valuable in reducing the overdesign of the equipment by optimizing the margins, thus saving space and weight.

8.4.2 Compressor antisurge system design

One of the key applications of dynamic simulation in LNG plant design is the development of an antisurge control system for the refrigeration and other compressors. The protection of compressors from surge is critical as reliable operation of these compressors is important for overall plant availability. The main objective of the antisurge control system is to prevent the compressors operating in the surge region. The operation of the machines in the surge region is damaging for the compressor impellers due to rapid flow reversals that can occur under these conditions.

There are two distinct aspects of antisurge control system development where dynamic simulation models are valuable. First is the ability of the system to maintain safe operation away from the surge region during deviations around the normal operating point. This is accomplished by the antisurge control system. The second is to prevent surge during shutdown scenarios, where the compressor rapidly coasts down to a stop. This machine protection during shutdown may be done by the safety instrumented system or by other means.

8.4.2.1 Verification of antisurge control performance under normal conditions

During normal plant operation, the process is affected by various disturbances that can move the compressor operating point toward surge. The antisurge control system will manipulate the antisurge valves to maintain the operating point to the right of the surge line. The antisurge valves need to be sized and selected to enable the operation to the right of the surge control line under these conditions. The design of this functionality is normally done by the antisurge control system vendor. Dynamic simulation is not needed to size the valve to prevent surge during normal operation. But it can still be used to provide adequate verification of the ability of the antisurge valve and the surge control algorithms to prevent surge under all possible scenarios.

The verification of antisurge control performance involves various operational scenarios. Some example scenarios follow. Other scenarios specific to the process under investigation can also be identified.

- Inadvertent closure of suction or discharge valves on the compressor
- Trip of one compressor affecting the other
- Failure of antisurge valve on one stage of a multistage compressor affecting the other stages.

These scenarios can be simulated to ensure adequacy of surge control margin, antisurge valve size, speed, and the antisurge control algorithms that are used.

The key to performing very accurate verification of antisurge system performance is having the right control algorithms in the model. In the past, the use of an actual control scheme in the model meant using the actual antisurge control hardware. Since this was a cumbersome task, most of the antisurge control schemes were emulated in the dynamic model. This could be adequate, but some fidelity is sacrificed. Recently, the soft emulators for the antisurge control system have become available, and can run in a computer. This can be connected to the dynamic model via OPC interface, whereby the process data is sent from the model to the control emulator and the valve outputs are sent back to the model. The various disturbances can be tested in this realistic environment (Wu et al., 2007). The same configuration files for the compressor control system that are used in the field can be used for this test. This way, the enhanced fidelity of the overall system can be used to develop reliable solutions to protect the compressors in the field.

8.4.2.2 Design of antisurge system for shutdown scenarios

The other application of dynamic simulation for antisurge control system development is the sizing of the antisurge valve to prevent surge during a machine emergency shutdown (ESD). The ESD of a compressor involves the sudden stop of fuel gas to the gas turbine (or loss of motor power). This results in a loss of shaft power within a short time (typically 100 milliseconds or similar time duration). For electric motor drives, the loss of power happens even faster. The compressor will coast down depending on the inertia of the compressor-turbine string. The rate of coast-down is determined by the energy balance that relates the compressor power, inertia, and rate of change of speed as given below:

$$W_{driver} = I_{string} N_{string} \frac{dN_{string}}{dt} + CompressionPower + Losses \tag{8-5}$$

where W_{driver} is the driver shaft power, I_{string} is the inertia of the compressor-driver string, N_{string} is the speed of the string, *CompressionPower* is the total compression power for all the compressor stages, and *Losses* denote the friction and windage losses in the system. For the strings where driver and compressor are at different speeds, appropriate string inertia can be calculated as:

$$I_{string} = I_{driver} + I_{comperrssor} \left(\frac{N_{compressor}}{N_{driver}} \right)^2 \tag{8-6}$$

where I_{driver} is the driver inertia, $I_{compressor}$ is the compressor inertia, and N_{driver} and $N_{compressor}$ are the speeds of driver and compressor, respectively. The string inertia in this case is with reference to the driver speed. If present, the gear and coupling inertias needs to be accounted for in a similar manner.

The antisurge valve and other system parameters need to be selected to protect the machine during coast-down from full power to a complete stop. This is especially important at high speeds, where the compression power is high and surge can be damaging to the impeller. A dynamic analysis can be performed for the compressors to ensure that the antisurge valves are sized adequately and that the stroke times of the antisurge valves are appropriate.

The input data to the dynamic simulation is very important in the simulation. The key inputs that determine the performance of this analysis are as follows.

- **Compressor performance curves and the rotational inertia.** The actual performance curves are required along with accurate values of rotational inertia for the string.

- **Piping layout.** The layout is important in determining the final size of the antisurge valves because it determines the suction and discharge volumes. Further, the hydraulic resistance is a determining factor in the overall system behavior.
- **Shaft power decay.** Electric motor drivers lose power instantly during trip. Gas turbine drivers have residual power even after the trip signal is initiated. This is due to the fuel gas valve actuation time and the holdup in the system that provides residual power.

The typical desired compressor coast-down response during ESD is shown in Figure 8-6. Here, the compressor initially moves toward the surge line due to the valve opening time and the time for the flow to reach the suction from discharge (Botros and Ganesan, 2008). In the subsequent phases, the compressor moves away from surge and coasts down safely.

Several parameters affect the dynamic response during the trip and can be used to design the system for effective compressor protection. Some of the considerations in the design of an effective antisurge system are listed as follows (Figure 8-7).

- **Antisurge valve sizes and valve timing.** The antisurge valve size needs to be large enough to provide adequate flow for machine protection. Also, they need to open fast to prevent trip surge. Normally, 1 second or similar is utilized for the opening time.
- **Suction and discharge valve timings.** The compressor may have to be isolated fast to prevent the need to pressurize/depressurize the additional system volume.
- **Volume of the discharge piping and suction piping.** It would be beneficial to minimize the piping volumes on the compressor discharge to the valve. This is especially true for cases where the air cooler is in the loop and can be located close to the compressor discharge in the layout.
- **Check valves to isolate volumes.** It may be beneficial to put additional check valves to isolate and reduce the volumes.

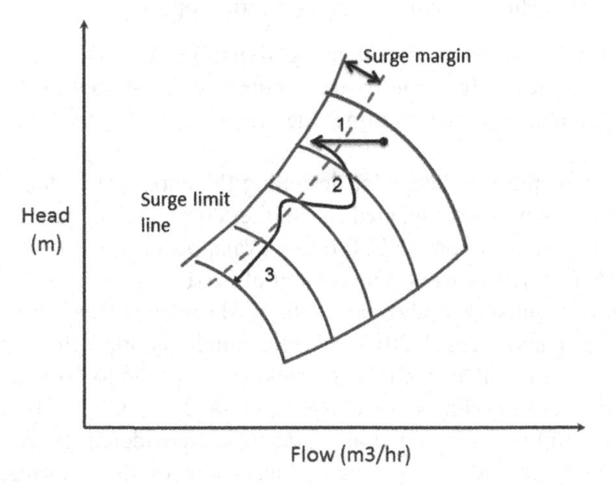

FIGURE 8-6

Desired compressor response during coast-down after an emergency shutdown.

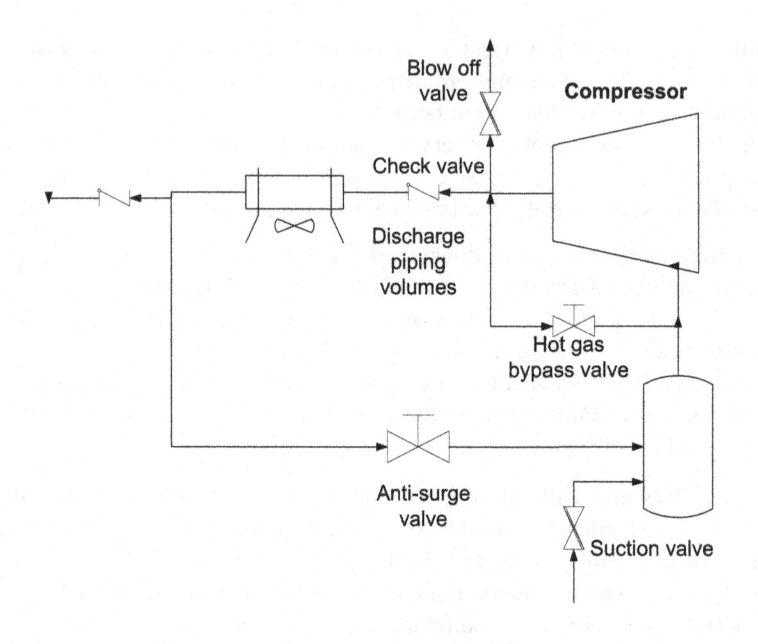

FIGURE 8-7

Various options for antisurge system design to handle a compressor ESD.

- **Hot-gas bypass for antisurge protection.** For systems with discharge cooler and large volumes, a hot gas bypass option can be beneficial, as this reduces the volume in the antisurge loop.
- **Blow-down valves.** For certain systems, blow-down valves can be used to reduce the discharge pressure right after trip. This is normally not a preferred option.

These various options can be analyzed using dynamic simulation to determine the effectiveness of a potential solution. In some cases, a combination of options may be required. Dynamic simulation is valuable in determining the most cost effective and robust solution for antisurge protection.

The application of dynamic simulation for designing the antisurge system for refrigeration compressors in an LNG plant is well documented in the literature. This includes the refrigeration compressors in the Cascade Refrigeration LNG Process (Valappil et al., 2004) and the MR and other compressors in the APCI C3MR process (Fantolini et al., 2012).

There are cases where antisurge valves sized for ESD may not be sufficient to provide surge protection during startup (Fantolini et al, 2012). In the example reported, the startup of the compressor with propane needed a control valve size (Cv) of 2600 (as opposed to 1600 required for ESD surge protection). Further, the startup with N_2 or defrost gas needed a Cv of 3000. In this case, it was necessary to make the startup Cv a function of the case considered by clamping the valve size appropriately. In addition, the various options in the design of the antisurge system for the MR compressor were studied here. This included using the bleed valve, hot gas bypass valve, and discharge valve to flare to optimize the design (Figure 8-8).

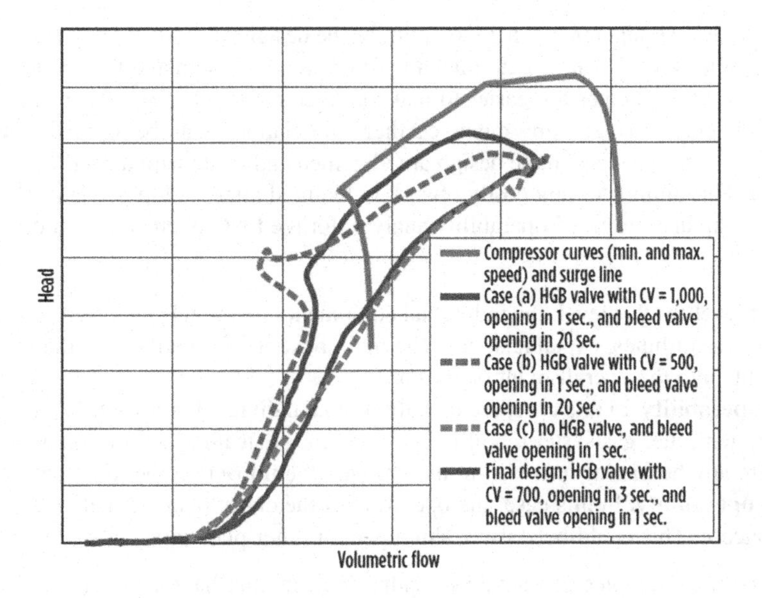

FIGURE 8-8

The various solutions to prevent surge during a compressor trip.

(Fantolini et al., 2012)

8.4.2.3 Other applications in compressor systems

There are other applications of dynamic simulation in the design and operation of compressor systems. One of them is for evaluating the effect of deviations in compressor curves between two parallel refrigeration compressors (Patel et al., 2007). This difference will have an impact on the load balancing between the two strings, especially for single-shaft machines with limited speed ranges. In this case, suction throttling valves may need to be used for load balancing. The impact of these deviations on the compressor operation can be tested prior to operation. In one case, it was found that a 5% deviation can cause one of the compressors to operate in recycle, thus reducing the overall efficiency.

Other applications of dynamic simulation in compressor systems include developing startup procedures (Section 8.4.4), developing compressor control strategies, and tuning the various control loops. The application of simulation in developing control strategy for an MR compressor was reported by Patel et al. (2007). In this case, the low pressure (LP) and medium pressure (MP) compressors, which are in the same string, experienced surge when the high pressure (HP) MR compressor in a different string was shut down. The normal surge detection and control was not sufficient in this case due to the delays in the valve stroke and actuator response. Feed-forward control actions that manipulate the antisurge valves for LP and MP compressors upon trip of the HP stage were developed to prevent surge in this application.

8.4.3 Operability and controllability analysis

Dynamic simulation is a valuable tool in evaluating the design and operability of the LNG plants in the early engineering stage. This helps to mitigate any risks related to operability during plant startup and operation. Further, the simulation can help to optimize the design by reducing any unnecessary margin. The operability of the entire plant is determined by both process design and the control system design. The changes in control schemes are easier to make at later stages. On the other hand, process design changes in later project stages are much costlier. Simulations can be carried out in the early design stages, where the changes in the design are identified and made with minimal adverse impact on the project cost. This eliminates any costly rework in the field after startup and commissioning.

The dynamic simulation-based operability analysis for the LNG plant can be of different scope, as follows.

- **Equipment level.** At this level, the individual equipment operation is analyzed. This could be the operation of gas turbines, expanders, etc. The main benefit is to verify how the individual equipment fits into the overall plant operation.
- **Unit level operability.** In this case, the operation of an individual unit is analyzed. This could be an amine system, fuel gas system, boil-off gas system, or steam system in the plant. The main concern here are the interactions within the unit and the impact on the unit controllability.
- **Plantwide operability.** In this case, the operation of the entire plant is analyzed to identify any areas of concern. This could be startup, shutdown, or other plantwide operational scenarios.

There are several examples of operability enhancements that have been accomplished for LNG plants using dynamic simulation. Some of these are listed here:

- One of the main concerns in LNG plant is smooth plant startup. The value of simulation analysis here is discussed separately in the next section.
- Another important issue related to LNG plant operability is the ability of parallel refrigeration compressor systems to operate reliably when one of the compressors trips. This has been studied for both APCI C3MR (Okasinski and Liu, 2010) and cascade refrigeration processes (Wilkes et al., 2007; Valappil et al., 2011). The main area of concern here is the possibility of overloading and tripping the parallel compressor train, when one of the compressor strings trips. Suitable control strategies to prevent this overload trip of parallel turbine were developed using dynamic simulation. In addition, trip of helper motors, wherever used, can be analyzed for any risk to the overall operation.
- For APCI C3MR plants, the failure of Main Cryogenic Heat Exchanger (MCHE) is also an important issue. The effect of this on the overall plant reliability is realistically analyzed using dynamic simulation.
- For systems where the electrical system performance can have an impact, the model can be updated to include the performance of the electrical system. In this case, the effect of electrical faults or trip of Load Commutated Inverter can be analyzed using a suitable dynamic simulation model (Ismail et al., 2004).
- The operability of fuel gas systems to ensure appropriate fuel gas header pressure and fuel properties during transients.
- Application of surge analysis for the LNG loading line (Gandhi and de la Vega, 1998). This involves analyzing the pressure surge in the loading line under various scenarios like valve closure. This can be used to identify and implement surge dampening devices in the system.
- The confirmation of operability of NGL recovery unit in the LNG plant as reported by Masuda et al. (2012). In this case, dynamic simulation was used to verify the operability and

controllability during startup and shutdown of the expander, turndown of the NGL recovery unit, failure of ASV valves in the residue gas compressor, and the trip of the residue gas compressor.

These are some examples of improvements in operability achieved using dynamic simulation. Other similar concerns related to LNG plant operability and controllability can be addressed using simulation.

8.4.4 LNG plant startup simulations

One of the key benefits of dynamic simulation is the ability to test startup of the LNG plant before field commissioning. Startup is inherently dynamic and the use of steady state process models is not appropriate for this exercise. This is even more important for the plants where design changes are introduced. This inherently brings in additional risks as to the ability for a smooth plant startup. Some of these risks can be addressed early in the design stage using dynamic simulation. The important aspects of the startup applications are detailed in the following.

8.4.4.1 Startup procedure development

One of the startup applications of dynamic simulation is the development of a startup procedure for the plant. This is very relevant in the case of LNG plants with design modifications, which may introduce issues with startup procedures. Identifying these issues early in the design can help to modify the design to handle issues. This also helps to automate certain startup steps in the DCS, which can help reduce the variability in the startup with different plant operators. Another very important benefit is having a well-defined startup procedure ready for the field startup. This procedure can further be refined in the field. The key to startup of the LNG plant is the liquefaction section, especially the refrigeration compressor systems. These can be verified and refined with dynamic simulation (Heckel et al., 1998; Simonetti et al., 2009).

An example of the use of dynamic simulation for startup enhancement is its use as a tool to tailor the cooldown procedure for the loading line and hence minimize the use of cold gas and LNG (Simonetti et al., 2009). Another example of enhancing startup using simulation is the solution to the issue of loading and balancing of two parallel trains. The development of a procedure to do this using dynamic simulation is reported for APCI C3MR plants (Okasinski et al., 2010) and cascade processes (Wilkes et al., 2007; Valappil et al., 2011). For the C3MR process, the suction drum for the machine in recycle needs to be precooled by an appropriate gas stream. The rate of cooldown is controlled by the antisurge valves. Once the temperatures are equilibrated, the block valves can be opened and the forward flow through the check valves can be established (Okasinski et al., 2010).

8.4.4.2 Verification of driver capability and startup conditions

The refrigeration compressors and the drivers are some of the key equipment in the LNG plant. These machines have to be started up without any operational issues. The drivers for these compressors could be gas turbines, steam turbines, or electric motors. It needs to be ensured that these drivers have enough power to bring the entire compressor string from a shutdown condition to normal operating speed. This could be an issue for single-shaft gas turbine and electric motors, which have limited torque available during startup at speeds lower than the normal operating speed. Dynamic simulation can be utilized to verify the driver torque adequacy during startup.

A related issue is the startup conditions that are used in the compressor loop. The initial pressure used for the startup has a big impact on the torque requirements during the ramp-up to normal speed. Hence, the selection of right starting pressure is critical, if it needs to be different from the settle-out pressure. These starting conditions can be fine-tuned and selected using simulation (Heckel et al., 1998). A related concern is the process conditions in the compressor loop during startup. The possibility of encountering vacuum conditions in the compressor suction can be verified for startup using low pressures. This information is valuable to ensure the right design pressures are used for the piping and equipment that can experience low pressure conditions. It may also be required to ensure that the antisurge valve openings are clamped during the startup to prevent excessive flow through the stages. The suitable valve openings can be developed using simulation analysis (Fantolini et. al., 2012).

The key input here is the driver speed-torque curve that defines the capability of the machine at various speeds. The torque available has to be sufficient to provide both the compression power and the acceleration power. A certain amount of margin also needs to be provided (Figure 8-9). The gas turbine has a predetermined speed acceleration rate that can be programmed into the model for simulation of startup. The model would then calculate the total torque based on the operating point on compressor curves and the rotational inertia of the string. Another important factor that should be included here is the possible seal gas leakage. This can add to the torque requirements during startup as mass is added to the compressor loop, resulting in higher torque demand.

8.4.5 Dynamic optimization of LNG plant design

The optimization of the plant design is normally carried out using steady state models, as these models are capable of evaluating multiple designs seamlessly. This is important for optimization with a large number of variables. There are benefits to be gained from using dynamic models for other optimization purposes during the design phase of the LNG projects. There are costs associated with equipment parameters like surge volumes, control valve sizes, and so on, which are provided to satisfy certain dynamic performance needs. These costs are determined by dynamic characteristics of the system. The ability to optimize these parameters using simulation and optimization tools is valuable for improving the profitability of LNG plant design.

One area of application of dynamic optimization are the offshore LNG plants, where the weight or size of equipment can be a constraint. Dynamic optimization is a valuable tool for reducing the weight by optimizing the size or selecting a different material (Feng et al., 2009). Another example is optimizing the size of propane condenser in the APCI C3MR process using simulation (Gandhi et al., 1998).

Dynamic optimization is best performed by utilizing an optimizer linked to the dynamic simulation model. The economic parameters for the optimization have to be input along with any other constraints. These could be limits on the process variables or constraints on optimization parameters. The optimization tool is utilized to determine the best design, maximizing an objective function (alternatively, minimizing cost). For this, repeated simulation is needed and the performance of each run collected and analyzed. The adjustable parameters like antisurge valve sizes can be varied to determine the optimum point and determine the exact benefits from reducing the capital cost of the plant via optimization.

Another important area of application is determining the maximum plant capacity and the best operating variables to optimize the plant production (Ismail et al., 2004). This is closely related to

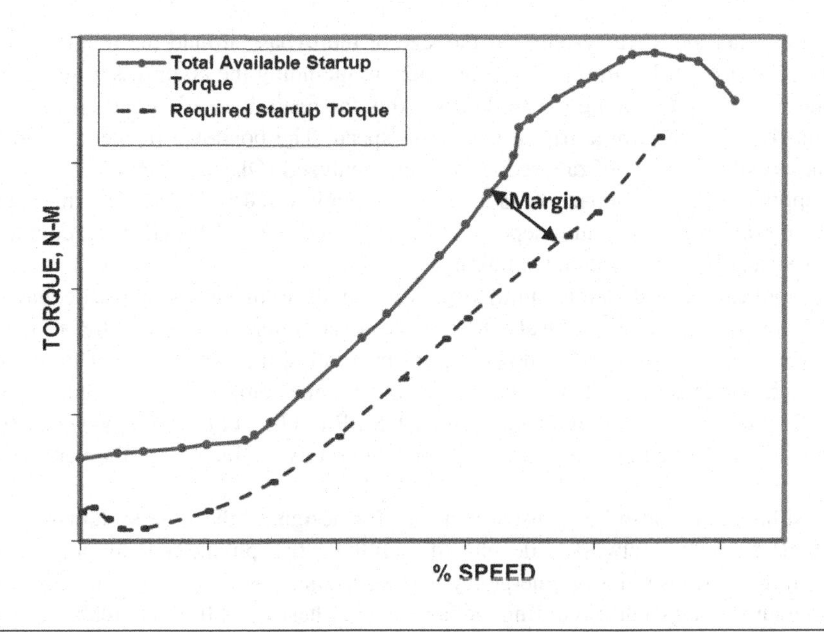

FIGURE 8-9

The comparison of available and the required torque during startup.

debottleneck and advanced control applications. The example reported by Ismail et al. (2004) optimizes the LNG production by varying key operating parameters like Inlet Guide Vane (IGV) angle of a LP MR compressor, Inlet pressure of a LP MR compressor, Outlet pressure of a MR expander and the MR composition. The optimization is done with the constraints on refrigeration driver power, outlet pressure of a heavy MR expander, IGV angle and the temperature approaches in MCHE. For the cascade refrigeration process, similar optimization can be accomplished by varying refrigeration compressor suction pressures, condensing pressure, and other load balancing (between the three refrigeration systems) handles in the process.

8.4.6 Relief system design and safety scenarios

A properly designed relief system is very important for the safe and reliable operation of the LNG plant (refer to Appendix 3). Hence, the proper sizing of the relief valves is a key part of the engineering design of the plant. The majority of the relief valves are adequately sized using the traditional calculation methods. There are certain scenarios like a compressor blocked outlet or general power failure that are better analyzed through dynamic simulation. In these cases, various competing and time-dependent events can become active, which can impact the relief flow rates. Also, dynamic simulation accounts for the actual volume and inventory of the refrigeration systems in the LNG process. This helps to provide more realistic relief rates during various transient scenarios.

One of the LNG process relief cases where dynamic simulation has found applicability is the compressor blocked outlet case. The use of dynamic simulation in analyzing the inadvertent closure of the compressor discharge block valve and the propane receiver outlet block valve for APCI C3MR process has been reported (Gandhi et al., 1998; Okasinski et al., 2007). This was utilized to verify the adequacy of the relief and flare system and the size of the bypass around the relief valve. During a blocked outlet scenario, different scenarios can happen, including the compressor going into recycle and the chiller pressures continuing to rise. Further, this can result in overloading of the turbine driver and bogdown, resulting in turbine trip at minimum speed. The bogdown of the gas turbine and its impact on the possibility of relief rate reduction were analyzed (Okasinski et al., 2007).

Another application is the scenario of General Power Failure (GPF) and the determination of relief flow rates during this event. Several independent events are triggered by a GPF, the cumulative effect of which is well analyzed by dynamic simulation.

A related application of dynamic simulation is the analysis of flare system adequacy to handle various relief scenarios. Simulation can also look at the actual timing of various relief scenarios and the ways to mitigate the maximum relief flows, hence preventing the overdesign of flare systems. An example of this is the relief due to loss of power for the propane compressor air-cooled exchanger fans in the APCI C3MR LNG process (Gandhi et al., 1998). Providing emergency power to some of the air-cooler fans was identified as a more cost effective alternative to increasing the flare header size in this case.

Dynamic simulation is also being used as a tool for designing the flare system itself. Here, the dynamic model of the flare network is developed with the actual piping and equipment data. This is used along with the various relief or emergency depressurizing scenarios to identify the conditions in the flare header and knockout drums during the flare event. These conditions are realistic estimates and can be used to optimize the design of the flare system (Urban et al., 2010).

Another area of application of dynamic simulation is the analysis of safety scenarios and HAZOP scenarios. These are scenarios where the plant could potentially reach an unsafe operating condition. The possibility of attaining such conditions can be evaluated to ensure that the proper preventive measures are built into the design.

8.4.7 Plant troubleshooting and operational support

A very key benefit of developing a high fidelity LNG plant dynamic model is its use in assisting plant operations. The model can be validated with plant data at frequent intervals to make it an adequately accurate representation of the LNG plant and used for "what-if" or other operational studies.

The main application of a validated plantwide dynamic model is plant troubleshooting, where the model is used to identify the root cause for certain plant events. These troubleshooting studies involve first replicating the plant scenario in the validated model. Then, the dynamic model is used to identify potential options for mitigation. A number of operational problems can be investigated and mitigated with this approach.

Another very important value of having an LNG dynamic model is providing operational support. This involves providing the right operating setpoints under various plant conditions like ambient temperature, feed compositions, and so on. The dynamic model that is representative of the plant can also be used to identify better operating strategies for the LNG plant.

8.4.8 Plant production enhancements and debottlenecking

There is a significant emphasis on utilizing the existing natural gas assets to their maximum potential. The first step in achieving this is operating the existing LNG plants at their maximum capacity. Beyond that, there are economic incentives in debottlenecking the existing LNG plants to increase their capacity or availability. There are various options available to perform the debottlenecking of existing plants. The evaluation and selection of the best options to increase the plant capacity is important for optimum return on the investment. An overall plantwide dynamic simulation model is a useful tool to perform the evaluation and optimization of the various debottlenecking options.

The main objective is to identify and remove the capacity limiting constraints in the plant. Once the existing constraints are identified, various options are considered to debottleneck the LNG plant and are studied with the simulation models. There are various options to perform the debottlenecking of existing plants. These include:

- Upgrades or modifications to existing equipment
- Replacement of existing equipment with ones that provide higher capacity
- Changes in process configuration or conversion to alternate technologies.

The dynamic model can be used as a tool to screen the various debottleneck options and to identify the most economical one before the detailed engineering design (Omori et al., 2001; Valappil et al., 2009). The dynamic model is an actual replica of the existing plant, and can be used for rapid screening of multiple debottleneck options. This is especially valuable when evaluating various combinations of options.

The simulation model needs to be first validated and baselined with the operating data from the LNG plant. This is essential to establish a valid baseline for further debottleneck simulations. The simulation models may be built during the engineering and construction phase of the plant to evaluate and predict the plant performance prior to the initial operation. This initial model needs to be updated to incorporate actual tested equipment performance data and baselined to reflect the actual plant behavior. The model parameters are tuned with the actual plant data that is reflective of the normal plant operation over time. The main focus of the baselining exercise was to validate the performance of the refrigeration systems (especially the compressors and gas turbines) and other constraints against the plant operation.

It is important to identify the existing plant constraints before performing any debottleneck evaluations. These constraints vary with the operating conditions like ambient temperature, feed composition, and so on. Some of the main constraints that limit production in LNG plants are related to the refrigeration compressor drivers. Also, various other equipment limitations and hydraulic constraints may be present in the plant. In addition to steady state models, a plantwide dynamic model provides a suitable tool to identify these existing plant constraints over the range of operating conditions. The percentage of time annually that each constraint is active can be determined from simulation to identify the debottleneck options to be selected (Figure 8-10).

Once the existing constraints are identified, various options to debottleneck the LNG plant can be studied. Of particular interest are plant changes that can be made with minimum downtime or within the downtime required for other purposes. For example, gas turbine drivers can be upgraded to those with higher power ratings.

The simulation models can be run with the actual field conditions to identify the plant performance with these debottleneck modifications individually or in selected combinations. The incremental LNG production is estimated at various times of the year to obtain an annualized production increase (Figure 8-11). This is important to optimize the selection of LNG plant debottleneck modifications that can be used in estimating the return on capital investment.

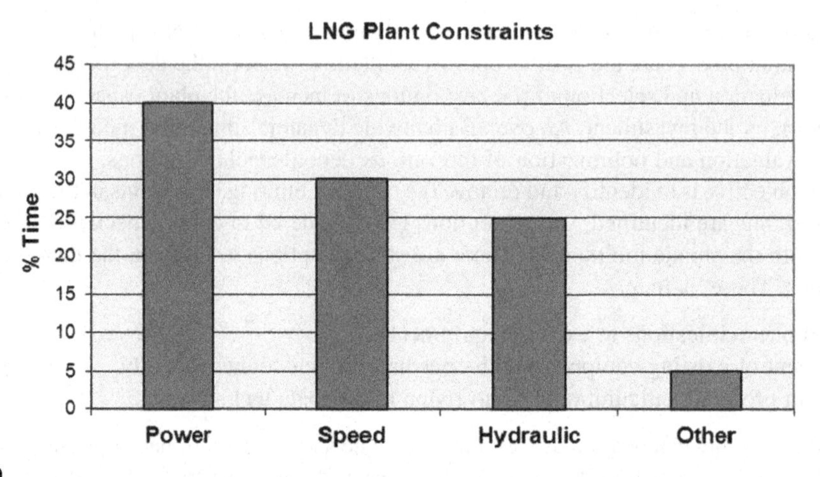

FIGURE 8-10

Typical debottleneck simulation constraint results.

8.4.9 Control system checkout

The dynamic simulation model can be used for a variety of control system related applications. One of the potential applications is the determination of the right control structure in the early stages of the project. Systematic ways of analyzing the process controllability using the dynamic model are available at present. This helps to establish the economic trade-off between steady state design and

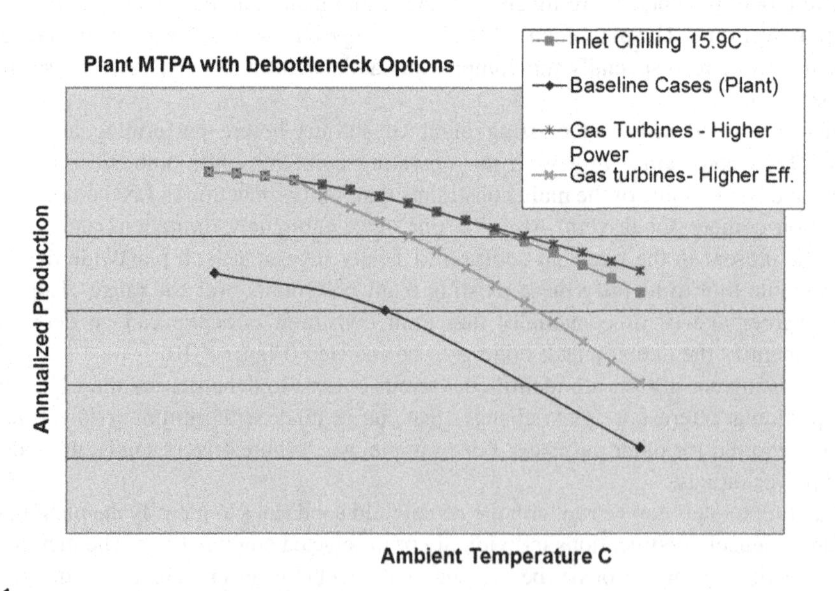

FIGURE 8-11

Typical debottleneck simulation results showing the achievable production against the existing plant baseline.

process controllability. These studies would have to be done in the early stages of the project, preferably in the FEED stage. Since these decisions are more qualitative in nature, the lack of detailed equipment information should not be a serious limiting factor in these applications.

The other important use of dynamic simulation is the selection of appropriate tuning parameters for the plant. Usually, the tuning parameters from a similar, previously constructed plant are used or the control loops are tuned on-site during the commissioning phase of the project. This prolongs the startup-commissioning time, thus making it economically very inefficient. The availability of a dynamic simulation model makes it possible to tune the control loops off-line in advance using established tuning procedures. These can then be fine-tuned quickly on-site during the commissioning. These studies can be performed during the EPC stage or precommissioning.

The other major application of dynamic simulation in control system studies is in the checkout of the plant control system prior to commissioning. In traditional practice, simple tie-back models are used for this purpose. This helps to confirm the input/output (I/O) integrity, but does not assist in other studies and does not include any process knowledge. A high fidelity dynamic simulation model provides an alternative to check the control system connections, the control logic, and the tuning parameters before the commissioning phase. This also helps to provide the framework for a future operator training simulator development. The LNG plant commissioning time saved by better checkout and operator training adds economic value to the project.

The process control system landscape has changed with the introduction of the digital fieldbus, as explained in Chapter 6. This has added further capability in performing model-driven control system checkout in a transparent manner. High fidelity dynamic simulation models can be linked to the emulation of the control system to perform the checkout in a single PC. Linking between the dynamic simulation models and control system emulations can be effectively performed using standards like OPC and CORBA (Figure 8-5). This avoids the need to purchase the control system hardware to do the checkout and operator training. Also, the hardware independence of the emulation tools makes it ideal for a system-independent approach (Mansy et al., 2002).

8.4.10 Operator training simulator

A high fidelity plantwide LNG dynamic model can be integrated with the control system emulations for use as an operator training simulator (OTS). An OTS based on such a model is a true replica of the plant, reflecting both the process and control elements. Control system emulators are useful to replicate the exact plant control strategy (most commercial DCS systems have an emulator that can be used for this purpose). The shutdown/interlock logic (SIS) of the plant is also implemented in the OTS. This logic is either implemented in separate script files or by using commercially available programmable logic control emulators. The OTS also utilizes the actual operator displays to capture the exact control room environment. The field operator station is replicated to enable the field operated valves, hand switches, and other functionalities. The instructor station is used to start and manage the training scenarios and perform operator evaluations as shown in Figure 8-12.

The benefits of using an OTS in LNG plants are several, including:

- Many of the LNG plants are located in remote areas, where experienced operators are not readily available. Hence, training the operator on an OTS is valuable in improving the plant availability and reliability. This training enables the operators to learn the functionality of the plant automation system and operating procedures in an offline environment.

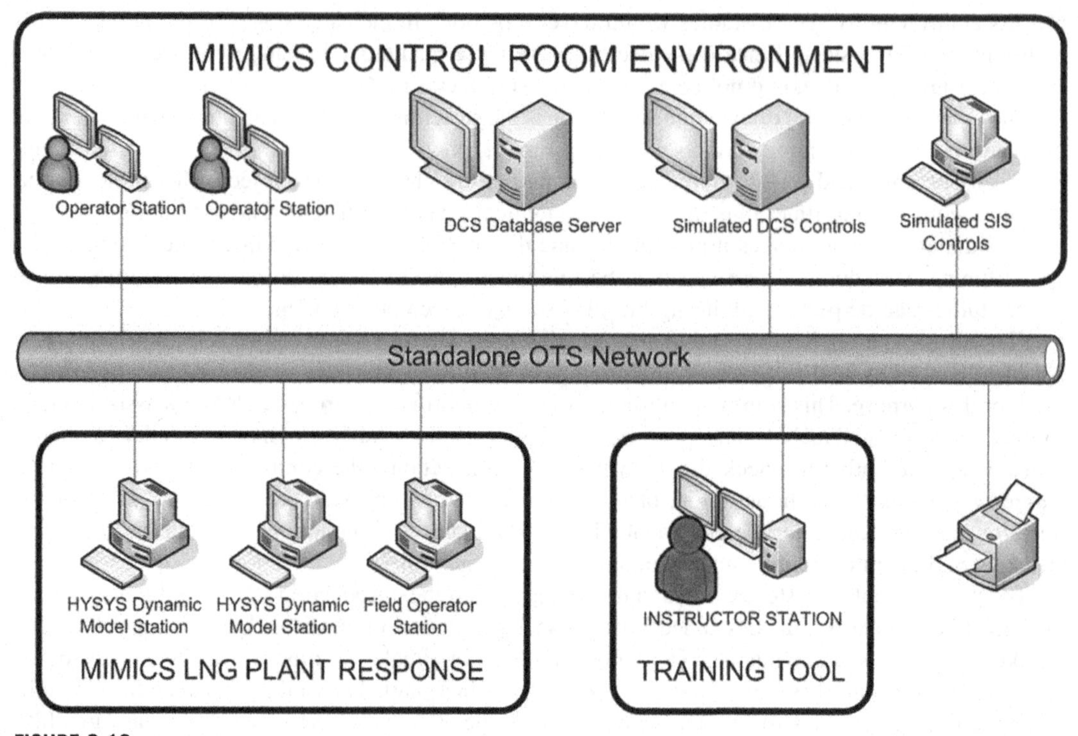

FIGURE 8-12

Components of LNG plant Operator Training System (OTS).

(Tekumalla et al., 2008)

- The OTS allows for complete control system checkout prior to plant commissioning. This enables engineers to resolve any issues regarding graphics or control logic coding early in the project.
- Operating personnel can also use the OTS for training purposes to ensure a smooth plant startup and commissioning. Feedback from operating personnel on the OTS also helps to provide the framework for the future OTS development.
- The OTS can be used for operational support once the LNG plant is up and running. This includes design changes, debottleneck analysis, troubleshooting, etc.

The training scenarios performed in an OTS include startups, normal operations, and abnormal situations. This abnormal situation training is very valuable, as these events occur rarely and the operators are not exposed to this in the plant. The selection of these malfunctions to train the LNG plant operator is the key to realizing maximum benefits from training. Some of the typical malfunctions that are simulated in an LNG plant OTS are:

- Ambient temperature changes
- Failure of air cooler
- Failure of the amine pump
- Foaming in the amine column

- Trip of refrigeration compressors
- Trip of Main Cryogenic Heat Exchanger
- LNG transfer pump failure
- Trip of boil-off gas compressor.

The relative performance of the operator in handling these malfunctions can be tracked using a Trainee Performance Monitoring (TPM) tool. This can be set up by the instructor using certain critical parameters and their limits that should be honored. This way, the operator can be trained to effectively handle these abnormal situations in addition to the normal operation of the plant (Dawson et al., 2006).

To realize the benefits of OTS fully, the training simulator should be developed during the engineering and construction stage of the project. The OTS ideally should be in the field anywhere from 6 months to a year before the startup. This way, the operators can be adequately trained and prepared for the plant startup and operation.

A novel approach to development of an OTS system for a Yemen LNG plant was described by Rabeau et al. (2010). A dynamic simulation model was used extensively in this case for the engineering design of the LNG plant. The same model was then used for development of OTS. Different phases of OTS development were used with a different OTS in this case, as shown:

- **Dynamic Engineering Simulator.** Simulation model used for engineering studies that does not include any DCS emulation or consoles.
- **Early Operator Training Simulator.** This is a simplified emulated OTS system with Human Machine Interface (HMI) that is used for process courses.
- **Extended Early Operator Training Simulator (EEOTS).** This is a more detailed OTS with emulated controls and DCS HMI.
- **Operator Training Simulator.** Full OTS with DCS and ESD stimulation and actual DCS consoles.

The Emulated OTS was used in this case for several process studies. The EEOTS was used for developing startup procedures. Since this was available several months before the actual plant startup, EEOTS was also used to train operators extensively before startup.

8.4.11 Advanced process control development

APC/optimization has established itself as a widely used technology in the process and other industries, including LNG plants. The implementation of APC/optimization is currently done after the plant has been operational for some time. Even in the grassroot plants, this is the normally accepted practice. There are some drawbacks to this implementation philosophy though. Recently, dynamic simulation is being used for developing APC applications before field implementation in LNG plants. This helps to minimize the impact on the plant operation during APC implementation (Alsop et al., 2006; Valappil et al., 2006; Sturm et al., 2009).

The dynamic models developed for various LNG project stages can be used for various APC applications (Figure 8-13). In the front-end stage, it can be utilized for design studies. The EPC stage dynamic model can be used to validate the regulatory control systems and to prototype the advanced controller. The dynamic model that is validated with plant data is a valuable asset in optimization and advanced control studies. This model can be used for estimating the potential benefits of advanced control, performing plant tests to obtain preliminary models, and testing the APC/optimization under various operational scenarios.

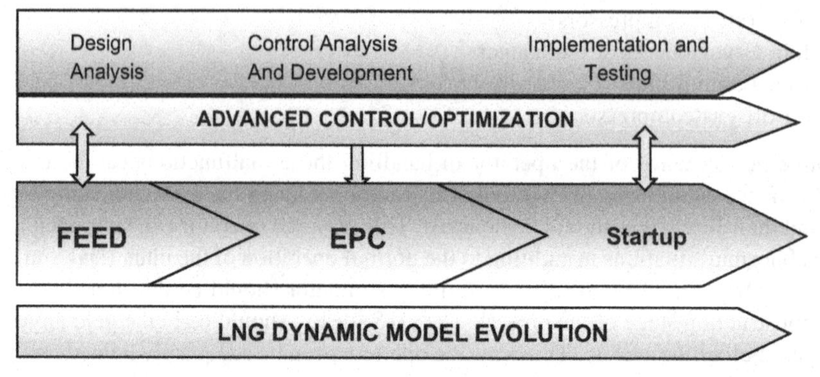

FIGURE 8-13

Use of dynamic simulation models in testing and implementing LNG plant advanced control.

- **Analyzing LNG process design for APC.** The analysis of process design in the early plant life cycle stages can verify the operability of the plant at optimal conditions after plant startup. This involves ensuring that the right operating constraints are considered and that no auxiliary variable becomes a limiting factor in optimization. Dynamic simulation can be used in the early project stage to identify the plant operating envelope after the plant startup. From a design point of view, it is important to consider the undesirable constraints that can become active during the operation. These can be identified during the operating envelope runs. This way, it can be ensured that the advanced control can operate the plant at proper constraints during the future plant operation.
- **Designing regulatory control systems.** The regulatory control layer of a typical plant is designed and finalized during the detailed engineering and construction phase. The instrumentation and control system design that is done during this stage is intended for normal plant operation. The implementation of advanced control requires additional instrumentation or changes in the regulatory control layer. Identifying these changes or additional control features in the engineering phase will prepare the plant for future advanced control implementation.
- **Estimating benefits from LNG plant APC.** A dynamic model that is validated with plant data can be used for estimating the potential benefits of advanced control, performing plant tests to obtain preliminary models, and testing the optimization under various operational scenarios. It is important to accurately estimate the benefits from optimization to justify its implementation. In most cases, purely plant data driven methods have been used to estimate the APC benefits. A different way to estimate the benefits is to actually simulate the operation of the plant with optimization and then compare it with a baseline operation.
- **Evaluating plant performance with APC.** The performance of the plant with optimization under various conditions can be studied with dynamic simulation. There are several advantages to doing this. One advantage is the fact that the constraints that the plant will operate normally are easily identified. Also, any changes in constraints with changing ambient or other variables can be detected using dynamic simulation. The plant operators can be trained to manage these changes.

8.5 Dynamic simulation of LNG import terminals

Dynamic simulation has found a range of applications in LNG import terminals, including safety studies, operational optimization, and rigorous hydraulic analyses. A brief description of the modeling of LNG import terminals and the various applications are given in the following sections.

8.5.1 Dynamic modeling of LNG import terminals

The methodology for the development of LNG import terminal dynamic models is similar to the LNG plants, as many of the unit operations are similar in nature. The main components of the import terminal dynamic model are the LNG storage tanks, BOG compressors, and LNG vaporizers.

The modeling methodology used for an LNG storage tank depends on the scope of the model and application that the model is used for. For process studies, where the interest is in the BOG generation and the tank level, the tanks can be modeled with the standard unit operation models. These consider the tank volume as a simple holdup at the same conditions in the entire tank. Assuming perfect equilibrium in the tank is not very accurate, and reasonable tank efficiency has to be specified. The heat leak into the tank can be calculated externally and specified or calculated based on the heat transfer parameters in the tank. The accurate heat loss calculation is important, as it determines the generation of boil-off gas. In addition, the heat gained from pumps would also contribute to the vapor generation. For more detailed calculations of the tank including scenarios like roll-over behavior, more rigorous models are needed. This is especially valuable to study the behavior of mixing of two different density fluids like roll-over (Maeda and Shirakawa, 2007). The tank model in this case needs to take into account the spatial distribution of fluid properties. This results in partial differential equations, with an accompanying turbulence model that can be solved using Computational Fluid Dynamics (CFD) tools.

The other important component of the LNG import terminal model are the vaporizers used to convert LNG to vapor. These vaporizers need to be modeled with adequate fidelity to handle the scenarios studied. Simple heat exchanger models, with parameters chosen to reflect the actual vaporizer, would be sufficient to study the operability of the system. For further exact modeling of the vaporizers, more rigorous heat transfer correlations can be used. This is especially valuable for modeling the ambient air vaporizers, used in some of the LNG vaporization processes.

8.5.2 Applications of dynamic simulation in LNG import terminals

8.5.2.1 Design and testing of safety and operational scenarios in LNG Terminals

One of the valuable applications of dynamic simulation is in performing safety analysis, particularly designing and testing the Emergency Shutdown systems for the LNG terminal (Contreras and Ferrer, 2005). The design of an ESD system for the LNG terminal is critical to ensure safety. The various safety hazards can be simulated to ensure that the ESD system design is adequate to deal with these situations. If needed, the system can be redesigned or fitted with ancillary equipment. Plant personnel can define more effective safety action plans and shutdown strategies.

In addition, other operational applications were reported by Contreras and Ferrer (2005). These include the following:

- **Startup and shutdown of pumps.** This involves studying the transients involved in shutdown and startup of various pumps in the regasification terminal.

- **Vaporizer shutdown scenario.** The shutdown of LNG vaporizers is a severe upset in the LNG import terminal operations. The effect of this upset and the ability of the system to accommodate this upset can be studied.
- **Failure in the BOG compressor.** The failure of one of the BOG compressors during ship unloading and the resulting pressure transients are studied here.

8.5.2.2 Determining optimal operational strategy and LNG terminal responses

The optimization of LNG terminals using a dynamic model was discussed by Lee et al. (2012), whereby various operational strategies were developed to minimize the total operating cost. This indirectly amounts to reducing the generation of boil-off gas in the system. The main emphasis is to evaluate the effect of bypass flow rate and recirculation flow rate on the various terminal characteristics to enhance the operating efficiency.

The other application of simulation is in optimizing the operation of boil-off gas compressors in the LNG regasification plant. This includes determining the operating points for the various compressors to minimize the overall energy consumption. An application of this using steady state plant models was discussed in Shin et al. (2007). This can be extended to use a dynamic model instead of steady state model.

The use of dynamic simulation in studying various operational scenarios and validating control systems for the Freeport LNG terminal was discussed by Nanda et al. (2010). The main concern here was the transition from summer to winter mode of operation and its effect on the heating tower. This transition involved a change of about 30°F for the most conservative case. A dynamic model for the entire facility was built and used to verify the plant operation during the transition. The results of the analysis were used to determine setpoints for the various control loops and to pretune the loops before commissioning. The model was also used to optimize the energy consumption by changing the setpoint of the main controllers.

8.5.2.3 Transient flow analysis in LNG flowlines

The safe and reliable operation of the LNG regasification plant is a key objective of the overall facility design. There will be a need to isolate the inventory in the system during emergency scenarios. Emergency shutdown valves are provided at various points in the process to accomplish this objective. The closing times of these valves need to be selected to ensure quick protection without adverse hydraulic transients like water hammer. The water hammer phenomena can be caused by startup or shutdown of pumps or by sudden closure of valves (Wink et al., 2007).

The first area of application of this analysis is the carrier unloading system, which includes the pipeline from the jetty to the LNG terminal. The closure of valves in this area can cause pressure peaks, the magnitude of which can be studied. The second area is the LNG flow from the tanks to the vaporizers, from the submerged pumps. The effect of valve closure in this area can also be evaluated. The other scenarios that can be examined include the high pressure sendout pump valve closure and LNG storage system valve closures.

8.5.2.4 Evaluating tank filling operations

The other application of simulation is to ensure that there are no rollover phenomena during tank filling operations (Maeda and Shirakawa, 2007). This can happen when LNG is added to a storage tank with LNG of different density, resulting in stratification and a sudden release of a large amount of boil-off gas. A model used to study this needs to have spatial details, which results in partial differential equations (continuity, momentum, and energy balance equations for the tank). Commercially available CFD codes can be used to solve this problem.

8.5.3 Case study: LNG import terminal dynamic simulation

A dynamic model was developed to study the transient performance of LNG terminals with Ambient Air Vaporizers (AAV). This model (shown in Figure 8-14) was used in calculating the physical properties and relieving performance of LNG at blocked in conditions. The model was also used to calculate the mass flow through the dump valve in order to prevent the system from overpressuring.

The specific objectives of the dynamic modeling exercise were as follows.

- Calculate the physical properties of relief valve discharge for abnormal heat input case.
- Calculate the physical properties of relief valve discharge for fire case.
- Estimate the withdrawal rate through the dump valve so that the system does not reach 95% of the design set pressure.
- Estimate the settle out pressure and mass of LNG remaining in the system when the system exit isolation valve is closed after 60 seconds of the inlet valve closing.

A dynamic model for the system was constructed with AAVs modeled using heat exchanger unit operation. The pumps and the isolation valves were also included in the dynamic simulation model (Figure 8-14).

8.5.3.1 Simulation runs

This section presents details of different cases studied using the dynamic model.

Blocked in condition with abnormal heat input

In this case, the inlet isolation valve (XV) and exit XV of the system are assumed to close in 3 seconds. During this period the flow from the pump decays to zero. As a result of the closing of the XVs, the system becomes isolated and heat transfer occurs from the ambient air to the AAVs. This transfer of

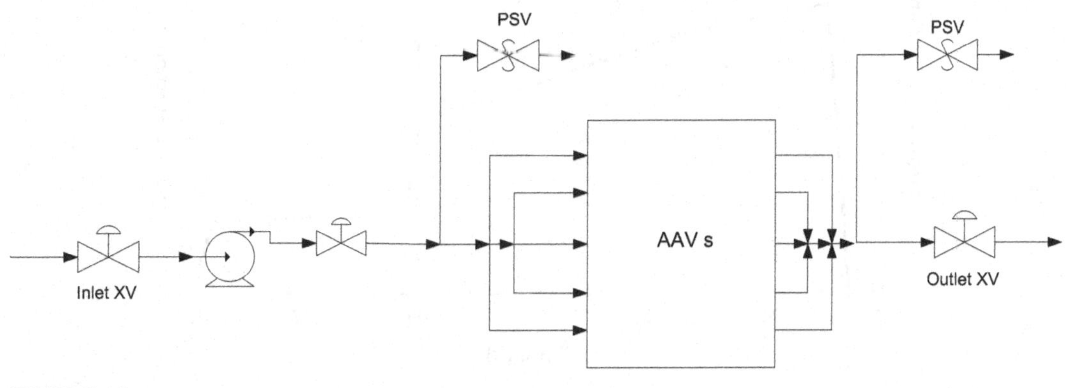

FIGURE 8-14

The schematic of the equipment in the dynamic model for LNG import terminal transient analysis.

energy leads to an increase in temperature of the fluid trapped inside the system and hence results in rapid pressure rise. When the set pressure is reached the relief valve opens and releases gas to the atmosphere. The pressure response at the inlet of the relief valve was determined from this simulation.

Blocked-in condition with fire

This case is similar to the previous one. In this case, a plant fire near one of the AAVs leads to an increased heat input to the system. For the sake of simplicity this radiant heat input is assumed to be constant and limited to the bottom quarter of the length of the tubes that faces the fire. The extra heat input causes the pressure of the system to increase more rapidly. The maximum flow rate of gas when the relief valve opens is more than that for Case 1.

Delayed dumping of LNG before exit XV closing

In this simulation, the possibility of preventing overpressuring of the system by removing fluid through a dump valve is investigated. The dump opens as soon as the inlet XV is closed and remains open until the closing of the exit XV in 60 seconds. The valve Cv is varied such that the final pressure of the system after closing the exit XV does not exceed 95% of the design pressure. The maximum flow rate through the dump valve is determined form this simulation. Plots of temperature, pressure, and fluid vapor fraction across the dump valve is given in Figure 8-15.

Delayed closing of exit XV

In this scenario the dump valve does not open. It is assumed that the inlet LNG valve shuts off in an ESD event, which stops the LNG flow. The AAV system outlet valve stays open for 60 seconds from the start of the event before closing. During this time the LNG in AAV gets heated. This causes the fluid to expand and flow out of the system as the downstream pressure is constant. After the exit is blocked the system pressure rises. The mass of LNG left in the system after isolation is calculated.

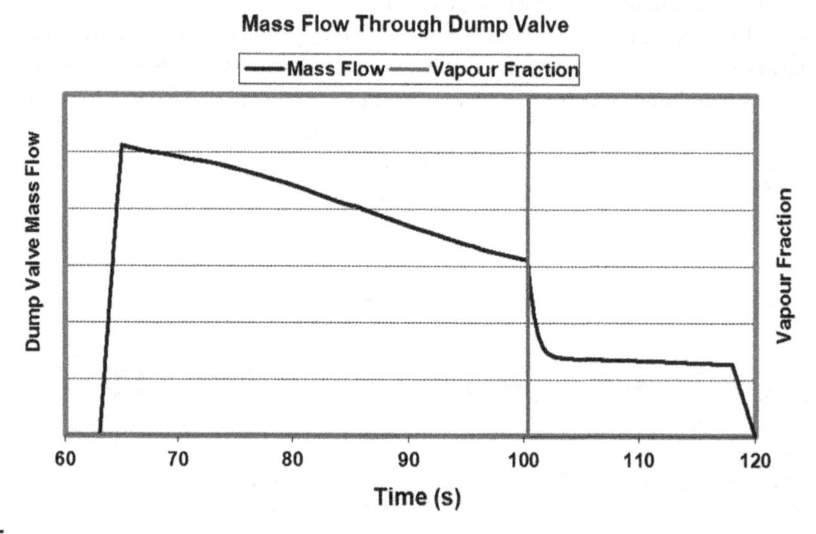

FIGURE 8-15

The mass flow and vapor fraction for the case where the dump valve is used.

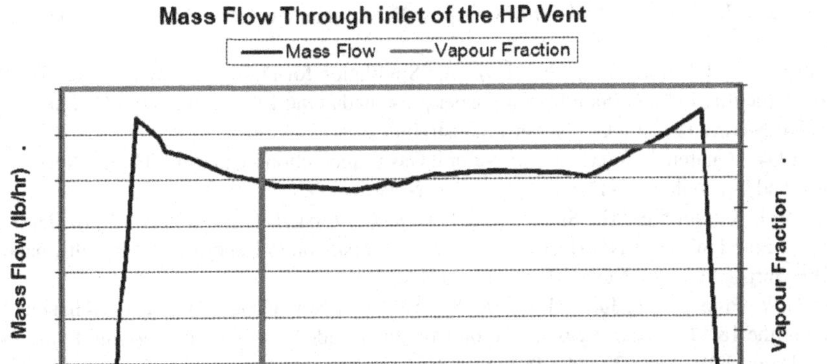

FIGURE 8-16

The mass flow and vapor fraction for the case where the HP vent is used.

Opening of HP vent for depressurizing

This scenario is similar to the previous case. An HP vent is located at the exit to the AAV arrays. The exit XV closes after 1 minute of the closing of the inlet XV. After this, the pressure in the system rises as the fluid inside gets heated. When the pressure reaches a set pressure, the HP vent discharges gas to the BOG header, which is at a very low pressure. The HP vent closes as soon as the system pressure goes below a predetermined pressure. In this simulation the maximum flow rate through the HP vent was determined. Plots of temperature, pressure, and fluid vapor fraction across the dump valve are given in Figure 8-16.

References

Alsop, N., Ferrer, J., April 23–27, 2006. Step-test Free APC Implementation Using Dynamic Simulation, paper presented at the AIChE Spring National Meeting, Orlando, FL, USA.

Botros, K.K., Ganesan, S.T., Sept. 8–11, 2008. Dynamic Instabilities in Industrial Compression Systems with Centrifugal Compressors, paper presented at the 37th Turbomachinery Symposium, Houston, TX, USA.

Boyce, M.P., 2003. Centrifugal Compressors: A Basic Guide, first ed. PennWell Publishing Company, Tulsa, OK, USA.

Broussard, M., May 2002. Maximizing Simulation's Reach. Chemical Engineering 50–55.

Contreras, J., Ferrer, J.M., 2005. Dynamic Simulation: A Case Study. Hydrocarbon Engineering 10 (5), 103–107.

Dawson, J.M., Pekediz, A., Womack, J.W., 2006. RasGas Makes Extensive Use of Process Operator Training Simulators in LNG Operations, paper presented at the 6th Topical Conference on Natural Gas Utilization, AIChE Spring Meeting, Orlando, FL, USA.

Fantolini, A.M., Pedone, L., D'Orazi, L., Prodan, R., Bhattad, D., Stvarkas, D., Harismiadis, V., 2012. Use Dynamic Simulation for Advanced LNG Plant Design. Hydrocarbon Processing 91 (8), 81–86.

Feng, J., Aggarwal, A., Dasgupta, S., Shariat, H., 2009. Using Dynamic Analysis to Reduce Weight of Offshore Installations. Offshore Magazine 69 (11), 72–74.

Fulton, S., Collie, J., May/Jun 1997. Confirm Complex Heat Exchanger Performance. Hydrocarbon Engineering 75–82.

Gandhi, S., de la Vega, F.F., May 4–7, 1998. Dynamic Simulation for Design Improvements, Cost Reduction and Operational Stability in LNG Plant Design, paper presented at the 12th International Conference & Exhibition on Liquefied Natural Gas (LNG 12), Perth, Australia.

Hammer, M., 2004. Dynamic Simulation of A Natural Gas Liquefaction Plant. PhD Thesis. Norwegian University of Science and Technology (NTNU), Trondheim, Norway.

Hammer, M., Vist, S., Nordhus, H., Sperle, I.L., Owren, G., Jorstad, O., March 31–April 03, 2003. Dynamic Modeling of Spiral Wound LNG Heat Exchangers - Comparison to Experimental Results, paper presented at the AIChE Spring Meeting, New Orleans, LA, USA.

Haarlemmer, G.W., Pigourier, J., June 01–04, 2008. Towards A New Generation Heat Exchanger Models, paper presented at the 18th European Symposium on Computer Aided Process Engineering, Lyon, France.

Heckel, B.G., Davis, F.W., Sept. 22–24, 1998. Starter Motor Sizing for Large Gas Turbine (Single Shaft) Driven LNG Strings, paper presented at the 27th Turbomachinery Symposium, College Station, TX, USA.

Ismail, E.N.H., Arrifin, M.S., Omori, H., Konishi, H., Ray, S.A., Durr, C.A., March 21–24, 2004. Application of Dynamic Simulation for Design and Commissioning of LNG Plants, paper presented at the 14th International Conference & Exhibition on Liquefied Natural Gas (LNG 14), Doha, Qatar.

Lee, C., Lim, Y., Han, C., 2012. Operational Strategy to Minimize Operating Costs in Liquefied Natural Gas Receiving Terminals Using Dynamic Simulation. Koran Journal of Chemical Engineering 29 (4), 444–451.

Maeda, M., Shirakawa, Y., April 24–27, 2007. Development of Fluid Dynamics Simulation for LNG Storage Tanks to Enhance Safety, Operational Flexibility, and Cost Performance, paper presented at the 15th International Conference & Exhibition on Liquefied Natural Gas (LNG 15), Barcelona, Spain.

Mansy, M.M., McMillan, G.K., Sowell, M.S., 2002. Step Into the Virtual Plant. Chemical Engineering Progress 98 (2), 56–61.

Masuda, K., Nakamura, M., Momose, T., Oct. 8–11, 2012. The Use of Advanced Dynamic Simulation Technology in the Engineering of Natural Gas Liquefaction Plants, paper presented at the Gastech 2012 Conference & Exhibition, London, UK.

Nanda, R., Lindahl, P., Eyermann, D., July 2010. Liquefied Natural Gas Terminal with Low Environmental Impact. Hydrocarbon Processing 89 (7), 51–58.

Okasinski, M., Schenk, M., April 24–27, 2007. Dynamics of Baseload Liquefied Natural Gas Plants – Advanced Modeling and Control Strategies, paper presented at the 15th International Conference & Exhibition on Liquefied Natural Gas (LNG 15), Barcelona, Spain.

Okasinski, M., Liu, Y., April 18–21, 2010. Dynamic Simulation of C3-MR LNG Plants with Parallel Compression Strings, paper presented at the 16th International Conference & Exhibition on Liquefied Natural Gas (LNG 16), Oran, Algeria.

Omori, H., Ray, S.A., de la Vega, F.F., Durr, C.A., May 14–17, 2001. A New Tool-Efficient and Accurate for LNG Plant Design and Debottlenecking, paper presented at the 13th International Conference & Exhibition on Liquefied Natural Gas (LNG 13), Seoul, Korea.

Patel, V., Feng, J., Dasgupta, S., Ramdoss, P., Wu, J., Sept. 10–13, 2007. Application of Dynamic Simulation in the Design, Operation, and Troubleshooting of Compressor Systems, paper presented at the 36th Turbomachinery Symposium, Houston, TX, USA.

Rabeau, P., Rieubon, R., Coupier, R., Thiabaud, P., Rey, F., April 18–21, 2010. Innovative Use of Emulated Operator Training Simulator in the Yemen LNG BAL HAF Plant, paper presented at the 16th International Conference & Exhibition on Liquefied Natural Gas (LNG 16), Oran, Algeria.

Ramesh, K., Aziz, N., Abd Shokur, S.M., Ramaswamy, M., 2007. Dynamic Rate-Based and Equilibrium Model Approaches for Continuous Tray Distillation Column. Journal of Applied Sciences Research 3 (12), 2030–2041.

Rowen, W.I., Oct.1983. Simplified Mathematical Representations of Heavy Duty Gas Turbines. Journal of Engineering for Power 105, 865–868.

Sarah, A., April 26–30, 2009. Performance Testing and Prediction of Cryogenic Liquefied Gas Reaction Turbines, paper presented at the 9th Topical Conference on Natural Gas Utilization, AIChE Spring Meeting, Tampa, FL, USA.

Shin, M.W., Shin, D., Choi, S.H., Yoon, E., Han, C., 2007. Optimization of the Operation of Boil-off Gas Compressors in Liquefied Natural Gas Plant. Ind. Eng. Chem. Res. 46 (20), 6540–6545.

Simonetti, T., Gadelle, D., Nanda, R., Oct. 5–9, 2009. Using CFD and Dynamic Simulation Tools for the Design and Optimization of LNG Plants, paper presented at the 24th World Gas Conference, Buenos Aires, Argentina.

Stephenson, G., Wang, L., 2010. Dynamic Simulation of Liquefied Natural Gas Processes. Hydrocarbon Processing 89 (7), 37–44.

Sturm, W., Parra-Calvache, M., Chantant, F., Opstal, J., Jan. 10–12, 2009. Unlocking the Potential of Modern Control and Optimization Strategies in LNG Production, paper presented at the 1st Annual Gas Processing Symposium, Doha, Qatar.

Szatny, M., Mullick, S., Autumn 2005. Modeling for the LNG Value Chain. LNG Industry, 69–72.

Tekumalla, R., Valappil, J., Dec. 2008. Innovation, Safety and Risk Mitigation via Simulation Technologies. Bechtel Technology Journal 1, 141–155.

Thomas, P., 1999. Simulation of Industrial Processes for Control Engineers, first ed. Butterworth Heinemann, Woburn, MA, USA.

Urban, Z., Matzopoulos, M., Marriott, J., Marshall, B., 2010. Consider New Analysis for Flares. Hydrocarbon Processing 89 (11), 41–47.

Valappil, J., Mehrotra, V., Messersmith, D., Bruner, P., Jan./Feb. 2004. Virtual Simulation of LNG Plant. LNG Journal 35–39.

Valappil, J., Mehrotra, V., Messersmith, D., Oct. 2005. LNG Lifecycle Simulation. Hydrocarbon Engineering 10 (10), 27–34.

Valappil, J., Messersmith, D., Wale, S., Mehrotra, V., May 8–10, 2006. Lifecycle Dynamic Simulation in the Design and Testing of Advanced Process Control, paper presented at the 2006 IEEE Advanced Process Control Applications for Industry Workshop (APC 2006), Vancouver, BC, Canada.

Valappil, J., Tekumalla, R.T., Hall, L., Anderson, K., Ortego, D., April 26–30, 2009. Evaluating the Debottleneck Options for Existing LNG Plants Using Simulation Models, paper presented at the 9th Topical Conference on Natural Gas Utilization, AIChE Spring National Meeting, Tampa, FL, USA.

Valappil, J., Tekumalla, R.T., Williamson, J., Wilkes, M., March 21–24, 2011. Angola LNG Plant – Enhancing Design and Operability Using Dynamic Simulation, paper presented at the Gastech 2011 Conference & Exhibition, Amsterdam, The Netherlands.

Vist, S., Hammer, M., Nordhus, H., Sperle, I.L., Owren, G., Jorstad, O., March . Dynamic Modeling of Spiral Wound LNG Heat Exchangers – Model Description, paper presented at the AIChE Spring Meeting, New Orleans, LA, USA.

Wedel, L.V., Bayer, B., Marquardt, W., July 18–23, 1999. Perspectives on Lifecycle Process Modeling, paper presented at the 5th International Conference on Computer Aided Process Design, Breckenridge, CO, USA.

Wilkes, M., Gandhi, S., Valappil, J., Messersmith, D., Bellomy, M., April 24–27, 2007. Large Capacity LNG Trains – Focus on Improving Operability During Design Stage, poster presented at the 15th International Conference & Exhibition on Liquefied Natural Gas (LNG 15), Barcelona, Spain.

Wink, B., Marcum, S., April 22–27, 2007. Transient Flow Analysis in LNG Regasification Terminal Design, paper presented at the 7th Topical Conference on Natural Gas Utilization, AIChE Spring Meeting, Houston, TX, USA.

Wu, J., Feng, J., Dasgupta, S., Keith, I., Oct. 2007. A Realistic Dynamic Modeling Approach to Support LNG Plant Compressor Operations. LNG Journal 27–30.

CHAPTER

LNG Safety and Security Aspects

9

9.1 Introduction

The LNG industry has an excellent safety record. This is due in large part to the combination of industry practice and regulations that are in place to prevent incidents from occurring and to reduce or mitigate the impacts of incidents if they occur. For decades, the LNG industry has also maintained secure operations around the world, including areas where terrorism is a concern. Even so, the safe and environmentally sound operation of the LNG facilities, both ships and terminals, is a concern and responsibility shared by operators. Hazards in handling LNG and natural gas do exist and it is important not to understate or exaggerate them.

This chapter explores hazards associated with and safety features designed for the unusual characteristics of LNG. Also summarized here are regulations governing LNG and a wide framework of various risk assessment methods applicable to LNG shipping and export/import terminals, both onshore and offshore. A compilation is included of accident frequencies for LNG carriers per ship year. Examples are provided of risk analysis for specific ports. Innovative applications of risk analysis are treated, along with current research findings. This chapter also discusses the growing concerns over LNG infrastructure security, some factors associated with security, and recent initiatives undertaken to analyze and improve security.

9.2 Hazards associated with LNG plants

LNG is a clean fuel and as such is considered environmentally favorable to other fuels. The main hazards handling LNG are fire and explosion, cryogenic freeze burns, embrittlement of metals and plastics, and confined spaces hazards. These are all well understood and can be well mitigated with a careful appreciation of the hazards.

9.2.1 Properties of spilled LNG

LNG is usually stored and transported at near ambient pressures, typically less than 5 psig, in well-insulated containers. Inevitable heat conduction causes liquid evaporation, and removal of the boil-off gas helps maintain the LNG in its liquid state by auto-refrigeration. For large-scale storage, the boil-off gas is compressed and recondensed to limit losses. The density of LNG is about half that of water. If it is spilled on the ground, it will boil rapidly at first and then more slowly as the ground cools. It vaporizes completely, without leaving a residue. If it is spilled on water it will float and vaporize

Handbook of Liquefied Natural Gas.

rapidly since even at water temperatures near freezing, the water is significantly warmer than the spilled LNG. Boiling LNG sets up convection currents in the water, so it will form ice only if the water is very shallow . Temperature sensors are commonly used to detect LNG ground spills, and provide a stronger signal than hydrocarbon gas detectors.

As is typical of flammable liquids, the liquid phase of LNG itself will not burn or explode. It must be vaporized and mixed with air in the flammable range prior to ignition. The flammable region of an LNG vapor cloud (typically between 4% and 15% concentration of gas in air) is usually visible as a white cloud of water vapor and ice crystals condensed out of the air by the cold LNG vapor. Vapor comes off the pool initially around –161.7°C (–259°F) and consequently is heavier than air. Since the vapor cloud hugs the ground it is more likely to encounter an ignition source such as a vehicle muffler. If not ignited, the vapor cloud spreads due to its negative buoyancy. As the cloud spreads it entrains air and warms toward the ambient air temperature and may cease to be visible. As it warms above about –110°C (–166°F), it becomes positively buoyant and "lifts off."

LNG vapors consist of low-boiling, light hydrocarbons (mainly methane, ethane, propane, with some nitrogen) that are flammable and odorless. Although the vapors are nontoxic, they can be an asphyxiant when they displace oxygen in a confined space.

9.2.2 Ignition and fires

Fire and explosion hazards at LNG facilities may result from leaks and spills, especially during transfers including loading and unloading activities. They also require the presence of ignition sources. Plant designers try to minimize ignition sources such as vehicles, sparks associated with the buildup of static electricity, heat sources such as process furnaces, and lightning but risk assessments can never rule out ignition. If there is no ignition, the LNG will vaporize rapidly, spread, and carry downwind with no injurious effects after diluting below flammable limits.

LNG vapors are difficult to ignite partly because clouds contain condensed humidity from water vapor. Tests have shown that several cloud fires extinguished on their own (MIACC, 2007). Natural gas has an auto-ignition temperature of 540°C, which is higher than most other fuels (e.g., diesel at 280°C, premium grade gasoline at 400°C, propane at 468°C).

If ignited, LNG presents four potential fire risk scenarios: vapor cloud flash fire, jet fire, pool fire, and vapor cloud explosion.

- **Vapor cloud flash fire.** By late ignition the entire cloud does not ignite at once. Only the portion of the vapor cloud that has concentrations in the flammable range burns. A transient fire can burn both forward to the cloud front and back to the release point where it produces either a pool fire or a jet fire. Experiments show that LNG flash fires propagate at a relatively slow speed of 10 to 20 m/s. They can also stall if burning into a head wind. It is accepted (MIACC, 2007) that flash fires could result in serious consequences for anyone within the flames, but they pose a low risk for public exposure outside of the cloud's flammable area.
- **Jet fire.** An ignited liquid, vapor, or two-phase mixture discharging under pressure through a hole in a container will form a jet fire. A jet fire can cause severe damage but would be confined to a local area. This type of fire is unlikely for an LNG storage tank since the product is *not stored under pressure* (except for the hydraulic head pressure). At base load import terminals, there is little storage of any pressurized liquids. However, jet fires can occur from pressurized LNG vaporizers or during unloading or transfer operations under pump pressure even after

isolation valves close a pressurized line. Livingston et al. (2009) tabulates the typical isolatable segments for an onshore LNG regasification terminal, listing pressures, temperatures, phase state, and volumes. They also tabulate hole size, release duration, and jet flame length consistent with design variables for isolated sections.

- **Pool fire.** Calculations and experiments agree that a spill of LNG breaks into an aerosol, and air entrained in a vertical spill has only enough heat capacity to partly evaporate the liquid. Thus, an LNG release from the height of a storage tank or transfer pipeline will partly evaporate while falling and the balance "rains out" to form a liquid pool. A pool fire tilts in a wind and a pool fire on land partially wraps around obstacles such as the front of a dike. If the spill occurs inside a properly designed and maintained diked area, the pool fire will remain contained inside and will continue to burn until the fuel is consumed. If the spill occurs outside a confined area, the burning pool fire is free to flow based on topography and the geometry of the spill.

For LNG spills on the sea without ignition, the pool will spread to a steady state pool size where the evaporation rate matches the liquid rainout into the pool. Upon ignition, the feedback of heat from the fire increases the evaporation rate, so the pool shrinks to a smaller steady-state pool size. A steady-state pool fire on water has been found appropriate for calculating long-term exposure contours. For evaluating short-duration exposure to contours of low heat flux (5 to 12.5 kW/m^2) the initial maximum unignited pool size is recommended (MIACC, 2007).

The preferred extinguishing agent for small LNG fires is a dry chemical such as potassium bicarbonate. However, in general, it is often poor strategy to extinguish an LNG fire since the resulting dispersing vapor cloud is more dangerous than a burning pool in an impoundment area.

Spraying water on an LNG pool only increases the vaporization rate and intensifies any fire; spraying a given volume of water onto LNG will vaporize about twice that volume of LNG. High expansion foams are not considered to be effective LNG fire extinguishing agents, but they are effective in controlling LNG pool fires in dikes because the foam blanket reduces the heat generated by the fire radiating back to the LNG pool and thereby reduces the evaporation rate. Foams can prove valuable in vapor control of unignited LNG, and are sometimes automatically activated by temperature sensors in the ground. An experiment by Yun et al. (2010) applied high expansion foam into a concrete pit (6.5 m x 10 m x 1.2 m). They obtained an initial increase in vaporization, then a gradual decrease of the 5 kW/m^2 fire radiation contour from 19.7 m to 5.5 m, and a 58% reduction in flame height.

Experience has shown that a fire impinging upon structural steel takes only a few minutes of exposure to threaten the steel's integrity. The heat flux associated with large LNG pool fires is about 280 kW/m^2 for fires larger than a nearby exposed object, and approximately 85 kW/m^2 for pool fires comparable in size to the exposed object. The heat flux from jet fires would be approximately 250 to 300 kW/m^2 (Livingston et al., 2009). According to experienced operators, for a typical onshore LNG facility, the time to detect, isolate, and shut down the facility in the event of an unplanned release or leak is typically greater than 15 minutes.

9.2.3 Vapor cloud explosions (VCE)

Pure methane (natural gas) has not been known to generate damaging overpressures if ignited in an unconfined area. Only if the flammable plume is in a confined or congested zone can flame speeds accelerate to form a deflagration type of explosion. A deflagration explosion has flame speeds less than sonic and lasts only within congested zones before flames decay to flash fire flame speed. Even so, a deflagration often has a large impact. A deflagration is a far less damaging explosion than a detonation. A detonation occurs with more reactive gases such as ethylene and burns all the fuel within the flammable range, including fuel outside of congested zones. This is because it perpetuates the combustion reaction in the advancing supersonic flame front and does not need the turbulence enhancement of obstacles.

Pressurized flammable liquids exposed to an external fire can become superheated. The unwetted metal over the vapor space in a container can become weakened and fail, releasing the liquid that partially flashes to vapor. This flashing vapor expands and ignites as a boiling liquid expanding vapor explosion (BLEVE). LNG is not pressurized (that is the purpose for liquefying the gas) so there is a very low possibility of a BLEVE. It is possible with a faulty design for an LNG truck to have pressure relief valves with setpoint pressures that are too high, allowing a pressurized, heated metal situation to develop in an engulfing fire. This occurred with a tank truck accident in Spain in 2002 (Planas-Cuchi et al., 2004).

9.2.4 Cryogenic effects

Storage and handling of LNG may expose personnel to contact with very low temperature liquid, vapors, or solid surfaces. The viscosity of cryogenic liquids is low, meaning that they penetrate through porous materials of clothing more quickly than water. Contact with a cryogenic can cause severe damage to the skin and eyes. It can also make ordinary metals, plastics, rubber, and some clothing materials subject to embrittlement and fracture; therefore, cryogenic operations require specialized containers, materials, and protective clothing. Training should always be provided to educate workers regarding the hazards of contact with cryogenic liquid and cold surfaces and the need for personal protective equipment (e.g., gloves, insulated clothing).

LNG containers are manufactured from high quality metals intended for cryogenic storage. LNG carriers and some storage tanks are designed with an inner and outer cryogenic shell that prevents the LNG from coming into contact with the outer hull at ambient temperature. International ship design rules require that areas where cargo tank leakage or spill during unloading might be expected (e.g., ship deck and tank covers) must be designed for contact with cryogenic LNG to prevent brittle facture. Since near the beginning of the LNG trade in 1969 there have been eight marine LNG incidents resulting in spillage with some hull damage due to cold fracture (Livingston et al., 2009). In an early experience at an export terminal, a valve failed, spraying a worker with LNG.

The brittle transition temperature range for most carbon steels is 200 to 250°K (–73 to –23°C or –132 to –42°F). Experiments have shown that immersion in LNG of 25.4 mm (1 inch) thick pieces of painted steel completely cools the steel to LNG temperatures in less than two minutes. With these high heat transfer fluxes structural steel sections reach nominal failure criteria in as little as one to five seconds (Livingston et al., 2009). Since cooling rates are so rapid, early leak detection and system isolation have little effect on managing stress hazards in the immediate release area. Thus, cryogenic protection requires appropriate materials of construction and/or adding insulation and shielding.

The cold vapors from the venting of pressure relief valves on an LNG line or tank are a possible source of exposure. Careful location of relief valve vents is needed. Other normal system design practices include using remotely operated isolation valves and a reliable system for gas detection that helps to isolate the source of a release.

9.2.5 Rollover

Storage of large quantities of LNG in tanks has lead to a phenomenon known as rollover. Rollover may occur if LNG stratifies within the storage tank into two layers of different densities over an extended period with inadequate mixing. The upper layer preferentially evaporates lighter components or "weathers" and becomes denser. Heat transfer affects both layers, but the pressure of the hydrostatic head of the upper layer allows the vapor pressure of the lower level to increase. As the density of the upper layer becomes slightly heavier than the lower level, suddenly the upper layer sinks or rolls over, bringing the lower layer to the surface. The lower layer, which has been superheated relative to atmospheric pressure, gives off a large amount of vapor. This can result in a rather sudden increase in the tank head pressure and could cause structural damage since pressure relief valves are not sized for rollover.

Two types of conditions typically bring on rollover conditions. The first is fill-induced stratification. This occurs when the added liquid (cargo) is less dense than the liquid in the tank (the heel), and is added through a top fill line, or, conversely, when cargo denser than the heel is added through a bottom fill line. The second is nitrogen-induced (Melhem, 2005). Liquid nitrogen is the most volatile component of LNG and when it is present at greater than 1% it boils off preferentially and leads to an increase in the bubble point temperature of the mixture and a reduction in the density of the top layer, hence to stratification. In contrast, when methane is the most volatile component (of a non-LNG mixture), its loss leads to a slight increase in saturation temperature without a significant change in the liquid density.

Rollover could occur on LNG carriers docked for extended periods, but ullage costs make extended stays unlikely. Rollover is not a problem for carriers at sea because of mixing by sea motion.

Recommended measures to prevent rollover include the following:

- Monitor LNG storage tanks for pressure, density, and temperature all along the liquid column.
- Maintain sufficient mixing. If necessary, recirculate the LNG in within the tank.
- Install multiple loading points at different tank levels to allow for the distribution of LNG with different densities within the tank to prevent stratification.

These measures have become common in the industry, and rollover is no longer considered credible in a well-managed facility.

9.2.6 Rapid phase transition

Some LNG spills on water have involved a nearly simultaneous transition from the liquid to vapor phase with an associated rapid pressure increase. The rapid phase transition (RPT) energy comes from a physical phase change and is much less than the energy available from a chemical combustion reaction. One set of investigators claim it is more likely to occur when the LNG contains high concentrations of the heavy hydrocarbons (C2 to C4 components), or after a time delay allowing the lighter methane to boil off leaving a heavier liquid with higher concentrations of those components. It may result in two effects: a localized overpressure resulting from rapid phase change, and enhanced dispersion as LNG is rapidly vaporized and expelled to the atmosphere. No known incidents of RPT have occurred in commercial transportation or handling of LNG, but experimentation has shown that potential for them to occur does exist. Attempts to model the energy of an RPT have been made for LNG (Bubbico and Salzano, 2009) and for events involving molten metal and water, for which there is a history of large damaging events. The history of LNG RPT events is also treated (Woodward and Pitblado, 2010).

The ignition of an LNG vapor cloud during the Falcon test series in the late 1990s that destroyed the test apparatus has generated a concern that an RPT could be responsible for ignitions (Nedelka et al., 2003).

9.2.7 Confined spaces

As in any other industry sector, confined space hazards are potentially fatal to workers. Confined spaces may include storage tanks, secondary containment areas, and stormwater/wastewater management infrastructure. Facilities should develop and implement confined space entry procedures as described in general EHS (Environmental, Health, and Safety) guidelines. Gas detection devices should also be used to authorize entry and to monitor operations into enclosed spaces.

Vapor leaks have also occurred along unexpected confined spaces. The Cove Point, MD LNG terminal experienced an explosion from vapors that flowed through 200 ft of electrical conduit (described in Woodward and Pitblado, 2010). This event resulted in three major changes to the National Fire code.

9.2.8 Chemical hazards

Common to any processing plant, the design of the onshore LNG facilities should reduce exposure of personnel to fuels and products containing hazardous chemical substances. Use of substances and products classified as very toxic, carcinogenic, allergenic, mutagenic, teratogenic, or strongly corrosive should be identified and substituted by less hazardous alternatives, wherever possible. For each chemical used, a Material Safety Data Sheet (MSDS) should be available and readily accessible at the facility. A general hierarchical approach to the prevention of impacts from chemical hazards is provided in the General EHS Guidelines.

9.3 Safety features of LNG facilities

Safety features of LNG facilities are an inherent part of each design. Only a brief overview can be provided here.

The broad categories of safety features at LNG facilities are (1) primary containment, (2) secondary containment, (3) plant safety systems, and (4) separation distance.

9.3.1 Safety of LNG storage tanks

Nearly all modern LNG tanks have double walls. *Primary containment* requires materials designed, tested, and selected for cryogenic service. Designs for removal of boil-off vapors, to prevent the ingress of air, to prevent frost heave, and to withstand a number of filling, emptying cycles and cool-down and warming operations are planned for the design life of tanks (ABS Consulting Inc, 2004; Er, 2007).

Secondary containment is effectively a second tank surrounding the onshore LNG storage tank. The secondary tank is designed for a capacity exceeding the volume of the primary container. An insulation system surrounds the inner wall, which contains the cryogenic liquid. The tanks are constructed of metals or alloys with low coefficients of thermal expansion that do not embrittle when in contact with cryogenic fluids (i.e., aluminium or 9% nickel steel). Internal pumps are used to pump out LNG. There are no bottom connections to leak or fail. Embankments, berms, bunds, or dikes surround the tanks and are scaled, in modern facilities, as a precaution to contain leakage up to 110% of the tank capacity.

Additional devices include:

- Cooldown temperature sensors on the storage tanks and base
- Leak detection in the annular space on tanks, e.g., low temperature alarms
- LNG tank gauging that provides remote readings, with high/low level alarms and shutdown systems
- Combined temperature and density sensors to detect rollover potential.

Explosion risk is minimized by storing LNG slightly above atmospheric pressure so that air cannot inadvertently leak into a tank, except by the opening of vacuum breaker valves. If LNG does however escape from a tank or is spilled during transfer it will mix with air and either ignite, forming a pool fire, or will rapidly vaporize, leaving no residue.

Some tanks are underground. In-ground LNG storage tanks are accredited with the European standard EN1473 as the safest way to store LNG. They also have the highest degree of security and some environmental benefits. Even in the event of the concrete roof being destroyed by a projectile in a terrorist attack, since LNG is stored below the ground surface the LNG would not leak onto the ground. In an earthquake, the seismic motion is not amplified for in-ground storage tanks in contrast to above-ground structures, thus making them safer in earthquake-prone regions. A potential remains for a leak from an earthquake, but this is likely self-sealing by freezing surrounding ground water. Ground water can pose problems such as generating buoyant forces.

9.3.2 Export and import plants: prevention and emergency systems

Plant safety systems are designed with two types of layers of protection:

- A Prevention System that prevents Loss of Containment (LOC), such as pressure relief valves.
- An Emergency System that mitigates loss after LOC, such as an Emergency Depressuring System (EDS), an Emergency Shutdown System (ESD) and a fire protection system that automatically activates fire suppressants.

ESD systems are required by the US codes NFPA-59A. By European codes prEN1473 4.5.6 EDS systems are optional.

Computerized Emergency Systems, among other automatic systems, are designated as Safety Instrumented Systems (SIS). An SIS is commonly designed to a Safety Integrity Level (SIL) by specific criteria for their design. The principles for conducting an SIL study are given by ISA-TR84.00.02 (2002), and supplemented by several International Electrotechnical Commission (IEC) documents including IEC-61508/61511 (Tanabe and Miyake, 2011).

The order of work is (de la Vega and Durr, 2004):

1. Establish acceptable risk for the plant (see Section 9.4.10).
2. Work back to find the reliability of the Safety Instrumented System (SIS) needed to meet the plant risk criteria.
3. Use the IEC methodology with process hazard and risk assessment studies to design the Safety Instrumented Functions (SIF) needed to achieve the required safety standards.

Table 9-1 Safety Integrity Levels (SIL) for Low Demand Safety Instrumented System

Safety Integrity Level (SIL)	Probability of Failure on Demand (PFD)
4	1E-4 – 1E-5
3	1E-3 – 1E-4
2	1E-2 – 1E-3
1	1E-1 – 1E-2

(Tanabe and Miyake, 2011)

Table 9-1 defines the SIL levels by the required reliability of a SIS that is calculated by the Probability of Failure on Demand (PFD) of SIS components. The order of magnitude increases in SIL going from one level to the next implying considerable effort to achieve. Measures are required such as increased inspection and test frequency, higher quality system components, and instrument redundancy.

9.3.3 Safety of LNG unloading facilities

Standard practices have been adopted worldwide to prevent leaks and their escalation, including (GIIGNL, 2011b; Wang, 2012):

- Compliance with national and internationally accepted codes and standards, as well as company guidelines, for siting, designing terminals, inspection, and maintenance
- Siting new terminals a safe distance from adjacent populations based on risk assessments
- Use of materials and systems designed to safely insulate and store LNG
- Impoundment areas; spills are contained in these areas to control spread and vaporization rate, as well as to minimize pool fire consequences
- Vapor reduction systems; foam generators reduce the rate of vapor formation and movement
- The LNG unloading process systems incorporate monitoring and control devices to detect deviation from acceptable parameters, facilitating corrective actions
- Specifically, Powered Emergency Release Couplings (PERC) on unloading lines have shutdown triggered by several signals (e.g., gas detection, low temperature, fire)
- Overpressure protection (pressure controllers and relief valves)
- Leak detection and spill control using temperature and gas detection probes
- Ignition source control
- UV/IR fire detectors and combustible vapors detection systems
- Closed-circuit TV monitoring
- Fire zoning
- Automatic emergency shutdown and depressurization systems and isolation valves
- Passive fire protection (fireproofing, barriers, and coatings)
- Active fire protection (firefighters, preinstalled monitors, etc.)
- Emergency release couplings on unloading lines
- Trained operators are always present; their response includes making emergency notifications to responders and broadcasts to the community

- Emergency shutdown buttons are at the pier, the control room, on board the LNG ship, and at field locations; this shutdown generally shuts off all pumps and closes off all piping so that the LNG stays either on the ship or in the storage tank
- Using manufacturer's service engineers for all vital machineries and systems
- Preemployment crew security screening, medical tests, training, and licensing
- Safety and operational inspection/audits of crews and ship managers are done with every arrival, annually by ship managers, and every 2.5 years by external inspectors during scheduled dry dock and maintenance
- The Home Doctor concept (a designated shipyard) includes standard specifications and pricing as part of a Master Maintenance plan for all vital machinery and systems.

9.3.4 Protection features for LNG facilities

Some normal plant protection practices are modified for LNG applications, as discussed next (Livingston et al., 2009).

9.3.4.1 Diking and sloping

Even if pooled LNG does not ignite, the bases of columns and equipment supports could fail by cryogenic exposure. Two principles applied to LNG plants are sloping and using insulating concrete. These minimize the area of an LNG pool and reduce heat conduction from the substrate. Sloped and paved troughs are under all LNG lines, draining to impoundment basins.

When designing LNG spill containment systems, it is necessary to consider the film boiling or Leidenfrost effect that leads to a vapor layer under boiling LNG. Flowing on a vapor layer reduces friction and produces higher liquid velocities when compared to flowing water. In turn, higher velocities could cause splashing around obstructions, through turns, and at changes in elevation. Structural supports within the curbing and drainage paths should be on a concrete base that prevents the exposure of steel to pooling, splashing, or draining liquid.

9.3.4.2 Coatings and insulation

A primary difficulty in designing for LNG release scenarios is that there could be a release that results in a cryogenic exposure, a fire exposure (jet or pool), or a combination of events. No industry standard tests have yet been developed for cryogenic exposure followed by fire exposure.

Standard fire-approved cementitious insulation provides economical protection of structural steel against fires and also against short-term cryogenic exposure. Unfortunately, not all potential insulation products have been tested for both types of exposure.

Intumescent and subliming coatings expand upon exposure to fire. These coatings have also been tested and found, in conjunction with a cryogenic insulating coating, to provide good protection from either cryogenic or jet fire exposure. They are more expensive to apply than cementitious insulation.

9.3.4.3 Instrument and electrical cabling

Protection of instrument and electrical cabling is not normally done because these systems are designed to be fail-safe. However, direct exposure from cryogenic spray to shutdown/blowdown valves or actuators could fail the isolation or deinventory process.

9.3.4.4 Cloud effect

An LNG spill could generate a large fog cloud that can impair employees' visual response. They may not be able to see where the LNG is pooled. Based on this, employees need more than one route to temporary refuge no matter what the wind direction or where an incident may occur.

9.3.5 Safety features of LNG trucks

LNG tank trucks have safety devices to prevent overfilling and overpressurization, as well as safety systems to prevent the LNG road tanker from driving away while still connected to the loading facilities. Pressure-relief setpoints should be set to the lowest practical value to reduce BLEVE potential. LNG road tankers must comply with country-specific codes for design and operation.

9.4 LNG risk analysis and controls

The safe processing, storage, and transportation of LNG is an essential condition for the continued existence, growth, and sustenance of the entire industry. Both marine transport and onshore LNG plants and transportation follow two basic paths for safe operations:

- All applicable codes and standards should be met with rigor (even voluntary ones).
- Each operation must establish their own Process Safety Management (PSM) system. The objectives are usually to establish and follow best industry practices, to use innovative measures, and to obtain the best risk/reward ratio for their safety budgets.

Risk is usually defined by consequence and probability or frequency. Mitigation measures can address either the reduction of consequences, the reduction of frequency, or both.

9.4.1 Risks to natural gas supply train

Table 9-2 lists a broad range of risk issues that can affect parts of the natural gas supply train (Nicholson, 2006). The risk assessment methods discussed subsequently have been applied to only some of the listed risks.

Risk assessments have various objectives such as (1) set insurance coverage, (2) justify risk mitigation measures, (3) develop contingency plans, and (4) make arrangements to provide coverage for business interruption or missed LNG cargos.

9.4.2 Government oversight

The construction of an LNG receiving terminal is subjected to a considerable number of design standards, local regulations, and national regulations, a complete discussion of which is beyond the scope of this book (more complete coverage in Woodward and Pitblado, 2010). Mainly the European and American standards are widely used throughout the world.

The LNG industry adheres to an international network of codes and standards that specify safe materials, designs, and generally approved technologies for import terminals. This network promotes sharing state-of-the-art technologies and research. An international work group, TC67 Work Group 10: "Standardization for Installations and Equipment for LNG, Excluding Product for Testing" was formed in 2006 under the International Organization for Standardization (ISO) organization. The group's

Table 9-2 Risk Issues for the Natural Gas Supply Train

Position in Supply Train	Risk Issues
Gas production	Age of facilities? Political stability of host government? Number of reservoir resources? Exposure to natural perils? Single or multiple gas production facilities (especially offshore)? Drilling risk in the production field? Quality of risk management of the gas producer? Security around gas production facilities?
Gas treatment and separation facilities	Number, location, and spacing of treatment and NGL (natural gas liquid) extraction plants? Could accident in gas treatment or NGL plant interrupt LNG production? Quality of risk management of the gas treatment company? Security of offshore gas treatment plants?
Gas transportation to the LNG plant	Single or multiple pipelines available? Length of gas pipelines and design at vulnerable points such as river crossing, mountain slopes, etc.? Dependence on gas compressors? Redundancy of gas compressors? Gas pipeline security and monitoring?
LNG liquefaction plant	Number of trains and capacity at LNG source? Liquefaction redundancy or contractual flexibility in the event of plant problems? Age of equipment and quality of maintenance/ inspection at the LNG facility? Quality of safety management? Quality of security management? Tankage flexibility and redundancy? Utilities redundancy? Number of jetties and ship schedule flexibility? Weather risk at jetties? Port access problems? Machinery spare parts?
LNG ship transportation	Ship/shore safety interface effective? Size and number of ships required? Age of ships? Experience of crew? Type of LNG tanks on ships? Quality of risk management of shipper? Quality of tug and berthing support? Schedule and weather problems en route?
LNG import terminal	Port access and security? Jetty flexibility? Leaks and breaks of unloading line? Startup, cooldown stress breaks? Storage size c.f. throughput? Send out flexibility? Single or double-walled storage tanks? Safety design of LNG storage tanks? Layout of plant? Vaporizer capacity and redundancy? Weather and natural perils (earthquakes)? Quality of safety and security management? Reliability and redundancy of power supply?

(Nicholson, 2006)

objective is compatibility and harmonization of LNG codes among countries (GIIGNL, 2011a). Another international trade association, Society of International Gas Tanker and Terminal Operators (SIGTTO), compiled a single publication that summarizes best practices in the LNG industry (SIGTTO, 2000).

The US government oversight of LNG facilities is provided by three federal agencies under an Interagency Agreement.

- The Federal Energy Regulatory Commission (FERC) grants approval for new onshore facilities and is the lead agency for review of environmental and safety concerns, including public comment meetings. Every two years, FERC staffs inspect LNG facilities to monitor the condition of the physical plant and inspect changes from the originally approved facility design or operations.
- The US Department of Homeland Security (DHS) exercises regulatory authority over LNG facilities that affect the safety of ports and navigable waterways. A key law governing the marine portion of an LNG terminal in the United States is 33 CFR Part 127, Waterfront Facilities Handling Liquefied Natural Gas and Liquefied Hazardous Gas. The USCG also establishes criteria for evaluating a proposed deepwater port. Terminals operate under site-specific USCG Operating Plans (OPLANS) that require prearrival boarding and inspection of ship certificates, crew licenses, safety equipment, ship condition, ship's log, and procedures.
- The US Department of Transportation (DOT) and specifically the Pipeline and Hazardous Materials Safety Administration (PHMSA) promulgates and enforces safety regulations and standards for transportation and storage of LNG for interstate and foreign commerce under the pipeline safety laws. PHMSA regulations are contained in US Federal Law 49 CFR Part 193 and cite NFPA standards. The Maritime Administration (MARAD), also within DOT, has licensing authority for the construction and operation of deepwater ports, including offshore (floating) import terminals.
- Under several Memorandums of Understanding (MOU), it is agreed that OSHA 1910 and EPA Risk Management Planning (RMP) obligations do not apply to LNG import, export, or peak shaving plants.

9.4.3 Codes and standards for LNG onshore (United States, Europe, Japan)

The European code EN1473 "Installation and Equipment for LNG Design on Onshore Installations" is risk-based, meaning focused on outcomes rather than specific ways to achieve a desired level of safety. The European Council Directive 96/82/EC (SEVESO II) aims at the prevention of major accidents involving dangerous substances, including LNG. The provisions in the Directive were developed following a review analyzing major accidents since the implementation of the SEVESO I directive. Failures of management systems were found to have contributed to the cause of over 85% of the accidents. Additional codes for all modifications of "Installation and Equipment for LNG" include EN1160, EN1474, EN1532, and EN 13645.

Canada requires compliance with US and European standards. In addition Canadian standard CSA Z276-01 requires underground unloading lines at import terminals, encased in a concrete caisson with a nitrogen inert atmosphere. Exclusion zones are set for:

- A 500 m radius around unloading arms at the head of a jetty
- A 100 m radius around the impoundment basin of onshore facilities
- A 400 m radius around LNG tanks and process facilities.

In Japan, LNG terminal siting and operation is regulated by the Ministry of Economy, Trade, and Industry (METI), which enforces the Gas Utility Industry Law, the Electric Utility Industry Law, and the High Pressure Gas Regulation Law. Gas utility companies must:

- Maintain a facility in accordance with a technical standard.
- Define, submit, and observe their companies' own security standards in order to ensure safe construction, maintenance, and operation of gas facilities.
- Assign a licensed engineer to ensure the safety of construction, maintenance, and operation of a gas facility.

For regulations in China, Korea, India, and Taiwan and a description of industry associations in the LNG industry, see GIIGNL (2011a).

Table 9-3 lists the US and European codes directed specifically to LNG facilities.

9.4.4 Technical feedback on codes

For LNG receiving terminals the governing US Federal Law 49 CFR Part 193 refers to sections of the National Fire Protection Association design standard NFPA 59A "Standard for the Production, Storage, and Handling of Liquefied Natural Gas." Uniquely, the US standards requires each LNG terminal, tank, and process area to have a thermal exclusion zone and a vapor exclusion zone within the owner's control. The thermal exclusion zone is the area with a fire flux of or below 5 kW/m^2 for exposure to the public. In addition, a 37.5 kW/m^2 threshold is set for the integrity of exposed structures. The vapor dispersion exclusion zone is the area within the contour to half LFL (lower flammable limit). The criteria for these exclusions zones have been the subject of technical criticism such as:

- The criteria do not take sufficiently into account more vulnerable individuals (children, the elderly, or mobility impaired; Mannan et al., 2005).
- Some of the concepts of Process Safety Management (PSM) described by the American Institute of Chemical Engineers (AIChE) and the Occupational Safety and Health Administration (OSHA) PSM regulation in 29 CFR 1910.119 should be included in codes (Mannan et al., 2005; Zinn, 2005).
- The effects of air dilution and wind "scooping" of LNG vapors from impoundment basins should be accounted for in modeling the distance to half LFL for vapor dispersion exclusion zones (Havens and Spicer, 2005, 2006, 2007).
- Apply vapor dispersion zones for a docked LNG carrier (Havens, 2005).
- Use the LFL instead of half LFL as the flammable vapor end point (Raj, 2008a and 2008b). This is disputed by Ivings and Webber (2007).
- Develop a procedure that permits more advanced consequence models to be used in addition to the DEGADIS, FEM3A, and LNGFIRE3 models, originally the only ones accepted by the DOT and FERC. The last point has been addressed, and PHMSA has approved additional models under the Model Evaluation Protocol (MEP) that was incorporated into the 2009 edition of NFPA59. Kohout (2012) provides a recent review of the application of the PHMSA protocol and of dispersion models for LNG siting applications.

The subject of wind "scooping" of LNG vapors from an impoundment basin was tested experimentally and modeled by Chan (1992). Figure 9-1 illustrates how Chan's modeling predicts that the vapors

Table 9-3 Pertinent Regulations for the Design, Construction, and Operation of LNG Facilities

Regulation	Description
49CFR Part 193	Liquefied Natural Gas Facilities: Covers siting requirements, design, construction, equipment, operations, maintenance, personnel qualification and training, fire protection, and security.
33CFR Part 127	Waterfront Facilities Handling Liquefied Natural Gas and Liquefied Hazardous Gas: Governs import and export LNG facilities or other waterfront facilities handling LNG. Its jurisdiction runs from the unloading arms to the first valve outside the LNG tank.
NFPA 59A	Standard for the Production, Storage, and Handling of Liquefied Natural Gas (LNG) (2006, rev. 2012): Covers general LNG facility considerations, process systems, stationary LNG storage containers, vaporization facilities, piping systems and components, instrumentation, electrical services, transfers of natural gas and refrigerants, fire protection, safety and security.
EN1473	Installation and Equipment for Liquefied Natural Gas—Design of Onshore Installations: Evolved from the British Standard BS 777742 in 1996. The standard is not prescriptive but promotes a risk-based approach for the design of onshore LNG terminals.
EN1160	Installation and Equipment for Liquefied Natural Gas—General Characteristics of Liquefied Natural Gas: Contains guidance on properties of materials commonly found in LNG facilities that may come into contact with LNG.
EEMUA 14743	Recommendations for the Design and Construction of Refrigerated Liquefied Gas Storage Tanks: Contains recommendations for the design and construction of single, double, and full-containment tanks for the storage of refrigerated liquefied gases down to −165°C with both metal and concrete materials.
33CFR 160.101	Ports and Waterways Safety; Control of Vessel and Facility Operations: Describes the authority exercised by District Commanders and Captains of the Ports to ensure the safety of vessels and waterfront facilities, navigable waters, and the resources therein. The controls described here are directed to specific situations and hazards.
33CFR 165.20	Regulated Navigation Areas and Limited Access Areas; Safety Zones: This section defines a safety zone as a water area, shore area, or water and shore area to which access is limited, for safety or environmental purposes, to authorized persons, vehicles, or vessels (stationary or moving). The safety zone is commonly used for ships carrying flammable or toxic cargoes, fireworks barges, long tows by tugs, or events like boat races.
33CFR 165.30	Regulated Navigation Areas and Limited Access Area; Security Zones: Defines a security zone as an area of land, water, or land and water that is designated by the Captain of the Port or District Commander for such time as it is necessary to prevent damage or injury to any vessel or waterfront facility, to safeguard ports, harbors, territories or waters of the United States from sabotage or other subversive acts, accidents, or causes of a similar nature.

(Alderman, 2005)

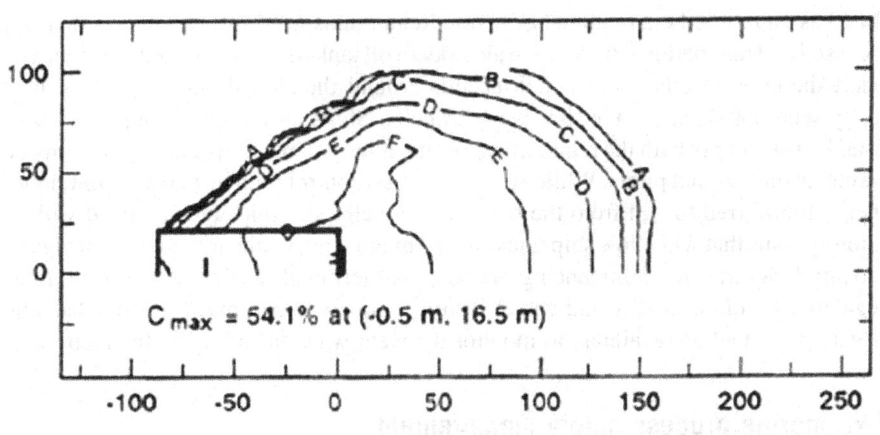

FIGURE 9-1

Concentration contours above vapor fence predicted for Falcon-1 test. The contour levels are (in mole %): A = 0.5, B = 1, C = 2, D = 5, E = 10, F = 15, G = 25, H = 35, I = 50.

(Chan, 1992)

overflow a structure similar to an impoundment zone. This test (Falcon 1) was conducted under stable atmospheric conditions at low wind speeds, and was notable for superheating the vapors by prolonged contact with the water in the walled-in area. Vapors at various dilution overflow the dike walls and do not gradually fill the confined volume before flowing over like a liquid would do.

Gavelli (2010) also modeled the scooping effect of LNG vapors over an impoundment sump using the CFD model FLACS. Upon varying the wind speed he found that the plume length to the LFL does not increase with wind speed. That is, while the "vapor scooping" increases with wind speed, turbulent mixing also increases at a faster rate, so the net effect is a reduction in vapor dispersion hazard distances.

9.4.5 Codes and standards for LNG marine operations

LNG ships must comply with all relevant local and international regulations including those of the International Maritime Organization (IMO, 2002), International Gas Carriers Code (IGC, 1975), and the US Coast Guard (USCG). Insurance companies "classify" the vessel designs and verify vessel integrity.

The regional Captain of the Port USCG marine safety unit reviews LNG ship management procedures and emergency plans. These procedures include requirements for prearrival notification, harbor transit, docking operations, cargo transfer, inspection, monitoring, and emergency procedures. Companies involved in LNG shipping work with the local Pilotage Authority and the USCG to develop optimum plans for safe transit in and out of port. This coordination helps manage port shipping traffic, similar to air traffic controllers, with the aim of protecting against collisions while facilitating movement of other traffic. If warranted, the USCG can assign sea marshals to escort LNG ships as they transit in and out of US ports to provide for harbor safety and security.

International regulation for the training of seafarers is covered by an International Maritime Organization (IMO) convention known as the Standards of Training, Certification and Watchkeeping (STCW) 1995, which has specific requirements for operations on gas carriers.

The IMO has established international Collision Regulations for ship navigation, which apply to all vessels in coastal and international trade. Like all modern oil tankers, sophisticated radar and positioning systems alert the crew to other traffic and hazards around the ship. Distress systems and beacons automatically send out signals if the ship is in difficulty. Ships also employ antipiracy and boarding measures, and must comply with the requirements of the International Ship and Port Security code. This has requirements for tugs and pilots. While at sea, the cargo control room is manned continuously when cargo is being transferred to and from the ship. Additionally, the ships are equipped with automatic identification systems that will allow ship tracking and monitoring while travelling on navigable waters.

Prior to any LNG transfer commencing, the ship and terminal staff meet to ensure all aspects of safety. Regulations require facility and vessel security officers to be present. At the discretion of the USCG, USCG personnel are available to monitor the waterway, the ship, and the facility.

9.4.6 LNG marine process safety management

LNG facilities and ships are viewed in the industry as the "top of the line" (GIIGNL, 2011b). LNG ships have operated worldwide for more than 45 years without major accidents or safety problems either in port or at sea. However, the current active fleet of LNG vessels is aging, with many built in the 1970s, operating with steam drives. An example of new issues involves a 138,000 m^3 LNG carrier delivered in June 2004. It had a problem with leakage of nitrogen injected into the interbarrier space of the membrane system for monitoring and inerting purposes (Nicholson, 2006). The delivery of an LNG carrier was delayed 18 months because of such permeation problems.

9.4.6.1 LNG ship design features

Currently there are two main ship types: single-wall self-supporting spheres (Moss spheres) and a dual membrane design by Gaz Transport or TechniGaz (GTT). The spherical tanks in the Moss design are usually constructed of aluminium with 49 to 57 mm thickness (Pitblado et. al. 2004). The sphere maintains its own structural integrity and the cargo load is transferred to the vessel through a continuous metal skirt attached to the equator of the sphere. The hull is a double hull and some vessels have an additional wall surrounding the spheres. The spheres are, on average, much further separated from the external environment than the tanks of a membrane-type LNG carrier.

The dual membrane design consists of thin stainless steel or high nickel steel membranes 0.7 to 1.2 mm thick, capable of containing the hydrostatic load of LNG but rely on the vessel structure to provide structural support. There is at least 2 m and often 3 to 4 m between the outer hull and the cargo. If a grounding or collision produced damage to the hull and secondary liquid barrier the design of the primary barrier would prevent leakage. Similarly if the primary barrier fails LNG is contained by the secondary liquid barrier and the outer hull, which is also insulated. The design prevents leakage of cryogenic liquid onto unprotected steel or other materials not designed for cryogenics. The insulation spaces are continuously monitored by sensors for any sign of leakage.

Carriers of both types have LNG capacity of more than 135,000 m^3. Vanem et al. (2008) reported that the LNG fleet consisted of 50% membrane, 40% Moss spheres, and 5% other types. The average size carrier in 2008 was 120,000 m^3 and the average size on order books was 156,000 m^3. The size of LNG carriers is increasing, recently by the design of the Q-Max type (capacities to 250,000 m^3). Even so, a recent risk analysis for the Rabaska Project (MIACC, 2007) found that with the larger ships the risk levels would not change much because the visit rate decreases, decreasing the frequency of potential accidents. Furthermore, the size increase in membrane-type LNG tankers is due to their larger width and number of tanks, and the liquid full level above water is only 0.2 m higher with the new Q-Max carriers (Oka, 2010). The liquid height turns out to be a main parameter in calculating consequences for various breach sizes.

Many of the features described in Section 9.3.3 for onshore tanks apply to LNG ships, and include extensive cargo safety systems. LNG in transit is maintained at near atmospheric pressure (< 5 psig) in insulated tanks (each ship usually has 4 to 6 separate tanks). Pressures, levels, and temperatures are monitored automatically. Deviations from predetermined limits sound alarms and require actions to assure safety. The cargo transfer system cannot be operated if all cargo-related safety systems are not fully functioning. Submerged internal pumps are used to pump out LNG. There are no bottom connections to leak or fail. A large coffer dam separates each LNG membrane tank, reducing the potential for an event in one tank to affect its neighbor.

These vessels are designed to withstand the impact of both collision and grounding without damage to the containment system. The double-hull design proved successful in the grounding at 19 knots on June 29, 1979 of the 125,000 m^3 El Paso Paul Kaiser on rocks near Gibraltar. Figure 9-2 shows

FIGURE 9-2

Damage from grounding of the El Paso Paul Kayser, 1979.

(Vaudalon, 2000; publicity release by El Paso Marine Co.)

considerable denting of the outside hull, and minor denting inside the LNG compartment. There was no penetration of the compartment or loss of cargo.

9.4.6.2 Risks to LNG carriers

Wang (2012) writing from experience from service in a marine and offshore surveying company compiled a list of risks that can occur to LNG carriers. A sample of this list is provided in Table 9-4. Vanem et al. (2008) compiled the number of accidents for the first eight categories that are also on Wang's list (e.g., 19 collisions). They also tabulated ship years of experience, and calculated accident frequency per ship year (e.g., 6.7 x 10^{-3}). The accident rate for LNG carriers was found to be slightly lower than that for LPG tankers, oil tankers, chemical tankers, and bulk carriers. Being aware of the possible risks is the first step in protecting against them.

9.4.6.3 Analyzing potential hole sizes in LNG carriers

Risk analysis studies of LNG carriers have postulated arbitrary hole sizes from ship collisions, typically in the range of 0.75 to 5.0 m diameter. The European Union (EU) sponsored a research project known as HARDER to establish design criteria for ship stability for vessels involved in collisions (Laubenstein et al., 2001). This study accumulated data from almost 3000 collision cases to establish probabilistic actual hole sizes for a wide range of scenarios: different vessel sizes, speeds, angle of approach, and striking bow shapes. Paik et al. (2001) identified critical speeds for the collision of two LNG carriers of 6.6 to 7.4 knots leading to tank spills and for VLCCs (Very Large Cargo Carriers) onto LNG carriers of 1.7 to 7.7 knots for light and heavy collision loadings, respectively. A complicating

Table 9-4 Experienced Risks to LNG Carriers

1. Collision (19)(a) 6.7E-3	10. Loss of electric power	19. Leak from loading arm
2. Grounding (8) 2.8E-3	11. Loss of hydraulic system	20. Lightning
3. Contact (8) 2.8E-3 (b)	12. Leak of nitrogen	21. Tsunami/earthquake
4. Fire and explosion (boiler, engine room from fuel gas, etc.) (10) 3.5E-3	13. Structural damage due to incorrect/unbalanced loading	22. Sabotage
5. Equipment and machinery failures (collective) (55) 1.9E-2	14. Crane operation accident	23. Piracy, hijacking
6. High wind/waves (9) 3.2E-2	15. Loss of navigational or maneuvering capability	24. Refueling leak of bunker fuel
7. Loading/unloading (22) (c) 7.8E-3	16. Illness and epidemics	25. Failure of mooring by tidal effects
8. Failure of cargo containment (27) 9.5E-3	17. Failure of emergency shutdown system	26. Mooring failure/anchor drag
9. Loss of instrumentation during loading	18. Overfilling/overpressure of tanks	

(a) The number of accidents are in parentheses, followed by the frequency of accidents per ship year.
(b) Striking or being struck by a fixed or floating object other than another ship or the sea bottom.
(c) 9 of the 22 incidents involved spills of LNG.
(Vanem et al., 2008; Wang, 2012)

factor is that many collisions will leave the vessels connected, so the effective hole size is not the entire damaged area measured later in the repair dock.

Pitblado et al. (2008) extended this analysis to consider a wider range of striking vessels (90, 140, and 230 m long) at 45° to 90° angles to both membrane and Moss sphere carriers. The ABAQUS finite element (FEM) code was used to obtain predictions of bow intrusion such as is illustrated in Figure 9-3. In this work, two colliding bow profiles (bulbous and raked) were forced into LNG vessels to determine the energy required for different resultant hole dimensions. Figure 9-3 illustrates the finding that while the LNG vessels are very strong, when sufficient energy exists to penetrate a hull and all the structural elements, there is little residual resistance offered by the tank to differentiate a small hole from a large one. This paper recommended a planning approach based on risk principles, rather than nominating a specific hole size, which may be too pessimistic or too optimistic. However, the paper suggested 750 mm as a maximum credible hole size for an operational accident.

The breach sizes selected for the Rabaska project (Pitblado et al., 2008), based on past events and upon discussion with experts were 250 mm for a tank puncture, 750 mm for a collision or grounding accident, and 1.5 m for an intentional act. In previous studies, Sandia National Laboratories (Hightower et al., 2004) used 1.1 and 1.6 m for accidental breaches and 2.5 m for an intentional breach. A study done by the ABS Consulting Inc. (2004) for the Federal Energy Regulatory Commission (FERC) used 1 m breach for long-lasting leaks and 5 m to obtain shorter-lasting peaks, specifically not attributing causes for these sizes or accounting for their probability. The subject of breach sizes is still unsettled (MIACC, 2007).

9.4.6.4 Location of LNG tanker penetration

Analysts have defined three categories of postulated LNG spills from carriers as shown in Figure 9-4. The types are basically Category I (above the water level), Category II (at the water level), and Category III (below the water level).

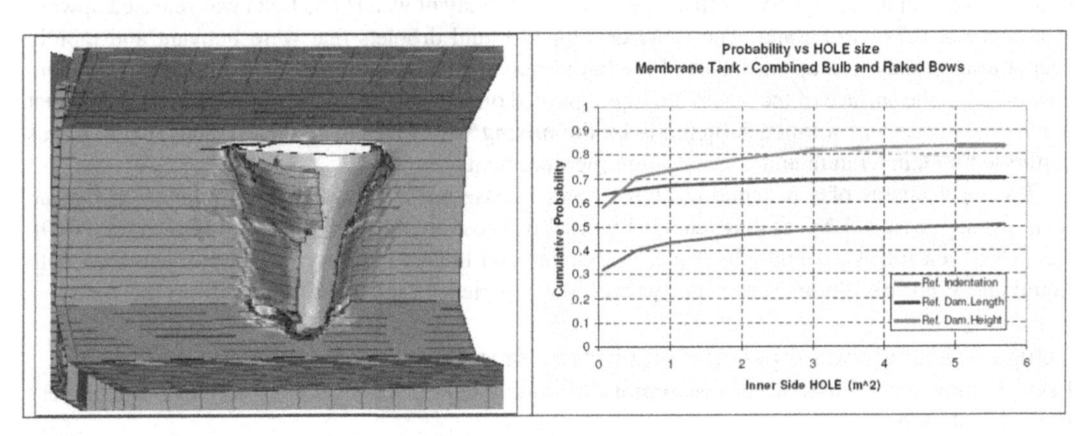

FIGURE 9-3

FEM prediction of bow intrusion into a membrane carrier and resultant hole sizes.

(Pitblado et al., 2008)

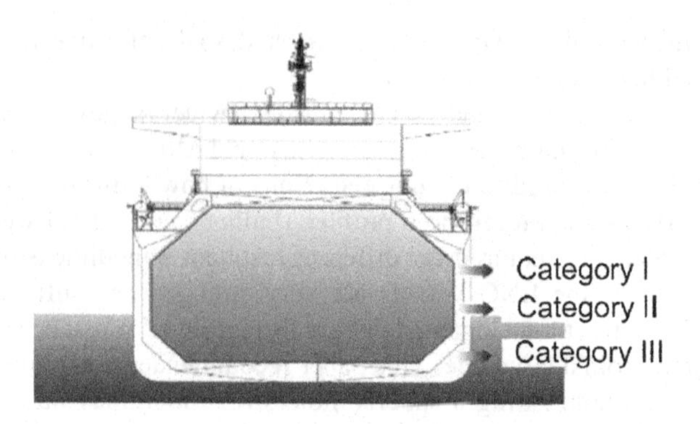

FIGURE 9-4

Types of LNG leak location.

(Luketa-Hanlin and Hightower, 2008a)

Category I, above water penetrations of membrane carriers

For penetrations of LNG carriers above the water level (Category I), the potential leak of LNG is obviously limited. High penetrations, just under the LNG level, do not have enough hydraulic head pressure to develop a jet that shoots beyond the double hull gap. With lower penetrations, still well above water level, the jet can shoot the double hull gap and spill into the sea. These plumes will partially evaporate before reaching the sea, and then penetrate into the sea water. Such penetrations into the water will consequently mix with water and rapidly evaporate, become buoyant, and "geyser" back upward.

This type of geyser behavior was observed from a release of LNG underwater as reported by Qi et al. (2011). The resulting visible plumes of condensed water vapor and partially evaporated LNG are shown in Figure 9-5 for (A) the initial plume, and (B) a later plume. The motivation for the tests was to find the effects of a leak of LNG from a pipe laid under a body of water. The LNG was released upward at a depth of 0.71 m of water. The LNG developed liquid droplets that were buoyant and rapidly evaporating as they rose upward. Part of the liquid remained unevaporated as the plume carried them upward from the surface of the water. No liquid pooled on the water surface. Figure 9-5(B) shows that the visible plume was warmed sufficiently by the mixing with water to be clearly buoyant. This was confirmed by temperature and concentration measurements.

The mechanisms of a penetrating above-water release are expected to be similar to that of an underwater release of LNG as illustrated in Figure 9-6. These are modeled by Raj and Bowdoin (2010). The LNG first forms an umbrella shape, evaporates and breaks into drops of liquid, and then both liquid and vapor rise through the water where they experience rapid heat transfer from the water.

Category III, underwater penetration of membrane carriers

Less obvious, is that underwater penetrations of double-hull LNG ships are likely to result in very limited LNG loss.

Underwater penetration of the outer hull only would produce a build-up of pressure in the double hull space as water inflow compresses the air space. This is shown by Woodward (2008) to produce an intermediate condition of pressure equilibrium between the water pressure at the breach and the air

FIGURE 9-5

Plumes from underwater release of LNG: (A) Initial plume; (B) Later plume.

(Qi et al., 2011)

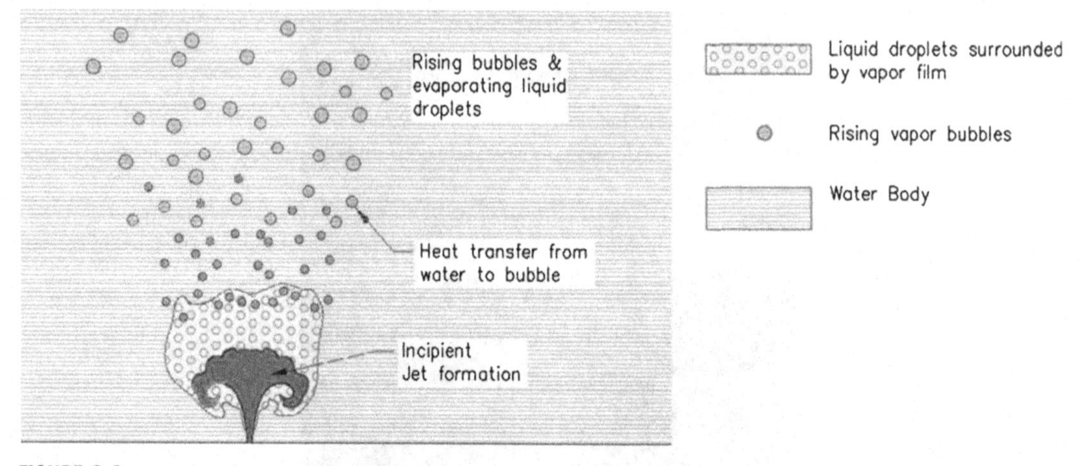

FIGURE 9-6

Schematic mechanism of underwater LNG release developing liquid and vapor bubbles by heat conduction with water.

(Raj and Bowdoin, 2010)

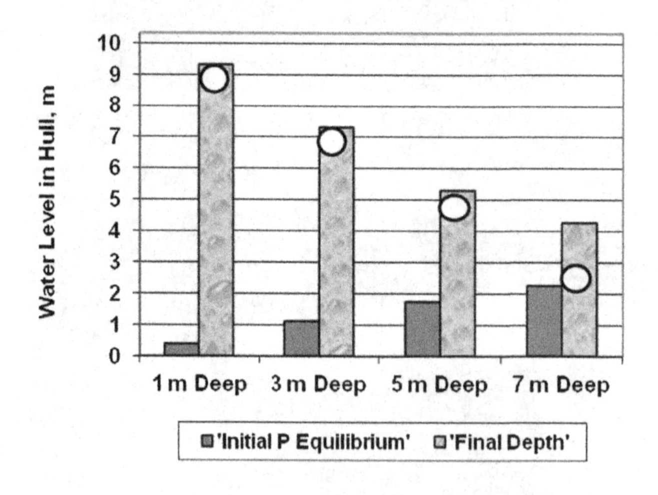

FIGURE 9-7

Water depth in hull with Type III outer hull penetration.

(Woodward, 2008)

pressure within the hull. Figure 9-7 illustrates a 0.5 m diameter hole at four water depths (white circles). The dark bars on the left indicate that this intermediate pseudo-equilibrium point occurs with increasing water depth as the hole occurs deeper beneath the sea level. After this intermediate pressure equilibrium point, water inflow continues at constant pressure by a mechanism of equal volume exchange of water and air, giving essentially constant water flow rate until a final water level is reached

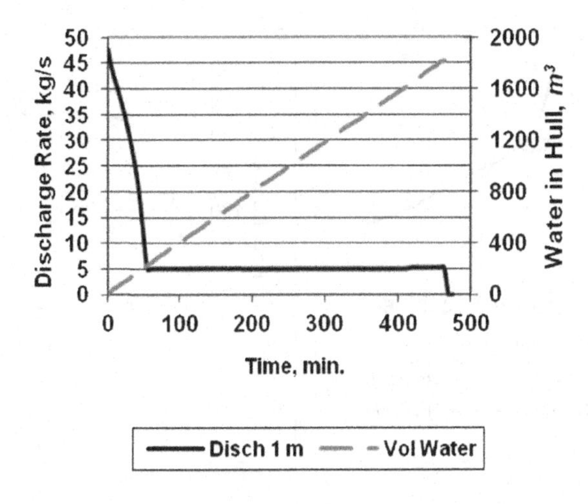

FIGURE 9-8

Predicted behavior of water inflow rate for 0.15 m hole at 1 m depth in outer hull of membrane tanker.

(Woodward, 2008)

at or above the top of the breach. This final level is indicated by the right-hand bar drawn behind the white circles.

The predicted ideal behavior of water inflow to a double hull membrane carrier from an underwater penetration is illustrated in Figure 9-8. This is for a 0.15 m diameter hole topped at 1 m water depth. The initial inflow rate drops off rapidly as compression builds the hull pressure and reduces flow. A period of constant flow rate by equal volume exchange follows, allowing the water level in the hull space to rise to just above the hole.

The increase of pressure in the hull space is further enhanced if the inner hull is also penetrated, and the evaporating LNG further increases the hull pressure, inhibiting the leak rate of LNG. This is predicted to have a strong mitigating effect making double hull release rates much lower than those of single hulled vessels.

Category II, penetration of membrane carriers at water level

Category II LNG spills, at the water level, must realistically be considered as falling partially below the water level. Thus, both seawater and LNG flow into the double hull area at first, until the double hull below the breach is filled. Depending on the relative size of the breach in the outer hull to the inner hull, any serious breach of the inner hull will leak enough LNG to freeze the water in the hull space. With high mixing of water and LNG there may also be complicating RPTs that could even be damaging. The complications of this situation are, as a first-pass approximation, ignored and analyses of Category II spills consider only the LNG leaking over the filled double hull and into the sea.

For a Category II release, the problem is inherently dynamic, since the discharge rate increases and the duration decreases as hole size increases. In addition, the discharge rate decreases over time as the liquid level in the LNG tank drops. Woodward (2007) coupled the dynamics of a decreasing discharge rate (blow down) with a pool spread and evaporation model. As shown in Figure 9-9, this method predicts that a pool from a 1 m hole rapidly reaches an equilibrium where pool evaporation equals

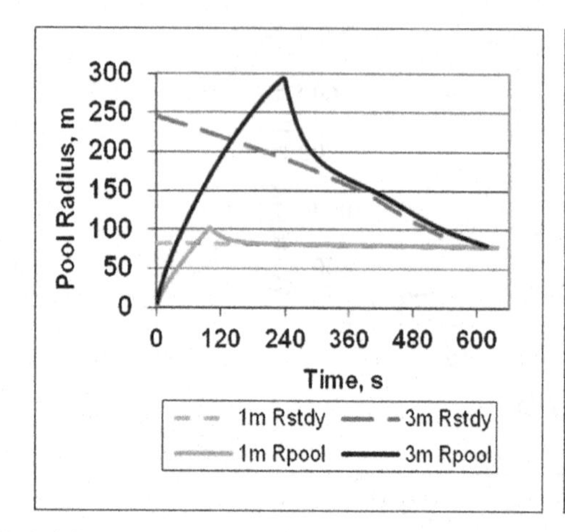

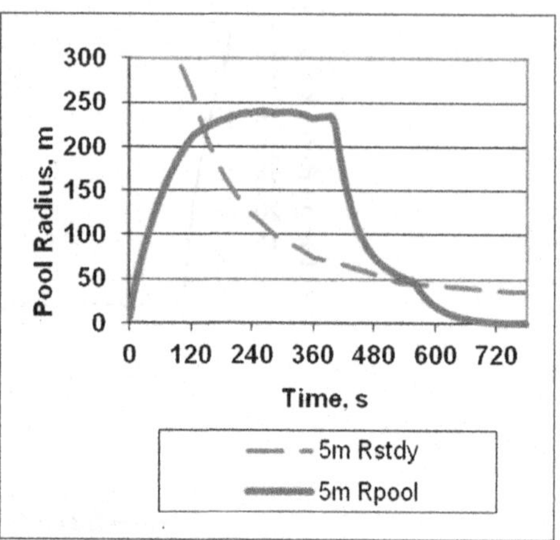

FIGURE 9-9

Time-varying source rate from blowdown coupled with pool radius for an unignited LNG pool for a 25,000 m^3 spill.

(Woodward, 2007)

discharge rate, labeled a "steady-state" curve. A 3 m hole would produce the largest pool extent, reaching a peak value just under 300 m only briefly and then dropping down to the equilibrium or steady-state curve. The predicted pool radius from a 5 m hole rises quickly and then flattens to a broader peak just under 250 m because the discharge rate drops more quickly. After the discharge from the tank stops, the pool radius drops so rapidly that the pool radius curve does not follow the equilibrium "steady-state" curve.

9.4.6.5 Steady-state predictions for hazard distances from LNG spills

With the understanding from Figure 9-9 that maximum pool sizes may remain only very briefly, next consider predictions using the equilibrium or steady-state pool assumption (pool area such that evaporation or burn rate equals initial discharge rate).

To obtain a feel for the expected scale of the hazards from a large LNG spill on water, two scenarios are evaluated, a pool with immediate ignition and a pool with no ignition. For the first of these scenarios, the analysis finds radiation contours from a pool fire to thermal flux levels of 37.5 kW/m^2 and 5 kW/m^2, commonly recognized for defining hazard distances for fires (NFPA 2001). The 37.5 kW/m^2 is a level suggesting severe structural damage and major injuries if continued for over 10 min. The 5 kW/m^2 level found to produce second-degree burns on bare skin exposed for over 20 sec, and is proposed as the protection standard for people in open spaces.

Several studies have developed hazard distances for Category 2 LNG breaches, both ignited and unignited, as summarized in Woodward and Pitblado (2010). The Sandia guidance report (Hightower et al., 2004) evaluates ignited and unignited scenarios for currently standard membrane LNG vessels holding 125,000 to 140,000 m^3 of LNG. The loss for a single tank out of the four to six tanks in a membrane carrier would be about 12,500 m^3. Qiao et al. (2006) investigated the influence of geometric difference between membrane and Moss spherical tanks on the LNG release rate and blow down, but did not carry out any further consequence analysis.

Table 9-5 Predicted Thermal Hazard Distances in Sandia Report

Hole Size (m^2)	Multiple of Base Burn Rate	Pool Diameter (m)	Burn Time (min)	Distance to 37.5 kW/m^2(m)	Distance to 5 kW/m^2 (m)
Accidental Events					
1	1	148	40	177	554
2	1	209	20	250	784
Intentional Events					
5	1	330	8.1	391	1305
5	2	220	8.1	253	810
12	1	512	3.4	602	1920

(Hightower et al., 2004)

Table 9-5 summarizes predicted distances for pool fires as a function of the size of penetration of the inner hull. The assumptions for Table 9-5 are:

- LNG composition = 100% methane (density at boiling point = 422.5 kg/m^3)
- Discharge coefficient = 0.6
- Burn rate = 0.30 mm/s (0.127 kg/m2.s)
- Surface emissive power = 220 kW/m^2
- Pool at steady-state (burn rate = initial discharge rate)
- Burn time = discharge time at initial discharge rate
- Pool shape = semicircle.

By comparison, a study by the Major Industrial Accidents Council of Canada (MIACC, 2007) found for a 750 mm breach (0.44 m^2) with very similar assumptions, radiation contours to 5 kW/m^2 range from 450 m to 480 m, consistent with Table 9-5 values.

For the second scenario, an unignited (or delayed ignition) pool, the dispersion hazard distance is the longest length of a (transient) flash fire. This is taken as the distance to the lower flammability limit (LFL), the lowest concentration at which LNG will burn.

Table 9-6 summarizes these distances from the Sandia report (Hightower et al., 2004) for stable atmospheric conditions at low wind speed. Example spills from larger vessels are given in later Sandia reports (Hightower et al., 2006; Luketa-Hanlin et al., 2008b).

A study by Oka (2009, 2010) uses the same models as the Sandia report (Hightower et al., 2004), and provides more detail of predictions as a function of hole size. Oka (2010) extends an earlier treatment (Oka and Ota, 2008) to the larger Q-Max carriers, using the assumptions summarized in Table 9-7.

Oka's modeling assumptions are:

- Category II breach centered at the water line
- LNG composition = 100% methane (density at boiling point = 422.5 kg/m^3)
- Discharge coefficient, $C_D = 0.65$
- Burn rate = 0.668 mm/s (0.282 kg/m^2.s)

Table 9-6 Predicted LFL Distances for Unignited LNG Spills in Sandia Report

Hole Size (m^2)	Tanks Breached	Pool Diameter (m)	Spill Duration (min)	Distance to LFL (m)
		Accidental Events		
1	1	181	40	1536
2	1	256	20	1710
		Intentional Events		
5	1	405	8.1	2450
5	3	701	8.1	3614

(Hightower et al., 2004)

Table 9-7 Release Assumptions of Oka (2010)

LNG Carrier	Conventional	Latest
Total cargo capacity	125,000 m^3	250,000 m^3
Single tank volume	25,000 m^3	50,000 m^3
Total spill volume	14,300 m^3	28,600 m^3
Initial LNG level above water	13.0 m	13.2 m
Breach equivalent diameter	0.5 to 15 m	0.5 to 15 m
Draft	11.8 m	

- Evaporation flux (not burning) = 0.17 kg/m^2s
- Surface emissive power = 265 kW/m^2
- Friction effects included in pool spread mode = yes
- Flame model = two-zone solid cylinder including tilt by wind
- Pool at steady-state (burn rate = initial discharge rate)
- Burn time = discharge time at initial discharge rate
- Pool shape = semicircle
- Averaging time for dispersion = few seconds (point values not averaged)
- Obstacles or terrain effects for dispersion = none.

The tank dimensions and spill volume used by Oka are based on Fay's study (Fay, 2003). The membrane tanker geometry is simplified to a rectangular box with a draft, d_r, and the height of the tank initially above the water level, h_t, including vapor ullage. The initial LNG level above the water line, h_O, is $1.1d_r$ for a conventional carrier. This gives the cargo surface area, A_t, in terms of the volume of cargo, V_{ct}, as:

$$A_t = 0.52\frac{V_{ct}}{d_r} \quad (9\text{-}1)$$

The discharge rate for a box with a circular hole of area A_{hole} driven by the pressure of the hydraulic head (ρgh) is:

$$\rho A_t \frac{dh}{dt} = C_D A_{hole}\, \rho (2gh)^{1/2} \tag{9-2}$$

which is integrated to give the drain time, t_s, as:

$$t_s = \sqrt{\frac{2h_0}{g}} \left(\frac{4A_t}{C_D \pi} \right) d^{-2} \tag{9-3}$$

Thus, on a log-log plot, the drain time plots linearly against hole diameter d with a slope of –2 as shown in Oka's results in Figures 9-10 and 9-11 for ignited and unignited pools, respectively. With the increase in the breach diameter, though, the curve representing the fire duration in Figure 9-10 or evaporation duration in Figure 9-11 begins to deviate from the straight line for the spill duration. The total spill duration is much shorter than the fire duration or the evaporation duration when breach diameters are larger than about 5 to 6 m for both sizes of LNG carrier (LNGC).

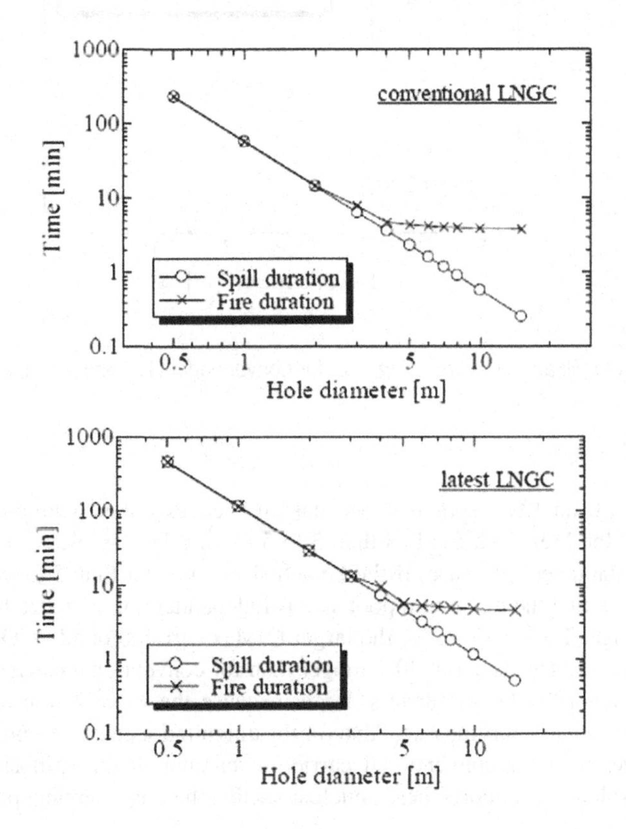

FIGURE 9-10

Predictions of duration of spill and of pool fire for conventional (150 km^3) and latest LNG carriers (250 km^3).

(Oka, 2010)

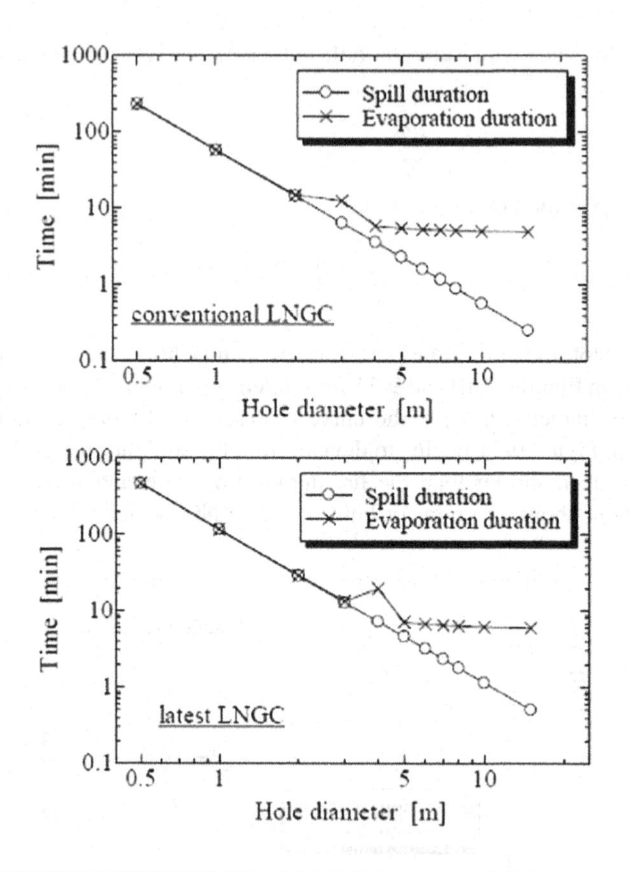

FIGURE 9-11

Predictions of duration of spill and of evaporating pool for conventional (150 km^3) and latest LNG carriers (250 km^3).

(Oka, 2010)

From these findings, an LNG spill can be characterized as either a long-duration release (or continuous release) if the breach size is less than 3 to 5 m, or a large-scale release of short duration (approximating an instantaneous release) if the breach size is greater than 5 to 6 m.

The implications are that the maximum pool size is independent of hole size for large holes in the instantaneous spill range. The pool size for the larger Q-Max carriers found by Oka is approximately 430 m for the ignited case. This is about 30% longer than for conventional carriers. For the unignited spill it is about 480 m for the larger carriers, larger because the vaporization rate in the unignited scenario is lower than the mass burning rate. Also for the unignited scenario the pool size for the newer carriers is 30% longer than for conventional carriers even though the spill size doubles. This is important if further evaluation supports these conclusions that there is a capping pool size for breaches larger than 5 m.

The main uncertainties that need to be addressed are whether environmental conditions such as waves and currents break up a single pool into multiple irregular-shaped pools.

9.4.7 Onshore and offshore plant differences

Offshore plants, including Floating LNG plants (FLNG), have close spacing because of the high cost of building on platforms. Accordingly, fire and cryogenic protection becomes an asset protection issue. Egress and safe refuge for operators is a high priority. High congestion requires attention to blast walls around the control area. Jet fire is often a design basis for fire protection. These considerations favor the use of modular designs (Tanabe and Miyake, 2010). For most onshore plants more generous spacing can be applied between equipment to limit escalation. Savings in fire and cryogenic protection can be achieved without compromising safety. Even so, pumping and piping costs result in still relatively congested process equipment. Relocating personnel to safe areas is usually not an issue. A major risk concern for onshore plants is the effects beyond the plant boundaries. Pool fire is the main design basis for risk and jet fire is considered a residual risk. Congestion can be moderated.

With an increasing number of onshore plant developments in remote locations where labor mobilization or extreme weather conditions are difficult and/or site area is minimized to protect sensitive environments, some of the considerations for offshore construction are pertinent. In these cases, modular construction is being applied onshore. An alternative approach for onshore modularized design mediates between offsite and onsite approaches (Tanabe and Miyake, 2010).

9.4.8 Onshore plants, process safety management systems

Safety is vital to the public acceptance of LNG and to the economic viability of the industry. Even though it is highly regulated, regulations do not require Process Safety Management (PSM) systems. Most commonly throughout the world, LNG liquefaction plants and LNG import terminals have implemented a Safety Management System (SMS) and an Environment Management System (EMS) based on the ISO 14000 standard. A survey by Working Committee 3 of the industry cooperative group Safety and Environmental Management in LNG Plants in 2002–3, reported that 81% of participating companies had an SMS (Goy, 2003). This percentage is likely higher now. In Europe SMSs are required by a European regulation. In Taiwan and Korea SMSs are required by local regulation. In the United States and the Caribbean, no such federal requirement exists, but companies have developed SMS in accordance with their internal company policy. In Japan Goy (2003) reported at that time there was neither regulation nor implementation by company policy.

The benefits of an SMS are reported to be a decrease in material damages and in the number of employee injuries as well as improvements in plant productivity, availability of sendout equipment, and decreased product losses.

Improvements are always possible in human endeavors. It is wise to consider lessons from the best available example of an SMS, in spite of the fact that it is in a different "industry" than LNG: the safety program of the US Navy for nuclear submarines. For over 50 years of nuclear plant operations with as many as 100 operational reactors, the Nuclear Navy has not had a single loss of life or major environmental accident because of process safety. As described by Paradies (2011), at the dawn of the nuclear age, Admiral Hyman G. Rickover (2011) realized that using nuclear reactions to make steam requires a different approach to operation, maintenance, and management than was the tradition of naval engineering. He created a set of standards and a safety culture that was, and still is, effective and truly unique. His three major elements and 18 detailed elements are outlined in Congressional testimony as Technical competence, Total responsibility, and Facing the facts.

These principles require first detailed technical knowledge of the process, not only for engineers designing processes but also for the senior managers, middle managers, supervisors, and operators of the technology. Rickover (2011) said

> *At Naval Reactors, I take good engineers and make them managers. ... "The Devil is in the Details" is especially true in technical work. If you ignore these details and try to rely on management techniques or gimmicks you will surely end up with a system that is unmanageable and problems will be immensely more difficult to solve.*

By Total Responsibility Admiral Rickover developed a policy that "unless you can point your finger at one person who is responsible when something goes wrong, then you have never really had anyone responsible." Anyone in his organization was given total responsibility to stop the job if something goes wrong. This policy does not tolerate finger-pointing. "The lack of a single person taking total responsibility is what happens with shared responsibility, i.e., no one is really responsible" (Rickover, 2011).

Facing the facts is terminology for making difficult decisions that favor process safety and quality despite the cost, effort, delay, or potential bad press. Rickover (2011) said that it is human inclination to "...hope that things will work out, despite evidence or suspicions to the contrary."

9.4.9 Risk analysis tools

At the heart of a good risk management system is a systematic method to identify hazards, assess their likelihood and consequences, control the process effectively, shut down safely in an emergency, and recover from any loss of containment with minimum consequences. Budget decisions are made each year considering alternative strategies to justify the most cost-effective measures to bring the risks "As Low As Reasonably Practicable, ALARP" (Chamberlain, 2006). Qualitative risk analysis tools can identify and prioritize hazards; quantitative methods can quantify the benefit side for risk/benefit analyses.

The following formal methods of Process Hazard Analysis (PHA), Hazard Identification (HAZID), Semiquantitative Risk Analysis, and Quantitative Risk Analysis (QRA) are introduced here (expanding upon Keong, 2012; see also HSE, 2002; ISO Standard 17776:2000).

- Qualitative Methodologies
  - Preliminary Risk Analysis
  - What-if Analysis
  - Hazard and Operability Analysis (HAZOP)
  - Failure Mode and Effects Analysis (FMEA/FMECA)
  - Multiple Attribute Utility Function Analysis
- Approximate Quantitative Methods
  - Layers of Protection Analysis (LOPA)
- Tree-Based Techniques
  - Fault Tree Analysis (FTA)
  - Event Tree Analysis (ETA)
  - Cause-Consequence Analysis (CCA)
  - Bow-tie method

    - Barrier-Systematic Cause Analysis Technique (BSCAT)
    - Management Oversight Risk Tree
    - Safety Management Organization Review Technique
- Techniques for Dynamic System Analysis
    - Go Method
    - Digraph/Fault Graph
    - Markov Modeling
    - Dynamic Event Logic Analytical Methodology
    - Dynamic Event Tree Analysis Method
- Individual and Societal Risk Methods
    - Quantitative Risk Analysis (QRA).

9.4.9.1 Qualitative risk analysis methodologies

Preliminary risk analysis

Preliminary Risk Analysis or hazard analysis is a qualitative technique that involves a disciplined analysis of the event sequences that could transform a potential hazard into an accident. In this technique, the possible undesirable events are identified first and then analyzed separately. For each undesirable event or hazard, possible improvements or preventive measures are then formulated.

This methodology provides a basis for determining which categories of hazard should be looked into more closely and which analysis methods are most suitable. Needed safety measures can also be readily identified. With the aid of a frequency/consequence diagram, the identified hazards can then be ranked and prioritized according to risk.

What-if analysis. What-if studies are qualitative, postulating a deviation and asking what would be the consequences of this deviation.

Hazard and operability studies (HAZOP)

The HAZOP technique was developed in the early 1970s by Imperial Chemical Industries Ltd. (Crowl and Louvar, 2002). HAZOP can be defined as the application of a formal systematic examination of the hazard potential from deviations in designed operations and the consequential effects on the facilities. HAZOP evaluations assemble a team of specialists in a given process, including designers and operators. The team reviews the design piping and instrumentation drawings concentrating on each piece of equipment in succession.

This technique is usually performed using a set of guidewords: NO/NOT, MORE/LESS OF, AS WELL AS, PART OF, REVERSE, and OTHER THAN. From these guidewords, scenarios are identified that may result in a hazard or an operational problem. For example, assessing possible process line flow problems, the guide word MORE OF will correspond to high flow rate, LESS THAN, for low flow rate. The consequences of the hazard and measures to reduce the frequency of occurrence is then discussed and documented by Key Causation, Controls, and Needs for Improvement. This technique had gained wide acceptance in process industries.

Failure modes and effects analysis (FMEA/FMECA)

This method was developed in the 1950s by reliability engineers to determine problems that could arise from malfunctions of military system (Stamatis, 1995). Failure mode and effects analysis is a procedure by which each potential failure mode in a system is analyzed to determine its effect and to classify it according to its severity.

When the FMEA is extended by a criticality analysis, the technique is then called failure modes and effects criticality analysis (FMECA). FMEA has gained wide acceptance by the aerospace and the military industries. In fact, the technique has been adapted to other forms such as misuse mode and effects analysis.

Multiple attribute utility risk analysis. The multiple attributes as described by (2010) refer to consequences to (1) crew (C), (2) third-party personnel (P), (3) environment (E), (4) ship(S), (5) downtime (D), (6) reputation (R), and (7) third-party material assets (M). The method basically obtains a weighted average ranking over the seven variables by assigning a ranking for likelihood and consequence. The likelihood scale is (scaled per vessel year) (1) improbable (< 0.0005), (2) remote (0.0005–0.005), (3) occasional (0.005–0.05), (4) probable (0.05–0.5), and (5) frequent (> 0.5).

The consequence scale is:

1. Minor or negligible effect for all six attributes
2. Major (e.g., serious injury to crew or minor injury to third-party personnel, moderate damage to ship, one day downtime, local effect on reputation, and minor damage to third-party assets)
3. Critical (e.g., single fatality to crew, serious injury to third-party personnel, major release reportable to regulatory authorities, major damage to ship, one week downtime, national effect on reputation, and major damage to third-party assets)
4. Catastrophic (e.g., several fatalities to crew, fatalities to third-party personnel, uncontrolled pollution, loss of ship, more than one week downtime, loss of company reputation, and extensive damage to third-party assets).

The strength and the weakness of this method is that it requires estimation of values for up to seven values for every scenario for both frequency and consequence. The method forces management to refine their relative valuations, but it remains entirely in the qualitative domain. It adds little value when all frequencies and consequences are blunt estimates.

Applications of qualitative methods

The techniques outlined earlier require involvement of *hardware familiar* personnel. FMEA tends to be more labor intensive, as the failure of each individual component and subsystem and overall system limits have to be considered. These qualitative techniques can be used in the design as well as the operational stage of a system.

These techniques have seen wide use in nuclear and chemical processing plants including offshore platforms. FMEA has been used in several industries to improve the reliability of their products.

9.4.9.2 Approximate quantitative methods

Layers of Protection Analysis (LOPA) analysis is based on the concept that if independent barriers can be put in place at various steps along the chain of events leading to an undesirable consequence, the probability of the event can be decreased (CCPS, 2001). A LOPA analysis draws diagrams with each independent protection layer (IPL) shown in series. Figure 9-12 illustrates the general layer of protection concept. Specific protection devices or practices are applied for specific cases. The method is approximate because the risk reduction is taken as an order of magnitude for each added barrier.

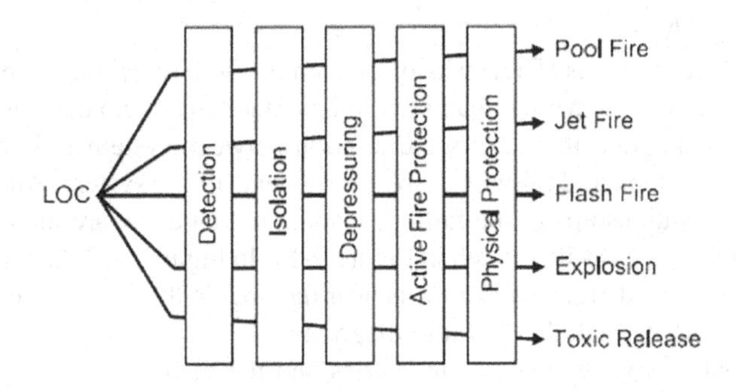

FIGURE 9-12

General layer of protection concept for emergency systems.

(Tanabe and Miyake, 2011)

Not all safeguards are IPLs because IPLs must meet seven requirements (CCPS, 2007; Dowell, 2011):

- **Independence.** Protection layer is not affected by the initiation event or by other protection layers.
- **Functionality.** The protection layer can prevent the consequence from occurring.
- **Integrity.** The protection layer performs at a specified low failure rate.
- **Reliability.** The protection layer will operate as intended under defined conditions for a specified time period.
- **Auditability.** Ability to inspect and demonstrate achievement of attributed feature.
- **Access security.** Administrative and physical means to reduce the potential for unintentional and unauthorized changes.
- **Management of change.** Formal process to review other than "replacement in kind."

If they are independent, then each layer must fail for the loss event to occur. The probability of failure does not need to be known precisely, but rather, an order of magnitude estimate is used. Enough layers are added until the tolerable risk criteria adopted by the facility is reached. LOPA can also be used to rank the estimated risk values and give priority to the mitigation measures that contribute most to risk reduction. Essentially, though, LOPA studies one barrier at a time and is not strong in finding the systemwide risk for an interacting system.

9.4.9.3 Tree-based techniques

Tree-based methods include Fault Tree Analysis (FTA), Event Tree Analysis (ETA), Cause–Consequence Analysis (CCA), the Bow-Tie method, Barrier-Systematic Cause Analysis Technique (B-SCAT), Management Oversight Risk Tree (MORT) and the Safety Management Organization Review technique (SMORT). The first four provide graphic description of accident sequences and include analysis of barriers to prevent accidents. B-SCAT and MORT use Bow-Tie methods, but are used more for accident investigation than risk assessment.

These methods were first developed to determine the reliability of electronic systems. They often involve substantial effort and cost.

Fault tree analysis (FTA)

The concept of Fault Tree Analysis (FTA) was originated by Bell Telephone Laboratories in 1962 to perform a safety evaluation of the Minuteman intercontinental ballistic missile launch control system. A fault tree is a logical diagram that shows the relation between system failure; that is, a specific undesirable event in the system, and failures of the components of the system. An undesirable event is first defined and causal relationships of the failures leading to that event are then identified as related through "AND" and "OR" gates as illustrated in Figure 9-13. In Figure 9-13 the top event, "Fire breaks out" is above an AND gate, and "Ignition source is near flammable fluid" is above an OR gate. Human error probabilities can also be included in these diagrams.

This method is used in a wide range of industries and the method is readily documented using software packages such as FT+. The NASA Handbook provides an excellent description of the method. FTA has a number of limitations, including the assumption that the causes are random and statistically independent, but certain common causes can lead to correlations in event probabilities. Such correlations violate the basic assumptions and could exaggerate the calculated likelihood of the top event. Missed or unrecorded causes may equally bias the calculated likelihood. The assumption that the sequence of events is not relevant can be a serious flaw. Markov-chain techniques are needed in this event.

Event tree analysis (ETA)

ETA diagrams the branches of consecutive events (fail or no-fail; Pate-Cornell, 1984). Each branch point has a probability, p_i for yes, and $(1-p_i)$ for no. For events that occur in series each probability along the path is multiplied to give the final probability for that path. ETA ideally identifies all possible failures. (What faults might we expect? What do they affect?) It is a good technique for working out the overall probability of a catastrophic event occurring.

A simplified event tree for an LNG spill onto the sea is shown in Figure 9-14. Pool ignition obviously applies to both pool fire and VCE. The diagram includes a potential RPT (Rapid Phase

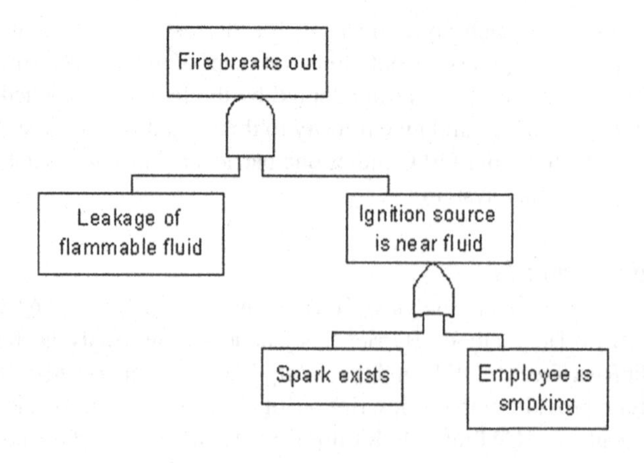

FIGURE 9-13

A fault tree depicting the event "Fire breaks out."

(Pacific.net web site, 2012)

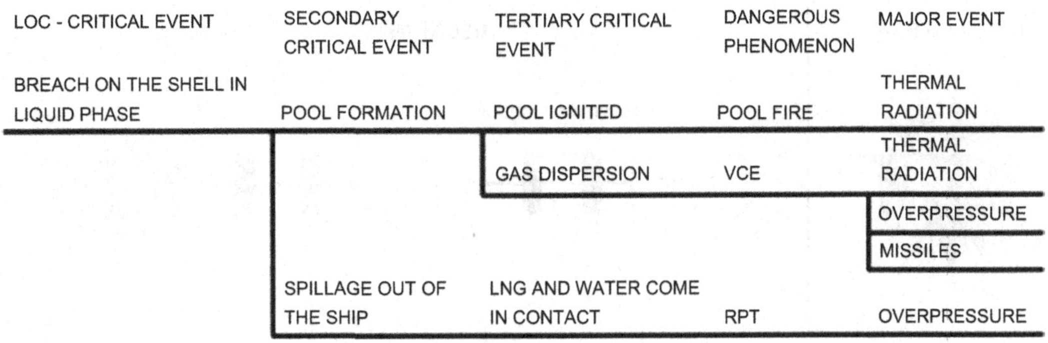

LOC - CRITICAL EVENT	SECONDARY CRITICAL EVENT	TERTIARY CRITICAL EVENT	DANGEROUS PHENOMENON	MAJOR EVENT
BREACH ON THE SHELL IN LIQUID PHASE	POOL FORMATION	POOL IGNITED	POOL FIRE	THERMAL RADIATION
		GAS DISPERSION	VCE	THERMAL RADIATION
				OVERPRESSURE
				MISSILES
	SPILLAGE OUT OF THE SHIP	LNG AND WATER COME IN CONTACT	RPT	OVERPRESSURE

FIGURE 9-14

Event tree for LNG spills on sea from an off-shore floating storage and regasification unit (FSRU).

(Tugnoli et al., 2010)

Transition) even though it is not yet considered to be a damaging event. The probability of each branch in the event tree can be estimated using failure rate tables. These probabilities can be decreased by Levels of Protection (LOP) and the analysis is used to justify these LOP.

Bow-tie analysis and BSCAT. Bow-tie analysis and cause–consequence analysis (CCA) combine cause analysis (described by fault trees) and consequence analysis (described by event trees), and hence employ both deductive and inductive analysis.

The bow-tie method expands logic trees on both sides of a "top event," hence the appearance of a bow tie. To the left is a fault tree representing failure modes that lead to the top event, and to the right are event trees representing possible consequences of the top event. The method adds barriers to the left and right of the top event, representing, respectively, Prevention Controls and Mitigation Controls. Thus, the bow-tie method highlights the direct link between hazard controls (barriers) and elements of the management system. It satisfies the UK Control of Major Hazards (COMAH) regulations to "Provide a clear link between the various accident scenarios identified and the measures which are in place to defend against them." By visually illustrating the hazard, its causes and consequences, and the controls to minimize the risk, the bow-tie can be readily understood at all levels from senior management to the public. It also provides greater ownership by stressing that people are responsible to keep barriers in working order. When people feel involved they tend to "buy-in" and take ownership. It can reduce the volume of safety analysis and lead to a reduction in unnecessary or low-importance barriers (Smith, 2010).

After being developed in Australia and The Netherlands in the 1980s it has grown in popularity. Shell Oil describes the method in a paper to the European Commission (Zuijderduijn, 2000). A further description is given by Delvosalle et al. (2006), Pitblado and Tahilramani (2009), and Smith (2010).

An extension of the method by Pitblado et al. (2011) as shown in Figure 9-15 is called Barrier-Systematic Cause Analysis Technique (BSCAT). The figure also compares BSCAT with a previous SCAT method developed primarily for accident investigation. The objective of accident investigation methods is to work backward from an accident event to find root causes, whereas the objective of prevention measures is to work forward from possible causes (faults) to accidents. Thus, the same logic readily applies to both objectives. Accident investigation methods are further developed in Section 9.4.12.

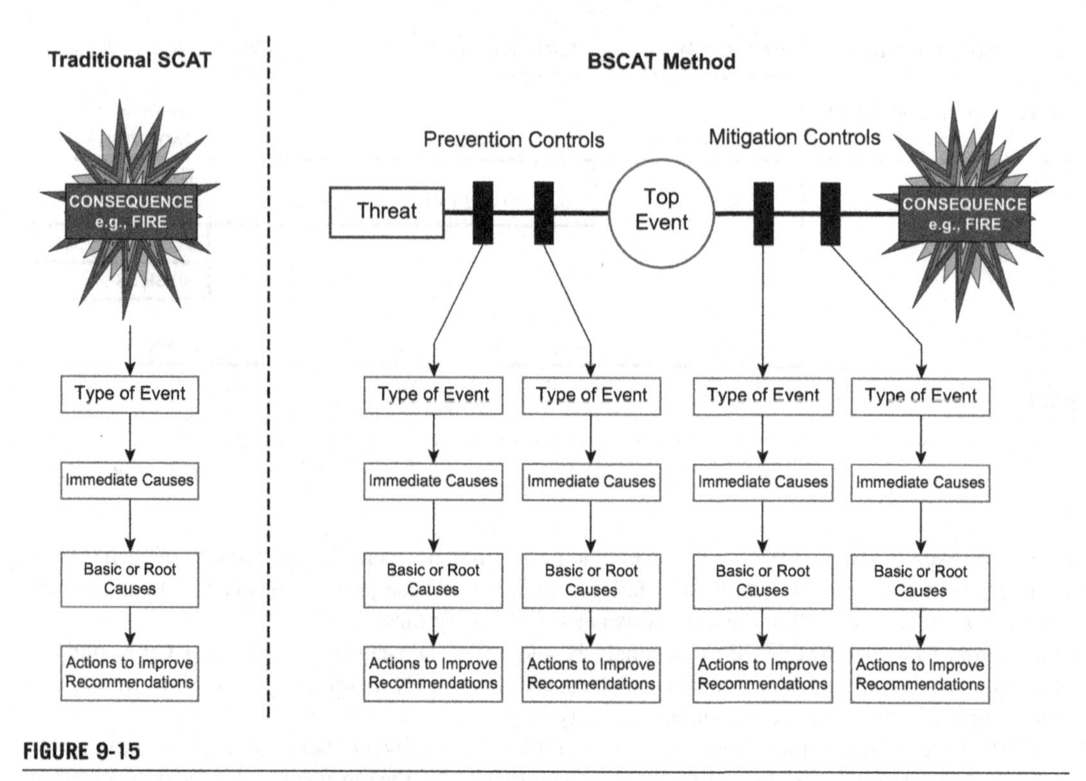

FIGURE 9-15

Generic BSCAT bow-tie diagram.

(Pitblado et al., 2011)

The premise of the BSCAT method is that for an incident to occur, at least some of the barriers must fail or become partly degraded. Since the nature of each barrier can be quite different from other barriers, a separate fault tree is drawn for the degradation of each barrier. The strength of the resulting method is in bringing clarity to a complex situation where, in an actual accident, up to eight barriers have been found to fail for the accident to occur (e.g., a damaging water hammer event, Berrera and Kamel [2010], the BP findings for the Macondo platform incident in the Gulf of Mexico, 2009 [Pitblado et al., 2011]).

The SCAT method and hence the BSCAT method makes use of a list of over 150 Basic Causes and 80 Immediate Causes as prompts for the accident investigators. These are listed in four categories of Process Safety events, arranged in declining severity: A-B-C-D.

Dynamic procedure for atypical scenarios identification (DyPASI). A consortium of Italian university professors addressed a concern that not all potential incidents are being addressed by other methods. As part of the ARAMIS project for identification of accident scenarios (Delvosalle et al., 2006) they developed the Dynamic Procedure for Atypical Scenarios Identification, DyPASI (Paltrinieri and Cozzani, 2010). The authors define atypical scenarios that are usually excluded in risk analysis for lack of data and modeling procedures. In particular, rapid phase transition (RPT) explosions fit into this category. The method proposed is not dynamic in the sense of treating a time-varying sequence of accident events, but rather in the sense of accepting continual model improvements.

Table 9-8 Specific External Threats for LNG Terminals

HAZOP Guideword	Offshore Threat (Or Hazard)	Onshore Threat (Or Hazard)
Man-made	Direct attack Third-party activity Dropped object Helicopter operation Human factor	Direct attack Third-party activity Human factor

(Paltrinieri and Cozzani, 2010)

After drawing up a list of hazardous materials in a plant and listing existing safety barriers the DyPASI method combines HAZOP, a bow-tie diagram, with event trees (ETA) and fault trees (FTA). The method then defines critical events, and postulates new safety barriers. As an extension of the HAZOP method to include intentional threats, the authors suggest a guide word applied to threats listed in Table 9-8.

Cause–consequence analysis

CCA was invented by RISO Laboratories in Denmark to be used in risk analysis of nuclear power stations. However, it can also be adapted by the other industries in the estimation of the safety of a protective or other system.

The purpose of CCA is to identify chains of events that can result in undesirable consequences. The probability of various events in the CCA diagram are found, leading to the probabilities of various consequences and the risk level of the system. Figure 9-16 shows a typical CCA.

Management oversight risk tree (MORT)

Management Oversight Risk Tree (MORT) was developed in the early 1970s, for the US Energy Research and Development Administration as a safety analysis method that would be compatible with complex, goal-oriented management systems. MORT is a diagram that arranges safety program elements in an orderly and logical manner. Its analysis is carried out by means of a predeveloped fault tree; that is, the investigator does not create his or her own fault tree, which would be a very large task for routine investigations. The top event is "Damage, destruction, other costs, lost production or reduced credibility of the enterprise in the eyes of society." The tree gives an overview of the causes of the top event from management oversights and omissions and/or from assumed risks.

The generic MORT tree has defined more than 1500 possible basic events compressed to 100 events applicable in the fields of accident prevention, administration, and management. MORT is used in the analysis or investigation of accidents and events, and evaluation of safety programs.

Safety management organization review technique (SMORT)

Safety management organization review technique (SMORT) is a simplified modification of MORT developed in Scandinavia. This technique is a structured analysis process that employs analysis levels with associated checklists, as distinguished from MORT, which is based on a comprehensive tree structure.

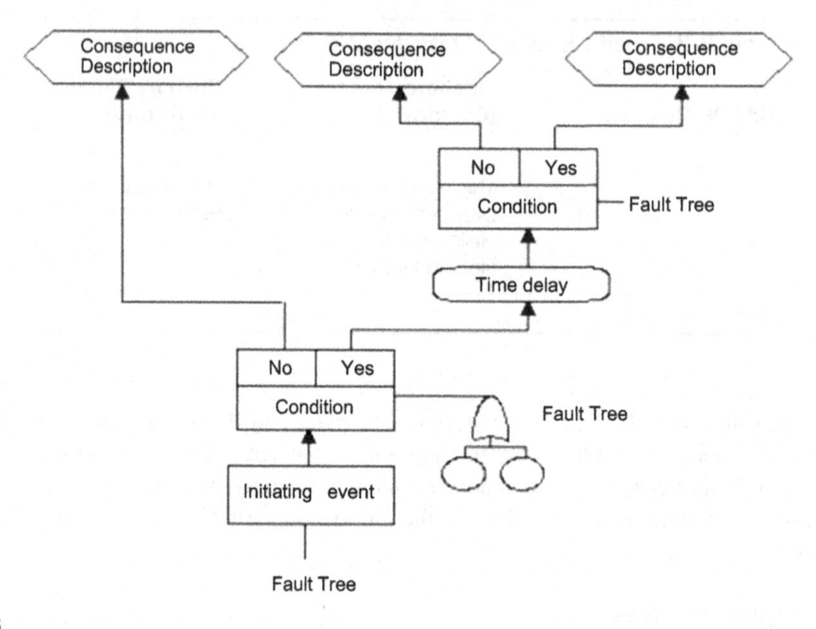

FIGURE 9-16

A typical cause-consequence analysis.

SMORT analysis begins with data collection based on the checklists and their associated questions, followed by evaluation of results. The information can be collected from interviews and studies of documents and investigations. This technique can be used to perform detailed investigation of accidents and near misses, safety audits, and planning of safety measures.

Application of tree methods

Tree-based methods are mainly used to find cut-sets or critical paths through the logic trees that lead to the undesired events. Event trees and fault trees have been widely used in probabilistic risk assessment. A strength of the methods is that hardware failures and human errors can be placed on the same tree. This requires some estimation because human behavior cannot be quantified explicitly. New techniques such as human cognitive reliability attempt to reconcile this deficiency.

9.4.9.4 Methodologies for analysis of dynamic systems

These methods do not incorporate time-varying dynamic analysis. Methods that incorporate dynamics include GO method, digraph/fault graph, event sequence diagrams, Markov behavior, dynamic event analytical methodology, and dynamic event tree analysis.

GO method

The GO method is a success-oriented system analysis that uses seventeen operators to aid in model construction. It was developed by Kaman Sciences Corporation during the 1960s for reliability analysis of electronics for the US Department of Defense.

The GO model can be constructed from engineering drawings by replacing system elements with one or more GO operators. Such operators are of three basic types: independent, dependent, and logic. Independent operators are used to model components requiring no inputs but at least one output. Dependent operators require inputs. Logic operators combine the other operators following the logic of the system being designed. After assigning a probability for success of each operator the probability of successful operation of the system can then be calculated.

The GO method is used where the boundary conditions for the system are well defined by a system schematic or other design documents. Since the failure modes are implicitly part of the GO structure, it is unsuitable for detailed analysis of failure modes. Furthermore, it does not treat common cause failures nor provide structural information (critical paths or cut sets) regarding the system.

Digraph/fault graph

The fault graph method/digraph matrix analysis uses the mathematics and language of graph theory such as path set (a set of models connected on a path) and reachability (the complete set of all possible paths between any two nodes).

This method is similar to a GO chart but uses AND/OR gates instead of GO operators. The connectivity matrix, derived from the adjacency matrix for the system, shows whether a fault node will lead to the top event. These matrices are then computer analyzed to give singletons (single components that can cause system failure) or doubletons (pairs of components that can cause system failure). The digraph method allows cycles and feedback loops that make it attractive for analyzing a dynamic system. Figure 9-17 shows a success-oriented system digraph of a simplified emergency nuclear reactor core cooling system.

Markov modeling

Markov behavior is a classic technique used for assessing the time-dependent behavior of dynamic systems. Essentially, the dynamic response is calculated again and again with different parameter values set randomly. The state probabilities of the system P(t) in a continuous system are obtained by the solution of a coupled set of first order, constant coefficient differential equations:

$$\frac{dP}{dt} = MP(t) \tag{9-4}$$

where M is the matrix of coefficients whose off-diagonal elements are the transition rates and whose diagonal elements are such that the matrix columns sum to zero. An application of Markov behavior to fire propagation on an offshore platform is discussed by Pate-Cornell (1983).

Dynamic event logic analytical methodology (DYLAM)

Dynamic event logic analytical methodology (DYLAM) provides an integrated framework to explicitly treat time, process variables, and system behavior (Siu, 1994). A DYLAM will usually be comprised of the following procedures: (1) component behavior, (2) system equation resolution algorithms, (3) setting of TOP conditions, and (4) event sequence generation and analysis.

DYLAM is useful for the description of dynamic incident scenarios and for reliability assessment of systems whose response is to be kept within certain limits (Mendola, 1988). This technique can also be used for identification of system behavior and thus as a design tool for testing proposed protective barriers and operator procedures.

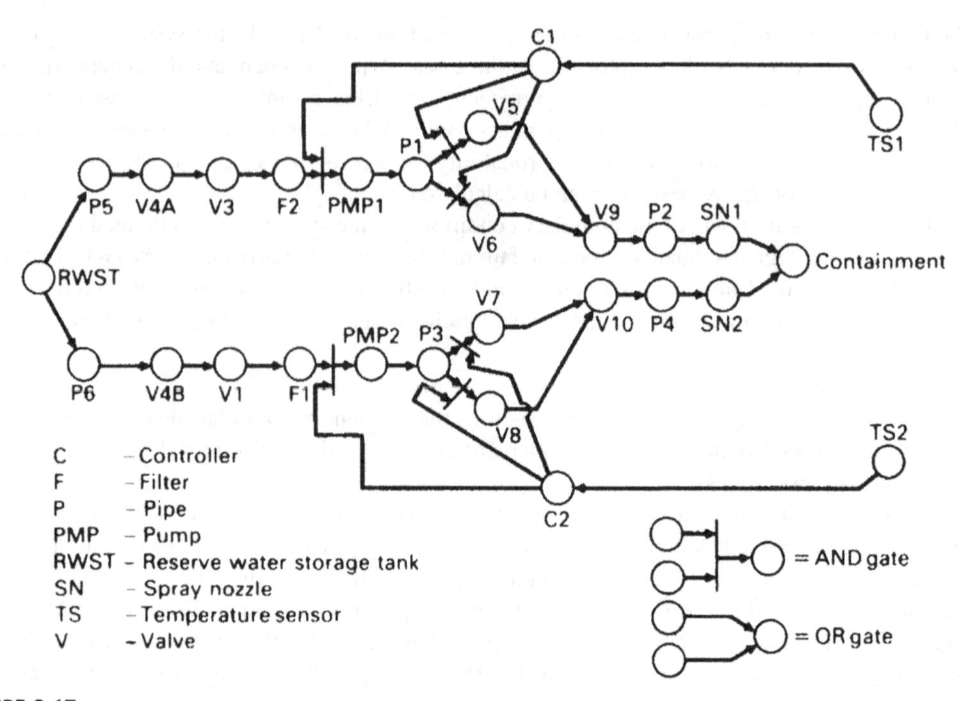

FIGURE 9-17

Success oriented system digraph of simplified emergency core cooling system in a nuclear power plant.

(Fullwood and Hall, 1988)

A system-specific DYLAM simulator must be created to analyze each particular problem. Furthermore, DYLAM requires considerable setup. It requires input data such as probabilities of a component being in certain initial states, independency of such probabilities, transition rates between different states, conditional probability matrices for dependencies among states, and values for the process variables.

Dynamic event tree analysis method (DETAM)

Dynamic event tree analysis method (DETAM) is an approach that treats time-dependent evolution of plant hardware states, process variable values, and operator states over the course of a scenario. In general, a dynamic event tree is an event tree in which varying behaviors are allowed at different points in time. This approach is defined by five characteristic sets: (1) branching set, (2) set of variables defining the system state, (3) branching rules, (4) sequence expansion rule, and (5) quantification tools. The branching sets refer to the set of variables that determine the space of possible branches at any node in the tree. Branching rules are used to determine when a branching should take place (a constant time step). Sequence expansion rules are used to limit the number of sequences.

This approach can be used to represent a wide variety of operator behavior, to model the consequences of operator actions, and as a framework to employ a causal model for errors of commission. Thus it allows the testing of emergency procedures and identifying where and how changes can be made to improve their effectiveness.

Applications of dynamic methods

The dynamic methods address an important deficiency found in fault/event tree methodologies. Even so, there are also limitations to their usage. The digraph and GO techniques model the system behavior and deal, to a limited extent, with changes in model structure over time. Markov behavior requires the explicit identification of possible system states and the transitions between these states. This is a problem as it is difficult to envision the entire set of possible states prior to scenario development. DYLAM and DETAM can solve the problem through the use of implicit state-transition definitions. Developing these definitions is no small task. With the large tree structure generated through the DYLAM and DETAM approaches, large computer resources are required along with a considerable amount of analyst effort in data gathering and model construction.

9.4.9.5 Summary of risk analysis methods

A total of 18 risk analysis techniques are reviewed here. Qualitative methodologies, though lacking the ability to account for dependencies between events, can identify potential hazards and failures within the system. Tree-based techniques address the dependencies between events. They quantify system failure frequency within the availability of operational data. Progress has been made with DYLAM and DETAM to study accident scenarios by treating time, process variables, system behavior, and operators action through an integrated framework. These techniques address the problem of having less than adequate models of conditions affecting control systems and operator behavior. However, the drawbacks for these techniques are the requirement for large computer resources and extensive data collection.

9.4.9.6 Multiple step systems

Many of the general tools described earlier have been successfully applied across many fields, including the area of maritime and port safety. The Formal Safety Assessment (FSA) is considered to be the most standardized framework of risk analysis in the regulated maritime environment (Bichou, 2008). The FSA was first developed by the UK Maritime and Coast Guard Agency (MCA) and later incorporated into the International Maritime Organization (IMO) interim guidelines for safety assessment (IMO, 1997). The FSA method consists of a five-step process: (1) hazard identification, (2) risk assessment, (3) risk management with alternative mitigation options, (4) cost-benefit analysis, and (5) decision-making (MCA, 1996).

An example FSA is provided by Vanem et al. (2008) directed toward the risk of fatalities to crew members and passengers on an LNG carrier. They used a 138,000 m^3 membrane carrier, 30 person crew, societal risk levels of 10^{-3} as intolerable and 10^{-6} as negligible, and economic benefit earned by each LNG carrier of 1.6×10^6 US$/ship year. For cost/benefit analysis they calculated a Gross Cost of Averting a Fatality (GCAF) and a Net Cost of Averting a Fatality (NCAF) defined in terms of the cost of a mitigation measure, ΔC, the risk reduction from that measure, ΔR, and the economic benefit from the measure, ΔB:

$$GCAF = \frac{\Delta C}{\Delta R} \tag{9-5}$$

$$NCAF = \frac{\Delta C - \Delta B}{\Delta R} \tag{9-6}$$

Using the frequencies shown earlier in Table 9-4, the individual risk for crew members on board for 182 days/year was found to be 1.6×10^{-4}/person year, in line with risks found by Hansen et al. (2002)

Table 9-9 Potential LOSS of Life (PLL) from LNG Carrier Operations per Ship Year

Accident Category	PLL (Crew)
Collision	4.42×10^{-3}
Grounding	2.93×10^{-3}
Contact	1.46×10^{-3}
Fire and/or explosion	6.72×10^{-4}
Loading/unloading	2.64×10^{-4}
Total	9.74×10^{-3}

(Vanem et al., 2008)

for crew on gas tankers of 4.9×10^{-4}/person year. The statistics justify safety improvement measures costing less than $3 million (GCAF value).

Table 9-9 lists the likely frequency for the number of crew lives lost from LNG fleet operations per ship year distributed by type of accident calculated in the FSA by Vanem et al. (2008).

The FSA risk analysis by Vanem et al. (2008) concludes that their results justify focusing further risk reduction efforts on:

- Navigational safety
- Maneuverability reliability
- Collision avoidance
- Cargo protections
- Damage stability
- Evacuation arrangements.

9.4.9.7 Risk modeling approach for LNG plants

An approach that makes use of predefined and general FTA and ETA events and structure is proposed by Rathnayaka et al. (2011). These authors observe that accident prevention barriers can be defined in five categories shown in Figure 9-18 as

- Release prevention barriers
- Dispersion prevention barriers
- Ignition prevention barriers
- Escalation prevention barriers
- Damage control, emergency management barriers.

Failure of all of these barriers is required before a catastrophic accident can occur.

All physical barriers are under the influence of a Management and Organizational (MO) Barrier and a Human Factors Barrier. A general fault tree diagram can be drawn for the MO barriers as shown in Figure 9-19. A similar, more complex fault tree is drawn for the Human Factor Barrier. The failure probabilities assigned to each of the numbered inputs in Figure 9-19 are listed in Table 9-10. The generality of Figure 9-19 is apparent since it could be applied to almost any risk-reduction application such as flying airplanes, construction, or operating LNG terminals, with slight adjustments to the failure probabilities in Table 9-10.

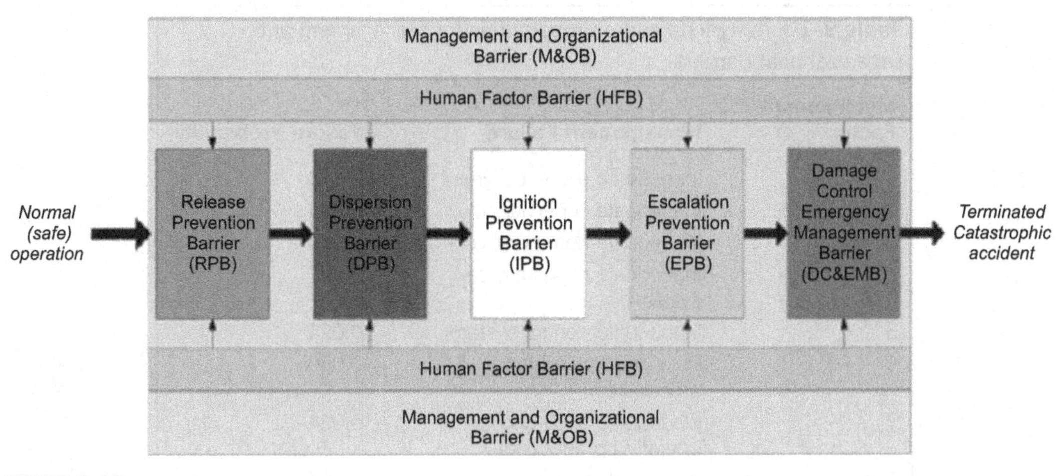

FIGURE 9-18

Conceptual framework for process accident model.

(Rathnayaka et al., 2011)

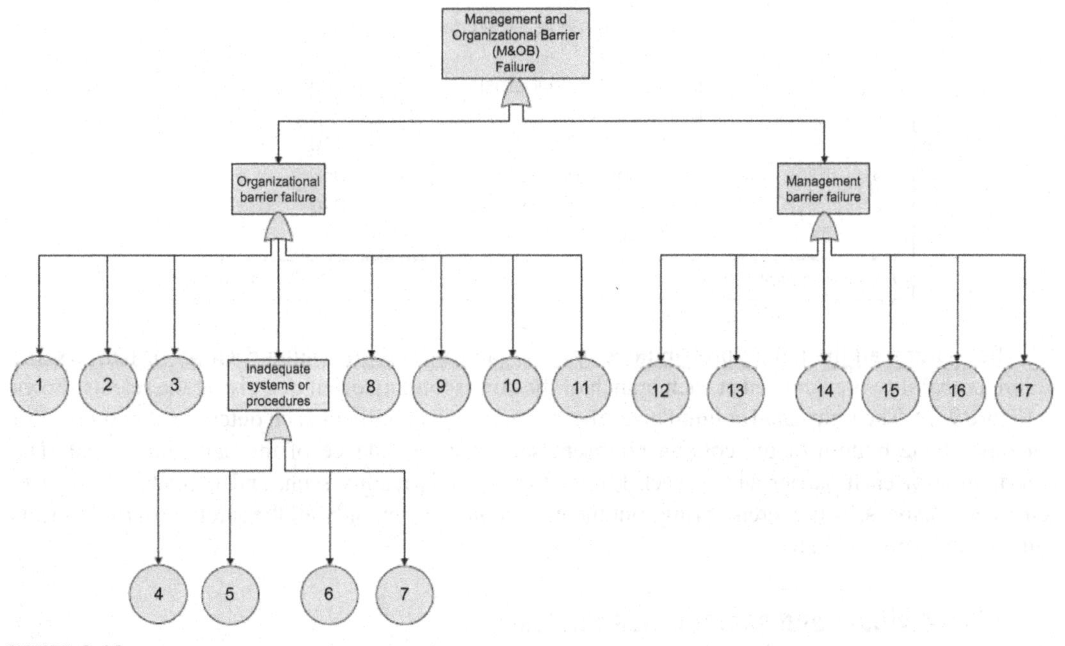

FIGURE 9-19

Fault tree analysis of human factor barriers.

(Rathnayaka et al., 2011)

Table 9-10 Assigned Failure Probabilities for Management and Organizational Barriers

Management Factor	Management Failure	Failure Probability
1	Inadequate safety program	0.010
2	Inadequate supervision	0.034
3	Inadequate communication	0.050
4	Inadequate maintenance system	0.020
5	Inadequate control system	0.025
6	Poor or no work permit procedures	0.050
7	Inadequate audit and operating procedures	0.034
8	Inadequate training	0.025
9	Inadequate company policies	0.020
10	Inadequate staff resources	0.020
11	Inadequate planning and organization	0.025
12	Poor decision making or failure to make	0.040
13	Inadequate management job knowledge	0.020
14	Inadequate management policies	0.025
15	Leadership failure	0.010
16	Poor communication	0.050
17	Incompetent or insufficient management	0.020

(Rathnayaka et al., 2011)

After putting all the failure probabilities into the fault trees representing each set of barriers, the failure probability is calculated at each branch of the corresponding event tree giving the results shown in Figure 9-20. The right-hand column gives the estimated probability of each outcome. All but the last outcome at the bottom of the column represent successful avoidance of the damaging event. The contribution of each barrier and of each failure frequency is readily seen, and is readily subject to sensitivity analysis. This exercise brings out the need to keep essentially all the safety program aspects and barriers working well.

9.4.10 Individual and societal risk analysis

Quantitative risk analysis (QRA) methods are applied widely to petrochemical plants, including LNG plants. QRA provides highly transparent and readily understandable results. Frequencies and modeled consequences are explicit inputs. The effect of mitigation measures is also modeled.

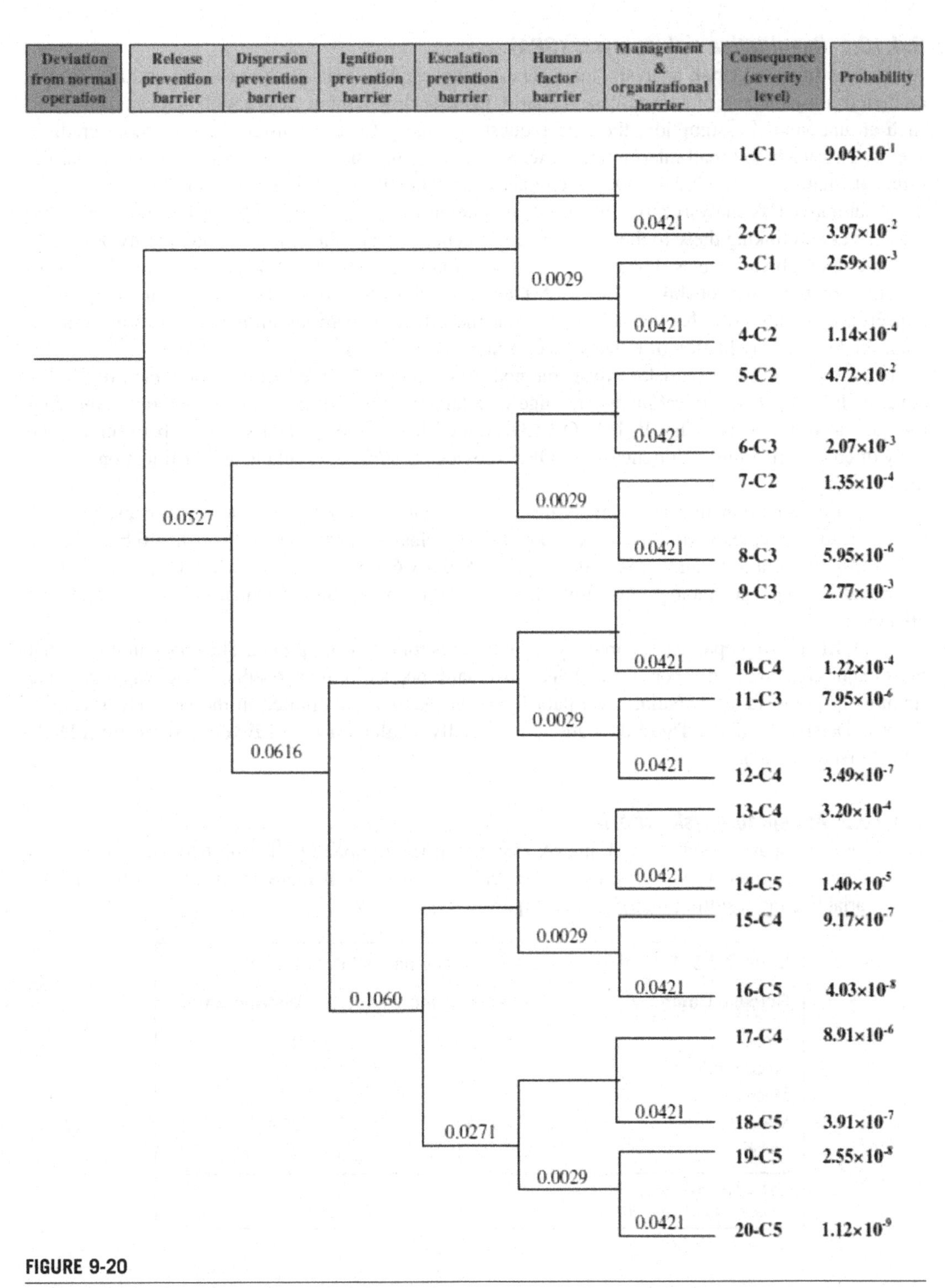

FIGURE 9-20

Predicted probabilities of each branch in the fault tree.

(Rathnayaka et al., 2011)

9.4.10.1 Quantitative risk analysis (QRA)

At the beginning of a risk analysis, managers usually want to bracket the problem and ask for the predicted consequences of a worst-case scenario. Upon seeing extreme destruction that can result with such an unbounded assumption, the next request is usually for the consequences of more credible events. This leads to broad interpretations where some, citing their own personal experience, set the limits at small-bore pipe breaks. Others cite actual events with much larger line breaks.

Quantitative risk analysis solves the problem of defining credible breaks by considering all possible break sizes and linking these to their estimated frequencies. Since the line break probability decreases with increasing break size as shown in Table 9-11 (Tanabe and Miyake, 2011) large events (holes) are weighted with a low probability. The QRA method displays the consequence and probability of all conceivable events. This has the advantage that the effects of various mitigation measures can be quantified by their reduction of either consequence or probability.

In addition, an adequate risk analysis method must account for the fact that some facilities in the general class of petrochemical plants are huge with large numbers of possible leak sources. Applying the qualitative methods (What-If, HAZOP, LOPA, etc.) leaves the issue of selecting between a large body of conceivable mitigation measures. QRA provides cost/benefit ranking and justification of such measures.

QRA is calculation intensive, involving hundreds and even thousands of scenario evaluations. Consider, for example, 8 wind speeds and directions, 6 classes of atmospheric stability, 6 break sizes, each for 8 different lines and already this requires 8 x 6 x 6 x 8 = 2304 modeling runs. Fortunately, improved computer technology and streamlined modeling make such calculations feasible and cost effective.

The GRI (1990) prepared a report on types of failures for LNG equipment. The report indicates that most major fires involved vaporizers and that most major breaks were the result of either vaporizer tube ruptures or pump failures. Failure rate data for LNG plants are compiled in the HSE Hydrocarbon Release Database (HRD). These data are commercially available as the LEAK database from DNV (Det Norske Veritas).

9.4.10.2 Acceptable risk criteria

Commonly, companies define a risk matrix that quantifies company policy on two axes; likelihood (probability or frequency) versus severity (consequence or cost). If there are five levels for each of these variables, the resulting matrix is like Figure 9-21.

Table 9-11 Typical Release Hole Size Normalized Distribution

Release Category	Hole Size, mm	Distribution
Small	~5	0.6
Small-medium	> 5–12	0.25
Medium	> 12–25	0.10
Medium-large	> 25–50	0.04
Large	50–100 (FBR)	0.01

FBR = full bore rupture
(Tanabe and Miyake, 2011)

CONSEQUENCE SEVERITY	LIKELIHOOD				
	A	B	C	D	E
5	Medium	Medium	High	High	High
4	Low	Medium	Medium	High	High
3	Low	Low	Medium	Medium	High
2	Low	Low	Low	Medium	Medium
1	Low	Low	Low	Low	Low
0	OK	OK	OK	OK	OK

FIGURE 9-21

Typical risk matrix.

The likelihood values for Figure 9-21 are:

A. < 0.0001, less than once in 10,000 years
B. 0.0001 to 0.001, once in 1,000 to 10,000 years
C. 0.001 to 0.01, once in 100 to 1,000 years
D. 0.01 to 0.1, once in 10 to 100 years
E. 0.1 to 1, averages once a year.

The divisions are not always decades. Guidance in designing an effective risk matrix is provided by Talbot (2011) and Ozog (2012).

The action levels are:

- Red = Corrections required to reduce risk to yellow area
- Yellow = Cost-effective measures should be used to reduce risk
- Green = No further mitigation required.

The consequence values and authority levels of responsibility are given in Table 9-12.

The matrix in Figure 9-21 can be compared to an FN (Fatality versus Number) chart shown in Figure 9-22. The action lines are often parallel and diagonal on a log-log plot, defining the same three actionable zones.

Table 9-12 Consequence Values and Authority Levels of Responsibility

Severity	People	Assets	Environment	Reputation	Responsibility
0	No injury	No damage	No effect	No impact	Shift Supervisor
1	Slight injury, first aid	Slight, < $10k	Slight	Slight	Group Manager
2	Minor injury, lost time	Minor $10k to $100k	Minor	Limited	Section Manager
3	Major injury	Localized $100k to $1M	Localized	Considerable, community	General Manager
4	Single fatality	Major $1M to $10M	Major	State	Plant Manager
5	Multiple fatalities	Extensive > $10M	Massive	National	Headquarters

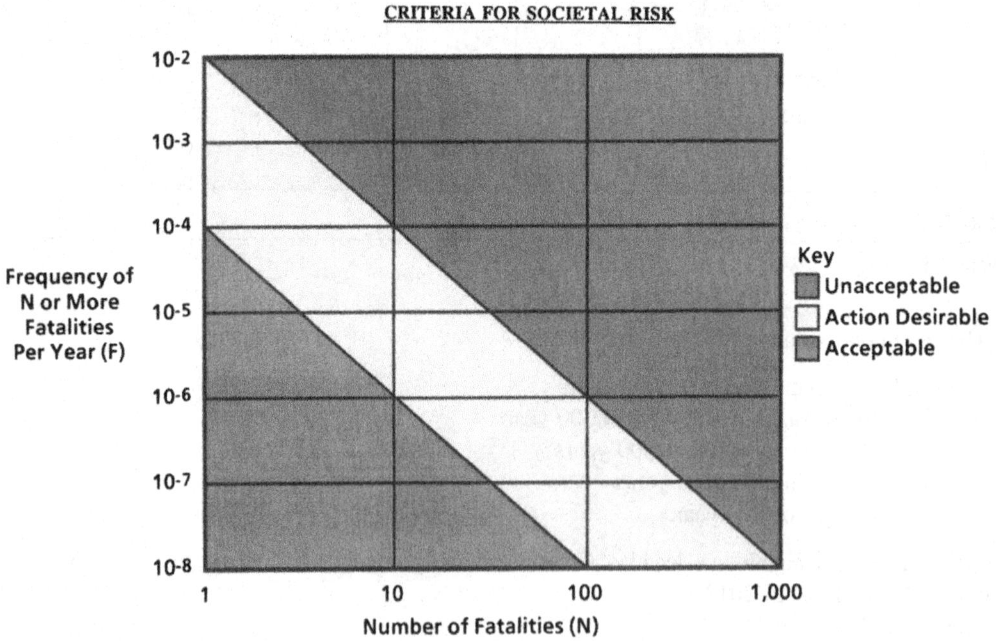

CRITERIA FOR SOCIETAL RISK

Frequency of N or More Fatalities Per Year (F)

10^{-2}
10^{-3}
10^{-4}
10^{-5}
10^{-6}
10^{-7}
10^{-8}

1
10
100
1,000

Number of Fatalities (N)

Key
Unacceptable
Action Desirable
Acceptable

FIGURE 9-22

FN curve with risk acceptance criteria.

A QRA typically provides societal risk around a plant in the form of FN curves and contour plots of individual risk on a plan-view of the plant and neighboring areas. Individual risk quantifies the risk of death for a person living in a fixed place near the plant for years. (This is an idealization colloquially described as the risk for someone tied to a stake at a fixed point the entire time.) A QRA also usually quantifies the benefits of prospective mitigation measures to obtain a cost/benefit ratio. This enables a ranking to obtain the best results for the safety budget.

Societal risk is defined as the relation between the occurrence frequency of each accident and the number of people that could be affected by the impact of each accident (normally considering only death). The calculation of societal risk considers the population density around a plant and accounts for movement patterns of the exposed population from night to day, weekdays to weekends.

An FN (frequency-number) curve sorts the scenario events in increasing order by N and plots the cumulative frequency for N or more number of deaths. Risk is normally categorized on an FN curve into one of three categories of tolerance or acceptability defined by two lines:

- A lower line below which the risk is acceptable
- An upper line above which mitigation measures are required to reduce the risk below this line
- Between the two lines mitigation measures are subject to cost/benefit analysis and may be required to bring the risk As Low as Reasonably Practical (ALARP).

These lines on an FN curve represent one highly visible form of risk tolerance or risk acceptance criteria. As would be expected, risk acceptance requires some degree of debate in the political arena, and considerable differences can be expected worldwide. Important summaries of risk acceptance criteria are provided by CCPS (2009) and by Pitblado et al. (2012), which cite an important summary for Europe by Trbojevic (2010).

For example, Figure 9-23 provides the societal risk criteria for Abu Dhabi as prescribed by the national oil company ADNOC. They have a Societal Risk criterion in their Health, Safety, and Environment (HSE) Management code of practice, ADNOT-CoP-V6-06, and also have an individual risk criterion.

Brazil regulates societal risk on both a federal and a state level. As of 2012, four states have formally established risk guidelines (Rio de Janeiro, Sao Paulo, Rio Grande do Sul, and Bahia State) and the risk criteria are different in each state. A comparison of the four state criteria is shown in Figure 9-24.

The RTC Guidelines (CCPS, 2009) include recent thinking by a large number of regulatory bodies and experts in risk analysis, as discussed by Frank and Jones (2010) and Frank (2011). To be precise, a risk assessment and a corresponding acceptance criterion should define the particular groups to which the criteria are intended to apply; for example, workers exposed to the risk, workers in the general area, public, vulnerable populations, and so on; and the level of harm addressed by the criteria (fatality or injury). The criteria should specify whether frequency represents an annual frequency or a fixed probability of injury or death. The CCPS CPQRA Guidelines (2000) presents 14 different risk measures, all derived from the same set of incident, likelihood, and consequence data.

The FN curve method is graphic and provides a superficially understandable visual comparison with risk tolerance criteria curves. In addition, an FN curve shows top contributing events as steps along the curve, but it is not the best way to indicate what drives risk or how to mitigate it. Other useful methods present societal risk as total onsite societal risk per process unit, per building, or per contribution per source, or PLL (Potential Loss of Life) favored for offshore platform assessments.

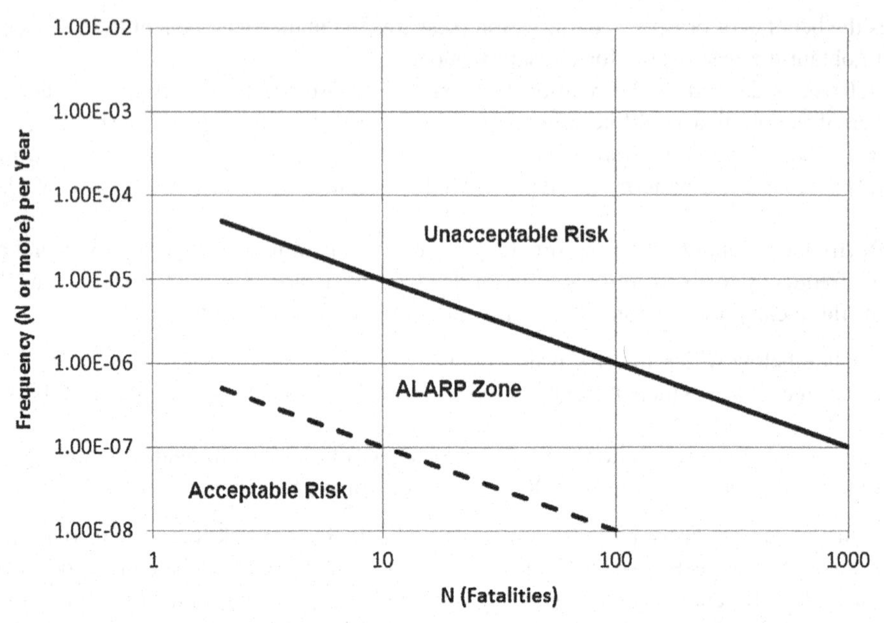

FIGURE 9-23

Societal risk criterion for single fixed installation in Abu Dhabi.

(Pitblado et al., 2012)

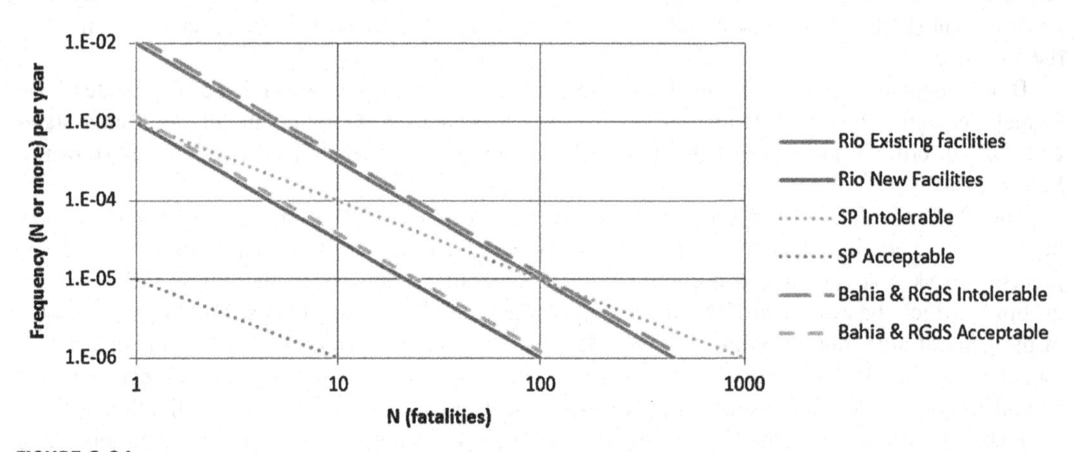

FIGURE 9-24

Societal risk criteria in Brazil.

(Pitblado et al., 2012)

As an example the Major Industrial Accidents Council of Canada (MIACC) specifies the following criteria for land use and occupancy as cited in MIACC (2007):

- For a risk of 10^{-4} per annum (or 100 deaths every million years): no land use other than industrial shall be allowed.

- For risks between 10^{-4} and 10^{-5} per annum (10 to 100 deaths in a million years): Uses that require permanent access, a limited number of people, and an easy and timely evacuation are allowed (manufacturing, warehouses, etc).
- For areas between the contours of 10^{-5} and 10^{-6} per annum (1 to 10 deaths in a million years): Uses that require permanent access, a limited number of people, premises easily evacuated, with a low population density.
- For areas of risk at or below 10^{-6} per annum: no limits are made for land occupancy.

The MIACC cited other sources for consistency, including the CSChE (2004), the UK Health and Safety Executive (HSE, 1992), and Pate'-Cornell (1994) who stated that the Norwegian Petroleum Directorate, among others, uses a maximum risk criterion for the collapse of offshore rigs to be 10^{-4} per annum.

9.4.11 Accident investigation techniques

As pointed out in Section 9.4.9.3 discussing the bow-tie and BSCAT methods, accident investigation techniques are essential the inverse of some risk analysis methods, since both employ an event tree path, only evaluated in forward or reverse directions. Broadribb (2003) identifies three main incident investigation approaches: Domino Theory of Causation, System Theory or Multiple-Causation Theory, and Hazard-Barrier-Target Theory.

The Domino Theory of Causation is one of the earliest systems, incorporated in the International Safety Rating System (ISRS), as documented by Bird et al. (2003). This theory postulates that a series of failures in barrier systems leads to an accident. The system approach postulates that multiple failures could independently occur "in parallel," as could be analyzed by FTA (fault tree analysis). The BSCAT method is based on ideas of the so-called Swiss cheese model (see Section 9.4.9.3) and can link incidents back to underpinning failures in management systems. It is claimed to be designed for less expert investigators such as process supervisors, who do the first level of accident investigation.

Some other accident investigation methods include (Pitblado et al., 2011):

- What-If method
- Fishbone diagrams
- Fault tree analysis
- MORT (Management Oversight and Risk Tree)
- Common List of Causes
- TapRoot (System Improvements Inc.)
- Tripod (Univ. Manchester and Univ. Leiden)
- System Dynamics (Mass. Institute of Technology).

Some of these techniques require training in quite different skill sets, and FTA, MORT, Tripod, and System Dynamics require specialist investigators. Tripod is designed to address safety culture deficiencies. It is unlikely that a single technique would meet the needs of every incident.

9.4.12 Innovative systems under development

Innovative approaches under development include the Fault Semantic Network for fault detection and Bayesian-LOPA Methodology for improving the database used in QRA for LNG plants.

9.4.12.1 Fault semantic networks (FSN)

A proposed fault detection approach uses software tied to a plant's instrumentation. The computer algorithms of Gabbar and Bedard (2010) and Gabbar and Khan (2010) use real-time process data. The approach compares certain patterns of deviations in values of operating variables with patterns from previous accidents. This approach introduces *diagnostic capability* to normal instrument logging to provide early warnings. It is limited, of course, to its database of previous accidents. The proposed software uses Process Object Oriented Modeling (POOM) to construct a "fault semantic network" to construct logic rules to automate the identification of failures by pattern recognition. The authors cite successful use of predictive maintenance to estimate remaining life using real-time monitoring and equipment failure records.

9.4.12.2 Bayesian-LOPA methodology

In the developmental stages at universities, the Bayesian-LOPA methodology (Yun et al., 2009; Pasman and Rogers, 2011) may improve Quantitative Risk Assessment (QRA) for LNG projects. It could also fill some voids in applying Levels of Protection Analysis (LOPA). Both QRA and LOPA rely on failure rate data of facilities and equipment that are in relatively short supply in the LNG industry, precisely because there are few incident reports and few failure rate compilations. Generic failure data from other industries such as the petrochemical and nuclear power industries have sufficient and longer-term historical records. However, these data may not provide appropriate risk results for the LNG industry because the operational conditions and environments differ so much. Bayesian methods were developed specifically to statistically produce updated failure data using the prior generic data from other industries (Wan, 2001) with likelihood information from a new source such as the LNG industry. For plant-specific likelihood information, the LNG plant failure database (Johnson and Welker, 1981) that was established from 27 LNG facilities is recommended.

The Bayesian methods are general enough that they have been developed into computer programs capable of treating very complex systems such as an air traffic controller program. The program allows any number of logical connections to be drawn between nodes, represented by ovals in the logic diagram. These are called Bayesian Belief Networks (BBN). This development, specifically at the University of Pittsburgh, and at Delft University in the Netherlands, provides free software with good capability. Pasman and Rogers (2011) are taking advantage of the BBN software by applying it to a LOPA calculation. The methods may prove useful in analyzing accidents, as well as making QRA analysis simpler and more transparent.

9.4.13 LNG risk analysis examples

9.4.13.1 LNG port risk analysis

Examples of risk analysis for an LNG port are provided by VanDoorn (2011, 2012) and MARIN[1] (2012) for the ports of Ferrol, Spain, and Rotterdam. The probabilities of a grounding or collision incident were calculated using MARIN's Safety Model for Shipping and Offshore in the North Sea (SAMSON). The probability of developing a hole in a cargo tank were calculated using an analytical Maritime Collision Model MARCOL. It determines the penetration probability of ship and cargo tanks

[1]Maritime Research Institute Netherlands

in a few seconds. A fast-time simulation model SHIPMA was used for training and to check the assumptions of proposed tug assistance.

Figure 9-25 shows for the Port of Rotterdam the five parts of the path of an arriving LNG carrier that were evaluated for risk: (1) anchorage area; (2) approach to port entrance; (3) approach to terminal; (4) just in front of the terminal; (5) moored at the jetty. Some findings were that the penetration probability for cargo tanks is much higher in the anchorage area than within the harbor because passing ships at sea sail at higher speeds than ships within the port. In port, because of the layout of the terminal, an LNG carrier cannot be hit in the side by another large ship.

The Rotterdam area study demonstrated that traffic regulation can strongly reduce the collision risk, such as stopping all other traffic during the arrival of LNG carriers, and scheduling carrier arrivals for a specified time slot at night, when other traffic density is low.

For the Port of Ferrol, Spain, risks from grounding are found by first plotting grounding boundaries along the LNG carrier route into the port. A navigational error rate is applied to the SAMSON model to calculate "ramming" probabilities at each point along the route as a function of the carrier speed. A mechanical failure rate is similarly applied all along the route to find "drifting" probabilities. Unexpected weather changes and emergencies during transit are also considered. These become inputs to the MARCOL model to calculate vessel penetration and possible leaks of LNG. These scenarios are inputs for a quantitative risk analysis (QRA) that models LNG leaks, fires, and such resulting in a conventional FN curve such as Figure 9-26. The three curves in Figure 9-26 are for the cases of high, nominal, and lower speeds of the carrier. The graph shows that reducing the ship speed by one knot increases the safety level by a factor of 10. The study also results in recommendations for nautical procedures, entrance rules, and limiting environmental conditions used to develop port procedures.

An additional example of a risk analysis for the Italian LNG terminal of Panigaglia near La Spezia, Italy is provided by Bubbico et al. (2009). In this case, accidental collision and grounding cases were not considered likely because of ship speed restrictions in the harbor. The authors evaluated the effects of a deliberate attack on an LNG carrier at two locations indicated in Figure 9-27 by the center of fire radiation contours of 5 kW/m^2 and 37.5 kW/m^2 drawn as circles accounting for all wind directions. The southern-most point is at the end of a breakwater on a required path for LNG ships where an attack could possibly block or reduce passage into the harbor. The northern-most point is at an anchorage point, India 2, which can be used only under special conditions and for a limited time (6 hours). A 5 m^2 breach at the water line is assumed caused by an explosive-laden small boat attack. The models ALOHA 5.4.1 and CHEMS-PLUS 2.0 were used to predict pool size, fire radiation contours, and dispersion from an unignited spill. The predictions are consistent with those of the Sandia study (Hightower et al., 2004).

Superimposed on the predicted fire radiation circles in Figure 9-27 is a plume representing a possible delayed-ignition flash fire. This is considered highly unlikely since it is not compatible with the assumption of an explosive attack (that would likely result in immediate ignition). A flash fire burning back would be an important threat mainly to people caught in the flash fire plume. Altogether, since the hazard plumes are largely at sea, little collateral damage is predicted.

9.4.13.2 Establishing onshore terminal vapor dispersion zones

Increasingly realistic simulations are being developed to set the required vapor dispersion zones for an onshore LNG import terminal. An example by Melton and Cornwell (2010) uses Computational Fluid

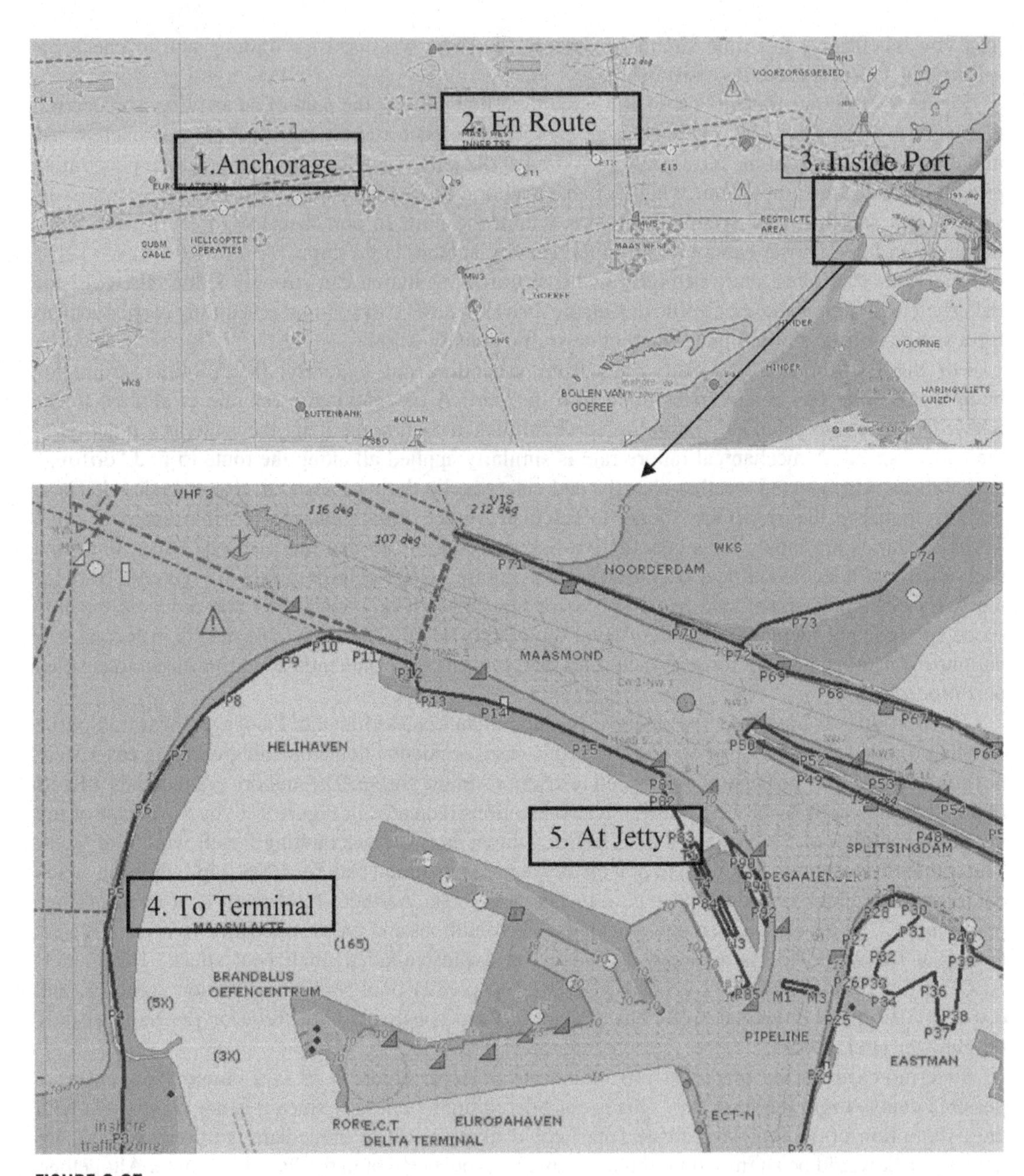

FIGURE 9-25

Rotterdam port with LNG carrier route to jetty showing points evaluated with risk analysis.

(MARIN, 2011)

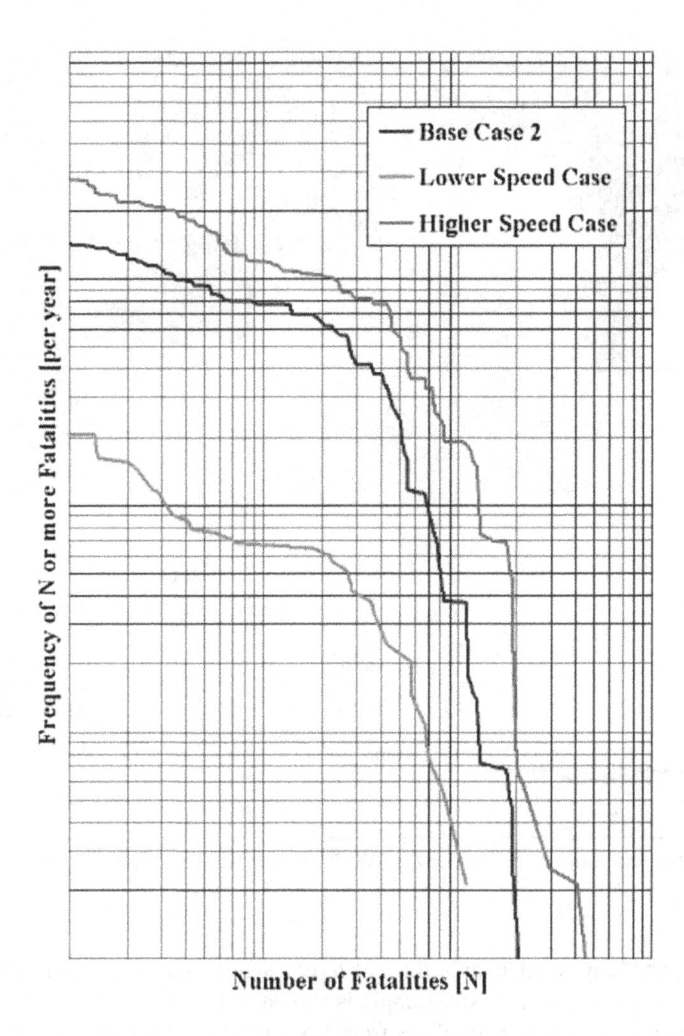

FIGURE 9-26

Example FN curve for LNG risk to port of Ferrol, Spain.

(vanDoorn, 2012)

Dynamics (CFD) to model LNG spills flowing into trenches that direct the spill to an impoundment basin. A free model available from the US government has been found appropriate for modeling buoyancy-dominated vapor dispersion and is relatively easy to use for routine modeling tasks. This is the Fire Dynamics Simulator (FDS) developed by the National Institute of Standards and Technology's Building Fire Research Laboratory. The model has been checked and validated in several studies conducted at the National Bureau of Standards (see Melton and Cornwell, 2010), and has been successfully used to model large-scale LNG tests Burro 8 and 9 by Clement (2000) and Chang and Meroney (2003).

CFD models have the ability to simulate a full range of trench layouts and drainage paths. In addition, the analyst can incorporate important features such as sloping terrain, vapor fences, escarpments,

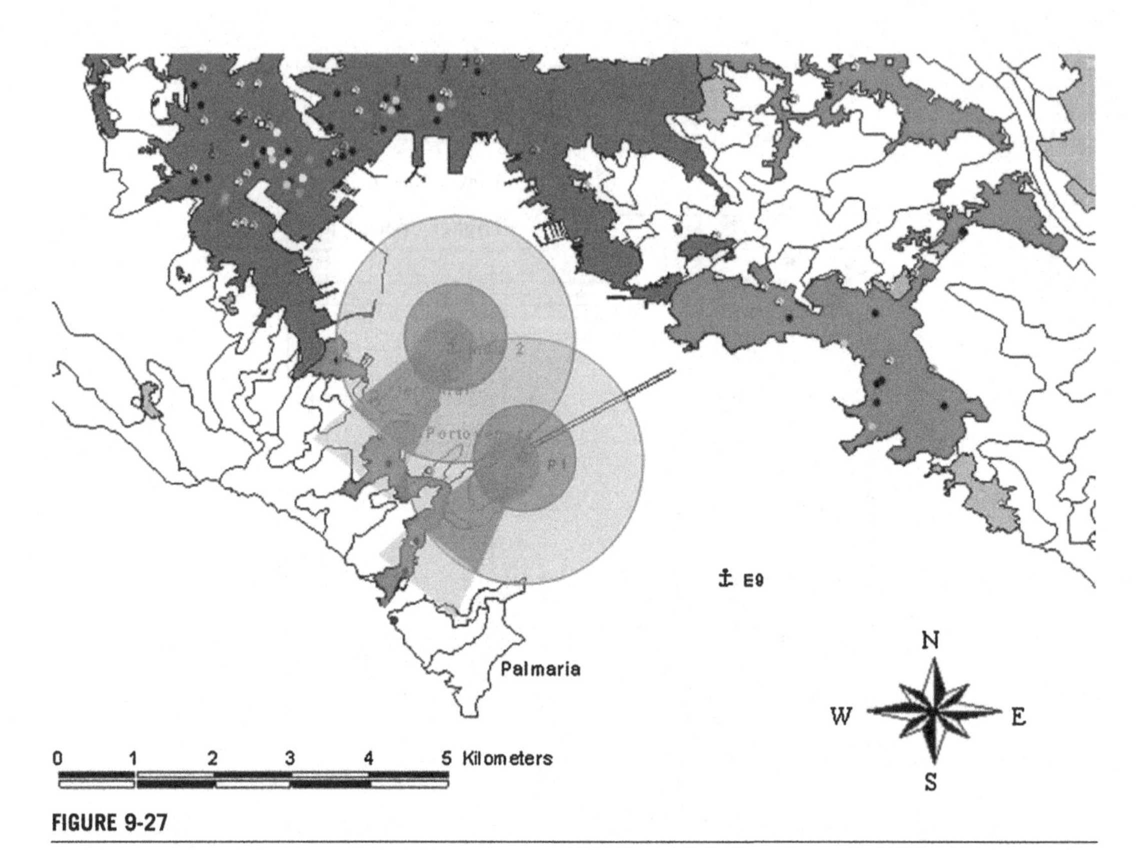

FIGURE 9-27

Predicted fire radiation contours to 5.0 and 37.5 kW/m^2 from an attack at two locations in the harbor at Panigaglia, Italy.

(Bubbico et al., 2009)

process equipment, and LNG storage tanks that can have a significant effect on the dispersion of vapors evaporating from drainage channels. An example is shown in Figures 9-28 and 9-29 of the transient vapor generation and dispersion of natural gas from LNG in a trench and an impoundment basin near three storage tanks. The event is a full-bore rupture of an unloading line spilling into a channel and

FIGURE 9-28

Predictions of dispersion of evaporating LNG vapors from channel and impoundment basin flowing over berm.

(Melton and Cornwell, 2010)

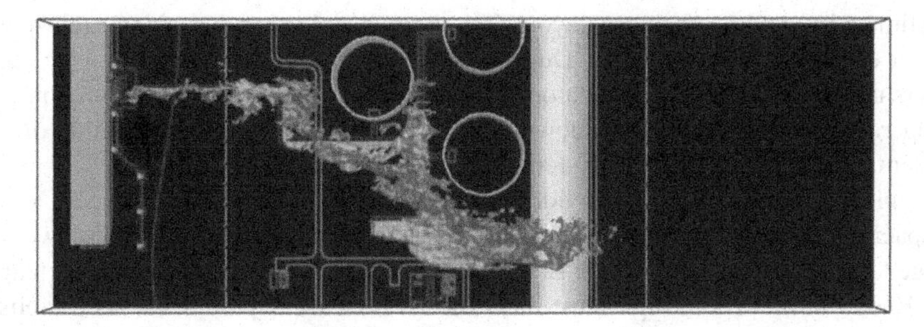

FIGURE 9-29

Plan-view predictions of evaporating LNG vapors from channel and impoundment basin flowing over berm.

(Melton and Cornwell, 2010)

impoundment basin made of medium-density concrete. A side view in Figure 9-28 shows vapors elevated by flowing over the impoundment berm. A plan view in Figure 9-29 shows a trench leading to the impoundment basin and evaporated vapors from these sources partially caught in the wake of a storage tank, and somewhat contained by the impoundment berm. Such simulations can be used to establish the vapor dispersion zone requirements of the US codes citing NFPA-59A.

9.4.13.3 Optimizing onshore terminal layout by risk analysis

Optimal design of an LNG terminal is usually done considering risk factors. Plant layout is a key passive measure for plant safety and avoidance of escalation of fires and explosions. Plant layout safety principles include (de le Vega and Durr, 2004):

- Separation should allow for effective firefighting and prevent fire from one area to propagate to others.
- Each area should have access from at least two different ways.
- Minimize liquid inventory.
- Equipment with high inventory of flammable material should be located downwind of the prevalent wind direction, away from community areas, control rooms, etc.
- Fire water systems should be looped around entire plants, so if one line is cut, there is another.
- Plant electric power should be provided through two separate circuits.
- Select safer tank designs.

Additional passive measures such as fire walls assist when additional space is impractical.

ESD (emergency shutdown) and EDS (emergency depressurization systems) divide the plant into possible fire zones. Each zone can be isolated at its boundaries by ESD valves before proceeding to depressurization. Depressurization philosophy is crucial to metallurgy selection in an LNG plant. Depressurization rapidly reduces pressure of process equipment by relieving its inventory to flare or vent to prevent vessel bursting, and removes process fluid from equipment to a safe destination.

Current best design practice involves studying different layouts using 3D software that allows easy rearrangement of the main equipment, followed by computer rerunning of the pipe rack design. A new cost estimate is found for each new layout. The thermal and vapor dispersion exclusion zones are calculated for each case (de le Vega and Durr, 2004).

Designers use dynamic process simulators to perform real-time studies of the unit under different modes of operation such as startup, shutdown, upsets, and so on to verify the stability of the process control system, and to verify the EDS system.

An example design approach is provided by Taylor (2007) for an LNG export terminal with nominal capacity of 3.0 million tonnes per annum (mtpa). The proposed plant includes two 140,000 m^3 LNG storage tanks, and the rate at which LNG will be transferred to the ships was set tentatively for more than 10,000 m^3/hr. The design must meet the NFPA 59A requirements, as established using consequence modeling. To do so, three impoundment areas are needed, with the design requirements shown in Table 9-13. The values in Table 9-13 are inputs to dispersion modeling to establish the thermal radiation exclusion zone, the vapor dispersion exclusion zone, and tolerable vapor cloud explosion (VCE) blast loads. The response times in Table 9-13 are justified by incorporating sophisticated leak detection and shutdown systems.

The model used was the CANARY model by Quest Consultants, Inc., which incorporates the Baker-Strehlow explosion model (Baker et al., 1997).

The results of modeling to find the separation distances required by NFPA 59A are shown in Table 9-14, with two sets of weather parameters, those specified in NFPA 59A and site-specific values, taken to be more credible.

With the required separation distances in Table 9-14, the designers can optimally arrange process units and equipment. Once the equipment arrangement and the site location plan have been finalized, further studies verify the location and design criteria for process buildings according to API RP 752.

9.4.13.4 Explosion considerations in terminal design

Modern theory for vapor cloud explosions requires modeling how the flammable vapor cloud overlaps congested zones such as zones of high piping density in a plant. Models have been developed for calculating this overlap, and some models account, as well, for how the burning gas vapors push unburned flammable gases into other nearby congested zones. Predictions of the SafeSite$_{3G}$ model of Baker Engineering and Risk Consultants are illustrated in Figures 9-30 and 9-31 for loss of propane refrigerant from a 2 in (50.8 mm) hole in a 4 in line discharged horizontally at 0.5 m elevation.

Table 9-13 Spill Impoundment Modeling Parameters

LNG Leak Scenario	Nominal Leak Rate, kg/s	Duration (min)	Impoundment Size (m)	Basis
Single loading arm break	600	1	15 x 29 x 0.1143	Half maximum loading rate (5000 m^3/hr)
Leak from liquefaction train to process channel	147	3	5 x 6 x 7	LNG liquefaction rate
Leak from storage tank pump-out line	600	10	10 x 14 x 5.5	Maximum pump-out rate from one tank

(Taylor, 2007)

Table 9-14 Separation Distances to Satisfy NFPA 59A Requirements

Flammable Vapor Exclusion Zone Distances		
Description	**Maximum Downwind Distance (m) to:**	
	LFL	**½ LFL**
10 min spill from LNG pump-out line	Within dike	Within dike
3 min spill from liquefaction process to impoundment	70	115

Fire Radiation Exclusion Zone Distances				
		Maximum Downwind Distance (m) from Center of Impoundment to Thermal Radiation Endpoint		
Description	**Weather**	**30 kW/m²**	**9 kW/m²**	**5 kW/m²**
Impoundment sump (14 m x 14 m) for 10 min LNG tank spill	(a)	20	45	60
	(b)	45	60	75
Impoundment for liquefaction process (5 m x 6 m)	(a)	6	15	22
	(b)	20	25	30
LNG tank impoundment (140m x 210m) fire	(a)	180	310	400
	(b)	240	360	435
Weather Case		**Wind, m/s**	**Air temp, °C**	**Relative hum, %**
(a) NFPA 59A required		0	21	50
(b) Site-Specific		7	24	90

(Taylor, 2007)

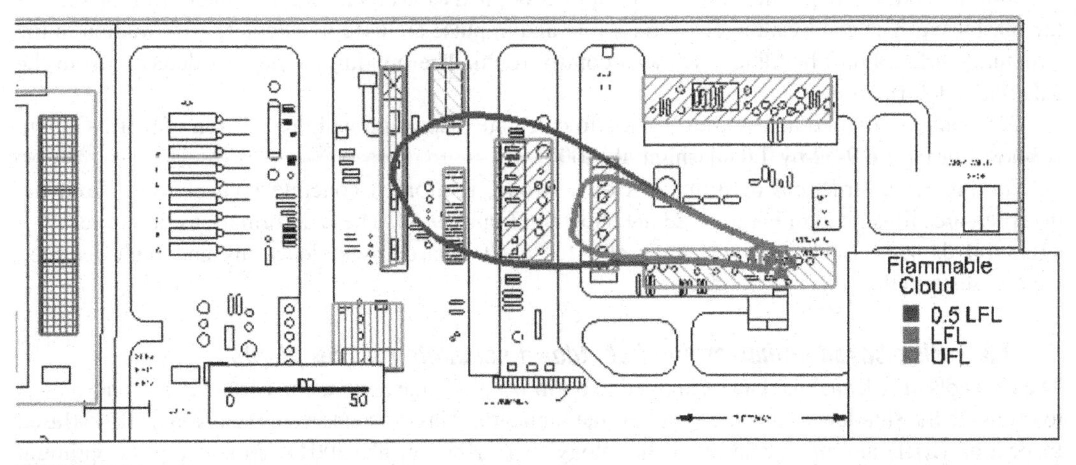

FIGURE 9-30

Contours of flammable vapor clouds in congested zones of a plant before ignition predicted by $SafeSite_{3G}$ model.

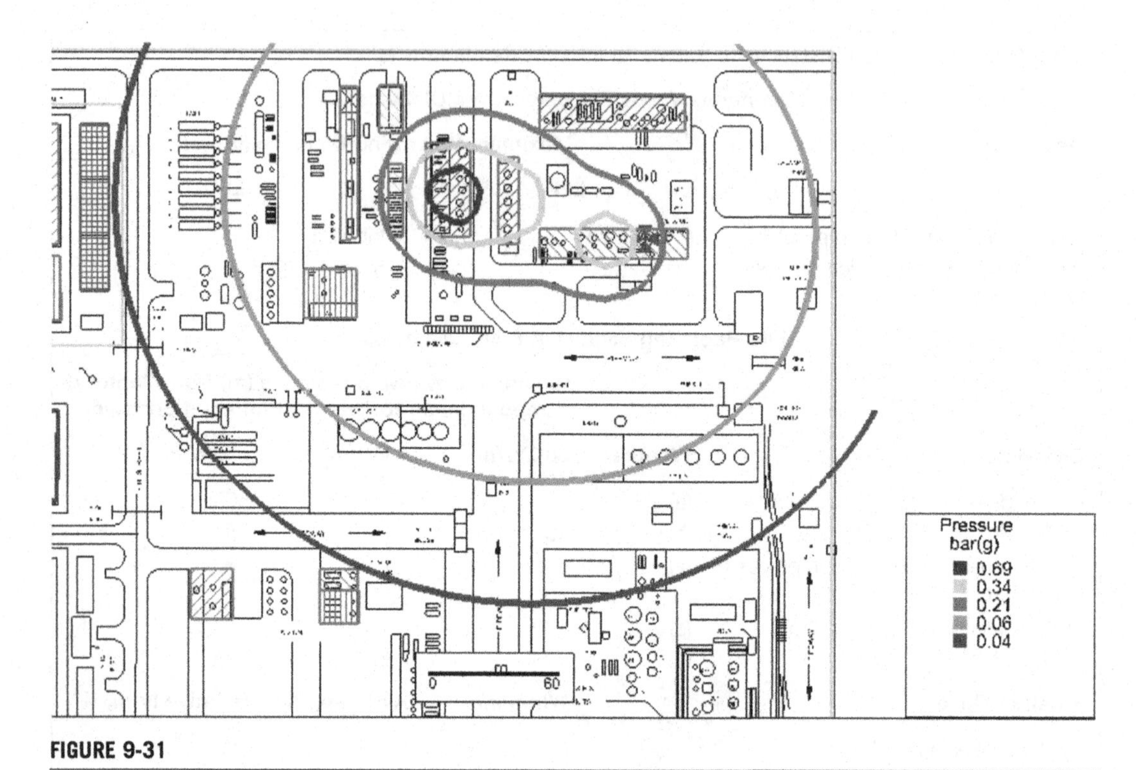

FIGURE 9-31

Contours of blast overpressure predicted for ignition of propane vapors depicted in Figure 9-30 (predictions of SafeSite$_{3G}$ model).

Similar contours are predicted for blast impulse, defined as overpressure integrated over the time of the blast wave. The combination of overpressure and impulse are used to calculate deformation of the structural surfaces of a building, such as a control room. The building damage is deduced from the calculated deformations.

CFD models are used in a similar fashion to calculate dispersion within a portion of an LNG plant as shown in Figure 9-32 by Takahashi et al (2007).

The predicted structural deformations of a strong, reinforced concrete control room from the overpressures illustrated in Figure 9-33 are shown in Figure 9-34. These deformations do not result in structural damage. Such analyses are useful to select control room locations and their required structural strength.

9.4.13.5 Risk-based optimization of shutdown schedule for LNG plants

The concepts of risk analysis can be applied to optimize plant management in certain areas. One area is to optimize the shutdown schedule for plant maintenance. This concept is an extension of Risk-Based Inspection (RBI) and maintenance methodology (e.g., Khan et al., 2004). To develop an optimum shutdown strategy, a set of scenarios are postulated that employ different combinations of the parameters: redundancy, standby, and shutdown periods. For each scenario, the failure probability and

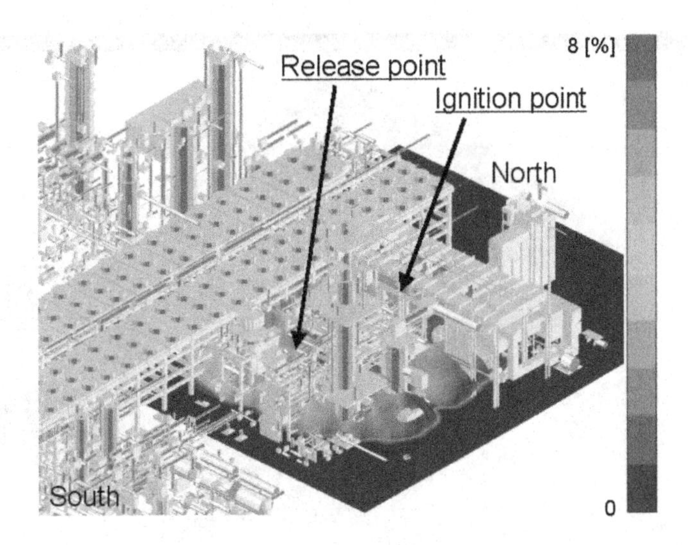

FIGURE 9-32

Dispersing flammable propane vapors predicted by CFD model.

(Takahashi et al., 2007)

consequence are calculated. The optimization proceeds by iterating on the parameters until the lowest cost is found (Keshavarz et al., 2012).

The failure rates of key equipment can be found from plant records, but is costly and time-consuming. Keshavarz et al. (2012) suggest, instead, to use life-testing data from the Offshore

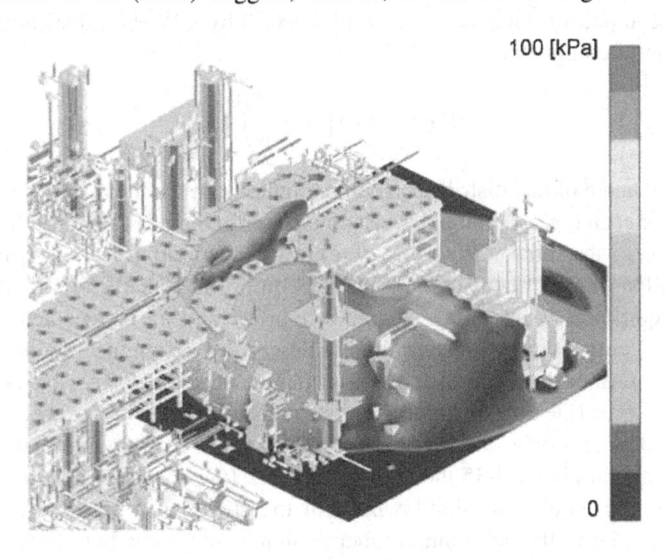

FIGURE 9-33

Explosion overpressure contours predicted by CFD model.

(Takahashi et al., 2007)

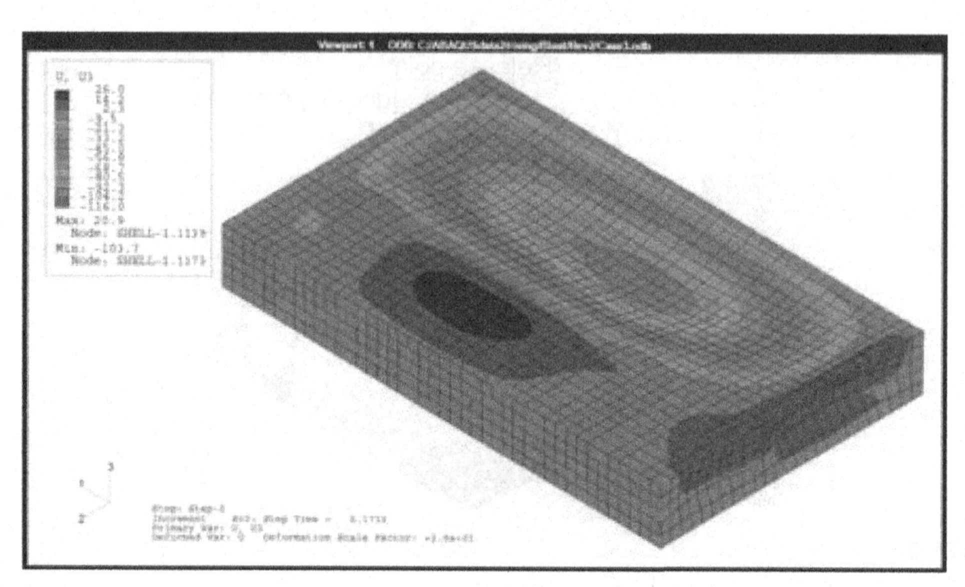

FIGURE 9-34

Structural response deformation profiles of control room at 1.3 s after ignition of the vapor cloud.

(Takahashi et al., 2007)

Reliability Data Handbook (OREDA, 2002). The major assumption is that these data can be well-represented (at least in part of the general "bathtub curve") by a Weibull distribution, *R(T)*, where *T* is the operation time of equipment:

$$R(T) = \exp\left(-\left(\frac{T}{\theta}\right)^{\beta}\right) \tag{9-7}$$

The parameters β and θ of this distribution can be found from the mean and variance of mean time to failure (MTTF) for each type of equipment. Examples are given in Table 9-15 that illustrate that the shape factor, β, differs substantially from that of a normal distribution for which β is 2.

When the operation time is much smaller than the equipment's characteristic life, Equation 9-7 is close to one, the equipment is operating at high reliability, and no further action is required. The reliability "bath-tub" curve shows higher failure rates at the beginning of operation and later as the system deteriorates. Short intervals of preventive maintenance may be needed at both extremes. As the average shutdown time (the time at which the system needs to be down for preventive maintenance) or goal time (the operational time of the plant) increases, the risk of an unintentional shutdown also increases. For example, in Figure 9-35 the authors plot risk (the cost of maintenance and of business interruption) against the number of shutdowns with the shutdown time as a parameter. For long shutdown time (e.g., 9 days), the optimum number of shutdowns is low but the optimum cost is high. For shorter shutdown times (5 days and 1 day), the optimum risk is decreased, and the optimum occurs at a higher number of shutdowns. By repeating the analysis by doubling the goal time, the authors show that the required preventive maintenance and the associated optimum risk more than doubles. The

Table 9-15 Weibull Distribution Parameters for Equipment

Equipment	Characteristic Life, θ (hrs)	Shape Parameter, β
Centrifugal compressor	113,930	2.54
Axial compressor	118,700	2.13
Gas turbine	182,470	1.17
Heat exchanger	116,580	4.84
Pump	94,270	2.47
Expander	58,630	1.43
Generator	118,770	2.5
Electric motor (helper)	94,900	2.28
Cooler	226,000	3.78

(Keshavarz et al., 2012)

effect of having more standby redundant equipment can be shown to also lower the risk and the number of shutdowns.

The optimized maintenance shutdown schedule may not represent the overall optimum as Keshavarz et al. (2012) point out since it does not focus on asset management. Asset management requires a comprehensive study to achieve minimum risk associated with the best plant performance.

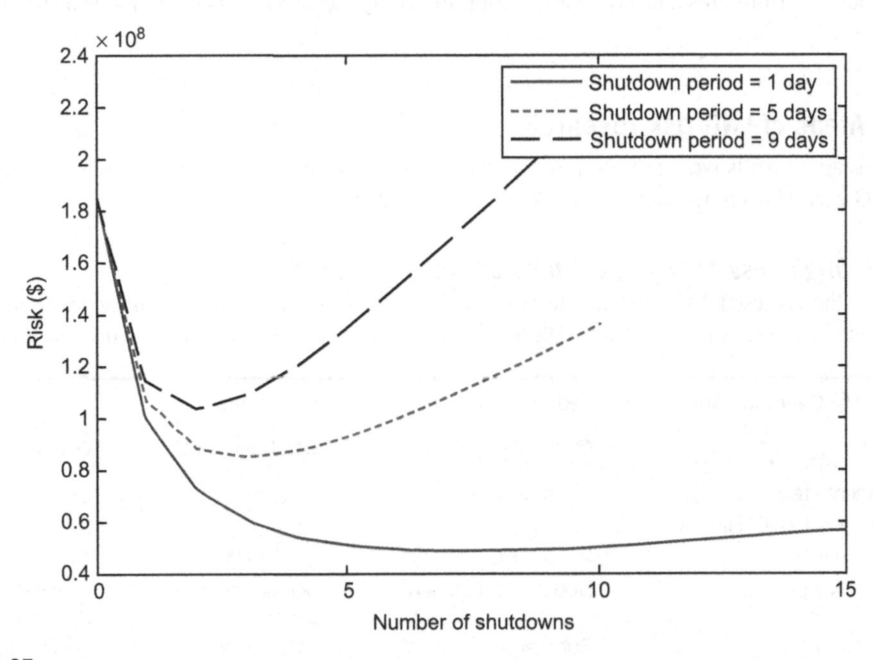

FIGURE 9-35

Risk as a function of the number of maintenance shutdowns for a goal interval of 10,000 hr (1.14 years).

(Keshavarz et al., 2012)

9.4.13.6 Comparing alternate LNG technologies for risk

An example of the application of risk analysis comparing three types of LNG regasification terminals for their inherent safety is provided by Tugnoli et al. (2010) for onshore terminals, offshore gravity-based structures (GBS), and offshore floating storage and regasification units (FSRU). Table 9-16 compares some current characteristics of these technologies.

A set of Loss of Containment (LOC) scenarios were selected for the risk comparison:

- Breach on the shell in the liquid or vapor phase
  - Large (e.g., 100 mm equivalent diameter)
  - Medium (e.g., 35 to 50 mm diameter)
  - Small (10% of the pipe diameter)
- Leak from liquid or gas pipe
  - Large (full-bore break)
  - Medium (22 to 44% of the pipe diameter)
  - Small (10% of the pipe diameter)
- Catastrophic rupture
- Vessel collapse.

For this example, a Performance Indicator (PI) is defined as the distance to 1% fatality found by applying a consequence model. Upon adding the PI values for each branch of the event tree, the authors conclude that all three technologies have a similar safety performance. This is because the extent of damage from fires and explosions is dominated by the LNG properties, and is nearly the same in each case.

9.4.14 Areas of LNG risk research

Interim research results were reported by Sandia Laboratories on two important issues related to risks to an LNG carrier from accident or attack (Hightower, 2012).

9.4.14.1 Brightness of large pool fires on water and land

Previously, the largest LNG pool fire test was on a 35 m diameter concrete pit at Montoir, France (Nedelka et al., 1989; Malvos and Raj, 2006). As the diameter increases for a pool fire, larger amounts

Table 9-16 Characteristics Considered of LNG Regasification Terminals

Type	Onshore	Offshore GBS	Offshore FSRU
Development stage	Operational	Starting up	Under design
Potential capacity (10^9 Nm^3/yr)	3.5	7.6	3.7
Storage size (m^3)	2 x 50,000	2 x 125,000	4 x 35,000
Storage tank type	Double containment	Self-supporting prismatic	Kvaerner/Moss-Rosenberg
Vaporizer type	Submerged combustion	ORV (Open rack vaporizer)	Intermediate fluid vaporizer

(Tugnoli et al., 2010)

of smoke occurred as would be expected from limited air diffusion into larger fires. The Montoir test fire was modeled with a surface emissive power (SEP) ranging from 300 kW/m^2 near the base of the fire to 100 kW/m^2 at the top of the visible flame (Raj, 2007). Raj fit a quadratic profile for SEP as a function of the fraction of the visible flame length.

Sandia Laboratories constructed a 120 m diameter pool, 2 m deep, for conducting LNG pool fire tests on water of 30 m, 70 m, and 100 m diameter. Sandia reported finding that water entrainment and low amounts of smoke created average flame SEP values higher than expected at 280 kW/m^2 (Hightower, 2012). This report also disclosed that the flame height/diameter ratio was less than expected from previous correlations extrapolated to large pool fires. The overall LNG pool fire hazard predicted distances were decreased by 3 to 7% from Sandia reports of 2004 and 2008.

9.4.14.2 Investigation of possible cascading damage on LNG carriers

The issue of cascading effects has been a concern for some time in LNG risk assessments. In particular, a breach releasing LNG into a double-hulled LNG carrier introduces the threat of metal stress on structural steel members from two temperature extremes, cryogenic cooling and fire heating. The interim Sandia report concludes:

- About 40% of LNG spilled can stay within the LNG vessel, causing cryogenic and fire thermal damage to the vessel's structure.
- The cargo tank insulation and pressure relief valve systems appear adequate to prevent overpressurization of the cargo tanks in an LNG fire.
- Simultaneous, multiple cargo tank spills (cascading failure) from an initial event seems unlikely.
- More detailed risk mitigation and management may be required, depending on site-specific conditions and operations (e.g., improved traffic control, lightering operations and capabilities, high-capacity firefighting tug escorts, etc.

9.5 LNG security

For decades, the LNG industry has maintained secure operations around the world, including areas where terrorism is a concern. However, since LNG infrastructure is highly visible and easily identified, it can be vulnerable to terrorist attack.

9.5.1 Codes for security

A number of international and national safety and design standards have been developed for LNG ships to prevent or mitigate spills of LNG over water. These standards are designed to prevent groundings, collisions, steering or propulsion failures, and attacks. They include traffic control, exclusion zones around a vessel while in transit within a port, escort by Coast Guard vessels, as well as early notice of a ship's arrival, investigation of crew backgrounds, at-sea boarding of LNG ships, special security sweeps, and positive control of an LNG ship during port transit.

Several of the provisions covered earlier for codes and standards have security provisions. The US codes 33CFR104 Part C—Vessel Security Assessment (VSA) and 33CFR105 Part C—Facility Security Assessment (FSA) give the US Coast Guard jurisdiction over maritime security on vessels and at facilities subject to 33CFR Parts 126, 127, and 154. The VSA is the responsibility of vessel owners and operators. The FSA is the responsibility of the facility owner/operator who must designate a Facility Security Officer (FSO) with background and experience specified in the regulation. The FSO is responsible to prepare a Facility Security Plan (FSP) based on an FSA. Paragraph 105.305 sets forth the minimum requirements for the FSA.

In addition, Canada requires LNG tankers and terminals to have an approved security plan under the Marine Transportation Security Act (SC 1994, c 40) and the Marine Transportation Security Regulations (SOR/2004-144) (MIACC, 2007).

9.5.2 Security vulnerability analysis (SVA)

Security risk is defined as a function of consequence, vulnerability, and threat. Consequence is a measure of a result if an item, process, or system is destroyed or interrupted. Vulnerability is a measure of how well a site is physically protected by barriers, electronics, people, and processes. Threat is a measure of how likely it is that a person or group has targeted the site for penetration (Walker et al., 2003).

A performance-based SVA method that is relevant to LNG facilities is available from the American Petroleum Institute (API) and the National Petrochemical & Refiners Association (NPRA). This guideline, "Security Vulnerability Assessment Methodology for the Petroleum and Petrochemical Industries," draws heavily on the AIChE Center for Chemical Process Safety (CCPS) SVA methodology (CCPS, 2002).

Essentially all SVA methods involve similar steps as illustrated in Figure 9-36 (Luketa-Hanlin and Hightower, 2006). The five steps in the method illustrated in Figure 9-36 are similar to the widely accepted Navigation Vessel Inspection Circular (NVIC) No. 11-02, "Recommended Security Guidelines for Facilities," published by the US Coast Guard.

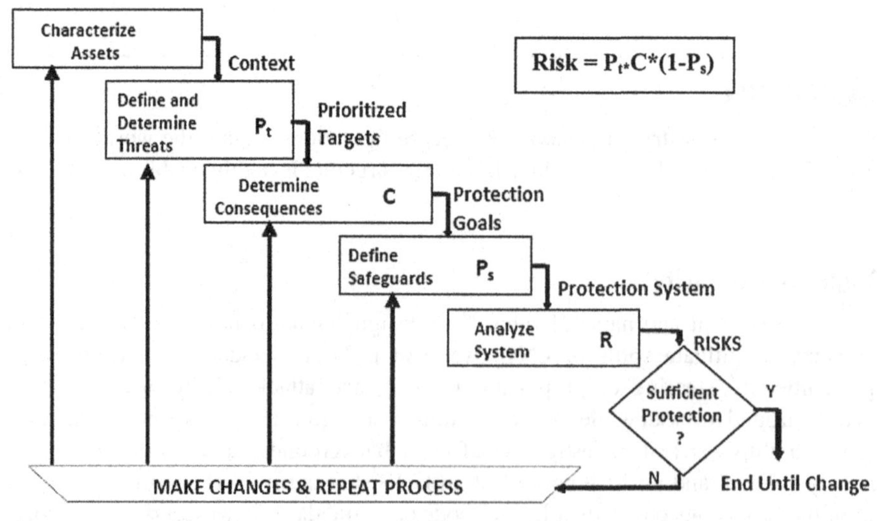

FIGURE 9-36

Security vulnerability analysis process.

(Luketa-Hanlin and Hightower, 2006)

Figure 9-36 is illustrated as an iterative process in which layers of safeguards (barriers or procedures) are added and the risk is reevaluated, very much like conventional risk analysis except that the threat probability is very difficult to quantify. The steps are:

- Characterize assets. Collect information on critical operations, hazardous materials, nearby populations and businesses, and security systems.
- Identify desirable target assets and prioritize their value.
- Identify possible threats and vulnerabilities. Rank the threats after determining the consequences and safeguards. Assign a probability P_T to quantify this ranking.
- Determine consequences for each threat scenario. Establish the potential damage to a ship cargo tank or onshore facility and the potential volume and rate of spills that could occur.
- Establish the environmental conditions (e.g., wind, currents, atmospheric stability) and calculate the volatilization, dispersion, fire radiation contours, and explosion potential and estimate the cost of each threatened event, C.
- List current safeguards, layers of protection, and response resources. Estimate the effectiveness in reducing the severity of an event, P_S.
- Calculate risk, R, as the product P_T (1–P_S)C for each threat and sum all threats.
- If the risk seems too high provide additional layers of protection to reduce risk of attack/sabotage.

Improvements to reduce security risk usually focus on reducing vulnerability by developing strategies to detect, delay, and respond to an adversary in the shortest amount of time.

Examples of layers of protection and areas for improvement include:

- Physical: fencing, trenches, motion detectors, trip wires, lighting, etc.
- Internal surveillance: security personnel, TV monitors, etc.
- Tank and insulation upgrades
- Tanker standoff protection systems
- Improved surveillance and searches of tugs, ship crews, and other vessels
- Redundant offshore mooring and offloading systems
- External surveillance: local and state law enforcement, port authorities, USCG, FBI
- Improved emergency response coordination, communication, drills, and training.

These measures reduce the likelihood that those with criminal intent will target their vessels or facilities.

9.5.3 Security vulnerability criticality index

Several approaches are used in security analysis with application to LNG facilities. One approach is to estimate an index that semiquantifies the impact or criticality, *C*, of an attack, or the value to a terrorist, as additive factors such as (Bernatik et al., 2011):

$$C = 2C_{type} + C_{casualties} + C_{econ\ \mathrm{impact}} + C_{outage\ \mathrm{time}} + C_{sectors\ impact} + C_{environmental} \quad (9\text{-}8)$$

where C_{TYPE} is the significance of a monument, $C_{CASUALTIES}$ is number of casualties, $C_{ECONOMIC\ IMPACT}$ is cost of replacement damage to equipment (or ship), $C_{OUTAGE\ TIME}$ is cost of lost business (business interruption), $C_{IMPACT\ ON\ OTHER\ SECTORS}$ is cost to restore the supply chain and to replace energy supply, and $C_{ENVIRONMENT}$ is cost to restore the damaged environment.

Table 9-17 Effectiveness of Protective Measures

Effectiveness of Protective Measures	Definition of Effectiveness	Value
Very high	Represents significant obstacle to achieving objectives of an attack.	0.9
High	Overcoming this obstacle will require great effort.	0.7
Somewhat effective	Overcoming this obstacle requires moderate effort.	0.5
Minimally effective	Overcoming this obstacle requires minimal effort.	0.3
Low	The system does not put obstacles that would prevent achieving the objectives of an attack.	0.1

(Senovsky, 2008)

There are also less quantifiable factors that could enter a terrorist's estimation of the value of a target such as the reaction that can be achieved from society, such as an impact on stock markets, or a depression of the economy.

Similarly, estimate the effectiveness of protective measures, E_P, or, inversely, the likelihood of a successful attack, L_A, where:

$$L_A = 1 - E_P \tag{9-9}$$

The effectiveness of protective measures can be given a value (probability) from Table 9-17.

Such security analyses have placed LNG facilities generally below a number of other potential targets for both impact and for the likelihood of a successful attack.

9.5.4 Deciding on sufficiency of protective measures

The approach described previously selects strategies based on site-specific conditions and the expected impact of a spill on public safety and property. Less intensive strategies could often be sufficient in areas where the impacts of a spill are low (Luketa-Hanlin and Hightower, 2006). A two-tiered approach is illustrated in Table 9-18. For any operation, some level of residual risk remains after continued risk reduction programs.

9.5.5 Security of ships and land-based LNG facilities

LNG ships may be physically attacked in a variety of ways to destroy their cargo, or commandeered for use as weapons against coastal targets. However, with their double-hull construction, robust cargo tanks with multiple layers of insulation, implementation of maritime security measures following the attack on the World Trade Center in New York in 2001, scrutiny from regulators, transit risk mitigation measures, and the training required for the crew, LNG ships are becoming less desirable targets.

Land-based LNG facilities may also be physically attacked with explosives or through other means. Alternatively, computer control systems may be "cyber-attacked," or both physical and cyber attack may happen at the same time. Some LNG facilities may also be indirectly disrupted by other types of terror strikes, such as attacks on regional electricity grids or communications networks, which in turn could affect dependent LNG control and safety systems (Skolnik, 2002).

Table 9-18 Examples of LNG Spill Risk Reduction Options

Impact on Public Safety	Event Damage Reduction (Prevention)	System Security and Safety (Mitigation)
Medium and high	Use early offshore interdiction Use ship inspection Control escorts, ships, or tugs Control vessel movement (safety/ security zones) One-way traffic Use LNG offloading system security interlocks	Use harbor pilots Upgrade ship and terminal security Expand emergency response to address firefighting, vapor clouds, and damaged vessels
Low	Use existing best risk management practices for traffic control, monitoring, and safety zones	Use existing best risk mitigation practices

(Luketa-Hanlin and Hightower, 2006)

LNG facilities have been the subject of many analyses and studies by government authorities, research centers, and large insurance companies. Security experts do not consider LNG facilities to be a priority terrorist target. Studies (OECD, 2003; Southwell, 2005; Tierney, 2006) concluded that full containment tanks were unattractive targets given the difficulty of undermining the structural integrity of these tanks. These facilities are identified as part of a country's critical infrastructure and enhanced security measures have been implemented. However, a report from the US Congressional Research Service noted that pipelines and oil facilities have already been the targets of attacks throughout the world (Parfomak and Frittelli, 2007).

9.5.6 Security initiatives, RAMCAP

The US Department of Homeland Security (DHS), Directorate of Information Analysis and Infrastructure Protection (JAIP), Protective Services Division (PSD) contracted with the American Society of Mechanical Engineers Innovative Technologies Institute, LLC (ASME: ITI) to develop guidance on Risk Analysis and Management of Critical Asset Protection (RAMCAP). The objectives are:

- Improve the framework for Security Vulnerability Analysis (SVA) by providing a common basis for developing SVAs and for making vulnerability assessments and risk-based decisions.
- Improve the screening process used by the DHS for understanding the assets that are important to protect against terrorist attacks and to prioritize risk management steps.

This effort recognized that initially in the security field there were differing approaches, terminology, criteria, scales, and outputs. The development of RAMCAP is expected to provide more consistent terminology, criteria, and such, and consistent, objective, and integrated application of risk analysis methods. A driving force is that resources are limited and allocating resources requires prioritization. A seven step process has been outlined as the RAMCAP framework for interaction between facility owners/operators and the DHS (Table 9-19).

Table 9-19 The Seven Steps in the RAMCAP Process

Step	Actions
1. Characterize assets	Identify assets Assess potential severity of consequences Screen out low consequence events
2. Characterize threats	Determine targets Characterize adversary capabilities: tactics, weapons Compare with threat characterized by DHS Owner/operator may choose lesser threats at their discretion
3. Analyze consequences	Find potential damage for each threat Find worst reasonable case consequences (C*)
4. Analyze vulnerability	Identify vulnerabilities to worst reasonable consequence event (V*) Assess likelihood of adversary success Evaluate existing countermeasures and mitigation capability
5. Assess threats	Assess attractiveness and deterrence DHS will determine adversary capability and determination DHS estimates threat (T*) as likelihood of attack as a function of attractiveness and adversary capability and intent
6. Assess risk	Risk= C*V*T from steps 3, 4, and 5
7. Manage risk	Consider risk goals and need for action Make recommendations Evaluate options and decide on enhancements

Values on a scale of 1 to 10.
(Moore et al., 2007)

9.5.7 Security of offshore and remote LNG facilities

Some have suggested that new LNG import terminals should be built only offshore to keep associated terrorism hazards away from populated areas. Such a strategy may indeed reduce terrorism risks to ports and coastal communities, but it may also increase the risks to the terminals and terminal operators themselves. Because offshore oil and gas facilities are remote, isolated, and often lightly manned, some experts believe they are more vulnerable to terror attacks than land-based facilities (Daughdril, 2003). Offshore oil and gas facilities have not been frequent terror targets, but they have been attacked in the past during wartime and in territorial disputes (George, 1993). Since September 11, 2001, international concern about terrorist attacks on these platforms has grown (McKenzie, 2003). Some experts believe terrorist attacks against offshore platforms and remote land-based facilities have been on the rise recently in countries with a history of terror and extortion activity like Nigeria, Colombia, Yemen, and Indonesia, although many of these attacks may be motivated by finance/extortion, rather than politics (Adams, 2003). Disruption of any single offshore LNG terminal would not likely have a great impact on natural gas supplies, but if several new offshore terminals were attacked in the future, the effects on natural gas availability and prices could have serious consequences for energy markets. Onshore versus offshore siting alternatives should, therefore, be considered in the context of exposure to security breaches, public security, and the security of national gas supplies.

Similarly, the greater availability of spot cargoes versus long-term supply contracts means that disruption of supply is now easier to accommodate. In the past, there were only long-term contracts, and if a ship were lost there was no replacement immediately available. With spot cargoes, if one vessel is no longer available, it is readily replaced by a spot cargo.

9.5.8 Policy issues in LNG security

The LNG industry has taken significant steps to secure the LNG infrastructure. But continued progress in implementing and sustaining LNG security faces several challenges.

Some LNG operating companies have resisted suggestions that they pay more for public security (McElhenny, 2001). Others have expressed a willingness to pay for "excess" security only if it exceeds the level of government agency service ordinarily provided from corporate tax payments (Dominion Resources, 2003). It is difficult to predict how the public component of LNG security costs will evolve as the LNG industry grows. The public component might be expected to decrease, particularly if security threats decline in the medium-term. Altogether, the potential increase in security costs from expanding LNG infrastructure and shipping warrants a review of these costs, the public share and sustainability.

While acknowledging the potential terrorist threat, many experts, including those in the LNG industry, believe that public concern about threats to LNG infrastructure is overstated. Industry representatives argue that deliberately attacking an LNG facility to cause disruption, terror, and injury might perhaps be attempted, but remains extremely difficult to execute effectively. LNG proponents also believe that LNG facilities are relatively secure compared to other hazardous chemical and hydrocarbon infrastructures, which also receive less public attention. However, it may be impossible for the LNG industry officials to ever prove that LNG infrastructure will not be targeted by terrorists. As the US FERC has remarked, "unlike accidental causes, historical experience provides little guidance in estimating the probability of a terrorist attack on an LNG vessel or onshore storage facility" (Foss et al., 2003). Because the probability and impact of a terrorist attack on LNG infrastructure cannot be known with certainty, policy makers, plant operators, emergency response specialists, and community leaders must ultimately rely on their judgment to decide on the adequacy of LNG security measures for a specific facility. Also what will adequately protect the public and what, if any, of the incremental security costs, personnel, and resources are to be provided from public funds.

References

ABS Consulting Inc, May 2004. Consequence Assessment Methods for Incidents Involving Release from Liquefied Natural Gas Carriers. Report 131–04, prepared for Federal Energy Regulatory Commission (FERC), Washington, DC, USA.

Adams, N., 2003. Terrorism in the Offshore Oil Field. Underwater 15 (2), 70–72.

Alderman, J.A., 2005. Introduction to LNG Safety. Process Safety Progress 24 (3), 144–151.

Baker, Q.A., Doolittle, C.M., Fitzgerald, G.A., Tang, M.J., March 9–12, 1997. Recent Developments in the Baker-Strehlow VCE Analysis Methodology, paper presented at the 31st Loss Prevention Symposium, Houston, TX, USA.

Bernatik, A., Senovsky, P., Pitt, M., 2011. LNG as a Potential Alternative Fuel—Safety and Security of Storage Facilities. Journal of Loss Prevention in the Process Industries 24 (1), 19–24.

Berrera, C.A., Kemal, A., March 22–24, 2010. Condensation-induced Water Hammer: Principles and Consequences, paper presented at the AIChE Spring Meeting, San Antonio, TX, USA.

Bichou, K., Dec. 11–12, 2008. Security and Risk-based Models in Shipping and Ports: Review and Critical Analysis. UK. Discussion Paper No. 2008–20, prepared for the OECD/ITF Round Table, Joint Transport Research Center, London.

Bird, F.E., Germain, G.L., Clark, M.D., 2003. Practical Loss Control Leader, third ed. Det Norske Veritas Inc, Georgia, GA, USA.

Broadribb, M.P., 2003. Guidelines for Investigation Chemical Process Incidents, second ed. Center for Chemical Process Safety (CCPS), AIChE, New York, NY, USA.

Bubbico, R., Salzano, E., 2009. Acoustic Analysis of Blast Waves Produced by Rapid Phase Transition of LNG Released on Water. Safety Science 47 (4), 515–521.

Bubbico, R., Di Cave, S., Mazzarotta, B., 2009. Preliminary Risk Analysis for LNG Tankers Approaching a Maritime Terminal. Journal of Loss Prevention in the Process Industries 22 (5), 634–638.

CCPS, 2000. Guidelines for Chemical Process Quantitative Risk Analysis. Center for Chemical Process Safety (CCPS), AIChE, New York, NY, USA.

CCPS, 2001. Layers of Protection Analysis—Simplified Process Risk Assessment. Center for Chemical Process Safety (CCPS), AIChE, New York, NY, USA.

CCPS, 2002. Guidelines for Managing and Analyzing the Security Vulnerabilities of Fixed Chemical Sites. Center for Chemical Process Safety (CCPS), AIChE, New York, NY, USA.

CCPS, 2007. Guidelines for Safe and Reliable Instrumented Protective Systems. Center for Chemical Process Safety (CCPS), AIChE, New York, NY, USA.

CCPS, 2009. Guidelines for Developing Quantitative Safety Risk Criteria (Risk Tolerance Criteria or RTC Guidelines). CCPS, AIChE, New York, NY, USA.

Chamberlain, G.A., June 5–9, 2006. Management of Large LNG Hazards, paper presented at the 23rd World Gas Conference, Amsterdam, The Netherlands.

Chan, S.T., 1992. Numerical Simulations of LNG Vapor Dispersion from a Fenced Storage Area. Journal of Hazardous Materials 30 (2), 195–224.

Chang, C., Meroney, R., 2003. Concentration and Flow Distributions in Urban Street Canyons: Wind Tunnel and Computational Data. Journal of Wind Engineering and Industrial Aerodynamics 91 (9), 1141–1154.

Clement, M., 2000. Experimental Verification of the Fire Dynamics Simulator (FDS) Hydrodynamic Model. PhD Thesis. University of Canterbury, Christchurch, New Zealand.

Crowl, D.A., Louvar, J.L., 2002. Chemical Process Safety- Fundamentals with Applications, second ed. Prentice Hall, Upper Saddle River, NJ, USA.

CSChE, 2004. Risk Assessment—Recommended Practices for Municipalities and Industry. Canadian Society for Chemical Engineering (CSChE), Ottawa, ON, Canada.

de la Vega, F.F., Durr, C., March 21–24, 2004. Designing Safety into LNG Export/Import Plants, paper presented at the 14th International Conference & Exhibition on Liquefied Natural Gas (LNG 14), Doha, Qatar.

Delvosalle, C., Fievez, C., Pipart, A., Debray, B., 2006. ARAMIS Project: A Comprehensive Methodology for the Identification of Reference Accident Scenarios in Process Industries. Journal of Hazardous Materials 130 (3), 200–219.

Dominion Resources, Aug.19, 2003. Corporate Security. Personal Communication, Richmond, VA, USA.

Dowell, A.W., 2011. Is It Really an Independent Protection Layer. Process Safety Progress 30 (2), 126–131.

ElSayed, T., 2010. Risk Assessment of Marine LNG Operations. In: Natural Gas Book. InTech Europe, Rijeka, Croatia.

Er, I.D., 2007. Safety and Environmental Concern Analysis for LNG Carriers. International Journal on Marine Navigation and Safety of Sea Transportation 1 (4), 421–426.

Fay, J., 2003. Model of Spills and Fires from LNG and Oil Tankers. Journal of Hazardous Materials 96, 2–3, 171–188.

Foss, M.M., Delano, F., Gülen, G., Makaryan, R., Oct. 2003. LNG Safety and Security. Research Report, Commercial Frameworks for LNG in North America consortium, Center for Energy Economics (CEE), Institute for Energy, Law & Enterprise, University of Houston Law Center, Houston, TX, USA.

Frank, W., 2011. Challenges in Developing and Implementing Safety Risk Tolerance Criteria. Process Safety Progress 30 (3), 232–239.

Frank, W., Jones, D., 2010. Choosing Appropriate Quantitative Safety Risk Criteria: Applications from the New CCPS Guidelines. Process Safety Progress 29 (4), 293–298.

Fullwood, R.R., Hall, R.E., 1988. Probabilistic Risk Assessment in the Nuclear Power Industry, first ed. Pergamon Press, Oxford, UK.

Gabbar, H.A., Bedard, R., June 25–26, 2010. Hazard Analysis and Accident Prediction for LNG Plants, paper presented at the International Workshop on Real Time Measurement, Instrumentation and Control (RTMIC), Oshawa, ON, Canada.

Gabbar, H.A., Khan, F.I., June 25–26, 2010. Design of Fault Semantic Networks to Integrate Fault, Failure, Hazard, and Accident Models for LNG Plants, paper presented at the International Workshop on Real Time Measurement, Instrumentation and Control (RTMIC), Oshawa, ON, Canada.

Gavelli, F., Davis, S., Hansen, O.R., March 21–25, 2010. A Unified Model for LNG Pool Spread and Vapor Dispersion: Is Wind Scooping Really a Factor?, paper presented at the AIChE Spring Meeting, San Antonio, TX, USA.

George, D., Oct. 1993. Piracy on Increase, Present Threat to Offshore Structures, Vessels. Offshore 25.

GIIGNL (French acronym for International Group of Liquefied Natural Gas Importers), Sept., 2011a. LNG Information Paper No. 4, Managing LNG Risks- Operational Integrity, Regulations, Codes, and Industry Organizations. http://www.giignl.org/.

GIIGNL (French acronym for International Group of Liquefied Natural Gas Importers), Sept. 2011a. LNG Information Paper No. 6, Managing LNG Risks- Industry Safeguard Systems. http://www.giignl.org/.

Goy, A., June 1–5, 2003. Liquefied Gases, Safety and Environmental Management in LNG Plants, report presented at the 22nd World Gas Conference, Tokyo, Japan.

GRI, April 20, 1990. Reduction of LNG Operator Error and Equipment Failure Rates. GRI 90/0008, Gas Research Institute (GRI), Chicago, IL, USA.

Hansen, H.I., Nielsen, D., Frydenberg, M., 2002. Occupational Accidents Aboard Merchant Ships. Occupational Environment Medicine 59 (2), 85–91.

Havens, J., 2005. Public Hazards of LNG Import Terminal Operations. Business Briefing: LNG Review 52–54.

Havens, J., Spicer, T., 2005. LNG Vapor Cloud Exclusion Zones for Spills into Impoundments. Process Safety Progress 24 (3), 181–186.

Havens, J., Spicer, T., Oct. 24–25, 2006. Error in FERC Environmental Impact Statement Determinations of LNG Vapor Cloud Exclusion Zones: Failure to Account for Air Mixing in Vapor Impoundments, paper presented at the 9th Annual International Symposium of the Mary Kay O'Connor Process Safety Center, Texas A&M University, College Station, TX, USA.

Havens, J., Spicer, T., 2007. United States Regulations for Siting LNG Terminals: Problems and Potential. Journal of Hazardous Materials 140 (3), 439–443.

Hightower, M., Gritzo, L., Luketa-Hanlin, A., Covan, J., Tieszen, S., Wellman, G., Irwin, M., Kaneshige, M., Melof, B., Morrow, C., Ragland, D., Dec. 2004. Guidance on Risk Analysis and Safety Implications of a Large Liquefied Natural Gas (LNG) Spill Over Water. Report SAND2004-6258, Sandia National Laboratories, Albuquerque, NM, USA.

Hightower, M., Luketa-Hanlin, A., Gritzo, L.A., Covan, J.M., Jan. 2006. Review of the Independent Risk Assessment of the Proposed Cabrillo Liquefied Natural Gas Deepwater Port Project. Report SAND2005-7329, Sandia National Laboratories, Albuquerque, NM, USA.

Hightower, M., May, 2012. DOE/Sandia National Laboratories LNG Safety Research. LNG Export Forum Safety Workshop, Houston, TX, USA.

HSE, 1992. The Tolerability of Risk from Nuclear Power Stations. Health and Safety Executive (HSE), London, UK.

HSE, 2002. Marine Risk Assessment. Offshore Technology Report 2001/063, Health and Safety Executive (HSE), London, UK.

IGC, 1975. Code for the Construction and Equipment of Ships Carrying Liquefied Gases in Bulk. International Gas Carriers (IGC). Now in updated amendments and published by International Maritime Organization (IMO), London, UK.

IMO, 1997. Interim Guidelines for Safety Assessment. International Maritime Organization (IMO), London, UK.

IMO, Dec. 17, 2002. IMO Adopts Comprehensive Maritime Security Measures. Press Release, International Maritime Organization (IMO), London, UK.

ISA –TR84.00.02, 2002. Safety Instrumented Functions (SIF)- Safety Integrity Level (SIL) Evaluation Techniques. The International Society of Automation (ISA), Research Triangle Park, NC, USA.

ISO Standard 17776:2000, Offshore Production Installations—Guidelines on Tools and Techniques for Hazard Identification and Risk Assessment, first ed. International Standards Organization (ISO) Catalogues (www.iso.org) (visited April 2012).

Ivings, M.J., Webber, D., 2007. On Defining Safety Criterion for Flammable Clouds. Health and Safety Laboratory (HSL) Report HSL/2007/30, Harpur Hill, Buxton, UK.

Johnson, D.W., Welker, J.R., 1981. Development of An Improved LNG Plant Failure Rate Data Base. Applied Technology Corp, Norman, OK, USA.

Keong, T.H., May, 2012. Risk analysis of engineering systems. MSc Thesis. National University of Singapore, Singapore.

Keshavarz, G., Thodi, P., Khan, F.I., 2012. Risk-based Shutdown Management of LNG Units. Journal of Loss Prevention in the Process Industries 25 (1), 159–165.

Khan, F.I., Sadiq, R., Haddara, M.M., 2004. Risk-based Inspection and Maintenance, Multi-attribute Decision-making with Aggregative Risk Analysis. Process Safety and Environmental Protection 82 (6), 398–411.

Kohout, A., April 1–5, 2012. Evaluation of Dispersion Models for LNG Siting Applications, paper presented at the 12th Topical Conference on Gas Utilization, AIChE Spring Meeting, Houston, TX, USA.

Laubenstein, L., Mains, C., Jost, A., Tagg, R., Bjørneboe, N.K., July 1–3, 2001. Updated Probabilistic Extents of Damage-based on Actual Collision Data, paper presented at the 2nd International Conference on Collision and Grounding of Ships, Copenhagen, Denmark.

Livingston, M., Gustafson, R., 2009. Minimize Risks from Cryogenic Exposure on LNG Facilities. Hydrocarbon Processing 88 (7), 51–58.

Luketa-Hanlin, A., Hightower, M., Nov. 30, 2006. Guidance on Safety and Risk Management of Large Liquefied Natural Gas (LNG) Spills Over Water, presented at the U.S Department of Energy LNG Forums, Houston, TX, USA.

Luketa-Hanlin, A., Hightower, M., 2008a. On the Evaluation of Models for Hazard Prediction for Liquefied Natural Gas (LNG) Spills on Water. Report by Sandia National Laboratories, Albuquerque, NM, USA.

Luketa-Hanlin, A., Hightower, M., Attaway, S., 2008b. Breach and Safety Analysis of Spills Over Water from Large Liquefied Natural Gas Carriers. Sandia Report SAND2008-3153, Sandia National Laboratories, Albuquerque, NM, USA.

Malvos, H., Raj, P.K., April 23–27, 2006. Details of 35 m Diameter LNG Fire Tests Conducted in Montoir, France in 1987 and Analysis of Fire Spectral and Other Data, paper presented at the AIChE Spring Meeting, Orlando, FL, USA.

Mannan, M.S., Wang, J.Y., West, H.H., 2005. LNG Safety—An Update on Recent Issues. Business Briefing: LNG Review 48–51.

MARIN, Jan. 17, 2012. Safety Assessment Studies for LNG Transport in Ports and Port Approaches. Newsletter, Maritime Research Institute Netherlands (MARIN), Wageningen, The Netherlands.

MCA, 1996. Guidelines for Formal Safety Assessment (FSA). Maritime and Coastguard Agency (MCA), Southampton, Hampshire, UK.

McElhenny, J., Nov. 2, 2001. State Says LNG Tanker Security Cost $20,500. Associated Press, New York, NY, USA.

McKenzie, S., Aug. 10, 2003. Submarine Terror Fear for North Sea Oil Rigs. Sunday Mail, Scotland, Glasgow.

Melhem, G.A., 2005. Managing LNG Rollover. IoMosaic Corporation Quarterly News 4 (1), 1.

Melton, T.A., Cornwell, J.B., 2010. LNG Trench Dispersion Modeling Using Computational Fluid Dynamics. Journal of Loss Prevention in the Process Industries 23 (6), 762–767.

Mendola, A., 1988. Accident Sequence Dynamic Simulation Versus Event Trees. Reliability Engineering and System Safety 22, 1–4, 3–25.

MIACC, May, 2007. Rabaska Project—Implementation of an LNG Terminal and Related Infrastructure. Report 241 (Translation), Major Industrial Accidents Council of Canada (MIACC), Ottawa, ON, Canada.

Moore, D.A., Fuller, B., Hazzan, M., Jones, J.W., 2007. Development of a Security Vulnerability Assessment Process for the RAMCAP Chemical Sector. Journal of Hazardous Materials 142 (3), 689–694.

Nedelka, D., Moorhouse, D.J., Tucker, R.F., Oct. 17–20, 1989. The Montoir 35 m Diameter LNG Pool Fire Experiments, paper presented at the 9th International Conference & Exhibition on Liquefied Natural Gas (LNG 9), Nice, France.

Nedelka, D., Sauter, V., Goanvic, J., Ohba, R., May 5–8, 2003. Last Developments in Rapid Phase Transition Knowledge and Modeling Techniques, paper presented at the Offshore Technology Conference, Houston, TX, USA.

Nicholson, P., Jan. 2006. Evolving Risks in the LNG Supply Chain. Public Utilities Fortnightly, 20–23. +62.

OECD, July 2003. Maritime Transport Committee, Security in Maritime Transport: Risk Factors and Economic Impact. Organization for Economic Cooperation and Development (OECD), Paris, France.

Oka, H., Ota, S., 2008. Evaluation of Consequence Assessment Methods for Pool Fires on Water Involving Large Spills from Liquefied Natural Gas Carriers. Journal of Marine Science and Technology 13 (2), 178–188.

Oka, H., 2009. Consequence Analysis of Pool Fire Hazards from Large-scale Liquefied Natural Gas Spills Over Water. Hydrocarbon World 4 (1), 90–93.

Oka, H., 2010. Consequence Analysis of Large-scale Liquefied Natural Gas Spills on Water. In: Natural Gas Book. InTech Europe, Rijeka, Croatia.

OREDA (Offshore REliability DAta) Handbook, 2002. fourth ed. DNV, Hovik, Norway.

Ozog, H., Designing an Effective Risk Matrix. Whitepaper, ioMosaic Corporation, Salem, NH, USA (accessed 12.02.2012).

Paik, J.K., Choe, I.H., Thayamballi, A.K., July 1–3, 2001. An Accidental Limit State Design of Spherical Type LNG Carrier Structures Against Ship Collisions, paper presented at the 2nd International Conference on Collision and Grounding of Ships, Copenhagen, Denmark.

Paltrinieri, N., Cozzani, V., Sept. 5–8, 2010. Atypical Major Hazard Scenarios and Their Inclusion in Risk Analysis and Safety Assessments, paper presented at the ESREL 2010 Annual Conference, Rhodes, Greece.

Paradies, M., 2011. Has Process Safety Management Missed the Boat? Process Safety Progress 30 (4), 310–314.

Parfomak, P.W., Frittelli, J., Jan. 9, 2007. Maritime Security: Potential Terrorist Attacks and Protection Priorities. CRS Report for Congress, Congressional Research Service (CRS), Washington, DC, USA.

Pasman, H., Rogers, W., Oct. 25–27, 2001. Bayesian Belief Networks (BBN), a Tool to Make LOPA More Effective, QRA More Transparent and Flexible and Therefore to Make Safety More Definable, paper presented at the 4th Annual International Symposium of the Mary Kay O'Connor Process Safety Center, Texas A&M University, College Station, TX, USA.

Paté-Cornell, M.E., 1983. Risk Analysis and Risk Management for Offshore Platforms: Lessons from Piper Alpha Accident. Journal of Offshore Mechanics and Arctic Engineering 115 (1), 179–190.

Paté-Cornell, M.E., 1984. Fault Tree Versus Event Tree in Reliability Analysis. Risk Analysis 4 (3), 177–186.

Paté-Cornell, M.E., 1994. Quantitative Safety Goals for Risk Management of Industrial Facilities. Structural Safety 13 (3), 145–157.

Pitblado, R.M., Baik, J., Hughes, G.J., Ferro, C., Shaw, S.J., June June–July 1–29, 2004. Consequences of LNG Marine Incidents, paper presented at the 19th Annual International Conference of Center for Chemical Process Safety (CCPS), Orlando, FL, USA.

Pitblado, R.M., Skramstad, E., Danielsen, H.K., Hysing, T., April 6–10, 2008. Energy-based Risk Methodology for Collision Protection of LNG Carriers, paper presented at the 8th Topical Conference on Natural Gas Utilization, AIChE Spring Conference, New Orleans, LA, USA.

Pitblado, R.M., Tahilramani, R., Oct. 27–28, 2009. Risk Communications; Websites for Barrier Diagrams and Process Safety, paper presented at the 12th Annual International Symposium of the Mary Kay O'Connor Process Safety Center, Texas A&M University, College Station, TX, USA.

Pitblado, R.M., Fisher, M., Benavides, A.J., Oct. 25–27, 2011. Linking Incident Investigation to Risk Assessment, paper presented at the 14th Annual International Symposium of the Mary Kay O'Connor Process Safety Center, Texas A&M University, College Station, TX, USA.

Pitblado, R.M., Bardy, M., Nalpanis, P., Crossthwaite, P., Molazemi, K., Bekaert, M., Raghunathan, V., April 1–4, 2012. International Comparison on the Application of Societal Risk Criteria, paper presented at the AIChE 8th Global Congress on Process Safety, Houston, TX, USA.

Planas-Cuchi, E., Gasulla, N., Ventosa, A., Casal, J., 2004. Explosion of a Road Tanker Containing Liquefied Natural Gas. Journal of Loss Prevention in the Process Industries 17 (4), 315–321.

Qi, R., Raj, P.K., Mannan, S., 2011. Underwater LNG Release Test Findings—Experimental Data and Model Results. Journal of Loss Prevention in the Process Industries 24 (4), 440–448.

Qiao, Y., West, H.H., Mannan, S., Johnson, D.W., Cornwell, J.B., 2006. Assessment of the Effects of Release Variables on the Consequences of LNG Spillage Onto Water Using FERC Models. Journal of Hazardous Materials 130, 1–2, 155-162.

Raj, P.K., 2007. LNG Fires: A Review of Experimental Results, Models and Hazard Prediction Challenges. Journal of Hazardous Materials 140 (3), 444–464.

Raj, P.K., Feb. 2008a. Substantiation Provided for Comment #59A–29", 2008 Annual Revision Cycle, Report on Comments. NFPA59A-8 Committee, National Fire Protection Association (NFPA), Quincy, MA, USA.

Raj, P.K., Oct. 28–29, 2008a. Use of Risk Analysis for LNG Facility Siting in the US: Changes in NFPA 59A and in Future Regulations, a Road Map and Required Actions, paper presented at the 11th Annual International Symposium of the Mary Kay O'Connor Process Safety Center, Texas A&M University, College Station, TX, USA.

Raj, P.K., Bowdoin, L.A., 2010. Underwater LNG Release: Does a Pool Form on the Water Surface? What are the Characteristics of the Vapor Released? Journal of Loss Prevention in the Process Industries 23 (6), 753–761.

Rathnayaka, S., Khan, F.I., Amyotte, P., 2011. Accident Modeling Approach for Safety Assessment in an LNG Processing Facility. Journal of Loss Prevention in the Process Industries 25 (2), 414–423.

Rickover, H.G., 2011. Congressional Testimony. Available at: http://www.taproot.com/wordpress/archives/16148.

Senovsky, P., 2008. Methods of the Critical Infrastructure Assets Risk Analysis. Spektrum SPBI 1, 8–10.

SIGTTO, 2000. Safety in Liquefied Gas Marine Transportation and Terminal Operations: A Guide for Self-Assessment. Society of International Gas Tanker and Terminal Operators (SIGTTO), London, UK.

Siu, N., 1994. Risk Assessment for Dynamic Systems: An Overview. Reliability Engineering and System Safety 43 (1), 43–73.

Skolnik, S., Sept. 2, 2002. Local Sites Potential Targets for Cyber-terror. Seattle Post-Intelligencer, Seattle, WA, USA.

Smith, K., Oct. 26–28, 2010. Lessons Learned from Real World Applications of the Bow-tie Method, paper presented at the 13th Annual International Symposium of the Mary Kay O'Connor Process Safety Center, Texas A&M University, College Station, TX, USA.

Southwell, C., Nov. 9, 2005. An Analysis of the Risks of a Terrorist Attack on LNG Receiving Facilities in the United States. CREATE Report, CREATE Homeland Security Center, University of Southern California, Los Angeles, CA, USA.

Stamatis, D.H., 1995. Failure Mode and Effect Analysis—FMEA from Theory to Execution. ASQC Quality Press, American Society for Quality (ASQ), Milwaukee, WI, USA.

Takahashi, K., Kado, K., Hara, N., April 24–27, 2007. Advanced Numerical Simulation of Gas Explosions for Assessing the Safety of LNG Plants, poster presented at the 15th International Conference & Exhibition on Liquefied Natural Gas (LNG 15), Barcelona, Spain.

Talbot, J.. What's right with risk matrices? A great tool for risk managers (accessed on 12.06.2012). http://31000risk.wordpress.com/article/what-s-rightwith-risk-matrices-3dksezemjiq54-4/.

Tanabe, M., Miyake, A., 2010. Safety Design Approach for Onshore Modularized LNG Liquefaction Plant. Journal of Loss Prevention in the Process Industries 23 (4), 507–514.

Tanabe, M., Miyake, A., 2011. Risk Reduction Concept to Provide Design Criteria for Emergency Systems for Onshore LNG Plants. Journal of Loss Prevention in the Process Industries 24 (4), 383–390.

Taylor, D.W., 2007. The Role of Consequence Modeling in LNG Facility Siting. Journal of Hazardous Materials 142 (3), 776–785.

Tierney, S.F., June 30, 2006. Report to the Massachusetts Special Commission Relative to Liquefied Natural Gas Facility Siting and Use. Analysis Group Inc, Boston, MA, USA.

Tugnoli, A., Paltrinieri, N., Landucci, G., Cozzani, V., March 14–17, 2010. LNG Regasification Terminals: Comparing the Inherent Safety Performance of Innovative Technologies, paper presented at the 4th International Conference on Safety & Environment in Process Industry, Florence, Italy.

Trbojevic, J., 2010. Risk Acceptance Criteria in Europe. Safety and Reliability of Industrial Products, Systems, and Structures. In: Soares, C.G. (Ed.). CRC Press, Boca Raton, FL, USA.

U.S Federal Law 49 CFR Part 193, Oct. 2010. Liquefied Natural Gas Facilities: Federal Safety Standards. US Government Printing Office, Washington, DC, USA.

Vanem, E., Antao, P., Ostvik, I., Del Castillo de Comas, F., 2008. Analyzing the Risk of LNG Carrier Operations. Reliability Engineering and System Safety 93 (9), 1328–1344.

Walker, A.H., Scholz, D., Boyd, J., Burns, G.H., April 6–10, 2003. LNG Transportation, Risk Management, and Maritime Security, paper presented at the International Oil Spill Conference, Vancouver, BC, Canada.

Wan, Z., 2001. Bayesian Primer, Piscataway, NJ, USA.

Wang, K.S., Loss Prevention Through Risk Assessment Surveys of LNG Carriers in Operation, Under Constructions and Repair, (www.bmtmarinerisk.com), BMT Marine & Offshore Surveys Ltd, Kobe, Japan (viewed May 25, 2012).

Woodward, J.L., 2007. Coupling Dynamic Blowdown and Pool Evaporation Model for LNG. Journal of Hazardous Materials 140 (3), 478–487.

Woodward, J.L., 2008. Modeling Underwater Penetration of LNG Carrier. Process Safety Progress 27 (4), 336–344.

Woodward, J.L., Pitblado, R.M., 2010. LNG Risk-Based Safety—Modeling and Consequence Analysis. John Wiley & Sons, New York, NY, USA.

Yun, G., Ng, D., Mannan, M.S., March 21–25, 2010. A Medium-scale Field Test on Expansion Foam Application—Key Findings on LNG Pool Fire Suppression on Land, paper presented at the AIChE Spring National Meeting, San Antonio, TX, USA.

Yun, G., Rogers, W.J., Mannan, M.S., 2009. Risk Assessment of LNG Importation Terminals Using the Bayesian-LOPA Methodology. Journal of Loss Prevention in the Process Industries 22 (1), 91–96.

Zinn, C.D., April 10–14, 2005. LNG Codes and Process Safety, paper presented at the AIChE Spring National Meeting, Atlanta, GA, USA.

Zuijderduijn, C., Nov. 10–12, 2000. Risk Management by Shell Refinery/Chemicals at Pernis, The Netherlands, paper presented at the Seveso 2000 European Conference, Athens, Greece.

CHAPTER 10

Advances and Innovations in LNG Industry

10.1 Introduction

There are increasing LNG demands from many parts of the world because LNG is now considered safe and is less polluting. The demands are to be met by an increase in new liquefaction plants that are based on the unconventional gas and new discoveries. The new liquefaction plants under consideration will be built with higher efficiency in an environmentally responsible manner. This means the whole LNG supply chain, from the well heads, gas treatment, liquefaction, transportation, and regasification must be designed and configured with low emissions and high efficiency. Additionally, there is also a drive to lower the cost of natural gas liquefaction and improve the value of the LNG regasification plant.

The earlier LNG production plants were based on proven technologies and operation tracking records. There have not been significant changes in the LNG plant design. This is understandable as the investment risk with natural gas liquefaction is large, and conservative designs are necessary. With the escalating energy costs, as natural gas prices are now pegged to oil price, the emphasis is on efficiency and lower production cost. However, recent cost overruns with large LNG plants makes investors rethink the project economics of large plants, most likely refocusing on the less expensive mid-sized plants.

On the liquefaction technology, there are recent developments in driver technologies that will change the liquefaction unit design and configuration, such as more efficient gas turbine drivers, aero-derivative turbine drivers, proven large motors, and the efficient combined cycle power plant. There are also innovations in gas treating technologies in processing sour gases.

On the regasification side, there are also advances and developments. Floating regasification and storage carriers appear to be less costly than land-based facilities. Emphasis is on LNG cold utilization to improve the overall LNG supply chain efficiency, such as cryogenic power generation, air separation plants, boil-off gas condensation, and integration to industrial complexes. The greenhouse gas emissions from fuel gas fired combustion vaporizers and the impact on marine life with cold water discharge from seawater vaporizers are environmental concerns. The use of ambient air vaporizers is now considered to be more environmentally friendly.

This chapter discusses the development, advances, innovations, and new ideas in the LNG supply chain. The focus will be on more efficient liquefaction technology, larger and higher efficiency drivers, improved liquefaction equipment design, NGL recovery, LNG regasification, and LNG cold utilization. The applications of some of these technologies may further improve the image of the LNG industries.

10.2 Innovations in LNG liquefaction

10.2.1 Larger trains

For economy of scale, the trend is to construct a larger train size. The LNG industry has taken a big step toward increasing train capacity with the implementation of the AP-X™ process in Qatar in 2008. Six trains were constructed, each with a nameplate capacity of 7.8 MTPA. Today's APX trains are becoming even larger with more drivers and improved liquefaction design, as shown in Figure 10-1.

The AP-X™ process is an extension of the C3MR process, maintaining its proven performance while increasing the train capacity. The configuration of the cycle is shown in Figure 10-2. In this process, propane is used for precooling, while natural gas is liquefied and partially subcooled in the main cryogenic heat exchanger (MCHE) with a mixed refrigerant. Final subcooling is accomplished by using a nitrogen expander loop. Nitrogen is compressed to a high pressure, cooled to ambient temperature, and further cooled by low pressure nitrogen before being expanded to a lower pressure. The nitrogen provides refrigeration for subcooling LNG. By using nitrogen to subcool LNG, the percentage of the total refrigeration load on the upstream C3MR section is reduced, allowing for greatly increased capacity in a single train without having to parallel major equipment. With today's larger gas turbine drivers, and the large heat exchanger designs, a train size over 10 MTPA is now feasible.

The AP-X™ process is claimed to achieve high efficiency and low production cost by using all three refrigerant cycles to their best advantage (Roberts et al., 2002). Bronfenbrenner et al. (2009) compared the thermodynamic efficiencies (power required/ton of LNG produced) of the main gas liquefaction process. On a relative basis, they claimed that the C3MR, AP-X and DMR processes offer the highest efficiencies, with POCLP (Phillips Optimized Cascade LNG Process) offering about 90%,

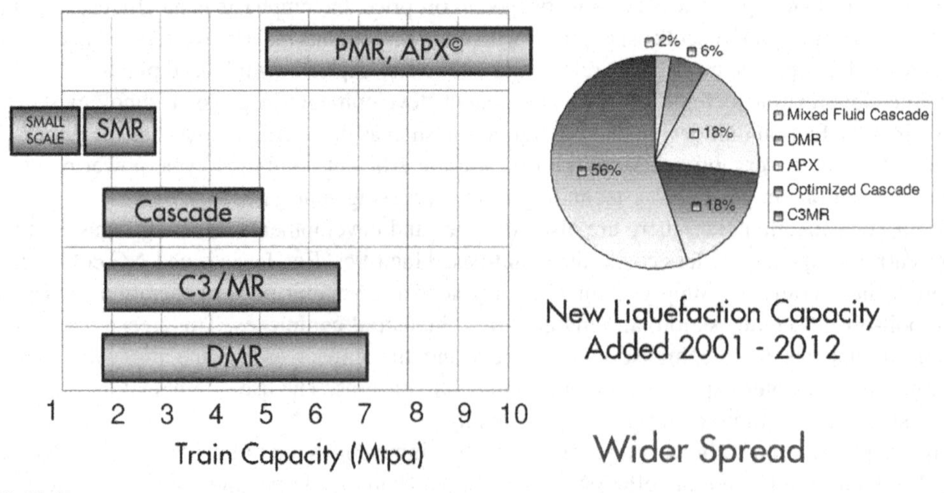

FIGURE 10-1

LNG plant size development.

(Nagelvoort, 2009)

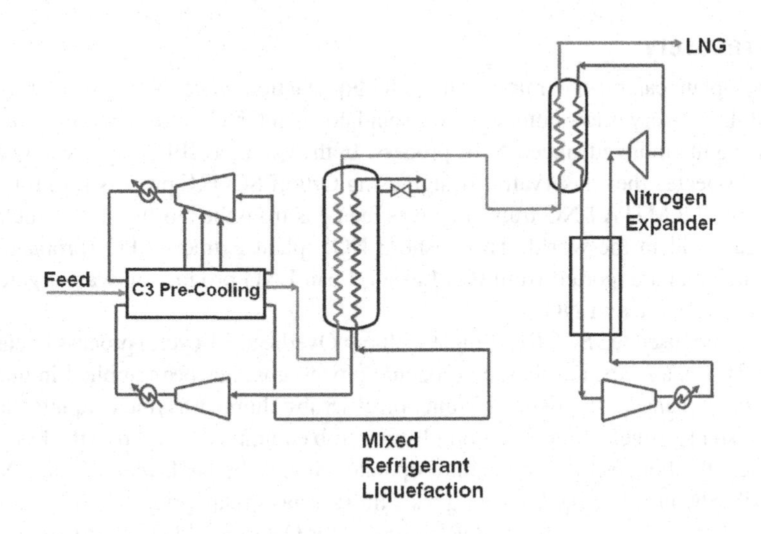

FIGURE 10-2

AP-X™ cycle.

(Bukowski et al., 2011)

SMR offering about 80%, and expander technologies offering about 70%, respectively, of the efficiencies achievable by C3MR and others. On the other hand, thermal efficiency studies conducted by ConocoPhillips on the POCLP disagree with such findings (Ransbarger, 2007).

Thermodynamic efficiencies of the refrigeration cycles are frequently compared among different processes. But the meaningful comparison should be made in consideration of the overall utility consumption instead of the cycle, particularly when high efficiency cycles are considered.

10.2.2 Main exchanger size

Several years ago, the maximum main cryogenic exchanger size was limited to a diameter of 4.6 m and a maximum exchanger weight of 310 tonnes. This was the maximum exchanger size necessary in order to achieve the required LNG train production, which had been limited by the power available from the compressor drivers. As the drivers keep increasing in size, larger train capacity has been developed to take advantage of the driver power. Improvements have been made to the manufacturing technique, and currently the maximum exchanger diameter increases to 5.0 m, and the exchanger weight increases to 430 tonnes. These larger exchangers play a key role in increasing capacity of the liquefaction train beyond 8 MTPA.

10.2.3 Liquefaction pressure

Higher tube-side design pressure allows for higher operating pressures, which results in increased production and lower specific power for the same size heat exchanger. An increase in liquefaction pressure from 65 barg to 80 barg can increase the production by about 6% and reduce the specific power by about 2% (Pilarella et al., 2007).

10.2.4 NGL recovery

An NGL recovery plant can be integrated to the LNG liquefaction process using a scrub column design. However, the NGL recovery is low compared to a standalone unit. NGL recovery can be integrated to the LNG process using the more advanced NGL process. In the Qatargas II LNG project, Qatar Petroleum and ExxonMobil selected the conservative design using Ortloff SCORE process for propane recovery.

Each of the two 7.8 MTPA LNG trains receives feed gas from the North Field, which is the largest nonassociated gas field in the world. The onshore NGL plant extracts LPG (propane and butane), which is fractionated and exported from Ras Laffan. Lean LNG production was targeted for the UK market, which requires a lean LNG.

The NGL recovery used the SCORE (Single Column Overhead REcycle) process licensed by Ortloff Engineers, Ltd. This is an expander-based cryogenic process that has been applied in numerous plants worldwide. The NGL plant accomplishes the function of scrub column in a typical liquefaction process-for removal of C_5+ prior to liquefaction. Accordingly, no scrub column is included in the LNG plant design. As a consequence, the liquefaction unit cannot operate without the NGL plant. Rich LNG production will still be possible by injecting the C_3 and C_4 back into the gas stream prior to liquefaction. Figure 10-3 shows a simplified process flow of the SCORE process for Qatargas II (Thompson et al., 2004).

Ortloff's SCORE process is primarily utilized as an efficient propane recovery technology. Propane recovery typically exceeds 97%, with 99% or higher easily achievable, while rejecting all the ethane. If ethane recovery is desired in addition to high propane recovery, SCORE can be operated in a partial ethane recovery mode by adjusting the amount of heat input to the column. This "incidental" ethane recovery mode is usually limited to about 40% ethane recovery (Thompson et al., 2004).

10.2.5 Liquid expander

Liquid expander has been used in the gas processing industry for decades for the recovery of power and production of refrigeration. In the traditional LNG process, pressure reduction is typically by a Joule-Thomson valve, in refrigeration units or in the last stage of the end-flash process. Only recently, liquid expanders have been introduced to the LNG industry and are now accepted in new liquefaction plants and in retrofit of existing facilities, starting with MLNG (Tiga Liquefaction Complex in Bintulu, Malaysia) in 1996. The liquid expanders are essentially pumps that are run backward that allow a subcooled liquid to be expanded. Work of the expansion is extracted from the fluid as shaft power that can be used to drive a pump or generate electricity (Figure 10-4). The recovery of power is valuable, but more importantly, when the fluid expands, it also lowers the fluid temperature. This is particularly important when the fluid, either refrigerant or LNG, can be chilled to a lower temperature, reducing the amount of vapor flashing at the lower pressure, lowering the vapor compression requirement.

According to Cryostar, with the cryogenic liquid turbine, the total plant savings are between 1.5 and 5%, depending on plant size and the process design, considering the additional production of liquid and the production of electricity.

10.2.6 Matching drivers

Optimizing the process/machinery configuration has played a major role in increasing LNG train capacity. Traditionally, the propane compressor and the mixed refrigerant compressors were driven by separate drivers (typically gas turbine drivers). Since operators prefer to use the same gas turbine

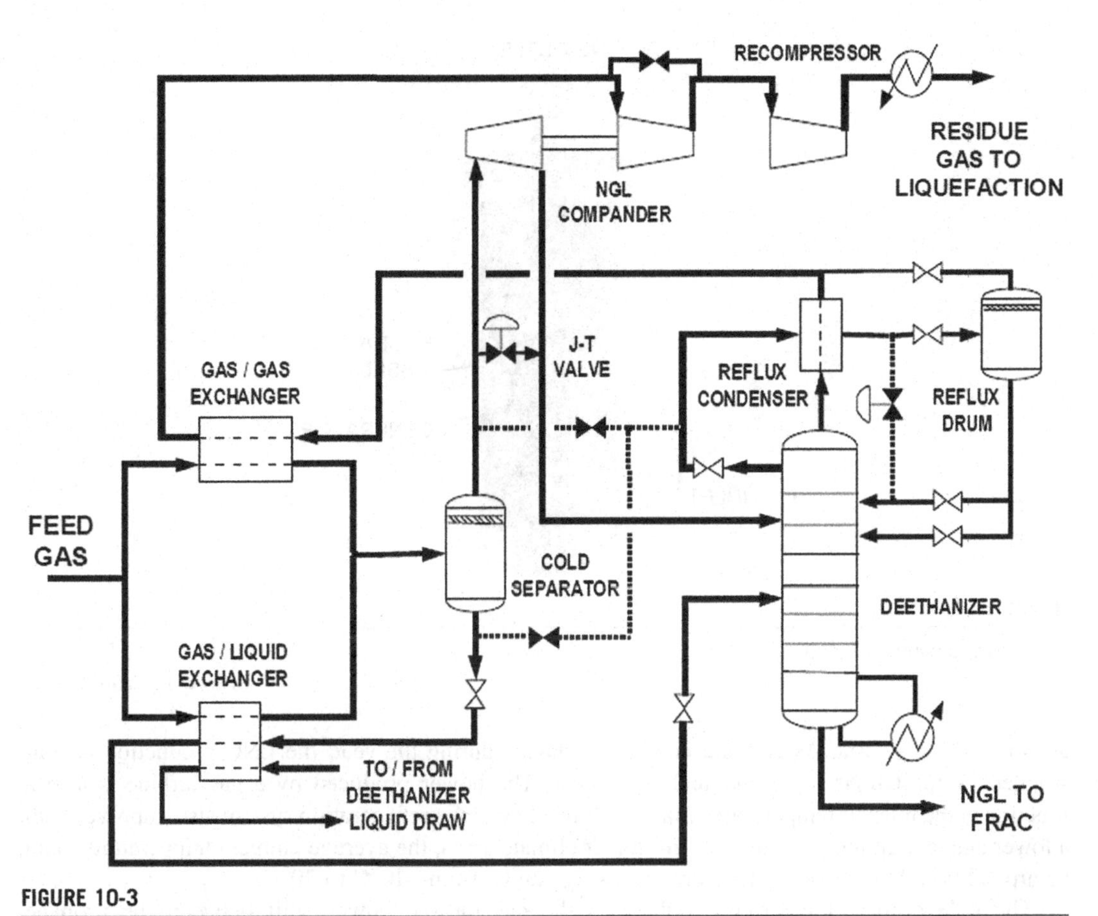

FIGURE 10-3

Ortloff SCORE process.

(Thompson et al., 2004)

drivers for maintainability, this driver configuration is not conducive to fully utilizing the available gas turbine power. In the recent liquefaction designs, such as RasGas Trains 3, 4, and 5 and Segas, the same driver is used to drive multiple refrigeration services and has improved the overall operability and cycle efficiency, making the process less dependent on driver selection.

With the advances of large motors, the gas turbine drivers can now be replaced with motors. Motors are available from different manufacturers and come in different sizes. This flexibility allows more optimum liquefaction cycle designs, and the thermal efficiency is expected to be higher, consequently producing less emissions.

10.2.7 Colder climate design

Ambient conditions, such as air and water temperature, have significant effects on plant design. Plants have been located mostly near the equator, but newer facilities are now located in Arctic climates such

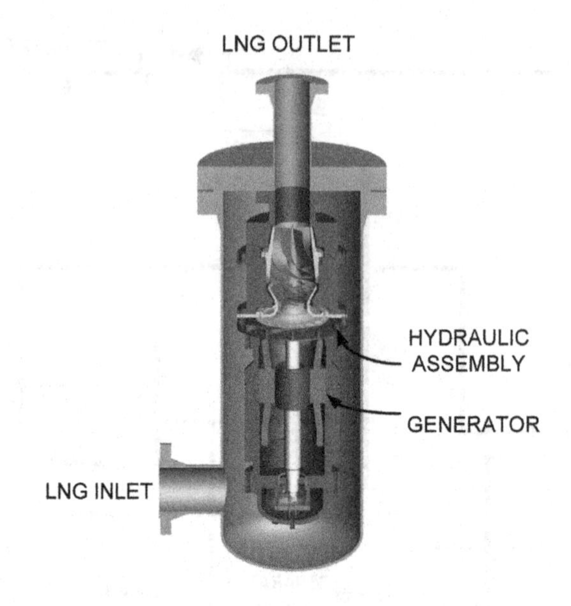

FIGURE 10-4

LNG liquid expander assembly.

(Kimmel and Cathery, 2010)

as Norway and Russia. As ambient conditions change during the year, the LNG production is constrained by limitations of mechanical equipment. The power produced by a gas turbine is also a function of the inlet air temperature, while the efficiency of the refrigeration system also improves with a lower ambient air temperature. For the colder climate areas, the average ambient temperature could be around 0°C, but the yearly temperature swing can be from –40°C to 30°C.

The cold climate temperature influences the gas turbine power split between the propane refrigeration section for precooling and the mixed refrigerant section for liquefaction and subcooling. Typically, the equipment is designed for high ambient temperature, say 60°F, and the power split is about one third for precooling and two thirds for liquefaction and subcooling. At 0°C ambient, the split will shift to about one fourth for precooling and three fourths for liquefaction and subcooling. The challenges are in designing the heat exchanger equipment to fully utilize the higher power output from the gas turbines as well as the deeper chilling during the cold ambient months (Pilarella et al., 2007).

10.2.8 Future LNG plants

Large trains in the range of 8 to 9 MTPA can be justified only in locations with very large gas reservoirs. It is likely that trains in the capacity range of 3 to 6 MTPA will become more important in the future as the LNG plants are likely to be located in more areas and offshore and there is more emphasis on flexibility, efficiency, and operating and capital costs.

Most of the LNG train designs built in the past decade use single cycle heavy-duty gas turbines both as refrigeration compressor drivers and in the power generation plant. The waste heat from the gas

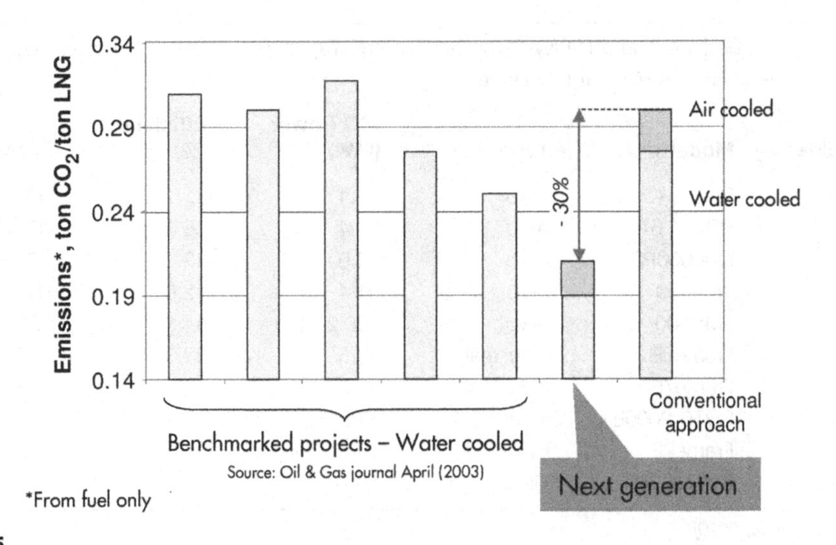

FIGURE 10-5

Future challenges—Reduced CO_2 emissions.

(Nagelvoort, 2009)

turbine exhaust is used in steam generation for power generation and process heating. Any excess waste heat, if not used by the process, is rejected to the atmosphere.

Most of the earlier processes also use seawater for cooling, which is now known to have significant impact on the ocean environment. A more environmentally friendly approach is to use air instead of seawater for cooling even though it will increase the energy consumption by liquefaction.

Today, the focus of innovation is shifted toward increasing energy efficiency, minimizing CO_2 emissions and lowering overall environmental impacts. Large equipment sizes and higher capacities may no longer be the solution. The next generation of LNG liquefaction plants will have a lower carbon footprint using the high efficiency gas turbines, combined cycle power plants, large motor drives, and more efficient NGL/LNG integration (see Chapter 2). Air cooling should be used where possible to minimize impact on the ocean environment. Nonproprietary and more efficient heat exchanger equipment will be readily available from manufacturers from different countries. More liquefaction process choices will be available as the LNG industry moves toward the less costly but efficient mid-scale LNG plants.

These advances in LNG production plant were predicted to be able to reduce CO_2 emissions in the next generation LNG plants by at least 30% as shown in Figure 10-5.

10.2.9 Gas turbines

Traditionally, industrial gas turbines are used as drivers for the refrigeration units. Aero-derivative gas turbines are thermally more efficient than industrial gas turbines and have been applied to power plant design. However, they have not yet been used as mechanical drives in LNG plants until recently. A list of the aero-derivative gas turbines and industrial gas turbines is shown in Table 10-1.

Table 10-1 Aero-Derivative and Heavy-Duty Gas Turbine Types, with Their ISO Powers and Energy Efficiencies When Used as Mechanical Drive

Gas Turbine	Model	Type	ISO Power (MW)	Efficiency (%)	Vendor
LM2500+	PGT25+	Aero	31.4	41.1	GE
RB211	RB211 6762	Aero	30.4	38.8	Rolls-Royce
LM6000	LM6000PD	Aero	43.9	43	GE
Trent 60	Trent 60	Aero	51.4	42.6	Rolls-Royce
LMS100	LMS100	Aero	100.2	44.1	GE
Frame 6	Frame6B	Industrial	43.5	33.3	GE
Frame 7	Frame7EA	Industrial	86.2	33	GE
SGT6-2000E	SGT6-2000E	Industrial	113	33.9	Siemens
Frame 9	Frame9E	Industrial	130.1	34.6	GE
SGT5-2000E	SGT5-2000E	Heavy duty	168	34.7	Siemens

(van de Lisdonk et al., 2010)

There are several advantages of the aero-derivative gas turbines. Aero turbines are lighter in weight and more compact than industrial gas turbines, which are favorable attributes for offshore installation. They are multishaft variable-speed machines, and can start under settle-out pressure without using starter/helper motors.

10.2.9.1 CO_2 emissions

The higher efficiencies can be translated into lower CO_2 emissions. Figure 10-6 shows the CO_2 emissions from industrial gas turbines and aero-derivative gas turbines. The aero-derivative gas

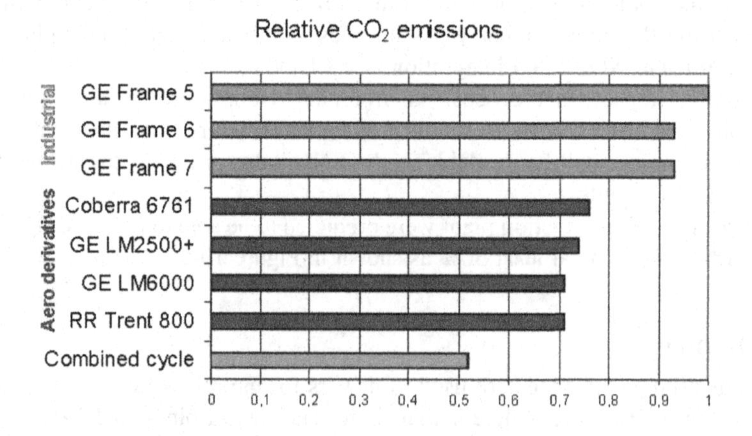

FIGURE 10-6

Relative CO_2 emissions from gas turbines.

(Neeraas and Marak, 2011)

turbines, being more efficient, produce lower CO_2 emissions than industrial-type gas turbines. However, when the gas turbine exhaust is recovered and used in power generation or heating, the combined cycle power generation efficiency can approach 60% (Section 10.2.12), which makes it the most efficient power plant in today's technology.

10.2.9.2 Performance impacts from ambient temperature changes

The power output of gas turbines varies with ambient temperature. Power output typically drops by about 0.3% to 0.5% for each °F increase in ambient temperature. Aero-derivative gas turbines are more sensitive to ambient temperature change than industrial gas turbines as shown in Figure 10-7.

Industrial gas turbines are less sensitive than aero-derivative turbines to ambient temperature changes. Typically, industrial gas turbines operate at lower pressure ratios than the aero-derivative counterparts. They also operate with a much higher air flow; consequently, temperature changes have less impact on power output. For example, on a 90°F day, the power output of a frame unit operating at a pressure ratio of around 10 to 1 will decline by 8 or 10% (from its standard 59°F nameplate rating) as compared to a 15% drop for an aero-derivative gas turbine operating at a 30 to 1 pressure ratio. In actuality, each gas turbine model has a unique power-temperature curve specific to its aerodynamic design that determines the power output, heat rate, and exhaust flow and conditions.

High ambient temperatures usually coincide with peak demand periods, which are especially detrimental during hot summer days when the reduction in power output is greatest. Inlet cooling offers a cost effective solution to offset power loss at high ambient temperatures. The effect of air cooling on power output is shown in Figure 10-7. As an example for a generic aero-derivative turbine, lowering the gas turbine inlet air temperature from 60°F to 40°F increases its power output by

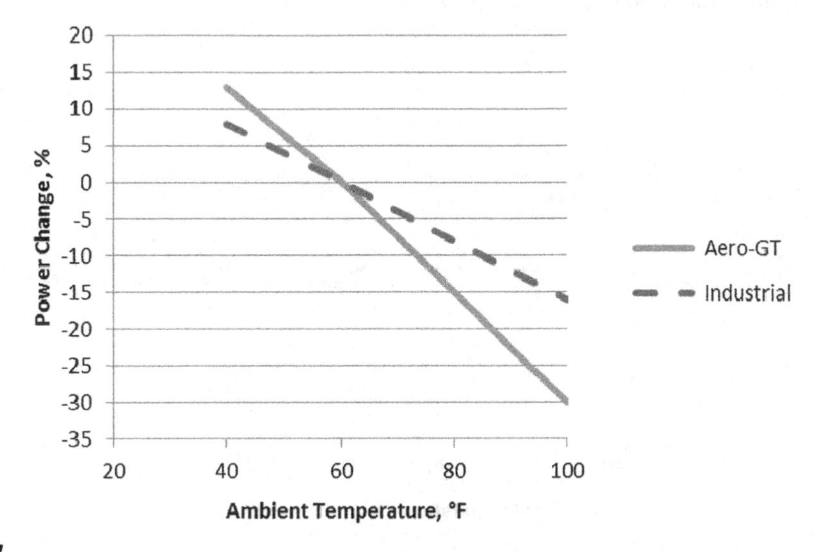

FIGURE 10-7

Generic performance of gas turbine power to ambient temperatures.

(Source: GE Energy Oil & Gas)

about 14%. On the other hand if the air temperature is raised to 100°F, the power output drops by about 30%. The impact in power output is less for the industrial turbines since their firing temperatures are lower and are less efficient. For this reason, modern power plants installed in hot climate areas are equipped with an ambient air cooler in order to maintain the power output.

Gas turbine air cooling also helps to reduce degradation in heat rate, as shown in Figure 10-8. The heat rates are lower at the lower temperatures, which means less fuel is consumed and less CO_2 emission is produced. For example, reducing the inlet temperature from 60°F to 40°F reduces the heat rate by about 2% or fuel efficiency by the same amount. At the same time, the exhaust flow from the gas turbine is increased by about 4%, which means more waste heat is available for steam generation in the HRSG (Heat Recovery Steam Generation), further increasing the power output from the steam turbines.

10.2.9.3 Application to LNG industries

At the present time, aero-derivative gas turbines can achieve thermal efficiencies over 25% higher than industrial gas turbines. This can result in a 3% or higher increase in overall plant thermal efficiencies. The use of aero-derivative gas turbines to improve plant output or reduce greenhouse is very attractive. The drawbacks are the maintenance costs for an aero-derivative gas turbine and their lower reliability than industrial gas turbines. But as time passes, with more in operation, the costs and reliability are expected to improve.

Aero-derivative gas turbines were successfully deployed in the optimized cascade liquefaction trains of the Darwin LNG facility (Meher-Homji et al., 2008). Positive aspects are the improved plant availability as a result of the ability to completely change out an aero-derivative gas turbine generation set in less than three days versus 14-plus days that could be required for a major overhaul on an industrial gas turbine (Yates, 2002).

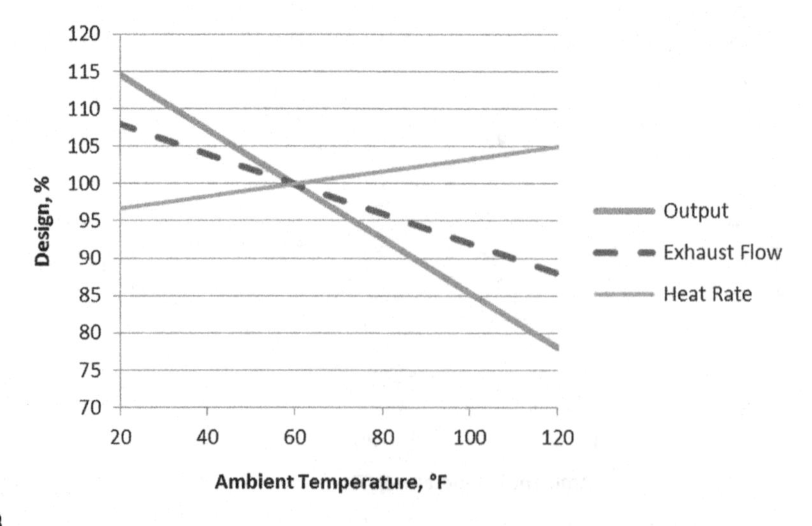

FIGURE 10-8

Impact of temperature on MS7001 power output, exhaust flow, and heat rate.

(Source: GE Energy Oil & Gas)

10.2.9.4 Emissions from ship propulsion system

The CO_2 emissions or energy efficiencies from the available engine choices for an LNG carrier propulsion system are given in Table 10-2. The use of steam turbines for propulsion is the least efficient method and emits the most pollutants. They were used in the earlier ships but are no longer used. The operation of a steam system also generates significant amount of waste water, which must be processed prior to discharge to the open sea.

The demand for larger and more energy efficient LNG carriers has resulted in rapidly increasing use of the diesel engine as the prime mover, replacing traditional steam turbine propulsion plants. Two alternative propulsion solutions have established themselves on the market to date: low speed, heavy fuel oil burning diesel engines that can be combined with a reliquefaction system for BOG recovery and medium speed, dual-fuel engines with electric propulsion.

A further low speed direct propulsion alternative, using a dual-fuel two-stroke engine, is also available: high thermal efficiency, flexible fuel/gas ratio, and low operational and installation costs are the major benefits of this alternative engine version. The engine utilizes a high pressure gas compression system to supply boil-off gas at pressures of 250 to 300 bar for injection into the cylinders.

For many years, the LNG market has not really valued the boil-off gas, as this has been considered a natural loss and typically is burned in the steam boiler for the steam turbine movers. Generally, emission considerations favor the selection of the dual-fuel engine over the HFO (Heavy Fuel Oil) engine and, with today's fuel prices, the dual-fuel gas injection engine is found to be an economical option. However, project specific factors, such as a requirement for a fixed amount of LNG to be delivered, need to be considered in specific cases, and may influence the balance between the gas injection engine alternative and the HFO engine alternative.

10.2.10 Electric motor drive

Traditionally, refrigeration system drivers are either by gas turbines or steam turbines. Motor drive is used for starter or helper motors in a large gas turbine driver. The helper motor drive for a GE Frame 7

Table 10-2 Typical Emissions from Different Engines for LNG Ship

Type	NO_x (g/kWh)	SO_x (g/kWh)	Particulates (g/kWh)	CO_2 (g/kWh)	Relative CO_2 emissions (0 0.2 0.4 0.6 0.8 1)
Low speed diesel	17	12.9	0.5	580	
Medium speed diesel	12	13.6	0.4	612	
Dual-fuel diesel electric	1.3	0.05	0.05	420	
Steam turbine	1	11.0	2.5	850*	
Gas turbine	2.5	0	0.01	480	

*Using 50% boil-off gas and 50% heavy fuel oil for steam generation
(Neeraas and Marak, 2011)

turbine is typically 20 MW. But as larger motor sizes are available, the natural progression is to use motor drives instead of turbine drives for refrigeration compressors.

In most applications, there are backup systems for the power supply and the system is highly reliable. As the motor sizes are approaching the 100 MW range, motor drives are becoming cost competitive with mechanical drivers. The electric driven design is particularly attractive in the 5 to 8 MTPA capacity range (Del Vecchio and de Jong, 2004). Electric motors of 65 MW have been developed for LNG service (see Figure 10-9), and motors up to 80 MW are now feasible (Kleiner and Kauffman, 2005).

Electric motor drives are more expensive than mechanical drives but have the following advantages.

- Electricity can be supplied from a combined cycle power plant, which is the most energy efficient power generation process, and can lower greenhouse gas emissions. The liquefaction process configuration is now independent from the gas turbine drive selection.
- Combined cycle power plants can be sourced from many manufacturers. Competitive design and equipment pricing are readily available.
- The dominance of a few liquefaction suppliers that ties to specific gas turbine drives is over. There are competitive designs that will lower the cost of the liquefaction plants.
- The plot plan of the refrigeration system is significantly reduced since the gas turbines and heat recovery equipment are located away in the power plant.
- The process safety is improved since there is no combustion equipment within the liquefaction plant.
- Motor drive is more reliable, eliminating downtime and shutdown/maintenance required by the gas turbine drive.

FIGURE 10-9

Siemens 65 MW E-drive motor for LNG.

(Courtesy of Siemens)

- Electric motor variable speed drive (VSD) provides turndown flexibility, reliability, and efficiency, which are important in LNG plants.
- Compared to gas turbine drives, motor drives are not sensitive to ambient temperature changes. Process design can be maintained as constant throughout the year.

10.2.11 Combined cycle power plant

The size of the combined cycle power plant is an important parameter in determining the LNG plant size. The larger plant sizes benefit from economies of scale (lower $/kW) and higher thermal efficiency. The combined cycle units can also be designed to produce extraction steam to supply steam to the process heating. A block diagram of an integrated combined cycle plant and LNG plant is shown in Figure 10-10. Figure 10-11 depicts a combined cycle power plant, courtesy of Siemens. The discharge from the gas turbines that are housed in a building is piped via a duct to the HRSG located in a structure on the left side.

There are several design configuration options for a combined cycle power plant:

- In a single shaft combined cycle plant, a gas turbine and a steam turbine can drive a common generator.
- In a multishaft combined cycle plant, each gas turbine and each steam turbine has its own generator.
- The multishaft design enables two or more gas turbines to operate in conjunction with a single steam turbine, which can be more economical than a number of single shaft units.
- A larger steam turbine also allows the use of higher steam pressures, more reheat stages, resulting in a more efficient steam cycle.

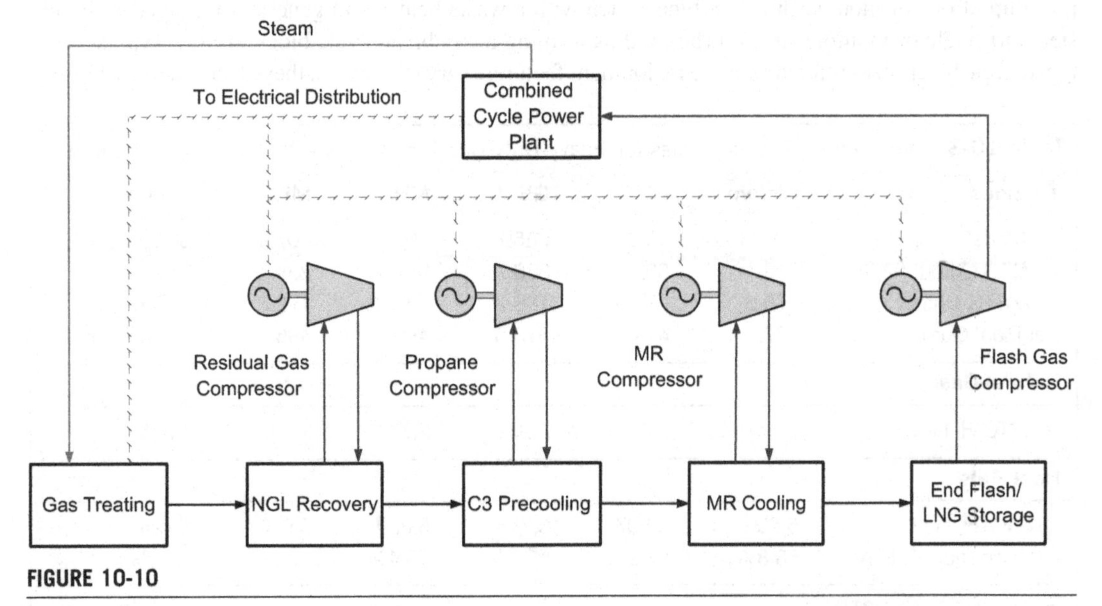

FIGURE 10-10

Typical setup of combined cycle/LNG plant.

FIGURE 10-11

A combined cycle power plant.

(Courtesy of Siemens)

Thus the overall plant size and the associated number of gas turbines have a major impact on whether a single shaft combined cycle power plant or a multiple shaft combined cycle power plant is more economical.

Large gas turbines of over 150 MW size are already available by at least four separate groups: General Electric and its licensees, Alstom, Siemens, and Westinghouse/Mitsubishi. Combined cycle units are made up of one or more such gas turbines, each with a waste heat steam generator arranged to supply steam to single or multiple-steam turbines, thus forming a combined cycle block or unit. Typical combined cycle block sizes offered by these major manufacturers vary slightly but they all can achieve close to

Table 10-3 Combined Cycle Efficiencies for Heavy-Duty Gas Turbine from Different Manufacturers

Turbines	Alstom	GE	GE	MHI	MHI	Siemens
GT Model	KA-26	109H	FE50	107F	107G	SGT5-8000H
2 - GTG Output (MW)	313.1	332.6	338.7	317.4	334	389.1
Total STG Output	175.6	173.5	186.2	163.2	172.3	205.3
Net Plant Output	477.7	494.9	513.5	469.9	495	581.4
Natural Gas:						
MMBTU/H, LHV	2,771	2,918	2,935	2,744	2,929	3,328
Heat Rate						
BTU/KWH, LHV	5,800	5,897	5,715	5,839	5,918	5,724
Net Efficiency %, LHV	58.8%	57.9%	59.7%	58.4%	57.7%	59.6%

(Source: Thermal Flow program)

60% thermal efficiency. Their performances are compared in Table 10-3 using the GTPRO program from Thermoflow Inc., based on the gas turbine's iso-conditions (60°F). Note that actual power plant outputs from the manufacturers are expected to be slightly different and are also dependent on site conditions.

More recent combined cycle designs, especially the larger units, have higher thermal efficiencies than the earlier designs. The development of high gas turbine combustion temperature 1370°C (2500°F) has led to higher combined cycle efficiency. The advances are mainly due to the development of a dry low NOx combustor, which allows the combustion temperature to increase, and the gas turbine exhaust can be produced at over 1300°F. The high gas turbine exhaust provides a higher heat sink temperature that boosts the efficiency of the steam cycle. The steam cycle is designed to operate at over 2200 psig steam pressure, and with steam turbine reheat, the steam cycle efficiency alone can exceed 35% thermal efficiency.

As published on GE's web site, GE has achieved 60% efficiency in the combined cycle unit in Baglan Bay, which uses a GE H-technology gas turbine with a NEM 3 pressure reheat boiler, utilizing steam from the HRSG to cool the turbine blades. Similarly, Siemens AG announced in May 2011 it had achieved a 60.75% net efficiency with a 578 megawatts SGT5-8000H gas turbine at the Irsching Power Station.

10.2.12 Gas turbine inlet air chilling

The efficiency and power output of the combined cycle power plant can be boosted by chilling the turbine intake air. The optimum choice of technologies is largely determined by site weather conditions. There are several methods for cooling inlet air and they differ in installation costs and efficiencies.

Turbine inlet air can flow through a continuously wetted honeycomb type fibrous material evaporating water off surrounding surfaces thereby cooling the air. In low humidity areas, the cooling can boost power output by up to 15%, while in humid areas, the boost is more likely to be under 10%.

Turbine air can also be cooled by fogging where fine droplets of water are sprayed into the inlet air. Fogging can cool the air up to 95% toward saturation, and is slightly more effective than the wetted media method.

In humid regions, more effective cooling can be achieved with refrigeration. Ambient air is cooled by cooling coils installed in the gas turbine inlet using a chilled heat transfer fluid. Mechanical compression type refrigeration chiller or Lithium Bromide absorption chiller (driven by steam or hot water) may be used to cool the heat transfer fluid. With refrigeration, it is possible to cool ambient air below its web bulb temperature, typically down to around 45°F, which will result in a significant increase in power output.

With the availability of refrigeration from LNG regasification, there is an opportunity to integrate the LNG and power plant to improve the power plant performance.

10.2.13 Gas turbine exhaust duct firing

With duct firing, the gas turbine exhaust temperature can be increased; subsequently increasing the steam production from the steam generator. During high ambient temperatures, the power plant output will drop and may not meet the power requirement of the LNG plant. The additional steam can be used to increase power to meet the power demands. The additional power may also be used to increase LNG production rate or liquefy a lower pressure feed gas. Alternatively, the additional steam can be used to supply heating to the process units, such as the amine regeneration and reboilers in the NGL fractionation unit.

10.2.14 Modularization

Modularization is a concept to design, build, and commission systems of equipment and piping in self-supporting transportable structures. Modularization is necessary when plant site is located in a remote area especially in an area with a harsh and extreme climate, such as the North Slope and Sakhalin Island. Modules are built in large fabrication yards and are transported to the plant site. The use of modules basically moves the costly site construction hours to the fabrication yards where cost and quality can be better controlled.

However, because of the complexity of an LNG plant, there is yet to be an LNG facility that is fully optimized for modular design. In the Woodside Train V Expansion Project in Karratha, Australia, modularization was based on the same plant layout from an existing design for Train IV. According to the Woodside web site, the major problem was the modularization of the complex piping at the plant, which was reported to require 75 different large structures and there were difficulties in fitting the pipe work among different modules. As a result, this approach uses more structural steel than the existing train IV project.

The liquefaction portion of the Statoil's Snohvit LNG project in Hammerfest, Norway was designed and constructed on a custom-built barge. Fabricated in Spain, the barge was towed to Norway and permanently fixed to the site. This strategy of prefabricating process units in an industrial location reduced the amount of construction hours that would have been required in an arctic location. However, the plant was reported to have huge cost overrun and schedule delays.

Modular design is quite different than a land-based design. The limitation of module size and weight require redistribution of the process equipment. The egress and safety design must be well planned because of the congested area. Consequently, the modular design will look and operate differently than a land-based plant. A successful module must be designed almost from the ground up. The delivery of modules and equipment must be coordinated with the on-site construction team. One drawback of a modular design is that revamping of the process is significantly more difficult than conventional design, due to the compact nature of the modules.

There are potentials of significant cost saving and advantages in modular design, but the application to LNG plants has not been successful in reducing the overall cost. However, innovation, development, and redefinition of the work flow are necessary for more cost effective modularization projects, including off-site modularization and prefabrication at all levels within the design, delivery, and installation activities. New tools and application models are now being developed (Haney et al., 2011) to produce a more sophisticated modular design that can pack more equipment and use the "Fit for Purpose" materials and designs. A successful modular project is expected to produce the following results.

- Improved labor productivity and transfer effort hours from site to shop
- Cost and schedule certainty
- Reduced total installed cost (TIC)
- Improved safety
- Improved quality
- Improved environmental footprint
- Enhanced operations and maintenance
- Reduce startup times after module delivery to begin production quicker.

A successful modularization project must also have buy-in and cooperation from all project team members on the new execution approaches and not rely on their past experience.

- Agreement on the basic concept of modularization driving the plant layout. The process and piping design will need to be redesigned to fit into modules. In the past projects, modularization meant packaging a land-based process unit to a module, which was proven not to be cost-effective.
- Patents and innovations are necessary to facilitate new generation modularization design and installation.
- Design takes longer, but quicker plug-and-play scenario at the site means a shorter project schedule.
- Complete owner and early project team buy-in. The design cannot be finalized without definition of the details and extent of modularization and prefabrication methodology.
- Owner operations personnel to be involved from the start to ensure module configuration accounts for operations and maintenance requirements.

Future LNG plants most likely will use a combination of the innovations described under this section—advanced efficient gas turbine drivers, combined cycle power plants, large electric motors, more efficient air cooling processes (Figure 10-12), and the application of the new approaches on modularization to reduce both capital and operating costs.

FIGURE 10-12

Combination of all innovations for LNG production.

(Nagelvoort, 2009)

10.3 LNG regasification

10.3.1 LNG industrial complex

The cold energy in LNG can be utilized to produce power using an organic Rankine power cycle (as described in the following sections), or distributed to various cold users in an industrial complex. Use of LNG directly as a cooling medium is a more efficient use of the cold energy since chilling requires only a heat exchanger and incurs less thermodynamic losses. Osaka Gas in Japan has been adopting such a concept in their terminal and has developed an LNG cryogenic energy cascade process that is designed to supply refrigeration to the various users in an industrial complex that consists of LPG storage, power plant, petrochemical plant, and oil refinery. The concept of this process is illustrated in Figure 10-13. The cold energy users include an air separation plant, carbon dioxide liquefaction plant, propane and butane product chiller units, chilled water system, and gas turbine inlet air chiller. The system uses an intermediate fluid to transfer the cold energy from LNG to the different users in the industrial complex. The chilled water system is an effective method for distributing the

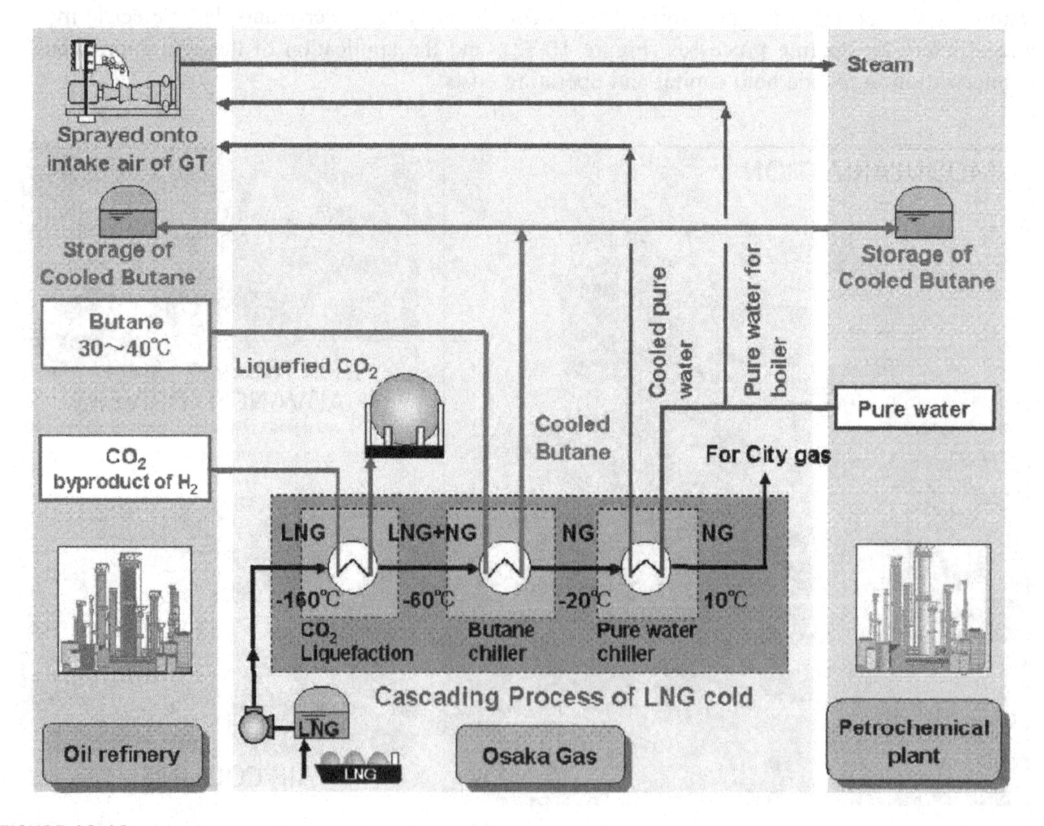

FIGURE 10-13

LNG industrial complex.

(Otsuka, 2006)

cold energy to the processing units in the petrochemical plant and oil refinery. With a low-temperature chilled water supply, fractionation can be more effective and product yields and purities can be increased, consequently reducing the energy consumption and emissions from the integrated complex.

10.3.2 LNG wobbe index

LNG import terminals typically receive LNG from different parts of the world and sometimes on a spot-market basis. The hydrocarbon content can vary depending on the gas fields and whether there is an upstream NGL recovery unit. As discussed in the section on LNG Quality and Gas Interchangeability in Chapter 1, the heating value and Wobbe Index of the LNG can vary widely, which may exceed the Wobbe Index specification of some countries, such as the United States and United Kingdom.

The most common method to reduce the Btu value of a gas is by diluting the vaporized LNG with nitrogen, up to the pipeline limit for inert content, usually 2 to 3%. Nitrogen is an effective diluent for lowering the heating value and Wobbe Index, but would require the investment of a nitrogen plant, either a membrane type or a cryogenic separation type, which would not add any revenues to the facility. In addition, if the LNG supply is very rich, nitrogen dilution may exceed the maximum 3 mole % inert limit for pipeline gas.

As an alternate to nitrogen injection, an LNG Btu conditioning unit can be installed that can be integrated to the LNG regasification plant as depicted in Figure 10-14. The conditioning unit can remove ethane and all the propane and heavier components from the LNG such that a lean LNG can be produced. The conditioning unit utilizes the cold energy from LNG for fractionation, and only a slip

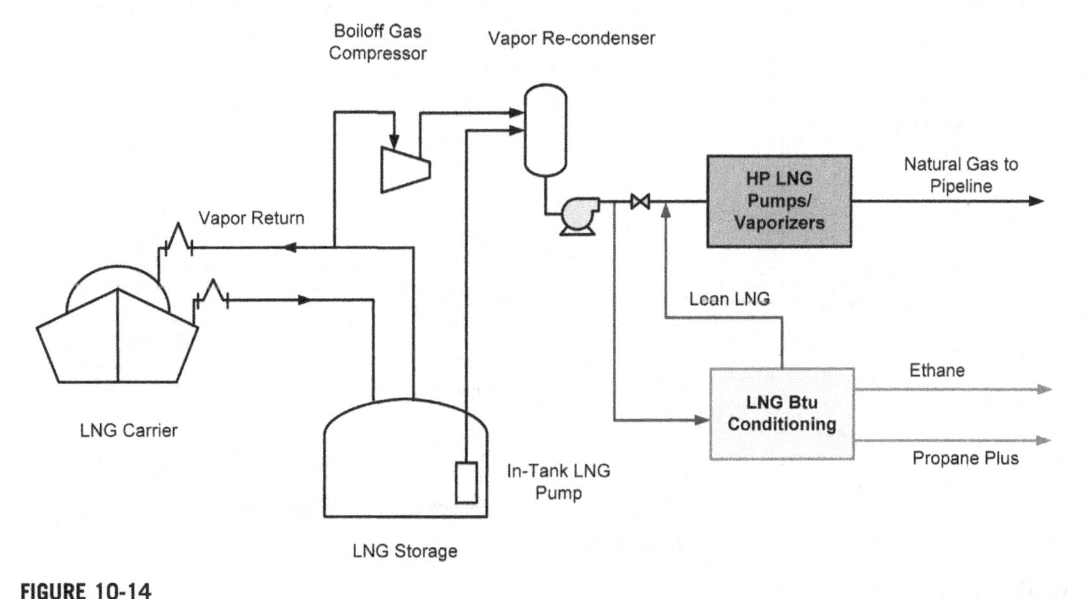

FIGURE 10-14

LNG terminal with Btu control.

(Mak et al., 2007)

stream is required due to the high LPG removal efficiency. The LNG conditioning unit can also be designed for ethane products for a petrochemical plant. The blended LNG stream is further pumped and heated by LNG vaporizers to meet pipeline specifications. Since the LNG is leaner and is at a higher temperature, lower vaporization duty is required.

There are many patented processes that can be used to remove ethane and LPG components in an LNG terminal using LNG cold. One of the processes is the Fluor LNG Btu conditioning process shown in Figure 10-15. In this process, the LNG cold energy is used for cooling and refluxing the demethanizer and deethanizer. In addition, LNG refrigeration is also used for reliquefaction of the demethanizer overhead gas, which allows the liquid to be pumped to the pipeline pressure, eliminating gas compression equipment.

The Fluor process can recover over 97% of the propane and 70% of the ethane from LNG, producing a lean gas with a lower Wobbe Index, as shown in Table 10-4. The design feature of the Fluor process is the feed exchanger, E-1, which consists of two heat exchanger passes. The first heat exchanger pass uses the inlet LNG to cool and condense the lean residue gas from the demethanizer reflux drum. The second pass uses the residual cold to partially condense the demethanizer overhead to produce a lean, methane-rich reflux to the demethanizer column.

LNG refrigeration is also used to provide column reflux to the deethanizer. With lower overhead temperature, the deethanizer can operate at a lower pressure than conventional design. Consequently, less reboiler duty and smaller column size are required due to the more efficient separation process. The deethanizer produces an ethane overhead product and an LPG bottom product.

As shown in Table 10-4, with a rich LNG as feed, this process can produce 40,700 BPD of LPG product, 31,200 BPD of ethane, and 1,094 MMscfd of 1,021 Btu/SCF HHV pipeline gas.

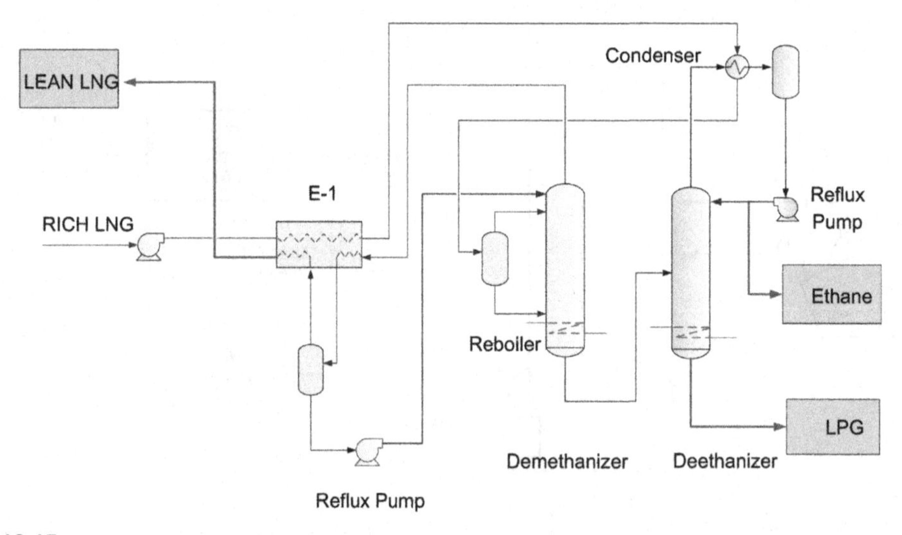

FIGURE 10-15

LNG conditioning plant with NGL production.

(Mak et al., 2007)

Table 10-4 Ethane and LPG Production Balance

Component, Mole Fraction	LNG Feed	LPG Fraction	Ethane	Pipeline Gas
N_2	0.0034	0.0000	0.0000	0.0036
C_1	0.8976	0.0000	0.0215	0.9439
C_2	0.0501	0.0100	0.9585	0.0517
C_3	0.0316	0.6277	0.0200	0.0009
IC_4	0.0069	0.1442	0.0000	0
NC_4	0.0103	0.2160	0.0000	0
NC_5	0.0001	0.0021	0.0000	0
Gross heat value, Btu/SCF	1,137	2,850	1,787	1,021
MMscfd	1,200	57	49	1,094
Barrels per day	NA	40,700	31,200	NA

(Mak, 2005)

10.3.3 LNG and CNG vehicle fuel production

With the increasing higher oil prices, the use of LNG as transportation fuel is becoming economically attractive. Most terminals today have incorporated vehicle fuel stations for distribution of natural gas as LNG fuel or CNG fuel.

However, the presence of ethane and heavier components in LNG limits its use as a vehicle fuel, as most LNG is too rich to meet the vehicle fuel specifications. For vehicle fuel, a methane number of 80 is required for engine performance, which would require most of the nonmethane components to be removed. Figure 10-16 shows the configuration that can be used to coproduce vehicle grade LNG and CNG. LNG and CNG can be produced at the required pressure for the fuel filling stations. In addition, a cryogenic power generation facility using an expander cycle (EP2) can be incorporated in the design to supply power to the facility.

Based on the rich LNG feed, this integrated plant can potentially produce about 200,000 gallons per day of vehicle fuel quality, 51,000 barrel per day of LPG, and 59 MMscfd of ethane as shown in Table 10-5 (Mak et al., 2004).

10.3.4 Integrated LNG regasification/power generation

The integration of an LNG regasification terminal with a power plant can significantly increase the power output and efficiency of power production while eliminating the heating requirement of an LNG facility. A conceptual design of an integrated plant is depicted in Figure 10-17. In this integrated plant, power is produced in three processes:

- Cryogenic power cycle
- Pressure letdown from the vaporized LNG
- Power increase from gas turbine inlet air cooling.

LNG cold is first used as a cold heat sink in power generation in an organic Rankine cycle using propane, butane, or mixed fluid as the working fluid. The working fluid is condensed in exchanger

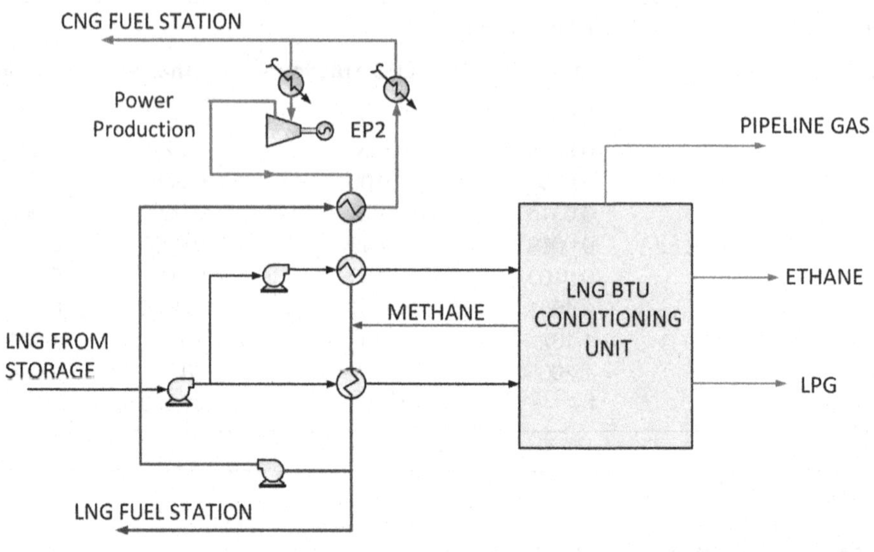

FIGURE 10-16

Integrated LNG regasification facility—LNG and CNG production.

(Mak et al., 2004)

E-1, pumped by working fluid pump P-2 to a higher pressure, heat exchanged with the expanded working fluid in E-4, and further heated in E-5 to a higher temperature using low pressure steam from the power plant. Power is produced by expanding the high pressure working fluid to a lower pressure in EP-1.

Table 10-5 Overall Balance of the Integrated Facility—LNG and CNG Production

Components Mole Fraction	LNG Feed	Ethane	LPG	Pipeline Gas	CNG Motor Fuel	LNG Motor Fuel
N_2	0.0065	0.0000	0.0000	0.0073	0.0073	0.0073
C_1	0.8816	0.0243	0.0000	0.9868	0.9868	0.9868
C_2	0.0522	0.9657	0.0053	0.0050	0.0050	0.0050
C_3	0.0328	0.0092	0.5399	0.0007	0.0007	0.0007
iC_4	0.0071	0.0000	0.1208	0.0001	0.0001	0.0001
NC_4	0.0107	0.0000	0.1821	0.0001	0.0001	0.0001
iC_5	0.0040	0.0000	0.0675	0.0000	0.0000	0.0000
NC_5	0.0020	0.0000	0.0338	0.0000	0.0000	0.0000
C_6+	0.0030	0.0000	0.0507	0.0000	0.0000	0.0000
Btu/SCF (HHV)	1,153	1,745	2,985	1000	1000	1000
Flow, MMscfd	1,200	59	70	1,046	12	12
Flow, BPD		37,100	51,000		4,900	5,000

(Mak et al., 2004)

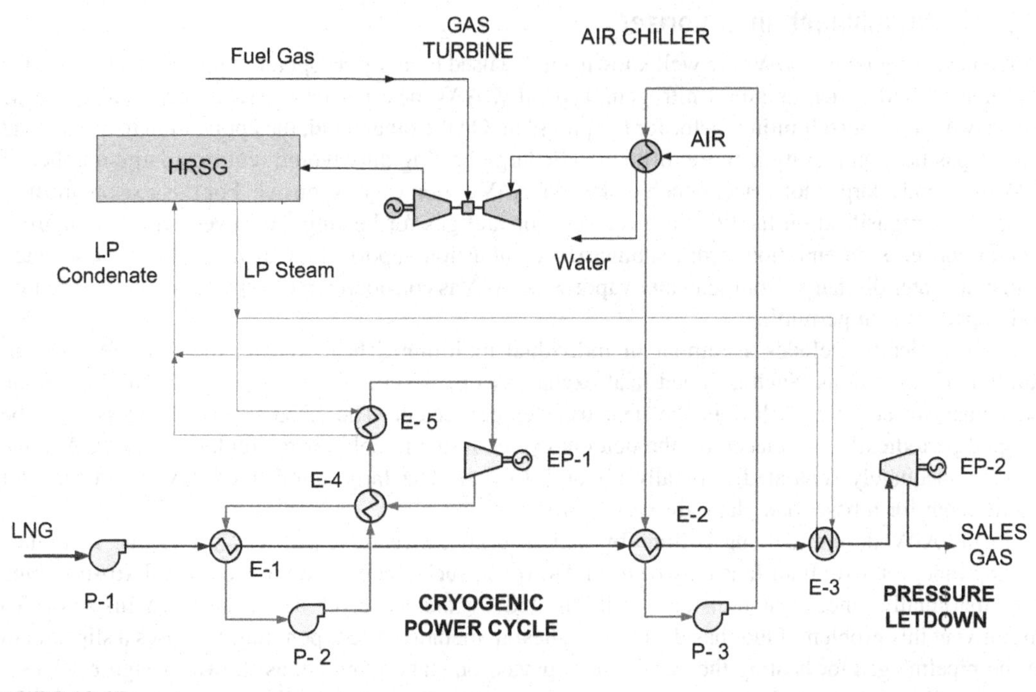

FIGURE 10-17

Integration of LNG regasification/power plant.

(Mak, 2009)

LNG exiting from the cryogenic power cycle is further utilized in exchanger E-2 for gas turbine inlet air chilling. The LNG cold is transferred to the gas turbine inlet using a heat transfer fluid such as ethylene glycol water mixture. The cold glycol from E-2 is pumped by glycol pump P-3 and heat exchanged with inlet air to the gas turbine. When ambient air is chilled to 40°F, almost all of its moisture content is condensed, and the air density is increased. The mass flow to gas turbine can then be increased resulting in a higher power output and efficiency. The condensate can be used as boiler feed water makeup to the power plant. The impact of lower gas turbine inlet temperatures on the gas turbine performance are further discussed in an earlier section.

Typically, a LNG high pressure pump is designed to deliver gas at 1500 psig or higher, while the fuel gas pressure requirement in most power plant facilities is about 400 psig. The sendout gas can be heated to a higher temperature using waste heat and expanded with a second expander EP-2 to produce more power. Low pressure steam from the power plant can be used for heating. The expanded gas is then sent directly to the sales gas pipeline.

Availability of waste heat to the LNG regasification facility and cooling to the power plant must be considered in an integrated facility. The variations in power demand and sales gas demand must be evaluated to ensure both plants can operate independently from each other. This will require back-up facilities such as duct firing in gas turbines and a seawater vaporizer or SCV as backup system in the integrated facility.

10.3.5 LNG ambient air vaporizer

Ambient air vaporizers (AAV) are well known and are used in many cryogenic liquid plants to vaporize cryogenic liquids, such as liquid nitrogen. Typically, AAV heating duty is relatively small in these plants, which are not a limiting factor for the plot plan. On the other hand, the application to a base load LNG regasification facility is more difficult. The large heating duty would require a large number of AAV units and a large plot space, which makes the AAV option very expensive. For this reason, most of the existing regasification facilities use seawater or fuel gas for heating. However, with the environmental concerns on emissions from submerged combustion vaporizers and the problems associated with cold water discharge from seawater vaporizers, AAV is considered more environmentally friendly and is preferred in permitting.

AAV typically includes a number of individual multifinned heat transfer elements in serial or parallel configurations. Such a finned heat exchanger operation is limited by the ice buildup on the exchanger surface, which lowers the heat transfer coefficient. The LNG heating process must be stopped periodically and placed on the deicing cycle. To avoid cold air recirculation, the AAV must also be adequately separated, typically about 5 to 6 ft. The large number of AAVs and the plot requirement for a base load plant are very costly.

In the AAV design, deicing is done by natural draft convection, which is very slow. To reduce deicing time, force draft air fans may be used. However, such an operation reduces the defrosting time only marginally since heat transfer is still limited by the ice layer that acts as an insulator. To circumvent this problem, Fluor has developed a design method (patent pending) that uses a slip stream of the pipeline gas for heating the AAV that is placed on standby mode, as shown in Figure 10-18.

In the Fluor proposed design, a small blower is used to direct the pipeline gas to the bottom of the standby AAV. The hot gas will melt the ice on the inside of the exchangers, which will fall freely into the basin below by gravity. This hot gas deicing method can almost eliminate the standby time, and consequently can reduce the number of AAVs required. Consequently the plot space requirement for the AAV is reduced, improving the viability of the AAV economics.

10.3.6 LNG new berthing designs: jetties and subsea pipelines

The LNG terminal cost of service can be reduced by using a subsea transfer pipeline in lieu of a traditional jetty, which also offers the opportunity to save dredging costs as well as the cost of a jetty.

This part of the facility represents a major portion of the overall capital costs when one considers dredging, jetty construction, and process piping costs. The environmental impact of any required dredging is also a consideration. The LNG berth and jetty is required for the unloading of tankers. For locations with sufficient deep water access close to the coast, terminals may consist of jetty structures and breakwaters, where tankers can be moored and offloading can take place via the standard loading arms. Many locations require channel dredging and berth dredging to obtain the required water depths at berth and in the turning basin.

As an alternate to the trestle supported piping, subsea pipelines may be considered. A subsea pipeline can be used to transport the LNG to and from an offshore terminal, thereby eliminating the need and cost for a connecting trestle. With current subsea cryogenic pipeline designs, LNG can be efficiently transferred over distances of up to 20 miles. Buried subsea pipelines that have the proper designed cover are inherently safer from leaks and damage than exposed pipes. In addition, having

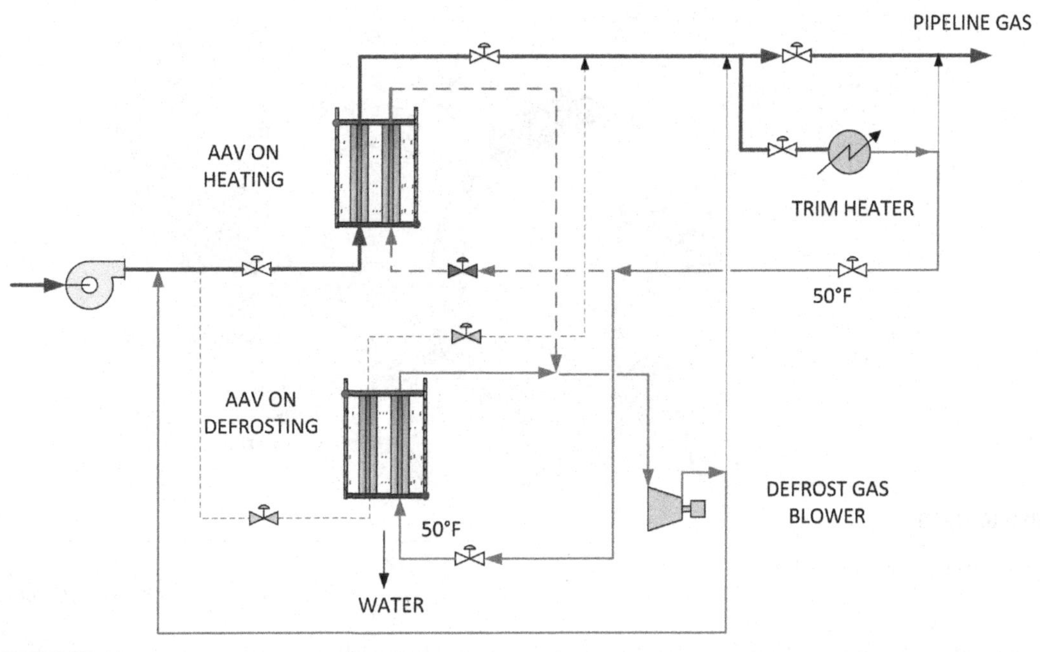

FIGURE 10-18

Ambient air vaporizer defrosting scheme.

(Mak, 2010)

buried cryogenic piping may improve the thermal performance of the pipeline in hot climates. By adding features such as real time monitoring of the performance of the pipeline for structural integrity, thermal performance and leaks using fiber optic technology, the safety of the subsea pipeline can be further enhanced.

Subsea cryogenic pipelines are emerging technologies that are essential for the new generation of offshore LNG loading and receiving terminals. A major issue for these systems is the pipe contraction due to the low temperature of the LNG. At present, there are mainly two methods to accommodate this contraction:

- Use of Invar or other alloys with an ultra-low thermal expansion coefficient
- Use of bellows, one in each segment (about 50 ft long) of the pipeline, which is a self-contained pipe-in-pipe segment with vacuum insulation.

Although technically feasible, both methods suffer major disadvantages in cost, reliability, durability, or maintenance requirement.

A new pipeline configuration (Patent Pending) has been developed to address these disadvantages. The configuration is the culmination of Fluor conceptual designs, which began in the 1970s with the design of a subsea LPG pipeline and continued into the 1980s with their first subsea LNG pipeline for an arctic LNG ship system. This new design, however, takes advantages of recent developments in insulation technology and uses a highly efficient thermal nano-porous insulation in the annular space between the

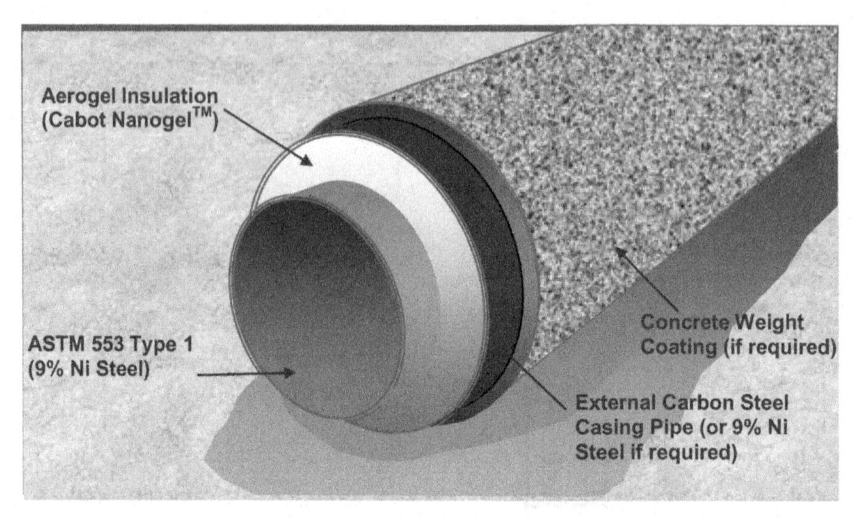

FIGURE 10-19

New piping design for offshore.

(Phalen et al., 2006)

inner and outer pipes (see Figure 10-19). This material is kept in an ambient pressure environment, which is produced by sealing metal or nonmetal bulkheads. The bulkheads transfer the contraction induced axial tension load on the inner cryogenic carrier pipe(s) to the external jacket pipe. The resulting pipeline bundle is a structural element, which accommodates the thermal contraction and expansion loads without resorting to expansion bellows or ultra-low thermal expansion alloys. As an example, the LNG carrier pipe would be 9% nickel steel, while the jacket pipe can be of carbon steel. The thermal insulation would be a high-performance nano-porous aerogel product, approximately 2" thick, in blanket or bead form installed within the annular space without vacuum and under ambient pressure.

10.3.7 Offshore loading and unloading hose

Open sea offloading of LNG cargoes to off-take carriers has been performed in a side-by-side configuration under benign environmental conditions. But because of potentially harsh sea environments, the ship-to-ship transfer of LNG must be done in a tandem offloading configuration.

SBM Offshore has achieved certification for its Cryogenic Offshore Offloading and Loading (COOL) offshore LNG transfer system as well as qualification of the manufacturing process for the COOL hose and its connectors. The COOL system comprises a flexible cryogenic floating LNG hose and connectors designed to allow the transfer of LNG between vessels in a tandem mooring configuration.

The SBM hose design is based on a patented system, and comprises a well-proven outer marine hose with an inner composite LNG hose. The space between these two hoses is filled with insulating materials with excellent properties over the full range of ambient to cryogenic temperatures. The connector consists of both a structural and a fluid connector. It is a cryogenic quick connect/disconnect system that handles and connects the hose to the LNG carrier bow manifold and therefore facilitates LNG loading and unloading.

References

Bronfenbrenner, J.C., Pillarella, M., Solomon, J., Summer 2009. Selecting a Suitable Process Technology for the Liquefaction of Natural Gas. LNG Industry, 17–25.

Bukowski, J., Liu, Y.N., Boccella, S., and Kowalski, L., "Innovations in Natural Gas Liquefaction Technology for Future LNG Plants and Floating LNG Facilities", paper presented at the International Gas Union Research Conference 2011, Seoul, Korea (Oct. 19–21, 2011).

Del Vecchio, J., and De Jong, E., "The Mariscal Sucre LNG Project—A Downstream Perspective", paper presented at the 4th Topical Conference on Natural Gas Utilization, AIChE Spring National Meeting, New Orleans, LA, USA (April 25–29, 2004).

Haney, F., Donovan, G., Roth, T., Lowrie, A., Morlidge, G., Lucchini, S., Halvorsen, S., June 23, 2011. Modular Processing Facility. U.S Patent Application US2011/0146164.

Kimmel, H.E., and Cathery, S., "Thermo-Fluid Dynamics and Design of Liquid-Vapour Two-Phase LNG Expanders", paper presented at the GPA Europe Technical Meeting, Paris, France (Feb. 24–26, 2010).

Kleiner, F., and Kauffman, S., "All Electric Driven Refrigeration Compressors in LNG Plants Offer Advantages", paper presented at the Gastech 2005 Conference and Exhibition, Bilbao, Spain (March 14–17, 2005).

Mak, J.Y., "LNG Regasification and Utilization", paper presented at the 5th Annual Atlantic Canada Oil and Gas Summit, Halifax, NS, Canada (May 30–31, 2005).

Mak, J.Y., "Configurations and Methods for Power Generation with Integrated LNG Regasification", U.S Patent No. 7,574,856 (Aug. 18, 2009).

Mak, J.Y., Feb.25, 2010. Ambient Air Vaporizers. U.S Patent Application 20100043453.

Mak, J.Y., Nielsen, D., and Graham, C., "A New 95% Ethane Recovery Process for LNG Receiving Terminal", paper presented at the GPA Europe Meeting, Paris, France (Feb. 22–23, 2007).

Mak, J.Y., Nielsen, D., Schulte D., and Graham C., "A New and Flexible LNG Regasification Plant", paper presented at the 83rd Annual GPA Convention, New Orleans, LA, USA (March 13–16, 2004).

Meher-Homji, C.B., Messersmith, D., Hattenbach, T., Rockwell, J., Weyermann, H., and Masani, K., "Aeroderivative Gas Turbines for LNG Liquefaction Plants—Part 1: The Importance of Thermal Efficiency and Part 2: World's First Application and Operating Experience", paper presented at the ASME Turbo Expo 2008 Congress, Berlin, Germany (June 9–13, 2008).

Nagelvoort, R.K., "LNG Innovations, Past, Present and Future", presented at the FPSO International 2009 Conference, Oslo, Norway (March 4, 2009).

Neeraas, B.O., and Marak, K.A., "Energy Efficiency and CO_2 Emissions in LNG Chains", presented at the 2nd Trondheim Gas Technology Conference, Trondheim, Norway (Nov. 2–3, 2011).

Otsuka, T., "Evolution of an LNG Terminal: Senboku Terminal of Osaka Gas", paper presented at the 23rd World Gas Conference, Amsterdam, The Netherlands (June 5–9, 2006).

Phalen, T., Mak, J.Y., and Prescott, N., "Innovations in LNG Receiving Terminals Design", paper presented at the 1st Asia LNG Summit, Shanghai, China (Sept. 20–21, 2006).

Pilarella, M., Liu, Y., Petrowski, J., and Bower, R., "The C3MR Liquefaction Cycle: Versatility for a Fast Growing, Ever Changing LNG Industry", paper presented at the 15th International Conference & Exhibition on Liquefied Natural Gas (LNG 15), Barcelona, Spain (April 24–27, 2007).

Ransbarger, W., Spring 2007. A Fresh Look at LNG Process Efficiency. LNG Industry, 73–80.

Roberts, M.J., Liu, Y.N., Petrowski, J.M., and Bronfenbrenner, J.C., "Large Capacity LNG Process—The AP-X Cycle", paper presented at the Gastech 2002 Conference and Exhibition, Doha, Qatar (Oct. 13–16, 2002).

Thompson, G.R., Adams, J.B., Hammadi, A.A., and Sibal, P.W., "Qatargas II: Full Supply Chain Overview", paper presented at the 14th International Conference and Exhibition on Liquefied Natural Gas (LNG 14) Conference, Doha, Qatar (March 21–24, 2004).

van de Lisdonk, S., van Rijmenam, C., Tanaeva, I., Di Nola, G., van Loon, M., and Nagelvoort, R.K., "Next Generation Onshore LNG Plant Designs", paper presented at the 16th International Conference and Exhibition on Liquefied Natural Gas (LNG 16), Oran, Algeria (April 18–21, 2010).

Yates, D., "Thermal Efficiency-Design, Lifecycle, and Environmental Considerations in LNG Plant Design", paper presented at the Gastech 2002 Conference and Exhibition, Doha, Qatar (Oct. 13–16, 2002).

CHAPTER

LNG Project Management

11

11.1 Introduction

Project management is of vital importance to LNG infrastructure projects because of the complex technology, large investments involved, and remote locations of liquefaction plants. There are also needs to source equipment and materials from different countries and coordinate with the supply chain components (e.g., feedgas pipeline, gas liquefaction plant, and marine export terminal). In addition LNG projects typically are developed by multiple contractors and subcontractors and are reliant on suppliers and service companies located around the world, requiring large multicultural labor teams and financial institutions providing large tranches of project finance. Even with strong project management teams and well-defined project management strategies it is difficult to deliver these projects on time and on budget. This chapter addresses the management of capital projects along the LNG supply chain, focusing particularly on the large-scale facilities associated with liquefaction and regasification plants. For the most part, best practice management for LNG facilities projects follows generic project management principles applicable to most industrial engineering and construction projects. These principles and best practice guidelines are well documented (e.g., PMBOK Guide, 2013) and are addressed here primarily to highlight their importance to the success of LNG projects.

11.2 Project management sequence

The sponsors and managers of an LNG facility construction project generally follow a sequence involving a series of defined steps:

1. **Project definition.** Define the plant capacity, feed and product specifications, site location data, and an overview of the conceptual design configurations.
2. **Project scope.** Identify the activities, services, and deliverables that must be provided to fulfill the project definition.
3. **Project budgeting.** Estimate the engineering efforts and costs, including contingencies required to meet the project scope.
4. **Project planning.** Define the engineering milestones that are required to meet the project objectives. The project is staffed with experienced engineers required to produce project deliverables specified in the project scope. Project risks are identified and mitigation plans developed.

5. **Project scheduling.** Estimate the start and completion dates for each activity and sequence of each activity to avoid any bottlenecks. Long lead equipment items are identified for early procurement to meet project schedule.
6. **Project execution and tracking.** Review completeness of the project deliverables, and measure work, time, and costs that have been expended to ensure that the project cost and schedule are on target.
7. **Project close-out.** Plant commission, performance tests, measurement against guarantee values, and final payment per contract terms. The plant is considered successful only when it meets the sponsor's expectation.

Successful projects require a focused, dedicated, and integrated team led by effective management, which requires clear objectives, a well-thought out plan, realistic stakeholder expectations, effective communication, integrated multicultural workforces, monitoring of project progress, and stakeholder support.

Project management in today's organizations demands multiskilled persons who can handle and manage far more than their predecessors. They must be capable of resolving project resource problems, interfacing with the clients, as well as making critical decisions in technical areas.

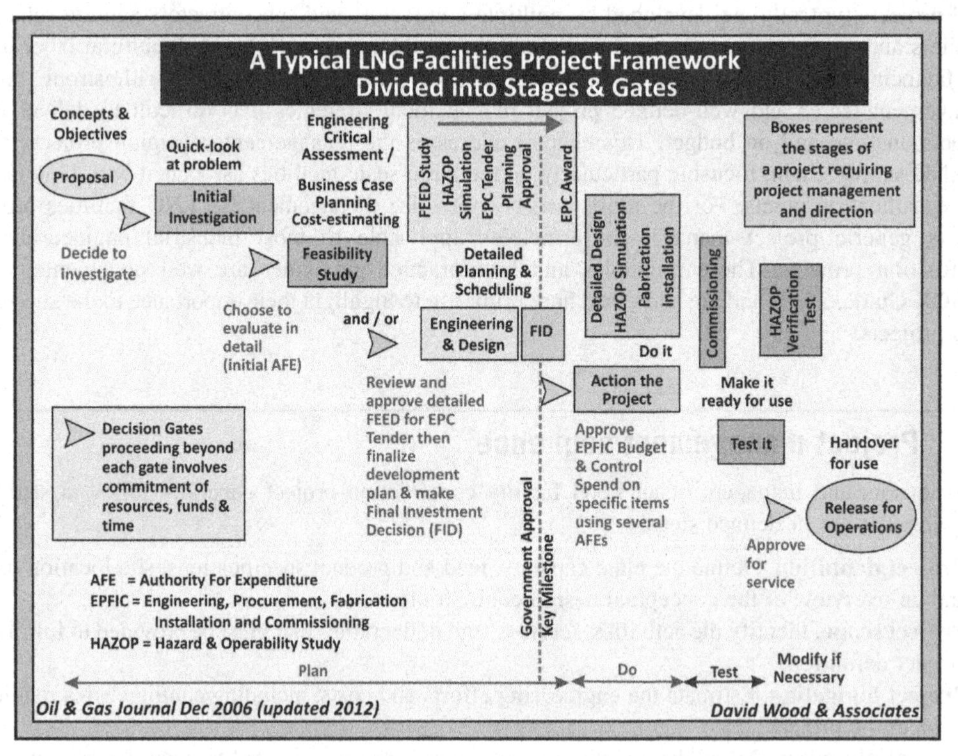

FIGURE 11-1

A stages and gates approach to the project management sequence of large LNG facilities projects.

(Diagram expanded from Wood & Mokhatab, 2006)

A stages or gates approach to manage a typical LNG project is illustrated in Figure 11-1. The planning stages (feasibility, pre-FEED, and FEED) must be completed satisfactorily to all stakeholders before the EPC phase that includes detailed engineering, procurement, EPC contracting, construction, fabrication, installation, commissioning, hazards and operability (HAZOP) verification test, and finally, an approval to operate.

To move from one stage to another requires decisions and approvals to be made on the design selection in respect to the design integrity, plant operability, and environmental impact. Project finance and funding must also be considered. The approval to proceed to a second stage requires AFE (authorities for expenditure) to be signed off by the project owners (and often other stakeholders, e.g., government authorities) and within the project budget. Although the diagram shown in Figure 11-1 appears to suggest a linear process development from one stage to another, in reality, there are often recycles and redesign of the work completed in earlier stages. However, efforts must be made to minimize the recycles as they may increase the project costs and delay the completion date.

11.3 Success of engineering and design

At the onset of a project, the project sponsor team will develop a business case for the project and provide justification to the board of directors for budget approval. Depending on the resources available to the sponsor and the complexity of the project, a consultant may be assigned the task in preparing a recommendation in a study report. Once the recommendation from the study is approved, funding will be released and project definition will be developed. The project sponsor will develop a Design Basis Memorandum (DBM), which typically includes a description of the facility, the feed and product specifications, and the technology selection.

For more complex, large projects that involve developments of infrastructure, acquisition of new technologies, and particularly installation in foreign countries, it is common practice to use consultants to develop a feasibility study and pre-FEED study to further validate the technology selection and plant configuration prior to the FEED phase. These activities must be conducted by competent consulting companies with experienced and knowledgeable staff. It is commonly recognized that the pre-FEED study phase is where most of the cost savings of the project can be realized with minimum man-hours spent.

The technology selection and plant configuration that are defined in the pre-FEED phase will be used as a starting point in the FEED phase. Once the FEED work is started, changes to technology and plant configuration are very costly and may potentially delay the project completion. Changes of the FEED content during the EPC phase are even more difficult, unless the FEED design is proven to be unsafe and inoperable. All these mean that the results from the pre-FEED or FEED phase must be sound, which requires a well-managed team ensuring the success of the project.

11.4 Sponsor–contractor relationships

The project sponsor, once satisfied with FEED design, will submit the FEED package to solicit bids from EPC contractors and will select the successful bidder(s). In some cases, project sponsors may enter into cooperation partnerships or joint ventures with an EPC contractor for certain types and sizes of projects. Pricing will be prenegotiated according to a range of possible contract models (fixed price or turnkey, cost-plus, risk-sharing, etc.) or for smaller projects perhaps performed on a reimbursable basis. Both parties, project sponsor and contractor, will set up project teams to do the work. Depending on the finance arrangement, a main EPC contractor, joint venture, or alliance with other EPC subcontractors can be awarded for all aspects of engineering, procurement, and construction or they may be used only to develop detailed engineering and some procurement with the construction carried out by other contractors. Expenditure programs and work package schedules will be set up by the main EPC contractor in consultation with the project sponsor.

The main EPC contractors for LNG facility projects are likely to be internationally known firms like Bechtel, Fluor, Foster Wheeler, KBR, Technip, Worley Parsons, etc. However, smaller contractors can be employed as subcontractors to the project. The main contractor typically has the applicable expertise, experience, and associations with particular technologies (e.g., Bechtel and the ConocoPhillips optimized cascade liquefaction process), but all tend to operate in very similar ways. Consultants and contract personnel are often employed to provide a "cold-eye" review of the design or to support a safety review or engineering studies to supplement the resources or expertise of the main EPC contractor. Most often, the project sponsors would assign their own staff to the contractor's home office to monitor the progress and quality of the design efforts.

A successful project requires the sponsor and EPC contractor to work closely together. Typically, the sponsor introduces the operation staff at an early stage to ensure that they will accept the design within the confines of the specifications in the design basis and are given the opportunity to review and modify the design as necessary. In this way, the EPC phase can proceed smoothly and would meet the plant operators' expectations, since they are the ones who will own and run the units. During the construction phase, project management must also address a diverse range of staffing issues, such as union or nonunion staffing, equipment delivery problems, logistic problems, and other unexpected problems that may arise throughout the project (Wood et al., 2011).

Most importantly during project execution, the project's design objectives, cost, schedule, deliverables, and expectations must be clearly communicated to the engineers, employees, subcontractors, and equipment suppliers.

11.5 Defining business and project objectives

The first step in project management is alignment of the business objectives with the project objectives. A project can be installed on time and on budget, but if it does not meet the business objectives, then the project cannot be deemed a success. Some of the questions that the project sponsors should ask include:

- How much gas is available for processing and for how long? Is the quantity sufficient to support the size of an LNG production plant that would satisfy the sales gas purchase agreements? What is the life cycle cost of the installation?
- What is the projected market value of LNG or gas when the plant is in operation? Has plant downtime been included in the economic analysis?
- What is a realistic schedule and startup date for the facility?
- What are the field production pressures, temperatures, and compositions?
- What are the predicted gas flow rates over the life of the project? Will there be a change in gas pressures, temperatures, and compositions? Will there be an increase in nitrogen content or acid gas content due to enhanced oil recovery upstream?

- What products can be sold from the facility in addition to LNG/gas (e.g., ethane, propane, butane, condensate, sulphur)? What are the short-term and long-term market value forecasts for these products?
- What are the specifications of the LNG product? Is there a specification on LNG heating value and/or Wobbe Index?
- Is there a disposal method that is environmentally acceptable in handling hazardous materials produced from the feedgas, such as mercury, organic sulfur, and other heavy components?
- How and where will the LNG products be sold (e.g., FOB, CIF, DES contracts)? If products are to be sold on a delivered basis, how will the shipping be conducted (e.g., new-build marine vessels owned/operated by or on behalf of the project sponsor or chartered from existing fleets)? Is the product sold on a BTU basis?
- Are there any local environmental regulations on emissions and effluents? How do these compare with industry best practice and the project sponsor's own policies?
- What are the local safety and security policies requirements? How do these compare with industry best practice and the project sponsor's own policies?
- What infrastructure is necessary? This may include roads, bridges, loading and unloading facilities, jetties, port access channels, off-sites and utilities, work force construction camp, permanent operating personnel housing, and recreation facilities.
- Are there any local skilled crafts available for construction, operation, or maintenance?
- What are the local procurement regulations and mandated requirements for locally hired personnel? Are their skills acceptable to work in LNG project?

Since most LNG facilities form part of a long supply chain stretching across continents with several owners ranging from international oil and gas companies (IOC), national oil and gas companies (NOCs), and utilities companies, the project sponsor may not be able to answer all of these questions, and consultation with the upstream personnel, engineers, permitting agency, government officials, legal entities, and the finance industries is necessary.

11.6 The project charter

The project charter is a document often adopted following negotiations between project sponsor and the project management team. It outlines and demonstrates clear project management support for the business and project objectives as outlined by the project sponsor. In particular, it authorizes the project manager or management team to lead the project and allocate resources to contractors/subcontractors and suppliers as required. The project charter is commonly distributed to all project stakeholders in order to encourage commitment and alignment with the project and its defined objectives. Indeed stakeholder consultations and negotiations may be required prior to the final issue of the project charter in order to ensure alignment with the project as planned. This process should occur early on and may require some compromises on the part of the project sponsor and the project management team in order, for example, to satisfy local community demands. A clearly drafted project charter usually helps with team building and motivation and encourages questions and concerns to be raised and clarifications established early in the project's schedule. The project charter also has the role of reinforcing the project manager's and project management team's authority in conducting the project.

The project sponsor may be a joint venture of organizations (IOCs, NOCs, utilities) with one designated as project operator and in some cases another as technical advisor. The project charter is usually signed off by all joint venture partners forming the project sponsor together with approval of the full project budget and authorities for expenditure (AFE) mandating the project operator to commence the initial stages of the project and the associated expenditures.

11.7 Project risk assessments and risk management plans

LNG project developments face many risks that can and frequently do lead to expensive delays. Late delivery by suppliers, contractors, and service providers, prolonged commissioning and plant start-up periods, and other logistical failures can all lead to time and cost overruns that negatively impact a project's long-term economic value. The complexity of the full LNG supply chain—comprising upstream facilities, liquefaction plants, a fleet of LNG carriers and, receiving terminals—means that delay or poor performance in any segment of that chain has significant consequences for the performance of the entire supply chain.

A typical 5 MTPA gas liquefaction facility for the whole chain requires a capital expenditure of $7 billion to $10 billion (2011 dollars), which may incur some $10 million expenditure, but generating product with a value, depending on prevailing prices of natural gas, of some $30 million to $40 million a week (Skramstad et al. 2011). Any delay in delivering the products to the customers represent in a significant loss in cash flow.

Recent analysis (Skramstad et al. 2011) has shown that only three of 22 LNG import terminals were delivered on time, while for export terminals, 50% of recent projects experienced significant delays. In addition, reduced performance, commissioning problems, and lengthy startup periods to reach full production have been reported for many gas liquefaction plants in recent years. Delays have ranged typically from 3 to 18 months. The worst performing gas liquefaction projects in this context are, in descending order of delay time with operator/technical leader in brackets: Hammerfest LNG Norway (Statoil), Sakhalin II Russia (Shell), Qatargas 2 (ExxonMobil/QP), Yemen LNG (Total), and Tangguh LNG Indonesia (BP). Note that these projects involve major companies that have had much past experience of delivering large projects. Some projects have been delivered ahead of their planned schedule, such as Egyptian LNG, Equatorial Guinea LNG, and Darwin LNG (Australia), so delay is not inevitable if the risks are well managed. The causes for the delays and cost overruns identified in that study suggested that six categories predominate:

- Political issues and circumstances
- Design faults
- LNG deliveries not meeting buyers' quality specifications
- Problems with new, unproved technologies
- Lack of management in addressing issues
- Unclear interfaces and poor communications between vendors, contractors, and sponsors.

Here we focus on just the following subset of project execution risks associated with specific facilities along the LNG supply chain:

- Pre-FEED or conceptual risks
- Technical risks (engineering and design)
- Execution risks (construction, installation, startup, plant commissioning, and operations).

However, in practice we recognize that execution risks cannot be considered in isolation from the full spectrum of asset risk and uncertainty (Wood et al., 2007).

There are benefits to be gained by identifying execution risks by major project phase and establishing a system that can be used to manage, document, and control risks throughout the project schedule. The aim is to communicate execution risks to management before the start of each major project phase. A definitive process is required in order to systematically and appropriately identify, assess, control, and reduce or mitigate risks. This should also not be confined only to mitigation of downside risks, but also should address the potential exploitation of opportunities (i.e., upside) during all phases of a project.

An integrated Risk Management System (RMS) should address all related project risks and their potential impacts on issues related to a project's technical quality and performance, cost, schedule, execution, safety, security, health, community, and environment. Risk should be addressed holistically for the project as a whole, including, to the extent that it is possible to predict impacts, all stakeholders, involved parties, all phases, and all types of project exposure and uncertainty. The RMS should be developed in order to properly address all actions to be undertaken to fulfill requirements contained in any and all contractual obligations, and the ongoing incorporation of the results of all assessments and risk mitigation/opportunity exploitation strategies concerned with uncertainties as they become better defined.

Specific equipment operational risks, such as control valve functionality, construction yard locations, or installation sequencing should be captured by HAZOP simulations incorporating detailed risk assessments at the equipment and control level. It is beneficial that such detail be specified in a project's Risk Management Plan (RMP), which will ultimately be passed on to plant operators once the facility is operational. However, in addition to the technical risk issues covered by the HAZOP, it is prudent for a project RMP to define a Project Execution Plan (PEP), which specifically includes the techniques, processes, plans, and procedures to be used or developed for risk assessment and management during the delivery of a construction project. The PEP should be project specific. Each project should develop a projectwide RMP that defines the specific assessments to be conducted, the timing of the assessments, and the roles and responsibilities of accountable personnel/contractors/suppliers involved. Risk management, though required for all projects, is scalable and each RMP and its component PEP should define the level of detail and analysis that is prudent in the project context.

An Initial Risk Management Plan (IRMP) should be developed to incorporate risk assessments to be conducted prior to detailed project design. This will usually be initiated during a project's conceptual, feasibility, or pre-FEED analysis. The IRMP should evolve into the RMP, which should be reevaluated and updated on an ongoing basis through to the start of detail design, and again revisited prior to the start of construction/installation activities. A key part of an RMP is that it evolves as a project unfolds through a process incorporating follow-up and/or close-out of all issues identified by the assessments of specific uncertainty events, covering both risks and opportunities.

All companies, small and large, whether they are gas companies, engineering firms, or EPC contractors, should have their own established, comprehensive, and systematic approaches to managing project risks. The gas company(s), as the project sponsor (and client of the contractors) and operator, should have its own established system to manage all facets of risks, which needs to be agreed to by its joint venture partners. In addition it should also have detailed coordination procedures that outline the goals and expectations of the contractors' Risk Management Plans. These coordination procedures should be clearly communicated to the contractors during the EPC tendering phase. Established EPC contractors are expected to have comprehensive risk management programs in place that meet the project sponsor's requirements and expectations and can be easily integrated with the project sponsor's RMP.

11.8 Pre-FEED or conceptual risks

During the pre-FEED or Conceptual phase an approach should be developed (as part of the IRMP), depending on the project complexity, that seeks to identify a comprehensive list of project issues, possible events, and their likely impacts. The analysis should be systematic, focusing on all aspects of the project, and might best be kicked off by a brainstorming exercise as part of a risk-identification workshop (e.g., Skramstad et al., 2011). The scope can be broad and include aspects such as regulatory, fiscal, logistics requirements, geographical locations, technical issues, and so on. This might lead into more detailed hazard identification (HAZID) exercises and some quantitative risk analysis (QRA) of specific issues of concern. See, for example, a discussion of HAZID and QRA results concerning safety drivers associated with FLNG facilities by Shell (Persaud et al. 2003).

The IRMP should address the resolution of key issues along with guidance on the levels of appropriate senior management involvement. It should also address and document transparently how issues are classified and assimilated. All significant issues (those with substantial negative or positive impacts) need to be brought to the attention of senior management and provisional mitigation strategies and/or contingency plans considered.

Risk assessments during the pre-FEED or Conceptual phase should be performed as set out in the IRMP. As the project progresses, new issues are likely to be identified and analyzed in the same manner. However, as illustrated in Figure 11-2, the earlier in a project that risk mitigation decisions are taken the more significant their impact is likely to be in controlling ultimate project costs. Late stage decisions focused on risk reduction during detailed design commonly result in cost and time escalations. A comprehensive conceptual risk assessment (CRA) should therefore be initiated prior to a FEED study and long before detailed design.

Pre-FEED risk assessment generally involves the appraisal of multiple concepts and multiple cases within each concept. The results of the assessments are utilized to make go or no-go decisions. Additionally, these assessments may simply be a review of a single concept. The purpose is to identify specific areas and further work that needs to be done in the project definition phase and is likely to influence specific equipment and technology selection.

Depending on complexity most LNG facilities projects benefit from a preliminary HAZOP review and simulation. This should be conducted on the facilities design using the design detail available in the draft basis of design (BOD), including layouts, preliminary planned numbered and initialed drawings (PNIDs), equipment data sheets, process flow diagrams, metallurgical flow diagrams, information on likely hazardous materials, and such, as well as the draft project design specifications. The purpose is to identify potential design hazards and special material of construction that need to be addressed in the detailed design phase.

Departures from required asset performances and reliability should be considered during the conceptual risk assessment (CRA). The CRA is based on preliminary drafts of the key early project documents, including the initial project execution plan (IPEP).

This early stage risk analysis enables the project team to move up the learning curve (Figure 11-3) and reduce the likelihood and potential impacts of identified risk events and in some cases eliminate

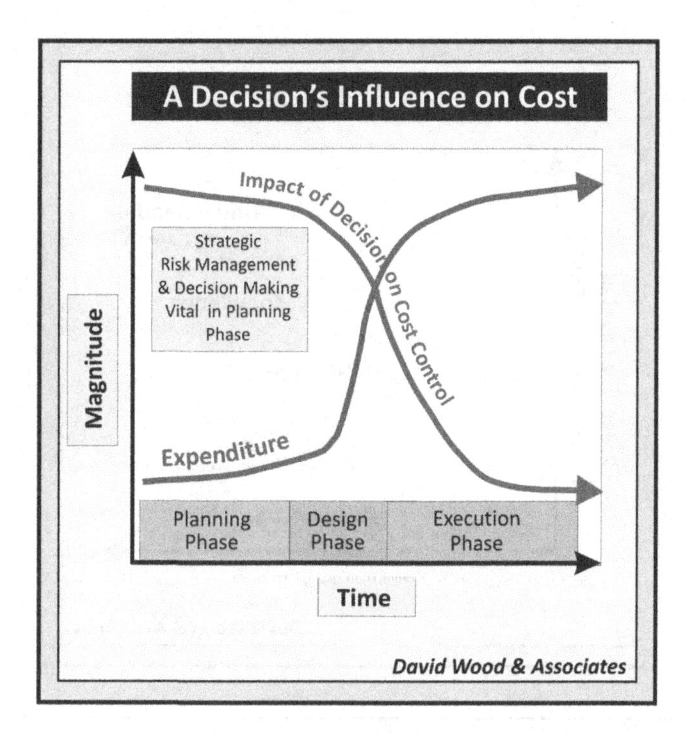

FIGURE 11-2

Cost and schedule control benefits of early planning phase risk assessment and risk mitigation decision-making is well established.

(Wood et al. 2007)

the causes of potential events. Figures 11-2 and 11-3 highlight the conceptual benefits of early stage risk assessment.

11.9 Key project management considerations for LNG projects

There are a number of critical issues for sponsors in the design and construction of LNG facilities projects. Some of these issues are technical in nature:

- Selection of technologies
- Selection of contractors (e.g., lead EPC contractor)
- Development of a project management team
- Well-considered and communicated strategies for design, contracting, and risk management
- Cost and budgeting
- Change orders and deviation to selected designs
- Scheduling and defining meaningful milestones and targets
- Quality of work and performance.

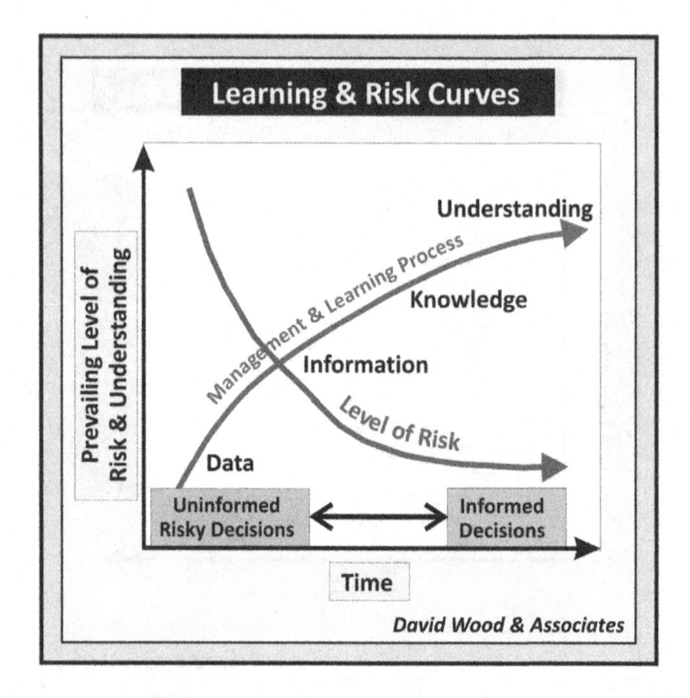

FIGURE 11-3

Early stage risk assessment aids learning and enables decisions to be made under conditions of less uncertainty.

(Wood et al. 2007)

Other critical issues that impact project performance and outcomes are of a contractual, commercial, and financial nature:

- EPC contract structure (e.g., lump sum turnkey or gain share)
- Liability limitations applied to contractors and suppliers
- Financial security to be posted by contractors (or retained for a period)
- Securing project finance and completion test terms for drawdown of loans
- Negotiating local content obligations and procurement rules with government
- Placing the appropriate insurance and export credit guarantees as part of financing.

11.10 Scenario-based risk assessment

A scenario-based assessment can be aided by a probability (likelihood) versus consequence (impact) matrix. Such matrices can take several formats varying from purely qualitative to quantitative, with different numbers of individual cells. Figure 11-4 shows a 5 × 4 matrix (i.e., 5 alternatives for likelihood of occurrence and 4 alternatives for impact consequences should an event occur). Each cell in the matrix represents a coded outcome with groups of outcomes color-coded, usually using variations of a traffic light approach (red for unacceptably high-risk events requiring mitigation actions and green for acceptably low-risk events requiring little or no control actions).

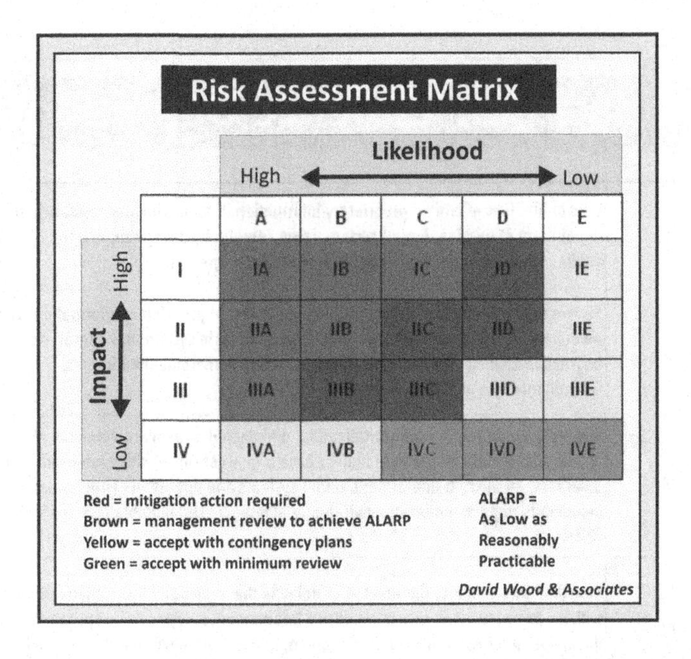

FIGURE 11-4

Example likelihood versus impact risk assessment matrix with cell coding and color classification of appropriate responses. See Figures 11-5 and 11-6 for definitions of impact and likelihood categories.

(Wood et al. 2007)

Figures 11-5 and 11-6 show example definitions for likelihood and consequence for the matrix illustrated in Figure 11-4. These definitions and quantitative magnitudes of the impacts are project-specific and it is important that they are defined, understood, and agreed by all involved in project risk assessment.

Scenario-based assessments utilizing a risk matrix approach are recommended in order to properly address, rank, and present assessment results for all project phases. It can often be complemented by the use of "bow-tie" and/or "butterfly" diagrams (Wood et al., 2007) that identify the links between events, their causes and potential consequences/outcomes. Such approaches help in the recognition that events can have multiple outcomes and multiple causes and to focus mitigation strategies on the ultimate causes rather than just the events or their potential outcomes.

11.11 Technical and extreme risks (engineering and design)

As a project moves into the definition phase the scenario-based approach can be further developed into an engineering and design risk assessment (EDRA), which should be conducted before initiating the detailed design work. Depending on the size and complexity of the project, the EDRA scope can be very broad and may cover—in addition to specific design aspects such as number of trains, operability, sparing philosophy, and so on—a broad range of project aspects such as regulatory approvals and third-

Impact Consequence Definitions

I	Loss of life, loss of critical proprietary information, loss of critical assests, significant impairment of mission, loss of system, large community disruption, major environmental impact. Quantifiable loss $10MM or greater.
II	Severe injury to employee or other individual, loss of proprietary information, severe asset loss, unacceptable mission delays, unacceptable system and operations interuptions, small community disruptions, serious environmental impact. Quantifiable loss of $1MM - $10MM
III	Minor injury not requiring hospitalization, undetected or delayed detection of unauthorized entry resulting in limited access to assets or sensitive information, minor asset loss, no mission impairment, minor systems and operation disruption, minor community impact, moderate environmental impact. Quantifiable loss of $100K - $1MM
IV	Less than minor injury, undetected or delay in the detection of unauthorized entry with no asset loss or access to sensitrive information, no systems or operation disruption, minimal to no public disruption, minor or no environmental impact. Quantifiable loss of $10K - $100K

David Wood & Associates

FIGURE 11-5

Event likelihood definitions that are project specific and in some systems involve a probability expressed as a percentage chance of occurrence.

(Wood et al. 2007)

party interfaces, contracting strategies, community impacts and relationships, system abandonment at the end of the asset's operational life, as well as construction. The EDRA may, but does not generally, address specific project details such as control valves or detailed procedures. These are normally covered during HAZOPs and subject to specific risk assessments clearly identified in the RMP. The PEP, work breakdown structure (WBS), BOD, and operating philosophy, all discussed already, are typical project documents that are reviewed and used as the basis of the EDRA.

In the wake of tougher corporate regulation introduced around the world in recent years most large organizations (e.g., gas companies and contractors) have established or expanded risk management procedures and systems integrated with companywide approaches, or enterprise risk management (ERM). The aim of such systems is to make companies more aware of and better prepared to deal with low likelihood but high impact (extreme risk) events should they occur. There is pressure on companies operating in the petroleum industry, from multilateral bodies, governments and some financial institutions to integrating their enterprise risk management frameworks with triple bottom line (3BL) analysis techniques (i.e., profit, people, and planet originally proposed by Elkington, 1997), or the like,

Likelihood & Probability Definitions

A	Possibility of repeated incidents. More than one per year.
B	Possibility of isolated incidents. More than one in 5 years.
C	Possibility of occurring sometime. More than one in 10 years.
D	Not likely to occur. More than one in 100 years.
E	Practically impossible. More than one in 1,000 years.

David Wood & Associates

FIGURE 11-6

Event consequence definitions for which the specific monetary magnitudes at least are project specific.

(Wood et al. 2007)

to broaden their risk analysis and investment decision processes with more emphasis on corporate social responsibilities (CSR) to improve their performance with respect to communities, staff, environment, health, security, and safety issues (Wood, 2011; Wood et al., 2011). Such developments now require LNG project sponsors and contractors to consider as part of their RMP how they will respond to low likelihood extreme risks and to develop credible contingency and emergency response plans for dealing with such eventualities should they occur. Such credible plans are now more frequently required and reviewed by regulators as part of the project approval and permitting process.

11.12 Hazards and operability (HAZOP) studies

HAZOPs are a key ingredient of sound project risk management. HAZOP studies and simulations are normally conducted by the EPC contractors after the design is frozen and/or prior to key deliverables, such as PFDs, PNIDs, and so on, being issued for construction (IFC). However, it is also common for provisional HAZOP studies and/or simulations to be conducted as part of the FEED studies (Figure 11-1) to confirm that the designs proposed are realistic and operable. All plot plans, layouts, data sheets, and vendor information must be available for review at the start of the HAZOP. The purpose of the HAZOP at the end of the detailed design phase is to surface and address all potential process and operational hazards prior to finalizing equipment and materials procurement and moving into the fabrication, construction, and installation phase. HAZOPs are also conducted to stress test a facility to verify that it performs as per its specified design under a range of possible operating conditions.

A successful HAZOP depends to a large extent on a well-documented, structured process where the complete design is systematically analyzed for deviations from design and operational intent by a team of experienced technical, construction, and operations personnel. The HAZOP team should be ideally independent of the project team or at least impartial when it comes to evaluating design and process performance. The presence and involvement of those individuals to be responsible for operating a facility after commissioning is highly recommended in order to capture all the anticipated operational issues (normal, cold, or hot start ups; normal or emergency shutdowns; normal, temporary, and emergency operations; etc.) and focus them upon responding to potential plant teething problems.

The results of the HAZOP will include a comprehensive list of issues along with recommendations covering—in addition to engineering and operations—safety, health, community, and environmental issues (i.e., in line with ERM and 3BL principles and regulatory standards as applicable; e.g., establishing acceptable quantities of gas flared and other greenhouse gas emissions as part of normal operations).

Responsibility for follow-up/close-out of HAZOP items (coordination of the "list" as well as for each individual item) is documented, understood, and followed up by the project management team to ensure issues are resolved satisfactorily and closed-out in a timely fashion. The HAZOP coordinator/ team leader should agree with actions to be taken and follow up after a specified period of time to ensure the HAZOP recommendations have been addressed.

After completion of the detailed design phase HAZOP, it is a prudent policy to mandate that no changes can be made in engineering, design, construction, or operation that materially affects the results of the HAZOP without a thorough reanalysis. This should be rigorously covered under the project management of change plan (MOCP), including appropriate approvals for initiation, evaluation, and acceptance of changes. All changes introduced after this stage are likely to have significant impacts on project cost and schedule, so ideally should be avoided unless absolutely essential.

All issues and action items stemming from the detailed design phase HAZOP must be ultimately resolved and all impacted project deliverables should be upgraded before the completion of detail design. For the precommissioning HAZOP verification test (Figure 11-1) all issues and action items stemming from that final HAZOP must be ultimately resolved and all impacted project deliverables should be upgraded before the facility is signed off for operations.

11.13 Project execution risks

The Construction Risk Assessment (CRA) is conducted prior to the start of facility construction. The CRA scope can be very broad and may cover all aspects of the construction phase of the project such as government/regulator approvals and interfaces, contractor/subcontractor performance, component integration, weather window implications, interfaces, and issues associated with the final construction plan. The CRA does not usually address very specific project details such as individual yard safety programs, welder qualification, or detailed contractor procedures. These detailed issues are covered by specific risk assessments as identified in the RMP.

Detailed design documents and drawings, construction contracts, supplier quotations, cost estimates, long-lead item schedules, and contractor plans are example input documents that contribute to the CRA. The purpose of the CRA is to identify project level issues and hazards and plans to mitigate the risks and/ or eliminate the causes of events that could have material projectwide impacts. The CRA also often develops into a valuable communication tool, in conjunction with its input to the project risk register, for identifying and documenting project risks to management before starting project construction.

11.14 Additional specific precommissioning risk assessments

Other areas of the project subject to specific risk assessments are usually defined in the PEP and RMP. These areas may include, but are not limited to, special design details, contracting strategies, fabrication and load-out activities, heavy lifts, installation activities, early construction work, hook-up and commissioning activities, precommissioning HAZOP, and so on.

All projects suffer exposure to vulnerabilities and uncertainties—these are often encompassed by a project's or company's definition of risk. Properly managing this risk exposure requires a structured systematic and holistic approach involving open communication, transparency, accountability from experienced personnel, and flexibility to adjust responses to risks and opportunities as they become identified and more clearly understood.

One of the keys to success is the early identification, assessment, evaluation, and resolution of risk in all phases and at all levels of an LNG project. A solid, tried, and tested method for achieving successful outcomes to project execution is to use the scenario-based approach. In all phases of a project, regardless of the source or which party has liability, there must be a concerted and coordinated effort to work together to manage the risk and share information related to it. If project risks, particularly project execution risks, are not managed through comprehensive assessment and measured responses the consequences frequently come back to haunt all the project stakeholders and can lead to human and financial catastrophes. The LNG sector has an excellent safety track record over the past five decades because close attention has been paid to project risk management at all stages of project development and operation.

11.15 Focusing planning stages on constructability

The Construction Industry Institute (CII) defines constructability as "the optimum use of construction knowledge and experience in planning, design, procurement, and field operations to achieve overall project objectives." This emphasizes the importance of putting construction knowledge to work early in the project development and integrating that knowledge with the conceptual, design, and procurement phases to allow for the most efficient, practical, and hence cost-efficient construction techniques to be used (Wood et al., 2008b). The CII claim that adopting a construction-focused methodology increases the chance of achieving:

- Reduction in total project costs
- Accelerated schedule to project completion
- Enhanced plant maintainability, reliability, and operability
- Selection of safest designs and construction methods.

Like the term "construction-driven," constructability is not "construction calling the shots." It is a recognition that the construction phase of a project can constitute up to 50% of the overall project costs. The construction stage of any facilities project is where the majority of project risks lie (schedule, cost, weather, impacts of new technology/implementation, labor skills, community disputes, etc.). As such, much thought and effort should be expended early in the project so that the most optimum construction methods, techniques, equipment, and materials are used or specified. The criticality and benefits of an early focus on constructability are highlighted in Figure 11-7.

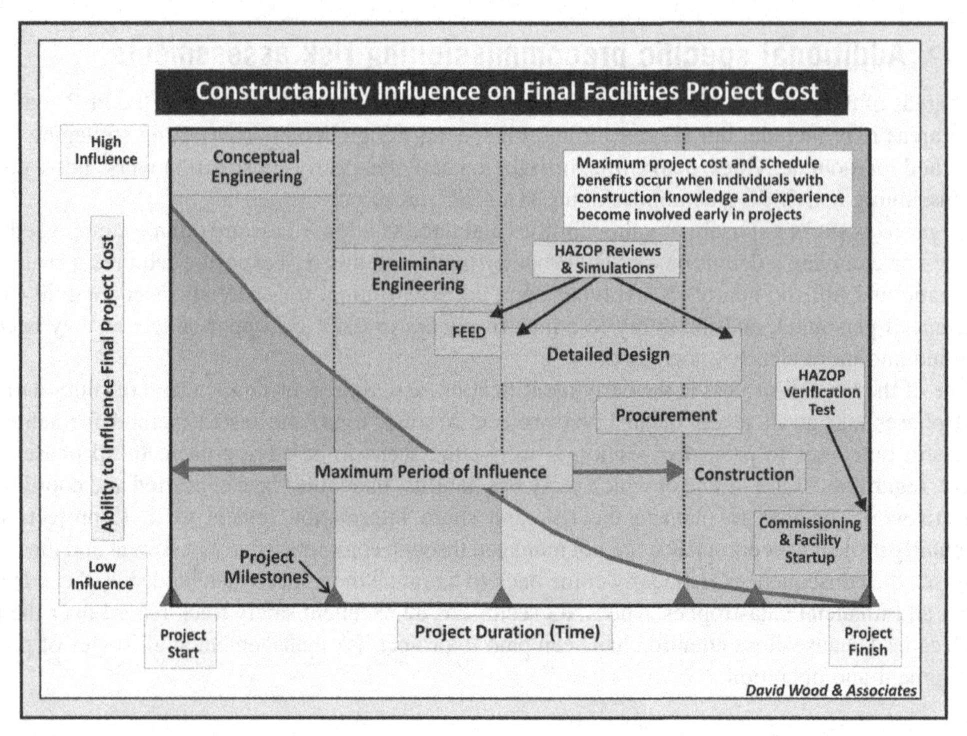

FIGURE 11-7

The maximum impact on the final project cost and schedule occurs in the planning stages of the project. Once procurement and construction have begun it is usually too late to have major impacts that can reduce cost and time expended.

(Diagram modified after Wood et al., 2008b)

Early stage constructability workshops for routine or repeat projects should include personnel from all participating companies (i.e., operator, joint venture partners, potential construction contractors, subcontractors, service companies, and suppliers) with expertise in construction operations, safety, and appropriate branches of LNG facilities engineering. On larger, more complex, and novel projects the range of skills involved in the workshops would be expanded to include environmental, regulatory, quality, procurement, logistics, health, public relations, legal, and security.

The basic objectives of constructability are to identify, define, and verify:

- Guidelines for engineers, designers, and planners to enhance the probability of a cost-efficient, buildable project
- Key engineering, design, and procurement deliverables that will have the greatest bearing on the ability to construct the project and to identify where early construction input is needed
- Construction-related enhancements to engineering, design, and planning that will improve overall safety, quality, costs, and schedule
- Major execution challenges, uncertainties (risks and opportunities)

- Project contracting strategy and plans for construction/installation considering the specific execution challenges faced
- Plans for the transfer of constructability information to the next phase of the project.

Regardless of the size of the project or the mix of the participants it is important to have a structured approach, which should be defined in the Constructability Program.

11.16 Developing a project constructability program

A constructability program can be developed to outline and document the project management team's policy regarding constructability. This should consider the minimum criteria required for defining constructability objectives, the general resources roles and responsibilities involved, and a feedback mechanism by which the contribution of the program might be assessed. A constructability program should include at least the following:

- Clearly defined and, where possible, quantified corporate objectives
- Constructability organization chart defining roles and responsibilities
- Constructability plans and implementation guidelines
- Methodology for key performance indicator (KPI) measurement and verification
- Analysis of KPIs and interpretation to provide project performance benchmarking
- Procedures for capturing feedback and learning outcomes
- Constructability tools: forms, agendas, checklists, and workshops
- Issue and change resolution methodologies
- Milestones for constructability workshops.

An overview of a typical constructability program timeline is shown in Figure 11-8.

As project construction kicks off, most of the significant constructability issues should have already been implemented into the construction plans. At this point, a constructability review follow-up action register is often useful to document actions and responsibilities that have been gathered from the formal constructability reviews ongoing through the planning and detailed design phases. A project constructability coordinator should be responsible for working with each manager or supervisor to ensure items are closed-out and properly documented as the project execution plan (PEP) progresses. The project manager or his or her designate should be responsible for ensuring lessons are captured from constructability reviews and discussed with the project constructability coordinator for dissemination to other project teams. In this way the constructability methodology facilitates an ongoing learning and improvement process built into the project strategies incorporated into agreed amendments to the project charter.

Constructability, if properly defined, structured, and used should ensure that construction knowledge and experience is used as early as possible in the planning and design process to ensure design and implementation plan practicality and cost effectiveness. Constructability is simply the utilization of construction knowledge, experience, and efficiency in a structured and systematic way throughout the design and execution planning stages of a facilities project. Constructability issues touch on all aspects of project execution, including safety, scheduling, detailed design, procurement, material/equipment delivery, contracting, temporary facilities/infrastructure needs, commissioning, and the project management team organization.

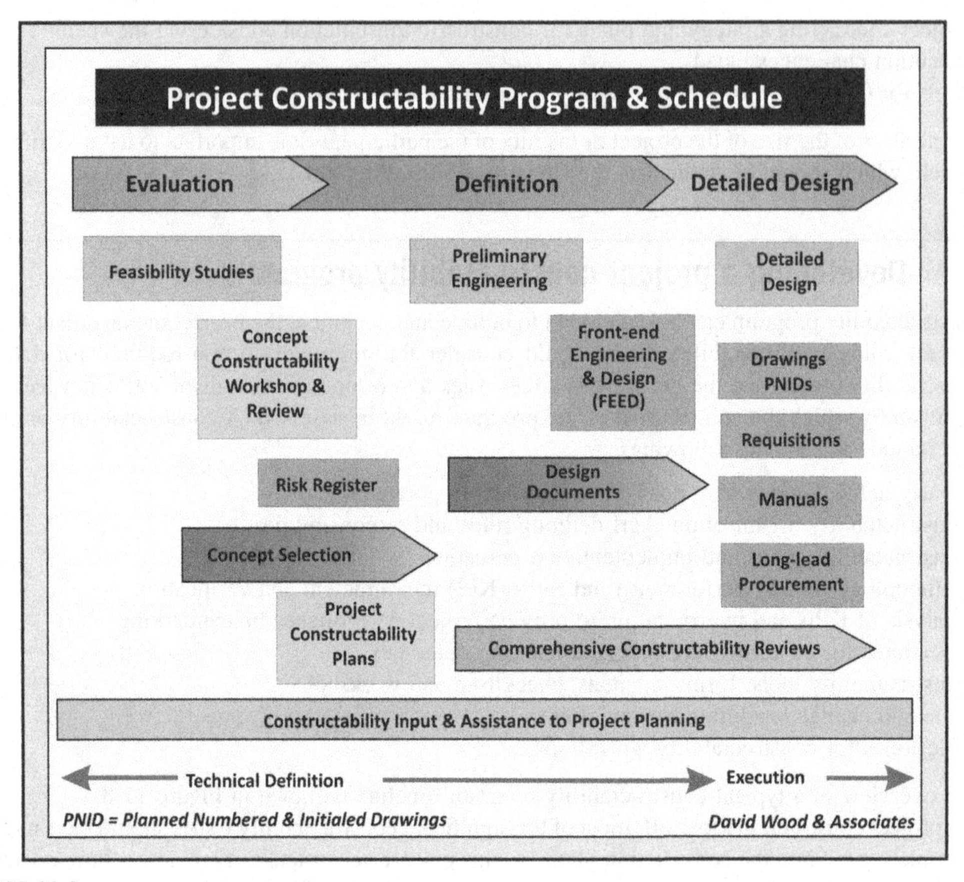

FIGURE 11-8

Typical constructability program timeline.

(After Wood et al., 2008b)

11.17 Project cost, schedule estimating, and modeling

As project definition improves through the design phases of the project so the uncertainties associated with cost estimates should decrease to a funding level of accuracy of approximately +15%/–7.5% with a similar expected range of contingency identified. A probabilistic approach to cost estimating identifying percentiles (P90, P50, and P10) has been used widely used for some time to define cost uncertainties in higher risk projects (e.g., McIntire, 2001; Yale and Knudsen, 2006).

Cost estimating and scheduling software exists to assist with conceptual cost and time estimates as well as detailed estimates and complex project networks involving the optimization of project networks with critical path analysis. Most major EPC contractors have their own custom tools project cost and schedule management tools. Smaller contractors and operating companies may use products supplied by a vendor specializing in these tools. For larger projects it is increasingly common for

Monte Carlo simulation analysis to be used in conjunction with project planning software including critical path identification to yield probabilistic estimates of cost and time associated with each project activity and for the project as a whole (e.g., Wood, 2001).

11.18 Quality assurance, control, and inspection

An important part in achieving the desired result and performance of a project is the specification and assurance of its quality. The establishment of quality-conscious construction practices is essential with quality control made a priority, ensuring that the required qualities are built into all the tasks. Most established EPC contractors operate integrated Quality, Health, Safety and Environmental (QHSE) management systems, which they apply generically across their operations. This is appropriate for LNG project teams and contractors as it makes clear that all four of its components are important to ensure successful project outcomes and that QHS&E all influence each other as well as the budgetary, schedule, and risk issues that drive project decisions. The adage that "*one should expect what one inspects*" applies to LNG facilities project management.

11.19 Technology verification and risk-based verification

Some facilities deploy new or novel uses of technologies, which can pose significant risks, such as the technology not working as expected based on pilot tests, or requiring modifications that jeopardize the project schedule. Systematic programs to identify, evaluate, and mitigate new-technology risks are available (Skramstad et al., 2011). These systems aim to provide evidence that a new technology being considered will function within specific limits with an acceptable level of confidence. Experience from such technology verification systems suggests:

- Improved end-user confidence in a new technology
- More efficient use of project resources
- Systematic identification of failure modes
- Identification of improvements and design changes at an early stage
- Optimization of qualification testing and analysis
- Recognition of interface issues between manufacturer and subvendors
- Increased likelihood of on-schedule delivery
- Reduced risk during operation, resulting from fewer uncertainties and increased reliability
- Traceability of qualification efforts
- Quicker requalification for new operating conditions.

Risk-based verification schemes (e.g., Skramstad et al., 2011) provide the ability to focus attention on the high-risk issues that have greatest impact on project cost and schedule and long-term operating performance. Independent verification of projects by competent and qualified engineers and risk managers provides sponsors with the confidence that their projects are designed, constructed, and installed in accordance with acceptable project objectives.

Systems critical to process and safety usually require special attention, particularly during the start-up and initial in-service phase. These systems may require independent verification to ensure integrity and compliance with industry standards. Process systems or equipment delivered from around the world typically need to be verified against project and regulatory requirements before delivery, ensuring unnecessary delays are avoided and items are used safely. Risk-based verification systems facilitate this and thereby aid project management.

11.20 Engineering, procurement, and construction (EPC) contracts

The relationship between project sponsors and the contractors they mandate through contractual agreements to engineer, build, and in most cases commission a facility are governed by the negotiated terms of the EPC contracts.

A typical EPC contract contains legal clauses defining the terms and usually a series of attachments and schedules that define the commercial and technical arrangements, such as:

- Scope of work and design basis
- Conceptual engineering documents
- FEED study documents
- Detailed list of contractor deliverables
- Payment schedule including milestone payments, retentions, and monthly billing process
- Change order procedures including agreed changes and unilateral changes
- Subcontractors' and suppliers' management and responsibilities
- Key performance measurement of progress (e.g., earned value)
- Start-up, commissioning, and performance testing
- Health safety and environment
- Risk management, including cost escalation, raw materials costs, labor disputes, currency exchange, and interest rates
- Warranties and performance guarantees
- Liquidated damages for failure to meet design guarantees
- Limitations of liability and indemnities (excluding misconduct and gross negligence)
- Insurance
- Financial security (e.g., parent guarantee, letter of credit, performance bond)
- Dispute resolution and arbitration procedures
- Project close out procedures.

For an LNG regasification plant for example separate EPC contracts are likely to be required for the following components of the project:

- Site access and basic civil work
- Soil stabilization
- Site preparation/civil engineering
- Constructing LNG storage tanks (main cost of most receiving terminals)
- Marine works—channel dredging
- Unloading jetty and unloading arms
- Pipework
- Insulation
- Buildings
- Control rooms and instrumentation.

Some issues that are high on the agenda of such projects and require careful attention from the EPC contractors with guidance from the project sponsors are:

- Importance and sensitivity of the project with respect to government and local community
- Regulatory compliance
- Safety, security, and environmental standards
- Health standards for labor force (policies on exposure to extreme conditions)
- Precautions against diseases, pests, and vermin
- Local staffing regulations and indigenous contractor and supplier priorities
- Nonmanual staffing strategies and interfaces with sponsor's project team
- Site access and supply logistics.

Change management requires particular attention and can be greatly helped by early warnings of potential problems and likely changes. The key performance measurements collected on an on-going basis throughout a project help in this early identification process. This facilitates informed and timely decisions on change orders with the sponsor made fully aware of the cost implications of any unilateral late-stage change orders imposed on the project. The aim is to minimize late stage surprises in order to avoid cost overruns and delays.

It pays to have the customer deliverables of projects clearly spelled out in an EPC contract addressing at least the following:

- What is actually to be delivered (a detailed list of items)
- In what format, specification, and condition is it to be delivered
- A review process to report and evaluate delivery as it progresses
- How is delivery to be approved and confirmed as being acceptable to the customer or sponsor
- How and when project and plant documents are to be handed over to the sponsor or plant operating team.

Relationships between the EPC contractors and project sponsor or customer require specific attention and joint team building initiatives from both contractors and sponsors.

Contractors prefer to have:

- A single point of contact with the sponsor or customer
- Formal lines of communication and notification regarding deliverables
- Timely and well-communicated decision-making from the sponsor
- Alignment within the project sponsor team.

Sponsor and customer expectations of EPC contractors include:

- Timely delivery to the project plan (within budget and on schedule)
- Clear understanding of what the project sponsor wants, not what the contractor would prefer to deliver
- Understanding of the project's goals and decisions as specified in the EPC contract
- Active listening to the project sponsor's management team and proactively taking on board responsibilities for issues as and when they arise.

Well-drafted EPC contracts greatly aid the integration and alignment of a project team.

11.21 EPC contractor selection and design competitions

EPC contractors are typically selected by a competitive bidding process or by bilateral negotiations between the sponsor and a single favored contractor. Procurement regulations in a specific country may specifically require a bidding process with a prequalification exercise to establish whether local companies are able to provide the desired deliverables. If it is through a bid process it is usual to award the FEED study to one company and to have that company excluded from the bid list for the EPC contracts. This avoids conflicts of interest or manipulation in the design specifications.

Competitive bidding takes time and may reduce flexibility in terms of the contractor ultimately being able to deliver the sponsor a tailored product. Nevertheless, competition does tend to drive down prices over time, but that depends on the conditions in the market. If the contractors' order books contain plenty of work and upcoming projects already awarded, some will elect not to compete in a time-consuming and costly (i.e., for the contractors) process with modest chances of success; others may just bid high on the basis that they will accept the work only if it offers them high margins.

Bilateral, sometimes open-book, negotiations can save time avoiding a bidding schedule, motivate the contractors to provide innovative solutions, and deliver a successful project. They are unlikely to deliver the lowest price budget, but if the sponsor and contractor have long-term relationships, or are trying to develop one, competition is not needed to motivate contractor performance. Contractor reputation counts for a lot regarding whether sponsors are prepared to deal with them on an open-book basis, leading ultimately to a lump-sum price for the contract.

There are in fact several alternative strategies that can be used to select EPC contractors. The most common options include one or more of the following:

- Negotiate a project with a preferred contractor
- Competitively bid every phase of the project
- Design competition for FEED followed by lump sum EPC bidding
- Reimbursable EPC payment terms
- Owner-contractor alliance-based execution (e.g., gain share contracts).

In bilaterally negotiated contracts the FEED and EPC work is typically done by the same tried and tested contractor. Additional advantages of this approach are that the sponsor does not have to pay bid compensation to multiple contractors and it gives the sponsor's team more time to focus on other project issues. Also the sponsor participates fully in the development of the EPC work program, cost estimate, schedule, budget, and contingencies. The sponsor should therefore gain a good insight into the uncertainties associated with those key items.

The most commonly used approach for selecting an EPC contractor for a gas liquefaction facility involves the development of a FEED package with one contractor followed by competitive bidding for the EPC contract (see Figure 11-9). Typically with this approach, a FEED contractor is competitively selected based on a qualitative evaluation, which includes past experience, evaluation of key personnel, price, and execution plans. The scope of deliverables for the FEED study are not usually determined competitively, although the liquefaction process technology may be preselected by the project sponsors with advice from a competent engineer or contractor during the pre-FEED phase. The FEED deliverables are then competitively bid in order to obtain lump sum EPC proposals from prequalified contractors.

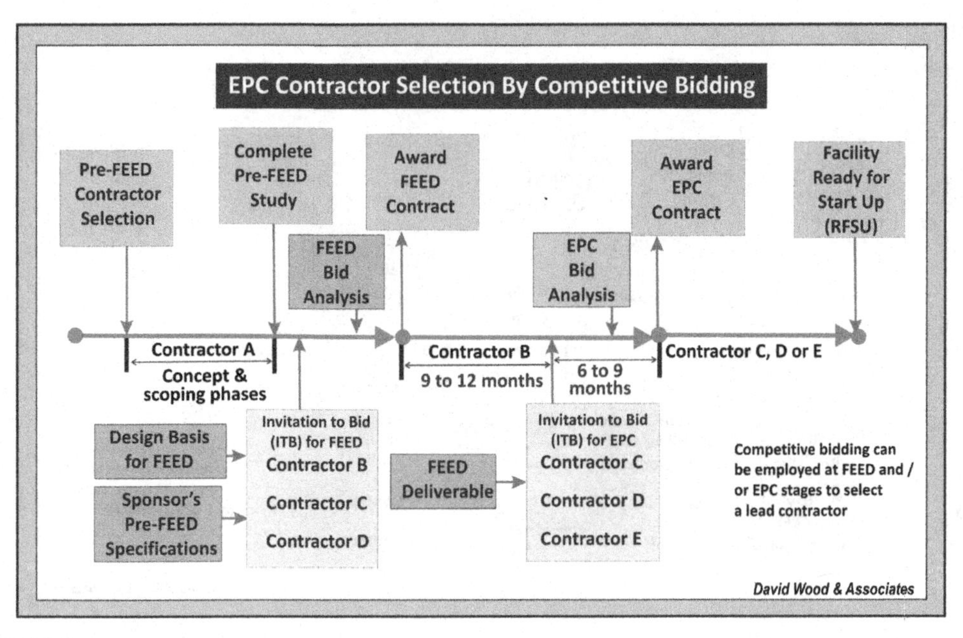

FIGURE 11-9

EPC contractor selection by competitive bidding, showing generic times for the bidding and contract award stages.

(Modified after Durr et al., 2004)

Another approach involves selecting more than one contractor to develop parallel FEED studies. Each FEED contractor commits to generate a lump sum bid for the EPC contract based on their respective FEED package. This process is referred to as a design competition (Figure 11-10). This strategy was adopted for the SEGAS liquefaction plant built in Egypt. There is no guarantee, however, that a design competition will lead to the lowest $/tonne of capacity, but as the FEED stage of the process is identified as the most important in influencing cost and identifying risks in time for them to be appropriately managed, having more than one FEED study conducted by different contractors has its appeal to project sponsors.

Project sponsors have to set rules and guidelines for the participants in conducting design competitions. To do so the following factors require careful consideration:

- Plant components for which the sponsor is prepared to entertain new or novel designs or configurations should be outlined at a high level in advance of the studies. Where ideas that would be considered new or novel are proposed, the onus should be on the design team to identify what steps would be taken to identify, minimize, or eliminate the associated risks.
- All the designs competing in the design competition should be based on process technologies that have been rigorously tested to at least "prototype" (a defined term in each study) level.
- All FEEDs should be required to use the same design basis to ensure a level playing field among the design teams. The design basis requires careful definition by the project sponsor. This can require quite onerous specification guidelines to be prepared in advance by the sponsor.

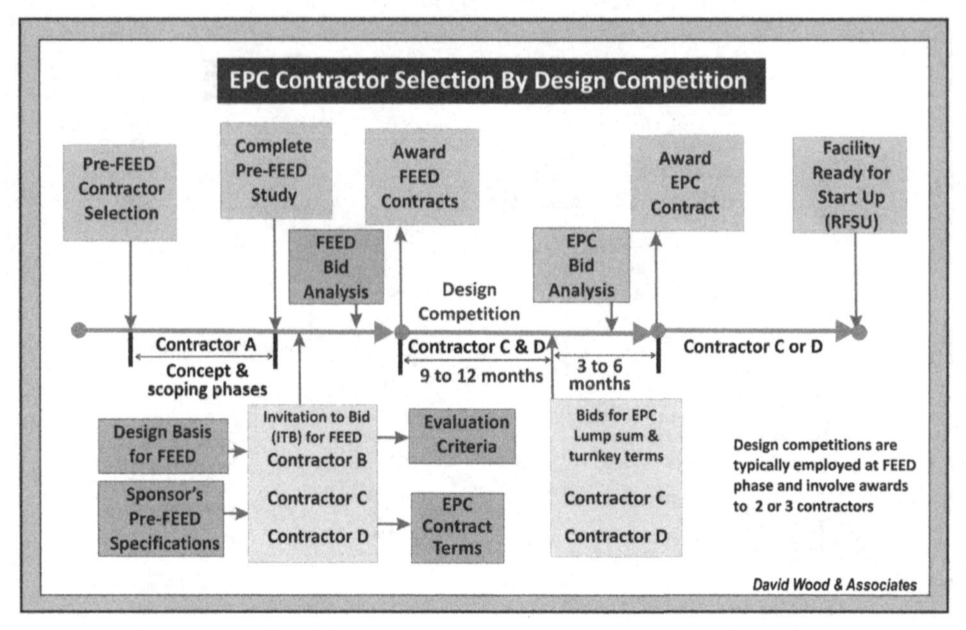

FIGURE 11-10

EPC contractor selection by a design competition, showing generic times for the bidding and contract award stages.

(Modified after Durr et al., 2004)

- The rules set at the beginning of design competitions should be closely followed to ensure fairness and integrity to the competitive process. Each team should be given the same level of information and guidance from the sponsor.
- The sponsor should define in advance the evaluation criteria prior to the start of the FEED. This allows the contractor teams to understand what is important to the owners and this ensures that the work done during the FEED is consistent with the owner requirements. These evaluation criteria provide the contractors with the basis to optimize the design and risk management.
- Sponsors should determine sparing philosophy, requirements for parallel (2-in-1) train designs that decide plant availability along with a list of acceptable equipment vendors as part of the design competition guidelines. This should result in bids that can be more easily compared.
- The sponsor should define in advance the key legal terms and conditions for the EPC contract as these will have an impact on the contractors' lump sum bids. For example, high liquidated damages for missing contractually guaranteed plant capacity values tend to increase the margins applied by the contractors when formulating their bids. If some tolerance is granted in achieving guaranteed specifications then contractors tend to reduce the margins added to the design or to project contingency lump sums, promoting a more fit-for-purpose design philosophy.
- Owners should ensure equal treatment for the competing teams in respect of design scope, compensation for the scope, schedule, and terms/conditions for the contract.

- All the competing EPC bids and FEEDs should be submitted at the same time. Sequential bidding and reverse auctions should be avoided. The contractors should be required to submit only one EPC estimate, on the basis that multiple EPC lump-sum estimates from each bidder compromises the competitive process.
- All contractors should be free to choose partners, suppliers, and subcontractors from an approved or prequalified list. Project sponsors should ensure key participants are free to compete in either design to avoid potential advantage or disadvantage of one design team over the other. If there are particular suppliers or subcontractors that have a definite commercial advantage, then they should be forced to work with each competing design team.

Even though design competitions have the potential to be very effective in achieving sponsor's EPC objectives and goals for the project, they have to be carefully conducted and managed. Trust and fairness ensure that all parties are aligned and work towards the benefit of the project. Structuring a design competition to achieve maximum benefit is not easy and requires the sponsor to balance their desire to specify the design versus encouraging the contractor to apply their knowledge and experience and be innovative.

11.22 Project management case study: Peru LNG

The Peru LNG Liquefaction Plant at Pampa Melchorita (Peru) was commissioned in 2009. A detailed case study describing the project and its evolution is provided by Bruce and Lopez-Piñon (2009). The information discussed here is drawn from that detailed study. The project is of interest because it was delivered on budget during a time of rapidly escalating costs, which impacted most other projects executed at the same time. The key stated project driver for the project management team was cost optimization and this worked in delivering the project at a unit EPC (and commissioning) cost of $420/ton of annual LNG production capacity (including both gas liquefaction plant and marine terminal). Mayer (2011) of CB&I quoted a $440/ton cost in 2008 dollars. This is at the lower end of a trend that extends up to some $1400/ton for other plants built at the same time elsewhere in the world. The Peru LNG project consists of the liquefaction plant, a marine terminal, and a gas supply pipeline. It was the largest single investment ever in Peru. At $3.8 billion, the total project cost including project finance costs constituted the largest project finance in Latin America, with loans obtained up to $ 2.2 billion.

The project was developed over eight years through a challenging period in which the project had to overcome:

- A steep global capital cost inflation trend
- Extended delivery times for equipment and materials
- Shortages of skilled labor and experienced contractors, due to competition from other projects around the world
- Stringent environmental and social conditions
- Opposition from some sectors of the community
- Difficult conditions for securing project financing.

Despite those difficulties PERU LNG project was completed as specified for production capacity, energy efficiency, and plant reliability; within the approved budget; and in accordance with the time schedule specified at FID.

Conceptual studies were conducted in 2001, which confirmed the technical feasibility and economic viability of the project. KBR performed a FEED study for the sponsors in 2002 based upon the proven and popular Air Product's C_3MR gas liquefaction technology. The FEED also confirmed that air cooling and gas turbine drivers were the most efficient and reliable solutions for the Peruvian plant.

A site for the liquefaction plant and marine export terminal was selected at Pampa Melchorita, 170 km south of Lima. In 2003 physical and geotechnical studies on the selected site—both onshore and offshore—and the Environmental Impact Assessment (EIA) was initiated and continued through 2004 and 2005 in order to obtain final approval by Peruvian authorities and potential lender agencies. During that period plant design optimization and technology improvement works were made to the FEED specifications.

While the EIA work was conducted PERU LNG negotiated the Commercial framework for the project. Three keenly negotiated agreements were established:

- A feedgas agreement for the supply of 620 MMscfd with the joint venture holding the Camisea gas fields. This involved a gas sales agreement signed in 2005 for gas to be acquired FOB the gas treatment plant at the gas fields some 600 km from the Peru LNG site, with initial supply of gas scheduled for March 2010.
- All LNG produced by the liquefaction plant was sold to Repsol—with a minimum annual quantity of 4.2 MTPA—through a FOB gas sales and purchase agreement (SPA) signed in June 2005, with the first cargo lifting date set for May 2010.
- The legal framework for the project, an agreed stable fiscal regime over the life of the project, investment requirements (local content), and some other supporting provisions, was executed between Peru LNG and the government of Peru, in the form of an Investment Agreement (contract) executed in January 2006.

These three critical agreements were supplemented by two further commercial agreements that bolstered the LNG offtake agreement:

- In September 2007 REPSOL entered an agreement with Mexican state-owned "Comision Federal de Electricidad" (CFE) to supply LNG to the Manzanillo regasification terminal, located on the Mexican Pacific coast. Initial regasification capacity of that terminal is 500 MMscfd. Construction of the Manzanillo terminal started in July 2008 with a completion date scheduled for the end of 2011 with an estimated investment of over $700 million (it was actually delayed by a few months in coming online).
- An agreement for the LNG tankers for this trade to be supplied by STREAM (50% Repsol, 50% Gas Natural of Spain) under long-term charters with ship owners. Three newly built dual fuel vessels are dedicated to this project, with an additional vessel available if necessary. Each vessel is capable of loading 173,400 m^3 of LNG. Expected delivery date for the first LNG tanker is March 2010.

Peru LNG executed extreme care when selecting and organizing its project management team. The emphasis was on a team of experienced and professional individuals within a simple team organization structure. Classic project management techniques were to be the backbone of the project management philosophy, namely:

- Project planning
- Task scheduling

- Cost budgeting
- Design optimization
- Contract strategy
- Risk profile definition
- Contractor selection.

Prioritizing these items played a fundamental role in the success of the PERU LNG project.

The most challenging difficulty faced by the gas liquefaction plant project was the high seismic activity exposure of the site location. As an example in August 2007, during initial construction works, the Peruvian coast experienced an earthquake registering 7.9 degrees on the Richter scale with its epicenter less than 30 km away from the site. Consequently the plant had to be designed to safely withstand an 8.6 degree earthquake with neither damage nor loss of containment. Specially designed tank foundations (e.g., including a tank base with special seismic isolators), thicker concrete rebar, increased sections of structural steel, and bracing of equipment and valves were some of the measures implemented to deal with the issue.

Other challenges included:

- Numerous geological and terrain hazards along the entire feedgas pipeline route varying from very steep slopes to unstable slopes and landslides, and from moving sand dunes to torrential rivers. This PERU LNG pipeline is one of the highest altitude pipelines ever built reaching an elevation of 4,901 m above sea level. The Peru LNG pipeline has a diameter of 34 in and is 408 km long. It is currently capable of delivering 1.2 billion scf/day at the Pampa Melchorita delivery point, of which approximately 700 MMscfd are dedicated to the LNG production and up to 500 MMscfd are for third-party transportation.
- The environmental impact commitments went far beyond the requirements of existing regulations and compliance with environmental standards in order to secure government and lenders' approvals. The additional requirements were specified especially on the social/economical side. In addition as work progressed local municipalities and local communities along the pipeline route expected the Peru LNG to fulfill their demands in terms of special taxes, monetary compensations, additional jobs for their people, or social investments. Managing these demands and requests in a way to minimize time delays and additional project costs required care and diplomacy skills.

The gas liquefaction plant site occupies 520 ha of desert land adjacent to the Peruvian coast 170 km south of Lima, at 135 m average altitude. It lies just off the Pan-American Highway, in between the towns of Cañete (30,000 inhabitants) and Chincha (over 100,000 inhabitants). The site is 40 km north of Pisco port where most equipment and materials for the project were unloaded during construction time. Climate is very mild with temperatures never exceeding 28°C nor dropping below 12°C, and practically no rain year round, though humidity is always well over 80%. Besides the highway no other infrastructure was available at the site.

Feed gas supplying the plant is very lean and dry (well above hydrocarbon and water dew points) with C_3+content below mol. 0.02 %. Consequently no NGL fractionation is necessary at the site. The plant is self-sufficient as it generates all the electricity it requires and produces all the water, nitrogen, and compressed air that are needed, as well as taking care of liquid effluents and solid wastes. A permanent community with housing and recreation facilities for all members of the operational staff is part of the project.

The current project consists of the installation of a single liquefaction train, though provisions for expansion to include and tie-in additional trains were part of the design. The LNG loading line slopes down 140 m from the tanks to the beach and then it extends for another 1.3 km up to the loading arms; PERU LNG adopted a special piping design for this line and incorporated quick opening valves and a surge drum at the loading platform.

The marine terminal is designed to handle LNG tankers ranging in capacity from 90,000 up to 175,000 m^3. It consists of following components:

- A 1.3 km long jetty into the Pacific Ocean
- The loading platform and ships dock
- Four breasting and six mooring dolphins
- 3.6 km dredged navigation channel 18 m deep
- 800 m long breakwater protecting the whole harbor
- Tug boat dock, with a 200 m long breakwater
- Mooring and navigation aids and ancillary equipment.

The 1.3 km long trestle consists of a two-lane road and a steel rack containing the LNG, with boil-off and other pipes running all the way to the loading platform. Both road and rack are supported by an extra strong steel structure that is anchored to the sea bottom by 550 steel piles that were driven into the soil down to the rock substratum. The main breakwater is located 1,600 m from the shore at more than 15 m of water depth, and stands up to 11 m over the lowest (at spring tide) water level in order to guarantee protection to the docked LNG tankers against the incoming ocean waves—up to the 100-year return design level.

The design criteria agreed upon was that the liquefaction plant should produce LNG at a rate above contracted sales (not less than 4.2 MTA net of losses) in an efficient manner (i.e., total fuel consumption + losses under 9%) with a high plant availability (i.e., above 94%/year) and ensuring that it would achieve that in a safe manner and free of environmental impacts (i.e., tightly controlled emissions and prevention of spills). A financing objective was to involve 40% equity and secure 60% debt finance to support the capital investment.

The conditions of the investment agreement with the government included:

- Creation of 12,000 direct jobs plus an estimate of 30,000 indirect in the country, during construction (at peak time).
- Contracts awarded and goods purchased from Peruvian companies for more than \$1 billion.
- A net contribution of 0.6% growth to the nation GDP.
- Job training for thousands of Peruvian nationals in the fields of civil engineering, pipe and structural welding, electrical and instrumentation, etc.

In addition it was estimated that some \$7.8 billion in royalties and taxes to the government would be generated by LNG sales throughout the life of the project.

Following EIA approvals and supply chain commercial agreements being secured, project implementation commenced in earnest in mid-2005. At that time several activities were launched, namely:

1. A project management organization was established and key positions filled.
2. Plant site preparation works (Phase I) were awarded.

3. Gas liquefaction plant EPC lump sum turnkey bids were requested from KBR (open book).
4. Bid packages were prepared for planned early awards as follows: Site preparation (Phase II) to several local bidders, MCHE supply (Air Products), and Refrigeration compressors supply (G.E. and Nuovo Pignone).
5. Marine terminal EPC package was prepared for launching a lump sum turnkey bid request (several international bidders).
6. A pipeline engineering bid package was prepared and sent to several bidders.

Almost two years were consumed in the bidding process for the gas liquefaction plant Engineering, Procurement, Construction and Commissioning (EPCC) contract, with lengthy negotiations being undertaken with KBR and Chicago Bridge & Iron (CB&I). Finally in December 2006 PERU LNG awarded the work to CB&I on a lump sum basis, with Notice to Proceed issued in January 2007, for a price of $1.6 billion (wholly including MCHE (Air Products) and Refrigeration compressors (Nuovo Pignone) early awards). CB&I then had 40 months EPCC execution time to First LNG Cargo.

Following detailed engineering at its London and Chicago offices, CB&I procured and installed following main equipment and liquefaction plant components:

- Process and pressure vessels: 44
- Heat exchangers (including MCHE): 19
- Air coolers: 228
- Pumps and compressors (with drivers): 57
- Electrical units: 295
- Tanks (including those built on site): 16
- Valves: 3,610
- Instruments: 1,870
- Buildings (process and nonprocess): 24, plus the permanent accommodation for the plant operating community.

Only four significant cost deviations occurred in the total project (i.e., pipeline, liquefaction plant, and marine terminal) from when it was approved in 2006, namely:

- An increase in marine EPC contract (CDB) price driven mainly by a geotechnical problem that was only evident during construction.
- An increase in pipeline construction contract (Techint) price due to unexpected work stoppages caused by local communities along the rights of way (ROW).
- An increase in staff and third-party services costs as a consequence of additional archaeological and social work in relation to the pipeline project.
- A decrease in financing costs mainly due to lower than predicted interest rates during construction; the savings in financing just offset the preceding three increases in project costs.

The gas liquefaction plant EPC lump sum price contract with CB&I did experience several change variations during project execution but these did not significantly impact the contract value from its award value (i.e., $1.6 billion budget; Dec, 2006). PERU LNG managed to deliver this gas liquefaction plant, capable of producing some 4.5 MTPA, by focusing on fundamental principles of project management (e.g., design optimization, anticipation in decision-making, competition in EPCs, and all other contracts and rigorous planning schedules). It also benefitted from full support and knowledgeable collaboration from all project stakeholders and exceptional performance from Peruvian construction workers.

Some key contributing factors to this successful outcome, as detailed by Bruce and Lopez-Piñon (2009) were:

- Cost reduction associated with a safe, reliable, and simple (1 train) design using proven technology.
- Establishing EPCC competition for the key contracts.
- Planning ahead; for example, timely land acquisition and securing rights of way (pipeline)
- Allocation of time and resources to commercial and contractual negotiations.
- Anticipation of time-consuming processes in the environmental impact studies, meeting corporate social responsibility obligations, and securing project finance agreements.
- Early selection of gas liquefaction plant site.
- Different fit-for-purpose contracting strategies adopted, based on previous experience, for the pipeline, liquefaction plant, and marine terminal components of the project. The initial open-book EPCC contract negotiations with KBR for the liquefaction plant failed to deliver acceptable commercial and contractual terms and were replaced promptly with open-book negotiated agreements with CB&I.
- Careful capital cost benchmarking against other comparable-scale liquefaction projects being specified in the 2003 to 2005 period (e.g., Damietta in Egypt and Skikda and Arzew plants in Algeria) proved vital in establishing a realistic project budget.

Taking into account the risks that contractors were taking in adopting lump sum turnkey contracts for EPC projects in an environment experiencing rapid inflation and rising prices for raw materials in 2005–2006, it is not surprising perhaps that EPC negotiations proved difficult for both sponsor and contractors.

Other published case studies providing significant insight to LNG facilities project management recommended for review are Technip's account of delivering mega-liquefaction projects in Qatar (Bamba et al., 2011) for Qatar Petroleum and ExxonMobil and the description of the ground-breaking Atlantic LNG train 1 project in Trinidad (Redding and Richardson, 1998), deploying the Phillips optimized cascade liquefaction technology for the first time for sponsors BP and partners.

11.23 Ensuring continuity from project construction to plant operations

Handing over from the development and construction workforce to a plant's operational team can be one of the most challenging aspects of managing any project (Gas Strategies, 2010). Capturing and transferring highly-detailed and specific knowledge of a new plant, its machinery, its suppliers, and its commercial contracts can be a complex, subtle process. It is important for ensuring continuity, safety, and long-term plant performance.

Training and retaining operations staff in remote facilities where local skills may be in short supply is not an easy task. This can be complicated by rules that require skilled staff to be primarily drawn from the local and national population. Good terms and conditions, long-term prospects and desirable housing and, potentially, schooling, onsite for employees and their families are required to retain skilled plant operators.

Operations personnel are often embedded into inspection and construction teams leading up to a project's startup. This ensures that final HAZOP testing, precommissioning, and commissioning are all witnessed, and in some cases, primarily handled by the core of the future operations team.

Operations management systems (OMS) are an essential part of ensuring that plant operations and the operating team are fit-for-purpose and able to operate the plant safely according to clearly specified guidelines (e.g., KBR, 2012a). Addressing the OMS early ensures that the system is ready to use during the critical facility-commissioning phase and helps avoid startup delays and incidents. A draft OMS should be one of the deliverables of the precommissioning activities and further developed by the future operating team. Indeed some projects recognize the need for an OMS and start to plan and design it as early as the FEED stage of a project. Such an approach ensures that the OMS is developed and available when required and provides accurate, real-time information to enhance decision-making.

An operator training simulator (OTS) can also be developed and maintained for training of new plant operators (e.g., KBR, 2012b) and analyzing plant performance under specific conditions relevant to planned operations. The long-term benefits of investing in an operator training simulator according to KBR are:

- Operator best practices are reinforced through refresher training
- Workforce flexibility is increased through cross-trained operators
- An opportunity to test trial modifications to the plant's control and automation systems and process configurations is provided
- Development and refinement of operating procedures is facilitated.

An OTS ensures plant operators receive plant specific and realistic hands-on training ahead of plant startup and throughout plant operation.

11.24 Integrated approaches to achieve high standards of performance

Achieving a smooth transition from construction to commissioning and startup, followed by ramp-up to full capacity and achievement of the operational performance set out in a project plan, is a challenge for any oil and gas project, but particularly so for large gas liquefaction plants. Lessons need to be learned from previous projects and other operators. Industry best practices need to be established, deployed, and benchmarked.

Preparation for commissioning, startup, and production process operations must be planned in an integrated fashion at an early stage of project planning. Some organizations rigorously implement "flawless start-up" or "perfect plant" methodologies through all project phases through to operations. The integrated implementation of operational-excellence programs before startup is useful in achieving specified design objectives without encountering problems at the commissioning and operations stages.

Central to a flawless startup program is the systematic application of lessons learned from earlier projects and earlier phases of projects as they progress. An extensive "flaw database" aids an integrated approach. This is built up from the experience of an organization or project team, containing lessons learned on all kinds of related projects, which helps to identify potential flaws. It can be complemented and added to by the experience of the contractors, subcontractors, and suppliers involved in the project. Flaws can then be systematically scanned for across the facility components. To be successful this approach needs to be embedded deep into the supply chains of the EPC contractors during construction as a quality assurance measure.

Figure 11-11 illustrates the functions, management, and reporting activities that require integration. No single team in the plant design, construction, and operational life cycle has control over all factors affecting functional and financial performance of a facility. However, the facility sponsor through its management system needs to ensure their integration and overlap, which will require appropriate information technology and database systems (Lauzon, 2008). Isolation between the many groups involved in designing, building, operating, and maintaining such facilities wastes time and money, jeopardizes plant safety, and limits the owner-operator's ability to optimize plant performance (Snitkin, 2007).

To prepare for an intended flawless startup and subsequent high-performing plant operations, all project and operations staff need to be trained and provided with tools to enable them to detect and root out defects and potential problems that are often concealed within the process designs.

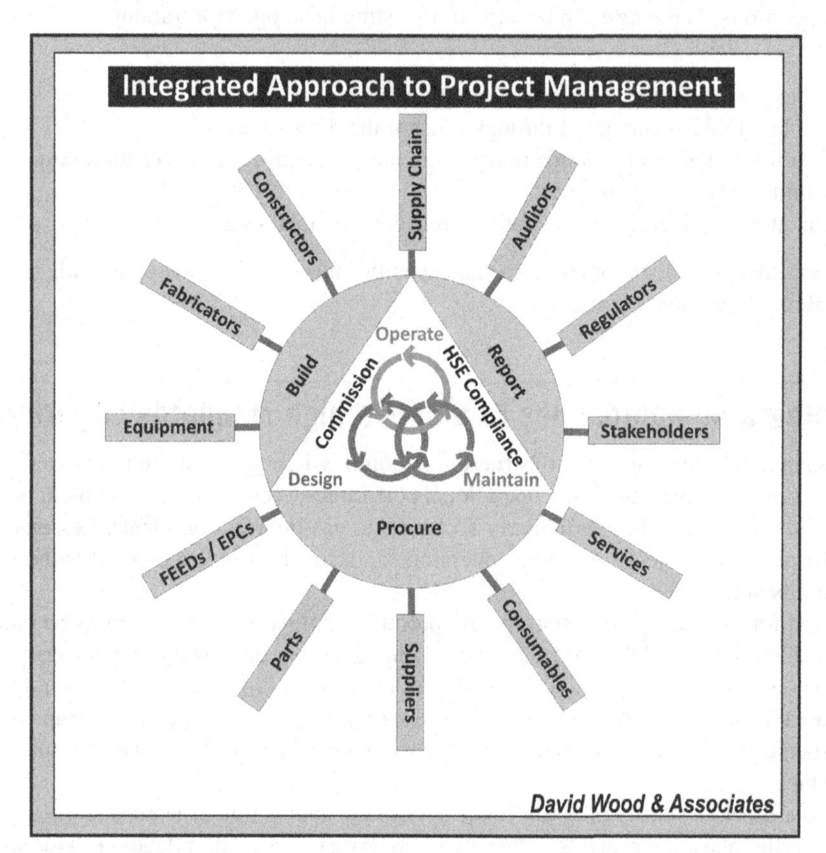

FIGURE 11-11

An integrated approach to design, construction, operations, and maintenance is more likely to result in high performing LNG facilities.

(Modified after Lauzon, 2008)

Recruiting and training quality staff is essential to the success of LNG plant projects and the long-term operations that need to follow on from a project's development. Because of the breadth and scale of activities in large projects it is important to be able to attract the required numbers of competent staff with relevant experience from recently completed, similar, or related projects. Continuity of core staff is essential and several project team members ideally need to remain in the permanent operating organization well into the operational phase of a new facility. Recruitment of experienced operational staff needs to begin early in project development (i.e., not long after a positive final investment decision (FID) is made). Key operators and technicians need to be recruited, trained, and mobilized to a project site long before the commissioning phase.

An integrated project team can influence long-term project value and return on capital invested in some key ways:

- Define responsibilities, accountabilities, and interfaces across the team at an early stage in the project
- Measure value and risk across all component parts of a project (using a comprehensive risk register) to focus attention
- Manage and consult with all project stakeholders with a professional project culture
- Align the vision of the project objectives across the project team and repeatedly reinforce that vision
- Focus on how value can be improved and risk reduced at all stages
- Rigorously and consistently deploy constructability principles to drive project execution.

11.25 Conclusions

Detailed, systematic, team-involved plans are the foundation for project success in the LNG industry. Project success means completing all project deliverables on time, within budget, and to a level of quality that is acceptable to the project sponsors and stakeholders. The project manager must keep the project team's attention focused on achieving these broad goals and the stakeholders aligned to the project objectives. Project risk assessment, mitigation strategies derived from scenario-based risk identifications, and attention to contingency plans for dealing with low-likelihood extreme risk exposures play a key role in the successful design and delivery of LNG projects.

Attention to constructability principles in the planning stages can also reduce the risks of cost, time, and quality problems during the project construction phase. LNG facility construction projects are not complete until the plant, unit, or equipment is placed in service and has demonstrated that it can deliver efficiency levels and uptime availability at or above its design specifications. It may take more than a year after a facility is commissioned and starts operations before acceptable full-capacity performance levels can be established and verified. At that stage the long-term operating team must to an extent live with what a project team has handed to them.

References

Bamba, M.O., Millot, N., Mahe, H., Sept. 21–23, 2011. The challenges of Qatar mega LNG Projects, paper presented at the GPA Europe Annual Conference, Prague, Czech Republic.

Bruce, B., Lopez-Piñon, C., Oct. 5–9, 2009. Peru LNG: A Grassroots Gas Liquefaction Project Optimized for Cost in Difficult Times, paper presented at the 24th World Gas Conference, Buenos Aires, Argentina.

Durr, C., Hill, D., Shah, P., March 21–24, 2004. LNG Project Design Competition—A Contractor's Viewpoint, paper presented at the 14th International Conference & Exhibition on Liquefied Natural Gas (LNG-14), Doha, Qatar.

Elkington, J., 1997. Cannibals with Forks: The Triple Bottom Line of 21st Century Business. Capstone Publishing, Oxford, UK.

Gas Strategies How to Build an LNG Operations Team. LNG Business Review, Jan. 2010.

KBR, 2012a. Operations Management Systems. http://www.kbr.com/Services/Advanced-Chemical-Engineering-Services/Operations-Management-Systems/.

KBR, 2012b. Operator Training Simulators. http://www.kbr.com/Services/Advanced-Chemical-Engineering-Services/Operator-Training-Simulators/.

Lauzon, S., 2008. The Path to the Perfect Plant. Chemical Engineering 115 (9), 36–42.

Mayer, M., March 21–24, 2011. The Challenge of Building South America's First LNG Production Facility, paper presented at the Gastech 2011 Conference & Exhibition, Amsterdam, The Netherlands.

McIntire, P., Aug. 13, 2001. Cost Estimating Challenges Face Frontier Projects. Oil & Gas Journal 99, 33.

Persaud, G.A., Chamberlain, S., Kauffman, S., March 30 –April 3, 2003. Safety Drivers in the Lay-out of Floating LNG Plants, paper presented at the 3rd Topical Conference on Natural Gas Utilization, AIChE Spring National Meeting, New Orleans, LA, USA.

PMBOK Guide, 2013. In: A Guide to the Project Management Body of Knowledge, fifth ed. Project Management Institute (PMI), Newtown Square, PA, USA.

Redding, P., Richardson, F., Nov/Dec, 1998. The Trinidad LNG Project: Back to the Future. LNG Journal.

Snitkin, S., Aug. 2007. Improve the Functional and Financial Performance of Your Plant through Better Plant Asset Information Management. ARC White Paper. ARC Advisory Group, Washington, DC, USA.

Skramstad, E., Blikom, L.P., Wredmark, J., July 1, 2011. Managing LNG Project Risk. Petroleum Economist.

Wood, D.A., 2001. Probabilistic Methods with Simulation Help Predict Timing, Costs of Projects. Oil & Gas Journal 99 (46), 79–83.

Wood, D.A., Oct. 2011. Is the Oil and Gas Industry Adequately Handling Exposure to Extreme Risks? World Oil 232 (10), 113–118.

Wood, D.A., Mokhatab, S., 2006. Challenges, Risks Can be Managed in Deepwater Oil and Gas Projects. Oil & Gas Journal 104 (44), 37–42.

Wood, D.A., Lamberson, G., Mokhatab, S., 2007. Project Risk: A Key Consideration for Upstream Project Management. World Oil 228 (9), 127–130.

Wood, D.A., Lamberson, G., Mokhatab, S., 2008a. Better Manage Risks of Gas-processing Projects. Hydrocarbon Processing 87 (6), 124–128.

Wood, D.A., Lamberson, G., Mokhatab, S., 2008b. Project Execution Risk Management for Addressing Constructability. Hydrocarbon Processing 87 (12), 35–42.

Wood, D.A., Lamberson, G., Mokhatab, S., 2011. Staffing Strategies for Large Projects Must Tackle Many Diverse Issues. Oil and Gas Financial Journal 8 (9), 35–40.

Yale, R., Knudsen, J.I., Jan 2006. Business-risk Management: An Essential Tool. World Oil 227 (1), 7–10.

APPENDIX

An LNG Primer: Basic Facts, Safety and Security Clarifications

1

This section is structured in the form of answers to frequently asked questions about LNG posed by those with little or no previous understanding of the LNG industry.

A1.1 Background of LNG

Liquefied Natural Gas (LNG) is the liquid form of natural gas. LNG is a natural gas, just like the gas produced and delivered by pipelines to energy markets around the world.

It is called liquefied natural gas because it is a liquid. When natural gas is cooled at atmospheric pressure to very low temperatures (about –162°C, or about –260°F, depending on the composition of the natural gas) it condenses into a liquid. The critical temperature and pressure of natural gas are around –82°C and 46 bar, again depending upon the exact composition of the gas.

Natural gas is composed primarily of methane (typically at least 90%), but may also contain some heavier hydrocarbons such as ethane, propane, or butane, and typically less than 1% nitrogen. Prior to liquefaction natural gas is typically conditioned to remove any oxygen, carbon dioxide, sulfur compounds, other trace impurities (such as mercury), and water. Removal of these impurities eases the handling of the LNG formed by minimizing corrosion or damage to materials used for its containment.

A1.1.1 Why LNG?

It typically takes about 600 m^3 of natural gas to yield 1 m^3 of LNG, with 1 tonne of LNG holding the energy equivalent of some 50,000 cubic ft of natural gas. Exact conversions depend upon the composition of the natural gas in question. It is the large contraction in volume from the gaseous state to the liquid state that makes LNG much easier and more economical to transport over large distances and store in large quantities.

A1.1.2 Is LNG explosive?

LNG is not at all explosive or flammable in its liquid state.

A1.1.3 What happens when LNG is warmed?

As LNG vapor warms above –160°F (–106.7°C), it becomes lighter than air and will rise and disperse rather than collect near the ground. However, it is not explosive unless flammable concentrations of gas occur in enclosed or otherwise confined spaces.

A1.1.4 How much energy does it take to make LNG from natural gas?

Typically about 10 to 20% of gas delivered into an LNG supply chain is consumed in the process and transportation facilities. This is comparable with long-distance high-pressure gas pipelines.

A1.1.5 What are the advantages of storing gas as LNG?

LNG facilities offer two clear advantages over alternative gas storage options:

- Because LNG facilities can be located above ground, operators and/or owners have many more opportunities for locating LNG facilities in comparison with traditional underground gas storage alternatives that depend on underground geological conditions such as depleted reservoirs, aquifers, and salt caverns.
- LNG facilities are often constructed with a higher degree of "deliverability" (the amount of gas the facility can send out under peak conditions relative to stock in inventory) than traditional underground storage facilities. This deliverability provides the opportunity to meet demand spikes, sometimes called "needle peaks."

A1.1.6 What is the history of LNG?

LNG has been used for more than 50 years, especially in Asia, Europe, and the United States. Improved technology is now making it more economical to produce, transport, and store LNG in large quantities. These economies of scale have opened up wider markets for its use where it can compete effectively on price with other sources of fuel.

LNG fed by gas from large gas fields remote from gas markets has become an increasingly attractive alternative to oil or piped gas (natural gas transported from its country of origin through pipelines). Natural gas as a "clean" source of energy is becoming the fuel of choice or preference in many regions if significant volumes can be brought reliably and at competitive prices into a market enabling it to compete with coal, oil, and petroleum products for power generation and industrial and commercial fuels.

There is an increasing need globally for diversification of energy supplies due to politics, economics, and reserves. LNG carriers can be diverted to a number of LNG consuming countries easily providing higher confidence in security of supply for major gas-importing nations.

A1.1.7 How can we keep LNG cold?

LNG is stored in large insulated tanks that are designed to minimize any heat ingress. The insulation of the tanks, as efficient as it is, does not keep the temperature of LNG at cryogenic temperatures by itself. LNG will stay at near constant temperature if kept at constant pressure. This phenomenon is called auto-refrigeration.

As long as the LNG vapor boil-off is allowed to leave the tank in a safe and controlled manner, the auto-refrigeration process will keep the temperature constant. This vaporization loss is typically collected as it leaves the tank and either reliquefied, sent to the gas line connecting to a gas distribution network, or used as fuel on the site or to power an LNG carrier ship.

A1.1.8 What are the differences between LNG and LPG or LNG and NGL?

All are light liquid hydrocarbons that can be used as fuel. To most nonengineers, the terminology is confusing. Liquefied petroleum gas (LPG) is composed primarily of propane (upward of 95%) and smaller quantities of butane. This is quite different from the primarily methane composition of LNG. LPG is used primarily as residential fuel, petrochemical feedstock and often is used as vehicle fuel. In fact LPG is a cleaner burning liquid fuel than gasoline.

Natural gas liquid (NGL) is a light hydrocarbon mixture that may also consist of ethane, propane, butane and traces of condensate (heavier gasoline range hydrocarbon) components. Ethane can be used for petrochemical production and the remaining portion can be sold as LPG.

LPG can be maintained as a liquid by means of elevated pressure alone or by chilling to temperatures to around –40°C. On the other hand, it is not possible to liquefy natural gas (methane) at ambient temperature even at elevated pressure.

Despite perceived safety concerns, LNG is also safer to handle than LPG in most circumstances. Because LPG is heavier than air, it "hangs" low to the ground if leaks occur in storage facilities. A leaked, low-lying cloud of LPG is more easily ignited than LNG. In contrast, LNG vapor (primarily methane) is lighter than air. As a result, the revaporized gas stream in an uncontained condition typically floats away into the atmosphere and poses a much lower threat of fire or explosion.

A1.1.9 What are the sources of LNG?

Qatar, Indonesia, Australia, Malaysia, Trinidad, Algeria and Nigeria are leading exporters of LNG. There are in total some nineteen countries that export LNG. LNG is imported by many countries and in particularly large quantities by Japan and South Korea and some European countries with China and India increasing their demand for LNG quite rapidly.

Growth in LNG applications depends on expansion of current facilities and new construction and infrastructure investment along the LNG supply chains. The industry has experienced growth of some 7.5% per year for the past 20 years and investment commitments suggest this rate of growth will continue for the next decade at least.

The continued expansion and diversification of LNG supply indicates that sources of LNG will become more readily available in gas consuming markets around the world. The emergence of new LNG import markets in South America (i.e., Argentina, Brazil, and Chile) and the Middle East (i.e., Dubai, Kuwait) in recent years is testament to the diversification of the industry.

Floating regasification and LNG storage units have enabled smaller markets to secure LNG imports quickly (i.e. in less than a year) without building land-based LNG receiving terminals. This approach has now become popular with countries seeking to import occasional or seasonal LNG cargoes.

Demand for LNG is expected to increase as emission restrictions favor gas over coal for power generation, and gas supply companies make inroads into niche markets such as road vehicular fuel, as a marine vessel fuel, and as LNG replaces propane as a fuel for facilities not connected to the pipeline gas grid.

A1.2 LNG supply chain

The LNG supply chain includes all the facility and equipment involved in taking natural gas from an underground reservoir, liquefying it, and transporting to an end-user customer of natural gas.

That supply chain is typically long in terms of distance and expensive in terms of the capital costs of the equipment and facility involved. The components of the supply chain typically include:

- Gas field production infrastructure
- Feed gas pipeline to gas processing and conditioning plant
- A large-scale refrigeration plant involving heat exchangers to liquefy the feed gas
- LNG storage and port loading facilities: everything must be kept cold
- LNG marine tankers
- LNG receiving terminal including port unloading, LNG storage, regasification, and gas sendout compression facilities
- Connection to a natural gas transmission and distribution network to deliver gas to customers
- In some cases distribution of LNG by truck to small, remote off-grid gas customers.

A1.2.1 Can LNG compete commercially with pipeline gas?

Yes. An LNG supply chain (i.e., gas field development, liquefaction plant, transportation by LNG carrier, LNG terminal) is generally set up when pipeline transmission is too expensive due to the long distances involved or the technical/political difficulties of pipeline construction, or to enable the gas to be delivered to more than one geographic market location. LNG supply chains are much more flexible than gas pipelines, being able to serve different markets at different times and to avoid the political and geopolitical instability of transit countries that transcontinental gas pipelines have to deal with.

A1.2.2 What are the commercial terms?

In traditional LNG markets, buyers and sellers are generally linked by long-term contracts (i.e., 10 to 25 years duration typically) for predefined quantities of LNG produced in a liquefaction plant and received at an LNG terminal specified in the contract.

There are usually penalties for the customer not taking the contracted quantities—take-or-pay. The contracts usually specify a price often linked to benchmark gas prices or to other fuel (crude oil) prices or inflated from an initial floor price.

Rapidly growing short-term markets are changing this as more uncontracted LNG carrier vessels enter the market. However, the LNG market will remain dominated by long-term contracts for the foreseeable future. Short-term LNG cargo sales amount to about 15% of the overall LNG supply.

A1.3 LNG regasification terminals

An LNG regasification terminal is where the LNG is delivered to the end users, which typically comprises the LNG unloading jetty and LNG storage and sendout facility, along with heating to reconvert LNG back to natural gas. The regasification process is a heating process typically using ambient temperature heat sources. Most terminals use seawater for heating, and in some terminals ambient air is also used. In cold climate regions, fuel gas is necessary to supplement heating during the winter months.

The LNG terminals typically unload the LNG shipment in 10 to 12 hours, in order to minimize the docking times for the ships and reduce the operating cost of the ship.

A1.3.1 How are terminals designed?

All LNG storage facility designs must comply with stringent regulations as required by national planning legislation codes (e.g., in the United States: the US Department of Transportation (DOT)'s safety standards in Title 49 Code of Federal Regulations (CFR) Part 193—Liquefied Natural Gas Facilities: Federal Safety Standards and National Fire Protection Association (NFPA) 59A—Standard for the Production, Storage, and Handling of LNG).

In accordance with safety standards, vapor dispersion distances must be calculated to determine how far downwind a natural gas cloud could travel from an onshore storage facility and still be flammable. As required by these regulations, these exclusion zones must not reach beyond a property line where other development could occur.

Since a fire would burn with intense heat, each onshore LNG container and LNG transfer system must also have thermal exclusion zones established in accordance with prevailing safety standards. These exclusion zones must be legally controlled by the LNG facility operator, or a government agency, to ensure adequate separation between members of the public and the heat from a fire.

A1.4 Seismic design requirements

LNG facilities must meet stringent standards to ensure public safety and plant reliability in the event of an earthquake. Extensive studies of the geological conditions and earth history of a proposed LNG site are required to determine appropriate design loads on the critical components of the LNG plant, such as the design of the LNG storage tanks.

A1.4.1 What gas markets do LNG regasification facilities serve?

Some LNG facilities have the flexibility to participate in several markets at once. For example, some fill both base load and peak shaving roles. Some also provide LNG for commercial vehicle fuel which is growing in demand since the LNG fuel cost is significantly less than petroleum fuel that has been escalating in price.

A1.4.2 How long can LNG be stored at LNG regasification terminals?

Scheduling for both the arrival of the LNG shipment and the dispatch of the regasified product generally is necessary to maintain an optimum operation of the LNG facility. A balance of multiple sources of LNG supply and storage capacity to match the variations in consumption is essential to minimize the inventory shortages that might be brought about by weather or tanker-scheduling problems.

Typically, the regasified LNG is sent out to customers on a routine schedule under a contract that calls for a set daily volume. Consequently, the LNG may be in storage at a marine import terminal for only a few days and, depending on the terms of individual contracts and the time of the year, is seldom held for more than a few months, unless it is held only for emergency back-up.

A1.4.3 What is an LNG peak shaving plant?

Such facilities typically involve a small liquefaction unit linked with a large LNG storage tank and gas sendout facilities to a gas distribution network capable of responding to gas demand peaks or supply crises. However, some peak shaving plants receive their LNG in liquid form by ship or by truck from other LNG facilities and do not have liquefaction facilities of their own.

Typically, gas is taken from a pipeline supply and liquefied and stored as LNG at the peak shaving plant. LNG remains in storage for several months and in most cases is only used to supply the extreme demand periods or needle peaks of just a few days each winter.

Peak shaving plants are used by gas utilities and regional pipeline companies as a means of storing gas in liquid form for peak periods and emergency backup. They are usually located at strategic points within the supply network to enable rapid delivery of gas to key markets.

A1.4.4 Are there air emissions from an LNG regasification terminal?

During the operation of an LNG regasification terminal, atmospheric emissions are mainly combustion emissions resulting from the burning sulfur-free natural gas.

LNG terminals are typically subject to regulations set by a government environment agency. Authorization for a specified level of emissions is typically granted by such agencies once they are satisfied that best available techniques are being employed in the operation and the terminal to eliminate, minimize, and render harmless any resultant emissions to the environment.

A1.5 LNG's safety records

The LNG industry has a long and excellent safety record, due to strict industrial safety standards applied worldwide. Up to 2012 there have been some 50,000 LNG carrier voyages, without a significant accident or safety problem (i.e., loss of containment) either in port or on the high seas.

Two major accidents have impacted the LNG industry: Cleveland LNG storage facility, located in the United States, in 1944; and, Skikda liquefaction plant, located in Algeria, in 2004. These incidents are described together in the following, with lessons learned.

In most jurisdictions "safety case" and "environmental impact study" reports are required by the government's regulatory authorities before consent for building an LNG facility is granted. The safety case considers all aspects of management, handling facilities, and operation of the plant, particularly on potential accidents and how major accidents would be prevented. Fires are more likely at liquefaction plants than regasification terminals, but are extremely rare; for example, a fire (August 2003) shut down the seventh train of the Malaysian MLNG Tiga plant (Bintulu), with a capacity of 3.4 MTPA, for seven months.

As part of the safety case and environmental impact studies, "credible" LNG spill incidents are the subject of risk analyses to review and assess the suitability of the site and the design of an LNG facility and any access waterway and road network. Modern LNG facilities are designed and operated such that persons not involved in the operation of the facility (including ships) that are outside the clearly designated safety and exclusion zones would not be at risk should these credible incidents occur. Rapid response planning to potential incidents by emergency services and training of responders is also part of the safety systems put in place for LNG facilities (e.g., National Association of State Fire Marshals, 2005; Shelley, 2007).

A1.5.1 What caused the Cleveland LNG tank failure in 1944?

The East Ohio Gas Company built the world's first LNG peak shaving facility in Cleveland in 1941. It consisted of three small spherical LNG storage tanks covered by cork insulation and a mild steel outer shell. The tanks were supported by uninsulated mild steel legs. The facility adjoined a residential neighborhood. The facility was run without incident until 1944, when a larger new tank was added. Since stainless steel alloys were scarce because of shortages resulting from World War II, the new tank was built using a toro-segmented design using low-nickel content (3.5%) alloy steel. Shortly after going into service, on October 20, 1944, the tank failed. LNG spilled into the street and storm sewer system where vaporizing gas ultimately met an ignition source and ignited. The resultant fire killed 128 people.

The US Bureau of Mines investigation (1946) that followed showed that the accident was due to the low-temperature embrittlement of the inner shell of the cylindrical tank. The inner tank was made of 3.5%-nickel steel, a material now known to be susceptible to brittle fracture at LNG storage temperature (−260°F). Had the fourth Cleveland tank been built using appropriate materials, the investigation concluded that the tragic accident would not have happened. LNG tanks constructed around the world of 9%-nickel steel have never had a brittle crack failure over decades of subsequent operations.

Clearly there is no reason why such an accident should be repeated.

A1.5.2 What caused the Skikda liquefaction train fire in 2004?

On January 19, 2004, a leak in the hydrocarbon refrigerant system at one of the natural gas liquefaction units (Train 40) in Skikda, Algeria formed a vapor cloud that was ingested into the inlet of the combustion fan of a steam boiler. The hydrocarbon acted as increased fuel to the boiler causing a rapidly rising pressure within the steam generating equipment. The rapidly rising pressure quickly exceeded the capacity of the boiler's safety valve and the steam drum ruptured, tearing apart the boiler fire box and housing. The flames from the boiler firebox ignited the leaked refrigerant gas, which was confined by the equipment and structures in the area producing an explosion and an ensuing fire.

The explosion, along with the shrapnel from the ruptured steam drum, caused further damage to the process piping and pressure vessels in the immediate area leading to additional flammable fluid release. The fire took eight hours to extinguish. The explosions and fire destroyed a portion of the LNG plant and caused 27 deaths and injury to 72 more. No one outside the plant was injured nor were the LNG storage tanks damaged by the hydrocarbon explosions.

A joint report issued by the US Federal Energy Regulatory Commission (FERC) and the US Department of Energy (DOE) was issued in April 2004. The findings in that report indicate that:

- There were ignition sources in the process area.
- There was a lack of "typical" automatic equipment shutdown devices required by modern design codes.
- There was a lack of hazard detection devices, which should have provided advanced warning of the refrigerant leak and helped to prevent the explosion.

While Skikda liquefaction trains 10, 20, and 30 had been upgraded in the late 1990s, Skikda train 40 was, in fact, of an obsolete design and scheduled for demolition at the time of the incident. Train 40 was originally built in 1981 and not well maintained. The poor maintenance, obsolete design, and poor general condition of Skikda train 40 suggest that such an incident should not be repeated in a modern liquefaction plant. Nevertheless, the incident highlights the need for comprehensive maintenance schedules and appropriate hazard detection systems at LNG facilities.

A1.5.3 What other serious incidents have occurred at LNG regasification terminals?

In addition to Cleveland, there have been two other incidents attributed to LNG.

A construction accident on Staten Island in 1973 has been cited by some parties as an "LNG accident" because the construction crew was working inside an (empty, warm) LNG tank.

In another case, the failure of an electrical seal on an LNG pump in 1979 permitted gas (not LNG) to enter an enclosed building. A spark of indeterminate origin caused an explosion in the building. As a result of that incident, the electrical code was revised for the design of electrical seals used with all flammable fluids under pressure.

This record suggests that LNG offers a safe and reliable source of natural gas and facilities can be located close to urban areas with confidence that they do not pose significant safety risks.

A1.6 LNG carriers

LNG carriers have double-hulled containment systems to limit loss of containment in cases of collision or grounding. LNG carrier safety equipment includes sophisticated radar and positioning systems that alert the crew to other traffic and hazards around the ship. A number of distress systems and beacons will automatically send out signals if the ship is in difficulty. The cargo system safety features include an extensive instrumentation package that safely shuts down the system if it starts to operate out of predetermined parameters.

Ships are also equipped with gas and fire detection systems. Crews are extensively trained to maintain high levels of onboard safety and how to handle emergency situations if they should arise.

Three LNG cargo tank types currently used in LNG carrier ships are self-supporting spherical, self-supporting prismatic SPB, and membrane. The membrane tanks are the most common in vessels built over the past decade and very few vessels have been built with self-supporting prismatic tanks, although there is renewed interest in such designs for floating liquefaction vessels.

For all cargo tank types, penetrating one or more LNG cargo tanks in a collision or grounding requires the penetration of all of the following:

- The ship's outer hull
- The 3-meter or so space between the outer and inner hulls (the water ballast tanks)
- The inner hull
- The insulation system around the LNG cargo tank
- The secondary containment of the individual LNG cargo tank
- The insulation system around the primary containment
- The primary containment vessel wall of the individual LNG cargo tank.

A1.6.1 What are the sizes of an LNG carrier and an LNG terminal?

LNG import terminals are equipped with storage tanks capable of holding at least one tanker load of LNG, and most modern facilities typically have a capacity of at least two tanker loads.

Modern large LNG carrier ships or tankers commonly hold some 145,000 cubic meters of LNG in liquid form, which equates to about 3 Bcf (about 80 mcm) in gaseous form. LNG vessels range in capacity from 19,000 to 265,000 m^3. The largest (Q-max) vessels of up to 265,000 m^3 have come into service since 2008 as part of Qatar's LNG supply chains.

LNG carriers are large double-hulled ships, several hundred meters in length, which travel at average speeds of 17 to 20 knots (18 knots is 33 km/h). It takes around 10 hours to fill an LNG tanker with a capacity of 120,000 m^3. Larger vessels require higher capacity loading infrastructure to enable rapid loading of cargoes.

Although the storage tanks at an LNG marine terminal often function as LNG storage facilities, the principal operation of an import terminal is not for gas storage, but rather for receiving the water-borne LNG imports and then regasifying LNG for shipment via pipelines to customers.

A1.6.2 Have LNG carrier groundings and collisions occurred?

Very few incidents involving LNG ships have occurred and those that have occurred have not resulted in loss of containment. The following example illustrates the robustness of these vessels and that the piercing or breaching an LNG ship tanks is extremely difficult (Beale, 2006).

In 1979, the El Paso Paul Kayser, loaded with about 125,000 m^3 of LNG, was steaming out of the Mediterranean Sea from an Algerian port. It was traveling at approximately 19 knots off the coast of Gibraltar when it struck a rock outcropping below surface and gouged a 750-foot long scar in its hull. This serious marine incident did not involve a loss of cargo, nor did it result in a breach of an LNG tank. The grounding did not even penetrate the outer hull. Another ship was brought alongside; the cargo was pumped out of the El Paso Paul Kayser into the second ship. The El Paso Paul Kayser was righted and sent to the shipyard for repairs and eventually returned to service.

A1.7 LNG spills

In order to be ignited, LNG must first be vaporized (heated and returned to a vapor state), mixed with between 5 to 15% air, and come in contact with an ignition source.

Were LNG to be released onto the ground as a consequence of a leak from a storage tank, the heat from the ground surface would initially cause very rapid boiling of the LNG. As the ground cools the boiling rate of the LNG would reduce. The amount of vapor formed would be in direct proportion to the amount of LNG released, the rate of release, and the surface area covered in the release.

In an unconstrained open-air environment the cold gas vapor will condense most of the water (humidity) in the surrounding air forming a white vapor cloud. If unhindered, the cloud will drift in the direction of the wind, further mixing with the air and picking up heat from both the ground and the air as it moves. As the vapor cloud warms up, it will become buoyant (lighter than air) and rise into the atmosphere where typically it will gradually disperse without ignition.

LNG released on water acts very similarly to the initial release on land. Assuming a large volume of water, the vapor formation rate will remain high as the surface water that is cooled by the LNG sinks and is replaced by warmer water.

Hence, in unconstrained scenarios spilled LNG should not typically ignite. Ignition is possible only in certain zones of an LNG vapor cloud where mixing with air would produce a flammable gas in the to

15% concentration in air range. At the center of the cloud the air quantity is too low for ignition; at the outer limits of the cloud the air quantity is too high for ignition. Natural gas has an auto-ignition temperature range of approximately 590°C to 650°C (1100°F to 1200°F), which is higher than LPG and significantly higher than gasoline. Those other fuels also ignite with lower concentrations in air. For example, gasoline has a lower flammability limit (LFL) of only 1.4% and propane is 2.1%, meaning that both can be ignited with significantly lower concentrations in air than natural gas (which has an LFL of 5%). If the limited flammable portion of a natural gas vapor cloud in an unconstrained environment met an ignition source, it would burn but not explode.

In order for the ignition of an LNG vapor cloud to result in an explosion, the gas must first be uniformly mixed with air in the 5 to 15% range, confined in an enclosed space, and then ignited through contact with an ignition source. It is extremely unlikely that such conditions could all occur together in a modern LNG facility.

A1.7.1 What are the likely impacts of large LNG spills?

Over the course of the past decade Sandia National Laboratories (New Mexico, USA) have conducted a series of detailed risk analysis studies and provided safety guidance associated with the consequences of a large LNG spill over water from a marine vessel (Hightower et al., 2004, 2006; Luketa et al., 2008; Kalan and Petti, 2011; Petti et al., 2011). The studies examine the vulnerability of LNG tankers and the impact of an intentional or accidental event that spills a large amount of LNG into a carrier or onto the water. As part of the research, Sandia in December 2009 intentionally set a very large LNG fire in a 120 meter-diameter pool in New Mexico that was made for that purpose. The test LNG spill was 83 m in diameter and created a 56 m diameter fire. The test also showed that the fire would likely stay attached to the ship instead of floating away.

Modeling associated with Sandia's 2011 studies showed that about 40% of spilled LNG could stay within the ship, causing cryogenic and heat damage. At high temperatures the strength of the steel on the ship is much reduced. On the other hand the extreme cold temperature of LNG is likely to cause fractures in all structural elements in and around the ship that come into contact with LNG. The largest intentional breach events modeled would cause significant damage and make the vessel unseaworthy. The study considered a wide range of events, including shoulder-fired weapons, stinger missiles, backpack explosives, underwater events, and small aircraft.

The estimated thermal hazard distances, even from a pool fire associated with the largest capacity LNG carriers in operation, involve significant impacts to public safety and property contained within approximately 500 m of a spill, with lower public health and safety impacts at distances beyond approximately 1600 m. However, the studies found that pool fire and vapor dispersion hazard distances are significantly influenced by site-specific environmental, topographical, climatic, and operational conditions including the breach and spill size.

A1.7.2 What should be done if there is spill from an LNG delivery truck?

Truck loading and transport of LNG carries a higher risk than normal pipeline delivery, but involves a relatively small, limited amount of LNG. First responders need to be trained to treat such a spill because pouring water on an LNG pool results in a large increase in LNG vaporization. A better approach is to confine a spill with quick, simple methods such as placing sand bags in ditches.

Application of firefighting foam is not too effective once an LNG pool fire is ignited, but it can limit the evaporation rate of a pool prior to ignition. Blocking traffic is important to reduce ignition sources from hot engines and mufflers. If ignition occurs, an explosion is very unlikely unless the vaporized LNG cloud drifts indoors into buildings.

A1.7.3 What are the safety concerns of LNG spills?

Because of its cryogenic temperature (atmospheric boiling point approximately −260°F), LNG poses exposure concerns to employees, facility structure, and equipment. The design and operation of LNG terminals minimize ignition sources; thus, cryogenic exposure is more likely than a fire incident. This is particularly true in the high-pressure processing areas where the fluid inventory is lower but where the higher pressure creates greater potential for cryogenic exposure.

Cryogenic exposure can cause freeze burns to employees. Gas from LNG vaporization is extremely cold and can produce irreparable burns on delicate tissues such as those of the eyes. Therefore unprotected parts of the body should not be allowed to touch uninsulated pipes or vessels containing LNG.

The cold LNG vapor can also cause embrittlement to carbon steel, thus possibly resulting in structural failure. Protection from cryogenic exposure with insulation, as well as from fire exposure, is installed in the facility. Protective measures should be chosen that are effective for both fire and cryogenic exposure.

The other safety concern is asphyxiation. The normal oxygen content of air is 20.9% in volume. Atmosphere containing less than 18% are potentially asphyxiant. In the case of gas leakage, the high concentration of gas can cause nausea or dizziness from anoxia. In an LNG facility, oxygen and hydrocarbon content of the atmosphere are constantly monitored to detect any gas leakage. For maintenance of equipment, the operator is equipped with instruments to ensure sufficient oxygen is present before entry to any equipment.

A1.7.4 How would an LNG facility be safeguarded against damages from an LNG spill?

Direct contact of LNG with structural steel can rapidly cool the material to below embrittlement temperature and may quickly cause failure in a short time. For this reason, the construction and protection system material is selected to withstand cold temperatures and must comply with the most stringent LNG safety standards. Insulation, shielding, and detection systems are provided in the facility to limit the volume of LNG release and mitigate the spread of LNG over a greater area. In the United States, NFPA-59A is one of the key design documents for the design of the LNG facilities. In Europe, BS EN-1473 is normally used. Both US NFPA-59A (2009) and BS EN-1473 (2007) require that equipment and structures whose failure would result in incident escalation must be protected from cryogenic embrittlement.

A1.8 Security for LNG facilities and ships

Each country has its own regulations and agencies responsible for marine vessel security. For example, in the United States FERC is among several federal agencies overseeing the security of LNG terminals and peak shaving plants. The Coast Guard has responsibility for LNG shipping and marine terminal security (FERC, 2005). DOT's Pipeline and Hazardous Materials Safety Administration (PHMSA) and the Department of Homeland Security's Transportation Security Administration (TSA) have security authority for LNG peak shaving facilities. In addition to federal agencies, state and local authorities provide security assistance at LNG facilities.

Security measures for both onshore and offshore portions of marine terminals are required by Coast Guard regulations under the Maritime Transportation Security Act. Requirements for maintaining security of LNG import terminals are in the Coast Guard regulations at 33 CFR Part 105. The Coast Guard keeps other ships and boats from getting near LNG vessels while in transit or docked by enforcing Regulated Navigation Areas and security zones. The Coast Guard performs a number of important security and safety checks before allowing an LNG tanker to enter a port and unload its LNG. Facilities are required to have a written security plan and an emergency response plan. FERC, DOT, and the Coast Guard require LNG companies to contact and coordinate procedures with local response organizations.

A1.9 Risk of terrorism adds new dimension to LNG safety risk

The LNG business has an admirable safety record overall but a whole new dimension has been introduced since the terrorist attacks of September 11, 2001.

By their nature, LNG import terminals are likely to be near centers of population, and issues of public protection and public acceptance of new terminal proposals have high community profiles.

How much LNG could conceivably be released in a major incident? How quickly and how far might the pool spread? How fast does it vaporize to gaseous methane? And, what is the maximum size and intensity of any resulting fire? These are the questions typically addressed in accessing the potential impact of terrorist attacks. These are questions typically addressed in the planning, siting and design of modern LNG import terminals.

References

Beale, J.P., June 9, 2006. The Facts About LNG. CH·IV International Document: RPT-06903–01.

BS EN-1473, 2007. Installation and Equipment for Liquefied Natural Gas—Design of Onshore Installations. British Standard Institute (BSI), UK.

FERC, April 29, 2005. A Guide to LNG—What all Citizens Should Know, pamphlet produced by Federal Energy Regulatory Commission (FERC). Office of Energy Projects, Washington, DC, USA.

Hightower, M., Gritzo, L.A., Luketa-Hanlin, A., Covan, J.M., Dec. 2004. Guidance on Risk Analysis and Safety Implications of a Large Liquefied Natural (LNG) Spill Over Water. Report SAND2004–6258. Sandia National Laboratories, New Mexico, USA.

Hightower, M., Luketa-Hanlin, A., Gritzo, L.A., Covan, J.M., 2006. Review of Independent Risk Assessment of the Proposed Cabrillo Liquefied Natural Gas Deepwater Port Project. Report SAND2005–7339. Sandia National Laboratories, New Mexico, USA.

Kalan, R.J., Petti, J.P., Dec. 2011. LNG Cascading Damage Study Volume I: Fracture Testing Report. Report SAND2011–3342. Sandia National Laboratories, New Mexico, USA.

Luketa, A., Hightower, M., Attaway, S., May 2008. Breach and Safety Analysis of Spills Over Water from Large Liquefied Natural Gas Carriers. Report SAND2008–3153. Sandia National Laboratories, New Mexico, USA.

National Association of State Fire Marshals, May 2005. Liquefied Natural Gas: An Overview of the LNG Industry for Fire Marshals and Emergency Responders. National Association of State Fire Marshals, Washington DC, USA.

Petti, J.P., Wellman, G.W., Villa, D., Lopez, C., Figueroa, V.G., Heinstein, M., 2011. LNG Cascading Damage Study Volume III: Vessel Structural and Thermal Analysis Report. SAND2011–6226. Sandia National Laboratories, New Mexico, USA.

Shelley, C.H., Nov. 2007. Liquefied Natural Gas: A New Urban Legend. Fire Engineering, 107–112.

U.S. Bureau of Mines, Feb. 1946. Report on the Investigation of the Fire at the Liquefaction, Storage, and Regasification Plant of the East Ohio Gas Co., Cleveland, Ohio, October 20, 1944. U.S. Bureau of Mines, Pittsburgh, PA, USA.

U.S. Department of Energy (DOE), April, 2004. Report of the U.S. Government Team Site Inspection of the Sonatrach Skikda LNG Plant in Skikda, Algeria, March 12–16, 2004. U.S. Department of Energy, Washington, DC, USA.

U.S. NFPA-59A, 2009. Standard for the Production, Storage and Handling of Liquefied Natural Gas (LNG). Quincy, MA, USA.

APPENDIX

Modeling the Phase Behavior of LNG Systems with Equations of State[1]

2

A2.1 Introduction

The success of the design and operation of separation processes in the oil and gas industry at low temperatures is critically dependent upon accurate descriptions of the complex phase behavior and thermodynamic properties of the multicomponent hydrocarbon mixtures with inorganic gases concerned. For many years vapor-liquid equilibria problems in natural gas systems have been successfully solved. Yet, at present, interest has shifted to systems containing not only species in the simple paraffinic homologous series, but also to water (H_2O), carbon dioxide (CO_2), hydrogen sulfide (H_2S), hydrogen (H_2), nitrogen (N_2), to mention a few, in which multiphase phenomena can occur. For example, the situation of interest in natural gas processing is often a methane-rich stream in which the presence of a second solvent might alter dramatically the pattern of the phase behavior of the customarily liquid-vapor mixture and cause problems.

In this aspect, a neglected field of LNG equilibria work has been that portion devoted to vapor-liquid-liquid behavior. Process inefficiencies are often a result of this multiphase behavior, because the additional phase can cause unexpected separation problems.

In the cryogenic processing of natural gas mixtures, species such as CO_2 and the heavier hydrocarbons can form solids. The latter might coat heat exchangers and foul expansion devices, leading to process shutdown and/or costly repairs. Thus, knowledge of the precipitation conditions for gas streams is essential in minimizing downtime for cleanup and repairs. However, the appearance of a solid phase in a process is not always a liability. Off-gas (primary nitrogen) either from power plants with light water reactors or from fuel processing plants can contain radioactive isotopes of krypton and xenon, which can be removed by solid precipitation. The phenomenological aspects of LLV/solid-liquid-vapor (SLV) behavior are hence also interesting because there is a need for a better understanding of the physical nature of the thermodynamic phase space.

The aim of this appendix is to reveal, examine, and analyze the challenges and difficulties encountered when modeling the complex phase behavior of LNG systems, and to compare the

[1]This work was partially supported by the Mexican Petroleum Institute under Projects D.00484 and I.00432. The authors of this appendix wish to thank Professor Daimler N. Justo-García of National Polytechnic Institute of Mexico for his interest in this work and for the helpful discussions. They also thank Germán A. Ávila-Méndez for his assistance in the calculations.

capabilities of two numerical techniques advocated for phase behavior predictions and calculations of complex multicomponent systems.

The appendix is organized as follows. First, the thermodynamics and topography of the phase behavior of LNG systems are briefly presented. Then, two computational techniques to predict and model complex phase behavior of nonideal mixtures and their application to LNG systems are described, followed by presentation of a numerical procedure for calculating critical points of multicomponent mixtures at constant composition. Localization of a mixture critical point is critical for the correct construction of its phase envelope. The next section of the appendix is devoted to the description of a modified new method to directly calculate critical end points (K- and L-points). Next, two representatives of the equations of state (EoSs) type thermodynamic models, which are usually the primary choice when modeling LNG systems, namely SRK EoS (Soave, 1972) and PC-SAFT EoS (Gross and Sadowski, 2001) are briefly outlined. Examples of application and discussion of the phase behavior modeling for selected binary, ternary, and multicomponent systems of particular interest to the natural gas industry are presented.

A2.2 Thermodynamics and topography of the phase behavior of LNG systems

Though only a limited number of immiscible binary systems (methane–*n*-hexane, methane–*n*-heptane, to name the most prominent ones) are relevant to natural gas processing, LLV behavior can and does occur under certain conditions in ternary and higher LNG systems even when none of the constituent binaries themselves exhibit such behavior. It is also known that the addition of nitrogen to miscible LNG systems can induce immiscibility, and this necessarily affects the process design for these systems.

The qualitative classification of natural gas systems and the topography of the multiphase equilibrium behavior of the systems in the thermodynamic phase space and the nature of the phase boundaries will be addressed in this section.

Kohn and Luks (1976, 1977, 1978), who carried out extensive experimental studies on the solubility of hydrocarbons in LNG and NGL (natural gas liquids) mixtures, qualitatively classify natural gas systems (or any system for that matter) as one of four types (Kohn and Luks, 1976). A first type system, for example, methane-*n*-octane (Kohn and Bradish, 1964), has a solid-liquid-vapor (S-L-V) locus that starts at the triple point of the solute (the less volatile component) and terminates at a Q-point (S_1-S_2-L-V) near the triple point of the solvent (the most volatile component) with an S-V gap in the locus. The S-L-V branches terminate with a K-point where the liquid becomes critical with vapor in the presence of a solid. A second type system, for example, methane-*n*-heptane (Kohn, 1961), has a Q-point (S-L_1-L_2-V) in the central portion of the SLV locus, at which point one loses the L_1 phase and gains the L_2 phase (solvent lean phase). These systems also have a L_1-L_2-V locus running from the Q-point to a K-point where the liquid L_2 becomes critical with vapor in the presence of L_1. The other two types of systems have no discontinuities between the triple points of the solvents; however, methane-*n*-hexane (Shim and Kohn, 1962), a typical third type system, has a L_1-L_2-V locus that is terminated by critical points, where the two liquid phases, or a liquid and vapor, become critically identical. A typical system of the fourth type is ethane-*n*-octane (Kohn et al., 1976). The topographical evolution of multiphase equilibria behavior is discussed in detail by Luks and Kohn (1978). Ternary ($N = 3$) and higher systems exhibit similar types of phenomena, but the loci and exact points of a binary become $N - 1$ and $N - 2$ dimensional spaces in those more complex systems.

It is known that systems rich in methane can display L_1-L_2-V behavior of the second and third type variety (e.g., methane-*n*-heptane and methane-*n*-hexane, respectively). However, investigations into ternary S-L-V and more complex systems revealed that L_1-L_2-V behavior can occur in systems whose binary pairs exhibit no immiscibility (Green et al., 1976; Orozco et al., 1977). Hottovy et al. (1981, 1982) observed that systems of the first type (methane-*n*-octane) could behave like a second type system when a second solvent is added (e.g., methane-ethane-*n*-octane, methane-propane-*n*-octane, methane-*n*-butane-*n*-octane, methane-carbon dioxide-*n*-octane). Other ternary systems that also exhibit L_1-L_2-V behavior are methane-*n*-pentane-*n*-octane, methane-*n*-hexane-*n*-octane (Merrill et al., 1983), to name just a few. Adding carbon dioxide to a typical third type system methane-*n*-hexane also induces L_1-L_2-V immiscibility (Merrill et al., 1983).

The onset and evolution of LLV behavior in mixtures is related to the evolution of SLV behavior in those systems. Thus, in natural gas it is often of interest to predict whether a methane-rich stream with one of the solutes (such as an *n*-paraffin of carbon number four or higher; or benzene, or carbon dioxide) will form a solid phase. The reason behind that is that the presence of a second solvent can considerably change the solubility of the solid solute, if the solute is a hydrocarbon; this also occurs to a lesser extent if the solute is carbon dioxide.

On the other hand, the presence of nitrogen as a second solvent reduces the solubility of solids in methane, ethane, and mixtures thereof. Furthermore, the addition of nitrogen to miscible LNG systems can induce immiscibility. The number of nitrogen binary systems relevant to LNG that exhibit LLV immiscibility is few—nitrogen-ethane and nitrogen-propane being among the most prominent ones. However, LLV phenomenon has been observed at certain conditions in many ternary and higher realistic nitrogen-rich LNG systems since LLV behavior can and does occur in multicomponent systems even when no LLV locus is reported for any of the constituent binaries themselves.

The type of the LLV region displayed by a system depends on whether it contains an immiscible binary or not. For a ternary system with no constituent binary LLV behavior present, the three-phase region is a "triangular" surface in the thermodynamic phase space with two degrees of freedom, while its boundaries have one. It is bounded from above by a K-point locus, from below by a LCST locus, and at low temperatures by a Q-point locus. The systems methane + *n*-butane + nitrogen (Merrill et al., 1984a) and methane + *n*-pentane + nitrogen (Merrill et al., 1984b) belong to this class.

Yet another interesting system of this class is methane + ethane + *n*-octane, which does not exhibit immiscibility in any of its binary pairs. Although immiscibility has been reported in binary systems of methane + *n*-hexane and methane + *n*-heptane, solutes such as *n*-octane and higher normal paraffins crystallize as temperature decreases before any immiscibility occurs. On the other hand, with ethane as solvent, solutes beginning with *n*-C_{19} and higher paraffins demonstrate LLV behavior. Apparently, the addition of modest amounts of ethane to methane creates a solvent mixture exhibiting immiscibility with *n*-octane.

A2.3 Numerical procedures for calculating the multiphase equilibrium

Phase equilibria calculations in natural gas systems simulation can take up to as much as 50% of the CPU time. In complicated problems it may take even more. Thus, it is important to develop a reliable thermodynamic modeling framework (TMF) that will be able to predict, describe, and validate robustly and efficiently the complex phase behavior of LNG mixtures.

The TMF has three main elements: a library of thermodynamic parameters pertaining to pure-substances and binary interactions, thermodynamic models for mixture properties, and algorithms for solving the equilibrium relations. Reliable pure-component data for the main constituents of LNG systems are available experimentally; an EoS is usually the primary choice for the thermodynamic model. Thus, the focal point of a TMF is the prediction of phase behavior in LNG systems, which involves the solution of two relevant thermodynamic problems: phase stability (PS) and phase equilibrium (PE) calculations. PS problems involve the determination of whether a thermodynamic system will remain at one phase at given operating conditions (i.e., temperature, pressure, and composition) or split into two or more phases. This type of problem usually precedes the PE problems, which involve the identification and determination of the quantity and composition of the phases at equilibrium at specified conditions (Zhang et al., 2011).

These calculations require the availability of robust methods for PS and of reliable efficient and effective flash routines for phase split calculations, and in what follows, two efficient techniques designed for the solution of those problems will be outlined briefly.

A2.3.1 Computational technique 1

The technique uses an efficient computational procedure for solving the isothermal multiphase problem by assuming that the system is initially monophasic. A stability test allows verifying whether the system is stable or not. In the latter case, it provides an estimation of the composition of an additional phase; the number of phases is then increased by one, and equilibrium is achieved by minimizing the Gibbs energy. This approach, advocated as a stagewise procedure (Michelsen, 1982b; Nghiem and Li, 1984) is continued until a stable solution is found.

In technique 1, the stability analysis of a homogeneous system of composition $\mathbf{z}$, based on the minimization of the distance separating the Gibbs energy surface from the tangent plane at $\mathbf{z}$, is considered (Baker et al., 1982; Michelsen, 1982a). In terms of fugacity coefficients, ϕ_i, this criterion for stability can be written as (Michelsen, 1982a)

$$F(\mathbf{y}) = \sum_{i=1}^{N} y_i[\ln y_i + \ln \varphi_i(\mathbf{y}) - h_i] \geq 0 \quad \forall\, \mathbf{y} \tag{A2-1}$$

where

$$h_i = \ln z_i + \ln \phi_i(\mathbf{z}) \quad i = 1, 2\ldots, N \tag{A2-2}$$

Equation A2-1 requires that the tangent plane, at no point, lies above the Gibbs energy surface; this is achieved when $F(\mathbf{y})$ is positive in all its minima. Consequently, a minimum of $F(\mathbf{y})$ should be considered in the interior of the permissible region $\sum_{i=1}^{N} y_i = 1$, $\forall\, \mathbf{y} \geq 0$. Since to test condition 1 for all trial compositions is not physically possible, then it is sufficient to test the stability at all stationary points of $F(\mathbf{y})$ since this function is not negative at all stationary points.

An equivalent stability criterion to that given by Equation A2-1 but based on variables ξ_i (formally interpreted as mole numbers with corresponding mole fractions as $y_i = \xi_i / \sum_{j=1}^{N} \xi_j$) can also be formulated as (Michelsen, 1982a)

$$F(\xi) = 1 + \sum_{i=1}^{N} \xi_i[\ln \xi_i + \ln \varphi_i(\xi) - h_i - 1] \geq 0 \quad \forall\, \xi > 0 \tag{A2-3}$$

To guarantee that the tangent plane, at no point, lies above the Gibbs energy surface, $F(\xi)$ must be positive in all its minima. The quasi-Newton BFGS minimization method (Fletcher, 1980) is applied to Equation A2-3 for determining the stability of a given system of composition $\mathbf{z}$ at specified temperature and pressure.

Once instability is detected with the solution at $\pi - 1$ phases, the PE problem is solved by minimization of the following function:

$$\underset{n_i^{(\phi)}}{Min}\ \Delta g = \sum_{\phi=1}^{\pi} \sum_{i=1}^{N} n_i^{(\phi)} \ln \left(\frac{x_i^{(\phi)} \varphi_i^{(\phi)} p}{p^\circ} \right) \tag{A2-4}$$

subject to the inequality constraints given by

$$\sum_{\phi=1}^{\pi-1} n_i^{(\phi)} \leq z_i \qquad i = 1, \ldots, N \tag{A2-5}$$

and

$$n_i^{(\phi)} \geq 0 \qquad i = 1, \ldots, N \ ; \ \phi = 1, \ldots, \pi - 1 \tag{A2-6}$$

where Δg is the dimensionless Gibbs energy of the system, z_i is the mole fraction of component i in the system, $n_i^{(\phi)}$ $(i = 1, \ldots, N;\ \phi = 1, \ldots, \pi - 1)$ is the mole number of component i in phase ϕ per mole of feed, $x_i^{(\phi)}$ is the mole fraction of component i in phase ϕ, T is the temperature, p is the pressure, and p° is the pressure at the standard state of 1 atm (101.325 kPa). In Equation A2-4, the variables $n_i^{(\pi)}$, $x_i^{(\pi)}$, and $\phi_i^{(\pi)}$ are considered functions of $n_i^{(\phi)}$.

Equation A2-4 is solved using an unconstrained minimization algorithm by keeping the variables $n_i^{(\phi)}$ inside the convex constraint domain given by Equations A2-5 and A2-6 during the search for the solution.

A2.3.2 Computational technique 2

The second approach to predict and calculate multiphase equilibria used here applies a rigorous thermodynamic stability analysis and a simple and effective method for identifying the phase configuration at equilibrium with the minimum Gibbs energy. The stability analysis is exercised once and on the initial system only. It is based on the well-known tangent plane criterion (Baker et al., 1982; Michelsen, 1982a), but uses a different objective function (Stateva and Tsvetkov, 1994). The key point is to locate all zeros ($\mathbf{y}^*$) of a function $\Phi(\mathbf{y})$ given as

$$\Phi(\mathbf{y}) = \sum_{i=1}^{N} [k_{i+1}(\mathbf{y}) - k_i(\mathbf{y})]^2 \tag{A2-7}$$

where

$$k_i(\mathbf{y}) = \ln \varphi_i(\mathbf{y}) + \ln y_i - h_i \qquad i = 1, \ldots, N \tag{A2-8}$$

with

$$h_i = \ln z_i + \ln \varphi_i(\mathbf{z}) \qquad i = 1, \ldots, N \tag{A2-9}$$

and assuming $k_{N+1}(\mathbf{y}) = k_1(\mathbf{y})$.

Therefore, from Equations A2-8 and A2-9, it follows that min $\Phi(\mathbf{y}) = 0$ when $k_1(\mathbf{y}^*) = k_2(\mathbf{y}^*) = \ldots = k_N(\mathbf{y}^*)$.

The zeros of $\Phi(\mathbf{y})$ conform to points on the Gibbs energy hypersurface, where the local tangent hyperplane is parallel to that at $\mathbf{z}$. To each zero $\mathbf{y}^*$, a number k^* (equal for each y_i^*, $i = 1, 2, \ldots, N$ of a zero of the functional) corresponds, such that

$$k_i^* = \ln y_i^* + \ln \varphi(\mathbf{y}^*) - h_i \quad i = 1, \ldots, N \tag{A2-10}$$

Furthermore, the number k^* can be either positive or negative. A positive k^* corresponds to a zero, which represents a more stable state of the system, in comparison to the initial one; a negative k^*, a more unstable one. When all calculated k^* are positive, the initial system is stable; otherwise, it is unstable.

The specific form of $\Phi(\mathbf{y})$ (its zeros are its minima) and the fact that it is easily differentiated analytically, allows the application of a nonlinear minimization technique for locating its stationary points; for example, the quasi-Newton BFGS (Fletcher, 1980) method with a line-search and a superlinear order of convergence. The implementation of any nonlinear minimization technique requires a set of "good" initial estimates, and the BFGS method is no exception.

Thus, as discussed by Wakeham and Stateva (2004), a method has been created that proved to be extremely reliable in locating *almost all* zeros of $\Phi(\mathbf{y})$ at a reasonable computational cost. The term "almost all" zeros is used because there is no theoretically based guarantee that the scheme will always find them all. If, however, a zero is missed, the method is self-recovering. Furthermore, the TPDF is minimized *once only*, which is a distinct difference from the approach that stagewise methods generally adopt. The PS routine provides good approximations to the compositions to which the subsequent flash calculations of the phase identification procedure converge.

The numerical implementation, practical application, versatility, and robustness of technique 1 and technique 2 have been demonstrated over the years on a very large number of case studies, to be quite different in nature—from highly difficult theoretical problems to large practical problems in petroleum and chemical engineering. Some of the systems examined were chosen primarily for the high degree of complexity, which provided a rigorous test for the techniques' capabilities. Further details on their performance can be found elsewhere (García-Sánchez et al., 1996a, 1997, 2001; Justo-García et al., 2008a; Stateva, 1995; Stateva and Wakeham, 1997; Stateva et al., 2004).

A2.4 Calculation of critical points at constant composition

Calculation of critical points of multicomponent mixtures is of high importance in the petroleum industry. One such example is the compositional simulation of oil reservoirs; the behavior of a reservoir fluid during production can be determined by the shape of its phase diagram and the position of its critical point.

Although the most direct way of determining the critical points of any system is through experimental measurements, the complexity of many mixtures makes it difficult to determine them correctly. Such a problem has motivated many researchers to develop rigorous and efficient numerical procedures for solving the criticality conditions in multicomponent mixtures from an EoS (see, for example, Peng and Robinson, 1977; Baker et al., 1980; Heidemann and Khalil, 1980; Michelsen and Heidemann, 1981; Michelsen, 1984; Nagarajan et al., 1991; among others).

Recently, Justo-García et al. (2008b) applied the simulated annealing algorithm for the calculation of critical points of binary and multicomponent mixtures by using appropriate cooling schedule parameters to minimize the computer time for solution of a critical point formulated as an optimization problem. The numerical procedure used for solving the criticality conditions with EoS models was based on the tangent plane criterion in Helmholtz energy and was similar to that presented by Michelsen and Heidemann (1988), but temperature and volume were used as the independent variables. A concise description of the criticality conditions and solution procedure for locating all the critical points of a given mixture follows.

A2.4.1 Criticality conditions

Based on the Helmholtz energy, Michelsen and Heidemann (1988) defined the tangent plane distance at fixed temperature and volume as

$$F = \sum_{i=1}^{N} n_i \ln\left(f_i / f_{i_0}\right) - V(P - P_0)/RT \tag{A2-11}$$

To develop the criteria for critical points, the tangent plane distance function F is expanded in a Taylor series around the test point as (Michelsen, 1984)

$$F = bs^2 + cs^3 + ds^4 + O\left(s^5\right) \tag{A2-12}$$

such that $F(0) = 0$ and $(dF/ds)_{s=0} = 0$ hold at the test point; s being a parameter that defines the distance in composition space from the test point at $s = 0$. As the sign of the tangent plane distance function F determines the stability of the test phase, it is necessary to find the minimum of this function. In this case, it is convenient to handle this minimization problem in terms of scaled mole numbers as

$$X_i = (n_i - z_i)/z_i^{1/2} \qquad i = 1, \ldots, N \tag{A2-13}$$

where z_i are mole fractions in the test phase and n_i are mole fractions in any alternate phase. At the test point ($\mathbf{n} = \mathbf{z}$), $\mathbf{X} = 0$ and $F = 0$, and the first and second derivatives of F with respect to $\mathbf{X}$ are

$$g_i = \left(\frac{\partial F}{\partial X_i}\right) = z_i^{1/2}[\ln f_i(\mathbf{n}, V) - \ln f_i(\mathbf{z}, V)] \qquad i = 1, \ldots, N \tag{A2-14}$$

and

$$B_{ij} = \left(\frac{\partial^2 F}{\partial X_i \partial X_j}\right) = \left(z_i z_j\right)^{1/2} \left(\frac{\partial \ln f_i}{\partial n_j}\right)_{\mathbf{n}=\mathbf{z}} \qquad i,j = 1, \ldots, N \tag{A2-15}$$

The function F is then minimized by varying $\mathbf{X}$ under the constraint that

$$\mathbf{u}^T \mathbf{X} = s \tag{A2-16}$$

where $\mathbf{u}$ is a vector of unit length.

By applying the method of Lagrange multipliers, it is found that coefficient b can be expressed as

$$b = (1/2)\mathbf{u}^T\mathbf{B}\mathbf{u} = (1/2)\sum_{i=1}^{N}\sum_{j=1}^{N} B_{ij}u_iu_j \tag{A2-17}$$

regardless of the choice of vector **u**. Michelsen (1984) showed that the least possible value of coefficient b is obtained by choosing **u** as the eigenvector of **B** corresponding to the smallest eigenvalue $\lambda_{\min}$,

$$\mathbf{B}\mathbf{u} = \lambda_{\min}\mathbf{u} \tag{A2-18}$$

At trial conditions of temperature and volume, matrix **B** is calculated and $(\lambda_{\min}, \mathbf{u})$ are determined by inverse iteration (Wilkinson, 1964; Peters and Wilkinson, 1979) using a triangular decomposition of **B** (Press et al., 1992). Then

$$b = \lambda_{\min}/2 \tag{A2-19}$$

If $b = 0$, the system is at the limit of intrinsic stability. At a critical point coefficients b and c in Equation A2-12 are zero for a given eigenvector **u** of **B** corresponding to the smallest eigenvalue $\lambda_{\min}$. For the evaluation of coefficient c, Michelsen (1984) showed that it can be determined efficiently from information already available as

$$c = \frac{1}{6\varepsilon^2}\left[\left(\frac{dF}{ds}\right)_{s=\varepsilon} + \left(\frac{dF}{ds}\right)_{s=-\varepsilon}\right] + \mathbf{O}(\varepsilon^2) \quad \varepsilon = 1 \times 10^{-3} \tag{A2-20}$$

where

$$\frac{dF}{ds} = \sum_{i=1}^{N} g_i\left(\frac{\partial X_i}{\partial s}\right) = \sum_{i=1}^{N} u_i g_i \tag{A2-21}$$

A2.4.2 Solution procedure

The critical point calculation formulated as an optimization problem can be written as

$$\text{Min } f(T, V) = b^2 + c^2 \tag{A2-22}$$

subject to the constraints

$$T_{\min} < T < T_{\max} \quad \text{and} \quad V_{\min} < V < V_{\max} \tag{A2-23}$$

where the coefficients b and c are those given by Equations A2-19 and A2-20, respectively.

Expressions of the fugacity derivatives of component i with respect to mole numbers required in Equation A2-15 for the SRK (Soave, 1982) and PR (Peng-Robinson, 1976) equations of state can be found elsewhere (García-Sánchez et al., 1996b). For the PC-SAFT equation of state, the expression of the fugacity derivatives of component i with respect to mole numbers can be calculated numerically according to the approach proposed by Arce and Aznar (2007). That is, if the perturbation of mole numbers of component j is Δn_j, we have

$$n' = \sum_{\substack{k=1 \\ j \neq k}}^{N} n_k + (n_j + \Delta n_j) \quad \rightarrow \quad x'_i = \frac{n_i}{n'}, \quad v' = \frac{V}{n'} \tag{A2-24}$$

$$n'' = \sum_{\substack{k=1 \\ j \neq k}}^{N} n_k + (n_j - \Delta n_j) \quad \rightarrow \quad x''_i = \frac{n_i}{n''}, \quad v'' = \frac{V}{n''} \tag{A2-25}$$

where, before the perturbation, $V = nv$ with $n = \sum_{i=1}^{N} n_i$. Therefore, the derivative of the fugacity of component i with respect to mole number of component j can be evaluated as

$$\left(\frac{\partial \ln f_i}{\partial n_j}\right) = \frac{\ln f_i(T, V', x') - \ln f_i(T, V'', x'')}{2\Delta n_j} \tag{A2-26}$$

where the pertinent expressions for evaluating the fugacity of component i in the mixture are given in Section A2.7 for both the SRK and PC-SAFT EoSs.

The optimization problem as formulated by Equation A2-22 can be solved with the simulated annealing algorithm of Goffe et al. (1994). This algorithm offers the advantage that only the calculation of the objective function is needed, thus avoiding the evaluation of the derivatives of the coefficients b and c with respect to temperature and volume. Further, this algorithm does not depend on the initial guesses of the independent variables and gives preference to global optima.

In the simulated annealing algorithm, the initial control parameter θ_0, the control parameter reduction factor r_θ, the number of iterations before the control parameter is reduced n_θ, the number of cycles before the step is adjusted n_S, and the error tolerance for termination ε are part of the input data known as cooling schedule. The choice of these parameters is a very important aspect in the implementation of the simulated annealing algorithm since it strongly affects the performance of the method. A detailed description of a set of appropriate cooling schedule parameters that offers a relative computational speed and either guarantees that all critical points of a given mixture are calculated or verifies the nonexistence of a critical point can be found elsewhere (Justo-García et al., 2008c).

A2.5 Calculation of K- and LCST-points at constant temperature

Ternary systems, which exhibit LLV behavior but do not exhibit such behavior in their constituent binaries, have the immiscibility region bounded by a K (L_1−L_2=V)-point locus, a LCST (L_1=L_2−V) locus, and a Q (S−L_1−L_2−V)-point locus. Ternary systems, which have immiscibility in a constituent binary, can have boundaries similar to those just mentioned , besides the intrusion of the binary LLV locus on the ternary LLV region. The K-point and LCST loci can intersect at a tricritical point where the three phases become critical; that is, L_1=L_2=V.

In a study on the modeling of the three-phase LLV region for ternary hydrocarbon mixtures with the SRK EoS, Gregorowicz and de Loos (1996) proposed a procedure for finding K- and LCST-points of ternary systems, based on the solution of thermodynamics conditions for the K- and LCST-points, by using the Newton iteration technique and carefully chosen starting points. The procedure was applied to calculate the K- and LCST-point loci for two ternary systems, namely, $C_2 + C_3 + C_{20}$ and $C_1 + C_2 + C_{20}$, in which the constituent binary $C_2 + C_{20}$ exhibit immiscibility. Consequently, the extension of the three-phase LLV region of these systems is bounded by the binary $L_1{-}L_2{-}V$ locus of the system $C_2 + C_{20}$ and the ternary K-point and L-point loci.

The strategy followed by these authors to find the critical end points was the following: (1) calculation of the critical line, the K-point, the three phase line, and the LCST-point for the system $C_2 + C_{20}$ using thermodynamic conditions, and (2) calculation of the K- and LCST-point loci for the ternary systems by using as starting points the coordinates of the K- and LCST-points found for the binary system $C_2 + C_{20}$. In this case, to obtain a K- and LCST-point for a ternary system, the following set of six nonlinear equations,

$$D = \begin{vmatrix} A_{VV} & A_{Vx_1} & A_{Vx_2} \\ A_{Vx_1} & A_{x_1x_1} & A_{x_1x_2} \\ A_{Vx_2} & A_{x_1x_2} & A_{x_2x_2} \end{vmatrix} = 0 \tag{A2-27}$$

$$D^* = \begin{vmatrix} D_V & D_{x_1} & D_{x_2} \\ A_{Vx_1} & A_{x_1x_1} & A_{x_1x_2} \\ A_{Vx_2} & A_{x_1x_2} & A_{x_2x_2} \end{vmatrix} = 0 \tag{A2-28}$$

$$\mu_i^c - \mu_i^\alpha = 0 \qquad i = 1, 2, 3 \tag{A2-29}$$

$$p^c - p^\alpha = 0 \tag{A2-30}$$

in seven variables, T, V^c, V^α, x_1^c, x_2^c, x_1^α, and x_2^α, have to be solved, where α designates either V for the L-point or L_2 for the K-point, D and D^* are the two determinants that must be satisfied at a critical point c, and μ_i is the chemical potential of component i. The ternary K- and LCST-points were calculated at chosen values of the temperature.

The critical criteria given by Equations A2-27 and A2-28 are based on the Helmholtz energy, which can be expressed as (Prausnitz et al., 1999)

$$A = -\int_V^\infty \left(p - \frac{nRT}{V}\right) dV - RT \sum_{i=1}^{N} n_i \ln\left(\frac{V}{n_i RT}\right) + \sum_{i=1}^{N} n_i \left(u_i^0 - T s_i^0\right) \tag{A2-31}$$

where p is the pressure, T is the temperature, V is the system volume, n_i is the mole number of component i, u_i^0 is the standard state molar internal energy, s_i^0 is the standard state molar entropy, and R is the gas constant. Derivatives of the Helmholtz energy are denoted by a subscript in Equations A2-27 and A2-28 (e.g., A_V, A_{x_i}) indicating the differentiation variable (volume V or mole fraction x_i of component i).

Recently, Ramírez-Jiménez et al. (2012) adopted the approach of Gregorowicz and de Loos (1996) to predict the ternary K- and LCST-points. The Newton iteration technique was applied to solve a set of only five nonlinear equations, namely, the conditions of criticality given by Baker and Luks (1980),

$$D_1 = \begin{vmatrix} A_{VV} & A_{Vn_1} & A_{Vn_2} \\ A_{Vn_1} & A_{n_1n_1} & A_{n_1n_2} \\ A_{Vn_2} & A_{n_1n_2} & A_{n_2n_2} \end{vmatrix} = 0 \tag{A2-32}$$

$$D_2 = \begin{vmatrix} A_{VV} & A_{Vn_1} & A_{Vn_2} \\ A_{Vn_1} & A_{n_1n_1} & A_{n_1n_2} \\ D_V & D_{n_1} & D_{n_2} \end{vmatrix} = 0 \tag{A2-33}$$

and the equilibrium conditions in terms of fugacities, f_i,

$$f_i^c - f_i^\alpha = 0 \qquad i = 1, 2, 3 \tag{A2-34}$$

in five variables, $p, x_1^c, x_2^c, x_1^\alpha$, and x_2^α.

In Equation A2-33, D_V and D_{n_1} denote the derivatives of determinant D_1 with respect to volume and number of moles of component i ($i = 1, 2$), respectively.

The difference between the procedure of Gregorowicz and de Loos (1996) and the one reported by Ramírez-Jiménez et al. (2012) for finding K- and LCST points of ternary systems at constant temperature is in the manner of calculating the values of the two determinants that must be satisfied at the critical point. The method used by the latter to calculate the determinants D_1 and D_2 was that proposed by Baker and Luks (1980). In this method, the row ordering of the array of determinants D_1 and D_2 were chosen in such a way that the arrays are identical, except for the last row.

The determinant D_1 is then evaluated by computing the cofactors of the last row, which are, in turn, used to evaluate the determinant D_2. In addition, Baker and Luks (1980) showed that the derivative of determinant D_1 with respect to a given variable (e.g., V) is the sum of the element-by-element products of two matrices: one matrix contains the cofactors of the original matrix, whereas the other contains the derivative of the original matrix elements with respect to the variable. This method allows a saving in computer time, in particular for multicomponent systems.

An additional important difference between the two procedures is that in the Gregorowicz and de Loos' (1996) procedure a set of six nonlinear equations are solved in six variables ($V^c, V^\alpha, x_1^c, x_2^c, x_1^\alpha$, and x_2^α), whereas in the procedure of Ramírez-Jiménez et al. (2012) a set of five nonlinear equations are solved in five variables ($p, x_1^c, x_2^c, x_1^\alpha$, and x_2^α), which makes the iterative process easier to perform and converge.

Ramírez-Jiménez et al. (2012) give a detailed description of their algorithm realizing the Newton iteration technique and demonstrate its capabilities on the examples of K- and LCST-points calculation of the ternary systems nitrogen + methane + ethane, nitrogen + methane + propane, and nitrogen + methane + *n*-butane at different temperatures.

A2.6 Calculation of phase envelopes

In the past, the generation of phase envelopes on pressure-temperature diagrams involved a series of bubble- and dew-point calculations. However, several difficulties are inherent to this approach: (1) it is not clear whether a bubble or a dew point should be calculated near the critical point, (2) the tracing of the phase envelope in the retrograde zone is tedious due to the presence of a lower and an upper dew point, and (3) the specification of the variables have to be done manually; that is, during the construction of the phase envelope, it is necessary to specify the pressure and calculate the dew point temperature near the cricondentherm (maximum temperature point of the phase envelope), and specify the temperature and calculate the bubble/dew point pressure near the cricondenbar (maximum pressure point of the phase envelope).

The first general algorithm for phase envelope construction was reported by Asselineau et al. (1979), which involved the simultaneous solution of $2N + 4$ equations for each point on the phase envelope, where N is the number of components.

Later, Michelsen (1980) presented a more efficient algorithm to construct pressure-temperature phase diagrams. The formulation of this author involved the simultaneous solution of only $N + 1$ equations for each point on the phase envelope. This algorithm selects internally the set of primary variables and the step size to the subsequent point on the phase envelope for which the initial guess is obtained by extrapolation. Thus, the bubble-point and dew-point curves are traced in a single pass, whereas the critical point, cricondentherm, and cricondenbar are estimated by interpolation.

Subsequently, Li and Nghiem (1982) extended the Michelsen's algorithm to the construction of pressure-composition, temperature-composition, and composition-composition phase diagrams, while Lindeloff et al. (1999) and Lindeloff and Michelsen (2003) presented an algorithm for the calculation of phase envelopes of systems exhibiting more than two phases, which follows the basic principles of the Michelsen's algorithm to the construction of vapor-liquid phase envelopes.

The procedure for calculating phase envelopes according to Michelsen's algorithm is as follows: Consider a mixture of N components and that this mixture of composition $(z_1, z_2, \ldots, z_N)$ is divided into liquid and vapor phases at a given pressure and temperature, where the liquid phase contains $(1 - F)$ moles of composition $(x_1, x_2, \ldots, x_N)$ and the vapor phase contains F moles of composition $(y_1, y_2, \ldots, y_N)$. Then the following conditions must be satisfied at equilibrium:

$$f_i^L = f_i^V \quad i = 1, \ldots, N \tag{A2-35}$$

where f_i is the fugacity of component i in the mixture. Equation A2-36 can also be written as

$$\ln K_i + \ln \varphi_i^V - \ln \varphi_i^L = 0 \quad i = 1, \ldots, N \tag{A2-36}$$

where K_i is the equilibrium ratio and φ_i^L and φ_i^V are the fugacity coefficients of component i for the liquid and vapor phases, respectively, which are defined as

$$K_i = \frac{y_i}{x_i} \quad i = 1, \ldots, N \tag{A2-37}$$

$$\phi_i^L = \frac{f_i^L}{x_i p} \quad i = 1, \ldots, N \tag{A2-38}$$

$$\varphi_i^V = \frac{f_i^V}{y_i p} \quad i = 1, \ldots, N \tag{A2-39}$$

The material balance for each component i is

$$z_i = (1 - F)x_i + F\,y_i \quad i = 1, \ldots, N \tag{A2-40}$$

with

$$\sum_{i=1}^{N} x_i = \sum_{i=1}^{N} y_i = 1 \tag{A2-41}$$

Introducing Equation A2-37 into Equation A2-40, the following expressions are obtained:

$$x_i = \frac{z_i}{1 + F(K_i - 1)} \quad i = 1, \ldots, N \tag{A2-42}$$

$$y_i = \frac{z_i K_i}{1 + F(K_i - 1)} \quad i = 1, \ldots, N \tag{A2-43}$$

which make that Equation A2-41, which can be written as

$$\sum_{i=1}^{N} \frac{z_i(K_i - 1)}{1 + F(K_i - 1)} = 0 \tag{A2-44}$$

Since compositions of the liquid, x_i, and vapor, y_i, phases can be calculated from Equations A2-42 and A2-43 once z_i, K_i, and F are known, these are treated as secondary variables. Hence, the $N+1$ equations governing the equilibrium calculation in $N+2$ variables are

$$g_i(\boldsymbol{\alpha}, \boldsymbol{\beta}) = \ln K_i + \ln \varphi_i^V - \ln \varphi_i^L = 0 \quad i = 1, \ldots, N \tag{A2-45}$$

$$g_{N+1}(\boldsymbol{\alpha}, \boldsymbol{\beta}) = \sum_{i=1}^{N} \frac{z_i(K_i - 1)}{1 + F(K_i - 1)} = 0 \tag{A2-46}$$

where $\boldsymbol{\alpha}$ is the vector of independent variables

$$\boldsymbol{\alpha}^T = (\ln K_1, \ldots, \ln K_N, \ln T, \ln p) \tag{A2-47}$$

and $\boldsymbol{\beta}$ is the vector of fixed variables

$$\boldsymbol{\beta}^T = (z_1, \ldots, z_N, F) \tag{A2-48}$$

The complete set of solutions $\boldsymbol{\alpha}$ is called phase envelope and a particular solution corresponds to one point on the phase envelope. To obtain a particular solution it is necessary to add a specification equation for one of the dependent variables,

$$g_{N+2}(\boldsymbol{\alpha}, \boldsymbol{\beta}) = \alpha_k - S = 0 \tag{A2-49}$$

and solve the resulting $N+2$ equations

$$\mathbf{g}\ (\boldsymbol{\alpha}, \boldsymbol{\beta}, S) = 0 \tag{A2-50}$$

with the Newton-Raphson; that is,

$$\mathbf{J}^{(k)} \Delta\boldsymbol{\alpha} + \mathbf{g}^{(k)} = 0 \tag{A2-51}$$

$$\boldsymbol{\alpha}^{(k+1)} = \boldsymbol{\alpha}^{(k)} + \Delta\boldsymbol{\alpha} \tag{A2-52}$$

where $\mathbf{J} = \partial \mathbf{g} / \partial \boldsymbol{\alpha}$ is the Jacobian matrix at $\boldsymbol{\alpha}^{(k)}$.

The expressions for calculating the derivatives of the fugacity coefficients with respect to mole numbers, temperature, and pressure can be found elsewhere (Michelsen and Mollerup, 2004).

The process is iterative and the solution is reached when the following convergence solution is satisfied:

$$\sum_{i=1}^{N+1}\left(\alpha_i^{(k+1)}-\alpha_i^{(k)}\right)^2 < 10^{-10} \tag{A2-53}$$

As pointed out by Michelsen (1980), the efficiency of the method strongly depends on the initial guess $\boldsymbol{\alpha}^{(0)}$ and on the specification variable S that leads to a unique solution of Equation A2-50.

To initiate the calculations, a specification pressure $\alpha_{N+2}=S_1$ (e.g., 10 bar) normally is used so that good estimates of the equilibrium ratios K_i can be obtained from the expression (Michelsen, 1980; Li and Nghiem, 1982),

$$K_i=\left(\frac{p_{c,i}}{p}\right)\exp\left[5.42\left(1-\frac{T_{c,i}}{T}\right)\right] \quad i=1,\ldots,N \tag{A2-54}$$

Solving Equation A2-44 for T at the bubble point ($F=0$) or the dew point ($F=1$) by using the expression in Equation A2-54 for the K_i factors, an initial estimate of the temperature and compositions of the liquid and vapor phases are obtained, whereas the subsequent approximations are obtained through the Newton method.

A2.7 Thermodynamic models

A2.7.1 The SRK equation of state

The explicit form of the SRK equation of state (Soave et al., 1972) can be written as

$$p=\frac{RT}{v-b}-\frac{a(T)}{v(v+b)} \tag{A2-55}$$

where v is the molar volume and constants a and b for pure components are related to

$$a=0.42747\frac{R^2T_c^2}{p_c}\alpha(T_r) \tag{A2-56}$$

$$b=0.08664\frac{RT_c}{p_c} \tag{A2-57}$$

and $\alpha(T_r)$ is expressed in terms of the acentric factor ω as

$$\alpha(T_r)=\left[1+\left(0.480+1.574\omega-0.176\omega^2\right)\left(1-T_r^{1/2}\right)\right]^2 \tag{A2-58}$$

For mixtures, constants a and b are given by

$$a=\sum_{i=1}^{N}\sum_{j=1}^{N}x_ix_ja_{ij} \tag{A2-59}$$

$$b = \sum_{i=1}^{N} x_i b_i \tag{A2-60}$$

and a_{ij} is defined as

$$a_{ij} = \left(1 - k_{ij}\right)\sqrt{a_i a_j} \qquad k_{ij} = k_{ji} \ ; \ \ k_{ii} = 0 \tag{A2-61}$$

where k_{ij} is an adjustable interaction parameter characterizing the interactions between components i and j.

Equation A2-55 can be written in terms of compressibility factor, $Z = pv/RT$, as

$$Z^3 - Z^2 + \left(A - B - B^2\right)Z - AB = 0 \tag{A2-62}$$

where $A = ap/(RT)^2$ and $B = bp/(RT)$.

The expression for the fugacity coefficient, $\varphi_i = f_i/y_i p$, is given by

$$\ln \varphi_i = \frac{b_i}{b}(Z - 1) - \ln(Z - B) - \frac{A}{B}\left(\frac{2\sum_{j=1}^{N} x_j a_{ij}}{a} - \frac{b_i}{b}\right)\ln\left(1 + \frac{B}{Z}\right) \tag{A2-63}$$

A2.7.2 The PC-SAFT equation of state

In the PC-SAFT equation of state (Gross and Sadowski, 2001), the molecules are conceived to be chains composed of spherical segments, in which the pair potential for the segment of a chain is given by a modified square-well potential (Chen and Kreglewski, 1977; Kreglewski, 1984). Nonassociating molecules are characterized by three pure component parameters: the temperature-independent segment diameter σ, the depth of the potential ε, and the number of segments per chain m.

The PC-SAFT equation of state written in terms of the Helmholtz energy for an N-component mixture of nonassociating chains consists of a hard-chain reference contribution and a perturbation contribution to account for the attractive interactions. In terms of reduced quantities, this equation can be expressed as

$$\tilde{a}^{res} = \tilde{a}^{hc} + \tilde{a}^{disp} \tag{A2-64}$$

The hard-chain reference contribution is given by

$$\tilde{a}^{hc} = \overline{m}\,\tilde{a}^{hs} - \sum_{i=1}^{N} x_i\left(m_i - 1\right)\ln g_{ii}^{hs}\left(\sigma_{ii}\right) \tag{A2-65}$$

where $\overline{m}$ is the mean segment number in the mixture

$$\overline{m} = \sum_{i=1}^{N} x_i m_i \tag{A2-66}$$

The Helmholtz energy of the hard-sphere fluid is given on a per-segment basis as

$$\tilde{a}^{hs} = \frac{1}{\zeta_0}\left[\frac{3\zeta_1\zeta_2}{(1-\zeta_3)} + \frac{\zeta_2{}^3}{\zeta_3(1-\zeta_3)^2} + \left(\frac{\zeta_2{}^3}{\zeta_3{}^2} - \zeta_0\right)\ln(1-\zeta_3)\right] \tag{A2-67}$$

and the radial distribution function of the hard-sphere fluid is

$$g_{ij}^{hs} = \frac{1}{(1-\zeta_3)} + \left(\frac{d_i d_j}{d_i + d_j}\right)\frac{3\zeta_2}{(1-\zeta_3)^2} + \left(\frac{d_i d_j}{d_i + d_j}\right)^2 \frac{2\zeta_2{}^2}{(1-\zeta_3)^3} \tag{A2-68}$$

with ζ_n defined as

$$\zeta_n = \frac{\pi}{6}\rho \sum_{i=1}^{N} x_i m_i d_i{}^n \quad n = 0, 1, 2, 3 \tag{A2-69}$$

The temperature-dependent segment diameter d_i of component i is given by

$$d_i = \sigma_i\left[1 - 0.12\exp\left(-3\frac{\varepsilon_i}{kT}\right)\right] \tag{A2-70}$$

where k is the Boltzmann constant and T is the absolute temperature.

The dispersion contribution to the Helmholtz energy is given by

$$\tilde{a}^{disp} = -2\pi\rho\, I_1(\eta, \overline{m})\, \overline{m^2\varepsilon\sigma^3} - \pi\rho\, \overline{m}\left(1 + Z^{hc} + \rho\frac{\partial Z^{hc}}{\partial\rho}\right)^{-1} I_2(\eta, \overline{m})\, \overline{m^2\varepsilon^2\sigma^3} \tag{A2-71}$$

where Z^{hc} is the compressibility factor of the hard-chain reference contribution, and

$$\overline{m^2\varepsilon\sigma^3} = \sum_{i=1}^{N}\sum_{j=1}^{N} x_i\, x_j\, m_i\, m_j \left(\frac{\varepsilon_{ij}}{k_B T}\right)\sigma_{ij}{}^3 \tag{A2-72}$$

$$\overline{m^2\varepsilon^2\sigma^3} = \sum_{i=1}^{N}\sum_{j=1}^{N} x_i\, x_j\, m_i\, m_j \left(\frac{\varepsilon_{ij}}{k_B T}\right)^2 \sigma_{ij}{}^3 \tag{A2-73}$$

The parameters for a pair of unlike segments are obtained by using conventional combining rules

$$\sigma_{ij} = \frac{1}{2}(\sigma_i + \sigma_j) \tag{A2-74}$$

$$\varepsilon_{ij} = \sqrt{\varepsilon_i\varepsilon_j}(1 - k_{ij}) \tag{A2-75}$$

where k_{ij} is a binary interaction parameter, which is introduced to correct the segment-segment interactions of unlike chains.

The terms $I_1(\eta, \overline{m})$ and $I_2(\eta, \overline{m})$ in Equation A2-71 are calculated by simple power series in density

$$I_1(\eta, \overline{m}) = \sum_{i=0}^{6} a_i(\overline{m})\eta^i \tag{A2-76}$$

$$I_2(\eta, \overline{m}) = \sum_{i=0}^{6} b_i(\overline{m})\eta^i \quad \text{(A2-77)}$$

where the coefficients a_i and b_i depend on the chain length as given in Gross and Sadowski (2001).

The density to a given system pressure p^{sys} is determined iteratively by adjusting the reduced density η until $p^{cal} = p^{sys}$. For a converged value of η, the number density of molecules ρ, given in Å^{-3}, is calculated from

$$\rho = \frac{6}{\pi}\eta\left(\sum_{i=1}^{N} x_i m_i d_i^3\right)^{-1} \quad \text{(A2-78)}$$

Using Avogadro's number and appropriate conversion factors, ρ produces the molar density in different units such as $\text{kmol}\cdot\text{m}^{-3}$.

The pressure can be calculated in units of $\text{Pa} = \text{N}\cdot\text{m}^{-2}$ by applying the relation

$$p = Z\,kT\rho\left(10^{10}\frac{\text{Å}}{m}\right)^3 \quad \text{(A2-79)}$$

from which the compressibility factor Z can be derived. The expression for the fugacity coefficient is given by

$$\ln \varphi_i = \tilde{a}^{res} + \left(\frac{\partial \tilde{a}^{res}}{\partial x_i}\right)_{T,v,x_{j\neq i}} - \sum_{k=1}^{N}\left[x_k\left(\frac{\partial \tilde{a}^{res}}{\partial x_k}\right)_{T,v,x_{j\neq k}}\right] + (Z-1) - \ln Z \quad \text{(A2-80)}$$

In Equation A2-80, the partial derivatives with respect to mole fractions are calculated regardless of the summation relation $\sum_{i=1}^{N} x_i = 1$.

A2.8 Examples of application

As part of an extensive program for investigating the nonideal behavior of mixtures of major constituents of natural gas at LNG temperatures, the authors have applied a TMF including the numerical procedures presented earlier to model the fluid phase equilibria of binary, ternary, quaternary, and quinary systems constituted by nitrogen, methane, ethane, propane, and *n*-butane, which are the main components of LNG systems.

To demonstrate the capabilities of the TMF the following calculations were performed: two- and three-phase isothermal flash calculations; phase envelope, critical- and critical-end point calculations (K- and LCST points). The target systems examined were binary (nitrogen + ethane and nitrogen + propane), three ternary (nitrogen + methane + ethane, nitrogen + methane + propane, and nitrogen + methane + *n*-butane), one quaternary (nitrogen + methane + ethane + propane), and one quinary (nitrogen + methane + ethane + propane + *n*-butane).

The components' physical properties required for the SRK EoS and the characteristic parameters for the PC-SAFT EoS, are given in Table A2-1. The former were taken from the DIPPR Data Compilation of Pure Chemical Properties (Rowley et al., 2006) whereas the three pure-component

Table A2-1 Physical Properties[a] and Pure-Component Parameters of the PC-SAFT Equation of State[b]

Component	*MW* g/mol	T_b K	T_c K	p_c MPa	ω	*m*	σ Å	ε/k K
N_2	28.01	77.35	126.2	3.39	0.039	1.2053	3.3130	90.96
C_1	16.04	111.63	190.58	4.604	0.012	1.0000	3.7039	150.03
C_2	30.07	184.55	305.42	4.880	0.099	1.6069	3.5206	191.42
C_3	44.10	231.05	369.82	4.250	0.153	2.0020	3.6184	208.11
n-C_4	58.12	272.65	425.18	3.797	0.199	2.3316	3.7086	222.88

[a]*Ambrose (1980)*
[b]*Gross and Sadowski (2001)*

parameters of those compounds for the PC-SAFT equation of state were taken from Gross and Sadowski (2001).

The binary interaction parameters used in all the phase equilibrium calculations for both equations of state are given in Table A2-2. For the systems studied the SRK EoS interaction parameters were taken from Knapp et al. (1982) and from Nagy and Shirkovskiy (1982); corresponding binary interaction parameters for the PC-SAFT equation are those of García-Sánchez et al. (2004) and Justo-García et al. (2008b).

A2.8.1 The nitrogen + ethane system

At certain low temperatures, mixtures of nitrogen and ethane containing from 5 to 80 mole % ethane form two immiscible liquid phases in equilibrium with a vapor phase. The locus of conditions under which the three phases coexist is shown on a pressure-temperature diagram in Figure A2-1. The figure shows the experimental VLL points for the binary reported by Eakin et al. (1955), Yu et al. (1969), Kremer and Knapp (1983), and Llave et al. (1985).

Different authors report slightly different values for the maximum point at which three phases coexist— 133.2K and 41.70 bar (Eakin et al., 1955); 132.38K and 40.71 bar (Llave et al., 1985); 133.3K and 40.74 bar (Yu et al., 1969). The system's quadruple point, at which two liquids, a vapor, and a solid coexist, is also shown in Figure A2-1 with values as reported by Eakin et al. (1955), namely 88.4K and 2.76 bar.

The three-phase experimental data for the system nitrogen + ethane reported by all these authors were used to compare the SRK EoS modeling results.

Figure A2-1 also shows the agreement between the SRK EoS modeling results and the experimental data of the preceding authors. As can be seen, the maximum three-phase coexisting point predicted with this model is 129.2K and 37.03 bar, which is lower both in pressure and temperature than that one determined experimentally.

Figure A2-2 shows the experimental VLL and SRK EoS predicted equilibrium data on a temperature-composition (in terms of nitrogen mole fraction) phase diagram. Though the K-point predicted is lower than the experimental one and the calculated concentrations in the liquid phase L_2 deviate from the experimental concentrations, still the agreement is acceptable. Of course, the liquid

Table A2-2 Binary Interaction Parameters for the PR and PC-SAFT Equations of State

	PR EoS					PC-SAFT EoS				
	N_2	C_1	C_2	C_3	n-C_4	N_2	C_1	C_2	C_3	n-C_4
N_2	0.0	0.0278	0.0407	0.0763	0.0700	0.0	0.0307	0.0458	0.0759	0.0570
C_1	0.0278	0.0	−0.0078	0.0090	0.0056	0.0307	0.0	0.0039	0.0019	0.0192
C_2	0.0407	−0.0078	0.0	−0.0022	0.0067	0.0458	0.0039	0.0	0.0089	0.0084
C_3	0.0763	0.0090	−0.0022	0.0	0.0000	0.0759	0.0019	0.0089	0.0	0.0034
n-C_4	0.0700	0.0056	0.0067	0.0000	0.0	0.0057	0.0192	0.0084	0.0034	0.0

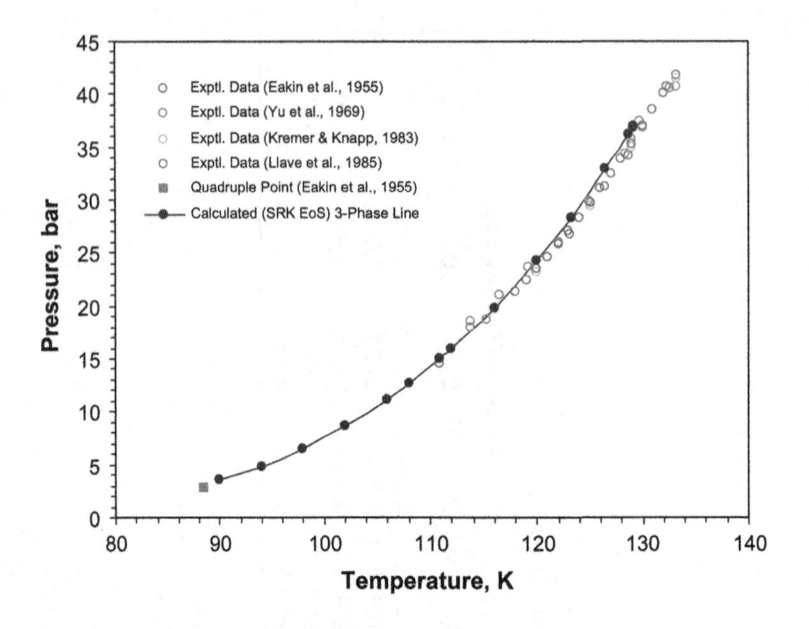

FIGURE A2-1

Experimental and calculated (SRK EoS) pressure-temperature phase diagram of the three-phase liquid-liquid-vapor equilibria in the binary system nitrogen + ethane.

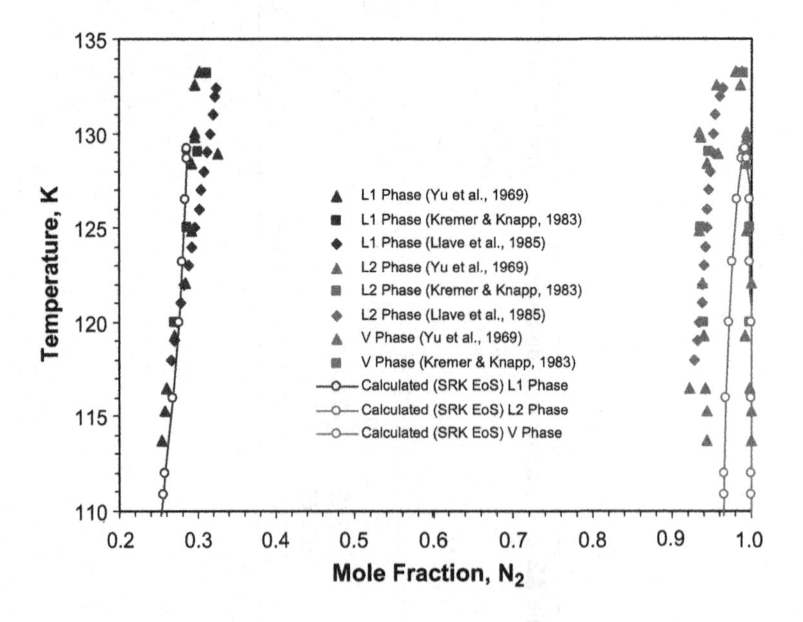

FIGURE A2-2

Experimental and calculated temperature-composition phase diagram for the binary system nitrogen + ethane.

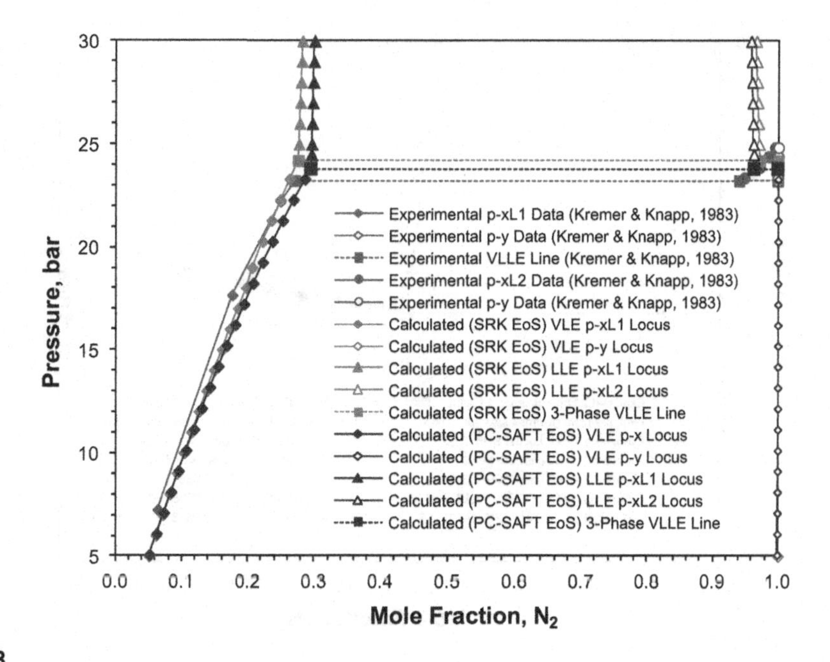

FIGURE A2-3

Experimental (Kremer and Knapp, 1983) and calculated pressure-composition phase diagram for the binary system nitrogen + ethane at 120K.

compositions calculated can be improved considerably by refitting the binary interaction parameter to the VLE data at each temperature level; that is, adjusting the k_{ij} parameter to the VLE data as a function of temperature.

Kremer and Knapp (1983) reported VLE and VLLE data for the nitrogen + ethane system at four temperatures between 120 and 133.15K. The experimental data reported by these authors at 120K were used to compare the PC-SAFT and SRK modeling results.

Figure A2-3 shows the experimental data (Kremer and Knapp, 1983) and data calculated with the PC-SAFT and SRK EoSs solubility limits for the nitrogen + ethane system at $T = 120$K. Both models can reliably predict the VL (V-L_1 and V-L_2) and V-L_1-L_2 solubility limits although the PC-SAFT results are slightly closer to the experimental data. Figure A2-3 also shows the predicted L_1-L_2 phase equilibria at higher pressures within the solubility limits using both thermodynamic models.

Figure A2-4 shows details of the nitrogen-rich liquid phase (liquid phase L_2) in equilibrium with the vapor phase on a pressure-temperature phase diagram. Again, the PC-SAFT performs better than the SRK: at $T = 120$K the experimental V-L_2 composition (in mole fraction) on the three phase line reported by Kremer and Knapp is 0.9398 N_2 (at 23.23 bar), whereas the calculated composition is 0.9632 N_2 (at 23.80 bar) and 0.9717 N_2 (at 24.20 bar), respectively.

Figure A2-5 shows the nitrogen + ethane experimental and calculated pressure-temperature phase diagram. The vapor pressure curves for nitrogen and ethane were calculated with the Wagner equation in the “3, 6” form using the parameters reported by Reid et al. (1987); the critical properties for these

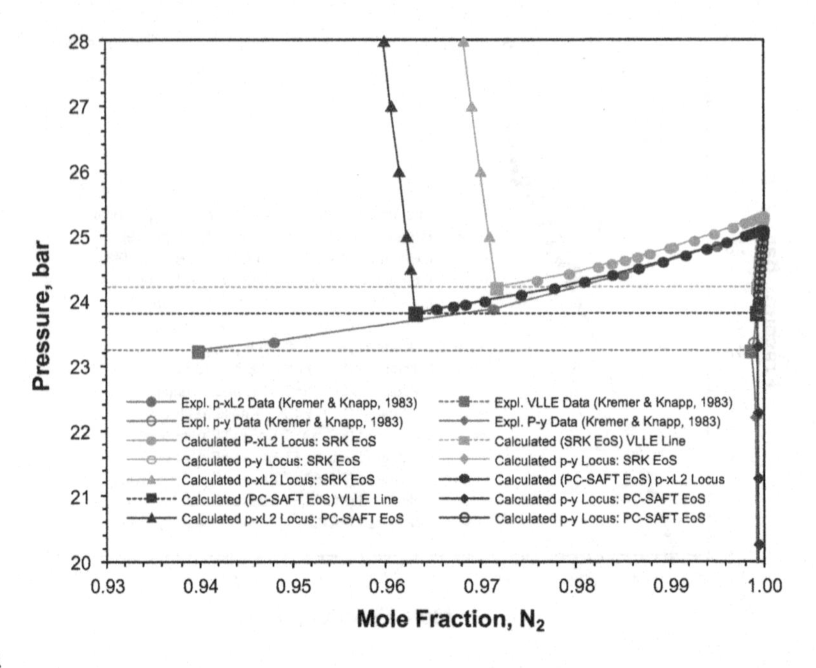

FIGURE A2-4

Experimental (Kremer and Knapp, 1983) and calculated V-L_2 (nitrogen-rich liquid phase) for the binary system nitrogen + ethane at 120K.

components are those given in Table A2-1, and the calculated three-phase VLL line is the one presented in Figure A2-1.

The mixture critical locus shown in Figure A2-5 was calculated using the procedure presented earlier for locating all the critical points of a given mixture. In this case, the feasible domain to locate all the critical points with the simulated annealing algorithm of Goffe et al. (1994) was defined as a close box in the $T - V$ plane in the form $[T_{\min}, T_{\max}] \times [V_{\min}, V_{\max}]$, where the lower and upper limits of temperature T and volume V are given by Equations A2-80 and A2-81 for one mole of mixture.

$$T_{\min} = 0.4 \sum_{i=1}^{N} z_i T_{c,i} \quad V_{\min} = 0.08R \sum_{i=1}^{N} z_i T_{c,i}/p_{c,i} \tag{A2-81}$$

$$T_{\max} = 4.0 \sum_{i=1}^{N} z_i T_{c,i} \quad V_{\max} = 0.8R \sum_{i=1}^{N} z_i T_{c,i}/p_{c,i} \tag{A2-82}$$

Figure A2-5 indicates that N_2 + ethane is a type III system, according to the classification scheme of van Konynenburg and Scott (1980). Different compositions were studied and it was found that for some the system exhibited more than one critical point. That is, two critical points were found over the N_2 mole fraction range of $[0.6944, 0.6800]$, one critical point was located within the N_2 mole fraction ranges of $[0.0010, 0.6750]$ and $[0.9995, 0.9600]$, and no critical point was found over the N_2 mole fraction range of $[0.9500, 0.9700]$. The latter case was also experimentally confirmed by Eakin et al.

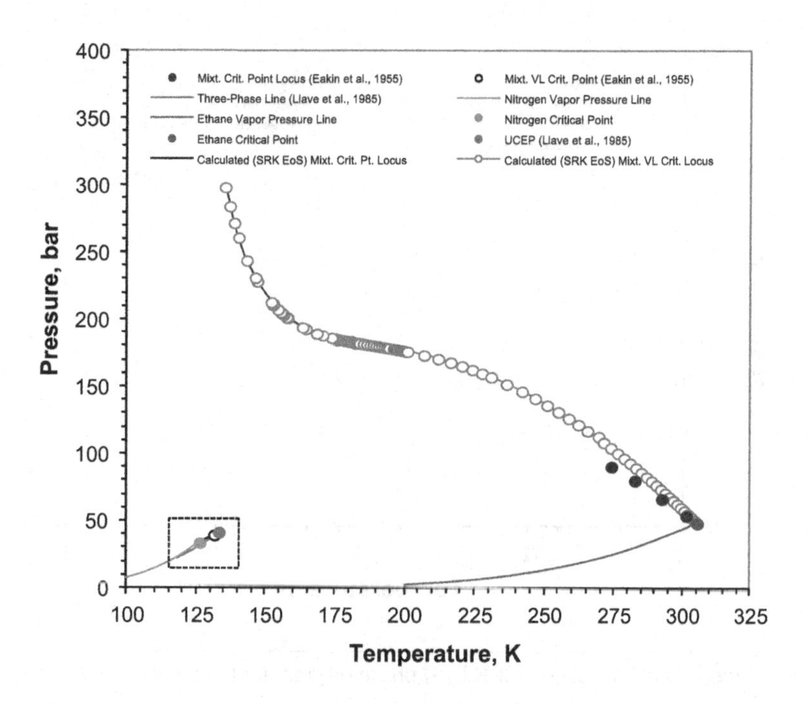

FIGURE A2-5

Experimental and calculated (SRK EoS) phase diagram for the binary system nitrogen + ethane.

(1955), although they did not report the N_2 mole fraction range in which there is not any critical point for this binary system.

Figure A2-6 shows details of the *P-T* phase diagram of Figure A2-5 at temperatures between 115 and 135K. For the sake of simplicity, this figure presents only the experimental and calculated three-phase VLL line and the UCEP (or K-point). The vapor pressure for nitrogen is given for reference and it can be seen that *both* the experimental and calculated VLL loci closely parallel it. It should be noted, however, that the calculated VLL locus is predicted at higher pressures than the experimental one and that the position of the UCEP is found at a temperature and pressure lower than the experimental one. Despite the deviations between the experimentally measured and calculated with the SRK EoS VLL locus and UCEP point, which can be attributed to the binary interaction parameter, this equation gives a satisfactory representation of the experimentally observed phase behavior of nitrogen + ethane system.

A2.8.2 The nitrogen + propane system

Schindler (1966) and Schindler et al. (1966) reported VL and LL equilibrium data of 13 isotherms for the nitrogen + propane system from 123.2 to 353.2K and pressures ranging up to 137.86 bar. The experimental data reported by these authors show that above 298.2K, the critical point of the system was encountered at pressures below 137.86 bar and that at temperatures below 126.7K two liquid

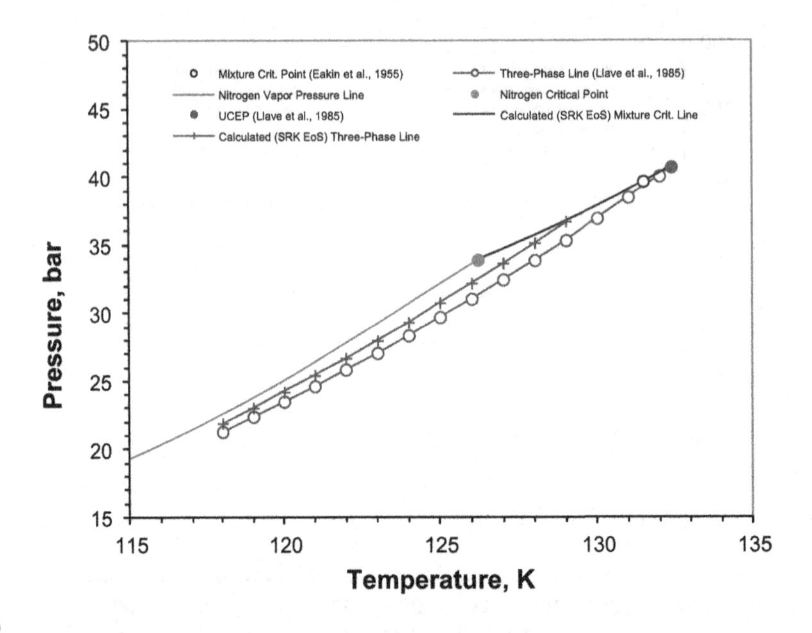

FIGURE A2-6

Details of the experimental and calculated (SRK EoS) phase diagram for the binary system nitrogen + ethane.

phases coexist. At a temperature below this value, the system is in vapor-liquid equilibrium up to pressure corresponding to the temperature in question on the three-phase locus.

This system was also studied by Kremer and Knapp (1983) at four isotherms from 120 to 127K and pressures up to 62.25 bar, and by Llave et al. (1985) over the temperature range from 117.0 to 126.62K and pressures ranging from 21.53 to 34.77 bar. The upper critical end point (K-point) for this binary system reported by Llave et al. (1985) is $T = 126.62$K and $p = 34.77$ bar, which is in a good agreement with the value reported by Schindler et al. (1966) of $T = 126.70$K and $p = 34.67$ bar.

Figure A2-7 shows on a *P-T* projection the experimental three-phase VLL locus of the system nitrogen + propane. The calculated three-phase line with the SRK EoS, which is in excellent agreement with the experimental data, is also shown. In this case, the estimated K-point obtained by interpolation of the calculated three-phase line is $T = 126.45$K and $p = 34.22$ bar, which is just slightly lower than the experimental data reported by Schindler et al. (1966) and by Llave et al. (1985). The experimental quadruple point (84.75K and 2.27 bar) reported by Schindler et al. (1966) is also shown Figure A2-7.

Figure A2-8 shows on a temperature-composition phase diagram the experimental and calculated (SRK EoS) mole fractions of nitrogen for the two liquid phases L_1 and L_2, and the vapor phase V. The calculated liquid phase L_1 mole fractions are in very good agreement with the experimental data of Llave et al. (1985). However, the calculated liquid phase L_2 mole fractions differ slightly from those reported by the same authors. The calculated mole fractions of the liquid phase L_2 and vapor phase V are very similar; these two phases are essentially pure nitrogen with concentrations higher than 99 mole %, which become identical at the K-point.

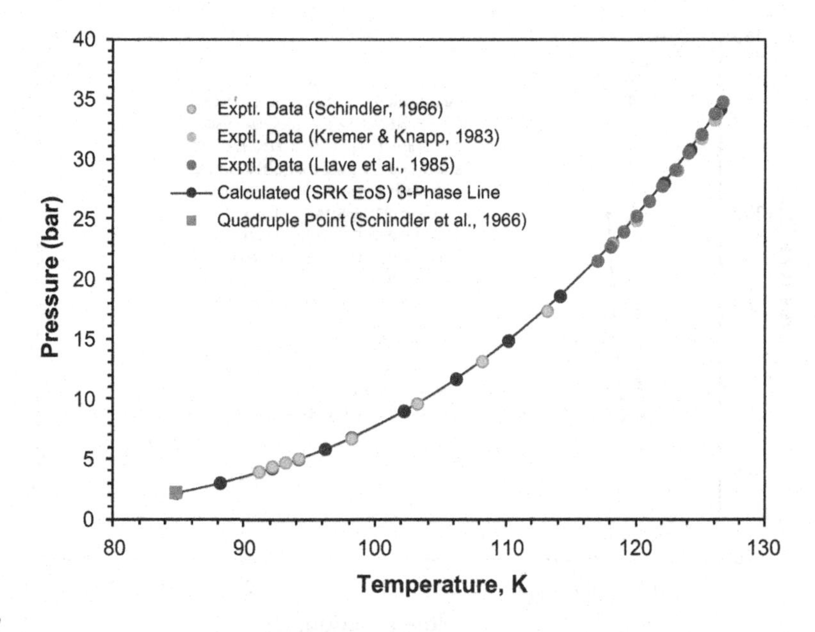

FIGURE A2-7

Experimental and calculated (SRK EoS) pressure-temperature phase diagram of the three-phase liquid-liquid-vapor equilibria in the binary system nitrogen + propane.

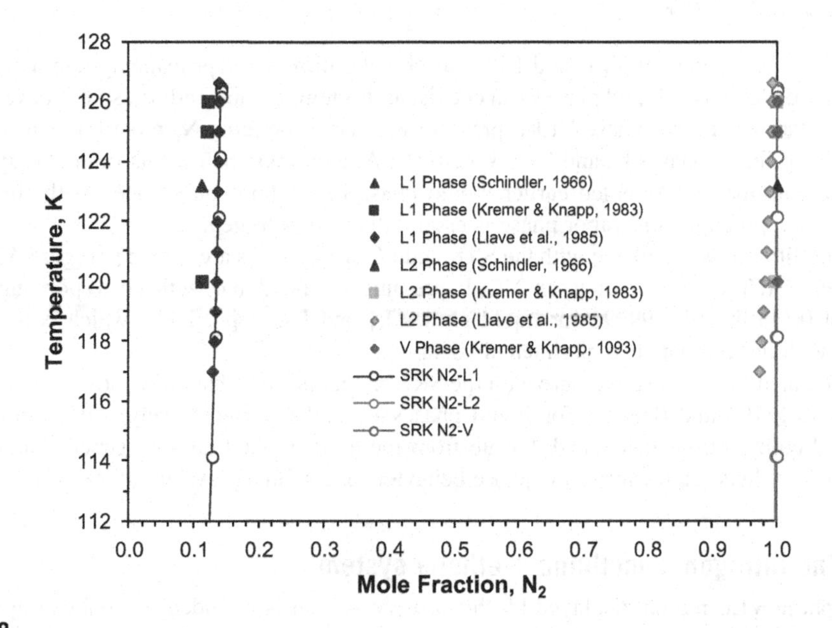

FIGURE A2-8

Experimental and calculated temperature-composition phase diagram for the binary system nitrogen + propane.

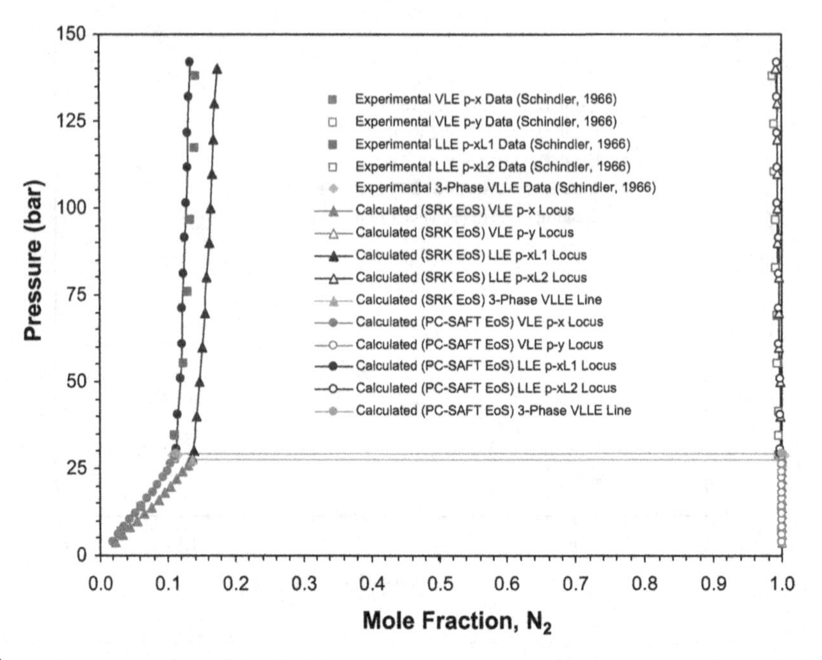

FIGURE A2-9

Experimental (Schindler, 1966) and calculated pressure-composition phase diagram for the binary system nitrogen + propane at 123.2K.

Schindler (1966) reported VLE and LLE data for the nitrogen + propane system at 123.2K. As shown in Figure A2-9, two liquid phases can coexist at this temperature and pressures above 29.09 bar, which is the estimated three-phase VLLE pressure with corresponding N_2 mole fractions of 0.11150 and 0.99962 for liquid phases L_1 and L_2, respectively. At a pressure below this value, the system is in vapor-liquid equilibrium. As noted earlier, liquid phase L_2 contains only traces of the high boiling component (i.e., propane) and vapor phase is essentially pure nitrogen.

The solubility limits calculated with the SRK and PC-SAFT EoSs are shown in Figure A2-9. In this case, the PC-SAFT model predicts an VLLE pressure of 29.13 bar with corresponding N_2 mole fractions of 0.11238 and 0.99996 for liquid phases L_1 and L_2, respectively, which is in very good agreement with the estimated "experimental" data.

The calculated three-phase pressure with the SRK model is 27.51 bar with corresponding N_2 mole fractions of 0.13702 and 0.99921 for liquid phases L_1 and L_2, respectively. Although the VLLE pressure and compositions calculated deviate from the experimental data, in general the thermodynamic model predicts satisfactorily the phase behavior of this binary system at low temperatures.

A2.8.3 The nitrogen + methane + ethane system

The three-phase VLL region displayed by this ternary system is bounded from above by a K-point locus, from below by a lower critical solution temperature (LCST) locus, while at low temperatures by a Q-point locus. Owing to the fact that the system contains a binary pair (nitrogen + ethane) that

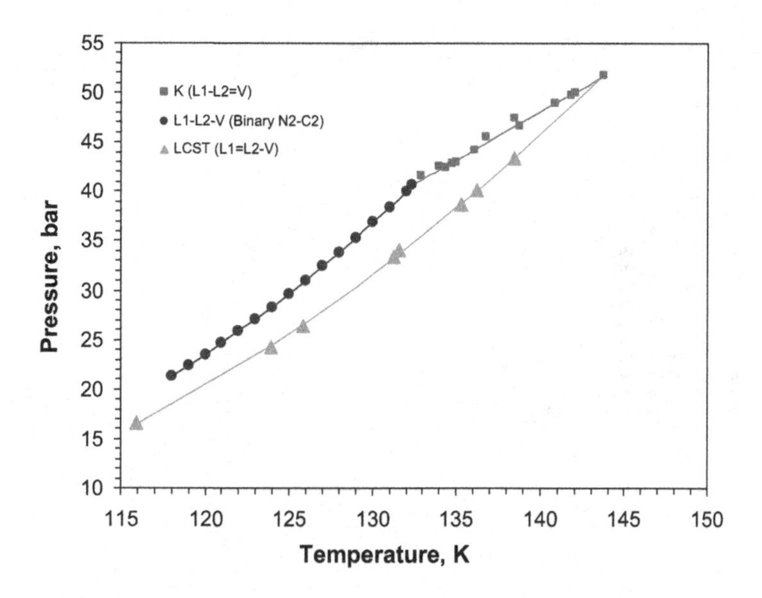

FIGURE A2-10

Experimental boundaries of the L_1-L_2-V immiscibility region for the system nitrogen +methane + ethane.

(Llave et al., 1985, 1987)

exhibits VLL behavior (Chang and Lu, 1967; Kremer and Knapp, 1983; Llave et al., 1985), its VLL space is truncated. In this case, the partially miscible pair nitrogen + ethane spans the VLL space from a position of the LCST locus to a position on the Q-point space. Because methane is of intermediate volatility compared with nitrogen and ethane, it creates a three-phase VLL space that extends from the binary VLL locus upward in temperature. The topographical nature of the regions of immiscibility for the system nitrogen + methane + ethane is shown in Figure A2-10. In this figure, it can be seen that the L-L=V and L=L-V critical end-point loci intersect at a tricritical point (L=L=V).

Figure A2-11 shows the good agreement between the experimental (Yu et al., 1969) and calculated with the PC-SAFT EoS L_1-L_2 phase behavior of the ternary system at $T = 113.7$K and different pressures.

An attempt to directly calculate the LCST points at $T = 113.7$K was carried out by using the procedure described earlier. As pointed out before, the use of good initial guesses, obtained from the VLL calculations close to the LCST, promotes the convergence of the Newton procedure. As a result, the SRK EoS LCST points at 113.7K (14.08 bar and 0.5461 N_2 mole fraction) predictions are completely acceptable (Figure A2-11). This figure also shows that at pressures away from the LCST point, the PC-SAFT model gives a good representation of the experimental compositions for both liquid phases.

Since there are not experimental data below 15.30 bar at 113.7K, a direct comparison of the model with experiment in the region of the LCST is not possible. However, by extrapolation of the experimental data, it was possible to determine the pressure of the "hypothetical" experimental LCST at $T = 113.7$ K; that is, $p \approx 15.24$ bar. The compositions (in mole fraction) at the LCST were obtained

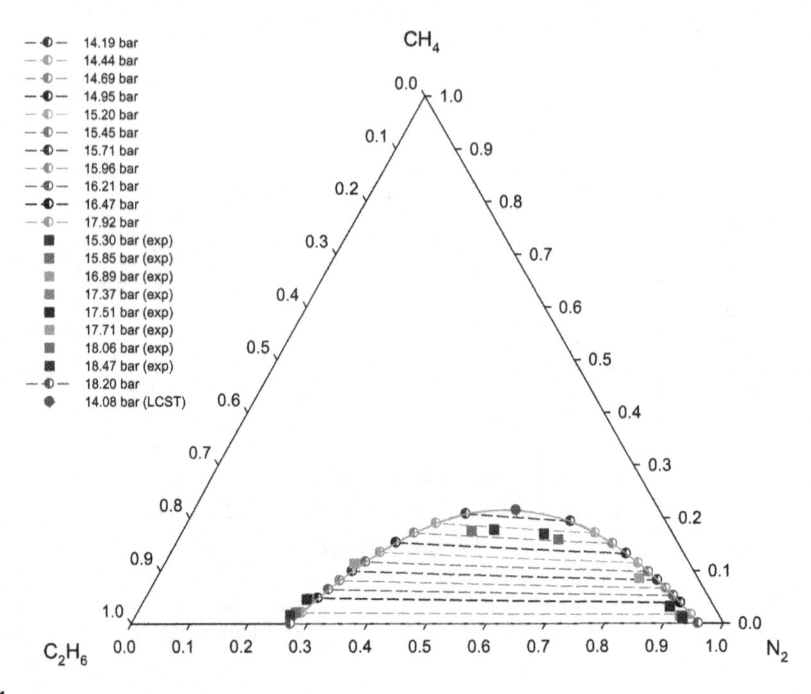

FIGURE A2-11

Experimental (Yu et al., 1969) and calculated (PC-SAFT EoS) liquid-liquid equilibrium diagram for the ternary system nitrogen + methane + ethane at 113.7K.

from the experimental tie-lines using the method of Treybal et al. (1946), and they are 0.5660 N_2, 0.1763 methane, and 0.2577 ethane.

A2.8.4 The nitrogen + methane + propane system

This system is topographically similar to the nitrogen + methane + ethane system, sharing the same sort of boundaries: K-point, Q-point, LCST, and binary (nitrogen + propane) VLL loci (Llave et al., 1985). Furthermore, the LCST and K-point loci intersect at a tricritical point. However, its three-phase VLL region extends over a much larger area and upward in temperature and pressure from the binary VLL locus in comparison to the nitrogen + methane + ethane system, as shown in Figure A2-12.

Figure A2-13 presents the experimental and calculated V-L_1-L_2 phase behavior (in terms of L_1-L_2 mole fraction data) for the nitrogen + methane + propane system at $T = 114.1$K and different pressures. The figure also shows the binodal curve with tie-lines and the LCST calculated. The SRK EoS was used as the thermodynamic model to represent the liquid and vapor phases. Overall, there is a good agreement between the experimental values of L_1 and L_2 phases, reported by Poon (1974) and Poon and Lu (1974), and those predicted with the model. It should be noted that the VLL calculations were performed up to almost the position of the binary VLL boundary ($p = 18.68$ bar) at $T = 114.1$ K, where methane mole fractions are practically zero in all three phases; that is, this ternary system exhibits LCST but not K-points at this temperature.

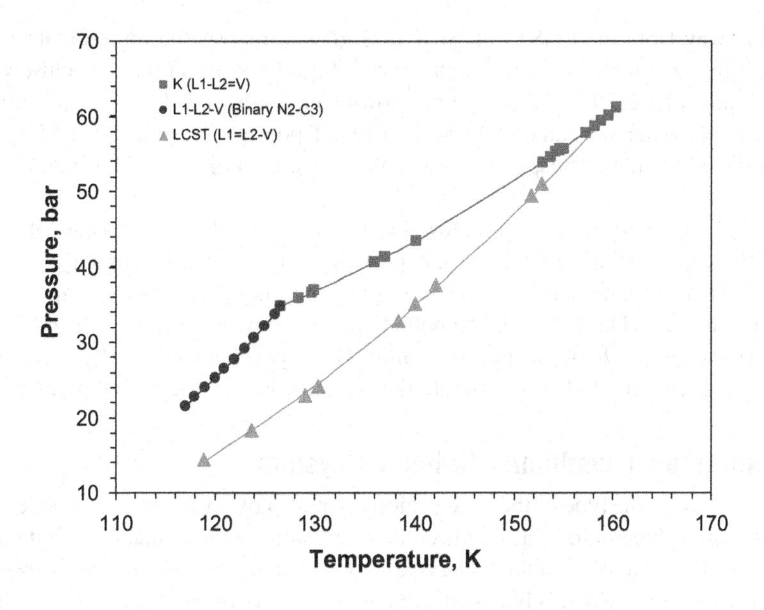

FIGURE A2-12

Experimental boundaries of the L1-L2-V immiscibility region for the system nitrogen + methane + propane.

(Llave et al., 1985, 1987)

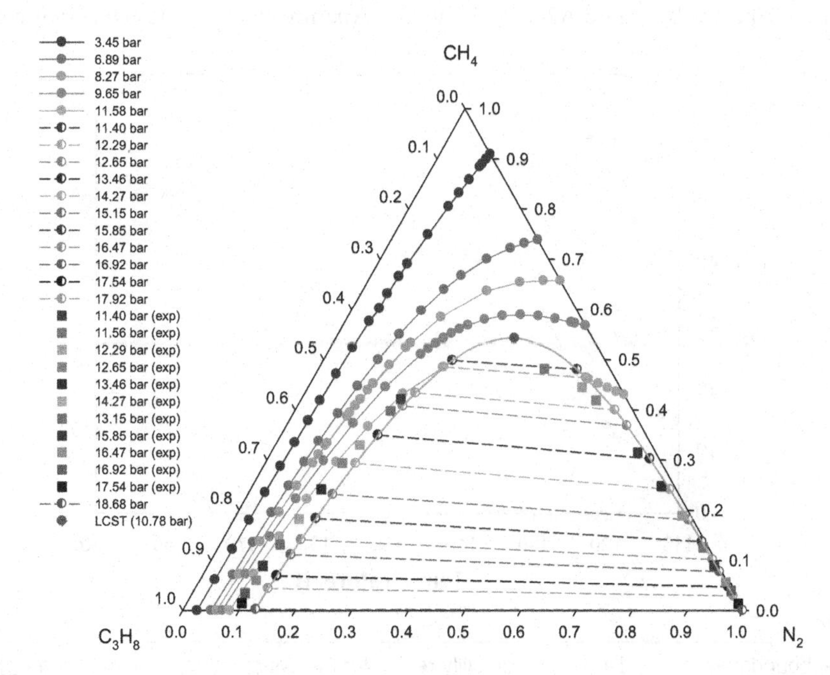

FIGURE A2-13

Experimental (Poon, 1974; Poon and Lu, 1974) and calculated (SRK EoS) liquid compositions and liquid-liquid equilibrium diagram for the ternary system nitrogen + methane + propane at 114.1K.

At pressures away from the LCST, the SRK EoS gives a reasonable representation of the experimental compositions for liquid phase L_1 (ethane-rich liquid phase) and agrees closely for L_2 phase (nitrogen-rich liquid phase). The LCST point showed in Figure A2-13 was determined using the procedure presented earlier for calculating K- and LCST-points. The calculated LCST at 114.1K is 10.78 bar with the following corresponding mole fractions: 0.3176 N_2, 0.5424 methane, and 0.1400 propane.

Since there are no experimental data below 11.40 bar at 114.1K, comparisons of the model with experiment in the region of the LCST are not possible. However, an estimate of the hypothetical experimental LCST was obtained by extrapolating the experimental pressure-composition data of the liquid phases L_1 and L_2. The estimated hypothetical experimental pressure at the LCST point is $p \approx 11.05$ bar. However, due to the scatter of the liquid L_2 experimental data, it was not possible to use the method of Treybal et al. (1946) to estimate the compositions at the LCST point.

A2.8.5 The nitrogen + methane + *n*-butane system

As mentioned previously, the type of the VLL region displayed by ternary systems depends on whether they contain an immiscible pair or not. In this context, the nitrogen + methane + *n*-butane system does not exhibit immiscibility in any of their binary pairs. Consequently, the three-phase region displayed by this system is triangular, which is bounded from above by a K-point locus (L-L=V), from below by a LCST (L=L-V) locus, and by a Q-point locus (S-L-L-V) at low temperatures. These three loci intersect at invariant points for this system: the K-point and LCST loci at a tricritical point, whereas the Q-point locus terminates at a point of the type S-L-L=V from above and at a point of the type S-L=L-V from below, respectively. Figure A2-14 presents the experimental pressure-temperature diagram of

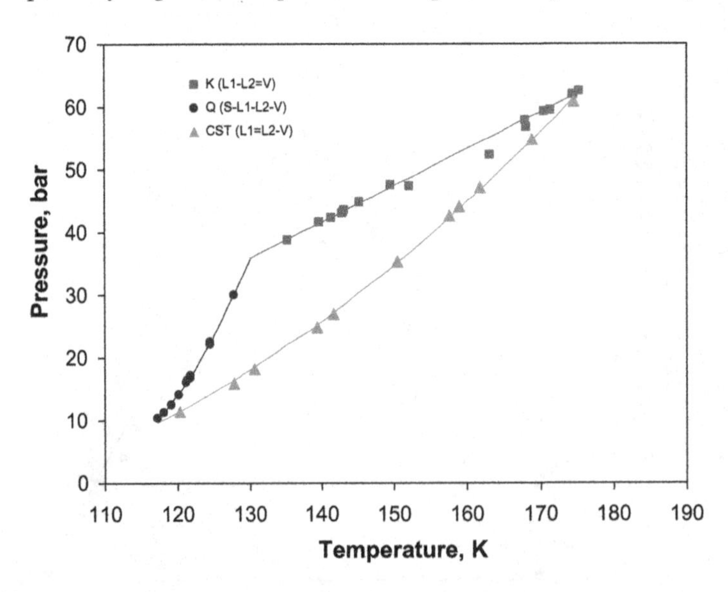

FIGURE A2-14

Experimental boundaries of the L1-L2-V immiscibility region for the system nitrogen + methane + *n*-butane.

(Merrill et al., 1984a)

the three-phase VLL space developed by this ternary system (Merrill et al., 1984a). This is a very interesting system because it is in contrast to the other ternary systems examined, which did not exhibit L_1-L_2-V behavior outside the boundary loci when projected on *P-T* space. That is, above 150K the critical solution temperature (CST) locus is comprised of LCST points, similar to those observed in other nitrogen-containing ternary systems (Merrill et al., 1984b; Llave et al., 1987; Chen et al., 1989) whereas below 150K the CST boundary changes to an UCST (upper critical solution temperature) locus, in a manner similar to that exhibited by the systems methane + 2-methylpentane + 2-ethyl-1-butene and methane + 3,3-dimethylpentane + 2-methylhexane reported by van Konynenburg (1968). Figure A2-14 shows the topographical nature of the regions of immiscibility for the system nitrogen + methane + *n*-butane. As can be seen, the L-L=V and L=L-V critical end-point loci intersect at a tricritical point (L=L=V).

VLL calculations have been performed at 159.5K with both the PC-SAFT and SRK models, where the obtained results are presented in Figure A2-15. As can be seen, the liquid L_1 and L_2 nitrogen mole fractions predicted with the SRK are in a better agreement with the experimental liquid L_1 data than those predicted with the PC-SAFT. However, the liquid L_2 nitrogen mole fractions predicted with both thermodynamic models do not agree well with the experimental data; that is, the SRK underpredicts the experimental data whereas the PC-SAFT overpredicted them. In fact, Figure A2-15 shows that the PC-SAFT overpredicts both liquid phases. This could be a result of the fact that the interaction parameter of the nitrogen + *n*-butane binary system for this model was determined at temperatures higher than those studied here.

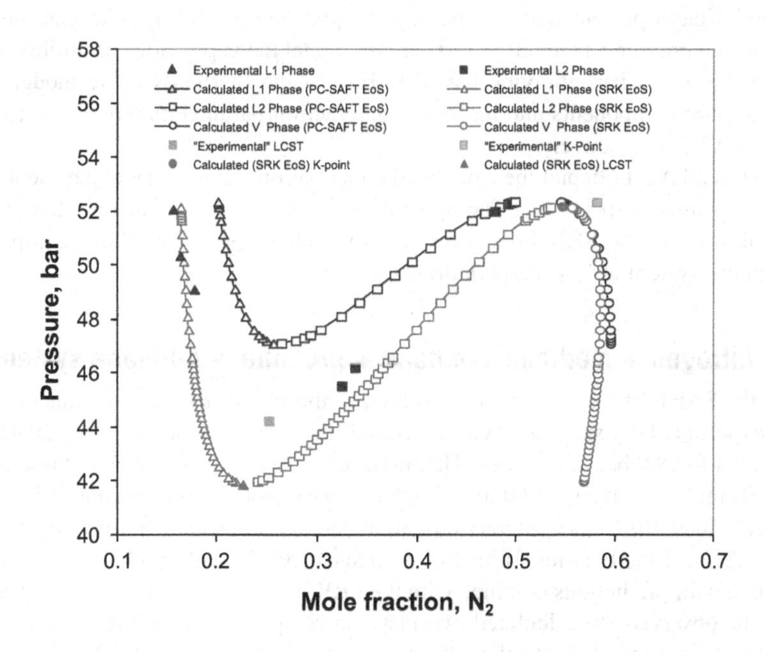

FIGURE A2-15

Experimental (Merrill et al., 1984a) and calculated N_2 mole fractions for the coexisting L_1 and L_2 phases of the system nitrogen + methane + *n*-butane at 159.5K.

Because at 159.2K two critical end points—a K-point and an LCST—exist on the pressure-temperature phase diagram, those were calculated with the SRK model using the aforementioned procedure, which gave a value of 52.20 bar (0.5463 N_2 mole fraction) and 41.88 bar (0.2256 N_2 mole fraction) for the K-point and LCST, respectively.

Although there is not any experimental K-point or LCST data for this isotherm, it was possible to estimate it by interpolation. The hypothetical experimental K-point and LCST values reported by Merrill (1983) are, respectively, 52.35 bar (0.5808 N_2 mole fraction) and 44.25 bar (0.2513 N_2 mole fraction), and they are also presented in Figure A2-15. In this case, the experimental K-point and LCST values are, respectively, 0.15 and 2.37 bar above the calculated ones with the SRK model. The estimated (by interpolation) K-point and LCST with the PC-SAFT model are 52.39 and 47.11 bar, respectively, which are 0.04 and 2.86 bar above the estimated experimental values, respectively.

A2.8.6 The nitrogen + methane + ethane + propane system

Though it has been well recognized that the design of natural gases' separation processes requires phase equilibrium data, most of the data available in the literature concerns ternary systems (see, for example, Cosway and Katz, 1959; Chang and Lu, 1967; Yu et al., 1969; Poon, 1974; Poon and Yu, 1974; Kremer and Knapp, 1983; Merrill et al., 1984a; Llave et al., 1987; to name just a few). To the best of our knowledge, quaternary vapor-liquid equilibrium data of systems including components of greatest interest to natural gas are limited to the experimental data reported by Trappehl and Knapp (1989).

Trappehl and Knapp present data on the vapor-liquid equilibrium of nitrogen-methane-ethane-propane at 200K and pressures from 20 to 120 bar. We model the vapor-liquid equilibria of this system at $T = 200$K and $p = 20$ bar, applying the SRK EoS as the thermodynamic model. The physical properties of the pure components and the binary interaction parameters required are those given in Tables A2-1 and A2-2, respectively.

Figures A2-16 and A2-17 depict the calculated compositions against the experimental ones for the liquid and vapor phases, respectively. The agreement, in most cases, is quantitative, which demonstrates the capabilities of the SRK EoS to represent the phase behavior of this complex quaternary nitrogen-containing system at low temperatures.

A2.8.7 The nitrogen + methane + ethane + propane + *n*-butane system

The SRK and PC-SAFT EOSs were applied to predict the phase behavior for this complex quinary LNG mixture experimentally studied by Gonzalez and Lee (1968) in the 116.5 to 220.4K temperature range and from 2.4 to 58.0 bar in pressure. The mixture composition reported by these authors is 1.60 mol % N_2, 94.50 mol % methane, 2.60 mol % ethane, 0.81 mol % propane, and 0.52 mol % butane.

For all calculations, the binary interaction parameters used for the SRK and PC-SAFT EoSs are those of Table A2-2 for the binaries formed of *n*-alkanes and N_2 with *n*-alkanes.

The phase behavior predictions obtained with the SRK and PC-SAFT EoSs are displayed in Figure A2-18. As can be observed the calculated two-phase envelopes exhibit a wide retrograde zone. This was verified with direct critical point calculations performed with the SRK (202.2K and 56.78 bar) and PC-SAFT (199.5K and 53.63 bar) models by using the procedures presented by Justo-García et al. (2008a,b). For the SRK model, the simulated annealing algorithm (Goffe et al., 1994) was used for

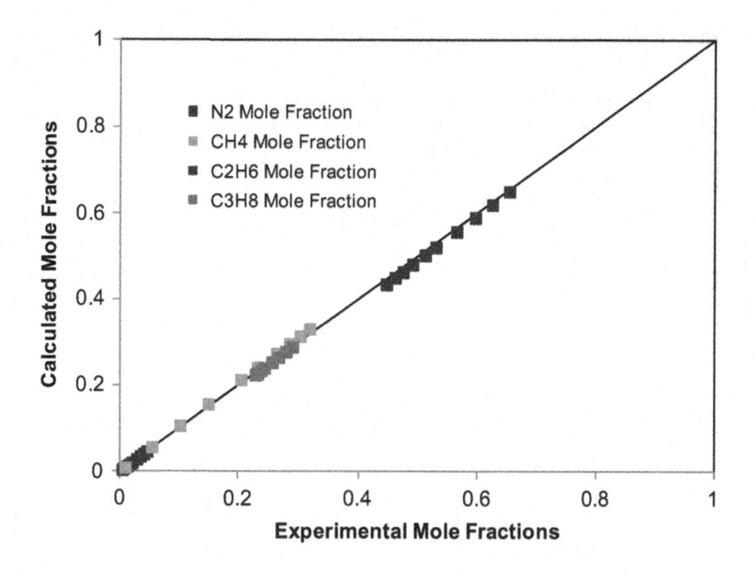

FIGURE A2-16

Experimental (Trappehl and Knapp, 1989) and calculated (SRK EoS) mole fractions of the liquid phase for the system nitrogen + methane + ethane + propane at 200K and 20 bar.

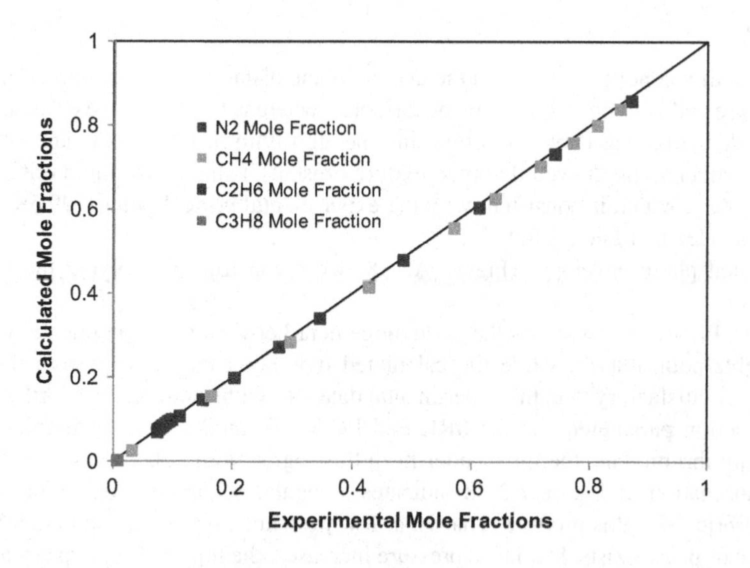

FIGURE A2-17

Experimental (Trappehl and Knapp, 1989) and calculated (SRK EoS) mole fractions of the vapor phase for the system nitrogen + methane + ethane + propane at 200K and 20 bar.

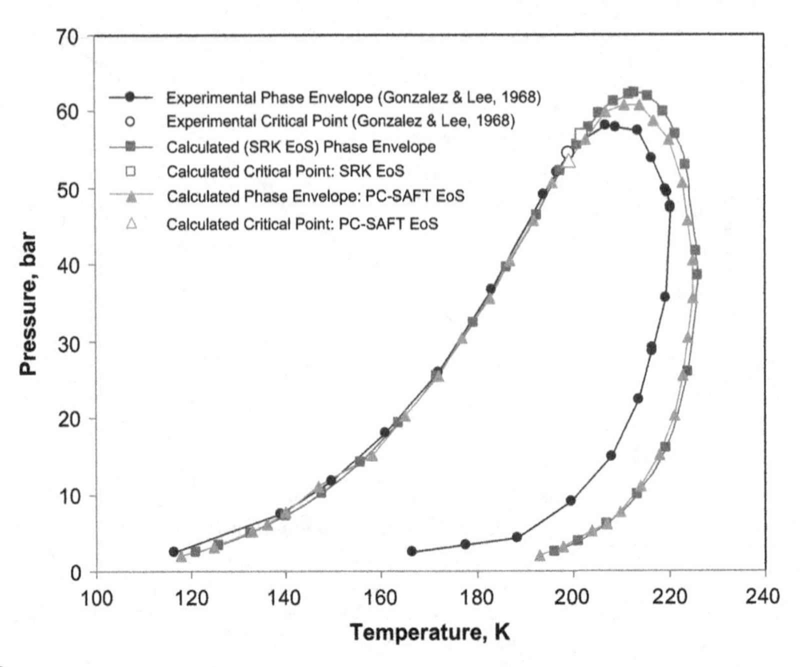

FIGURE A2-18

Experimental (Gonzalez and Lee, 1968) and calculated phase envelope for the system nitrogen + methane + ethane + propane + *n*-butane.

solving the criticality conditions based on the tangent plane distance in terms of the Helmholtz energy using temperature and volume as independent variables, whereas for the PC-SAFT model, the mixture critical point calculation was carried out by using the algorithm of Heidemann and Khalil (1980). In both cases, the calculations showed that this mixture presents a single gas-liquid critical point on the two-phase boundary, which is consistent with the experimental critical point (199.5K and 54.57 bar) reported by Gonzalez and Lee (1968).

The calculated phase envelopes (Figure A2-18) were constructed applying the two procedures presented earlier.

Furthermore, Figure A2-18 shows the good agreement between the experimentally measured and calculated bubble-point curves, while the calculated dew-point curves with both models are very similar to but less satisfactory than the experimental data. Notwithstanding, it is worth mentioning that since the interaction parameters for the SRK and PC-SAFT models were determined from binary vapor-liquid equilibrium data, the rather poor fit in this region with either model is not unexpected.

Further examination of Figure A2-18 indicates a regular sequential consistent behavior of the phases at equilibrium for this mixture. That is, at low pressures, on the dew-point curve side, only a single stable vapor phase exists. Insofar as pressure increases, the liquid phase appears and the mixture becomes stable as vapor-liquid. At higher pressures, the liquid disappears and the mixture becomes stable as vapor.

References

Ambrose, D., 1980. Vapour-Liquid Critical Properties. NPL Report Chem. 107, National Physical Laboratory, Teddington.

Arce, P., Aznar, M., 2007. Modeling of Critical Lines and Regions for Binary and Ternary Mixtures Using Non-Cubic and Cubic Equations of State. Journal of Supercritical Fluids 42, 1–26.

Asselineau, L., Bogdanic, G., Vidal, J., 1979. A Versatile Algorithm for Calculation of Vapour-Liquid Equilibria. Fluid Phase Equilibria 3, 273–290.

Baker, L.E., Luks, K.D., 1980. Critical Point and Saturation Pressure Calculations for Multicomponent Systems. SPE Journal 20, 15–24.

Baker, L.E., Pierce, A.C., Luks, K.D., 1982. Gibbs Energy Analysis of Phase Equilibria. SPE Journal 22, 731–742.

Chang, S.-D., Lu, B.C.-Y., 1967. Vapor-Liquid Equilibria in the Nitrogen-Methane-Ethane. Chem. Eng. Progr. Symp. Ser. 63 (81) 18–27.

Chen, S.S., Kreglewski, A., 1977. Applications of the Augmented Van der Waals Theory of Fluids. I. Pure Fluids. Ber. Bunsenges. Phys. Chem. 81, 1048–1052.

Chen, W.-L., Luks, K.D., Kohn, J.P., 1989. Three-Phase Liquid-Liquid-Vapor Equilibria in the Nitrogen + Methane + n-Heptane System. J. Chem. Eng. Data 34, 312–314.

Cosway, H.F., Katz, D.L., 1959. Low-Temperature Vapor-Liquid Equilibria in Ternary and Quaternary Systems Containing Hydrogen, Nitrogen, Methane, and Ethane. AIChE Journal 5, 47–50.

Eakin, B.E., Ellington, R.T., Gami, D.C., 1955. "Physical-Chemical Properties of Ethane-Nitrogen Mixtures", Institute of Gas Technology Research Bulletin No. 26. Chicago, IL, USA.

Fletcher, R., 1980. Practical Methods for Optimization- Vol. 1: Unconstrained Optimization. John Wiley & Sons, New York, NY, USA.

García-Sánchez, F., Schwartzentruber, J., Ammar, N.M., Renon, H., 1996a. Modeling of Multiphase Liquid Equilibria in Multicomponent Systems. Fluid Phase Equilibria 121, 207–225.

García-Sánchez, F., Hernández-Garduza, O., García-Flores, B.E., Eliosa-Jiménez, G., 1996b. Prediction of the Azeotropic Behaviour of Binary Mixtures from Gas-Liquid P-T-x Critical Data with Cubic Equations of State. Can. J. Chem. Eng. 74, 967–976.

García-Sánchez, F., Schwartzentruber, J., Ammar, N.M., Renon, H., 1997. Thermodynamics of Multiphase Liquid Equilibria for Multicomponent Systems. Rev. Mex. Fís. 43, 59–92.

García-Sánchez, F., Eliosa-Jiménez, G., Salas-Padrón, A., Hernández-Garduza, O., Ápam-Martínez, D., 2001. Modeling of Microemulsion Phase Diagrams from Excess Gibbs Energy Models. Chem. Eng. J. 84, 257–274.

García-Sánchez, F., Eliosa-Jiménez, G., Silva-Oliver, G., Vázquez-Román, R., 2004. Vapor-Liquid Equilibria of Nitrogen-Hydrocarbon Systems Using the PC-SAFT Equation of State. Fluid Phase Equilibria 217, 241–253.

Goffe, W.L., Ferrier, G.D., Rogers, J., 1994. Global Optimization of Statistical Functions with Simulated Annealing. J. Econometrics 60, 65–99.

Gonzalez, M.H., Lee, A.L., 1968. Dew and Bubble Points of Simulated Natural Gases. J. Chem. Eng. Data 13, 172–176.

Green, K.A., Tiffin, D.L., Luks, K.D., Kohn, J.P., 1976. Solubility of Hydrocarbons in LNG, NGL. Hydrocarbon Processing 56, 251–256.

Gregorowicz, J., de Loos, Th.W., 1996. Modelling of the Three Phase LLV Region for Ternary Mixtures with the Soave-Redlich-Wong Equation of State. Fluid Phase Equilibria 118, 121–132.

Gross, J., Sadowski, G., 2001. Perturbed-Chain SAFT: An Equation of State Based on Perturbation Theory for Chain Molecules. Ind. Eng. Chem. Res. 40, 1244–1260.

Heidemann, R.A., Khalil, A.M., 1980. The Calculation of Critical Points. AIChE Journal 26, 769–779.

Hottovy, J.D., Kohn, J.P., Luks, K.D., 1981. Partial Miscibility of the Methane-Ethane-n-Octane. J. Chem. Eng. Data 26, 135–137.

Hottovy, J.D., Kohn, J.P., Luks, K.D., 1982. Partial Miscibility Behavior of the Ternary Systems Methane-Propane-n-Octane, Methane-n-Butane-n-Octane, and Methane-Carbon Dioxide-n-Octane. J. Chem. Eng. Data 27, 298–302.

Justo-García, D.N., García-Sánchez, F., Romero-Martínez, A., 2008a. Isothermal Multiphase Flash Calculations with the PC-SAFT Equation of State. Am. Inst. Phys. Conf. Proc. 979, 195–214.

Justo-García, D.N., García-Sánchez, F., Díaz-Ramírez, N.L., Romero-Martínez, A., 2008b. Calculation of Critical Points for Multicomponent Mixtures Containing Hydrocarbon and Nonhydrocarbon Components with the PC-SAFT Equation of State. Fluid Phase Equilibria 265, 192–204.

Justo-García, D.N., García-Sánchez, F., Águila-Hernández, J., Eustaquio-Rincón, R., 2008c. Application of the Simulated Annealing Technique to the Calculation of Critical Points of Multicomponent Mixtures with Cubic Equations of State. Fluid Phase Equilibria 264, 195–214.

Knapp, H.R., Döring, R., Oellrich, L., Plöcker, U., Prausnitz, J.M., 1982. Vapor-Liquid Equilibria for Mixtures of Low Boiling Substances. DECHEMA Chemistry Data Series Vol. VI. Frankfurt, Germany.

Kohn, J.P., 1961. Heterogeneous Phase and Volumetric Behavior of the Methane-n- Heptane System at Low Temperatures. AIChE Journal 7, 514–518.

Kohn, J.P., Bradish, W.F., 1964. Multiphase and Volumetric Equilibria of the Methane-n-Octane System at Temperatures between –10° and 150°C. J. Chem. Eng. Data 9, 5–8.

Kohn, J.P., Luks, K.D., 1976. "Solubility of Hydrocarbons in Cryogenic LNG and NGL Mixtures", Research Report RR-22. Gas Processors Association, Tulsa, OK, USA.

Kohn, J.P., Luks, K.D., 1977. "Solubility of Hydrocarbons in Cryogenic LNG and NGL Mixtures", Research Report RR-27. Gas Processors Association, Tulsa, OK, USA.

Kohn, J.P., Luks, K.D., 1978. "Solubility of Hydrocarbons in Cryogenic LNG and NGL Mixtures", Research Report RR-33. Gas Processors Association, Tulsa, OK, USA.

Kohn, J.P., Luks, K.D., Liu, P.H., 1976. Three-Phase Solid-Liquid-Vapor Equilibria of Binary-n-Alkane Systems (Ethane-n-Octane, Ethane-n-Decane, Ethane-n-Dodecane). J. Chem. Eng. Data 21, 360–362.

Kreglewski, A., 1984. Equilibrium Properties of Fluids and Fluid Mixtures. Texas A&M University Press, College Station, TX, USA.

Kremer, H., Knapp, H., 1983. Three-Phase Conditions Are Predictable: VLE and VLLE in Mixtures of N_2 or CO with C_2H_6 or C_3H_8 Correlate with Generalized Equations of State with Acceptable Accuracy. Hydrocarbon Processing 62, 79–83.

Li, Y.-K., Nghiem, L.X., 1982. The Development of a General Phase Envelope Construction Algorithm for Reservoir Fluid Studies. SPE 11198, paper presented at the 57th Annual Fall Technical Conference and Exhibition of the Society Petroleum Engineers of AIME, New Orleans, LA, USA. Sept.26–29.

Lindeloff, N., Andersen, S.I., Stenby, E.H., Heidemann, R.A., 1999. Phase-Boundary Calculations in Systems Involving More Than Two Phases with Application to Hydrocarbon Mixtures. Ind. Eng. Chem. Res. 38, 1107–1113.

Lindeloff, N., Michelsen, M.L., 2003. Phase Envelope Calculations for Hydrocarbon-Water Mixtures. SPE Journal 8, 298–303.

Llave, F.M., Luks, K.D., Kohn, J.P., 1985. Three-Phase Liquid-Liquid-Vapor Equilibria in the Binary Systems Nitrogen + Ethane and Nitrogen + Propane. J. Chem. Eng. Data 30, 435–438.

Llave, F.M., Luks, K.D., Kohn, J.P., 1987. Three-Phase Liquid-Liquid-Vapor Equilibria in the Nitrogen-Methane-Ethane and Nitrogen-Methane-Propane. J. Chem. Eng. Data 32, 14–17.

Luks, K.D., Kohn, J.P., 1978. The Topography of Multiphase Equilibria Behavior: What Can it Tell the Design Engineer. Proc. 63rd Annual Convention. Gas Processors Association, Tulsa, OK, USA.

Merrill, R.C., 1983. Liquid-Liquid-Vapor in Cryogenic Liquefied Natural Gas Systems. PhD Dissertation. University of Notre Dame, Notre Dame, IN, USA.

Merrill, R.C., Luks, K.D., Kohn, J.P., 1983. Three-Phase Liquid-Liquid-Vapor Equilibria in the Methane-n-Pentane-n-Octane, Methane-n-Hexane-n-Octane, and Methane-n-Hexane-Carbon Dioxide. J. Chem. Eng. Data 28, 210–215.

Merrill, R.C., Luks, K.D., Kohn, J.P., 1984a. Three-Phase Liquid-Liquid-Vapor Equilibria in the Methane-n-Butane-Nitrogen System. Adv. Cryog. Eng. 29, 949–955.

Merrill, R.C., Luks, K.D., Kohn, J.P., 1984b. Three-Phase Liquid-Liquid-Vapor Equilibria in the Methane-n-Hexane-Nitrogen and Methane-n-Pentane-Nitrogen Systems. J. Chem. Eng. Data 29, 272–276.

Michelsen, M.L., 1980. Calculation of Phase Envelopes and Critical Points for Multicomponent Mixtures. Fluid Phase Equilibria 4, 1–10.

Michelsen, M.L., 1982a. The Isothermal Flash Problem. Part I. Stability. Fluid Phase Equilibria 9, 1–19.

Michelsen, M.L., 1982b. The Isothermal Flash Problem. Part II. Phase-Split Calculation. Fluid Phase Equilibria 9, 21–40.

Michelsen, M.L., 1984. Calculation of Critical Points and Phase Boundaries in the Critical Region. Fluid Phase Equilibria 16, 57–76.

Michelsen, M.L., Heidemann, R.A., 1981. Calculation of Critical Points from Cubic Two-Constant Equations of State. AIChE Journal 27, 521–523.

Michelsen, M.L., Heidemann, R.A., 1988. Calculation of Tri-Critical Points. Fluid Phase Equilibria 39, 53–74.

Michelsen, M.L., Mollerup, J.M., 2004. Thermodynamic Models: Fundamentals & Computational Aspects. Tie-Line Publications, Denmark.

Nagarajan, N.R., Cullick, A.S., Griewank, A., 1991. New Strategy for Phase Equilibria and Critical Point Calculations by Thermodynamic Energy Analysis- Part II. Critical Point Calculations. Fluid Phase Equilibria 62, 211–223.

Nagy, Z., Shirkovskiy, I.A., 1982. Mathematical Simulation of Natural Gas Condensation Processes Using the Peng-Robinson Equation of State. SPE 10982, paper presented at the 57th Annual Fall Technical Conference and Exhibition of the Society of Petroleum Engineers of AIME, New Orleans, LA, USA. Sept. 26–29.

Nghiem, L.X., Li, Y.-K., 1984. Computation of Multiphase Equilibrium Phenomena with an Equation of State. Fluid Phase Equilibria 17, 77–95.

Orozco, C.E., Tiffin, D.L., Luks, K.D., Kohn, J.P., 1977. Solids Fouling in LNG Systems. Hydrocarbon Processing 56, 325–328.

Peng, D.-Y., Robinson, D.B., 1976. A New Two-Constant Equation of State. Ind. Eng. Chem. Fundam 15, 59–64.

Peng, D.-Y., Robinson, D.B., 1977. A Rigorous Method for Predicting the Critical Properties of Multicomponent Systems from an Equation of State. AIChE Journal 23, 137–144.

Peters, G., Wilkinson, J.H., 1979. "Inverse Iteration, Ill-Conditioned Equations and Newton's Method", SIAM. Rev. 21, 339–360.

Poon, D.P.L., 1974. Phase Behavior of Systems Containing Nitrogen, Methane, and Propane. PhD Thesis. University of Ottawa, Ottawa, ON, Canada.

Poon, D.P.L., Lu, B.C.-Y., 1974. Phase Equilibria for Systems Containing Nitrogen, Methane, and Propane. Adv. Cryog. Eng. 19, 292–299.

Prausnitz, J.M., Lichtenthaler, R.N., Gomes de Acevedo, E., 1999. Molecular Thermodynamics of Fluids-Phase Equilibria, third ed. Prentice Hall PTR: Upper Saddle River, New Jersey, NJ, USA.

Press, W.H., Teukolsky, S.A., Vetterling, W.T., Flannery, B.P., 1992. Numerical Recipes: The Art of Scientific Computing, 2nd Edition. Cambridge University Press, New York, NY, USA.

Ramírez-Jiménez, E., Justo-García, D.N., García-Sánchez, F., Stateva, R.P., 2012. VLL Equilibria and Critical End Points Calculation of Nitrogen-Containing LNG Systems: Application of SRK and PC-SAFT Equations of State. Ind. Eng. Chem. Res. 51, 9409–9418.

Reid, R.C., Prausnitz, J.M., Poling, B.E., 1987. The Properties of Gases and Liquids, 4th Edition. McGraw-Hill, New York, NY, USA.

Rowley, R.L., Wilding, W.V., Oscarson, J.L., Yang, Y., Zundel, N.A., 2006. DIPPR Data Compilation of Pure Chemical Properties. Design Institute for Physical Properties. Brigham Young University, Provo, UT, USA.

Schindler, D.L., 1966. The Heterogeneous Phase Behavior of the Helium-Propane, Nitrogen-Propane, and Helium-Nitrogen-Propane Systems. PhD Dissertation. The University of Kansas, Lawrence, KS, USA.

Schindler, D.L., Swift, G.W., Kurata, F., 1966. More Low Temperature V-L Design Data. Hydrocarbon Processing 45, 205–210.

Shim, J., Kohn, J.P., 1962. Multiphase and Volumetric Equilibria of the Methane-n-Hexane System at Temperatures between −10° and 150°C. J. Chem. Eng. Data 7, 2–8.

Soave, G., 1972. Equilibrium Constants from a Modified Redlich-Kwong Equation of State. Chem. Eng. Sci. 27, 1197–1203.

Stateva, R.P., Tsvetkov, S.G., 1994. A Diverse Approach for the Solution of the Isothermal Multiphase Flash Problem. Application to Vapor-Liquid-Liquid Systems. Can. J. Chem. Eng. 72, 722–734.

Stateva, R.P., 1995. Predicting Calculating Complex Phase Equilibrium in Supercritical Fluid Systems with a New Technique. Collect. Czech. Chem. Commun. 60, 188–210.

Stateva, R.P., Cholakov, G.S., Galushko, A.A., Wakeham, W.A., 2004. A Powerful Algorithm for Liquid-Liquid-Liquid Equilibria Predictions and Calculations. Chem. Eng. Sci. 55, 2121–2129.

Stateva, R.P., Wakeham, W.A., 1997. Phase Equilibrium Calculations for Chemically Reacting Systems. Ind. Eng. Chem. Res. 36, 5474–5482.

Trappehl, G., Knapp, H., 1989. Vapour-Liquid Equilibria in the Ternary Mixture N_2-CH_4-CO_2 and the Quaternary Mixture N_2-CH_4-C_2H_6-C_3H_8. Cryogenics 29, 42–50.

Treybal, R.E., Weber, L.D., Daley, J.F., 1946. The System Acetone-Water-1,1,2-Trichloroethane. Ternary Liquid and Binary Vapor Equilibria. Ind. Eng. Chem. 38, 817–821.

Van Konynenburg, P.H., 1968. Critical Lines Phase Equilibria in Binary Mixtures. PhD Dissertation. University of California, Los Angeles, CA, USA.

Van Konynenburg, P.H., Scott, R.L., 1980. Critical Lines and Phase Equilibria in Binary Van der Waals Mixtures", Phil. Trans. Roy. Soc. London. Ser. A 298, 495–540.

Wakeham, W.A., Stateva, R.P., 2004. Numerical Solution of the Isothermal Multiphase Flash Problem", Rev. Chem. Eng. J. 20, 1–56.

Wilkinson, J.H., 1965. The Algebraic Eigenvalue Problem. Clarendon, Oxford, UK.

Yu, P., Elshayal, I.M., Lu, B.C.-Y., 1969. Liquid-Liquid-Vapor Equilibria in the Nitrogen-Methane-Ethane System. Can. J. Chem. Eng. 47, 495–498.

Zhang, H., Bonilla-Petriciolet, A., Rangaiah, G.P., 2011. A Review on Global Optimization Methods for Phase Equilibrium Modeling and Calculations. The Open Thermodynamics J. 5, 71–92.

APPENDIX

3 LNG Plant Relief and Flare Systems Design Guideline

A3.1 Introduction

The LNG plant process units are designed with controls that ensure equipment is operating within their design pressure and temperature limits, and maintain a stable operation. During startup, plant upset, or emergency situations, pressure will build up such that excess pressure or flow must be relieved to the flare systems to maintain a safe operation. The objectives of a flare and relief system are:

- Providing adequate safety for personnel and equipment, complying with all safety laws, design codes, and standards
- Minimizing atmospheric discharge, reducing environmental impacts from the relief
- Recovering boil-off vapors and liquids for reuse where economically viable
- Minimizing community impacts by reduction in emissions, flare luminescence, noise, and smoke.

The following sections describe the design and operation considerations in a flare and relief system, addressing the emergency scenarios as well as the normal operational requirements of the flare system.

A3.2 Flare and relief design considerations

The relief and flare system designs should accommodate the maximum relief loads, during emergency and off-design operations from the process units that may be required to continue operation.

The design of the flare system begins with an assembly of engineering documents that define the basis for the flare design. Typically, this would include:

- Heat and material balances of process units and utility systems
- A flare and blowdown distribution diagram
- Power distribution and one-line electrical diagram
- A plot plan with equipment location details.

A3.2.1 Plot plan consideration

Review the plot plan with equipment locations, especially vessels containing light hydrocarbons. If the equipment is grouped too closely together, a very large relief load may result during a fire, resulting in a large relief system. This activity should be completed early in the project to avoid expensive changes later in the project.

A3.2.2 Process and control system considerations

- Higher equipment design pressures to reduce relief loads
- More reliable power distribution system to avoid or reduce the impact of a power failure
- Highly reliable, double-lead electrical systems fed from dual power grids
- Instrumentation to automatically remove the source of pressure or heat that are the cause of overpressure
- Reliable driver with spares for reflux pumps and cooling water pumps.

A3.2.3 System optimization

- Operating and capital cost estimates for the flare system components
- The expected frequency of normal operational upsets and emergency situations, which will impact the flare system
- Recovery of vent gases from different sources, such as storage tanks, equipment and instrument leakage, purge gas, and compressor distance piece vents
- Segregation of the flare headers according to process services, such as high and low pressure headers, wet sour and dry flare headers, cold and warm flare headers
- Avoiding release of flammable liquid or two-phase mixtures, or of high molecular weight condensable vapors that may create unacceptable hazards
- Multiple combustion systems to handle various flare gas or liquid quantities, such as open pit combustion, ground flares, and thermal oxidizers
- Compliance with the latest safety standards and assessment of the impacts from the flare operation on the surroundings. Review relief material disposal methods. Are they safe, economical, and environmentally acceptable?

With the exception of low pressure (less than 15 psig) relief devices, the design and certification of almost all pressure relief devices is governed by ASME Boiler Pressure Vessel Code. The ASME Code specifies basic requirements for construction, set pressure, performance, and capacity testing and certification. Other standards also exist, which govern pressure relief device application, but these documents usually are based upon use of ASME-approved devices. See Chapter 8 for more detailed discussions on codes and standards.

A3.2.4 Administrative procedures

In addition to sound engineering and design of a flare and relief system, a strict administrative procedure must be enforced on plant operation, which plays an important role in the safety of pressure relief systems. Administrative procedures must be clearly defined, communicated to unit operators, and strictly enforced. Some of the administrative procedures are as follows:

- Lock (or car seal) procedures for block valves associated with pressure relief valves or associated equipment, such as separators, towers and suction drums. The procedures should

include a list of all block valves that are required to be locked in position, procedures for maintaining logs of locked block valve positions, and periodic inspection of the relevant valves.
- Requirements that equipment be continuously attended during certain operations, such as steam out, which is known to be problematic in generating excessive pressure.
- Limitations on modification of equipment, piping, and instrumentation. Any changes must be approved with the plant's engineering manager. Examples are pump impeller size changes, turbine driver speed adjustments, control valve trim alteration, and removal of control valve limit stops.
- Venting, draining, purging, and cleaning procedures for equipment maintenance.

A3.2.5 Design guidelines

A flare system typically consists of a piping system, liquid-vapor knockout drums, and a flare that includes flame igniters, pilots, and flare tips. The flare system also serves as a disposal system for excess hydrocarbon vapor release from pressure control valves during off-design operation.

The flare systems are primarily designed to dispose of vapor releases, but are also capable of handling liquids that may be present during unit upsets. Liquids may form in the flare header due to condensation of heavy hydrocarbons. The liquid content must be separated from the vapor in a flare knockout drum and processed in incinerators and burn pits. Incinerator is the preferred option for liquid disposal rather than a burn-pit because of its lower emissions.

An alternative to flare is to install a vapor recovery system that allows for recovery of valuable hydrocarbons; however, this would not reduce the flare size since the flare system must operate when the vapor recovery system is down.

A3.2.6 Flare systems configurations

Because of the wide ranges of operating pressures and temperatures of the different process units in an LNG production plant, the flare and relief systems typically are equipped with at least two independent flare systems, a warm/wet flare and a cold/dry flare.

The warm/wet flare is for vapor release from the front sections of the plant, including the condensate unit, acid gas removal unit, dehydration unit, NGL fractionation unit, and various utility systems. The cold/dry flare is for the liquefaction plant and refrigeration units, NGL recovery unit, and the cryogenic liquid storage tanks.

For the cold flare, it can be segregated according to operating conditions:

- Low pressure relief from the NGL fractionation columns, relief from the shell side of the liquefaction main exchanger, and for depressurization during plant upset.
- High pressure flare for refrigeration compressors blocked discharge cases and other high pressure relief scenarios.
- Liquid flare for disposal of liquids during the plant start-up or shutdown.

For the wet and warm systems, it can also be segregated according to operating pressure:

- Low pressure flare for the low pressure equipment, such as amine regenerator, the sulfur recovery unit, and the tail gas unit.
- High pressure flare for the feed gas inlet equipment including the dehydration unit.

Liquids that are released during startup or shutdown or other transient periods are routed to the flare knock out drums by liquid blowdown headers separately from the main flare headers. Liquids are disposed of in a liquid burn pit or incinerators specifically designed for liquid combustion.

A3.2.7 Material of construction

Low temperature fluids require special consideration, particularly if there is a possibility of cryogenic vapor release or low boiling liquids entering the disposal system that may cause low temperature due to auto-refrigeration as liquid depressurizes. In such cases, the system piping and drums will need to be fabricated of materials suitable for low temperature, such as stainless steel, to eliminate the risk of brittle fracture failures. For these reasons, the cold discharges are segregated in the cold/dry flare, which is constructed of stainless steel.

A3.2.8 Maintenance consideration

Operation of the LNG facility requires a reliable flare system. A flare system should be spared to some extent that would allow flare maintenance and replacement of individual components of the flare without shutdown of the facility. The maintenance of flare tip, riser, and knockout drums must be carried out without LNG production losses. Elevated flares, if they are installed on the same structure, should be of a dismountable and serviceable type while the other flares are on stream.

The flare capacity of an LNG plant is typically controlled by the blocked discharge case and is generally very large; consequently, the flare structure is very high and can be over 100 m in height. The flare must be inspected periodically and maintained when necessary to avoid potential hazards, which may take weeks to repair. In most installations, shutdown of an LNG plant because of flare repair would result in significant loss in revenue and is unacceptable. Therefore, special considerations should be made in the design selection and the sparing philosophy, such as a single spare flare versus multiple smaller flares.

A3.3 Relief scenarios

The design of each flare system is unique and shall be based on the process calculations and should not be copied from an existing plant. The relief scenarios can be broadly divided into emergency and operational.

Emergency cases are the events that result from plant upsets and emergencies that may exceed the system design capacity and beyond the controllable ranges of the control devices. Operational cases include the controllable events such as startup, shutdown, venting, nitrogen purging, blowdown of equipment, and piping, which are parts of the operation and maintenance procedure.

In the analysis of the relief loads, simultaneous occurrence of separate events that can lead to relief, is considered double jeopardy and generally is not considered. Exceptions to this rule occur when major events with unacceptable consequences of failure have been identified in HAZOPs or design reviews. In some cases, multiple causes of relief may be added together to form the design basis for a single relief case, such as in systems operating at very high pressures, potential of explosion, and material failure.

A3.4 Plant failure causes

There are two types of plant failure causes. The first one is due to operator error and the second one is due to utility failure.

A3.4.1 Operator error

Operator error is a potential factor causing plant upsets and overpressure. The most common error is the inadvertent closing or opening of block valves. Proper training, safe plant practices, posting of legible instructions and warning signs, and provisions for valves locked or sealed in the open or closed position are among the some of the preventative measures that can reduce the occurrence.

A3.4.2 Loss of utilities

The consequences that may develop from the loss of any utility service, whether plantwide or local, may result in partial or total shutdown of the equipment and associated system including the following.

A3.4.2.1 Electricity failure

- Pumps for circulating cooling water, boiler feed water, or reflux in columns
- Fans for air-cooled exchangers, cooling towers, or combustion air
- Compressors for process gas, instrument air, vacuum, or refrigeration
- Instrumentation
- Motor-driven valves.

A3.4.2.2 Cooling water failure

- Reflux condensers for fractionation columns
- Coolers for process coolers, equipment cooling fluids, lubricating oil, or seal oil
- Jacket coolers on rotating or reciprocating equipment.

A3.4.2.3 Instrument air failure

- Transmitters and controllers
- Alarm and shutdown systems
- Process control valves.

A3.4.2.4 Steam failure

- Steam turbine drivers for pumps, compressors, blowers, combustion air fans, or generators
- Heat to reboilers
- Stripping steam for strippers
- Steam to steam eductors.

A3.4.2.5 Fuel oil/fuel gas failure

- Fuel to operate heaters and boilers

- Fuel to engine drivers for pumps or electric generators
- Gas turbine drivers for compressors.

A3.4.2.6 Inert gas failure

- Loss of seal gas to pumps, compressors, and expanders
- Loss of tank pressure control on nitrogen blanketing
- Purge for instruments and equipment.

A3.5 Equipment or operation failure

A3.5.1 Reflux failure

Reflux failure is one of the more common causes for overpressure. The loss of reflux usually results in overhead condenser flooding. Relief devices are sized to handle relief loads based on total loss of cooling. If the operator fails to respond in time, the condenser can flood and the upset condition is likely to occur.

Steady state analysis of tower relief loads typically requires conservative assumptions based on normal tower compositions and temperature profiles. In reality, tower relieves result from dramatic changes in flows, pressure waves, heat and mass transfer, and other phenomenon that can only be assessed via dynamic analysis. Tower dynamic analysis is usually used to calculate an actual tower relieving load, which can be much lower than the steady state assumption. This is particularly true when the column contains a high boiling point material, which does not produce much vapor during this scenario.

A3.5.2 Reboiler failure

Excessive heat input or steam to a reboiler may cause column overpressure when vapor generation capacity exceeds the condensing capacity. This typically occurs when the reboiler temperature control system fails; steam pressure is higher than design, particularly with oversized heat exchangers.

In addition steam failure may trigger a series of events that could result in overpressure. Overpressure may result from the loss of steam to any steam turbine driven reflux pump, cooling water pump, or compressor. In a NGL fractionation train, where steam is the source of reboiler heat, loss of steam to the reboiler of the deethanizer will send light NGL to the depropanizer, which will produce excessive vapor that can overpressure the depropanizer. This effect may also be cascaded to the subsequent columns, and eventually the complete fractionation train will be overpressure.

A3.5.3 Heat exchanger tube failure

Heat exchanger tubes are subject to failure from a number of causes. The normal mode of failure is small leakage from a heat exchanger tube. However, complete rupture of a single tube should also be considered. A pressure relief valve or rupture disk may not be necessary for protection if the piping and downstream equipment on the low pressure side can handle the fluid from the high pressure source without exceeding the hydrostatic test pressure.

There is a safety caution on tube rupture. The sudden loss of a tube can be followed by acceleration of the fluid on the low pressure side and pressurization of that system. The duration and magnitude of these short-term spikes and their impact on the system can be assessed by a dynamic analysis. This type of analysis is recommended, in addition to the steady state approach, where there is a wide difference in design pressure between the two exchangers, especially where the low-pressure side is liquid-full and the high-pressure side contains a gas or a fluid that flashes across the ruptured tube.

A3.5.4 Air fan condenser failure

Air condenser failure may cause overpressure in upstream equipment. For air cooled condensers the relieving quantity may be reduced due to cooling by natural convection. A value of 25% of normal duty can typically be credited because of natural convection heat transfer. If necessary, a more vigorous simulation can be performed on the heat transfer. If the piping system is unusually large and uninsulated, the effect of heat loss from the piping to the surroundings can also be considered to reduce the relief load.

A3.5.5 Cooling water failure

Complete or partial loss of cooling water supply may be caused by power failure or equipment breakdown. The layout of the cooling water piping and system must be analyzed, as it may be too conservative to assume a complete loss of cooling water flow to all units. The standby pump driver usually has an alternate energy source using steam or separate power bus. The maximum quantity relieved under these circumstances can be calculated by heat and material balances around the system. Usually, loss of cooling water to an overhead cooler/condenser system also results in reflux failure.

On detection of low-low cooling water flow, all the trains should be shut down via the DCS. There may be some relief due to the heating up of the train and the loss of reflux in the fractionators.

The main impact in an LNG plant is on the supply of cooling to the propane refrigeration system. The propane compressor discharge pressure will quickly increase, which could lead to the opening of the PSVs of all the trains. To mitigate this problem, several control options can be considered in the design:

- Shutdown of the propane compressor driver on low-low cooling water flow
- Shutdown of the propane compressor driver on high-high discharge pressure
- HIPPS on the propane discharge to avoid overpressure.

The purpose is to avoid all the propane compressors of multiple trains relieving at the same time. Assuming that all devices mentioned earlier have been provided, it can generally be assumed that only one propane compressor trip is responsible for the relief, which must be carefully confirmed in the safety and design review.

A3.5.6 Loss of inert gas

Loss of inert gas will stop blanketing gas supply to the product storage tanks, which may eventually lead to vacuum relief. It will also cut off the supply of seal gas to rotating equipment, resulting in an unsafe operation. Alarms and a shutdown system to detect loss of seal gas pressure should be provided in the DCS and used for equipment shutdown to stop the hazardous operation.

A3.5.7 Instrument air failure

Instrument air failure may be local or total. In the total instrument air failure case, all ESD valves reach their fail safe position and the trains are shutdown and isolated. The main concern may be on the depressurization valves, which, if not supplied with air, will switch to the open position, possibly depressurizing all the trains at once. To resolve with this problem, these shutdown valves should be equipped with instrument air bottles.

A3.5.8 Loss of electric power

Loss of electric power can cause a wide range of upsets resulting in overpressure. Power failure may be further classified as local, intermediate, or total.

Local power failure can affect the operation of individual equipment items, such as air fans, reflux pumps, and solenoid valves that are not on uninterruptable power supply (UPS).

Intermediate power failure affects the operation of one electrical distribution center, one motor control center, or one bus. Depending upon how a group of motors are connected to the power source, multiple failure conditions may occur. For example, assume that the reflux pumps of a fractionator are motor driven and connected to the same bus line. If a bus line failure occurs, the accumulator is flooded, eventually resulting in a flooded condenser. The preferred practice is to place motors in complementary service on separate bus bars to minimize the chance of this type of failure.

Total power failure is assumed to be 100% loss of electric power except for the UPS. Backup electric power sources should be considered in the analysis of the effects of electric power failure. Emergency power to key equipment, lighting, or buildings is typically supplied by diesel generators that are supposed to start on power failure. The emergency loads are then picked up based on pre-established priorities. There is a time lag between loss of power and having equipment on emergency power start. Critical electronic instrumentation is generally connected to the UPS to ensure safe and orderly shutdown upon loss of normal power supply. This system provides battery backup to assure that electronic instrument functionality is maintained throughout the power failure.

Two possible scenarios may be envisaged for sizing the flare for power failure:

- For the case when the train inlets are closed, the relieves are from within the train, such as loss of reflux in fractionation.
- If the train inlets are not closed, pressure continues to rise in the liquefaction train and the pressure control valves from the feed gas section will start opening. In this scenario, several trains can be relieved to the flare system at the same time.

The UPS is typically designed to last for about 45 minutes. After that, if the emergency diesel generator has not taken over, the solenoids will be deenergized and the depressurization valves open. Therefore, in theory, all depressure valves of all trains may open at the same time, leading to the depressurization of all trains simultaneously, resulting in a large relief.

A3.5.9 External fire

Fire provides an unanticipated energy input to a system, which results in overpressure by thermal expansion or vaporization of the retained fluid. Depressurization can be used to reduce the pressure of vessels, equipment, and piping below normal operating pressure to prevent rupture caused by localized fire heating.

The procedures given in API standards or recommended practices (API ST 521 or API ST 2000) are to be utilized, unless otherwise ruled by local (or country) regulations. In some countries different calculation methods than API (for example, those countries that have adopted Japanese standards) are used. In the United States, some states dictate the use of methods in the NFPA codes. Therefore, it is important to define the fire load calculation methods at the beginning of any project.

Protection of process plant systems from the impact of a fire involves a combination of approaches such as fireproofing, fire monitors, equipment location, and equipment depressurizing, in addition to the use of pressure relief devices. The following basic assumptions are generally used in determining the fire case relief loads:

- The process is assumed to be shut down and isolated from other vessels, sources of process fluids, or other relief paths.
- Liquid inventories are assumed to be at maximum operating level at the high level alarm set point.
- Heat input to process equipment is calculated based on empirical equations required by codes and standards based on installation and environments.
- All fire heat input is assumed to be used for vaporizing the vessel contents instantaneously.

A3.5.10 Depressurization

Depressurizing can be used to reduce the pressure of vessels, equipment, and piping below normal operating pressure to prevent rupture caused by localized fire heating. To cope with this problem, the depressurizing sequence can be configured using a timer that allows depressurizing one train at a time in a timely ordered manner.

Emergency depressurizing of a high pressure system can impact the design of both the unit and the flare system. Internal to the unit, equipment temperatures will be lowered by the JT effect cooling during depressurizing. Flare capacity may also be impacted by high pressure drop and low temperature during depressurizing. Depressurization should be performed under emergency conditions only after a major incident. After depressurization, sufficient time should be allowed for the equipment to warm up to avoid temperature stress on the equipment. It is not envisaged to restart, hence repressurize, the train just after a depressurization has occurred.

Depressurization should be manually actuated from the Central Control Room with hard wired connections. A depressurization system actuated from the Central Control Room is mandatory for a piece of equipment or a process system isolated by ESD valves. The depressurization is achieved through a blowdown valve (on/off air actuated ball valve) and a restriction orifice, which set the depressurization rate. To determine the flow rate the following assumptions can be made:

- The initial pressure is the design pressure of the system.
- The system is exposed to an outside pool-type fire.
- The pressure of the section to be reached within the time set below is typically half the design pressure or as required by local codes and standards.
- The time allowed for depressurization is not to exceed 15 minutes.

A3.5.11 Thermal expansion

Thermal expansion or hydraulic expansion is the increase in liquid volume caused by an increase or decrease in temperature. It can result from several causes, the most common of which are the piping or vessels blocked in while they are filled with cold liquid and are subsequently heated, or an exchanger that is blocked in on the cold side with continuous flow on the hot side.

In certain installations such as in cooling circuits, equipment arrangements and operating procedures may eliminate the case for an expansion relieving scenario. For example, when one block valve on the cold side of a multiple shell and tube exchanger is locked open, it will eliminate the possibility of blocking and expanding the enclosed liquid in the entire train.

A3.5.12 Thermal stress

For Brazed Aluminum Exchanger, the maximum operating temperature is limited to 150°F. Heat source with high temperature must be avoided by a shutdown system that is necessary to avoid material failure of the aluminum exchanger.

A3.5.13 Vacuum relief

Under different circumstances, some of the same factors that lead to overpressure can lead to a drop in operating pressure to the extent that vacuum relief may be required to prevent the equipment from collapsing under the vacuum. Such occurrence includes the quenching of steam in a vessel during startup, and the withdrawal of liquid from a low pressure tank without makeup gas.

A3.5.14 Feed inlet control valves open

Every control valve must be considered as subject to inadvertent operation from the fully open to fully closed position independent of the failure position. Some control devices are designed to remain stationary in the last controlled position when the control signal or operating power fails. Since predicting the position of the valve at the time of failure is not possible, such devices could be considered either open or closed; therefore, no reduction in relief capacity should be considered when such devices are used.

The cause of control valve failure may be due to one of the following incidents:

- Loss of transmission signal to the valve positioner
- Failure of control valve operating medium (air, hydraulic oil, electricity, etc.)
- Process measuring element failure
- Process measuring element transmitter failure.

For these cases, the required relief capacity is the difference between the maximum inlet flow and the normal outlet flow at relieving conditions, assuming that the other valves in the system are still in normal operating position. If one or more of the outlet valves are closed by the same failure that caused the inlet valve to open, the required relief capacity is the difference between the maximum inlet flow and the minimum flow from the outlet valves that remain open.

If the gas is dry, the inlet gas can be routed to the cold/dry flare and in that case the flow is most likely less than that from the blocked discharge case from one of the propane compressors. The block feed inlet and the compressor block outlet should not be considered coincidental.

If the gas is wet, it is likely to be the sizing case for the wet/warm flare. Note that due to the JT cooling effect, the temperature may drop below the hydrate temperature of the wet gas, which may cause plugging problems in the flare system.

In the event of loss of upstream liquid level, vapor may pass into the downstream system at high rates determined by the differential pressure between the systems. In some cases the maximum vapor rate may cause the upstream system to gradually depressure and result in gradually declining flow.

A3.5.15 Blocked discharge

Blocked discharge situations apply to compressors, pumps, and other process systems. Generally, omission of block valves on vessels or locked open valves and equipment in series can simplify and avoid the pressure relieving scenarios.

In applying administrative controls, such as locks or car seals and associated procedures, to minimize the potential for block valve closure, consideration should be given to the potential consequences if the administrative procedure is circumvented and the block valve is accidentally closed.

For compressor block discharge cases, such as the propane compressor and MR compressor, the maximum flow should be based on vendor's compressor curves and the turbine driver operating at its maximum speed. For multiple compressors operating in parallel, two compressor outlet valves closed at the same time is typically not a credible scenario. The block compressor discharge case is generally the sizing case for the high pressure dry/cold flare.

A3.5.16 Amine absorber level control valve failure

When the amine absorber valve fails open, the feed gas flow can be sent directly to the amine flash drum through the level valve, potentially overloading the flare system. The relief rate depends mainly on the size of the level and the valve characteristics (Cv), and the system hydraulics and piping sizes. In most cases, the high pressure differential may result in a large vapor breakthrough, which can be one of the sizing cases for the wet /warm flare.

A3.5.17 Control valve bypasses

Control valve bypasses are provided for the purpose of maintenance of the main control valve and are subject to inadvertent opening due to misoperation.

Simultaneous full opening of a control valve and its bypass normally is considered to be double jeopardy. Either full opening of the control valve with the bypass in the normally closed position or full opening of the bypass with the control valve in its normal operating condition is generally considered as a credible case. In some instances, control valve bypasses have greater capacity than the control valve itself and may constitute a controlling factor in overpressuring of the downstream equipment. In such cases, installation of a reduced size bypass valve or restrictor can reduce the relief flow.

A3.5.18 JT valves open during MCHE and MR compressor shutdown

In the case of liquefaction unit trips, the MR compressor will be shut down, the suction valves closed, and JT valves closed. The discharge stays at a high pressure. If the JT valves are reset, a large quantity of MR will flow to the shell, vaporize, and the pressure might reach the design pressure of the shell. The relief flow is routed to the low pressure dry/cold flare.

A3.5.19 Warm LNG excursion

If the temperature control valve downstream of the main exchanger opens widely by error some warm LNG would be directed to the LNG storage and generate a significant amount of gas.

Since the ship loading flares are not designed for this large flow, a HIPPS system should be designed specifically for such a case. The HIPPS system consists of several independent temperature detectors, hard wiring, voting systems, several ESD valves, and triplicate redundant modular control system.

A3.6 Operator intervention

The decision to take credit for operator response must be carefully evaluated. Credits are given only after determining the relieving conditions considering the complexity of the process and controls, speediness to respond to the emergency situation, technically qualified personnel, and the risk associated with the failure of operator response. The minimum response time for operators typically ranges from 10 to 30 minutes. Among the factors that must be considered are:

- Operator response time is considered additional to the time required for an operator to recognize the emergency situation.
- Operating personnel must be able to correct or stop the event with a simple response before relief is required.

Special caution should be observed when using operator response to mitigate liquid overfill, due to the occurrence of several major industrial incidents that have resulted from liquid overfill. The effectiveness of operator response must be evaluated. Note that a correct response may reduce but may not totally eliminate the safety problem.

A3.7 Operation relieves

A3.7.1 Scrub column startup

During the startup and cooldown of the liquefaction plant, the pressure of the scrub column overhead must be maintained by venting the overhead gas from the scrub column reflux drum to the cold dry flare. At the same time, before the NGL fractionation is placed on stream, the scrub column bottoms are also routed to the cold flare for disposal.

A3.7.2 Liquefaction train startup

During liquefaction plant startup, the warm LNG (gas liquid mixture) at the outlet of the MCHE is sent to the cold flare till the LNG product temperature is cooled to the design condition that can be accepted by the LNG storage tanks.

A3.7.3 NGL fractionation startup

When the scrub column is stabilized, the NGL can be routed to the NGL fractionation train. Off-spec liquid products from the depropanizer and debutanizer are routed to flare.

A3.7.4 Vapor return from LNG tankers

During ship loading, the boil-off vapor is routed to a tank boil-off vapor header that feeds the boil-off compressor. The compressed gas vapor can be used as fuel gas; any excess is disposed of in the flare system.

A3.7.5 Vapor return from LPG tankers

Most of the tankers have their own on-board system to dispose of excess gas and some of the tankers have a reliquefaction unit. Therefore relief from the LPG tankers is infrequent but emergency flow needs to be considered. Overpressure can also be avoided by reducing the ship loading rate until the system is in balance.

A3.7.6 Ship loading dock relief

The followings are some of the loads that may contribute to the relief sizing during ship loading:

- LNG carrier is being loaded, generating vapors from pump heat
- LNG pumps operating and on recycle, generating excessive vapors
- LNG boil-off compressor out of service for maintenance
- Heat leaks from storage tank, rundown lines, and loading lines
- Maximum barometric pressure drop during a hurricane
- Instrument air failure
- Vacuum break gas valves failure
- LNG tankers vapor return blowers malfunction
- Total plant power failure.

A3.8 Design pressure

The design pressure for the equipment that will be relieving to the flare defines the service requirements for the relief system. Often, a small increase in the design pressure can reduce the cost and complexity of the relief system or even eliminate the need for pressure relief for particular contingencies. If the equipment design pressure is low or the anticipated relieving rates are high, additional care should be taken in this selection. It should be kept in mind that higher design pressure selection may reduce or eliminate frequent venting to the flare system.

The flare seal drum, knockout drum, and the flare piping are typically designed for 50 psig. However, higher pressures shall be specified if required by hydraulic evaluation.

A3.8.1 Vacuum pressure

It is possible to develop a vacuum condition in the relief header due to cooling and condensation of hot relieving vapors through ambient heat loss from the relief header, although the resulting event is the introduction of atmospheric air through the flare tip. Since this is undesirable because of potential explosion hazards, the flare design should provide an emergency purge (inert) gas addition through a control valve that opens when a hot flare relieving condition is detected in the flare header.

In any event, the relief header, flare seal drum, and flare K.O drum should be designed for at least half vacuum condition for these cases and for potential steam-out conditions during startup and maintenance. This should normally not impact the vessel and piping mechanical design because this equipment is typically designed for 50 psig.

A3.8.2 Maximum allowable working pressure

Design pressure and design temperature are used as the basis for design of ASME Section VIII pressure vessels. A required thickness for the walls is calculated and the next commercially available size is selected. Since this typically results in the actual vessel wall thickness being higher than the calculated wall thickness, the vessel will be able to withstand a higher pressure than the specified design pressure. The higher pressure is referred to as the vessel maximum allowable working pressure (MAWP). The higher pressure margin can be used to optimize the flare design by increasing relief valve set pressures and relief capacity, provided all other components in the system are also suitable for the higher design pressure.

A3.9 Design temperature

Since all relieving scenarios are considered to be under abnormal operation (i.e., either startup/shutdown or emergency operation), the normal condition for the flare system is with no flow other than the purge gas. Therefore, the normal operating temperature is considered to be the temperature of the purge gas even though the flow is so small that the physical reality is that the pipe and equipment wall temperatures will be at the ambient temperature.

A3.9.1 Minimum design temperature

The application of auto-refrigeration temperatures in setting the minimum design temperature for the flare system should consider the relevant relief rate and duration, as well as the system metal mass and ambient temperature.

Maximum Design Temperature (MDT): All the potential flare relieving scenarios to determine those with the highest associated temperatures should be analyzed. Most scenarios, such as fire, cooling water failure, and power failure would be considered short-term upsets. The application of these short-term temperatures excursion in setting the maximum design temperature for the flare system should be considered.

Longer duration scenarios such as startup and shutdown should also be considered in selecting the flare system design temperature.

> *Caution*: Selecting an unnecessarily high design temperature for the flare header will result in significant cost and plot plan impact as numerous large expansion loops will need to be included in the design to account for the thermal stresses.

A3.10 Other design considerations

A3.10.1 Use of restriction orifice

A restriction orifice or reduced trim valve can be used to limit the flow, such as that from amine absorber level valve to the flash drum. The use of restriction orifice can be used on other letdown stations to limit the vapor or liquid flow that may be due to operator errors or equipment failure.

A3.10.2 High integrity pressure protection systems

High integrity pressure protection systems (HIPPS) may be considered for the following duties:

- Propane compressor blocked discharge
- MR compressor blocked discharge
- High feed gas pressure when the upstream operation can be higher than the LNG plant design pressure
- LNG tank overflow.

A3.10.3 Purge gas supply to stack and pilots

It has to be stressed that the source of purge gas should be fully reliable, redundant, and with backup to avoid any disaster. If the purge gas has a density lower than air, which is generally the case in LNG plants, there is a risk of having a vacuum situation in the flare header in case of low purge gas flow, hence a risk of air ingress.

A3.10.4 Computer modeling

Many hydraulic programs have been developed and can be used specifically for relief system networks hydraulic design and evaluation. The following are some of the latest programs that are available in the industry.

- FLARENET is created for Flare Networks. It is a design and analysis tool with a steady state model of the flare network from PSV to flare tip.
- VISUAL FLOW ranges from line sizing and vessel depressurizing to the rating of complex relief systems. Process and Safety engineers can design, rate, and analyze processes with this rigorous steady-state simulator.

Index

Note: Page numbers with "f" denote figures; "t" tables.

A

B

C

D

E

F

G

H

I

M

N

O

P

S

T

U

V

W

Z

Printed in the United States
By Bookmasters